The Receptors

For other titles published in this series, go to
www.springer.com/series/7668

To my wife Samantha,
For her abundant support, for her patience
and understanding, and for her love

G. Di Giovanni

Preface

Accepting the challenge of editing the first book on a subject is a risky business, particularly if you attempt to produce an overview in the research of 5-HT$_{2C}$ receptors, an area now going back over 25 years with so productive results. Nevertheless, I have embarked on this unique editorial enterprise with the aim of giving pleasure to the readers and researchers and not least to myself, confident that it will be instrumental in future research in 5-HT$_{2C}$ pathophysiology. A few years ago Bryan Roth edited, for the same series "The Receptors," a very fine volume on serotonin receptors. The present volume is a further development of this and is a thorough examination of the 5-HT$_{2C}$ receptor subclass, needed to cover the extraordinary amount of research into nearly every aspect of 5-HT$_{2C}$ receptor function that has recently emerged. This is not surprising, considering that the 5-HT$_{2C}$ receptor is a prominent central serotonin receptor subtype, widely expressed within the central and the peripheral nervous system and is thought to play a major role in the regulation of a plethora of behaviors. Therefore, it has been shown by experimental and clinical observation that it may represent a possible therapeutic target for the development of drugs for a range of CNS disorders such as schizophrenia, depression, drug abuse, eating disorders, Parkinson's disease, and epilepsy, to cite but a few. The book, a result of the efforts of an international group of authors, has the aim of providing an update of the functional status of the 5-HT$_{2C}$ receptor, covering molecular, cellular, anatomical, biochemical, and behavioral aspects, to highlight its distinctive regulatory properties, the emerging functional significance of constitutive activity and RNA-editing in vivo, and the therapeutic potentiality in different diseases that are singled-out in different chapters.

While covering the latest research, for obvious reasons, this volume cannot be exhaustive and it has been impossible to include a number of authors of obvious merit. I hope that more volumes on the subject will be possible in the future.

I want to thank all the authors who have responded very willingly and contributed their time and expertise in preparing their individual contribution to a consistently high standard. My thanks go to Vincenzo Di Matteo and Ennio Esposito who have contributed to the realization of this book. I am indebted to Philippe De Deurwaerdère, who has unselfishly dedicated his time and expertise to insightful and helpful reading of this text and Dr Clare Austen for reviewing the English style of these manuscripts.

Finally, I would like to express my sincere appreciation to Kime Neve, series editor, and Matthew Giampoala, Springer publishing editor, for their help in driving the book's development and eventual publication.

I hope that the contents of this volume will further inspire and stimulate discussions and new interdisciplinary research on the 5-HT$_{2C}$ receptor.

May 2010

Giuseppe Di Giovanni
University of Malta
Msida, Malta

Contents

Chapter 1
The Making of the 5-HT$_{2C}$ Receptor

Jose M. Palacios, Angel Pazos, and Daniel Hoyer

1.1 Introduction

1.1.1 5-HT$_{2C}$ Receptors in the Context of Neurotransmitter Receptor Multiplicity: 1983–1988

1.1.1.1 The Concept of Receptor

This chapter is a personal account of how the 5-HT$_{2C}$ receptor was discovered and characterized. The discovery of 5-HT$_{2C}$ receptor took place more than 25 years ago, in the middle of a revolutionary period of change in the receptor concept: the realization that single messengers (such as neurotransmitters) were acting through a multiplicity of receptors. The concept of receptor was initially postulated more than 100 years ago by Paul Ehrlich as "selected binding sites for chemotherapeutic agents" and evolved though the work of J.N. Langley who formulated the concept of "receptive substances" (Bennett 2000).

For many decades, receptors were defined mainly through functional assays based on responses in isolated tissues to chemical series and translated in pharmacological effects in animal models and from there extended to the therapeutic use of such new molecular entities. This concept is the basis for the development of more than 45% of the about 500 known therapeutic targets and for the discovery of numberless drugs.

Despite many decades of receptor research, whether the receptors existed as actual molecular entities was still debated in the 1970s. It was the introduction of radioligand and affinity labeling that led to the solubilization and purification of receptor proteins opening the "molecular era of receptor research" and to the cloning of the first receptors, the cholinergic nicotinic and the β$_2$-adrenergic receptors

J.M. Palacios (✉)
BrainCo Biopharma, Parque Tecnológico de Vizcaya, Edificio 801,
48160 Derio (Bilbao), Spain
e-mail: jmpalacios@brainco.es

G. Di Giovanni et al. (eds.), *5-HT$_{2C}$ Receptors in the Pathophysiology of CNS Disease*,
The Receptors 22, DOI 10.1007/978-1-60761-941-3_1,
© Springer Science+Business Media, LLC 2011

representing the two big families of receptors, i.e., ligand gated channels and GPCRs (Lefkowitz 2004).

The development of the identification of receptors in cell-free preparations using radiolabeled ligands, a simple tool for the characterization of the recognition sites with the properties of the receptors, in a simple and fast manner, opened new possibilities to study reception. Radioligand binding was expanded later to intact tissue sections for the localization at the light microscope level. These new methodologies revealed many unexpected features of drug and neurotransmitter receptors.

1.1.2 The Multiplicity of Receptors for a Single Neurotransmitter

One concept that raised tremendous opposition in the beginning was the multiplicity of receptors for a single neurotransmitter. It was a generally accepted concept that the number of receptors for a given neurotransmitter was very limited. The concept moved from the extreme – "one transmitter-one receptor" – to the "liberal" – "one transmitter-two receptors." Examples include acetylcholine-nicotinic and muscarinic, noradrenalin-alpha and β-adrenergic receptors, and histamine-H_1 and histamine-H_2. Proposing additional or more than two receptors was viewed as a heresy, and if the proposal was coming from results based on radioligand binding, it was viewed as doubly heretical. In other words, radioligand binding was considered to represent a rather obscure side product of biochemical pharmacology, as were the effects of guanine nucleotides on agonist binding (Laduron 1984), which were merely associated with detergent-like effects of the GTP and the likes, rather than on active versus inactive states of the receptor (about 20 years later, the Nobel Prize was attributed for the discovery of G proteins). However, as Galileo said *e pur si muove*, more and more results were accumulating making it difficult to constrain the size of receptor families.

1.1.3 5-HT-M and 5-HT-D Versus 5-HT$_1$ and 5-HT$_2$ Receptors

With respect to serotonin receptors, it should be remembered that in those days (1983/1984 – this is when we started working together on this subject: José María (Chema) Palacios [JMP], Angel Pazos [AP] and Danny Hoyer [DH]), 5-HT receptors were considered to form a fairly simple family, and since the cloning era was to start a few years later, there was little structural evidence for further complexity. In the brain, one could distinguish between 5-HT1 and 5-HT$_2$ sites (Peroutka and Snyder 1979), labeled by [^{3}H]5-HT and [^{3}H]spiperone/[^{3}H]ketanserin, respectively. In the periphery, 5-HT-M and 5-HT-D receptors had been known for some time (the effects of 5-HT on this receptor were first reported in the guinea-pig ileum), but they were not labeled by anything, primarily since this was the field of physiologists working in the intestine, and electro-physiologists in the peripheral nervous system (Bradley et al. 1986; Gaddum and Picarelli 1957; Hoyer et al. 1994). As it was becoming clear that 5-HT$_2$ sites and 5-HT-D receptors were very closely related (Engel et al. 1984), it was decided to reconsider the whole nomenclature, and there was agreement to name these receptors 5-HT$_1$, 5-HT$_2$ (5-HT-D) and 5-HT$_3$ (5-HT-M): The Bradley scheme was born and the first receptor nomenclature party was in its infancy (Bradley et al. 1986).

1.2 [^{3}H]Mesulergine: An Interesting Molecule

1.2.1 CU32085: A Semisynthetic Ergot Compound

It is in this conceptual environment that the discovery of the 5-HT$_{1C}$ (later 5-HT$_{2C}$) receptors took place at Sandoz (now Novartis) in Basel, Switzerland. Sandoz had a long tradition of working with ergot derivatives with multiple useful pharmacological activities. Semisynthetic ergot compounds were being developed at the end of the seventies in many therapeutic areas. One of those was CU32085, also known as *mesulergine*. This compound presented interesting dopaminomimetic activities in animal models and was being developed as an antiparkinsonian drug. Binding studies carried out by Annemarie Closse at Daniel Hauser's group, with the technical help of Armand Wanner, using [^{3}H]mesulergine, surprisingly showed a potent binding of this compound to 5-HT$_2$ receptors in the rat brain (Closse 1983).

1.2.2 Receptor Localization Using [^{3}H]mesulergine

In 1983, having just set up the autoradiographic methodology for receptor localization developed in Mike Kuhar's laboratory at the Johns Hopkins University to label brain tissue sections with radioligands, Jose M. Palacios decided to investigate the localization of [^{3}H]mesulergine. At his lab, Rolf Lenherr incubated rat brain tissue sections with the radiolabeled ligand and generated some autoradiograms. When we developed the films, we were disappointed – at first glance the film appeared "empty." A closer look revealed some repeated dark lines in the film; after a while, we realized that those heavy labeled thready-looking spots corresponded to the choroid plexus (Fig. 1.1) and were filling the brain ventricles in our tissues sections. This labeling was selectively blocked by coincubation with unlabeled serotonin. We knew from our own experience that radiolabeled serotonin also labeled intensely the choroid plexuses in the rat brain, as well as many other brain areas, but no special attention had been paid to them.

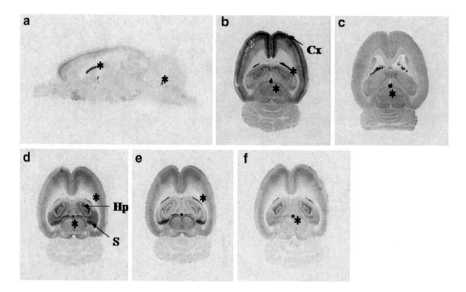

Fig. 1.1 Images from early autoradiograms of sagittal and horizontal rat brain sections incubated with [³H]mesulergine [(**a**), (**b**), and (**c**)] or ³H-5-HT [(**d**), (**e**), and (**f**)] and different unlabeled compounds. (**a**) Short exposure revealing intense labeling in choroid plexus. (**b**) and (**c**) are longer exposures showing binding of [³H]mesulergine to 5-HT$_{2A}$ and 5-HT$_{2C}$; (**b**) and (**c**) show the blockade of 5-HT$_{2A}$ binding by coincubation with ketanserin. (**d**) shows [³H]5-HT binding to 5-HT$_{1A}$, 5-HT$_{1B}$ and 5-HT$_{2C}$. (**e**) Binding to 5-HT$_{1A}$ has been blocked with 8-OH-DPAT. (**f**) 5-HT$_{1B}$ binding was blocked with compound SDZ 21009 (see text for details)

1.2.3   New 5-HT-Related Site

Thus, while one could expect some 5-HT$_2$ labeling as well as dopamine D$_2$ labeling with [³H]mesulergine, our first autoradiograms (probably developed in haste) were pointing to a very different pattern. Simple blockade studies suggested that we were in the presence of a "new 5-HT-related site."

Those preliminary autoradiographic studies were further confirmed by Angel Pazos (AP), a postdoctoral fellow who had just joined the JMP group, together with Roser Cortés, a Ph.D. student at the same laboratory. The striking results were first presented at a meeting of the British Pharmacological Society in 1984 in Birmingham and received with amused commentaries from the establishment. It is noteworthy that these were also the days where comments about "the binding of pepperoni to pizza" were made in Trends in Pharmacological Sciences (Guth 1982). Nevertheless, a full description of the pharmacological profile and characteristics of 5-HT$_{1C}$ receptors was published at the end of the year in the European Journal of Pharmacology, and their detailed anatomical distribution compared with 5-HT$_{1A}$ and 5-HT$_{1B}$ and 5-HT$_2$ receptors was published 1 year later in Brain Research. Those studies have been cited extensively (Pazos et al. 1984a, 1985; Pazos and Palacios 1985).

1.3   Collecting Pig Choroid Plexus: The Characterization of 5-HT$_{1C}$ Sites by means of Membrane Binding

1.3.1   Danny Hoyer Joined Sandoz

Classical membrane binding studies are simple to carry out if one wants to study the kinetic characteristics as well as the pharmacological profile with a large number of drugs to fully characterize a new binding site. Daniel Hoyer (DH), who had just joined Sandoz in 1983 from his postdoc at Perry Molinoff´s lab (Philadelphia), came into the picture and, due to his previous experience in membrane binding methodology, led the radioligand characterization of these sites in several species. Angel did not have a previous background in binding procedures – his PhD work had been mainly on in vivo biological recordings – but under the supervision of Danny, he soon became familiar with all aspects of binding assays. In order to characterize the pharmacological profile of a new binding site, one needs a tissue rich in these sites that can be easily obtained in sufficient amounts.

1.3.2   Collecting Pig Choroid Plexus

Since the putative 5-HT receptor had been clearly identified in the choroid plexus (we later demonstrated its presence in other brain regions but to a much lesser extent), we decided to study its pharmacology by applying radioligand binding to membranes of choroid plexus. As the rat choroid plexus are not easy to collect in large quantities in a reasonable manner, we turned to pig brains – it was cheaper to collect the choroid plexus from brains obtained from the local slaughterhouse. However, dealing with tissue preparation was not simple, since the choroid plexus is primarily a tight mix of vessels and harsh connective tissue, but with the initial help of Monique Rigo, a resourceful technician at JMP laboratory, we managed to obtain good enough membrane fractions, perform regular membrane labeling studies, and get decent quality binding data (Pazos et al. 1984a, b; Hoyer et al. 1985a).

From there, we started parallel autoradiographic studies to establish what came to be the 5-HT$_{2C}$ receptor, as will be discussed later in section 1.4 in this chapter. The project in fact followed a research plan carried out in parallel in two different buildings: Autoradiographic experiments were done at building 360 (JMP lab), while membrane assays were carried out at building 386 (DH lab). During the autumn of 1983 and the whole year 1984, it become quite usual to Angel to walk daily from one place to the other, bearing tissue samples, autoradiograms and counter prints in his hands, and frequently, a certain level of anxiety in his mind, since we were using old-generation liquid scintillation counters, with rather modest throughput, when we were preparing hundreds of samples. This meant that they had to be loaded and unloaded one by one, by hand with two or three loading sessions per day, weekends included; every sample was labeled by hand in case of counter failure and to allow recounting.

1.3.3   The Profiles of [³H]mesulergine, [³H]5-HT and [³H]LSD in the Choroid Plexus Membranes

Thus, we compared the profiles of [³H]mesulergine, [³H]5-HT, and [³H]LSD in the plexus membranes. The very first studies with [³H]mesulergine added an additional surprise to the story. We found that mesulergine had a peculiar and unique binding profile in the choroid plexus but that its pharmacology in the brain across various species was not overlapping. Species differences, a new feature in receptor studies, which was to appear repeatedly afterwards in the serotonin receptor field, made its first appearance adding further complexity to the field until it was recognized that such differences are real and dictated by the gene structure of these receptors (Hartig et al. 1996; Hoyer and Middlemiss 1989). After obtaining the initial data with [³H]mesulergine, the three of us spent many hours (in the labs and also at the canteen) talking about what kind of experiments (at that phase, focusing on binding to membranes and sections) had to be carried out, and which compounds – both labeled and unlabeled – were the best ones to define the new site, especially with respect to the classical 5-HT$_2$ site, which was disturbingly close to 5-HT$_{1C}$, depending on the tools used.

1.3.4   Discovery of 5-HT$_{1C}$

In rat cortex [³H]mesulergine-labeled sites with a profile indistinguishable to that of [³H]ketanserin-labeled sites (thus 5-HT$_2$), whereas in porcine and human cortex, [³H]mesulergine binding sites were different from [³H]ketanserin-labeled sites. On the other hand, in the choroid plexus of the three species, [³H]mesulergine binding was different from 5-HT$_2$ but equivalent to [³H]5-HT- and [³H]LSD-labeled sites. These sites were named 5-HT$_{1C}$ since our results clearly demonstrated that they were also different from the 5-HT$_{1A}$ and 5-HT$_{1B}$ subtypes, already well characterized as separate entities at that time in our labs, and since they were labeled by [³H]5-HT, the prototypical 5-HT$_1$ radioligand at that time. Obviously, as more receptor families appeared, it became clear that [³H]5-HT was by no means specific for the 5-HT$_1$ family (Hoyer et al. 1994).

1.4   Visualizing 5-HT Receptors at the Microscopic Level: The Power of the Anatomical Resolution

1.4.1   5-HT$_{1C}$ Autoradiographic Characterization

In parallel to the studies carried out in membranes, we performed a detailed auto-radiographic characterization. After gaining enough expertise with Chema´s advice,

Angel was dealing for a long time with the challenges and possibilities of brain anatomy across species. Studies were first designed to identify brain areas specially enriched in the different subtypes of 5-HT$_1$ receptors proposed in order to reinforce the singularity of 5-HT$_{1C}$ sites as a separate entity from the others. At that time, there was already quite some evidence that 5-HT1 binding was divided into 5-HT$_{1A}$ and 5-HT$_{1B}$ (Pedigo et al. 1981) and 8-OH-DPAT had just been reported as a 5-HT$_{1A}$ selective agonist (Gozlan et al. 1983; Hjorth et al. 1982; Middlemiss and Fozard 1983).

1.4.2   The Independence of 5-HT$_{1C}$ Sites with Respect to 5-HT$_{1A}$ and 5-HT$_{1B}$

In addition, [^{3}H]OH-DPAT was shown to label a more restricted population of sites in the brain than [^{3}H]5-HT (Pazos and Palacios 1985; Hoyer et al. 1986a). By using [^{3}H]5-HT and [^{3}H]mesulergine as radioligands and 8-OH-DPAT, the beta-blocker (SDZ 21-009, a very useful compound picked by Guenter Engel), and mesulergine itself as the main displacers, we succeeded in identifying anatomical areas particularly enriched in each 5-HT$_1$ subtype: Some examples are the dentate gyrus of the hippocampus for the 5-HT$_{1A}$, the substantia nigra (reticular) for the 5-HT$_{1B}$, and of course, the choroid plexus for the "new" 5-HT$_{1C}$. These studies contributed to establishing the independence of 5-HT$_{1C}$ sites with respect to 5-HT$_{1A}$ and 5-HT$_{1B}$.

1.4.3   The Pharmacological Profile of the New 5-HT$_{1C}$ Receptor

Subsequently, our work addressed the characterization of the pharmacological profile of the new receptor at the microscopic level, by constructing autoradiographic competition curves. These studies revealed a pharmacological profile fully comparable to the one found in parallel in membranes (see section 1.5). They also showed the first picture of the distribution of this subtype throughout the rat brain. In addition to the choroid plexuses that presented the highest autoradiographic density by far, clearly detectable levels of the new 5-HT sites were mainly localized over the olfactory system, hippocampus (CA1 field), thalamic nuclei, substantia nigra, and spinal cord (external); lower levels were detected in neocortex (piriform, cingulate, frontal), putamen, globus pallidus, hypothalamus (ventromedial), and several nuclei at brainstem (i.e., spinal trigeminal nucleus). In general, our autoradiographic studies illustrated very well the power of adding the anatomical dimension to the binding strategies at that time. In fact, without the initial identification of a region enriched in only one class of receptor site, it would be rather unlikely to go for the detailed studies in choroid plexuses membranes that eventually led to the full description of the renamed 5-HT$_{2C}$ receptors.

1.5 More Binding, More Fun

1.5.1 Using Beta-Blockers

We subsequently published evidence (Hoyer et al. 1985a) that brain 5-HT$_1$/
[^{3}H]5-HT binding could be displaced in a triphasic manner by a beta-blocker (SDZ
21-009), which was characterized by high affinity for 5-HT$_{1B}$ sites, intermediate for
5-HT$_{1A}$ sites, and very low affinity for 5-HT$_{1C}$ sites. Our autoradiographic data had
also demonstrated the ability of this compound to selectively bind to 5-HT$_{1B}$ sites
in specific brain areas (Hoyer et al. 1985a, b). Some further indole beta-blockers
turned out to be very important tools for the delineation of 5-HT$_1$ receptor subtypes.
Indeed, we noticed that ICYP, [^{125}I]iodocyanopindolol, a very potent antagonist at
β-adrenoceptors, which is still the most popular radioligand for labeling these
receptors (Engel et al. 1981; Hoyer et al. 1982), was also binding in the brain to
sites sensitive to 5-HT and other serotoninergic ligands. Although the radioligand
was initially described (in 1981) as highly specific for beta-receptors, we had to
admit that it was perfectly suitable to label 5-HT$_{1B}$ receptors in rodents but not in
other species.

1.5.2 Using Other Compounds

Thus, using an iterative process, we started to put bits and pieces together by comparing the binding profiles of radioligands known to interact with 5-HT/Dopamine
D$_1$-D$_2$/5-HT$_2$ or mixed 5-HT$_2$/D$_2$, beta-receptor ligands, and the one that seems to
label pretty much everything, [^{3}H]LSD and its derivative [^{125}I]LSD. In fact, [^{125}I]
SCH23982 and [^{125}I]LSD were characterized in the choroid plexus and found to
label 5-HT$_{1C}$ sites (Hoyer et al. 1986b). We also used [^{3}H]OH-DPAT, which at the
time was (and still is) an exquisite tool to define 5-HT$_{1A}$ sites, whereas [^{3}H]ketan-
serin turned out to be another valuable tool for labeling 5-HT$_2$ receptors. Both
membrane and autoradiographic studies allowed us to define the pharmacological
profile of affinities on the new receptor, with the drugs available at that time – in
addition to 5-HT and mesulergine, LSD, methysergide, and mianserin were among
the compounds showing high or very high affinity for 5-HT$_{1C}$ receptors; ketanser-
ine, pirenperone, and methergine bound to the new site with an intermediate affinity; finally, 8-OH-DPAT (a 5-HT$_{1A}$ drug), the beta-blocker and 5-HT$_{1B}$ compound
(−) 21009 and spiperone demonstrated low or very low affinity for the new recognition site (Hoyer et al. 1985a).

The iterations were multiple: Radioligand binding was performed in brain membranes, including choroid plexus, receptor autoradiography in brain slices of various species, and whichever other functional models could be used, e.g., contractions
of the guinea-pig ileum (Engel et al. 1984; Kalkman et al. 1986), inhibition of 5-HT
release in the cerebral cortex (Engel et al. 1986), stimulation or inhibition of cAMP

production in hippocampus (Markstein et al. 1986; Schoeffter and Hoyer 1989a) in the substantia nigra (Schoeffter and Hoyer 1989b), stimulation of PLC production in choroid plexus (Hoyer et al. 1989) or in smooth muscle cells, and contraction or relaxation of various vessels (Hoyer 1988a).

1.6   Labeling 5-HT Receptors in Human Brain

1.6.1   5-HT$_{IC}$ Receptors in Postmortem Human Brain Tissue

After characterizing the properties of the new site in animal (rat, pig) brain, we then went on with the characterization and localization of 5-HT$_{IC}$ receptors in postmortem human brain tissue. Taking advantage of the fact that Dr. Alphonse Probst, a pathologist at the University of Basel, shared a collaboration with Chema´s group aimed to visualize and analyze neurotransmitter receptors in human tissue, we could use similar experimental procedures both in membranes and sections from a series of human brains, although confronting the specific limitations associated to the work with postmortem material. The 360–386 buildings connection worked again fairly well and by the end of 1985 the anatomical distribution and pharmacological profile of 5-HT$_1$ and 5-HT$_2$ receptors in the human brain were obtained, proving to be relatively similar to that reported in animals (choroid plexuses starring again), although species differences were evident – in fact, the marked species-dependent differences in 5-HT$_{1B}$ pharmacology complicated for quite a while the exact delineation of the distribution of 5-HT$_{IC}$ sites in non-choroid plexus areas (Hoyer et al. 1986c, d; Pazos et al. 1987a, b).

1.7   Looking for a Physiological Role for 5-HT Receptors in the Choroid Plexus

The enormous concentration of 5-HT$_{IC}$ receptors in the choroid plexus of all the species investigated was calling for an examination of the role of these receptors in controlling one of the functions of this tissue, namely, the volume and composition of cerebrospinal fluid. Monique Rigo cannulated rat cerebral ventricles, examined using perfusion with artificial CSF, and radiolabeled inulin alterations in the volume of CSF. We found effects of 5-HT and other drugs, but the system was too complex to carry out detailed pharmacological studies, and we did not progress enough as to get publishable data. Along the same lines in 1988, Lindvall-Axelsson and colleagues (Lindvall-Axelsson et al. 1988) from the University of Lund in Sweden published a detailed study in the rabbit, a better suited model, described the inhibition of CSF production by 5-HT and its blockade by ketanserin.

The role of 5-HT$_{1C}$ (now 5-HT$_{2C}$) receptors in choroid plexus function has been the object of intense research, even in outer space. In the NASA Website, one can find a description of experiments proposed by Dr. J. Gabrion (Université Pierre et Marie Curie, Paris, France) aimed at studying the role of several molecules including 5-HT$_{2C}$ receptor in choroidal regulation in space/light conditions.

1.8 The Cloning of 5-HT$_{2C}$ Receptor: Bye Bye 5-HT$_{1C}$

1.8.1 The 5-HT$_{2C}$ Receptor Signal Transduction

Around that time, it became clear that 5-HT$_2$ receptors were acting via the PLC/PKC/calcium pathway (whereas 5-HT$_1$ receptors were modulating cAMP production), and since there was no evidence for cAMP modulation in the choroid plexus (Palacios et al. 1986), we searched and found that modulation: PLC, which was indeed positive for 5-HT$_{1C}$, as did others (Hoyer et al. 1989; Conn et al. 1986). We also suggested that 5-HT$_{1C}$ receptors were present in the stomach fundus and attempted to correlate both activities (5-HT$_{1C}$ binding and 5-HT-mediated contraction of the fundus). However, this is another story since it was the 5-HT$_{2B}$ receptor that is expressed in the fundus (Foguet et al. 1992a, b).

1.8.2 The Cloning of 5-HT$_{2C}$ Receptor

Since our initial studies, the pharmacological similarity between 5-HT$_{1C}$ and 5-HT$_2$ in terms of pharmacological profile was rather striking for us (Pazos et al. 1984a, b; Hoyer et al. 1985a, b). Eventually these similarities suggested these receptors to be closely linked (Hoyer 1988b); this was indeed the case when both receptors were finally cloned: 5-HT$_{1C}$ receptor was cloned first (Lubbert et al. 1987) although the full length sequence came a bit later (Julius et al. 1988), and almost simultaneously the 5-HT$_2$ receptor was also cloned (Pritchett et al. 1988).

The 5-HT$_2$-receptor family was then extended when the fundus receptor was also cloned, first named 5-HT$_{2F}$, but then renamed 5-HT$_{2B}$. Finally, the serotonin receptor nomenclature committee agreed that subtypes existed for 5-HT$_1$ and 5-HT$_2$ receptors (cloning had of course strongly supported these views)(Humphrey et al. 1993). It was then logically decided to allocate 5-HT$_{1C}$ to the 5-HT$_2$ family and 5-HT$_{1C}$ was renamed 5-HT$_{2C}$, which was least disruptive.

1.8.3 In situ Hybridization Histochemistry in Brain

The cloning of 5-HT$_{2C}$ receptors allowed for the first visualization of mRNA coding for this receptor by in situ hybridization histochemistry in brain (Mengod et al.

1990), showing that 5-HT$_{2C}$ binding and mRNA distributions were largely overlapping (Pompeiano et al. 1994) and selective 5-HT$_{2C}$ agonists and antagonists were synthesized, some of which are in clinical development (see reference Millan et al. 2003), although none of them has been demonstrated to bind to pizza (Guth 1982)! The progress in the field has been compelling, due to major advances in molecular biology, the availability of selective tools, and their judicious, occasionally "out of the box" thinking.

1.8.4   Gene Encoding the 5-HT$_{2C}$ Receptor

The gene encoding the 5-HT$_{2C}$ receptor is extraordinarily complex, which explains why it has taken quite some time to obtain the full sequence (see reference Lubbert et al. 1987). There are three splice variants of the 5-HT$_{2C}$ receptor: the full length receptor and two severely truncated forms (Canton et al. 1996), thought to be inactive. In addition, it has become clear that the primary transcript undergoes RNA editing (Burns et al. 1997), something that is unheard of on the GPCR field but was described for ionotropic glutamate receptors of the AMPA type. In rodents there are four editing sites within the coding region, whereas in humans a fifth editing site is present; together they may produce up to 32 different mRNAs and 24 different proteins. The 5-HT$_{2C}$ receptor is characterized by constitutive activity, the level of which decreases as editing increases (Herrick-Davis et al. 1999). Editing also leads to a loss of the active state of the receptor (Niswender et al. 1999) and a delay in agonist-stimulated calcium release in the fully edited isoforms (Price and Sanders-Bush 2000). The fully edited receptor couples to both Gq/11 and G13, whereas editing reduces or eliminates coupling to G13 (Price et al. 2001). Thus, editing may serve to stop constitutive activity by reducing coupling to G proteins.

1.9   A New Player in the Serotonin Field

1.9.1   Summary

We had initially noticed that [^{3}H]mesulergine, a ligand with high affinity for 5-HT$_2$ and dopamine D$_2$ receptors (Closse 1983; Enz et al. 1984; Markstein 1983), did strongly label the choroid plexus. By using binding techniques to both tissue membranes and sections, we had also shown the presence of a [^{3}H]5-HT binding site in the choroid plexus which was neither 5-HT$_{1A}$, 5-HT$_{1B}$, nor 5-HT$_{2A}$. Indeed, the choroid plexus was not labeled by [^{3}H]8-OH-DPAT (5-HT$_{1A}$) or [^{125}I]ICYP (5-HT$_{1B}$), nor was it labeled by [^{3}H]ketanserin, [^{3}H]spiperone, or classical 5-HT$_2$ or 5-HT$_2$/D$_2$ ligands. In contrast, [^{3}H]LSD and [^{125}I]LSD did strongly label the choroid plexus. We named this site 5-HT$_{1C}$, since it was also labeled by [^{3}H]5-HT: by definition [^{3}H]5-HT binding was defining 5-HT$_1$ sites (later however, it was renamed 5-HT$_{2C}$, for operational, transductional, and structural reasons as mentioned in section 1.8).

1.9.2   Posttransductional Changes

As previously mentioned, to believe that the 5-HT$_{1C}$ receptor was greeted with general acceptance is a vast overstatement. The area was complex, new, and rather hotly debated. It actually took years to be acknowledged by the 5-HT "experts": A famous 5-HT researcher at the time suggested that we go back to the books, since "there were only two 5-HT receptors (S1 and S2)," as it was well know that "there were only two dopamine receptors (D$_1$ and D$_2$)"; therefore, any additional complexity was considered superfluous. Needless to say, the statements made by eminent experts had also an influence on internal acceptance of the value of that type of research. For them the worst was still to come, and it arrived with the irruption of the molecular biology techniques. In fact, the family of serotonin receptors ended up with 14 children. The cloning of GPCRs, including that of multiple 5-HT receptors, the use of receptor KOs, and the development of selective tools/drugs facilitated the eventual acceptance of these strange binding sites as receptors. However, cloning was only to start only 3–4 years later, a long time in the career of a young scientist. The first member of the 5-HT$_2$ family to be cloned was the 5-HT$_{1C}$ receptor. Due to its transductional and structural characteristics, and according to the agreement of the Serotonin Receptor Nomenclature Committee, the site was then renamed 5-HT$_{2C}$.

1.10   A Receptor Looking for a Therapeutic Use

The clinical relevance of 5-HT$_{2C}$ receptor editing has been linked in association studies to suicidality (Niswender et al. 2001), schizophrenia (Sodhi et al. 2001), anxiety (Hackler et al. 2006), depression (Iwamoto et al. 2005), and spatial memory (Du et al. 2007). However, these data need confirmation, as is common in the field. 5-HT$_{2C}$ receptors have been shown to modulate mesolimbic dopaminergic function, where they exert a tonic inhibitory influence over dopamine neurotransmission (Di Giovanni et al. 1999; Bubar and Cunningham 2007) and, therefore, the interest in this receptor as a therapeutic target for treating abuse (Bubar and Cunningham 2006). The 5-HT$_{2C}$ receptor is also believed to mediate, in part, the effects of antidepressants, e.g., mirtazapine or agomelatine (Cremers et al. 2007), possibly by stimulating neurogenesis, as well as that of atypical antipsychotics (Herrick-Davis et al. 2000). 5-HT$_{2C}$ receptors are expressed in the amygdala, and functional magnetic resonance imaging fMRI data have demonstrated that 5-HT$_{2C}$ receptor agonists produce its neuronal activation (Hackler et al. 2007). Other potential indications relate to obesity and epilepsy (Tecott et al. 1995; Tecott and Abdallah 2003), but it will be a challenge to produce selective agonists for the 5-HT$_{2C}$ receptor (Bonhaus et al. 1997) that do not interact with the other 5-HT$_2$ subtypes, especially 5-HT$_{2B}$, which has serious liabilities (see reference Fitzgerald et al. 2000).

   In any case, 5-HT$_{2C}$ receptors have already made a relevant contribution to the modern pharmacology: The development of selective 5-HT$_{2C}$ receptor drugs has been pioneer in the demonstration of the capability of certain ligands to differentially activate different signal transduction pathways (mainly inositol phosphate accumulation versus arachidonic acid release) (Berg et al. 2003). This evidence has been instrumental for the concept of "ligand-dependent functional selectivity." In addition to challenging the dogma of classical pharmacology, this concept will have a clear impact on drug discovery (Millan et al. 2003).

Acknowledgments   We thank the members of the JMP and DH labs (M. Rigo, R. Lehnherr, M. Girod, E. Schuepbach, D. Fehlmann), who were instrumental in training and supporting our PhD students and postdocs active in the labs in the 5-HT$_{1C}$ times (C. Waeber, R. Cortés, M. Dietl, A. Karpf, H. Davies, S. Srivatsa, A. Bruinvels, P. Schoeffter, H. Neijt, V. Doyle, J. Creba, I. Sahin-Erdemli), which in turn contributed in discussions, shared lab and counter spaces, and helped in membrane preparations and some other atypical experiments. We also need to thank our colleagues (G. Engel, K.H. Buchheit, R. Markstein, H. Kalkman, H.G.W.M. Boddeke, A. Enz, A. Fargin, A.K. Dixon, A. Closse, E. Mueller-Schweinitzer, U. Ruegg, K.H. Wiederhold, B.P. Richardson, J.R. Fozard, M.P. Seiler, G. Mengod, E. Schlicker, M. Goethert, A. Probst, D. Middlemiss as well as various members of the 5-HT receptor nomenclature committee) for help and discussing our data in a critical and constructive manner. Finally, we thank Daniel Hauser and Manfred Karobath heads of Preclinical Research at Sandoz, which supported us all along those years.

References

Bennett MR (2000) The concept of transmitter receptors: 100 years on. Neuropharmacology 39:523–546.

Berg KA, Cropper JD, King BD, et al (2003) Effector pathway-dependence of ligand-independent 5-HT$_{2C}$ receptor activity. FASEB J 17:A1021.

Bonhaus DW, Weinhardt KK, Taylor M, et al (1997) RS-102221: a novel high affinity and selective, 5-HT$_{2C}$ receptor antagonist. Neuropharmacology 36:621–629.

Bradley PB, Engel G, Feniuk W, et al (1986) Proposals for the classification and nomenclature of functional receptors for 5-hydroxytryptamine. Neuropharmacology 25:563–576.

Bubar MJ, Cunningham KA (2006) Serotonin 5-HT2A and 5-HT$_{2C}$ receptors as potential targets for modulation of psychostimulant use and dependence. Curr Top Med Chem 6:1971–1985.

Bubar MJ, Cunningham KA (2007) Distribution of serotonin 5-HT$_{2C}$ receptors in the ventral tegmental area. Neuroscience 146:286–297.

Burns CM, Chu H, Rueter SM, et al (1997) Regulation of 5-HT$_{2C}$ receptor G-protein coupling by RNA editing [see comments]. Nature 387:303–308.

Canton H, Emeson RB, Barker EL, et al (1996) Identification, molecular cloning, and distribution of a short variant of the 5-hydroxytryptamine2C receptor produced by alternative splicing. Mol Pharmacol 50:799–807.

Closse A (1983) [^{3}H]Mesulergine, a selective ligand for serotonin-2 receptors. Life Sci 32:2485–2495.

Conn PJ, Sanders-Bush E, Hoffman BJ, et al (1986) A unique serotonin receptor in choroid plexus is linked to phosphatidylinositol turnover. Proc Natl Acad Sci USA 83:4086–4088.

Cremers TI, Rea K, Bosker FJ, et al (2007) Augmentation of SSRI effects on serotonin by 5-HT$_{2C}$ antagonists: mechanistic studies. Neuropsychopharmacology 32:1550–1557.

Di Giovanni G, De Deurwaerdère P, Di Mascio M, et al (1999) Selective blockade of serotonin2C/2B receptors enhances mesolimbic and mesostriatal dopaminergic function: a combined in vivo electrophysiology and microdialysis study. Neuroscience 91:587–597.

Du Y, Stasko M, Costa AC, et al (2007) Editing of the serotonin 2C receptor pre-mRNA: effects of the Morris Water Maze. Gene 391:186–197.

Engel G, Hoyer D, Berthold R, et al (1981) (+/−)[[125]Iodo] cyanopindolol, a new ligand for beta-adrenoceptors: identification and quantitation of subclasses of beta-adrenoceptors in guinea pig. Naunyn Schmiedebergs Arch Pharmacol 317:277–285.

Engel G, Hoyer D, Kalkman HO, et al (1984) Identification of 5HT2-receptors on longitudinal muscle of the guinea pig ileum. J Recept Res 4:113–126.

Engel G, Göthert M, Hoyer D, et al (1986) Identity of inhibitory presynaptic 5-hydroxytryptamine (5-HT) autoreceptors in the rat brain cortex with 5-HT1B binding sites. Naunyn Schmiedebergs Arch Pharmacol 332:1–7.

Enz A, Donatsch P, Nordmann R (1984) Dopaminergic properties of mesulergine (CU 32-085) and its metabolites. J Neural Transm 60:225–238.

Fitzgerald LW, Burn TC, Brown BS, et al (2000) Possible role of valvular serotonin 5-HT(2B) receptors in the cardiopathy associated with fenfluramine. Mol Pharmacol 57:75–81.

Foguet M, Hoyer D, Pardo LA, et al (1992) Cloning and functional characterization of the rat stomach fundus serotonin receptor. EMBO J 11:3481–3487.

Foguet M, Nguyen H, Le H, Lubbert H (1992) Structure of the mouse 5-HT1C, 5-HT2 and stomach fundus serotonin receptor genes. NeuroReport 3:345–348.

Gaddum JH, Picarelli ZP (1957) Two kinds of tryptamine receptor. Br J Pharmacol 12:323–328.

Gozlan H, el Mestikawy S, Pichat L, et al (1983) Identification of presynaptic serotonin autoreceptors using a new ligand: 3H-PAT. Nature 305:140–142

Guth PS (1982) The structurally specific, stereospecific, saturable binding of pepperoni to pizza. Trends Pharmacol Sci 3:467.

Hackler EA, Airey DC, Shannon CC, et al (2006) 5-HT(2C) receptor RNA editing in the amygdala of C57BL/6 J, DBA/2 J, and BALB/cJ mice. Neurosci Res 55:96–104.

Hackler EA, Turner GH, Gresch PJ, et al (2007) 5-Hydroxytryptamine2C receptor contribution to m-chlorophenylpiperazine and N-methyl-beta-carboline-3-carboxamide-induced anxiety-like behavior and limbic brain activation. J Pharmacol Exp Ther 320:1023–1029.

Hartig PR, Hoyer D, Humphrey PP, et al (1996) Alignement of receptor nomenclature with the human genome: classification of 5-HT1B and 5-HT1D receptor subtypes. Trends Pharmacol Sci 17:103–105.

Herrick-Davis K, Grinde E, Niswender CM (1999) Serotonin 5-HT$_{2C}$ receptor RNA editing alters receptor basal activity: implications for serotonergic signal transduction. J Neurochem 73:1711–1717.

Herrick-Davis K, Grinde E, Teitler M (2000) Inverse agonist activity of atypical antipsychotic drugs at human 5-hydroxytryptamine2C receptors. J Pharmacol Exp Ther 295:226–232.

Hjorth S, Carlsson A, Lindberg P, et al (1982) 8-hydroxy-2-(di-n-propylamino)tetralin, 8-OH-DPAT, a potent and selective simplified ergot congener with central 5-HT-receptor stimulating activity. J Neural Transm 55:169–188.

Hoyer D (1988) Molecular pharmacology and biology of 5-HT1C receptors. Trends Pharmacol Sci 9:89–94.

Hoyer D (1988) Molecular pharmacology and biology of 5-HT1C receptors. Trends Pharmacol Sci 9:89–94.

Hoyer D, Middlemiss DN (1989) Species differences in the pharmacology of terminal 5-HT autoreceptors in mammalian brain. Trends Pharmacol Sci 10:130–132.

Hoyer D, Engel G, Berthold R (1982) Binding characteristics of (+)-, (+/−)- and (−)-[[125]iodo] cyanopindolol to guinea-pig left ventricle membranes. Naunyn Schmiedebergs Arch Pharmacol 318:319–329.

Hoyer D, Engel G, Kalkman HO (1985) Molecular pharmacology of 5-HT1 and 5-HT2 recognition sites in rat and pig brain membranes: radioligand binding studies with [^{3}H]5-HT, [^{3}H]8-OH-DPAT, (−)[^{125}I]iodocyanopindolol, [^{3}H]mesulergine and [^{3}H]ketanserin. Eur J Pharmacol 118:13–23.

Hoyer D, Engel G, Kalkman HO (1985) Characterization of the 5-HT1B recognition site in rat brain: binding studies with ^{125}I-iodocyanopindolol. Eur J Pharmacol 118:1–12.

Hoyer D, Pazos A, Probst A, et al (1986) Serotonin receptors in the human brain. II. Characterization and autoradiographic localization of 5-HT1C and 5-HT2 recognition sites. Brain Res 376:97–107.

Hoyer D, Srivatsa S, Pazos A, et al (1986) [^{125}I]LSD labels 5-HT1C recognition sites in pig choroid plexus membranes. Comparison with [^{3}H]mesulergine and [^{3}H]5-HT binding. Neurosci Lett 69:269–274.

Hoyer D, Pazos A, Probst A, et al (1986) Serotonin receptors in the human brain. I. Characterization and autoradiographic localization of 5-HT1A recognition sites. Apparent absence of 5-HT1B recognition sites. Brain Res 376:85–96.

Hoyer D, Pazos A, Probst A, et al (1986) Serotonin receptors in the human brain. II. Characterization and autoradiographic localization of 5-HT1C and 5-HT2 recognition sites. Brain Res 376:97–107.

Hoyer D, Waeber C, Schoeffter P, et al (1989) 5-HT1C receptor-mediated stimulation of inositol phosphate production in pig choroid plexus. A pharmacological characterization. Naunyn Schmiedebergs Arch Pharmacol 339:252–258.

Hoyer D, Clarke DE, Fozard JR, et al (1994) International Union of Pharmacology classification of receptors for 5- hydroxytryptamine (Serotonin). Pharmacol Rev 46:157–203.

Humphrey PPA, Hartig P, Hoyer D (1993) A proposed new nomenclature for 5-HT receptors. Trends Pharmacol Sci 14:233–236.

Iwamoto K, Nakatani N, Bundo M, et al (2005) Altered RNA editing of serotonin 2C receptor in a rat model of depression. Neurosci Res 53:69–76.

Julius D, MacDermott AB, Axel R, et al (1988) Molecular characterization of a functional cDNA encoding the serotonin 1c receptor. Science 241:558–564.

Kalkman HO, Engel G, Hoyer D (1986) Inhibition of 5-carboxamidotryptamine-induced relaxation of guinea-pig ileum correlates with [^{125}I]LSD binding. Eur J Pharmacol 129:139–145.

Laduron PM (1984) Criteria for receptor sites in binding studies. Biochem Pharmacol 33:833–839.

Lefkowitz RJ (2004) Historical review: a brief history and personal retrospective of seven-transmembrane receptors. Trends Pharmacol Sci 25:413–422.

Lindvall-Axelsson M, Mathew C, Nilsson C, et al (1988) Effect of 5-hydroxytryptamine on the rate of cerebrospinal fluid production in rabbit. Exp Neurol 99:362–368.

Lubbert H, Hoffman BJ, Snutch TP, et al (1987) cDNA cloning of a serotonin 5-HT1C receptor by electrophysiological assays of mRNA-injected Xenopus oocytes. Proc Natl Acad Sci USA 84:4332–4336.

Markstein R (1983) Mesulergine and its 1,20-N,N-bidemethylated metabolite interact directly with D1- and D2-receptors. Eur J Pharmacol 95:101–107.

Markstein R, Hoyer D, Engel G (1986) 5-HT1A-receptors mediate stimulation of adenylate cyclase in rat hippocampus. Naunyn Schmiedebergs Arch Pharmacol 333:335–341.

Mengod G, Nguyen H, Le H, et al (1990) The distribution and cellular localization of the serotonin 1C receptor mRNA in the rodent brain examined by in situ hybridization histochemistry. Comparison with receptor binding distribution. Neuroscience 35:577–591.

Middlemiss DN, Fozard JR (1983) 8-Hydroxy-2-(di-n-propylamino)-tetralin discriminates between subtypes of the 5-HT1 recognition site. Eur J Pharmacol 90:151–153.

Millan MJ, Gobert A, Lejeune F, et al (2003) The novel melatonin agonist agomelatine (S20098) is an antagonist at 5-hydroxytryptamine2C receptors, blockade of which enhances the activity of frontocortical dopaminergic and adrenergic pathways. J Pharmacol Exp Ther 306:954–964.

Niswender CM, Copeland SC, Herrick-Davis K, et al (1999) RNA editing of the human serotonin 5-hydroxytryptamine 2C receptor silences constitutive activity. J Biol Chem 274:9472–9478.

Niswender CM, Herrick-Davis K, Dilley GE, et al (2001) RNA editing of the human serotonin 5-HT$_{2C}$ receptor. Alterations in suicide and implications for serotonergic pharmacotherapy. Neuropsychopharmacology 24:478–491.

Palacios JM, Markstein R, Pazos A (1986) Serotonin-1C sites in the choroid plexus are not linked in a stimulatory or inhibitory way to adenylate cyclase. Brain Res 380:151–154.

Pazos A, Palacios JM (1985) Quantitative autoradiographic mapping of serotonin receptors in the rat brain. I. Serotonin-1 receptors. Brain Res 346:205–230.

Pazos A, Hoyer D, Palacios JM (1984) The binding of serotonergic ligands to the porcine choroid plexus: characterization of a new type of serotonin recognition site. Eur J Pharmacol 106:539–546.

Pazos A, Hoyer D, Palacios JM (1984a) Mesulergine, a selective serotonin-2 ligand in the rat cortex, does not label these receptors in porcine and human cortex: evidence for species differences in brain serotonin-2 receptors. Eur J Pharmacol 106:531–538.

Pazos A, Cortés R, Palacios JM (1985) Quantitative autoradiographic mapping of serotonin receptors in the rat brain. II. Serotonin-2 receptors. Brain Res 346:231–249.

Pazos A, Probst A, Palacios JM (1987) Serotonin receptors in the human brain. III. Autoradiographic mapping of serotonin-1 receptors. Neuroscience 21:97–122.

Pazos A, Probst A, Palacios JM (1987) Serotonin receptors in the human brain. IV. Autoradiographic mapping of serotonin-2 receptors. Neuroscience 21:123–139.

Pedigo NW, Yamamura HI, Nelson DL (1981) Discrimination of multiple [^{3}H]5-hydroxytriptamine-binding sites by the neuroleptic spiperone in rat brain. J Neurochem 36:220–226.

Peroutka SJ, Snyder SH (1979) Multiple serotonin receptors: differential binding of [^{3}H]5-hydroxytryptamine, [^{3}H]lysergic acid die ethylamide and [^{3}H]spiroperidol. Mol Pharmacol 16:687–699.

Pompeiano M, Palacios JM, Mengod G (1994) Distribution of the serotonin 5-HT2 receptor family mRNAs: comparison between 5-HT2A and 5-HT$_{2C}$ receptors. Brain Res Mol Brain Res 23:163–178.

Price RD, Sanders-Bush E (2000) RNA editing of the human serotonin 5-HT(2C) receptor delays agonist-stimulated calcium release. Mol Pharmacol 58:859–862.

Price RD, Weiner DM, Chang MS, et al (2001) RNA editing of the human serotonin 5-HT$_{2C}$ receptor alters receptor-mediated activation of G13 protein. J Biol Chem 276:44663–44668.

Pritchett DB, Bach AW, Wozny M, et al (1988) Structure and functional expression of cloned rat serotonin 5HT-2 receptor. EMBO J 7:4135–4140.

Schoeffter P, Hoyer D (1989) 5-Hydroxytryptamine 5-HT1B and 5-HT1D receptors mediating inhibition of adenylate cyclase activity. Pharmacological comparison with special reference to the effects of yohimbine, rauwolscine and some beta-adrenoceptor antagonists. Naunyn Schmiedebergs Arch Pharmacol 340:285–292.

Schoeffter P, Hoyer D (1989) 5-Hydroxytryptamine 5-HT1B and 5-HT1D receptors mediating inhibition of adenylate cyclase activity. Pharmacological comparison with special reference to the effects of yohimbine, rauwolscine and some beta-adrenoceptor antagonists. Naunyn Schmiedebergs Arch Pharmacol 340:285–292.

Sodhi MS, Burnet PW, Makoff AJ, et al (2001) RNA editing of the 5-HT(2C) receptor is reduced in schizophrenia. Mol Psychiatry 6:373–379.

Tecott LH, Abdallah L (2003) Mouse genetic approaches to feeding regulation: serotonin 5-HT$_{2C}$ receptor mutant mice. CNS Spectr 8:584–588.

Tecott LH, Sun LM, Akana SF, et al (1995) Eating disorder and epilepsy in mice lacking 5-HT$_{2c}$ serotonin receptors [see comments]. Nature 374:542–546.

Chapter 2
Serotonin 5-HT$_{2C}$ Receptors: Chemical Neuronatomy in the Mammalian Brain

Guadalupe Mengod

2.1   Introduction

Serotonin 5-HT$_{2C}$ receptors belong to the 5-HT$_2$ family, which includes 5-HT$_{2A}$ and 5-HT$_{2B}$ receptors. All three share similarities in their molecular structure, pharmacology, and signal transduction pathways (Barnes and Sharp 1999; Hoyer et al. 1994). Initially named 5-HT$_{1C}$ based on the conventions for naming serotonin receptors at the time of its discovery, it was later renamed 5-HT$_{2C}$ receptors after the cloning of its gene and that of 5-HT$_{2A}$ (Pazos et al. 1984a; Prichett et al. 1988; Julius et al. 1988, 1990; Hoyer et al. 1985; Lubbert et al. 1987; Pazos and Palacios 1985) (see also Palacios et al., Chap. 1, this volume). Chemical neuroanatomical techniques were pivotal in the discovery of 5-HT$_{2C}$ receptors, which were first identified by autoradiography after labeling of rat brain sections with the 5-HT$_{2A}$ and dopamine D$_2$ receptor ligand mesulergine. This was followed by a thorough pharmacological characterization performed in pig brain choroid plexus, and the binding site was differentiated from the 5-HT$_{2A}$ receptor. The combined use of membrane receptor binding/pharmacology and brain slice autoradiography allowed its differentiation from 5-HT$_{1A}$, 5-HT$_{1B}$, and 5-HT$_{2A}$ receptors; it was then named 5-HT$_{1C}$. The pharmacology and distribution of the new site differed from the existing knowledge about other receptors (Pazos et al. 1984a, b; Hoyer et al. 1985). The 5-HT$_{1C}$ receptor was cloned soon thereafter (Julius et al. 1988). The knowledge of the messenger ribonucleic acid (mRNA) sequence and the derived protein sequence of these receptors allowed the development of new important tools for the study of their chemical neuroanatomy.

The autoradiographic localization of 5-HT$_{2C}$ receptors had suffered from the lack of selective ligands. While mesulergine remains the ligand of choice, 5HT$_{2C}$ sites can also be labeled by other ligands such as 5-HT self, LSD (lysergic acid diethylamide), and DOI (1-(2,5-dimethoxy-4-iodophenyl-2-aminopropane), but

G. Mengod (✉)
Departamento Neuroquímica y Neurofarmacología, IIBB/CSIC-IDIBAPS-CIBERNED, Barcelona, Spain
e-mail: guadalupe.mengod@iibb.csic.es

G. Di Giovanni et al. (eds.), *5-HT$_{2C}$ Receptors in the Pathophysiology of CNS Disease*, The Receptors 22, DOI 10.1007/978-1-60761-941-3_2,
© Springer Science+Business Media, LLC 2011

always in combination with unlabeled ligands to block additional sites labeled by these ligands. The ability to combine in situ hybridization with radioligand binding autoradiography has allowed the establishment of the anatomical distribution of 5-HT$_{2C}$ receptors in the brain of many animal species by different anatomical and pharmacological manipulations (Pazos and Palacios 1985; Eberle-Wang et al. 1997; Hoffman and Mezey 1989; Mengod et al. 1990a, b; Molineaux et al. 1989; Pompeiano et al. 1994). Immunohistochemistry with antibodies against parts of 5-HT$_{2C}$ receptors has also been used to visualize the receptor protein (Clemett et al. 2000; Abramowski et al. 1995).

In this chapter I will review some of the main findings concerning the anatomical and cellular distribution of 5-HT$_{2C}$ receptors in the brain of rodents, primates, and humans with special emphasis on recent studies on the characterization of the phenotype of the brain cells expressing these receptors and the significance of these findings for the understanding the role of the 5-HT$_{2C}$ receptors in brain function.

2.2   5-HT$_{2C}$ Receptors: Neuroanatomical Localization by Radioligand Binding Autoradiography, In situ Hybridization, and Immunohistochemistry in the Rodent Brain

5-HT$_{2C}$ receptor localization is restricted to the central nervous system (CNS), unlike that of 5-HT$_{2A}$ and 5-HT$_{2B}$ receptors. Radioligands that label 5-HT$_{2C}$ receptors are: [^{3}H]mesulergine (in the presence of a selective 5-HT$_{2A}$ antagonist), [^{3}H]5-HT (with adequate protection with a cocktail of 5-HT$_1$ ligands), [^{125}I]SCH23982 (also dopamine D1), and [^{125}I]LSD (in the presence of adequate 5-HT$_{2A}$ selective drugs). The remarkable concentration of 5-HT$_{2C}$ receptors in the mammalian choroid plexus somehow obscures its presence throughout the CNS. Autoradiographic studies have identified this receptor in anterior olfactory nucleus, olfactory tubercle, lateral amygdaloid nucleus, cortex, nucleus accumbens, hippocampus, amygdala, caudate, and substantia nigra in addition to the choroid plexus in rat brain (Pazos and Palacios 1985). 5-HT$_{2C}$ receptor mRNA is very abundant in the pyramidal cell layer of the ventral and posterior part of CA (cornu ammonis) 1 and CA2 fields of the hippocampus and of the anterior part of the CA3, while it is very low in the pyramidal cell layer of the dorsal and anterior region of CA1 and CA2, posterior part of CA3, and the granule cell layer of the dentate gyrus. 5-HT$_{2C}$ binding sites show a regional distribution in agreement with that of the mRNA, being concentrated on the pyramidal cell layer of CA1 and CA2 at ventral levels and the granule cell layer of the dentate gyrus. However, receptors are also seen in the stratum lacunosum molecular of the CA1 and CA3 fields at anterior and dorsal level (Palacios et al. 1991). These localizations are illustrated in Fig. 2.1.

The distribution of [^{3}H]mesulergine binding sites in rat brain (Pazos and Palacios 1985) are very similar to those found in mouse brain (Mengod et al. 1990a). In mouse brain, [^{3}H]mesulergine binding sites (in the presence of spiperone to block

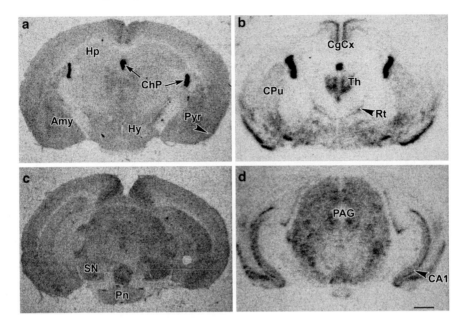

Fig. 2.1   5-HT$_{2C}$ receptors in the mouse brain. Regional distribution of [³H]mesulergine binding sites and 5-HT$_{2C}$ receptor mRNA in adjacent sections of the mouse brain. (**a**) and (**c**) Receptor binding sites labeled by 5 nmol/L [³H]mesulergine. (**b**) and (**d**) Hybridization signal obtained with a ³³P-labeled oligonucleotide probe complementary to the mRNA encoding 5-HT$_{2C}$ receptors. Pictures are digital photographs from film autoradiograms. *Amy* indicates amygdala, *CA1* CA1 field of the hippocampus, *CgCx* cingulate cortex, *ChP* choroid plexus, *CPu* caudate-putamen, *Hp* hippocampus, *Hy* hypothalamus, *PAG* periaqueductal gray, *Pn* pontine nuclei, *Pyr* pyriform cortex, *Rt* reticular nucleus of the hypothalamus, *SN* substantia nigra. Scale bar: 1 mm

binding of the radioligand to 5-HT$_{2A}$ receptors) (Mengod et al. 1990a) are present at high densities in choroid plexus, where it is predominant, although the presence of low/very low specific [³H]mesulergine signals can be detected in nucleus accumbens, patches of the caudate putamen, olfactory tubercle, claustrum, septum, cingular cortex, amygdala, dentate gyrus, periaqueductal gray, entorhinal cortex, and several brainstem motor nuclei. This binding is not detected in the 5-HT$_{2C}$ receptor knockout (KO) mouse brain (López-Giménez et al. 2002), indicating that 5-HT$_{2C}$ receptors are indeed present in the brain although at much lower densities than 5-HT$_{2A}$ receptors (with the remarkable exception of the choroid plexus). The distribution of mRNA is very similar to that of protein or binding sites, except for high levels in the habenular nucleus; where binding site levels are very low (Mengod et al. 1990a; López-Giménez et al. 2001a). There are multiple splice and editing variants of 5-HT$_{2C}$ receptors (Fitzgerald et al. 1999; Niswender et al. 1998), which are beyond the scope of this chapter; they are not discriminated, so far as is known, by the antagonist radioligands used in autoradiographic studies.

In monkey brain, 5-HT$_{2C}$ mRNA is present in choroid plexus, in layer V of most cortical regions, and in nucleus accumbens, ventral anterior caudate and putamen, septal nuclei, diagonal band, ventral striatum, and extended amygdala (López-Giménez et al. 2001a). Several thalamic, midbrain, and brainstem nuclei also contain 5-HT$_{2C}$ mRNA. [^{3}H]Mesulergine binding and mRNA show a good correlation across the brain supporting a predominant somatodendritic localization of 5-HT$_{2C}$ receptors. However, in a few instances, a lack of correlation between both patterns of signal suggests a possible location of 5-HT$_{2C}$ receptors on axon terminals. Examples of poor correlation are the septal nuclei and horizontal limb of the diagonal band (presence of mRNA with apparent absence of binding sites) and interpeduncular nucleus (presence of binding sites with apparent absence of mRNA).

2.3 Species Differences in 5-HT$_{2C}$ Receptors Distribution: Rodent Versus Human and Nonhuman Primates

The pharmacological profile for 5-HT$_{2C}$ receptors is very similar for human, pig and rat (Hoyer et al. 1985, 1986; Pazos et al. 1984b, 1987) both radioligand binding and autoradiographic procedures in frontal cortex, hippocampus, and choroid plexus of these species using [^{3}H]mesulergine. The distribution of 5-HT$_{2C}$ receptor binding sites in the human brain is somewhat different from that found in the rat brain (Pazos and Palacios 1985; Hoyer et al. 1986). In the rat hippocampus, 5-HT$_{2C}$ receptor binding sites are located in the stratum lacunosum molecular, whereas in the human hippocampus the pyramidal layer is enriched in these receptors.

There are also differences in [^{3}H]mesulergine binding sites in human brain when compared with monkey (*Macaca fascicularis*) brain. High densities of binding sites are observed in the human globus pallidus and substantia nigra (Pazos et al. 1987), whereas they are absent in monkey globus pallidus and low in substantia nigra (López-Giménez et al. 2001a). In monkey neocortex, low levels of [^{3}H]mesulergine binding sites are detected on layer V, whereas in human cortical areas binding sites are located predominately in layer III.

Although the distribution of 5-HT$_{2C}$ receptor mRNA in monkey brain (López-Giménez et al. 2001a) is very similar to that in rat (Eberle-Wang et al. 1997; Pompeiano et al. 1994; Wright et al. 1995), mouse (Mengod et al. 1990a), and human brain (Pasqualetti et al. 1999), there are some differences. In the neocortex of mouse and rat the 5-HT$_{2C}$ receptor mRNA are found at detectable levels only in prefrontal, cingulate, and retrosplenial cortices, whereas in monkey, mRNA is present in all neocortical areas except in the calcarine sulcus within the occipital cortex. The CA3 subfield of the hippocampus is another region of divergence: rat, mouse, and human brain contain this mRNA, and no signal is found in the monkey CA3. The rat entopeduncular nucleus contains 5-HT$_{2C}$ receptor mRNA, whereas its equivalent in primate, the internal segment of the globus pallidus, is devoid of it. In the striatum, the hybridization signal is uniformly intense in human, rat, and mouse brain, whereas in monkey brain this signal is not uniform and is restricted to ventral aspects of the anterior striatum. The substantia nigra is another brain region that presents differences in the distribution of 5-HT$_{2C}$ receptor mRNA among the species.

In human brain the presence of this mRNA is detected predominantly in the pars compacta of this nucleus, whereas monkey brain cells showing the hybridization signal were confined in the lateral part of substantia nigra. As in the rodent brain, there is in general a good correlation between mRNA and binding sites in human and monkey brains with the exceptions identified above. Some of these exceptions are now discussed in relationship to the cellular localization of these receptors.

2.4   Phenotype of Cells Expressing 5-HT$_{2C}$ Receptors

The widespread distribution of 5-HT$_{2C}$ receptors in the brain of the mammalian species studied until now suggests, as is the case for other neurotransmitter receptors, that 5-HT$_{2C}$ receptors are expressed by neurons with different neurotransmitter phenotypes. In this section evidence is presented that suggests the presence of these receptors in neuropeptidergic, cholinergic, serotonergic, and GABAergic neurons, as well as recent studies showing the interaction of 5-HT$_{2C}$ receptors with the cannabinoid and the dopaminergic systems.

2.4.1   5-HT$_{2C}$ Receptors and Neuropeptidergic Neurons

In nucleus accumbens and striatum, 5-HT$_{2C}$ receptor mRNA was found localized with each of the neuropeptides (enkephalin, substance P, and dynorphin) as shown by Ward and Dorsa (Ward and Dorsa 1996). The level of colocalization was similar among the three neuropeptides but varied by region: high levels of colocalization were observed (from 64% to 89%) ventrally, medially and scattered in patches with high expression of the receptor in the striatum, whereas lower levels of colocalization (43–54%) were observed in matrix-like areas of lower receptor expression. According to the authors this colocalization could provide an anatomical basis for earlier observations that alterations in serotonergic input can lead to changes in the levels of striatal neuropeptides (Kondo et al. 1993).

2.4.2   5-HT$_{2C}$ Receptors and Serotonergic or GABAergic Neurons

The cellular localization of 5-HT$_{2C}$ receptor mRNA in relation to serotonergic and GABAergic neurons has been studied in the anterior raphe nuclei of the rat (Serrats et al. 2005). In the dorsal and median raphe nuclei, 5-HT$_{2C}$ receptor mRNA is not detected in serotonergic cells identified as those expressing serotonin (5-HT) transporter mRNA. In contrast, 5-HT$_{2C}$ receptor mRNA is found in the majority of GABAergic cells of the anterior raphe nuclei, mainly located in the lateral and inter-mediolateral parts of the dorsal raphe and lateral part of the median raphe, supporting previous hypotheses that proposed a negative-feedback loop involving reciprocal connections between GABAergic interneurons bearing 5-HT$_{2A/2C}$ receptors and 5-HT neurons in the dorsal raphe and surrounding areas. According to this model, the excitation of GABAergic interneurons through these 5-HT$_{2C}$ (and also 5-HT$_{2A}$) receptors would result in the suppression of 5-HT cell firing.

The finding of 5-HT$_{2C}$-immunoreactive cells in the raphe nuclei has led to the proposal that some 5-HT neurons might express these receptors (Clemett et al. 2000). In contrast, electrophysiological data suggest that 5-HT$_{2C}$ receptors are located on local GABAergic neurons, inside or close to the dorsal raphe (Liu et al. 2000), being part of a local negative-feedback circuit that would involve reciprocal connections between GABAergic and 5-HT neurons. This model has been proposed to explain the increases in the frequency of inhibitory postsynaptic currents (IPSCs) induced by 5-HT and the 5-HT$_{2A/2C}$ agonist DOI [1-(2,5)-dimethoxy-4-io-dophenyl-2-aminopropane] when applied to rat brain slices containing the dorsal raphe.

The localization of the 5-HT$_{2C}$ receptor in GABAergic cells has been also described in other brain areas. Immunohistochemical analyses on the 5-HT$_{2C}$ receptor reveals that this receptor is mainly expressed in deep layers of the rat medial prefrontal cortex (Liu et al. 2007) and cortex (Abramowski et al. 1995) in agreement with the presence of the mRNA coding for this receptor in layers IV and V of PFC of mice (Mengod et al. 1990a), rat (Pompeiano et al. 1994), monkey (López-Giménez et al. 2001a), and human (Pasqualetti et al. 1999). Around 50% of the neurons expressing 5-HT$_{2C}$ receptor immunoreactivity in the prelimbic region of the medial prefrontal cortex also expressed GAD67 immunoreactivity (Liu et al. 2007), a marker of GABAergic interneurons. We have detected abundant expression of 5-HT$_{2C}$ receptor mRNA in layer V cells of the mouse cingulate cortex that were not GABAergic. A detail of these results is shown in Fig. 2.2.

2.4.3   5-HT$_{2C}$ Receptors and Cholinergic Neurons

In our studies on the distribution of 5-HT$_{2C}$ receptor mRNA in the Macaca brain (López-Giménez et al. 2001a) we remarked that several regions where cholinergic cell groups are located also contained mRNA for 5-HT$_{2C}$ receptor. These regions of codistribution include several forebrain areas [medial septal nucleus (cholinergic group Ch1), vertical nucleus of diagonal band (Ch2), horizontal nucleus of diagonal band (Ch3), and nucleus basalis of Meynert (Ch4)], several mesencephalic nuclei [pedunculopontine nucleus (Ch5), laterodorsal tegmental nucleus (Ch6), parabigeminal nucleus (Ch8), oculomotor and trochlear nuclei], and motor nuclei of the brainstem cranial nerve. This correspondence is also observed between the distribution of 5-HT$_{2A}$ receptor mRNA (López-Giménez et al. 2001b) and several mesencephalic and brainstem cholinergic cell groups, particularly in the latter region where the different cranial nerve motor nuclei are highly enriched in both 5-HT$_{2A}$ receptor mRNA and ChAT mRNA.

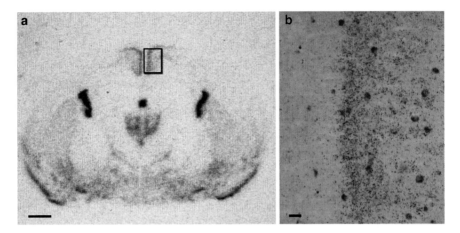

Fig. 2.2 Cellular visualization of 5-HT$_{2C}$ receptors mRNA in the cingulate cortex. (**a**) Macroscopic visualization of 5-HT$_{2C}$ receptor mRNA in the mouse coronal section. The inset in (**a**), corresponding to the cingulate cortex, is shown at higher magnification in (**b**). (**b**) Cellular localization of 5-HT$_{2C}$ receptors mRNA (labeled with ^{33}P, black silver grains) in the cells of cingulate cortical layers. GABAergic cells (GAD mRNA expressing cells seen as a brown precipitate) do not display 5-HT$_{2C}$ receptor mRNA hybridization signal. Scale bars: (**a**) 1 mm; (**b**) 20 μm

In fact, the interaction of cholinergic and serotonergic systems has been extensively studied, especially those aspects relating to the modulation of central cholinergic function by serotonin and its possible cognitive implications (See the review in Cassel and Jeltsch (1995)). Regarding 5-HT$_{2C}$ receptors and cholinergic function, several microdialysis studies carried out in rat brain showed the effect of 1-(3-chlorophenyl)piperazine (mCPP) (an unselective 5-HT$_{2C}$ receptor agonist) on the release of acetylcholine in rat cortex (Zhelyazkova-Savova et al. 1997) and hippocampus (Zhelyazkova-Savova et al. 1999). This effect consisted of an increase of acetylcholine release, which was shown to be mediated by 5-HT$_{2C}$ receptors, especially in the case of the cortex, providing corroborating evidence that the effect was produced particularly via the nucleus basalis magnocellularis (Zhelyazkova-Savova et al. 1997).

2.4.4   5-HT$_{2C}$ Receptors and the Cannabinoid Receptors System

The interaction of 5-HT$_{2C}$ receptors with the cannabinoid system through the CB$_1$ receptor has been recently studied (Aso et al. 2009). CB$_1$ KO mice exhibited a reduction in the expression of the 5-HT$_{2C}$ receptor in dorsal raphe, nucleus accumbens, and paraventricular nucleus, among other brain areas. In contrast, 5-HT$_{2C}$ receptor expression was higher in the CA3 field of the ventral hippocampus of CB$_1$ KO mice, suggesting different roles of this receptor in these brain areas. The decreased expression of the 5-HT$_{2C}$ receptor in the dorsal raphe of CB$_1$ mutant mice could lead to a reduction in the inhibitory effect exerted by the 5-HT$_{2C}$ receptor on 5-HT neurons through the GABAergic mechanism (Boothman et al. 2006), which supports the increased 5-HT extracellular levels in the brain areas receiving projections from the dorsal raphe observed in CB$_1$ KO mice. Likewise, the decreased levels of 5-HT$_{2C}$ mRNA in CB$_1$ KO mice, observed within the nucleus accumbens and the paraventricular nucleus of the hypothalamus, could indicate a diminished capacity of this receptor to inhibit dopamine activity (Dremencov et al. 2006) and to stimulate corticotrophin-releasing factor release (Heisler et al. 2007), respectively.

2.4.5   5-HT$_{2C}$ Receptors and the Dopaminergic Receptors System

Dopaminergic nuclei, such as retrorubral area, substantia nigra pars compacta, ventral tegmental area and periaqueductal gray, and dorsal striatum and nucleus accumbens, express 5-HT$_{2C}$ receptor mRNA (Eberle-Wang et al. 1997; Mengod et al. 1990a; Pompeiano et al. 1994; Ward and Dorsa 1996). Pharmacological activation of 5-HT$_{2C}$ receptors inhibits firing rates of ventral tegmental area neurons and dopamine release within the nucleus accumbens (Prisco et al. 1994; Di Giovanni et al. 1999; Di Matteo et al. 1998). The implication of 5-HT$_{2C}$ receptors in the regulation of nigrostriatal dopaminergic function has been a subject of debate mainly due to controversial results obtained with different 5-HT$_{2C}$ receptor acting molecules (Di Matteo et al. 2001; Porras et al. 2002; De Deurwaerdère et al. 2004; Navailles et al. 2004). Very recently (Abdallah et al. 2009) by using the 5-HT$_{2C}$ receptor null mutant mice, previously generated by Tecott and collaborators (Tecott et al. 1995), it has been studied in a more direct manner the influence of this receptor subtype on functions mediated by the nigrostriatal dopaminergic pathway. Based on results generated by the combination of electrophysiological, pharmacological, neurochemical, and behavioral methods, Abdallah and coworkers have recently described that 5-HT$_{2C}$ receptor null mutant mice displayed (1) an increment in the activity of the dopaminergic neurons of substantia nigra pars compacta, (2) an increment in the extracellular dopamine in the dorsal striatum and nucleus accumbens, (3) increased syntactic grooming chain failures and altered grooming behaviors, and (4) increased sensitivity to the stereotypic behavioral effects following selective dopamine transporter blockade. All these responses occur without phenotypic differences in the elevation of drug-induced striatal extracellular dopamine concentration, suggesting that the loss of 5-HT$_{2C}$ receptor function may be accompanied by enhanced behavioral responses to released dopamine. The phenotypic differences these authors observe in stereotypic behavior following selective stimulation of dopamine D$_1$ receptors with an agonist support this hypothesis. All these findings suggest that 5-HT$_{2C}$ receptors play a significant role in the control of nigrostriatal physiology and behavior.

2.5   Conclusions

The studies reviewed here show that 5-HT$_{2C}$ receptors are extensive but distributed heterogeneously in the mammalian brain. Although this distribution shows a remarkable similarity among the species, there are nevertheless significant differences that have been identified. Many different neuronal populations including neuropeptidergic, cholinergic, serotonergic and GABAergic as well as the cannabinoid and dopaminergic systems have been shown to express 5-HT$_{2C}$ receptor mRNA and/or protein. In addition, the distribution suggests the involvement of other transmitter systems such as the glutamatergic and dopaminergic, although colocalization data are still missing. All these chemical neuroanatomical studies clearly point to the role of these receptors in the functions of many brain pathways. The modulation through selective agonists or antagonists of 5-HT$_{2C}$ receptors, an area of intensive research, will reveal the importance of these receptors as therapeutic tools.

References

Abdallah L, Bonasera SJ, Hopf FW, et al (2009) Impact of serotonin 2C receptor null mutation on physiology and behavior associated with nigrostriatal dopamine pathway function. J Neurosci 29:8156–8165.

Abramowski D, Rigo M, Duc D, Hoyer D, Staufenbiel M (1995) Localization of the 5-hydroxytryptamine2C receptor protein in human and rat brain using specific antisera. Neuropharmacology 34:1635–1645.

Aso E, Renoir T, Mengod G, et al (2009) Lack of CB(1) receptor activity impairs serotonergic negative feedback. J Neurochem 109:935–944.

Barnes NM and Sharp T (1999) A review of central 5-HT receptors and their function. Neuropharmacology 38:1083–1152.

Boothman L, Raley J, Denk F, Hirani E, Sharp T (2006) In vivo evidence that 5-HT(2C) receptors inhibit 5-HT neuronal activity via a GABAergic mechanism. Br J Pharmacol 149:861–869.

Cassel JC, Jeltsch H (1995) Serotonergic modulation of cholinergic function in the central nervous system: cognitive implications. Neuroscience 69:1–41.

Clemett DA, Punhani T, Duxon MS, Blackburn TP, Fone KCF (2000) Immunohistochemical localisation of the 5-HT$_{2C}$ receptor protein in the rat CNS. Neuropharmacology 39:123–132.

De Deurwaerdère P, Navailles S, Berg KA, Clarke WP, Spampinato U (2004) Constitutive activity of serotonin2C receptor inhibits in vivo dopamine release in the rat striatum and nucleus accumbens. J Neurosci 24:3235–3241.

Di Giovanni G, De Deurwaerdère P, Di Matteo V, Di Mascio M, Esposito E, Spampinato U (1999) Selective blockade of serotonin2C/2B receptors enhances mesolimbic and mesostriatal dopaminergic function: a combined in vivo electrphysiological and microdialysis study. Neuroscience 91:587–597.

Di Matteo V, Di Giovanni G, Di Mascio M, Esposito E (1998) Selective blockade of serotonin2C/2B receptors enhances dopamine release in the rat nucleus accumbens. Neuropharmacology 38:1195–1205.

Di Matteo V, De Blasi A, Di Giulio C, Esposito E (2001) Role of 5-HT$_{2C}$ receptors in the control of central dopamine function. Trends Pharmacol Sci 22:229–232.

Dremencov E, Weizmann Y, Kinor N, Gispan-Herman I, Yadid G (2006) Modulation of dopamine transmission by 5HT$_{2C}$ 5HT3 receptors: a role in the antidepressant response. Curr Drug Targets 7:165–175.

Eberle-Wang K, Mikeladze Z, Uryu K, Chesselet MF (1997) Pattern of expression of the serotonin2C receptor messenger RNA in the basal ganglia of adult rats. J Comp Neurol 384:233–247.

Fitzgerald LW, Iyer G, Conklin DS, et al (1999) Messenger RNA editing of the human serotonin 5-HT$_{2C}$ receptor. Neuropsychopharmacology 21:82S–90S.

Heisler LK, Pronchuk N, Nonogaki K, et al (2007) Serotonin activates the hypothalamic-pituitary-adrenal axis via serotonin 2C receptor stimulation. J Neurosci 27:6956–6964.

Hoffman BJ, Mezey E (1989) Distribution of serotonin 5-HT1C receptor mRNA in adult rat brain. FEBS Lett 247:453–462.

Hoyer D, Engel G, Kalkman HO (1985) Molecular pharmacology of 5-HT$_1$ and 5-HT$_2$ recognition sites in rat and pig brain membranes: radioligand binding studies with [^{3}H]5-HT, [^{3}H]8-OH-DPAT, (–) [^{125}I]iodocyanopindolol, [^{3}H]mesulergine and [^{3}H]ketanserin. Eur J Pharmacol 118:13–23.

Hoyer D, Pazos A, Probst A, Palacios JM (1986) Serotonin receptors in the human brain. II. Characterization and autoradiographic localization of 5-HT1C and 5-HT2 recognition sites. Brain Research 376:97–107.

Hoyer D, Clarke DE, Fozard JR, et al (1994) International Union of Pharmacology classification of receptors for 5- hydroxytryptamine (Serotonin). Pharmacol Rev 46:157–203.

Julius D, MacDermott AB, Axel R, Jessell TM (1988) Molecular characterization of a functional cDNA encoding the serotonin 1c receptor. Science 241:558–564.

Julius D, Huang KN, Livelli TJ, Axel R, Jessell TM (1990) The 5HT2 receptor defines a family of structurally distinct but functionally conserved serotonin receptors. Proc Natl Acad Sci USA 87:928–932.

Kondo Y, Ogawa N, Asanuma M, et al (1993) Regional changes in neuropeptide levels after 5,7-dihydroxytryptamine-induced serotonin depletion in the rat brain. J Neural Transm Gen Sect 92:151–157.

Liu R, Jolas T, Aghajanian G (2000) Serotonin 5-HT$_2$ receptors activate local GABA inhibitory inputs to serotonergic neurons of the dorsal raphe nucleus. Brain Res. 873:34–45.

Liu S, Bubar MJ, Lanfranco MF, Hillman GR, Cunningham KA (2007) Serotonin2C receptor localization in GABA neurons of the rat medial prefrontal cortex: implications for understanding the neurobiology of addiction. Neuroscience 146:1677–1688.

López-Giménez JF, Mengod G, Palacios JM, Vilaró MT (2001) Regional distribution and cellular localization of 5-HT$_{2C}$ receptor mRNA in monkey brain: comparison with [^{3}H]mesulergine binding sites and choline acetyltransferase mRNA. Synapse 42:12–26.

López-Giménez JF, Vilaró MT, Palacios JM, Mengod G (2001) Mapping of 5-HT$_{2A}$ receptors and their mRNA in monkey brain: [3H]MDL100,907 autoradiography and in situ hybridization studies. J Comp Neurol 429:571–589.

López-Giménez JF, Tecott LH, Palacios JM, Mengod G, Vilaró MT (2002) Serotonin 5- HT (2C) receptor knockout mice: autoradiographic analysis of multiple serotonin receptors. J Neurosci Res 67:69–85.

Lubbert H, Hoffman BJ, Snutch TP, et al (1987) cDNA cloning of a serotonin 5-HT1C receptor by electrophysiological assays of mRNA-injected Xenopus oocytes. Proc Natl Acad Sci USA 84:4332–4336.

Mengod G, Nguyen H, Le H, Waeber C, Lubbert H, Palacios JM (1990) The distribution and cellular localization of the serotonin 1C receptor mRNA in the rodent brain examined by in situ hybridization histochemistry. Comparison with receptor binding distribution. Neuroscience 35:577–591.

Mengod G, Pompeiano M, Martinez-Mir MI, Palacios JM (1990) Localization of the mRNA for the 5-HT2 receptor by in situ hybridization histochemistry. Correlation with the distribution of receptor sites. Brain Res 524:139–143.

Molineaux SM, Jessell TM, Axel R, Julius D (1989) 5-HT1c receptor is a prominent serotonin receptor subtype in the central nervous system. Proc Natl Acad Sci USA 86:6793–6797.

Navailles S, De Deurwaerdère P, Porras G, Spampinato U (2004) In vivo evidence that 5-HT$_{2C}$ antagonist but not agonist modulates cocaine-induced dopamine outflow in the rat nucleus accumbens and striatum. Neuropsychopharmacology 29:319–326.

Niswender CM, Sanders-Bush E, Emeson RB (1998) Identification and characterization of RNA editing events within the 5-HT$_{2C}$ receptor. Ann NY Acad Sci 861:38–48.

Palacios JM, Waeber C, Mengod G, Pompeiano M (1991) Molecular neuroanatomy of 5-HT receptors. In: Fozard JR, Saxena PR, eds. Serotonin: Molecular Biology, Receptors and Functional Effects, Basel: Birkhäuser Verlag, 5–20.

Pasqualetti M, Ori M, Castagna M, Marazziti D, Cassano GB, Nardi I (1999) Distribution and cellular localization of the serotonin type 2C receptor messenger RNA in human brain. Neuroscience 92:601–611.

Pazos A, Palacios JM (1985) Quantitative autoradiographic mapping of serotonin receptors in the rat brain. I. Serotonin-1 receptors. Brain Res 346:205–230.

Pazos A, Hoyer D, and Palacios JM (1984) The binding of serotonergic ligands to the porcine choroid plexus: characterization of a new type of serotonin recognition site. Eur J Pharmacol 106:539–546.

Pazos A, Hoyer D, Palacios JM (1984) Mesulergine, a selective serotonin-2 ligand in the rat cortex, does not label these receptors in porcine and human cortex: evidence for species differences in brain serotonin-2 receptors. Eur J Pharmacol 106:531–538.

Pazos A, Probst A, Palacios JM (1987) Serotonin receptors in the human brain. III. Autoradiographic mapping of serotonin-1 receptors. Neuroscience 21:97–122.

Pompeiano M, Palacios JM, Mengod G (1994) Distribution of the serotonin 5-HT2 receptor family mRNAs: comparison between 5-HT$_{2A}$ and 5-HT$_{2C}$ receptors. Brain Res Mol Brain Res 23:163–178.

Porras G, Di Matteo V, Fracasso C, et al (2002) 5-HT$_{2A}$ and 5-HT$_{2C/2B}$ receptor subtypes modulate dopamine release induced in vivo by amphetamine and morphine in both rat nucleus accumbens and striatum. Neuropsychopharmacology 26:311–324.

Prichett DB, Bach AWJ, Wozny M, et al (1988) Structure and functional expression of cloned rat serotonin 5-HT$_2$ receptor. EMBO J. 7:4135–4140.

Prisco S, Pagannone S, Esposito E (1994) Serotonin-dopamine interaction in the rat ventral tegmental area: an electrophysiological study in vivo. J Pharmacol Exp Ther 271:83–90.

Serrats J, Mengod G, Cortes R (2005) Expression of serotonin 5-HT$_{2C}$ receptors in GABAergic cells of the anterior raphe nuclei. J Chem Neuroanat 29:83–91.

Tecott LH, Sun LM, Akana SF, et al (1995) Eating disorder and epilepsy in mice lacking 5-HT$_{2C}$ serotonin receptors. Nature 374:542–546.

Ward RP, Dorsa DM (1996) Colocalization of serotonin receptor subtypes 5-HT$_{2A}$, 5-HT$_{2C}$, and 5-HT6 with neuropeptides in rat striatum. J Comp Neurol 370:405–414.

Wright DE, Seroogy KB, Lundgren KH, Davis BM, Jennes L (1995) Comparative localization of serotonin1A, 1C, and 2 receptor subtype mRNAs in rat brain. J Comp Neurol 351:357–373.

Zhelyazkova-Savova M, Giovannini MG, Pepeu G (1997) Increase of cortical acetylcholine release after systemic administration of chlorophenylpiperazine in the rat: an in vivo microdialysis study. Neurosci Lett 236:151–154.

Zhelyazkova-Savova M, Giovannini MG, Pepeu G (1999) Systemic chlorophenylpiperazine increases acetylcholine release from rat hippocampus-implication of 5-HT$_{2C}$ receptors. Pharmacol Res 40:165–170.

Chapter 3
The Medicinal Chemistry of 5-HT$_{2C}$ Receptor Ligands

Marcello Leopoldo, Enza Lacivita, Paola De Giorgio, Francesco Berardi, and Roberto Perrone

3.1 Introduction

5-HT$_2$ receptors are major targets for a wide array of psychoactive drugs, ranging from nonclassical antipsychotic drugs, anxiolytics and antidepressants, which have a 5-HT$_2$ antagonistic action, to hallucinogens, which are agonists of the 5-HT$_2$ receptors. The 5-HT$_2$ receptors form a closely related subgroup of G-protein-coupled receptors and show the typical heptahelical structure of an integral membrane protein monomer. They are currently classified as 5-HT$_{2A}$, 5-HT$_{2B}$, and 5-HT$_{2C}$ subtypes, based on their close structural homology, pharmacology and signal transduction pathways. The amino acid sequence of the 5-HT$_2$ receptors shares a high degree (>70%) of identity within the transmembrane segments and consequently, it is not surprising that many compounds bind with high affinity to all these three receptor subtypes. The high degree of sequence homology has considerably hampered the identification of selective 5-HT$_{2C}$ antagonists and especially, agonists (Di Giovanni et al. 2006).

3.2 5-HT$_{2C}$ Receptor Antagonists

3.2.1 Indoline-Containing Structures

The search for selective 5-HT$_{2C}$ receptor antagonists has been extensively pursued by researchers at GlaxoSmithKline. The evolution of those studies is graphically summarized by the Structures I to VI depicted in Fig. 3.1. Early studies were aimed to obtain 5-HT$_{2C}$ agents with selectivity over 5-HT$_{2A}$ receptors, because both 5-HT$_2$

M. Leopoldo (✉)
Dipartimento Farmaco-Chimico, Università degli Studi di Bari,
via Orabona, 4, 70125, Bari, Italy
e-mail: leopoldo@farmchim.uniba.it

G. Di Giovanni et al. (eds.), *5-HT$_{2C}$ Receptors in the Pathophysiology of CNS Disease*,
The Receptors 22, DOI 10.1007/978-1-60761-941-3_3,
© Springer Science + Business Media, LLC 2011

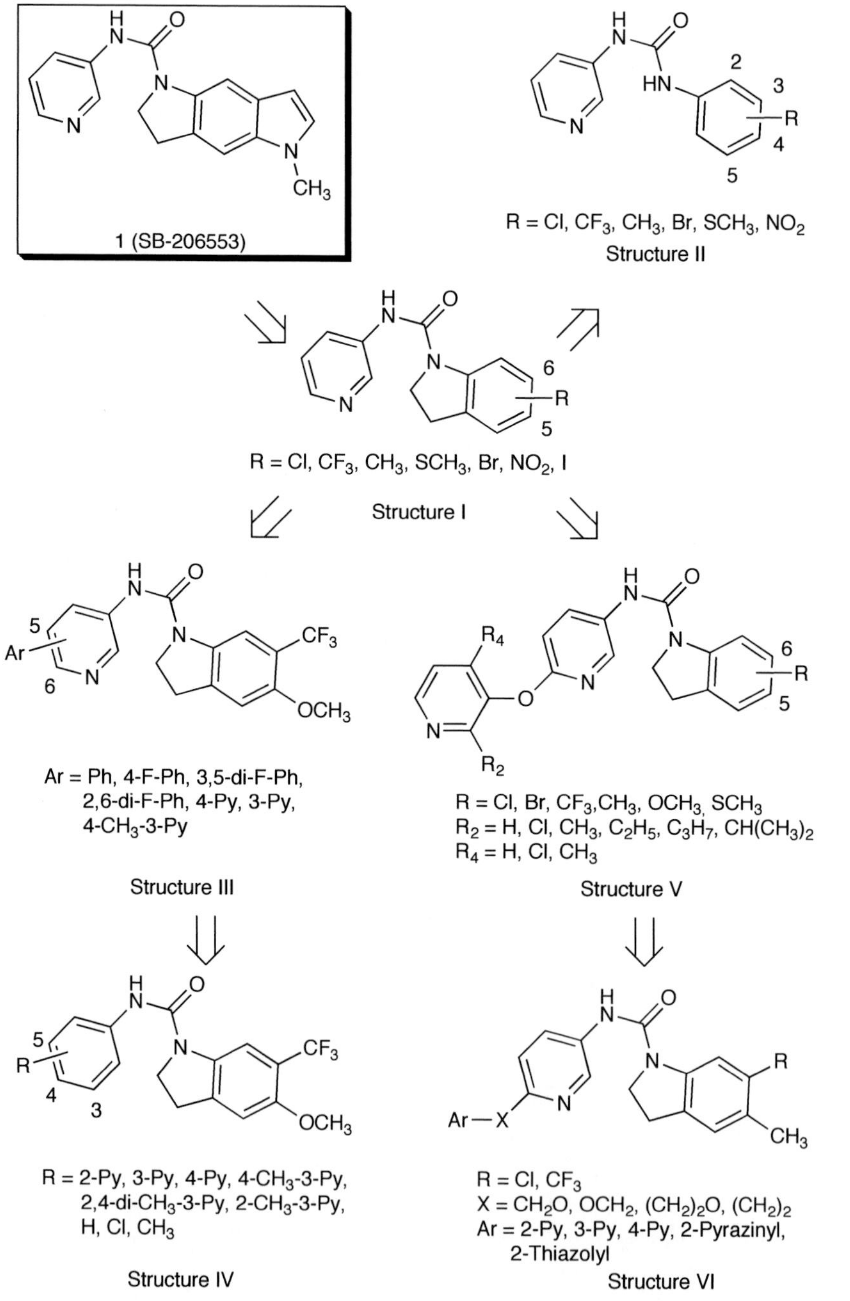

1 (SB-206553)

R = Cl, CF$_3$, CH$_3$, Br, SCH$_3$, NO$_2$

Structure II

R = Cl, CF$_3$, CH$_3$, SCH$_3$, Br, NO$_2$, I

Structure I

Ar = Ph, 4-F-Ph, 3,5-di-F-Ph,
2,6-di-F-Ph, 4-Py, 3-Py,
4-CH$_3$-3-Py

Structure III

R = Cl, Br, CF$_3$,CH$_3$, OCH$_3$, SCH$_3$
R$_2$ = H, Cl, CH$_3$, C$_2$H$_5$, C$_3$H$_7$, CH(CH$_3$)$_2$
R$_4$ = H, Cl, CH$_3$

Structure V

R = 2-Py, 3-Py, 4-Py, 4-CH$_3$-3-Py,
2,4-di-CH$_3$-3-Py, 2-CH$_3$-3-Py,
H, Cl, CH$_3$

Structure IV

R = Cl, CF$_3$
X = CH$_2$O, OCH$_2$, (CH$_2$)$_2$O, (CH$_2$)$_2$
Ar = 2-Py, 3-Py, 4-Py, 2-Pyrazinyl,
2-Thiazolyl

Structure VI

Fig. 3.1 The evolution of the first template of 5-HT$_{2C}$ receptor antagonists developed by GlaxoSmithKline

receptor subtypes are localized within the central nervous system, in contrast to the 5-HT$_{2B}$ subtype, which is localized peripherally. Later, the selectivity over 5-HT$_{2B}$ receptor was considered important because this receptor is notably expressed in heart valves and pulmonary arteries and could be implicated in cardiopulmonary toxicity. Particular attention has been devoted to the pharmacokinetic properties of the new chemical entities, and therefore, structural features that were not strictly necessary for the interaction with the receptor have been incorporated. The starting point was the 5-HT$_{2C}$ receptor antagonist **1** (SB-206553) (pK_i=7.9) with 160-fold selectivity over 5-HT$_{2A}$ receptor. This compound suffered from metabolic demethylation to a nonselective compound and therefore, to circumvent its metabolic liability, the replacement of the pyrroloindole ring was investigated. Modeling studies on a range of isomeric *N*-substituted pyrrole analogues of **1**, in which the pyrrole ring was fused across each of the 4-5, 5-6, and 6-7 bonds of the indoline, provided a definition of an "allowed" volume for 5-HT$_{2C}$ receptor affinity, which included a volume that was "disallowed" at the 5-HT$_{2A}$ receptor. These findings were rationalized by considering key differences in the sequences of the 5-HT$_{2C}$ and 5-HT$_{2A}$ receptors in a region adjacent to the indole *N*-methyl group in the proposed binding mode of **1**. These sequence differences would be expected to give rise in the 5-HT$_{2A}$ receptor to a smaller binding pocket which can less easily accommodate the indole *N*-methyl group of **1**, thus leading to the observed selectivity. On such basis, a series of mono- and disubstituted indolines were studied (Fig. 3.1, Structure I). 5,6-Disubstitution pattern gave the best results in term of affinity at 5-HT$_{2C}$ receptor (pK_i values >8) and selectivity over 5-HT$_{2A}$ receptor (>30-fold). In particular, an electron-withdrawing group at the 6 position was preferred and a correlation between 5-HT$_{2C}$ affinity and increasing lipophilicity was apparent at the 5 position. The properties of the substituent in 5 position have been evaluated in depth (compounds **2–17**, Table 3.1). In particular, the size and shape of the 5-substituent was varied in order to probe the crucial 5-HT$_{2C}$-allowed/5-HT$_{2A}$-disallowed region in combination with the small, lipophilic, electron-withdrawing Cl or CF$_3$ substituent in 6 position. The most interesting results were obtained in the case of the 5-alkylthio- and 5-alkyloxy-disubstituted compounds. Several of these compounds combined high 5-HT$_{2C}$ affinity (pK_i>8) with >100-fold selectivities over 5-HT$_{2A}$. Moreover, it was evidenced, among the 6-CF$_3$-substituted analogues, that the size and shape of the 5 substituent was crucial to the selectivity of the compounds. The best substituents were the thioethyl (**11**), thiopropyl (**12**), and *i*-propoxy (**17**). A further increase of the size of the substituent led to a drop in affinity and selectivity (compounds **13** and **14**). Molecular modeling on the 5-CH$_3$S-6-CF$_3$ derivative **10** (SB-221284) revealed that these substituents optimally interact/occupy the crucial 5-HT$_{2C}$-allowed/5-HT$_{2A}$-disallowed volume. The introduction of a 6-substituent also has the added beneficial effect of restricting the rotation of the 5-substituent to favor those conformations in which the alkyl group occupies the crucial region. Compound **10** was manually docked into a model of the 5-HT$_{2C}$ receptor.

The proposed binding mode was very similar to that proposed for **1** with the urea carbonyl oxygen double-hydrogen bonding to the hydroxyl side chains of Ser-312 and Ser-315. The 3-pyridyl ring occupies a lipophilic pocket defined by the side

Table 3.1 Binding affinities of 5-HT$_{2C}$ receptor antagonists[a]

Compound	R$_5$	R$_6$	pK$_i$		
			5-HT$_{2C}$	5-HT$_{2A}$	5-HT$_{2B}$
2	CH$_3$	Cl	8.2	6.8	–
3	CH$_2$CH$_3$	Cl	8.3	6.4	–
4	CH$_2$CH$_2$CH$_3$	Cl	7.7	5.9	–
5	CH(CH$_3$)$_2$	Cl	7.7	5.7	–
6	C(CH$_3$)$_3$	Cl	7.7	6.2	6.8
7	OCH$_2$CH$_3$	Cl	7.6	5.4	–
8	OCH(CH$_3$)$_2$	Cl	7.8	<5.2	–
9	SCH$_3$	Cl	8.2	5.6	–
10 (SB-221284)	SCH$_3$	CF$_3$	8.6	6.4	7.9
11	SCH$_2$CH$_3$	CF$_3$	8.5	5.5	8.0
12	SCH$_2$CH$_2$CH$_3$	CF$_3$	8.2	<5.2	7.8
13	SCH(CH$_3$)$_2$	CF$_3$	7.5	5.3	–
14	CH(CH$_3$)$_2$	CF$_3$	7.6	<5.2	–
15	OCH$_3$	CF$_3$	8.0	6.0	–
16	OCH$_2$CH$_3$	CF$_3$	8.2	5.8	–
17	OCH(CH$_3$)$_2$	CF$_3$	8.5	5.8	8.4

[a]Receptor binding assays were performed by using HEK-293 cell lines expressing recombinant human 5-HT$_{2C}$, 5-HT$_{2A}$, and 5-HT$_{2B}$ receptors. Determination of affinities was made using [³H] mesulergine for 5-HT$_{2C}$ receptors, [³H]ketanserin for 5-HT$_{2A}$ receptors, and [³H]-5-HT for 5-HT$_{2B}$ receptors

chains of the aromatic residues Phe-508, Trp-613, Phe-616, and Phe-617. Within this pocket the 3-pyridyl ring is able to form both π–π stacking and edge-to-face aromatic interactions with several of the aromatic residues lining the pocket. These interactions probably contribute significantly to the overall binding of these ligands to the receptor. The substituted indoline is placed in another pocket, the boundary of which is defined by residues Val-212, Phe-311, Val-608, Phe-609, Met-612, and Tyr-715. This pocket is also very lipophilic in nature, although less aromatic than the 3-pyridyl binding pocket. In the 5-HT$_{2A}$ receptor sequence both the 212 and 608 residues are Leu. These differences would be expected to lead to binding pockets of reduced size, and it was proposed that these steric differences in the receptors might account for the observed 5-HT$_{2C}$ specificity (Bromidge et al. 1998).

Due to the synthetic complexity of indolines with Structure I (Fig. 3.1), the more accessible equivalently substituted phenyl ureas were evaluated (Fig. 3.1, Structure II). However, this structural simplification was detrimental for affinity (pK$_i$ <7.8) (Bromidge et al. 1999).

Although indolines **2**, **10**, and **15** displayed high affinity and specificity for the 5-HT$_{2C}$ receptor, they potently inhibited a number of human cytochrome P450 enzymes, in particular, the CYP1A2 isoform, and this precluded further development. Introduction of steric hindrance around the pyridine nitrogen of **10** revealed that the unhindered nitrogen was responsible for the P450 inhibitory activity. Modeling studies revealed that the lipophilic pocket that was occupied by the pyridine ring was quite deep. Therefore, the substitution of the pyridyl ring with a variety of aryl groups was investigated with the twofold aim to reduce the cytochrome P450 inhibitory activity and to increase the 5-HT$_{2C}$ affinity (Fig. 3.1, Structure III). Although this modification was generally tolerated for the 5-HT$_{2C}$ affinity, none of the compounds displayed acceptable CYP1A2 activity. Alternatively, the replacement of the pyridyl with phenyl ring was accomplished (Fig. 3.1, Structure IV). In particular, the phenyl ring was decorated with various pyridyl substituents in order to retain good solubility and efficacious interaction with the 5-HT$_{2C}$ receptor. Compounds **18–26** (Table 3.2) generally retained good 5-HT$_{2C}$ affinity and selectivity over 5-HT$_{2A}$ receptors combined with reduced P450 liability. In particular, increasing the torsion angle between the two aromatic rings greatly increased the selectivity over 5-HT$_{2A}$ receptors. The most notable example was compound **23**, which displayed sub-nanomolar 5-HT$_{2C}$ affinity along with >20,000-fold selectivity over 5-HT$_{2A}$ receptor. Modeling studies of **23** showed an almost orthogonal relationship between the two aryl rings. However, this group of compounds yet displayed poor oral activity or inhibitory activity of other P450 enzymes.

In order to improve duration of action in vivo, a series of analogues of compound **19** (Table 3.2, R$_3$ = 3-Py, R$_4$ = R$_5$ = H) was studied. A variety of substituents capable of blocking the metabolism of the electron-rich phenyl ring (R$_4$ or R$_5$ = Cl, F, Br, CH$_3$, OCH$_3$) were well tolerated, affording good to excellent 5-HT$_{2C}$ affinity and increased selectivity over 5-HT$_{2A}$. Incorporating 4,5-disubstitution produced an additive effect on activity leading to very high 5-HT$_{2C}$ affinities and selectivities over 5-HT$_{2A}$ (compounds **27** and **28**, Table 3.2). These compounds demonstrated low inhibition of CYP1A2 and potent oral activity in the rat.

Many of the compounds described in Tables 3.1 and 3.2 possessed excellent 5-HT$_{2C}$ affinity and selectivity over 5-HT$_{2A}$ but lacked selectivity over the 5-HT$_{2B}$ receptor. Receptor–ligand modeling work suggested that the lipophilic pocket within the 5-HT$_{2C}$ receptor was still not fully exploited. This model suggested that introducing a linker group between the aromatic rings (Fig. 3.1, Structure V) would more optimally occupy this hydrophobic region (compounds **29–30**, Table 3.3). The bispyridyl ether **29** showed excellent 5-HT$_{2C}$ affinity and almost 100-fold selectivity over 5-HT$_{2A}$ receptor. Increasing the torsion angle between the two aryl rings by introduction of a 2-methyl substituent into the terminal pyridyl ring was beneficial for 5-HT$_{2C}$ affinity and increased tenfold the selectivity over 5-HT$_{2A}$ receptor (compound **30**). In addition, 80-fold selectivity over the 5-HT$_{2B}$ receptor was achieved. Modification of the substitution pattern of the indoline ring afforded compounds **31** (SB-243213) and **32** (SB-242084), which displayed optimal affinity for 5-HT$_{2C}$ receptor, >100-fold selectivity over 5-HT$_{2A}$ and 5-HT$_{2B}$ receptors, and suitable properties for in vivo evaluation (Bromidge et al. 2000a).

The structure–activity relationships of compounds **31** (SB-243213) and **32** (SB-242084) have been studied further (Fig. 3.1, Structure V). The results from the indoline substitution optimization overlapped those obtained for the compounds

Table 3.2 Binding affinities of 5-HT$_{2C}$ receptor antagonists[a]

Compound	R$_3$	R$_4$	R$_5$	pK$_i$ 5-HT$_{2C}$	5-HT$_{2A}$	5-HT$_{2B}$	CYP1A2 IC$_{50}$ (μmol/L)
18	2-Py	H	H	7.7	6.3	–	3%
19	3-Py	H	H	9.0	6.7	8.1	4
20	4-CH$_3$-3-Py	H	H	9.2	6.6	–	8%
21	2-CH$_3$-3-Py	H	H	8.3	6.3	7.8	2%
22	2,4-di-CH$_3$-3-Py	H	H	8.6	6.7	7.9	–
23	4-CH$_3$-3-Py	CH$_3$	H	9.5	<5.2	8.4	>100
24	4-CH$_3$-3-Py	Cl	H	9.2	<6.0	–	11%
25	H	3-Py	H	7.8	5.4	8.0	0%
26	H	4-Py	H	8.3	<5.0	7.9	–
27	3-Py	CH$_3$	F	9.2	5.8	8.1	–
28	3-Py	4,5-OCH$_2$CH$_2$		9.3	6.6	8.1	2%

[a]Receptor binding assays were performed by using HEK-293 cell lines expressing recombinant human 5-HT$_{2C}$, 5-HT$_{2A}$, and 5-HT$_{2B}$ receptors. Determination of affinities was made using [³H] mesulergine for 5-HT$_{2C}$ receptors, [³H]ketanserin for 5-HT$_{2A}$ receptors, and [³H]-5-HT for 5-HT$_{2B}$ receptors. The cytochrome P450 inhibitory potential was determined using isoform-selective assays and heterologously expressed human CYP1A2

Table 3.3 Binding affinities of 5-HT$_{2C}$ receptor antagonists[a]

Compound	R$_1$	R$_2$	R$_3$	pK$_i$ 5-HT$_{2C}$	5-HT$_{2A}$	5-HT$_{2B}$	CYP1A2 IC$_{50}$ (μmol/L)
29	H	CF$_3$	OCH$_3$	8.9	7.0	7.7	4%
30	CH$_3$	CF$_3$	OCH$_3$	9.2	6.1	7.3	>100
31 (SB-243213)	CH$_3$	CF$_3$	CH$_3$	9.0	6.8	7.0	>100
32 (SB-242084)	CH$_3$	Cl	CH$_3$	9.0	6.8	7.0	>100

[a]Receptor binding assays were performed by using HEK-293 cell lines expressing recombinant human 5-HT$_{2C}$, 5-HT$_{2A}$, and 5-HT$_{2B}$ receptors. Determination of affinities was made using [³H] mesulergine for 5-HT$_{2C}$ receptors, [³H]ketanserin for 5-HT$_{2A}$ receptors, and [³H]-5-HT for 5-HT$_{2B}$ receptors. The cytochrome P450 inhibitory potential was determined using isoform-selective assays and heterologously expressed human CYP1A2

with Structure I (Fig. 3.1): an electron-withdrawing group in R$_6$ and an alkyl group in R$_5$ were preferred. Optimization of the terminal aryl ring was focused on 2-and 4-positions because of its role on selectivity. Considering the R$_2$ substituent, the replacement of the methyl with ethyl or *n*-propyl or chloro was well tolerated with respect to the affinity, whereas the introduction of an *i*-propyl group was detrimental for affinity. Shifting of the methyl in 4 position (R$_2$= H, R$_4$= CH$_3$) or its replacement with a Cl maintained excellent 5-HT$_{2C}$ affinity (pK_i=9.3). However, these modifications had a negative effect on selectivity. As for the central pyridine ring, introduction of a methyl or its replacement by various diazines was well tolerated for 5-HT$_{2C}$ receptor affinity (pK_i=8.5–8.7), but produced a loss in 5-HT$_{2B}$ selectivity (Bromidge et al. 2000b).

As already discussed, the *bis*-pyridyl ether moiety in Structure V (Fig. 3.1) occupies a lipophilic pocket defined by side-chain aromatic residues on transmembrane helices 5 and 6. The size and shape of this binding pocket was further explored by incorporating longer linker groups between the aromatic rings and by evaluation of various terminal heteroaryl groups (Fig. 3.1, Structure VI). Affinity data indicated that the unsubstituted 2-pyridylmethoxy substituent (Ar-X) optimally fills the lipophilic binding pocket (R=CF$_3$: pK_i=9.3; R=Cl: pK_i=8.6). Finally, modifications of both length and nature of the linker were detrimental of affinity and selectivity (Bromidge et al. 2000c).

3.2.2   Piperidine-Containing Structures

With the aim of identifying additional new chemical entities capable of 5-HT$_{2C}$ receptor blockade, GlaxoSmithKline developed a pharmacophore model that was used to synthesize different chemical classes of compounds (Fig. 3.2). The model was validated by means of structure–activity relationships derived from available data of a wide range of antagonists that interacted with the receptor in the same

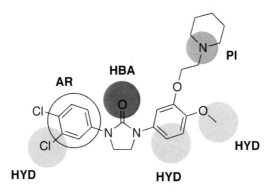

Fig. 3.2 Pharmacophore model for the 5-HT$_{2C}$ antagonists. Pharmacophoric features: *HBA* H-bond acceptor, *PI* positive ionizable, *HYD* hydrophobic, *AR* aromatic ring

manner. It demonstrated the key role played by both the positive ionizable group and the H-bond acceptor functionality to achieve high 5-HT$_{2C}$ affinity.

A first group of bisaryl imidazolidin-2-ones is exemplified by the structure of compound **33** (Fig. 3.3) which showed high 5-HT$_{2C}$ receptor affinity (pK_i=9.1) and >400-fold selectivity over 5-HT$_{2A}$ and 5-HT$_{2B}$ receptors. However, the compounds belonging to this class presented high in vivo clearance. The structure–activity relationships of these compounds indicated that the dihalosubstitution on the phenyl ring was preferred over the monosubstitution; the replacement of the piperidine ring by a morpholine or by a dimethylamino was well tolerated; the shifting of the ethoxyamino chain caused a marked loss in affinity; and the introduction of a double bond in the five-membered ring caused a loss in affinity (Goodacre et al. 2005).

A series of diaryl substituted pyrrolidinones and pyrrolones, which fitted the pharmacophore model described above, has been studied. The most relevant compound was **34** (Fig. 3.3), which showed the best compromise between pharmacological and pharmacokinetic properties (pK_is: 5-HT$_{2C}$=8.5; 5-HT$_{2A}$=6.1; 5-HT$_{2B}$<6.1). The different decoration of the scaffold led to different values in potency. In particular, a methoxy group on the phenyl ring was favored; substitution of the left hand side aromatic portion was well tolerated with respect to the potency and selectivity; the piperidine was the best basic moiety; hydrogenation of the lactamic double bond affected negatively both affinity and metabolic stability (Micheli et al. 2006).

Conformational analysis of the template **34** (Fig. 3.3) showed a rather small dihedral angle between the lactam and the adjacent left hand side phenyl ring for the global minimum conformer. Therefore, it was suggested that, if the conformation most relevant for binding at the target receptor was similar to the minimum energy conformer, tricyclic indole derivatives with a general formula exemplified by **35** (Fig. 3.3) might be able to bind to the 5-HT$_{2C}$ receptor. Moreover, inclusion of the carbon–carbon double bond of the lactam in an aromatic ring could modulate the overall properties of the resulting compounds by profoundly changing the chemical nature of this fragment. The new derivatives retained the same binding profile as the opened counterparts. The change from vinylic to aromatic nature of the carbon–carbon double bond of the lactam ring was compatible with a favorable profile at both receptor and in vivo level. For example, compound **35** showed pK_i=8.5 at 5-HT$_{2C}$ receptors, pK_i=6.9 at 5-HT$_{2A}$ and pK_i=7.4 at 5-HT$_{2B}$. This compound also possessed good oral bioavailability. The structural modifications summarized in Fig. 3.3 left unchanged the pharmacological profile of the compounds (Hamprecht et al. 2007a).

Rather than using an indole system as in **35**, phenyl fusion was considered as a further option to include the double bond of the pyrrolidone ring into a rigid framework. Following this criterion, a series of isoindolones was evaluated and the most notable example of this series was the compound **36** (Fig. 3.3). The affinity to the 5-HT$_{2C}$ receptor was high (pK_i=8.6) and the selectivities over the 5-HT$_{2A}$ and 5-HT$_{2B}$ subtypes higher than 100-fold. As far as the structure–activity relationships are concerned, evaluation of mono- and disubstitution pattern on the isoindolinone ring indicated that the 6,7-dichloro substitution was preferred. Variation of the basic

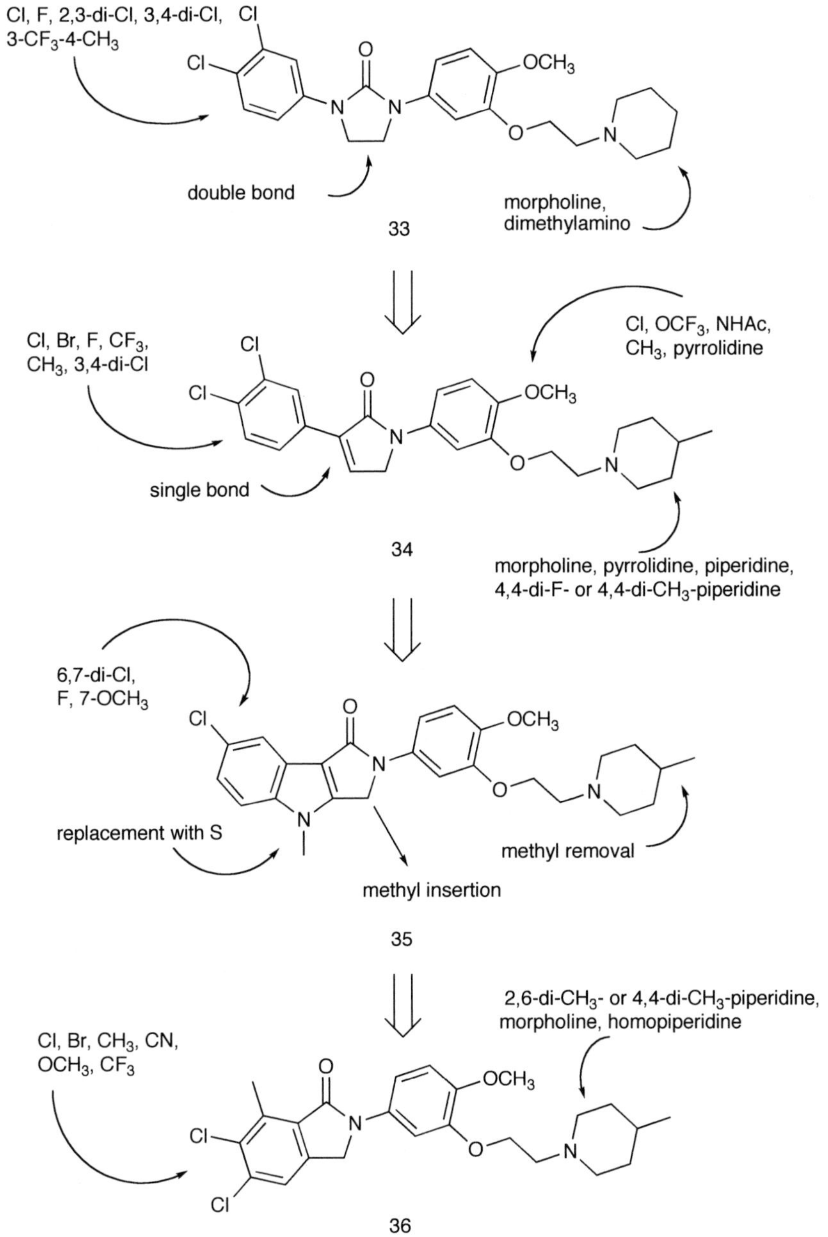

Fig. 3.3 The evolution of a set of 5-HT$_{2C}$ receptor antagonists developed by GlaxoSmithKline of the basis of pharmacophore model reported in Fig. 3.2

heterocycle indicated that various substituted piperidines were tolerated, as well as ring enlargement. α-Methylation to the basic nitrogen also led to slightly reduced target affinities in the case of replacement with morpholine. In line with previous findings, 4-methylpiperidine derivatives demonstrated the most favorable pharmacokinetic properties (Hamprecht et al. 2007b).

3.3 5-HT$_{2C}$ Receptor Agonists

It is quite difficult to describe detailed structure–activity relationships for 5-HT$_{2C}$ receptor agonists because a large part of the data has been published in patent applications and refers to compounds that are not directly related structurally. Moreover, interlaboratory comparison of agonist potency and relative efficacy can be difficult (Nilsson 2006). Apart from possible species differences, functional data are often dependent on the in vitro test system used. For example, the potency and efficacy of agonists in functional assays are highly dependent on receptor expression levels. Both receptor–effector coupling and receptor reserve can show large variations from one system and/or tissue to the other. Additionally, the signal transduction characteristics of transfected receptors may be dependent on the identity of the host cell line. Clearly, all these aspects may have a profound impact on the observed functional selectivities over 5-HT$_{2A}$ and 5-HT$_{2B}$ receptors. Generally, functional data are based on measurements of a second messenger response, such as phosphoinositide hydrolysis or increase in intracellular calcium at 5-HT$_{2C}$, 5-HT$_{2A}$, and 5-HT$_{2B}$ receptors expressed in the same host cell line. However, several other diverse functional in vitro models (biochemical assays or tissue preparations) have been employed such as phosphoinositide hydrolysis in rat or pig choroid plexus (5-HT$_{2C}$), phosphoinositide hydrolysis in rat cortex (5-HT$_{2A}$), effect on human platelet aggregation (5-HT$_{2A}$), contraction of rat jugular vein (5-HT$_{2A}$), contraction of the isolated rat tail artery (5-HT$_{2A}$), and contraction of isolated rat stomach fundus muscle strips (5-HT$_{2B}$). Yet, another complicating factor is that the 5-HT$_{2C}$ receptor undergoes post-transcriptional mRNA editing. Such editing is associated with different G-protein-coupling efficiencies of the isoforms. In this respect, it has been shown that 5-HT is less potent in stimulating phosphatidylinositol hydrolysis at edited versions of the 5-HT$_{2C}$ receptor than at the nonedited 5-HT$_{2C}$ receptor isoform INI.

All these considerations should be kept in mind when considering the literature data on 5-HT$_{2C}$ receptor agonists.

3.3.1 Arylpiperazine-Based Structures

The unsubstituted arylpiperazines mCPP (**37**) and TFMPP (**38**) (Fig. 3.4) were among the first nontryptamine 5-HT$_{2C}$ receptor agonists to be identified and still are commonly used as tools to probe 5-HT$_{2C}$ receptor function (Porter et al. 1999). Over the years,

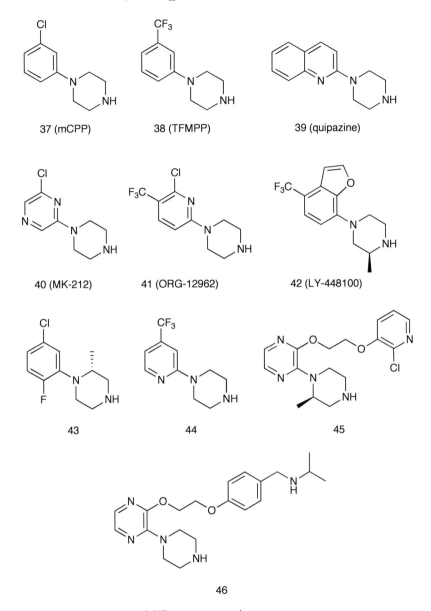

Fig. 3.4 Arylpiperazine-based 5-HT$_{2C}$ receptor agonists

many arylpiperazine based 5-HT$_{2C}$ receptor agonists have been reported (Fig. 3.4). However, the paucity of data does not allow the description of structure–activity relationships. As example, quipazine (**39**) and MK-212 (**40**) are nonselective agonists (Porter et al. 1999). ORG-12962 (**41**) behaves as a partial agonist at human 5-HT$_{2C}$ receptors with low binding selectivity over human 5-HT$_{2A}$ receptors (K_i=12 and 65 nmol/L, respectively). In functional in vitro assays measuring calcium release, **41** displayed pEC_{50} values of 7.01, 6.38, and 6.28, along with relative efficacies of 62%, 54%, and 41%, at the human 5-HT_{2C}, 5-HT_{2A}, and 5-HT_{2B} receptors, respectively (Leysen and Kelder 1998). Eli Lilly have reported the arylpiperazine-based 5-HT_{2C} receptor agonist LY-448100 (**42**), which showed high binding affinity ($K_i = 9$ nmol/L) and was claimed to be 15-fold selective over other 5-HT receptors. LY-448100 (**42**) behaved as a 5-HT_{2C} receptor agonist with high potency and efficacy ($EC_{50} = 8$ nmol/L, relative efficacy of 110%) (Briner et al. 2001). Also Arena Pharmaceuticals reported a series of arylpiperazine-based 5-HT_{2C} receptor agonists structurally related to mCPP (**37**). Among these, compound **43** displayed EC_{50} values of 8 and 529 nmol/L at the 5-HT_{2C} and 5-HT_{2A} receptors, respectively, while it was claimed to be essentially functionally inactive at the 5-HT_{2B} receptor, as determined by measurements of phosphoinositide hydrolysis (Smith et al. 2005). Other arylpiperazine 5-HT_{2C} agonists are **44–46** that possess moderate-to-high binding affinities (K_i values of 40, 3, and 5 nmol/L, respectively) (Leysen and Kelder 1998; Nilsson and Scobie 2002; Nilsson et al. 2002).

3.3.2 Indole-Derived Structures

The indole derivative RO600175 (Fig. 3.5, compound **47**) has been largely used as template for the development of 5-HT_{2C} receptor agonists. This compound displayed high affinity for the 5-HT_{2C} receptor, reasonably good selectivity over the 5-HT_{2A} receptor (5-HT_{2C} $K_i = 2.3$ nmol/L; 5-HT_{2A} $K_i = 14$ nmol/L; 5-HT_{2B} $K_i = 5.1$ nmol/L), and considerable potency ($EC_{50} = 18$ nmol/L, 93% relative efficacy). Structural modifications of compound **47** have initially been focused on the optimization of the substitution pattern of the indole ring (Fig. 3.5, Structure I). In particular, the monosubstituted halogen derivatives in 4-, 5-, and 6-position of the indole ring showed higher 5-HT_{2C} receptor affinity than the analogues bearing an electron-donating substituent (OCH_3, CH_3). The corresponding dihalogenated indoles (R = 5-F, 6-Cl) showed the highest 5-HT_{2C} receptor affinities within this series ($8 < pK_i < 9$). In particular, derivative **48** (Fig. 3.5, Structure I, R = 5,6-di-F) showed a better pharmacological profile than **47** (5-HT_{2C} $pK_i = 9.0$; 5-HT_{2A} $pK_i = 7.0$; $pEC_{50} = 6.7$, 100% relative efficacy). In general, the S-enantiomers displayed higher affinity and selectivity than their optical antipodes (Bös et al. 1997a).

Other 5-HT_{2C} receptor agonists are 1,4-dihydroindeno[1,2-b]pyrroles with Structure II (Fig. 3.5). Optimization of the substitution pattern of the indole ring indicated that, differently from the indoles with Structure I (Fig. 3.5), methoxy-substituted 1,4-dihydroindeno[1,2-b]pyrroles showed higher affinities than the halogenated counterparts. In this series, the 7-position was optimal for aromatic substitution. The introduction of a *gem*-dimethyl feature in the 1,4-dihydroindeno [1,2-b]pyrrole framework gave compound **49** (Fig. 3.5, Structure II, $R_1 = R_6 = CH_3$, absolute configuration S). This derivative showed binding affinities for the human 5-HT_{2C} and 5-HT_{2A} receptors similar to RO600175 (**47**) (pK_i values of 8.5 and 7.0, respectively), while displaying lower agonist potency in the 5-HT_{2B} rat fundus assay ($pD_2 = 6.1$ vs 7.9). Moreover, at the human 5-HT_{2B} receptor, **49** was about 100-fold less potent than **47** ($pEC_{50} = 7.26$, relative efficacy of 91%) (Bös et al. 1997b).

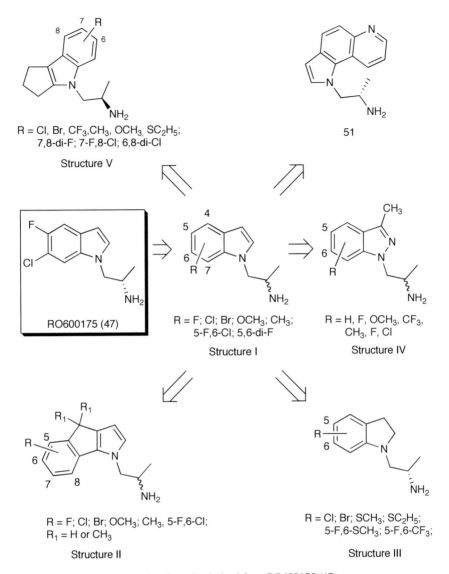

Fig. 3.5   5-HT$_{2C}$ receptor agonists formally derived from RO600175 (47)

Several indoline derivatives were formally derived from **47** by reduction of the pyrrole ring (Fig. 3.5, Structure III). These compounds are potent partial agonists at the 5-HT$_{2C}$ receptor, since their affinities are broadly in the same range as **47**, irrespective to the substitution pattern of the indole ring (3.2 nmol/L $< K_i <$ 26 nmol/L). The indoline **50** (Fig. 3.5, Structure III, R$_5$=F, R$_6$=Cl) showed similar selectivity (5-HT$_{2A}$/5-HT$_{2C}$ K_i ratio = 12; 5-HT$_{2A}$/5-HT$_{2C}$ K_i ratio = 2.2) as compared with the parent indole **47**, but significantly lower binding affinity (Bentley et al. 2004).

A series of indazole derivatives have also been studied (Fig. 3.5, Structure IV). These compounds demonstrated moderate affinities ($K_i > 37$ nmol/L) and partial agonist properties at the 5-HT$_{2C}$ receptor but lacked binding and functional selectivity over 5-HT$_{2A}$ and 5-HT$_{2B}$ receptors (Adams et al. 2000).

Several 5-HT$_{2C}$ receptor agonists with a tricyclic core were designed based on **47**. A first group of pyrroloquinoline derivatives originated from the fusion of the indole of **47** with a pyridine ring. An example is compound **51** (Fig. 3.5), which showed high 5-HT$_{2C}$ receptor affinity and an appreciable level of functional selectivity (5-HT$_{2C}$ $K_i = 9$ nmol/L; 5-HT$_{2A}$ $K_i = 45$ nmol/L; 5-HT$_{2B}$ $K_i = 12$ nmol/L; 5-HT$_{2C}$ EC$_{50} = 12$ nmol/L; 5-HT$_{2A}$ EC$_{50} = 360$ nmol/L). Other pyrroloquinolines possessed 5-HT$_{2C}$ affinities ranging between 22 and 156 nmol/L. These compounds acted as partial agonists and demonstrated relatively good functional selectivity (Adams et al. 2006).

In a second group, fusion of the indole of **47** with a cyclopentane ring gave the compounds with Structure V (Fig. 3.5). Although the pharmacological data were not exhaustive, it emerged that the S enantiomers displayed higher binding and functional selectivity over the 5-HT$_{2A}$ receptor. Variation of the indole ring substitution pattern resulted in moderate 5-HT$_{2C}$ receptor affinities (60 nmol/L $< K_i <$ 474 nmol/L) and low binding selectivity over the 5-HT$_{2A}$ receptor. Conversely, some of these agonists were endowed with remarkable functional selectivity as in the case derivative **52** (Fig. 3.5, Structure V, R$_7$ = F; R$_8$ = Cl; 5-HT$_{2C}$ EC$_{50} = 36$ nmol/L; 5-HT$_{2A}$ EC$_{50} = 2,102$ nmol/L) (Bentley et al. 2001).

The pyrazino[1,2a]indole framework (Fig. 3.6) has been extensively used for the design of 5-HT$_{2C}$ receptor agonists because it combines the structural features of mCPP (**33**) and of the indole derivative RO600175 (**47**).

The unsubstituted pyrazino[1,2a]indole **53** (Fig. 3.6) possessed low affinity for the 5-HT$_{2C}$ receptor (p$K_i = 5.7$), whereas the 10-methoxy substituted analog **54** (Fig. 3.6, Structure I, R = H) demonstrated much higher affinity (p$K_i = 7.1$). For this reason, a series of 10-methoxypyrazino[1,2-a]indoles bearing various substituents on the aromatic ring were evaluated (Fig. 3.6, Structure I). The affinity for the 5-HT$_{2C}$ receptor depended only upon the size of the substituent – neither the electronic properties nor the position on the aromatic ring were relevant. In fact, when R = H, 9-F, 9-CH$_3$ low affinities were found (7.1 < pK_i < 7.6). By contrast, compounds with R = 6-Br, 8-Br, 8-Cl, 9-CH(CH$_3$)$_2$ showed high 5-HT$_{2C}$ receptor affinities (8.0 < pK_i < 8.3) and partial agonist properties (Bös et al. 1997b).

A group of compounds formally derived from **53** by saturation of pyrazino[1,2a] indole double bond (Fig. 3.6, Structure II) displayed acceptable agonistic properties at 5-HT$_{2C}$ receptors (13 nmol/L < EC$_{50}$ < 162 nmol/L). Of the two enantiomers of **55** (Structure II, R = 7-Cl) the R- was more potent than the S- (EC$_{50}$ = 3 and 161 nmol/L, respectively).

The introduction of a CH$_3$ group in 4-position of the Structure II (Fig. 3.6) generated an additional chiral centre. The new compounds have two stereogenic centers and a maximum of four enantiomers. The role of stereochemistry on 5-HT$_{2C}$ activity has been studied in the case of the four stereoisomers **56–59** (Table 3.4). Compound **56** (stereochemistry $4R,10aR$) acted as full agonist with the highest affinity for all 5-HT$_2$ receptors. The corresponding enantiomer **57** was about 100-fold less potent and acted as partial agonist. Compound **58** (stereochemistry

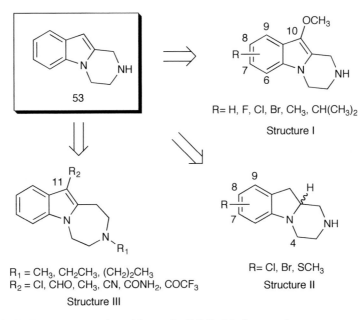

Fig. 3.6 5-HT$_{2C}$ receptor agonists with pyrazino[1,2a]indole framework

4S,10aR) and its enantiomer **59** were essentially full agonists and showed appreciable affinities for the 5-HT$_{2C}$ receptor with some selectivities against 5-HT$_{2B}$ receptors (7–16-fold). Based on these data, **56** was modified further (compounds **60–65**, Table 3.4). Small lipophilic substituents at the 7 or 6 position (Cl, CN, CF$_3$, and CH$_3$) led to high affinity for the 5-HT$_{2C}$ receptor, whereas polar substituents (NHAc, CH$_2$OH) led to a loss in affinity. Introduction of a 6-CH$_3$ substituent in compounds **56** and **60** led to **66** and **67**, respectively (Table 3.4). Derivative **67** showed remarkable subnanomolar agonist property and **66** was also selective over 5-HT$_{2A}$ and 5-HT$_{2B}$ receptors (≥10-fold) (Bentley et al. 2002; Rover et al. 2005).

Homologation of the nitrogen containing ring of pyrazino[1,2a]indole **53** gave rise to a series of azepino[1,7-a]indoles with Structure III (Fig. 3.6). Introduction of various alkyl groups on the basic nitrogen or of a range of substituents in 11-position gave compounds with modest 5-HT$_{2C}$ receptor affinity. The most active compound of the series was **68** (Fig. 3.6, Structure III, R$_1$=R$_2$=H) with K_i values of 4.8 and 18 nmol/L at 5-HT$_{2C}$ and 5-HT$_{2A}$ receptors, respectively (Ennis et al. 2003).

3.3.3 Arylpiperazine Isosters

As pointed out above, arylpiperazines are poorly selective for 5-HT$_{2C}$ receptors. The search for basic moieties alternative to arylpiperazine and specifically acting at the 5-HT$_{2C}$ receptor has been pursued by researchers at Athersys Inc

Table 3.4 Binding affinities and relative efficacies of 5-HT$_{2C}$ receptor agonists[a]

Compound		Stereochemistry	5-HT$_{2C}$	5-HT$_{2A}$	5-HT$_{2B}$	Relative efficacy 5-HT$_{2C}$ (%)
			K_i, nmol/L			
56	7-Cl	4R,10aR	2.2	1.9	15	97
57	7-Cl	4S,10aS	180	430	2,600	62
58	7-Cl	4S,10aR	13	55	210	94
59	7-Cl	4R,10aS	22	61	160	87
60	7-CH$_3$	4R,10aR	1.3	22	21	100
61	7-CF$_3$	4R,10aR	1.4	7.5	11	94
62	7-CN	4R,10aR	3.8	110	90	77
63	7-NHAc	4R,10aR	680	1,100	960	24
64	7-CH$_2$OH	4R,10aR	34	290	250	97
65	6-CH$_3$	4R,10aR	2.6	43	59	100
66	6-CH$_3$, 7-Cl	4R,10aR	0.3	2.6	3.2	98
67	6,7-di-CH$_3$	4R,10aR	1.7	34	13	96

[a]Receptor binding assays were performed by using CHO cell line expressing recombinant human 5-HT$_{2C}$, 5-HT$_{2A}$ and 5-HT$_{2B}$ receptors. Determination of affinities were made using [^{3}H]-5-HT for 5-HT$_{2C}$ and 5-HT$_{2B}$ receptors, and [^{125}I]-DOI for 5-HT$_{2A}$ receptors. The functional activity was determined in CHO cells using a Fluorometric Imaging Plate Reader. The maximum fluorescent signal was measured with the response produced by 10 μmol/L 5-HT

(Fig. 3.7). Initially, compound **69** was identified (EC$_{50}$ = 0.1 μmol/L), then the homopiperazine ring was bioisosterically replaced with the 2,7-diazabicyclo[3.3.0]octane system (compound **70**, EC$_{50}$ = 180 nmol/L). Of the possible enantiomers of **70**, the best was the S,S isomer (EC$_{50}$ = 23 nmol/L). The replacement of the pyrimidine ring with a phenyl gave the compound **71** that showed acceptable activity (EC$_{50}$ = 103 nmol/L) but poor selectivity. Next, removal of the nonbasic nitrogen afforded the moderately potent agonist **72** (5-HT$_{2C}$: EC$_{50}$ = 420 nmol/L; 5-HT$_{2A}$: EC$_{50}$ = 1,080 nmol/L; 5-HT$_{2B}$: EC$_{50}$ = 187 nmol/L) (Huck et al. 2006a). However, **72** presented unsuitable pharmacokinetic properties. In the search of better drug-like properties, the novel single-nitrogen motif was incorporated into a new tricyclic scaffold (Fig. 3.7, Structure I). The presence of CH$_3$ in 8-position (absolute configuration S) was beneficial for the specificity and two substituents on the aromatic ring (especially a halogen atom) enhanced activity. The most active compounds of this series (**73–80**) are listed in Table 3.5 (Huck et al. 2006b).

Another tricyclic system that retained the three-dimensional orientation of the basic amine in relation to the core phenyl ring in a similar manner as the compounds

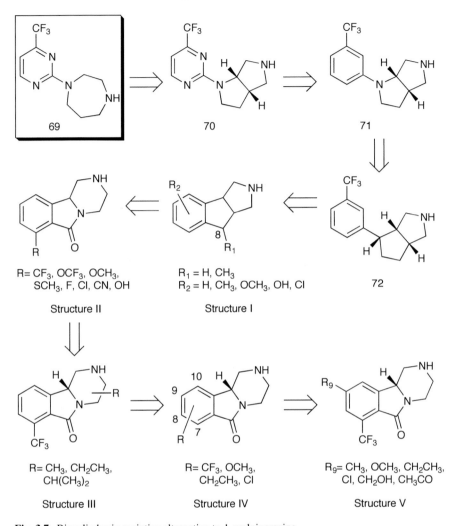

Fig. 3.7 Bicyclic basic moieties alternative to 1-arylpiperazine

73–80 is depicted in Fig. 3.7 (Structure II). Various substituents in 7 position of the racemic pyrazinoisoindolone system were evaluated. CF$_3$, OCF$_3$, and SCH$_3$ substituents were preferred due to their reduced electron-donating properties rather than the hydrophobic ones. From the optical resolution of the racemic mixture of the derivative **81** (Fig. 3.7, Structure II, R=CF$_3$), it emerged that the *R*-enantiomer was significantly more potent than the *S*-enantiomer at 5-HT$_{2C}$ receptor (EC$_{50}$=7 and 1,840 nmol/L, respectively). Therefore, for the subsequent modifications, only the *R*-enantiomers were considered. Insertion of an alkyl on various positions of the fused piperazine ring of *R*-**81** was detrimental for activity, especially when the steric hindrance was introduced near the basic nitrogen

Table 3.5 Binding affinities of 5-HT$_{2C}$ receptor agonists[a]

Compound	R$_5$	R$_6$	R$_7$	K_i, nmol/L	Selectivity vs (fold)	
				5-HT$_{2C}$	5-HT$_{2A}$	5-HT$_{2B}$
73	OCH$_3$	Cl	H	14	73	235
74	OCH$_3$	Br	H	27	12	81
75	OC$_2$H$_5$	Cl	H	231	9	31
76	OH	Cl	H	27	27	63
77	CH$_3$	Cl	H	155	8	32
78	Cl	CH$_3$	H	134	11	14
79	Cl	Cl	H	155	8	32
80	H	Cl	Cl	5	50	73

[a]Experimental details were not presented in the original publication

(Fig. 3.7, Structure III). Investigation of the substitution pattern of the aryl ring (Fig. 3.7, Structure IV) indicated that 7,9-disubstituted derivatives presented the most interesting properties, because although their potency was not extremely high (49 nmol/L < EC$_{50}$ < 86 nmol/L), they displayed >100-fold functional selectivity over 5-HT$_{2A}$ and 5-HT$_{2B}$ receptors. Conversely, 7,10-disubstituted derivatives were significantly less potent, whereas 7,8- and 8,9-disubstituted compounds displayed high potency but little or no selectivity. Particular attention has been devoted to the 7,9-disubstituted derivatives bearing CF$_3$ in 7 position (Fig. 3.7, Structure V). The presence in 9 position of -Cl or polar functionalities (Ac, CH$_2$OH) decreased activity, whereas 9-OCH$_3$ and 9-OCH$_2$CH$_3$ gave acceptable functional potency at 5-HT$_{2C}$ (EC$_{50}$ = 11 and 63 nmol/L, respectively). The 9-CH$_3$-substituted derivative was a very potent 5-HT$_{2C}$ agonist (EC$_{50}$ = 6 nmol/L) that was 27- and 47-fold functionally selective over 5-HT$_{2A}$ and 5-HT$_{2B}$, respectively. Increasing the size of the C9 substituent resulted in a decrease in 5-HT$_{2C}$ potency; the propyl, butyl, and isopropyl derivatives no longer exhibited high 5-HT$_{2C}$ functional efficacy but showed high excellent functional selectivity. The 9-ethyl derivative possessed good 5-HT$_{2C}$ functional potency (EC$_{50}$ = 16 nmol/L) and was >300-fold functionally selective over 5-HT$_{2B}$ and 73-fold selective over 5-HT$_{2A}$ (Wacker et al. 2007).

3.3.4 Benzazepine Derivatives

The arylethylamine motif, which is present in a number of nonselective 5-HT$_{2C}$ agonists, including 5-HT and norfenfluramine (**82**, Fig. 3.8) has been the starting point for the design of a series of 3-benzazepine derivatives

norfenfluramine (82)   83   X = H, F, Cl, Br, CF$_3$, OCH$_3$

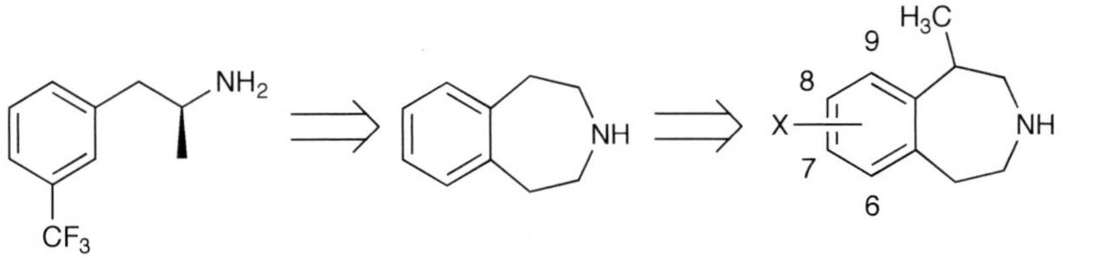

Structure I

Fig. 3.8 Development of 5-HT$_{2C}$ receptor agonists with 3-benzazepine structure

(Fig. 3.8, Structure I). In this structure type, the arylethylamine motif is constrained into a bicyclic system, which would reduce the number of available conformations to such a degree that target selectivity could be altered. During the development of this series only the functional activity of the target compounds at the h5-HT$_{2C}$ (INI isoform), h5-HT$_{2A}$, and h5-HT$_{2B}$ receptors was assessed by measuring the [^{3}H]IP turnover. The unsubstituted 3-benzazepine (**83**) displayed low potency at the 5-HT$_{2C}$ receptor (pEC$_{50}$ = 5.5), whereas the introduction of a methyl in 1-position (compound **84**: Fig. 3.8, Structure I, X = H) resulted in an increase of potency at the 5-HT$_{2C}$ receptor (pEC$_{50}$ = 6.5). The search for the optimal aromatic substitution pattern on **84** was accomplished. Monosubstitution at the 8-position (X = Cl, Br, CF$_3$) or at the 7-position (X = Cl) gave good 5-HT$_{2C}$ potency (7.9 < pEC$_{50}$ < 8.1), whereas the presence of OCH$_3$ or F at 7 or 8 position led to a reduction in potency (5.9 < pEC$_{50}$ < 6.7). 6- or 9-Cl substituted compounds also showed low potency (pEC$_{50}$ = 6.1). Introduction of an additional substituent in 7 or 9 position of the 8-chloro-1-methylbenzazepine (compound **85**, Table 3.6) resulted in potent compounds (8.1 < pEC$_{50}$ < 8.4), regardless of the nature of the substituent (Cl, F, OCH$_3$). The role of C1 stereochemistry was investigated for compounds **85–90**. Biological data revealed some stereospecificity but not all the eutomers showed the same absolute configuration (Table 3.6). The most selective compounds were the benzazepines S-**89** and S-**90** that were at least 40- and 400-fold selective over 5-HT$_{2A}$ and 5-HT$_{2B}$ receptors, respectively (Smith et al. 2008).

3.4   Conclusions

Among the 14 receptor subtypes of serotonin, the 5-HT$_{2C}$ subtype has received much attention because of its possible involvement in clinically relevant disease conditions such as schizophrenia and obesity. The studies performed by researchers at GlaxoSmithKline have led to the identification of several potent and selective 5-HT$_{2C}$ receptor antagonists. The large body of data on antagonists has allowed a detailed description of a pharmacophore model for 5-HT$_{2C}$ receptor antagonists as

Table 3.6 Intrinsic activities of 5-HT$_{2C}$ receptor agonists with 3-benzazepine structure[a]

Compound	X	5-HT$_{2C}$	5-HT$_{2A}$	5-HT$_{2B}$
			pEC$_{50}$	
85 (racemic)	8-Cl	7.9	6.7	6.0
R-85 (lorcaserin)	8-Cl	8.1	6.8	6.1
S-85	8-Cl	7.8	6.6	5.9
R-86	8-CF$_3$	8.1	6.9	6.3
S-86	8-CF$_3$	8.0	7.0	6.1
R-87	7,8-di-Cl	8.4	8.0	7.4
S-87	7,8-di-Cl	8.1	7.0	6.6
R-88	8-Cl, 7-OCH$_3$	8.1	6.7	6.4
S-88	8-Cl, 7-OCH$_3$	8.2	7.5	7.4
R-89	8,9-di-Cl	6.5	5.6	<5
S-89	8,9-di-Cl	8.5	6.9	5.9
S-90	8-Cl, 9-F	8.4	6.0	<5

[a]The functional activities were determined in HEK293 cells overexpressing human 5-HT$_{2A}$, 5-HT$_{2B}$, or 5-HT$_{2C}$ receptor with the intracellular inositol triphosphate (IP$_3$) accumulation assay

well as the description of the topography of the antagonist binding site of the receptor. On the other hand, studies from different laboratories have led to the identification of functionally selective 5-HT$_{2C}$ receptor agonists. This was a difficult task because the 5-HT$_{2C}$ receptor structure is closely related to that of the 5-HT$_{2A}$ and 5-HT$_{2B}$ receptor subtypes. Although information is available to describe the structure–activity relationships of the agonist, no pharmacophore model has been proposed to date. In recent years two 5-HT$_{2C}$ agonists [lorcaserin (R-85) and ATHX-105] entered phase II clinical trials for treatment of obesity. The outcome of these studies will reveal the validity of the 5-HT$_{2C}$ receptor as a pharmacologically treatable target and will likely fuel the search for newer selective agents for 5-HT$_{2C}$ receptors.

References

Adams DR, Bentley JM, Roffey JRA, et al (2000) Preparation of indazolylpropylamines as serotonin 5-HT2B and/or 5-HT2C agonists. PCT Int Appl WO 2000012481, 9 Mar 2000.

Adams DR, Bentley JM, Benwell KR, et al (2006) Pyrrolo(iso)quinoline derivatives as 5-HT2C receptor agonists Bioorg Med Chem Lett 16:677–680

Bös M, Jenck F, Martin JR, et al (1997a) Synthesis and biological evaluation of substituted 2-(indol-1-yl)-1-methylethylamines and 2-(indeno[1,2-b]pyrrol-1-yl)-1-methylethylamines. Improved therapeutics for obsessive compulsive disorder. J Med Chem 40:2762–2769.

Bös M, Jenck, F, Martin JR, et al (1997b) Synthesis, pharmacology and therapeutic potential of 10-methoxypyrazino[1,2-a]indoles, partial agonists at the 5HT2C receptor. Eur J Med Chem 32:253–261.

Bromidge SM, Dabbs S, Davies DT, et al (1998) Novel and selective 5-HT2C/2B receptor antagonists as potential anxiolytic agents: synthesis, quantitative structure-activity relationships, and molecular modeling of substituted 1-(3-pyridylcarbamoyl)indolines. J Med Chem 41:1598–1612.

Bromidge SM, Dabbs S, Davies DT, et al (1999) Model studies on a synthetically facile series of N-substituted phenyl-N'-pyridin-3-yl ureas leading to 1-(3-pyridylcarbamoyl) indolines that are potent and selective 5-HT2C/2B receptor antagonists. Bioorg Med Chem 7: 2767–2773.

Bromidge SM, Dabbs S, Davies DT, et al (2000a) Biarylcarbamoylindolines are novel and selective 5-HT2C receptor inverse agonists: identification of 5-methyl-1-[[2-[(2-methyl-3-pyridyl)oxy]-5-pyridyl]carbamoyl]-6-trifluoromethylindoline (SB-243213) as a potential antidepressant/anxiolytic agent. J Med Chem 43:1123–1134.

Bromidge SM, Dabbs S, Davies S, et al (2000b) 1-[2-[(Heteroaryloxy)heteroaryl]carbamoyl]indolines: novel and selective 5-HT2C receptor inverse agonists with potential as antidepressant/ anxiolytic agents. Bioorg Med Chem Lett 10:1863–1866.

Bromidge SM, Davies S, Duckworth DM, et al (2000c) 1-[2-[(Heteroarylmethoxy)aryl]carbamoyl] indolines are selective and orally active 5-HT2C receptor inverse agonists. Bioorg Med Chem Lett 10:1867–1870.

Briner K, Burkholder TP, Heiman ML, et al (2001) Benzofurylpiperazine serotonin agonists. PCT Int Appl WO 2001009123, 8 Feb 2001.

Bentley JM, Roffey JRA, Davidson JEP, et al (2001) Preparation of indole derivatives as agonists or antagonists of a 5-HT receptor, particularly a 5-HT2C receptor. PCT Int. Appl. WO 2001012603, 22 Feb 2001.

Bentley JM, Hebeisen P, Muller M, et al (2002) Preparation of 1,2,3,4,10,10a-hexahydro-1H-pyrazino[1,2-a]indoles and analogs and 5-HT receptor agonists for treatment of CNS diseases, cardiovascular disorders, gastrointestinal disorders, and obesity. PCT Int. Appl. WO 2002010169, 7 Feb 2002.

Bentley JM, Adams DR, Bebbington D, et al (2004) Indoline derivatives as 5-HT2C receptor agonists. Bioorg Med Chem Lett 14:2367–2370.

Di Giovanni G, Di Matteo V, Pierucci M, et al (2006) Central serotonin2C receptor: from physiology to pathology. Curr Top Med Chem 6:1909–1925.

Ennis MD, Hoffman RL, Ghazal NB (2003) 2,3,4,5-Tetrahydro- and 2,3,4,5,11,11a-hexahydro-1H-[1,4]diazepino[1,7-a]indoles: new templates for 5-HT2C agonists. Bioorg Med Chem Lett 13:2369–2372.

Goodacre CJ, Bromidge SM, Clapham D, et al (2005) A series of bisaryl imidazolidin-2-ones has shown to be selective and orally active 5-HT2C receptor antagonists. Bioorg Med Chem Lett 15:4989–4993.

Huck BR, Llamas L, Robarge MJ, et al (2006a) The identification of pyrimidine-diazabicyclo[3.3.0] octane derivatives as 5-HT$_{2C}$ receptor agonists. Bioorg Med Chem Lett 16:2891–2894.

Huck BR, Llamas L, Robarge MJ, et al (2006b) The design and synthesis of a tricyclic single-nitrogen scaffold that serves as a 5-HT$_{2C}$ receptor agonist. Bioorg Med Chem Lett 16:4130–4134.

Hamprecht D, Micheli F, Tedesco G, et al (2007a) 5-HT2C antagonists based on fused heterotricyclic templates: design, synthesis and biological evaluation. Bioorg Med Chem Lett 17:424–427.

Hamprecht D, Micheli F, Tedesco G, et al (2007b) Isoindolone derivatives, a new class of 5-HT2C antagonists: synthesis and biological evaluation. Bioorg Med Chem Lett 17:428–433.

Leysen D, Kelder J (1998) Ligands for the 5-HT2C receptor as potential antidepressants and anxiolytics. In: van der Goot H (Ed) Trends in Drug Research II, 11th Noordwijkerhout-Camerino Symposium; vol 29. Elsevier, Amsterdam, p 49.

Micheli F, Pasquarello A, Tedesco G, et al (2006) Diaryl substituted pyrrolidinones and pyrrolones as 5-HT2C inhibitors: synthesis and biological evaluation. Bioorg Med Chem Lett 16:3906–3912.

Nilsson BM (2006) 5-Hydroxytryptamine 2C (5-HT2C) receptor agonists as potential antiobesity agents. J Med Chem 49:4023–4034.

Nilsson B, Scobie M (2002) Preparation of piperazinylpyrazines as antagonists of serotonin 5-HT2 receptor. PCT Int Appl WO 2002040457, 23 May 2002.

Nilsson B, Tejbrant J, Pelcman B, et al (2002) Preparation of piperazinylpyrazinyl aryloxyalkyl ethers as 5-HT2C receptor agonists. US Patent 6465467, 15 Oct 2002.

Porter RHP, Benwell KR, Lamb H, et al (1999) Functional characterization of agonists at recombinant human 5-HT2A, 5-HT2B and 5-HT2C receptors in CHO-K1 cells. Br J Pharmacol 128:13–20.

Smith B, Tsai J, Chen R (2005) Preparation of N-phenyl-piperazine derivatives and methods of prophylaxis or treatment of 5-HT2C receptor associated diseases. PCT Int Appl WO 2005016902, 24 Feb 2005.

Smith BM, Smith JM, Tsai JH, et al (2008) (1R)-8-Chloro-2,3,4,5-tetrahydro-1-methyl-1H-3-benzazepine (Lorcaserin), a selective serotonin 5-HT$_{2C}$ receptor agonist for the treatment of obesity. J Med Chem 51:205–313.

Rover S, Adams DR, Benardeau A, et al (2005) Identification of 4-methyl-1,2,3,4,10,10a-hexahydropyrazino[1,2-a]indoles as 5-HT2C receptor agonists. Bioorg Med Chem Lett 15:3604–3608.

Wacker DA, Varnes JG, Malmstrom SE, et al (2007) Discovery of (R)-9-ethyl-1,3,4,10b-tetrahydro-7-trifluoromethylpyrazino[2,1-a]isoindol-6(2H)-one, a selective, orally active agonist of the 5-HT$_{2C}$ receptor. J Med Chem 50:1365–1379.

Chapter 4
Insights into 5-HT$_{2C}$ Receptor Function Gained from Transgenic Mouse Models

Stephen J. Bonasera

4.1 Introduction: Genetically Engineered Mouse Models – Strengths and Caveats

Reliable methods to generate specific DNA constructs targeted to specific regions of the mouse genome, in combination with advances in mouse embryonic stem cell technology, make it now possible to create lesions of any mouse gene locus of interest (for reviews, see Bonasera and Tecott 2000; Gaveriaux-Ruff and Kieffer 2007; Miyoshi and Fishell 2006). Mouse models generated using these kinds of genetic "dissecting tools" provide powerful and unique insights into the functional roles of specific genes at many different scales of organization. In fact, recently developed approaches allow the investigator to limit transgene expression to a specific developmental line of cells or switch transgene expression in response to drugs or introduction of viral vectors. These techniques offer a nearly limitless palette of tools for mechanistic studies associating gene expression with gene function. As long as one keeps in mind the limitations of these approaches, transgenic mouse models will remain important tools to elucidate the role that specific molecules contribute in organism development, cellular and tissue physiology, whole organ function, and organization of animal behavior.

Our current understanding of how serotonin-2C receptors (5-HT$_{2C}$Rs) influence central nervous system (CNS) development, neuronal physiology, and whole animal behavior owes much to the above-mentioned transgenic mouse technology. Currently, intense study has focused on the first 5-HT$_{2C}$R transgenic mouse model initially developed in the laboratory of Laurence H. Tecott (Tecott et al. 1995). In this model, a targeting vector complementary to the *htr2c* 5th-exon coding region containing an in-frame stop mutation was introduced into mouse embryonic stem cells. Mice derived from these embryonic stem cells were screened for animals that had the mutation present in germ line cells. These chimeras were the founders of the

S.J. Bonasera (✉)
Department of Medicine/Division of Geriatrics, University of Nebraska
Medical Center, Omaha, NE, 68190-5039, USA
e-mail: sbonasera@unmc.edu

G. Di. Giovanni et al. (eds.), *5-HT$_{2C}$ Receptors in the Pathophysiology of CNS Disease*,
The Receptors 22, DOI 10.1007/978-1-60761-941-3_4,
© Springer Science+Business Media, LLC 2011

htr2c⁻ mutant mouse line. Chimeras in turn were initially bred onto a hybrid mouse background B6D2 (an F1 hybrid between C57BL/6 and DBA/2) to take advantage of the hybrid's better breeding success. Later mice arising from this colony were bred to a pure C57BL/6 background, with the first mice congenic to C57BL/6 being produced around 2001. Since the mutation introduced into these transgenic mice produced a truncated 5-HT$_{2C}$R protein with no biological function, mutant mice descending from this line are characterized by having a complete lack of 5-HT$_{2C}$R function throughout their entire life span.

To fully appreciate the strengths and limitations of data coming from transgenic 5-HT$_{2C}$R mouse models, a few aspects of important mouse biology should be mentioned. Since the *htr2c* locus is X linked, the only genotypes possible when breeding a heterozygous female to a wild type male will be either wild-type or 5-HT$_{2C}$R mutant (in male mice) and wild-type or 5-HT$_{2C}$R heterozygotes (in female mice). To obtain female 5-HT$_{2C}$R mutant mice, one must cross 5-HT$_{2C}$R heterozygous female mice with 5-HT$_{2C}$R mutant males (this cross will also yield 5-HT$_{2C}$R mutant males and 5-HT$_{2C}$R heterozygous females). Thus, standard mouse breeding practices will readily produce populations of wild-type and littermate 5-HT$_{2C}$R mutant male mice, wild-type and littermate 5-HT$_{2C}$R heterozygous female mice, and heterozygous and 5-HT$_{2C}$R mutant female mice. However, no single cross can produce both wild-type and 5-HT$_{2C}$R mutant female mice. Thus, studies of 5-HT$_{2C}$R receptor function in female mice must address potential issues of either differential maternal care or cross-fostering of mouse pups. Additionally, inactivation of X-linked genes in females occurs in a mostly stochastic manner (Ohlsson et al. 2001). This leads to mouse-to-mouse differences in the precise 5-HT$_{2C}$R dose in female heterozygotes, and it is thus an additional source of experimental variation to be addressed in studies of 5-HT$_{2C}$R function in females.

Finally, the mouse genetic "background" chosen to carry the mutant allele may be a particularly important factor in interpreting experimental data. Currently available 5-HT$_{2C}$R mutant mice have undergone a rigorous backcrossing to a C57BL/6 genetic background, and most of the studies published regarding this mouse employ animals more than 20 backcrosses from the original F1 hybrids. While this degree of backcrossing is ample to remove any residual genetic background contributed by ES cells, DNA sequences flanking the transgene and included in the initial recombination event may remain even after 20 backcrosses. Controlling for this issue requires obtaining a second independent transgenic mouse line (that would certainly not have integrated the transgene in exactly the same way as the first line), and repeating behavioral assessments in this line. This highly rigorous control is often not available for a variety of reasons.

Alternatively, concerns about flanking DNA effects behavioral phenotypes can be allayed by comparing observed mutant phenotypes with those obtained by pharmacological inhibition or activation of the relevant gene target. Thus, most of the literature describing the phenotypes of 5-HT$_{2C}$R mutant mice evaluate animals that are otherwise mostly congenic to C57BL/6. C57BL/6 is a mouse strain particularly appropriate for studies of mouse behavior. Compared with most strains, C57BL/6 demonstrates better learning and memory task acquisition on available cognitive batteries, lower indices of trait anxiety-related behaviors, greater propensity for ethanol and cocaine ingestion, and less sensitivity to factors that change sensorimotor gating (Crawley et al. 1997; Holmes et al. 2002). C57BL/6 mice also have a relatively high seizure threshold. However, with age these mice demonstrate sensorineural hearing loss, diet-related obesity, type II diabetes, and atherosclerosis.

4.2   Insights into G-Protein-Coupled Receptor Sensitization/ Desensitization Dynamics from 5-HT$_{2C}$R Mutant Mouse Models

Like many G-protein-coupled transmembrane receptors, 5-HT$_{2C}$Rs contain a carboxy-terminal phosphorylation domain (PDZ) recognition site that participates in regulation of receptor ligand-binding sensitivity. In mutant mice bearing functional 5-HT$_{2C}$Rs without this PDZ recognition site (5-HT$_{2C}$RΔPDZ), as anticipated, no differences in serotonin binding kinetics, maximal agonist evoked inositol phosphate formation, or intracellular calcium dynamics were appreciated. However, 5-HT$_{2C}$Rs from 5-HT$_{2C}$RΔPDZ [+]/Y mice were unable to incorporate ^{32}P after ligand binding, and displayed attenuated responses to a second round of agonist stimulation given 2.5 min after an initial stimulation (Backstrom et al. 2000). These results provide genetic evidence that the loss of the PDZ binding site impairs G-protein-coupled receptor sensitization/desensitization processes, and clearly indicate that this process plays an important role in the molecular function of 5-HT$_{2C}$Rs. How this difference in receptor desensitization affects behavior remains unexplored, and is an important future avenue for further studies.

4.3   5-HT$_{2C}$R-Mediated Regulation of Ascending Serotonergic Neurotransmission

Serotonergic neurotransmission effects CNS systems through the binding of serotonin to one of 14 currently described receptors. Discussion of the known properties of these receptors is beyond the scope of this text, and the reader may enjoy many excellent reviews for further information (Gordon and Hen 2004; Bockaert et al. 2008; Roth 2006; Di Giovanni et al. 2006; Hedlund and Sutcliffe 2004; Thomas 2006; Woolley et al. 2004; Barnes and Sharp 1999). Mouse models engineered to lack a specific serotonin receptor are ideal for studying that receptor's effect on serotonergic function spanning molecular to behavioral scales. This statement is particularly true if parallel studies treating intact mice with specific antagonists or inhibitors of the targeted receptor replicate the mutant phenotype. In this manner, studies using 5-HT$_{2C}$R transgenic mouse models have provided important insights regarding how these receptors interact with other components of the serotonergic system.

Despite the key regulatory role of 5-HT$_{2C}$Rs in many serotonin related behaviors, constitutive loss of 5-HT$_{2C}$R function evokes no compensatory increase in the expression of 5-HT$_{1A}$, 5-HT$_{1B/1D}$, 5-HT$_{1F}$, 5-HT$_{2A}$, 5-HT$_4$, and 5-HT$_7$ receptors or in 5-HT transporter protein expression (as measured by quantitative receptor autoradiography; see Lopez-Gimenez et al. 2002). Consistent with these findings was the observation that microiontophoretic application of serotonin to neurons within either the orbitofrontal cortex or the striatum led to similar inhibition of neuronal firing rates in both wild-type and constitutive 5-HT$_{2C}$R $^-$/Y mice (Rueter et al. 2000). However, loss of 5-HT$_{2C}$Rs evoked a more global change in regulation of pyramidal neuron excitability in these two regions, as suggested by the finding that one could activate neurons from 5-HT$_{2C}$R $^-$/Y mice with less quisqualate-evoked excitation. Pyramidal cells from 5-HT$_{2C}$R $^-$/Y mice also fired at higher frequencies compared with cells from wild-type mice when excited with equivalent doses of quisqualate.

These findings suggest that loss of 5-HT$_{2C}$R function increases the excitability of cortical pyramidal cells in a currently undetermined manner independent of serotonergic neurotransmission. As will be discussed throughout this chapter, this global change in 5-HT$_{2C}$R-mediated neuron excitability may underlie a number of specific behavioral phenotypes observed in 5-HT$_{2C}$R mutant mice.

4.4   5-HT$_{2C}$R-Mediated Contributions to Cognitive Behaviors

Behaviors traditionally associated with cognitive function, including different aspects of long and short-term memory, behavioral planning, and behavioral inhibition are organized across a broad extent of the neuraxis and encompass regions of the cortex, basal ganglia, hippocampus, amygdala, and extended amygdala. 5-HT$_{2C}$R expression has been demonstrated in all of these regions (Eberle-Wang et al. 1997; Filip and Cunningham 2002; Huang et al. 2004; Krishnakumar et al. 2009; Pompeiano et al. 1994; Serrats et al. 2005). Thus, 5-HT$_{2C}$Rs may potentially regulate a large number of cognitive behaviors.

How 5-HT$_{2C}$Rs modulate hippocampal function is currently best understood. Prior studies have demonstrated that constitutive 5-HT$_{2C}$R $^-$/Y mutant mice possess a specific deficit in long term potentiation at the first major synaptic integration point of the hippocampus, the medial perforant pathway synapse onto dentate gyrus pyramidal cells (Tecott et al. 1998). This LTP deficit is not present in any of the other downstream hippocampal synaptic circuits, including the dentate to CA3 synapse, the CA3 to CA1 synapse, or the CA1 to subiculum synapse. 5-HT$_{2C}$R mutant mice have normal hippocampal cytoarchitecture when assessed by light microscopy of stained tissue, and demonstrate no significant differences in either hippocampal 5-HT$_{2A}$R expression or overall hippocampal catecholamine content compared with wild-type mice.

This specific deficit in hippocampal LTP had a testable behavioral correlate in the Morris water maze assay of spatial memory. While both wild type and constitutive

5-HT$_{2C}$R $^-$/Y mice demonstrated similar performances in learning trials to find the location of the submerged platform, constitutive 5-HT$_{2C}$R $^-$/Y mutant mice did not focus their search within the trained quadrant on performance of the "probe" trial. This deficit in place memory is often observed in animals with deficits in information processing at the perforant path to dentate gyrus synapse (Xavier and Costa 2009). Of note, no genotypic differences in "freezing" behavior evoked by exposure to an environment where the mice had previously received a mild aversive stimulus were observed. Thus, loss of 5-HT$_{2C}$R function leads to a very selective deficit in hippocampal learning.

4.5   5-HT$_{2C}$R-Mediated Contributions to Affective Behaviors

The prominent expression of 5-HT$_{2C}$Rs in deeper cortical structures, including the bed nucleus of stria terminalis (BNSt) and amygdalar nuclei (Huang et al. 2007; Pickering et al. 2007; Campbell and Merchant 2003), suggest that 5-HT$_{2C}$R function may modulate responses to affective stimuli. Additionally, many drugs used to treat affective disorders, including newer generation "atypical" antipsychotic medications, demonstrate significant binding affinity at the 5-HT$_{2C}$R (Di Matteo et al. 2002). As summarized below, recent research has demonstrated that 5-HT$_{2C}$Rs play an important role in modulating responses to anxiogenic and aversive stimuli.

Mice constitutively lacking 5-HT$_{2C}$Rs show multiple measures of diminished anxiety-related behaviors in many exploratory tasks. Constitutive 5-HT$_{2C}$R $^-$/Y mice spent more time in the central region of an open field, crossed into the open arm of an elevated zero maze more frequently (and spent more time in this open arm), spent more time exploring a novel object placed in the home cage, and spent more time in a mirrored chamber compared with wild-type littermates (Heisler et al. 2007a). Taken together, these results suggest that constitutive loss of 5-HT$_{2C}$Rs is mildly anxiolytic. Furthermore, mice demonstrating this anxiolytic phenotype were noted to activate fewer cells in the BNSt (when staining for the early gene product c-fos) compared with wild-type littermates.

Constitutive 5-HT$_{2C}$R $^-$/Y mice also show markedly heightened responses to mild, unconditioned aversive stimuli, such as brief foot shocks. In a fear-sensitized startle test (where mouse responses to brief acoustic startle stimuli before and 5 min after a series of mildly aversive footshocks are compared; see Davis 1989), constitutive 5-HT$_{2C}$R $^-$/Y mice showed a marked increase in poststimulus startle responses, particularly at the lowest startle sound intensities (Bonasera and Tecott 2004). This finding was true even at minimal footshock intensities (that provoked no response in wild-type littermate controls) and was not associated with changes in nociceptive thresholds to heat or skin irritants (Basbaum, personal communication). Responses to the aversive stimulus could be attenuated in constitutive 5-HT$_{2C}$R $^-$/Y mice following treatment with the selective D$_2$ receptor antagonist raclopride but not with the selective D$_1$ receptor antagonist SCH 23390. Additionally, constitutive 5-HT$_{2C}$R $^-$/Y mice activated fewer cells in the BNSt (as assayed by c-fos immunocytochemistry) following footshock when compared with wild-type littermates (Bonasera et al. 2005). These studies suggest that 5-HT$_{2C}$Rs on populations of BNSt projection neurons are required to appropriately activate circuits within the extended amygdala that inhibit the expression of motor responses to unconditioned anxiogenic and aversive stimuli. These 5-HT$_{2C}$R-expressing extended amygdala neuronal populations may prove to be important targets of therapeutic interventions designed to treat highly prevalent and disabling anxiety conditions, such as generalized anxiety and posttraumatic stress disorder.

While initial studies of mice constitutively lacking 5-HT$_{2C}$Rs demonstrated no significant phenotypic differences in immobility responses to tail suspension or water immersion (Tecott, personal communication), recent studies suggest that antagonists of 5-HT$_{2C}$Rs may be promising adjuncts for selective serotonin inhibitor (SSRI)-based treatment plans for affective disorders. In vivo microdialysis studies demonstrate a marked phenotypic increase in hippocampal extracellular serotonin levels in constitutive 5-HT$_{2C}$R^{-}/Y mice treated with common clinically available SSRIs (Cremers et al. 2004). Furthermore, this finding was also observed in mice receiving combination therapy with SSRIs and selective 5-HT$_{2C}$R antagonists. These studies suggest that 5-HT$_{2C}$R functional antagonism may prove to be a useful adjunct to treat patients with depression resistant to single SSRI therapy.

4.6   5-HT$_{2C}$R-Mediated Regulation of Hypothalamic–Pituitary–Adrenal (HPA) Axis

Shortly after constitutive 5-HT$_{2C}$R^{-}/Y mice were made available to other investigators, it was appreciated that these animals were highly sensitive to environmental perturbations. For example, mutant mice housed in a room nearby an ongoing building renovation project did not eat as much or gain body weight at the same trajectory as same-aged littermates housed in a nearby, quiet room (Dallman et al. 1999). Anecdotally, many of the postdoctoral scholars and graduate students breeding and working with constitutive 5-HT$_{2C}$R^{-}/Y mice note that mutant mice are consistently more difficult to handle (and more reactive to handling) compared with same-aged wild-type littermates. Investigation of this phenotype has led to an appreciation of the role that 5-HT$_{2C}$Rs contribute toward regulation of the hypothalamic–pituitary–adrenal (HPA) stress response axis.

Using constitutive 5-HT$_{2C}$R^{-}/Y and wild-type mice, Heisler and colleagues (Heisler et al. 2007b) demonstrated that 5-HT$_{2C}$Rs controlled the serotonin-evoked release of cortisol-releasing hormone (CRH) from the hypothalamic paraventricular nucleus (PVN). In hypothalamic slice preparations, infusion of the serotonin transporter inhibitor d-fenfluramine (which will increase extracellular serotonin concentrations) as well as m-chlorophenylpiperazine (mCPP), a nonselective 5-HT$_{2C/1B}$R agonist, increases CRH release in slices taken from intact mice. However, neither drug evokes increased CRH release in slices taken from constitutive 5-HT$_{2C}$R^{-}/Y mice. Furthermore, baseline PVN CRH expression (as measured by in situ hybridization studies against CRH message) is decreased in mutant mice compared with wild-type controls. Interestingly, there were no phenotypic differences observed in the diurnal variation of plasma cortisol, while d-fenfluramine and mCPP increased plasma cortisol only in intact animals. These data demonstrate that serotonin-evoked changes in the HPA axis can be dissociated from processes regulating HPA circadian variation.

Mice constitutively lacking 5-HT$_{2C}$Rs also show a heightened sensitivity to mild repeated environmental stressors (Chou-Green et al. 2003a). After a 6-day battery of repeated chronic mild stressors, including bedding changes and restraint, middle-aged constitutive 5-HT$_{2C}$R $^{-}$/Y mice displayed significantly greater weight loss and decreased caloric efficiency compared with same-aged wild-type mice or young (4 month old) mice of either genotype. Further, the above changes in energy balance in these middle-aged mutant mice were accompanied by increased serum concentrations of adrenocorticotrophic hormone (ACTH), corticosterone, and insulin. A similar increase in plasma ACTH (but not corticosterone) was also noted in 10-day-old constitutive 5-HT$_{2C}$R $^{-}$/Y mouse pups following the acute stress of being placed in the center of a 25-cm^2 open field for 5 min (Akana 2008).

The above findings highlight the complexity of the process whereby differing environmental stimuli (including representations of ambient noise, handling, restraint, and nest disruptions) influence serotonergic neurotransmission and lead to alterations in hypothalamic activity that are in part transduced through 5-HT$_{2C}$Rs. Breaking this signal flow at the 5-HT$_{2C}$R in turn leads to hypoactivation of hypothalamic CRH signaling and ultimately significant disruptions of behavior that impair the appropriate responses to these environmental factors.

4.7 Disrupted Maternal Behavior in 5-HT$_{2C}$R Mutant Mice

The prominent expression of 5-HT$_{2C}$Rs in multiple hypothalamic nuclei and over the extended amygdala also suggests that these receptors may play a key role in coordinating maternal–pup interactions. To study this question, Storm and colleagues (Storm et al. 2003) bred female constitutive 5-HT$_{2C}$R $^{-/-}$ mice from crosses of heterozygote 5-HT$_{2C}$R $^{-}$/X female and 5-HT$_{2C}$R $^{-}$/Y male mice. These females were noted to have normal pup gestation and normal litter size. However, litters raised by constitutive 5-HT$_{2C}$R $^{-/-}$ mothers demonstrated much diminished survival to weaning compared with litters raised by 5-HT$_{2C}$R $^{-}$/X mothers. Multiple potential etiologies were identified that may explain the increased mortality of pups born to constitutive 5-HT$_{2C}$R $^{-/-}$ mothers. Compared with both heterozygote 5-HT$_{2C}$R $^{-}$/X and wild-type mothers, constitutive 5-HT$_{2C}$R $^{-/-}$ mothers demonstrated decreased placentophagy. Nests constructed by constitutive 5-HT$_{2C}$R $^{-/-}$ mothers were smaller and less elaborate. In assays of pup retrieval, constitutive 5-HT$_{2C}$R $^{-/-}$ mothers were far less likely to successfully bring a pup back to the nest. No deficits were noted in mammary gland histology of constitutive 5-HT$_{2C}$R $^{-/-}$ mice or in their milk production; however, these mothers spent significantly less time nursing their pups. Overall, these deficits suggest that constitutive 5-HT$_{2C}$R $^{-/-}$ mothers may have begun nursing their litters in a less nutritionally replete state compared with heterozygous or wild-type females. Additionally, deficits in nest building, nursing time, and pup retrieval suggest an elevated pup stress level that may have contributed to their increased mortality.

4.8   Dysregulated Function of Mesoaccumbal and Nigrostriatal Dopamine Systems in 5-HT$_{2C}$R Mutant Mice

5-HT$_{2C}$R mRNA is prominently expressed in the striatum (Str) and nucleus accumbens (NAc) (Eberle-Wang et al. 1997), two subcortical regions vital for coordinating many important behaviors such as motor program selection and responses to appetitive stimuli. The Str and NAc also play vital roles in how the phenotypes of Parkinson's disease and substance abuse, two highly morbid and difficult-to-treat clinical conditions, are expressed. Studies of basal ganglia physiology and behavior in 5-HT$_{2C}$R mutant mouse models have provided important insights into potential strategies to address these devastating problems (Fig. 4.1).

Neither substantia nigra pars compacta (SNc) or ventral tegmental area (VTA) dopaminergic neurons, the projection neurons for the nigrostriatal and mesoaccumbal tracts, respectively, express 5-HT$_{2C}$Rs. However, neurons of the substantia nigra pars reticulata (SNr), a significant source of input to both the SNc and VTA, highly express 5-HT$_{2C}$Rs (Di Giovanni et al. 2001; Invernizzi et al. 2007). There is also an extensive literature suggesting that in both mouse and rat models, pharmacological inhibition of the 5-HT$_{2C}$R evokes a marked increase in VTA firing rates (Di Giovanni et al. 1999; Di Giovanni et al. 2000), and a smaller, but still significant increase in SNc firing rates (Di Giovanni et al. 1999). Burst firing is also more prominent in VTA and SNc neurons following 5-HT$_{2C}$R pharmacological blockade (Di Giovanni et al. 1999; Gobert et al. 2000). It was thus interesting to observe that chloral hydrate anesthetized constitutive 5-HT$_{2C}$R $^-$/Y mice do not show a marked increase in VTA neuronal firing rate or bursting compared with wild-type littermates (Abdallah et al. 2009). By contrast, this same study demonstrated significant increases in baseline SNc neuronal firing and number of action potentials in bursts from SNc neurons of constitutive 5-HT$_{2C}$R $^-$/Y mice.

Although constitutive 5-HT$_{2C}$R $^-$/Y mutant mice show only a modest elevation of baseline firing rate in SNc projection neurons, there is a much more pronounced effect appreciated in extracellular dopamine concentrations at these neuron terminal fields. Using no net flux microdialysis, the most sensitive assay for detecting baseline differences in extracellular neurotransmitter concentrations, we noted higher extracellular dopamine concentrations in both Str and NAc of constitutive 5-HT$_{2C}$R $^-$/Y mice (Abdallah et al. 2009).

The above-described differences in ascending dopaminergic system regulation between intact and constitutive 5-HT$_{2C}$R $^-$/Y mice are associated with observable changes in how mice perform a number of important behaviors organized through the Str and NAc. Regarding striatal function, constitutive 5-HT$_{2C}$R $^-$/Y mice are more likely to inappropriately terminate a bout of syntactic grooming compared with

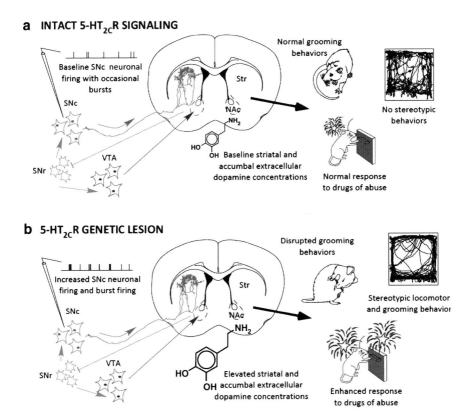

Fig. 4.1 Serotonin$_{2C}$-receptor-mediated actions on ascending dopaminergic function. (**a**) Under baseline conditions, dopaminergic neurotransmission in the striatum and nucleus accumbens is regulated through activity of the nigrostriatal and mesoaccumbal projections. Important target behaviors, including grooming, exploration of novel environments, and responses to psychstimulant drugs demonstrate characteristic normal phenotypes. (**b**) With constitutive loss of 5-HT$_{2C}$R function, there is a significant increase in nigrostriatal dopaminergic activity, as measured by both increased neuronal firing rates and increased burst firing of SNc neurons. This change in turn leads to a significant increase in extracellular dopamine concentrations in both the striatum and nucleus accumbens. 5-HT$_{2C}$R loss in basal ganglia nuclei distal to the striatum in turn contributes to observed dysruption of a number of target behaviors, including aberrant grooming, increased stereotypic motor patterns, and markedly enhanced motivation to self-administer psychostimulants

wild-type mice (Abdallah et al. 2009). Further, constitutive loss of 5-HT$_{2C}$R function effects syntactic grooming in a manner similar to drugs that increase dopaminergic tone. Constitutive 5-HT$_{2C}$R $^-$/Y mice also display compulsive behaviors. For example, mutant mice will chew more nonnutritive clay, neatly trim plastic circular mesh, and engage in frequent head-dipping behaviors compared with wild-type mice (Chou-Green et al. 2003b). Correlates of increased dopaminergic tone in the NAc include delayed habituation to a novel environment (Rocha et al. 2002) and increased interactions with a novel object placed in the home cage (Heisler et al. 2007a).

Constitutive 5-HT$_{2C}$R $^-$/Y mice also clearly demonstrate that serotonergic neurotransmission is an important regulator of overall mouse locomotor responses. For example, rodents receiving mCPP demonstrate a modest dose-dependent suppression of locomotion in a home cage or arena environment (Kennett and Curzon 1988; Lucki et al. 1989). However, administration of mCPP to 5-HT$_{2C}$R mutant mice evokes a dramatic hyperlocomotor response that can be blocked using antagonists of 5-HT$_{1B}$ (Heisler and Tecott 2000) (see also Dalton et al. 2004; Schlussman et al. 1998), 5-HT$_{2A}$, and 5-HT$_{2B}$ receptors (Dalton et al. 2004). Thus, serotonergic stimulation of global basal ganglia circuits produces both stimulatory and inhibitory locomotor responses, with responses transduced through 5-HT$_{2A}$, 5-HT$_{2B}$, and 5-HT$_{2C}$ receptors (which are dominant) causing decreased activity while responses transduced through 5-HT$_{1A}$ and 5-HT$_{1B}$ receptors cause increased locomotor activity (Dalton et al. 2004).

Constitutive 5-HT$_{2C}$R $^-$/Y mice also demonstrate altered responses to drugs of abuse. In wild-type mice, administration of either cocaine (a promiscuous inhibitor of serotonin, dopamine, and norepinephrine reuptake) or d-amphetamine (an inverse agonist of these catecholaminergic reuptake systems) leads to a behavioral response characterized by increased locomotion, locomotor route tracing, and at higher doses, long periods of behavioral stereotypies (Schlussman et al. 1998; Simpson and Iversen 1971). Psychostimulant-evoked locomotion is thought to reflect increasing NAc activation (Gold et al. 1988), while psychostimulant-evoked stereotypies are associated with increasing imbalance in ratio of striasomal to matrix striatal activation (Canales and Graybiel 2000). Constitutive 5-HT$_{2C}$R $^-$/Y mice receiving cocaine demonstrate more locomotor activity at similar doses and time points compared with wild types (Rocha et al. 2002). Concordantly, cocaine-evoked efflux of dopamine into the NAc is much exaggerated in constitutive 5-HT$_{2C}$R $^-$/Y mice compared with wild types. Furthermore, mutant mice receiving d-amphetamine, the selective dopamine reuptake inhibitor GBR 12909 (vanoxerine), and the dopamine D$_1$ receptor agonist SKF-81297 are more sensitive to both the locomotor-activating and stereotypy-development effects of drug treatment (Abdallah et al. 2009). Interestingly, complete loss of 5-HT$_{2C}$R function did not lead to phenotypic changes in Str extracellular dopamine concentrations following treatment with either cocaine, d-amphetamine, or vanoxerine. Since constitutive 5-HT$_{2C}$R $^-$/Y mice do not display any significant changes in Str or NAc expression of D$_1$ and/or D$_2$ receptors, nor any phenotypic changes in Str medium spiny neurons, it is likely that basal ganglia systems downstream of the Str and NAc are functionally effected by alterations in 5-HT$_{2C}$R activity.

Finally, constitutive 5-HT$_{2C}$R $^-$/Y mice demonstrate an altered drive to ingest cocaine. While both wild-type and mutant mice learn lever pressing to acquire a cocaine reward at a 1:1 reinforcement, constitutive 5-HT$_{2C}$R $^-$/Y mice will work harder for the same dose of cocaine (as measured by their performance in a progressive ratio reward situation; see Rocha et al. 2002). In fact, constitutive 5-HT$_{2C}$R $^-$/Y mice will make 25% more lever presses for cocaine (over multiple trial sessions) compared with wild types. No phenotypic differences were noted in extinction of operant responses to cocaine, or in reinstatement of lever pressing responses for a nondrug reward.

4.9   Dysregulated Organization of Proopiomelanocortin (POMC)-Mediated Feeding Behavior in 5-HT$_{2C}$ R Mutant Mice

Central serotonergic systems have long been appreciated to have key roles in the regulation of feeding and body weight (for review, see Lam and Heisler 2007; Nelson and Gehlert 2006; Heisler et al. 2003). Studies of mutant mice with functional alterations of 5-HT$_{2C}$R activity have elucidated some of the mechanisms underlying this important regulation.

In the hypothalamus, 5-HT$_{2C}$Rs are prominently expressed on POMC-containing neurons of the ventrolateral arcuate nucleus (Wright et al. 1995; Pasqualetti et al. 1999; Heisler et al. 2002), and serotonin binding to these receptors is thought to contribute to neuronal activation (Qiu et al. 2007). POMC neurons in the arcuate are also activated by leptin signaling through the leptin receptor (Cheung et al. 1997; Elias et al. 1999; Cowley et al. 2001) and serotonin signaling through the 5-HT$_{1B}$ receptor (Ho et al. 2007; Nonogaki et al. 2007). POMC neurons in turn signal the ventromedial hypothalamus (VMH) via melanocortin neurotransmission; binding of melanocortin to the melanocortin 4 receptor (MC$_4$R) inhibits expression of feeding behavior (Cowley et al. 1999; Adage et al. 2001). This anorexigenic pathway is in opposition to signaling through anorexigenic pathway consisting of neuropeptide Y (NPY)/agouti-related protein (AgRP) neurons (which are inhibited by leptin signaling, see Yokosuka et al. 1998) projecting to VMH via NPY neurotransmission (which stimulates feeding behavior) and agouti-related protein (AgRP) neurotransmission (which is a functional antagonist of the MC$_4$R, and thus blocks the inhibition of feeding behavior, see Morton and Schwartz 2001). This "canonical" circuit is depicted in Fig. 4.2.

Multiple studies have examined the effects of 5-HT$_{2C}$R loss on overall mouse weight, and consensus results suggest that while 5-HT$_{2C}$R loss does not initially influence body weight of newborn mouse pups, weanlings, or young adults (<3 months old), constitutive 5-HT$_{2C}$R $^-$/Y mice develop a progressive middle-aged onset obesity. This obese phenotype becomes more exaggerated when comparing body-weight trajectories of mice fed a high fat diet (Nonogaki et al. 1998). Of note, this obesity phenotype occurs despite observing only modest hyperphagia (particularly during the light cycle, when the animals are expected to be inactive) in young constitutive 5-HT$_{2C}$R $^-$/Y mice (Goulding et al. 2008).

As depicted in the Fig. 4.2, current models of ingestive behavior neurophysiology suggest that 5-HT$_{2C}$R related feeding regulation occurs through stimulation of POMC neurons, which in turn excite MC$_4$Rs and inhibit expression of feeding behaviors. Supporting this hypothesis, recent studies have demonstrated decreased expression of both POMC and a nucleobindin 2 (NUCB2, a satiety-related molecule) in the hypothalamus of constitutive 5-HT$_{2C}$R $^-$/Y mice compared with wild types (Nonogaki et al. 2008). Additionally, stimulation of 5-HT$_{2C}$Rs using mCPP leads to dose-dependent increases in hypothalamic NUCB2 expression that are not

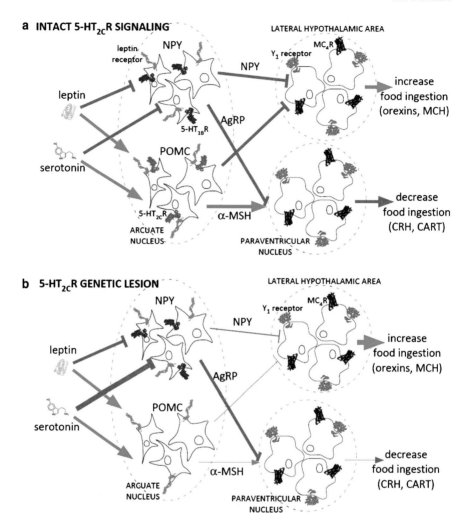

Fig. 4.2 (**a**) Integration of leptinergic and serotonergic signaling at the hypothalamic arcuate nucleus regulates both orexigenic and anorexigenic neuronal pathways controlling feeding behavior. (**b**) Loss of 5-HT$_{2C}$R function leads to decreased signaling within the pro-opiomelanocortin (POMC) neuronal population, decreased activation of anorexigenic signaling within the hypothalamus, and an imbalance between activities of the orixigenic and anorexigenic pathways favoring increased food consumption. 5-HT$_{2C}$R indicates serotonin$_{2C}$ receptor, 5-HT$_{1B}$R serotonin$_{1B}$ receptor; MC4R melanocortin 4 receptor, Y1 neuropeptide Y receptor 1, POMC pro-opiomelanocortin, NYP neuropeptide Y, AgRP agouti-related protein, CRH cortisol-releasing hormone, CART cocaine- and/or amphetamine-regulated transcript, MCH melanin-concentrating hormone

observed in constitutive 5-HT$_{2C}$R $^-$/Y mice. These results suggest that serotonergic neurotransmission onto these ventrolateral arcuate POMC neurons is a major factor driving the expression of both an anorexigenic-related transcription factor and the melanocortin transcript.

The functional status of this above-described pathway appears to be necessary for expression of SSRI-evoked anorexia. For example, fluvoxamine does not mediate its usual anorectic effect in mice with intact melanocortin signaling but constitutively lacking functional 5-HT$_{2C}$Rs (Nonogaki et al. 2009). However, fluvoxamine again has anorectic action (potentially through 5-HT$_{1B}$ receptor signaling) in mice with a combined homozygous deletion of 5-HT$_{2C}$Rs and a hemizygous mutation of β-endorphin. Thus, SSRI-evoked anorexia appears to require intact signaling at both the 5-HT$_{2C}$R/ POMC neuron and MC$_4$R/arcuate neuron. Mice with constitutive 5-HT$_{2C}$R knockout also show a similar resistance to the anorectic effects of d-fenfluramine, a potent stimulator of serotonin release and inhibitor of serotonin reuptake (Vickers et al. 1999). This finding was seen in analyses of behavioral satiety sequences, where it was noted that intermediate doses of d-fenfluramine had a much-blunted effect of suppressing feeding in constitutive 5-HT$_{2C}$R $^{-}$/Y mice compared with its effect in wild types.

Double mutant mice constitutively lacking both 5-HT$_{2C}$Rs and the adipocyte hormone leptin (*htr2c*$^{-}$/Y, *ob^{-}/ob^{-}*) provide particularly convincing evidence that signal integration at the POMC neuron plays a vital role in control of ingestive behaviors. Young double mutant mice had many hallmarks of significant diabetes, including hyperphagia and marked polydipsia, obesity, hyperglycemia, aberrant glucose tolerance tests, glucosuria, and hyperinsulinemia (Wade et al. 2008). Double mutant mice also had markedly elevated serum concentrations of the anti-insulin counterregulatory hormones corticosterone and glucagon. While there were no phenotypic differences appreciated in the above outcomes for young mice carrying a single constitutive 5-HT$_{2C}$R deletion, addition of the 5-HT$_{2C}$R mutation on top of the *ob^{-}/ob^{-}* genetic background markedly increased the size of each of these deficits well above the baseline *ob^{-}/ob^{-}* phenotype.

Similarly, mice lacking 5-HT$_{2C}$Rs demonstrate a complex interplay regarding how 5-HT$_{2C}$Rs and 5-HT$_{1B}$Rs regulate food intake. Wild-type mice receiving the 5-HT$_{1B}$R agonist CP-94,253 demonstrate a modest decrease in proportion of time spent feeding and performing other behaviors as evaluated by behavioral satiety sequences (over a 5-min bin resolution) (Dalton et al. 2006). However, performing this same treatment on 5-HT$_{2C}$R mutant mice led to a dramatic decrease in feeding and other activities. These studies suggested that 5-HT$_{1B}$R mediated hypophagia occurs independently of functioning 5-HT$_{2C}$Rs. Since 5-HT$_{2C}$R mutation does not lead to significant changes in expression of other serotonin receptors (Lopez-Gimenez et al. 2002), these results also suggest that persistent loss of 5-HT$_{2C}$R function effects cellular responses to 5-HT$_{1B}$R stimulation. Furthermore, contrary to expectation, middle-aged mice carrying double mutations for both the 5-HT$_{2C}$R and 5-HT$_{1B}$R are less heavy than 5-HT$_{2C}$R-mutant, 5-HT$_{1B}$R intact littermates (Abdallah, personal communication).

Interestingly, *htr2c* transcripts are one of the relatively few gene products that undergo posttranscriptional editing, an unusual RNA regulatory process where individual transcript nucleotides are enzymatically modified (usually to inositol, a process that changes their complementary binding to tRNAs and thus their peptide sequence). This process occurs through the action of multiple enzymes, predominantly through adenosine deaminase acting on RNA-1 (ADAR1, which selectively edits adenosines on residues 157, 159, and 161, see Hartner et al. 2004), as well as ADAR2 and ADAR3. Multiple, different posttranscriptional 5-HT$_{2C}$R variants have been described, with the main difference being the overall level of constitutive signal transduction activity in the absence of bound serotonin. The INI variant results from an unedited *htr2c* transcript, while the VGV variant results from a fully edited *htr2c* transcript. INI variant 5-HT$_{2C}$Rs demonstrate the greatest constitutive activity and native ligand affinity; conversely, VGV variant 5-HT$_{2C}$Rs demonstrate the least constitutive activity and weakest native ligand affinity (Herrick-Davis and Niswender 1999). Recent studies using transgenic mice expressing solely the INI or VGV 5-HT$_{2C}$R variants demonstrate that how *htr2cs* undergo posttranscriptional editing has a dramatic effect on mouse phenotype and feeding regulation. While mice solely expressing the INI 5-HT$_{2C}$R variant grew normally compared with wild-type littermates, mice expressing the VGV 5-HT$_{2C}$R variants demonstrated dramatically decreased fat mass, marked hyperphagia, increased energy expenditure and sympathetic tone, and increased overall 5-HT$_{2C}$R expression (Kawahara et al. 2008). Of note, the VGV phenotype was not suppressed in mice that expressed only the VGV 5-HT$_{2C}$R variant *and* a homozygous *MC$_4$R-/ MC$_4$R-* mutation. This data suggests that 5-HT$_{2C}$R mediated control of food ingestion is mediated through both MC$_4$R-dependent and -independent pathways.

It is intriguing to note that many of the differences in feeding behavior observed between constitutive 5-HT$_{2C}$R -/Y mice and their wild-type counterparts can be fully reversed by selective reintroduction of 5-HT$_{2C}$R expression into hypothalamic POMC neurons (Xu et al. 2008). In this study, a 5-HT$_{2C}$R construct was generated containing a transcription blocker flanked by loxP sites between exons 3 and 4 (loxTB 5-HT$_{2C}$R, as per authors). This construct was substituted for the native 5-HT$_{2C}$R allele, and mice containing this construct were bred to transgenic mice expressing *Cre* recombinase under control of the POMC promoter (POMC-*Cre*). In loxTB 5-HT$_{2C}$R/POMC-*Cre* mice, expression of *Cre* recombinase is restricted to POMC containing neurons. In these POMC neurons, *Cre* recombinase expression leads to excision of the transcription blocker site, thus restoring intact 5-HT$_{2C}$R expression in these cells. Dramatically, restoring 5-HT$_{2C}$R expression to POMC neurons alone normalized many of the above-described deficits. For example, loxTB 5-HT$_{2C}$R/POMC-*Cre* mice do not display middle-aged obesity, hyperadiposity, or hyperleptinemia. Additionally (and unexpectedly), both the daily and dark cycle locomotor hyperactivity seen in mice constitutively lacking 5-HT$_{2C}$R function is not observed in loxTB 5-HT$_{2C}$R/POMC-*Cre* mice.

4.10   5-HT$_{2C}$R Mutant Mice Demonstrate a Metabolic Syndrome Consistent with Type II Diabetes

Given the modest hyperphagia observed in middle-aged, obese, constitutive 5-HT$_{2C}$R -/Y mice, it is likely that other factors affecting overall animal energy balance are altered by loss of 5-HT$_{2C}$R function. Interestingly, both young and middle-aged constitutive 5-HT$_{2C}$R -/Y mice demonstrate greater home cage locomotor activity than wild-type littermates (Goulding et al. 2008; Nonogaki et al. 2003). This increased activity is a particularly robust finding in constitutive 5-HT$_{2C}$R $^{-}$/Y mice and has been appreciated by multiple investigators using different measuring systems. During these studies, much of the locomotor behavior observed in both wild-type and constitutive 5-HT$_{2C}$R $^{-}$/Y mice was associated with repeat visits to a known food source, and no genotypic differences were found in this behavioral aspect (Nonogaki et al. 2003).

Prior to the development of obesity (but in the context of ongoing hyperphagia), young constitutive 5-HT$_{2C}$R $^{-}$/Y mice have normal serum glucose, insulin, triglyceride, free fatty acid, corticosterone, and leptin concentrations. A separate study revealed no phenotypic differences in hepatic expression of peroxisome proliferator-activated receptor (PPAR) α and γ, two transcription factors with key roles regulating lipid metabolism and energy balance (Memon et al. 2000). By 9 months of age, constitutive 5-HT$_{2C}$R $^{-}$/Y mice are significantly more obese than wild-type littermates and demonstrate increased hepatic expression of both PPAR-α and PPAR-γ. In these mice, no differences were noted in serum glucose, insulin, triglyceride, free fatty acid, corticosterone, or leptin concentrations. However, constitutive 5-HT$_{2C}$R $^{-}$/Y mice older than 9 months are much less responsive to 3 days of leptin treatment, showing less blunting of food intake and less short-term decline in body weight. These middle-aged mice were also noted to have impaired glucose tolerance over a 2-h test infusion. While there were no phenotypic differences noted in glucose tolerance in young (2–3 months old) 5-HT$_{2C}$R $^{-}$/Y mice, middle-aged mutant mice display less of a hypoglycemic response to d-fenfluramine (Wade et al. 2008). This finding suggests that intact 5-HT$_{2C}$R function is a significant contributor to counter-regulatory processes set in effect by sudden serum glucose loads.

Additionally, middle-aged 5-HT$_{2C}$R $^{-}$/Y mice demonstrate significantly decreased oxygen consumption (although there is no phenotypic difference in minimum oxygen consumption, both as measured by indirect calorimetry, see Nonogaki et al. 2003; Nonogaki et al. 2002). The combination of increased locomotor activity combined with decreased oxygen consumption suggests that constitutive loss of 5-HT$_{2C}$Rs reduced the energy cost of physical activity. Phenotypic differences in skeletal muscle fiber type were not observed, but 5-HT$_{2C}$R $^{-}$/Y mice do display increased uncoupling protein 2 (UCP2) expression in white adipose tissue, liver, and skeletal muscle and decreased β_3-adrenergic receptor expression in white adipose tissue. Altogether, these findings indicate an altered regulation of sympathetic outflow onto white adipose tissue, altered thermogenesis, and altered regulation of energy balance in constitutive 5-HT$_{2C}$R $^{-}$/Y mice.

4.11   5-HT$_{2C}$R-Mediated Effects on Sleep and Entrainable Rhythms

As the above discussion suggests, 5-HT$_{2C}$Rs participate in the dynamic regulation of energy balance. To this effect, these receptors exert direct control over the hypothalamic generation of feeding behavior and indirectly influence important metabolic regulators of energy expenditure, including responsiveness to insulin and leptin signaling, tissue energy reserves, and baseline sympathetic tone. Recent studies also demonstrate that deficits in 5-HT$_{2C}$R function alter both light and food entrainable circadian rhythms and change the quality of many sleep parameters.

Light signaling from the retina initially converges onto the suprachiasmatic nucleus (SCN) core, a specialized neuronal structure characterized by strong light-evoked patterns of gene expression (for review, see (Aton and Herzog 2005; Reppert and Weaver 2002). The SCN core is densely innervated by both NPY-ergic and serotonergic fibers (Abrahamson and Moore 2001) and strongly expresses 5-HT$_{2C}$Rs (in a manner largely complementary to arginine vasopressin [AVP] expression) (Hsu and Tecott LH in preparation). By contrast, neurons of the SCN shell do not receive significant input from the retina, demonstrate significant periodic changes in circadian gene expression, and do not express 5-HT$_{2C}$Rs. Simply put, the SCN shell is the endogenous circadian oscillator whose intrinsic periodicity is modulated by light input initially processed in the SCN core. Consistent with this interpretation, mice constitutively lacking 5-HT$_{2C}$R expression show no phenotypic changes in circadian day length. Furthermore, constitutive 5-HT$_{2C}$R $^-$/Y mice do not retard their clock phase in response to a brief light pulse at the beginning of the dark cycle (Hsu and Tecott LH in preparation). Thus, 5-HT$_{2C}$R function is required for normal entraining of circadian clocks to changing day lengths.

Light is not the only environmental condition capable of entraining circadian clocks. Recent studies have also demonstrated the presence of an independent, food-entrainable oscillator that can modulate organism activity in cycle with the availability of food resources (Abe et al. 1989; Storch and Weitz 2009). While the molecular networks required for light-entrainable circadian rhythms have been well studied in the past, much of the structure of the food-entrainable network remains unknown (Davidson 2006). In this context, it is interesting to note that mice constitutively lacking 5-HT$_{2C}$R function demonstrate pronounced deficits in their ability to entrain their activity levels to new feeding schedules (Hsu 2009). Specifically, it was observed that while intact C57BL/6 mice increase their home cage locomotor activity in the 30 min preceding food availability, constitutive 5-HT$_{2C}$R $^-$/Y mice lack food anticipatory increases in locomotion.

At a global level of behavioral organization, the changes in regulation of ingestive behavior, metabolism, and circadian entrainment give rise to observable phenotypic differences in sleep architecture (Frank et al. 2002). Over the entire day, constitutive 5-HT$_{2C}$R $^-$/Y mice show less non-REM sleep (particularly during the active phase) and increased waking bouts (again, most prominent during the active phase) compared with wild-type littermates. Constitutive 5-HT$_{2C}$R mutant mice also showed a greater homeostatic drive to sleep after 6 h of sleep deprivation. By EEG studies, no phenotypic differences in spectral power density were observed during REM sleep, non-REM sleep, or wakefulness.

4.12   5-HT$_{2C}$R Mutant Mice as a Model of Audiogenic Seizure Disorder

Audiogenic seizures were a prominent phenotype observed in constitutive 5-HT$_{2C}$R $^-$/Y mice following derivation of the initial breeding line. For example, brief exposure (2–18 s) to modulated tones of 5–19 kHz (at the low range of mouse hearing) at 108 dB could reliably evoke seizure activity in constitutive 5-HT$_{2C}$R $^-$/Y mice but not wild-type littermate mice (Brennan et al. 1997). These seizures were characterized as tonic–clonic in nature, and were often preceded by a period of repetitive snout grooming. Seizures occasionally recurred within a few minutes. Mice were also noted to have two to three seizure episodes per day (Tecott et al. 1995). Significantly, these seizures occasionally evolved to a tonic-extension phenotype, leading to respiratory arrest and mouse death. Seizure penetrance increased as a function of age. Exposure to seizure-evoking sound stimuli was relatively ineffective in 21- and 60-day-old constitutive 5-HT$_{2C}$R $^-$/Y mice; however, this same stimulus evoked seizures in 100% of mice tested at 120 and 180 days old (Brennan et al. 1997). Mutant mice were also noted to have significantly lower seizure thresholds in response to IV administration of the pro-convulsant drug pentamethylenetetrazol (metrazol).

Two aspects complicate interpretation of this seizure phenotype. First, the constitutive 5-HT$_{2C}$R $^-$/Y mice studied all retain some DBA/2 genetic loci. Mice of strain DBA/2 are known to have increased susceptibility to seizures. However, DBA/2 mice show this tendency early in their life, and seizure susceptibility loci identified within this strain are autosomally located. Furthermore, constitutive 5-HT$_{2C}$R $^-$/Y mice backcrossed five to eight times onto a pure C57BL/6 background continued to demonstrate phenotypic differences in susceptibility to olfactory bulb kindling, chemoconvulsant, or corneal electroshock evoked seizures (Applegate and Tecott 1998). These findings suggest that contributions from the DBA/2 strain were not significant factors influencing the seizure phenotype observed in constitutive 5-HT$_{2C}$R mutant mice. Additionally, C57BL/6 J mice are known to have a well-described phenotype of middle-aged high frequency hearing loss. Microelectrode recordings of inferior colliculus cells in young animals demonstrated no phenotypic differences in neuronal auditory response. It is intriguing to note that the increase in constitutive 5-HT$_{2C}$R $^-$/Y mouse seizure susceptibility corresponds with the development of this high frequency hearing loss and remapping of auditory responses to the lower frequency sounds that are epileptogenic. The role of age-related high frequency hearing loss in C57BL/6 J mice was evaluated by breeding the 5-HT$_{2C}$R mutation into B6.CAST-$Cdh23^{Ahl+}$/Kjn mice, a strain congenic to C57BL/6 J mice in all ways except for having a wild-type Ahl (age-related hearing loss) gene that prevents these mice from developing age-related hearing loss. In these mice, audiogenic seizure frequencies remained low in mutants without functional 5-HT$_{2C}$Rs, and no age-related increase in seizure incidence was appreciated (Tecott, personal communication).

Audiogenic seizures in constitutive 5-HT$_{2C}$R $^-$/Y mice evoked a widespread activation of CNS regions implicated in sound processing. c-fos expression indicative of cellular activity was noted in the external and internal cortex regions of the inferior colliculus, the dorsal lateral lemniscal nucleus, the medial geniculate nucleus of the thalamus, and the posterior intralaminar thalamic nucleus (Brennan et al. 1997). Intracellular recordings from the inferior colliculus, however, did not demonstrate significant phenotypic differences in neuronal responses to either sound frequency or volume. Constitutive 5-HT$_{2C}$R $^{-}$/Y mice were noted to have slightly increased response latencies to stimulation, a finding that has been correlated with a decreased ability to respond to repetitive auditory stimuli (Langner et al. 1987). Overall, these findings suggest that 5-HT$_{2C}$Rs have their most powerful effects on animal seizure susceptibility in the context of auditory sensory deprivation. Further studies are warranted to determine if this phenotype reflects disuse hypersensitivity of the inferior colliculus.

4.13   Conclusions

It is clear that transgenic mice modeling different aspects of 5-HT$_{2C}$R function have provided important and otherwise unobtainable knowledge of how serotonergic neurotransmission affects cellular function and whole organism behavior. In fact, there is still much future work to be accomplished. For example, the generation of currently unavailable mouse models with tailored 5-HT$_{2C}$R function (e.g., overexpression, lesions, or specific edited receptor isoforms) restricted to important CNS regions, including but not limited to the striatum, nucleus accumbens, bed nucleus of stria terminalis, and amygdala, will further clarify how 5-HT$_{2C}$Rs regulate the behavioral response to environmental stressors and appetitive/aversive stimuli. Temporal control of 5-HT$_{2C}$R function would similarly allow investigators to determine if later-life loss of 5-HT$_{2C}$R function is ameliorated in the presence of intact function in neonate and young mice, and conversely, if reestablishment of 5-HT$_{2C}$R function in later adulthood can avert the development of middle-aged obesity and type II diabetes in animals born with genetic lesions of 5-HT$_{2C}$R expression. It is also intriguing to note that many of the therapeutic agents currently used to treat highly morbid psychiatric diseases (including schizophrenia, bipolar disorder, major depression, anxiety disorders, and substance abuse disorders) in part function through their actions on 5-HT$_{2C}$Rs. Thus, further elucidation of how 5-HT$_{2C}$Rs specifically affect the function of individual neuronal ensembles, and how these effects change the overall regulation of organism behavior, may provide important therapeutic insights to treat these major psychiatric illnesses.

References

Abdallah L, Bonasera SJ, Hopf FW, et al (2009) Impact of serotonin 2C receptor null mutation on physiology and behavior associated with nigrostriatal dopamine pathway function. J Neurosci 29:8156–8165.

Abe H, Kida M, Tsuji K, et al (1989) Feeding cycles entrain circadian rhythms of locomotor activity in CS mice but not in C57BL/6 J mice. Physiol Behav 45:397–401.

Abrahamson EE, Moore RY (2001) Suprachiasmatic nucleus in the mouse: retinal innervation, intrinsic organization and efferent projections. Mol Brain Res 916:172–191.

Adage T, Scheurink AJ, de Boer SF, et al (2001) Hypothalamic, metabolic, and behavioral responses to pharmacological inhibition of CNS melanocortin signaling in rats. J Neurosci 21:3639–3645.

Akana SF (2008) Feeding and stress interact through the serotonin 2C receptor in developing mice. Physiol Behav 94:569–579.

Applegate CD, Tecott LH (1998) Global increases in seizure susceptibility in mice lacking 5-HT2C receptors: a behavioral analysis. Exp Neurol 154:522–530.

Aton SJ, Herzog ED (2005) Come together, right...now: synchronization of rhythms in a mammalian circadian clock. Neuron 48:531–534.

Backstrom JR, Price RD, Reasoner DT, et al (2000) Deletion of the serotonin 5-HT2C receptor PDZ recognition motif prevents receptor phosphorylation and delays resensitization of receptor responses. J Biol Chem 275:23620–23626.

Barnes NM, Sharp T (1999) A review of central 5-HT receptors and their function. Neuropharmacology 38:1083–1152.

Bockaert J, Claeysen S, Compan V, et al (2008) 5-HT(4) receptors: history, molecular pharmacology and brain functions. Neuropharmacology 55:922–931.

Bonasera SJ, Tecott LH (2000) Mouse models of serotonin receptor function: toward a genetic dissection of serotonin systems. Pharmacol Ther 88:133–142.

Bonasera SJ, Tecott LH (2004) Increased fear-sensitized acoustic startle in serotonin 2C receptor mutant mice. In: Neuroscience 2004. San Diego: Society for Neuroscience.

Bonasera SJ, Schenk AK, Tecott LH (2005) Greater affective responses to aversive stimuli in mice lacking the serotonin 2C receptor. In: Neuroscience 2005. Washington, DC: Society for Neuroscience.

Brennan TJ, Seeley WW, Kilgard M, et al (1997) Sound-induced seizures in serotonin 5-HT2c receptor mutant mice. Nat Genet 16:387–390.

Campbell BM, Merchant KM (2003) Serotonin 2C receptors within the basolateral amygdala induce acute fear-like responses in an open-field environment. Brain Res Mol Brain Res 993:1–9.

Canales JJ, Graybiel AM (2000) A measure of striatal function predicts motor stereotypy. Nat Neurosci 3:377–383.

Cheung CC, Clifton DK, Steiner RA (1997) Proopiomelanocortin neurons are direct targets for leptin in the hypothalamus. Endocrinology 138:4489–4492.

Chou-Green JM, Holscher TD, Dallman MF, et al (2003) Repeated stress in young and old 5-HT(2C) receptor knockout mice. Physiol Behav 79:217–226.

Chou-Green JM, Holscher TD, Dallman MF, et al (2003) Compulsive behavior in the 5-HT2C receptor knockout mouse. Physiol Behav 78:641–649.

Cowley MA, Pronchuk N, Fan W, et al (1999) Integration of NPY, AGRP, and melanocortin signals in the hypothalamic paraventricular nucleus: evidence of a cellular basis for the adipostat. Neuron 24:155–163.

Cowley MA, Smart JL, Rubinstein M, et al (2001) Leptin activates anorexigenic POMC neurons through a neural network in the arcuate nucleus. Nature 411:480–484.

Crawley JN, Belknap JK, Collins A, et al (1997) Behavioral phenotypes of inbred mouse strains: implications and recommendations for molecular studies. Psychopharmacology (Berl) 132:107–124.

Cremers TI, Giorgetti M, Bosker FJ, et al (2004) Inactivation of 5-HT(2C) receptors potentiates consequences of serotonin reuptake blockade. Neuropsychopharmacology 29:1782–1789.

Dallman MF, Akana SF, Bell ME, et al (1999) Warning! Nearby construction can profoundly affect your experiments. Endocrine 11:111–113.

Dalton GL, Lee MD, Kennett GA, et al (2004) mCPP-induced hyperactivity in 5-HT2C receptor mutant mice is mediated by activation of multiple 5-HT receptor subtypes. Neuropharmacology 46:663–671.

Dalton GL, Lee MD, Kennett GA, et al (2006) Serotonin 1B and 2C receptor interactions in the modulation of feeding behaviour in the mouse. Psychopharmacology (Berl) 185:45–57.

Davidson AJ (2006) Search for the feeding-entrainable circadian oscillator: a complex proposition. Am J Physiol Regul Integr Comp Physiol 290:R1524–1526.

Davis M (1989) Sensitization of the acoustic startle reflex by footshock. Behav Neurosci 103:495–503.

Di Giovanni G, De Deurwaerdere P, Di Mascio M, et al (1999) Selective blockade of serotonin-2C/2B receptors enhances mesolimbic and mesostriatal dopaminergic function: a combined in vivo electrophysiological and microdialysis study. Neuroscience 91:587–597.

Di Giovanni G, Di Matteo V, Di Mascio M, et al (2000) Preferential modulation of mesolimbic vs. nigrostriatal dopaminergic function by serotonin(2C/2B) receptor agonists: a combined in vivo electrophysiological and microdialysis study. Synapse 35:53–61.

Di Giovanni G, Di Matteo V, La Grutta V, et al (2001) m-Chlorophenylpiperazine excites non-dopaminergic neurons in the rat substantia nigra and ventral tegmental area by activating serotonin-2C receptors. Neuroscience 103:111–116.

Di Giovanni G, Di Matteo V, Pierucci M, et al (2006) Central serotonin2C receptor: from physiology to pathology. Curr Top Med Chem 6:1909–1925.

Di Matteo V, Cacchio M, Di Giulio C, et al (2002) Biochemical evidence that the atypical antipsychotic drugs clozapine and risperidone block 5-HT(2C) receptors in vivo. Pharmacol Biochem Behav 71:607–613.

Eberle-Wang K, Mikeladze Z, Uryu K, et al (1997) Pattern of expression of the serotonin2C receptor messenger RNA in the basal ganglia of adult rats. J Comp Neurol 384:233–247.

Elias CF, Aschkenasi C, Lee C, et al (1999) Leptin differentially regulates NPY and POMC neurons projecting to the lateral hypothalamic area. Neuron 23:775–786.

Filip M, Cunningham KA (2002) Serotonin 5-HT(2C) receptors in nucleus accumbens regulate expression of the hyperlocomotive and discriminative stimulus effects of cocaine. Pharmacol Biochem Behav 71:745–756.

Frank MG, Stryker MP, Tecott LH (2002) Sleep and sleep homeostasis in mice lacking the 5-HT2c receptor. Neuropsychopharmacology 27:869–873.

Gaveriaux-Ruff C, Kieffer BL (2007) Conditional gene targeting in the mouse nervous system: insights into brain function and diseases. Pharmacol Ther 113:619–634.

Gobert A, Rivet JM, Lejeune F, et al (2000) Serotonin(2C) receptors tonically suppress the activity of mesocortical dopaminergic and adrenergic, but not serotonergic, pathways: a combined dialysis and electrophysiological analysis in the rat. Synapse 36:205–221.

Gold LH, Swerdlow NR, Koob GF (1988) The role of mesolimbic dopamine in conditioned locomotion produced by amphetamine. Behav Neurosci 102:544–552.

Gordon JA, Hen R (2004) The serotonergic system and anxiety. Neuromolecular Med 5:27–40.

Goulding EH, Schenk AK, Juneja P, et al (2008) A robust automated system elucidates mouse home cage behavioral structure. Proc Natl Acad Sci USA 105:20575–20582.

Hartner JC, Schmittwolf C, Kispert A, et al (2004) Liver disintegration in the mouse embryo caused by deficiency in the RNA-editing enzyme ADAR1. J Biol Chem 279:4894–4902.

Hedlund PB, Sutcliffe JG (2004) Functional, molecular and pharmacological advances in 5-HT7 receptor research. Trends Pharmacol Sci 25:481–486.

Heisler LK, Tecott LH (2000) A paradoxical locomotor response in serotonin 5-HT(2C) receptor mutant mice. J Neurosci 20:RC71.

Heisler LK, Cowley MA, Tecott LH, et al (2002) Activation of central melanocortin pathways by fenfluramine. Science 297:609–611.

Heisler LK, Cowley MA, Kishi T, et al (2003) Central serotonin and melanocortin pathways regulating energy homeostasis. Ann N Y Acad Sci 994:169–174.

Heisler LK, Zhou L, Bajwa P, et al (2007) Serotonin 5-HT(2C) receptors regulate anxiety-like behavior. Genes Brain Behav 6:491–496.

Heisler LK, Pronchuk N, Nonogaki K, et al (2007) Serotonin activates the hypothalamic-pituitary-adrenal axis via serotonin 2C receptor stimulation. J Neuroscience 27:6956–6964.

Herrick-Davis K, Grinde E, Niswender CM (1999) Serotonin 5-HT2C receptor RNA editing alters receptor basal activity: implications for serotonergic signal transduction. J Neurochem 73:1711–1717

Ho SS, Chow BK, Yung WH (2007) Serotonin increases the excitability of the hypothalamic paraventricular nucleus magnocellular neurons. Eur J Neurosci 25:2991–3000.

Holmes A, Wrenn CC, Harris AP, et al (2002) Behavioral profiles of inbred strains on novel olfactory, spatial and emotional tests for reference memory in mice. Genes Brain Behav 1:55–69.

Hsu J (2009) Contribution of serotonin2C receptors to the regulation of circadian rhythm entrainment, in Neuroscience. Ph.D. Thesis. Neuroscience Program, University of California, San Francisco.

Hsu JL, Yu L, Tecott LH (in preparation) Involvement of serotonin$_{2c}$ receptors in the regulation of circadian phase shifts to light.

Huang XF, Han M, Storlien LH (2004) Differential expression of 5-HT(2A) and 5-HT(2C) receptor mRNAs in mice prone, or resistant, to chronic high-fat diet-induced obesity. Brain Res Mol Brain Res 127:39–47.

Huang XF, Tan YY, Huang X, et al (2007) Effect of chronic treatment with clozapine and haloperidol on 5-HT(2A and 2C) receptor mRNA expression in the rat brain. Neurosci Res 59:314–321.

Invernizzi RW, Pierucci M, Calcagno E, et al (2007) Selective activation of 5-HT(2C) receptors stimulates GABA-ergic function in the rat substantia nigra pars reticulata: a combined in vivo electrophysiological and neurochemical study. Neuroscience 144:1523–1535.

Kawahara Y, Grimberg A, Teegarden S, et al (2008) Dysregulated editing of serotonin 2C receptor mRNAs results in energy dissipation and loss of fat mass. J Neurosci 28:12834–12844.

Kennett GA, Curzon G (1988) Evidence that mCPP may have behavioural effects mediated by central 5-HT1C receptors. Br J Pharmacol 94:137–147.

Krishnakumar A, Abraham PM, Paul J, et al (2009) Down-regulation of cerebellar 5-HT(2C) receptors in pilocarpine-induced epilepsy in rats: therapeutic role of Bacopa monnieri extract. J Neurol Sci 284:124–128.

Lam DD, Heisler LK (2007) Serotonin and energy balance: molecular mechanisms and implications for type 2 diabetes. Expert Rev Mol Med 9:1–24.

Langner G, Schreiner C, Merzenich MM (1987) Covariation of latency and temporal resolution in the inferior colliculus of the cat. Hear Res 31:197–201.

Lopez-Gimenez JF, Tecott LH, Palacios JM, et al (2002) Serotonin 5- HT (2C) receptor knockout mice: autoradiographic analysis of multiple serotonin receptors. J Neurosci Res 67:69–85.

Lucki I, Ward HR, Frazer A (1989) Effect of 1-(m-chlorophenyl)piperazine and 1-(m-trifluoromethylphenyl)piperazine on locomotor activity. J Pharmacol Exp Ther 249:155–164.

Memon RA, Tecott LH, Nonogaki K, et al (2000) Up-regulation of peroxisome proliferator-activated receptors (PPAR-alpha) and PPAR-gamma messenger ribonucleic acid expression in the liver in murine obesity: troglitazone induces expression of PPAR-gamma-responsive adipose tissue-specific genes in the liver of obese diabetic mice. Endocrinology 141:4021–4031.

Miyoshi G, Fishell G (2006) Directing neuron-specific transgene expression in the mouse CNS. Curr Opin Neurobiol 16:577–584.

Morton GJ, Schwartz MW (2001) The NPY/AgRP neuron and energy homeostasis. Int J Obes Relat Metab Disord 25(Suppl 5):S56–S62.

Nelson DL, Gehlert DR (2006) Central nervous system biogenic amine targets for control of appetite and energy expenditure. Endocrine 29:49–60.

Nonogaki K, Strack AM, Dallman MF, et al (1998) Leptin-independent hyperphagia and type 2 diabetes in mice with a mutated serotonin 5-HT2C receptor gene. Nat Med 4:1152–1156.

Nonogaki K, Memon RA, Grunfeld C, et al (2002) Altered gene expressions involved in energy expenditure in 5-HT(2C) receptor mutant mice. Biochem Biophys Res Commun 295:249–254.

Nonogaki K, Abdallah L, Goulding EH, et al (2003) Hyperactivity and reduced energy cost of physical activity in serotonin 5-HT(2C) receptor mutant mice. Diabetes 52:315–320.

Nonogaki K, Nozue K, Takahashi Y, et al (2007) Fluvoxamine, a selective serotonin reuptake inhibitor, and 5-HT2C receptor inactivation induce appetite-suppressing effects in mice via 5-HT1B receptors. Int J Neuropsychopharmacol 10:675–681.

Nonogaki K, Ohba Y, Sumii M, et al (2008) Serotonin systems upregulate the expression of hypothalamic NUCB2 via 5-HT2C receptors and induce anorexia via a leptin-independent pathway in mice. Biochem Biophys Res Commun 372:186–190.

Nonogaki K, Ohba Y, Wakameda M, et al (2009) Fluvoxamine exerts anorexic effect in 5-HT2C receptor mutant mice with heterozygous mutation of beta-endorphin gene. Int J Neuropsychopharmacol 12:547–552.

Ohlsson R, Paldi A, Graves JA (2001) Did genomic imprinting and X chromosome inactivation arise from stochastic expression? Trends Genet 17:136–141.

Pasqualetti M, Ori M, Castagna M, et al (1999) Distribution and cellular localization of the serotonin type 2C receptor messenger RNA in human brain. Neuroscience 92:601–611.

Pickering C, Avesson L, Lindblom J, et al (2007) Identification of neurotransmitter receptor genes involved in alcohol self-administration in the rat prefrontal cortex, hippocampus and amygdala. Prog Neuropsychopharmacol Biol Psychiatry 31:53–64.

Pompeiano M, Palacios JM, Mengod G (1994) Distribution of the serotonin 5-HT2 receptor family mRNAs: comparison between 5-HT2A and 5-HT2C receptors. Brain Res Mol Brain Res 23:163–178.

Qiu J, Xue C, Bosch MA, et al (2007) Serotonin 5-hydroxytryptamine2C receptor signaling in hypothalamic proopiomelanocortin neurons: role in energy homeostasis in females. Mol Pharmacol 72:885–896.

Reppert SM, Weaver DR (2002) Coordination of circadian timing in mammals. Nature 418:935–941.

Rocha BA, Goulding EH, O'Dell LE, et al (2002) Enhanced locomotor, reinforcing, and neurochemical effects of cocaine in serotonin 5-hydroxytryptamine 2C receptor mutant mice. J Neurosci 22:10039–10045.

Roth B (2006) The serotonin receptors: from molecular pharmacology to human therapeutics, Totowa, NJ: Humana.

Rueter LE, Tecott LH, Blier P (2000) In vivo electrophysiological examination of 5-HT2 responses in 5-HT2C receptor mutant mice. Naunyn Schmiedebergs Arch Pharmacol 361:484–491.

Schlussman SD, Ho A, Zhou Y, et al (1998) Effects of "binge" pattern cocaine on stereotypy and locomotor activity in C57BL/6 J and 129/J mice. Pharmacol Biochem Behav 60:593–599.

Serrats J, Mengod G, Cortes R (2005) Expression of serotonin 5-HT2C receptors in GABAergic cells of the anterior raphe nuclei. J Chem Neuroanat 29:83–91.

Simpson BA, Iversen SD (1971) Effects of substantia nigra lesions on the locomotor and stereotypy responses to amphetamine. Nat New Biol 230:30–32.

Storch KF, Weitz CJ (2009) Daily rhythms of food-anticipatory behavioral activity do not require the known circadian clock. Proc Natl Acad Sci USA 106:6808–6813.

Storm EE, Carra S, Holt M, et al (2003) Maternal behavior is disrupted in serotonin 2C receptor mutant mice. In: Society for Neuroscience 2003. New Orleans, LA: The Society for Neuroscience

Tecott LH, Sun LM, Akana SF, et al (1995) Eating disorder and epilepsy in mice lacking 5-HT2c serotonin receptors. Nature 374:542–546.

Tecott LH, Logue SF, Wehner JM, et al (1998) Perturbed dentate gyrus function in serotonin 5-HT2C receptor mutant mice. Proc Natl Acad Sci USA 95:15026–15031.

Thomas DR (2006) 5-ht5A receptors as a therapeutic target. Pharmacol Ther 111:707–714.

Vickers SP, Clifton PG, Dourish CT, et al (1999) Reduced satiating effect of d-fenfluramine in serotonin 5-HT(2C) receptor mutant mice. Psychopharmacology (Berl) 143:309–314.

Wade JM, Juneja P, MacKay AW, et al (2008) Synergistic impairment of glucose homeostasis in ob/ob mice lacking functional serotonin 2C receptors. Endocrinology 149:955–961.

Woolley ML, Marsden CA, Fone KC (2004) 5-ht6 receptors. Curr Drug Targets CNS Neurol Disord 3:59–79

Wright DE, Seroogy KB, Lundgren KH, et al (1995) Comparative localization of serotonin1A, 1C, and 2 receptor subtype mRNAs in rat brain. J Comp Neurol 351:357–373.

Xavier GF, Costa VC (2009) Dentate gyrus and spatial behaviour. Prog Neuropsychopharmacol Biol Psychiatry 33:762–773.

Xu Y, Jones JE, Kohno D, et al (2008) 5-HT2CRs expressed by pro-opiomelanocortin neurons regulate energy homeostasis. Neuron 60:582–589.

Yokosuka M, Xu B, Pu S, et al (1998) Neural substrates for leptin and neuropeptide Y (NPY) interaction: hypothalamic sites associated with inhibition of NPY-induced food intake. Physiol Behav 64:331–338.

Chapter 5
Serotonin 5-HT$_{2C}$ Receptor Signal Transduction

Maria N. Garnovskaya and John R. Raymond

5.1 Introduction

The 5-HT$_{2C}$ receptor was identified as a high-affinity [^{3}H]5-HT-binding site in epithelial cells of the choroid plexus (Berg et al. 2008; Heisler et al. 1998; Miller 2005). Originally, it was classified as a member of the 5-HT$_1$ receptor family and was named the 5-HT$_{1C}$ receptor (Pazos et al. 1985). Later, this receptor was reclassified as 5-HT$_{2C}$ due to its structural and signaling similarities with the 5-HT$_2$ receptor subclass (Hoyer et al. 1994). The partial cloning of the mouse 5-HT$_{2C}$ receptor (Lubbert et al. 1987) was followed by the sequencing of the full-length clone in the rat (Julius et al. 1988), human (Saltzman et al. 1991), and mouse (Foguet et al. 1992). Comparison of the cDNA-deduced amino acid sequences from rat, mouse, and human reveals strong sequence homology. The gene for the 5-HT$_{2C}$ receptor is located on the human X chromosome at position q24 (Xq24), and on mouse X chromosome at region D-F4 (Milatovich et al. 1992). Mice lacking the 5-HT$_{2C}$ receptor have been produced and show an eating disorder and epilepsy (Tecott et al. 1995). Xie et al. isolated the complete 4775-nt cDNA encoding the human 5-HT$_{2C}$ receptor and characterized its gene structure (Xie et al. 1996). The entire 5-HT$_{2C}$ receptor gene from exon I to exon VI spans at least 230 kb of DNA, encompassing six exons and five introns (Xie et al. 1996). The coding sequence of the human 5-HT$_{2C}$ receptor gene is interrupted by three introns (rather than two as in the case of the 5-HT$_{2A}$ and 5-HT$_{2B}$ receptors) (Stam et al. 1994), and the positions of the intron–exon junctions are conserved between the human and the mouse 5-HT$_{2C}$ receptor genes (Xie et al. 1996).

This complex gene structure allows for the existence of variants of the 5-HT$_{2C}$ receptor. An alternatively spliced variant of the 5-HT$_{2C}$ receptor RNA that contains a 95-nt deletion in the region coding for the second intracellular loop and the fourth

M.N. Garnovskaya (✉)
Medical and Research Services of the Ralph H. Johnson Veterans Affairs Medical Center and the Department of Medicine (Nephrology Division), Medical University of South Carolina, 96 Jonathan Lucas Street, MSC 629 Charleston, SC 29425, USA
e-mail: garnovsk@musc.edu

G. Di Giovanni et al. (eds.), *5-HT$_{2C}$ Receptors in the Pathophysiology of CNS Disease*, The Receptors 22, DOI 10.1007/978-1-60761-941-3_5, © Springer Science + Business Media, LLC 2011

transmembrane domain of the receptor generates a truncated, nonfunctional isoform of the 5-HT$_{2C}$ receptor (Xie et al. 1996; Canton et al. 1996). In addition, the 5-HT$_{2C}$ receptor exhibits a unique mechanism of generating multiple functional receptor variants through a process called mRNA editing (Burns et al. 1997) recently reviewed by Werry et al. (2008a), which will be discussed in Sect. 5.1.1.

The 5-HT$_{2C}$ receptor is expressed primarily in the central nervous system (Roth et al. 1998). High levels of 5-HT$_{2C}$ receptor expression have been detected by ligand autoradiography, immunocytochemistry, and in situ hybridization in the choroid plexus, the cortex, the nucleus accumbens, the amygdala, the hippocampus, caudate nucleus, and substantia nigra (Pazos et al. 1985; Mengod et al. 1990; Abramowski et al. 1995). The highest density of 5-HT$_{2C}$ receptors is found in epithelial cells of the choroid plexus, where these receptors control spinal fluid production (Leysen 2004). Pre- and postsynaptic localizations of 5-HT$_{2C}$ receptors have been confirmed by electron-microscope studies performed in the mouse (Bécamel et al. 2004). 5-HT$_{2C}$ receptors also are present in hypothalamic pro-opiomelanocortin (POMC)/ cocaine amphetamine-regulated transcript (CART) neurons within the arcuate nucleus (Heisler et al. 2002) and in rat retinal neurons (Pérez-León et al. 2004).

With regard to signaling, 5-HT$_{2C}$ receptors are positively coupled via G$_{\alpha q}$ to phospholipase Cβ (PLC) protein in several brain regions including choroid plexus. In addition to PLC, 5-HT$_{2C}$ receptors couple to multiple cellular effector systems including phospholipase A$_2$ (PLA$_2$), phospholipase D (PLD), extracellular signal-regulated kinases (ERK1/2), pertussis toxin-sensitive G proteins, PDZ domain-containing proteins, and to regulation of different channels and transport processes (for reviews, see Leysen 2004; Raymond et al. 2001, 2006; Millan et al. 2008). Consequently, the net cellular effect of activation of 5-HT$_{2C}$ receptors results from concurrent regulation of several effector pathways within cells.

5.1.1 RNA Editing of the 5-HT$_{2C}$ Receptor

RNA editing is a posttranscriptional modification resulting in an alteration of the primary nucleotide sequence of RNA transcripts by mechanisms other than splicing. The enzymatic conversion of adenosine to inosine by RNA editing has been identified within an increasing number of RNA transcripts, indicating that this modification represents an important mechanism for the generation of molecular diversity. Editing of the 5-HT$_{2C}$ receptor was discovered by comparing sequences from genomic DNA and cDNAs from the rat striatum when four A-to-G discrepancies were identified in rat brain (Burns et al. 1997). It was proposed that these four adenosine residues in the 5-HT$_{2C}$ receptor pre-mRNA were converted into inosines in the mature mRNA by enzymes called adenosine deaminases that act on RNA (ADAR). The ADAR family consists of three members: ubiquitously expressed ADAR1 and ADAR 2, and ADAR3, which expression appears to be restricted to brain (Chen et al. 2000; Sanders-Bush et al. 2003). ADAR enzymes bind to double-stranded RNA and catalyze the conversion of adenosine to inosine by hydrolytic deamination. Because inosines have base-pairing properties similar to that of guanosines, they are recognized and translated as guanosine by the ribosomes. RNA transcripts encoding the human 5-HT$_{2C}$ receptor undergo adenosine–inosine (A–I) editing events at five positions, A, B, C, D, and E (Fig. 5.1), altering the amino acid sequence in the putative second intracellular loop of the protein and resulting in at least 21 discrete mRNA species encoding 14 different editing variants of the 5-HT$_{2C}$ receptor (Burns et al. 1997; Niswender et al. 1998). In human brain, the genomic (unedited) receptor expresses amino acids isoleucine, asparagine, and isoleucine (INI) at positions 156, 158, and 160, respectively (in the sequence IRNPI in the intracellular loop 2), whereas conver-

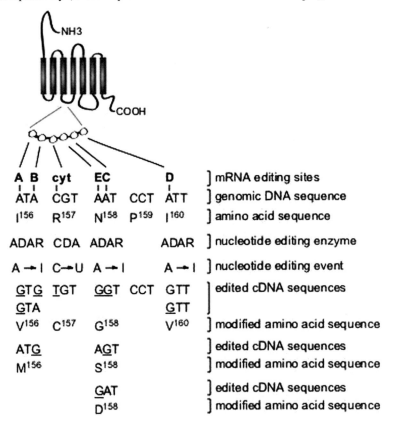

Fig. 5.1 RNA editing sites of 5-HT$_{2C}$ receptor mRNA. This figure depicts the human 5-HT$_{2C}$ receptor with seven membrane-spanning domains and an amino terminus facing the extracellular space. The genomic DNA encoding the second intracellular loop contains six mRNA editing sites (white oblongs). Five of those mRNA editing sites (A, B, E, C, and D) are modified by adenosine deaminase (ADAR1 and/or ADAR2), and one (cyt) is modified by cytidine deaminase (CDA). ADAR edits the mRNA sequences from adenosine to inosine (A → I), and CDA edits the mRNA sequences from cytidine to uridine (C → U), resulting in cDNA sequences that are modified from the genomic sequences. This results in the translation of amino acids that are distinct from those coded by genomic DNA. Amino acids are designated by their one-letter codes (I for isoleucine, R for arginine, etc.) followed by the superscripted amino acid number in the sequence of the 5-HT$_{2C}$ receptor protein (Adapted from Niswender et al. 1998; Tohda et al. 2006)

sion of all three amino acids results in valine, glycine, and valine (VGV), valine, serine, and valine (VSV), valine, asparagine, and valine (VNV), or valine, aspartic acid, and valine (VDV) (Niswender et al. 1998; Backstrom et al. 1999; Fitzgerald et al. 1999). These are referred to as $5\text{-HT}_{2C\text{-INI}}$, $5\text{-HT}_{2C\text{-VNV}}$, $5\text{-HT}_{2C\text{-VSV}}$, $5\text{-HT}_{2C\text{-VGV}}$, and $5\text{-HT}_{2C\text{-VDV}}$ receptors. Other partially edited human receptor variants include $5\text{-HT}_{2C\text{-VNI}}$, $5\text{-HT}_{2C\text{-VSI}}$, $5\text{-HT}_{2C\text{-ISI}}$, $5\text{-HT}_{2C\text{-INV}}$, $5\text{-HT}_{2C\text{-ISV}}$, $5\text{-HT}_{2C\text{-IGI}}$, $5\text{-HT}_{2C\text{-VGI}}$, $5\text{-HT}_{2C\text{-IDI}}$, and $5\text{-HT}_{2C\text{-IDV}}$ receptors. In addition to the five A-to-I RNA editing sites, a novel sixth site has recently been reported, which is deaminated from cytidine to uracil by citidine deaminase; this variant is thought to plays role in neuronal differentiation (Flomen et al. 2004; Tohda et al. 2006).

In the rat 5-HT_{2C} receptor, only four (A–I) conversion sites (termed *sites A–D*) have been identified, resulting in 11 discrete mRNAs and seven different receptor protein isoforms (Fitzgerald et al. 1999; Niswender et al. 1999). Because the second intracellular loop of the protein is involved in the interaction of 5-HT_{2C} receptors with G proteins, specific isoforms of 5-HT_{2C} receptors show different coupling profiles, agonist potency, and ligand binding affinity (Niswender et al. 1999; Herrick-Davis et al. 1999). They also show varying degrees of constitutive activity, ranging from pronounced for the wild-type $5\text{-HT}_{2C\text{-INI}}$, to intermediate for partially edited isoforms such as $5\text{-HT}_{2C\text{-VSV}}$, to minimal for the fully edited $5\text{-HT}_{2C\text{-VGV}}$ isoforms (Niswender et al. 1999; Herrick-Davis et al. 1999). The variants also exhibit differential abilities to mobilize intracellular Ca^{2+} and to stimulate accumulation of inositol phosphates (Backstrom et al. 1999; Fitzgerald et al. 1999).

The 5-HT_{2C} receptor is the only G-protein-coupled receptor known to be edited, and even mRNAs encoding closely related 5-HT_{2A} and 5-HT_{2B} receptors have not been shown to undergo RNA editing (Niswender et al. 1998). RNA editing of the 5-HT_{2C} receptor is prominent in the central nervous system; indeed, most of the receptors in the brain exist as edited isoforms (Burns et al. 1997; Fitzgerald et al. 1999; Niswender et al. 2001). Because the 5-HT_{2C} receptor is involved in the pathogenesis of various psychological disorders, RNA editing of the 5-HT_{2C} receptor has been proposed to play role in the etiology of schizophrenia (Sodhi et al. 2001), depression (Tohda et al. 2006), and affective disorders (Niswender et al. 2001; Gurevich et al. 2002).

5.2   The 5-HT_{2C} Receptor Couples to Multiple G_α Subunits

Although 5-HT_{2C} receptors are "classically" considered to couple to $G_{\alpha q}$, only a few studies have directly demonstrated this. Thus, stimulation of $[^{35}S]GTP\gamma S$ binding was observed in membranes of Sf9 insect cells expressing high levels of 5-HT_{2C} receptors reconstituted with exogenous $G_{\alpha q}$ proteins (Hartman and Northup 1996), and cell-permeable peptides mimicking the C-terminal component of $G_{\alpha q}$ were shown to block activation of native 5-HT_{2C} receptors in the choroid plexus (Chang et al. 2000). Several studies that used pertussis toxin (PTX), which uncouples G-protein-coupled receptors from $G_{\alpha i/o}$ protein subtypes, have demonstrated that in addition to $G_{\alpha q}$, 5-HT_{2C} receptors also interact with PTX-sensitive $G_{\alpha i/o}$ proteins controlling endogenous Cl^- membrane currents in *Xenopus laevis* oocytes (Quick et al. 1994), DNA synthesis and proliferation in NIH-3 T3 cells (Westphal and Sanders-Bush 1996), and adenylyl cyclase activity in the AV12 cell line (Lucaites et al. 1996). Direct functional coupling of 5-HT$_{2C}$ receptors expressed in *Xenopus laevis* oocytes to G$_{\alpha o}$ and G$_{\alpha i1}$ that resulted in phospholipase C activation was shown by an antisense strategy (Quick et al. 1994; Chen et al. 1994). Similarly, 5-HT$_{2C}$ receptors overexpressed in human embryonic kidney (HEK) 293 cells mediated [^{35}S]GTPγ S binding to G$_{\alpha i}$ proteins, and PTX-sensitive G$_{\alpha i}$ subtypes contributed to stimulation of phospholipase C, along with PTX-insensitive G$_{\alpha q/11}$ subtypes (Alberts et al. 1999). Finally, Cussac et al. demonstrated that the partially edited 5-HT$_{2C\text{-VSV}}$ receptor isoform stably expressed in Chinese hamster ovary (CHO) cells couples to both G$_{\alpha q/11}$ and G$_{\alpha i3}$, which are recruited in an agonist-dependent and receptor reserve–dependent manner, and suggested that the differential influence of agonists on G-protein coupling to the 5-HT$_{2C}$ receptor variants may be relevant to their functional profiles in vivo (Cussac et al. 2002).

However, these studies were conducted in artificial systems in which the receptors may have promiscuous interactions with various heterotrimeric G proteins. At the same time, studies in primary cultures of choroid plexus epithelial (CPE) cells demonstrated that PLCβ signaling of endogenously expressed 5-HT$_{2C}$ receptors is mediated by G$_{\alpha q/11}$ heterotrimeric proteins, specifically by the G$_{\alpha q}$ subunit, and that G$_{\beta \gamma}$ subunits released from Gq/11 heterotrimers did not contribute to the activation of PI hydrolysis signal in natural systems in which the stoichiometry of signaling molecules is undisturbed. In this study, peptide mimics of G$_{\alpha q}$ and the G$_{\alpha q}$ interaction domain of PLCβ1 prevented 5-HT$_{2C}$ receptor-mediated accumulation of inositol phosphates, whereas peptides derived from $\beta\gamma$-binding sites of PLCβ2 did not (Chang et al. 2000). Other studies in CPE cells have shown that the endogenous 5-HT$_{2C}$ receptor also is able to activate phospholipase D via G$_{\alpha 13}$ heterotrimers, and that both G$_{\alpha 13}$ subunit and free $\beta\gamma$ subunits are responsible for this activation (McGrew et al. 2002). In addition, 5-HT$_{2C}$ receptor has been shown to activate the small GTPase RhoA to induce stress fiber formation (Gohla et al. 1999).

5.3   The 5-HT$_{2C}$ Receptor Activates Phospholipases

Phospholipases form a ubiquitous class of enzymes that hydrolyze specific ester bonds in phosphoglycerides or glycerophosphatidates, converting the phospholipids into fatty acids and other lipophilic substances (reviewed in Roberts 1996). The 5-HT$_{2C}$ receptor can activate phospholipase C, phospholipase A$_2$, and phospholipase D (Fig. 5.2).

5.3.1   *Phospholipase C*

Perhaps the best-studied effector system to which the 5-HT$_{2C}$ receptor couples is the phospholipase Cβ (PLC) pathway (for reviews, see Roth et al. 1998; Leysen 2004).

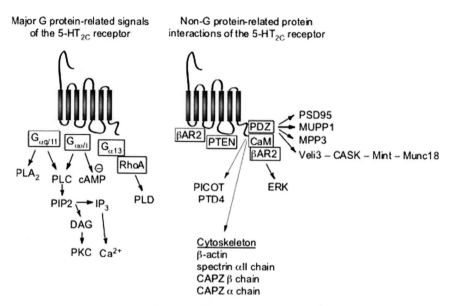

Fig. 5.2 G-protein-signaling and non-G-protein-coupling partners of the 5-HT$_{2C}$ receptor. The *left panel* depicts the three major heterotrimeric G-protein families (G$_{\alpha q/11}$, G$_{\alpha o/i}$, and G$_{\alpha 13}$) through which the 5-HT$_{2C}$ receptor activates phospholipase C (PLC), phospholipase A$_2$ (PLA$_2$), and phospholipase D (PLD) and inhibits the accumulation of cyclic adenosine monophosphate (cAMP). Activation of phospholipase C leads to the hydrolysis of PIP$_2$, resulting in the generation of diacylglycerol (DAG) and inositol trisphosphate (IP$_3$), which lead to activation of protein kinase C (PKC) and elevation of intracellular calcium levels, respectively. The receptor also modulates many ion channels and transporters, nitric oxide synthase, and other effectors through G proteins (see text for details). The *right panel* depicts various proteins that physically associate with the 5-HT$_{2C}$ receptor and that might mediate non-G-protein-mediated signals. Beta-arrestin2 (βAR2) binds to both the second intracellular loop and carboxyl terminus of the receptor. In the carboxyl terminus, βAR2 physically associates with the receptor and calmodulin (CaM), resulting in activation of the mitogenic kinase, extracellular signal-regulated kinases (ERK). The third intracellular loop binds to phosphatase and tensin homolog (PTEN), which is a lipid and protein phosphatase. The carboxyl terminus also binds to cytoskeletal elements such as actin, spectrin, and CAPZ and to PTD4 (a putative GTP-binding protein) and PICOT (PKC θ-interacting protein). Through a PDZ-ligand interaction, the carboxyl terminus of the receptor also interacts with PDZ-containing proteins such as MUPP1, PSD95, MPP3, and the Veli-CASK-Mint-Munc18 complex (see text for details). For the most part, the functional effects of those interactions remain to be elucidated

Stimulation of PLC by the 5-HT$_{2C}$ receptor leads to the production of diacylglycerol and inositol triphosphate (IP$_3$) and subsequent intracellular calcium release. This signaling pathway has been demonstrated in several brain regions including choroid plexus and in recombinant cell lines expressing the 5-HT$_{2C}$ receptor (Chang et al. 2000; Conn et al. 1986; Berg et al. 1994). The activation of PLC by endogenously expressed 5-HT$_{2C}$ receptor most likely is mediated through G$_{\alpha q/11}$-proteins, although coupling of 5-HT$_{2C}$ receptor to PLC activation through PTX-sensitive G$_{\alpha i}$/G$_{\alpha o}$ subtypes has been shown in *Xenopus laevis* oocytes (Chen et al. 1994) and in HEK 293 cells overexpressing 5-HT$_{2C}$ receptors (Alberts et al. 1999). In addition, involvement of botulinum ADP-ribosyltransferase-sensitive low-molecular-weight Gproteins in IP$_3$ accumulation induced by the 5-HT$_{2C}$ receptors transfected into COS-7 cells also has been proposed (Tohda et al. 1995).

RNA-edited isoforms of the 5-HT$_{2C}$ receptor, 5-HT$_{2C-VGV}$, and 5-HT$_{2C-VSV}$, which have different amino acids in the second intracellular loop from the nonedited 5-HT$_{2C-INI}$ receptor, have reduced capacity to activate PLC signaling pathway compared with the nonedited receptor (Burns et al. 1997; Fitzgerald et al. 1999; Niswender et al. 1999; Berg et al. 2001).

5.3.2   Phospholipase A$_2$

The 5-HT$_2$ receptor subclass including 5-HT$_{2C}$ receptors, also has been shown to couple to the phospholipase A$_2$ (PLA$_2$) signaling cascade, which leads to the release of arachidonic acid (AA) and the subsequent production of a myriad of AA metabolites (Leysen 2004; Felder et al. 1990; Berg et al. 1996).

Berg et al. tested the capacity of a series of serotonergic ligands to differentially activate PLC and PLA$_2$ signal transduction pathways associated with human 5-HT$_{2A}$ and 5-HT$_{2C}$ receptors transfected into CHO cells (Berg et al. 1998). The authors measured agonist-induced PLC-mediated accumulation of IP$_3$ and PLA$_2$-mediated release of AA simultaneously from the same population of cells and observed that the relative efficacy of a series of ligands for each of the receptors differed depending upon assessed signal transduction pathway. For the 5-HT$_{2A}$ receptor all tested agonists had greater relative efficacy for PLA$_2$ than for PLC. In contrast, for the 5-HT$_{2C}$ receptor some agonists preferentially activated the PLC pathway, whereas others favored PLA$_2$ when relative efficacies were referenced to 5-HT (Berg et al. 1998). These data strongly support the "agonist-directed trafficking of receptor stimulus (ADTRS)" hypothesis, which suggests that a single receptor subtype may couple with different efficacies to multiple signaling pathways depending upon the nature of the agonist to which the receptor is exposed (Kenakin 1995).

RNA-edited isoforms of the 5-HT$_{2C}$ receptor, 5-HT$_{2C-VGV}$,and 5-HT$_{2C-VSV}$, also couple to PLA$_2$, but the potency for 5-HT to stimulate AA release was less for the edited receptors than that for the nonedited receptor (5-HT$_{2C-INI}$) when receptors were expressed at similar densities (Berg et al. 2001).

5.3.3   Phospholipase D

Native 5-HT$_{2C}$ receptors in choroid plexus epithelial cells also activate phospholipase D (PLD), an enzyme that catalyzes the hydrolysis of phosphatidylcholine to form phosphatidic acid and a released choline head group. 5-HT$_{2C}$ receptor-induced PLD activation involves the G$_{\alpha 13}$ subunit and free G$_{\beta\gamma}$ subunits, as well as transactivation of the small GTPase RhoA (McGrew et al. 2002). These events are regulated by RNA editing of the 5-HT$_{2C}$ receptor. The RNA-edited isoform of the 5-HT$_{2C}$ receptor, 5-HT$_{2C-VGV}$, has a greatly attenuated ability to functionally couple to G$_{\alpha 13}$ subunit compared with the nonedited 5-HT$_{2C-INI}$ receptor, and is unable to promote Rho GTPase activity. Indeed, the PLD signal was undetectable in NIH-3 T3 cells heterologously expressing the 5-HT$_{2C-VGV}$ isoform (McGrew et al. 2004).

5.4 The 5-HT$_{2C}$ Receptor Can Modulate Cyclic Adenosine Monophosphate (AMP) Accumulation

When expressed at high density (12 pmol/mg membrane protein) in stably transformed AV12 cells, the 5-HT$_{2C}$ receptor has been shown to inhibit forskolin-stimulated cAMP production with an IC$_{50} \approx 50$ nmol/L. This effect was sensitive to PTX, suggesting the involvement of G$_{\alpha i/o}$ proteins. In contrast, the 5-HT$_{2A}$ and 5-HT$_{2B}$ receptors expressed at similar levels did not decrease adenylate cyclase (AC) activity in those cells, although all three receptor subtypes were coupled to PLC activation. Pretreatment of the cells expressing high levels of 5-HT$_{2C}$ receptor with PTX also unmasked a small stimulatory effect of the 5-HT$_{2C}$ receptor on cAMP accumulation. When expressed at low density (~150 fmol/mg of membrane protein), the 5-HT$_{2C}$ receptor could increase forskolin-stimulated cAMP production by twofold (Lucaites et al. 1996).

These results demonstrate that 5-HT$_{2C}$ receptor density may determine the signaling pathway(s) to which the receptor couples. At the same time, the 5-HT$_{2C}$ receptor endogenously expressed in pig choroids plexus has not been linked either to stimulation of AC or to inhibition of forskolin-stimulated cyclase activity (Palacios et al. 1986). Thus, it is not clear whether the 5-HT$_{2C}$ receptor can modulate cAMP levels in native tissues.

5.5 The 5-HT$_{2C}$ Receptor Couples to Cyclic Guanosine Monophosphate (GMP) Formation

The 5-HT$_{2C}$ receptor has been shown to trigger the formation of cyclic GMP (cGMP) in porcine choroid plexus tissue slices through a pathway that requires PTX-insensitive G proteins, Ca^{2+}, PLA$_2$, and lipoxygenase activity (Kaufman et al. 1995). Further, an inhibitory effect of cGMP on phosphoinositide turnover in choroid plexux has been demonstrated suggesting a possibility of negative interaction between second messenger pathways concurrently activated by 5-HT$_{2C}$ receptor (Kaufman and Hirata 1996). At the same time, the 5-HT$_{2C}$ receptor has been shown to prevent the nitric oxide (NO)-dependent increase of cGMP elicited by activation of N-methyl-D-aspartate (NMDA) receptors in rat cerebellum (Marcoli et al. 1997) and in human neocortical slices (Maura et al. 2000).

5.6   The 5-HT$_{2C}$ Receptor Regulates Channels

The ability of the 5-HT$_{2C}$ receptor to regulate potassium channels and Ca^{2+}-activated chloride channels has been shown by several studies, both in native and artificial systems.

5.6.1   K$^+$ Channels

The 5-HT$_{2C}$ receptor has been shown to modulate the activities of various K$^+$ channels; for example, the receptor closes K$^+$ channels in the apical membrane of the mouse choroid plexus epithelium (Hung et al. 1993). When the 5-HT$_{2C}$ receptor was coexpressed with rat cerebral cortex K$^+$ channels in *Xenopus laevis* oocytes, activation of the receptor also resulted in closing of the K$^+$ channels in a Ca^{2+} -independent manner. The coupling between the receptors and channels appears to be mediated by the inositol phospholipid second messenger pathway (Panicker et al. 1991). Another study has suggested that the suppression of K$^+$ currents by the 5-HT$_{2C}$ receptor coexpressed with a cloned mouse brain K$^+$ channel in *Xenopus laevis* oocytes, occurs through a novel mechanism, which is independent of A- or C-type protein kinases, but which involves a Ca^{2+}/calmodulin-activated phosphatase (Hoger et al. 1991).

DiMagno et al. showed that the 5-HT$_{2C}$ receptor, when coexpressed in *Xenopus* oocytes with rat brain mRNA, could inhibit an inwardly rectified, Ba^{2+} -sensitive K$^+$ conductance through a PKC-dependent pathway (DiMagno et al. 1996). When coexpressed in *Xenopus* oocytes, the 5-HT$_{2C}$ receptor also suppresses the activity of the Shaker-related K$^+$ gene Kv1.3, which encodes the type n K$^+$ channel. This effect on Kv1.3 currents occurs via activation of a PTX-sensitive G protein and a subsequent rise in intracellular Ca^{2+} but not via PKC, Ca^{2+}/calmodulin, or phosphatases (Aiyar et al. 1993). Timpe and Fantl used a similar system to show that the 5-HT$_{2C}$ receptor suppressed the activity of the voltage-activated K$^+$ channel Kv1.5 through a PLC-dependent pathway (Timpe and Fantl 1994). Speake et al. provided functional and immunocytochemical evidence for the presence of Kv1.1 and Kv1.3 K$^+$ channels in choroid plexus epithelial cells and demonstrated that the 5-HT$_{2C}$ receptor inhibits the Kv conductance by stimulating PKC (Speake et al. 2004). This regulation of K$^+$ channels by 5-HT may be important in inhibiting the cerebrospinal fluid secretion and/or increasing K$^+$ absorption by the choroid plexus epithelium.

5.6.2   Cl$^-$ Channels

The 5-HT$_{2C}$ receptor also has been shown to stimulate an apical Cl$^-$ conductance in mouse choroid plexus (Hung et al. 1993). 5-HT$_{2C}$ receptors expressed in

Xenopus laevis oocytes can stimulate Ca^{2+} release from IP_3-sensitive intracellular stores, which results in the opening of Ca^{2+}-gated Cl^- channels (DiMagno et al. 1996). This effect occurs through endogenous $G_{\alpha o}$ proteins. The same Cl^- current could be also stimulated by coexpression of 5-HT_{2C} receptors with PTX-sensitive ($G_{\alpha o}$) and PTX-insensitive ($G_{\alpha q}$) G protein subunits, suggesting that 5-HT_{2C} receptor couples within the same cell type to several different heterologously expressed G protein α subunits to activate endogenous *Xenopus laevis* oocyte Ca^{2+}-gated Cl^- channels. This effect was mediated by an endogenous PLC-β (Quick et al. 1994).

In addition to 5-HT_{2C} receptor–dependent regulation of K^+ and Cl^- channels, studies in the prefrontal cortex (PFC) have described effects of serotonin treatment on sodium and Ca^{2+} currents (Carr et al. 2002; Day et al. 2002), and on N-methyl-D-aspartate (NMDA) receptor channels (Yuen et al. 2008). PFC pyramidal neurons express detectable levels of 5-HT_{2A} and/or 5-HT_{2C} receptor mRNA, with half of the cells expressing both mRNAs. However since 5-HT_{2A} and 5-HT_{2C} receptors have similar pharmacological profiles and intracellular signaling cascades (Sanders-Bush et al. 2003), these studies did not address the relative contribution of the 5-HT_{2A} and 5-HT_{2C} receptors to 5-HT effects. $5\text{-HT}_{2A/C}$ receptor stimulation decreased rapidly inactivating Na^+ currents by reducing maximal current amplitude, and by producing a negative shift in the voltage dependence of fast inactivation. These effects of 5-HT on voltage-dependent Na^+ channels in PFC pyramidal neurons were mediated by $G_{\alpha q}$- dependent activation of PLC and subsequent activation of PKC, and resulted in reduced dendritic excitability (Carr et al. 2002). Voltage-clamp and single-cell RT-PCR studies in acutely isolated deep-layer prefrontal pyramidal neurons show that $5\text{-HT}_{2A/C}$ receptor activation reduces L-type Cav1.2 Ca^{2+} channel currents via a signaling cascade initiated by $G_{\alpha q}$ protein stimulation of PLCβ, leading to the mobilization of IP_3-sensitive intracellular Ca^{2+} stores and activation of the Ca^{2+}-dependent phosphatase calcineurin (Day et al. 2002). $5\text{-HT}_{2A/C}$ receptor activation regulates N-methyl-D-aspartate (NMDA) receptor currents in PFC pyramidal neurons by opposing the 5-HT_{1A} receptor-induced disruption of microtubule-based transport of NMDA receptors (Yuen et al. 2008).

Furthermore 5-HT_{2C} receptors have been shown to inhibit γ-aminobutyric acid (GABA)-A receptor channels by a Ca^{2+}-dependent, phosphorylation-independent mechanism in *Xenopus* oocytes (Huidobro-Toro et al. 1996). Activation of the 5-HT_{2C} receptor in hypothalamic pro-opiomelanocortin (POMC) neurons desensitizes the $GABA_B$ response in neurons through inhibition of G-protein-coupled inwardly rectifying potassium (GIRK) channels. The mechanism involves activation of PLC by $G\alpha_{q/11}$ subunits and hydrolysis of plasma membrane PIP_2 to IP_3 and DAG, which leads to the IP_3-induced release of Ca^{2+} from intracellular stores and DAG-dependent activation of PKC. Next, PKC phosphorylates and up-regulates adenylyl cyclase VII activity, leading to cAMP-dependent activation of PKA, which can rapidly uncouple $GABA_B$ from effector system through phosphorylation of GIRK (Qiu et al. 2007).

5.7 The 5-HT$_{2C}$ Receptor Regulates Transport Processes

The 5-HT$_{2C}$ receptor modulates a number of distinct transport processes. The 5-HT$_{2C}$ receptor has been shown to activate an electrogenic Na$^+$/Ca^{2+} exchanger in histaminergic neurons in the rat hypothalamic tuberomammillary nucleus leading to depolarization of the neurons (Eriksson et al. 2001); whereas the cloned 5-HT$_{2C}$ receptor has been linked to activation of Na$^+$/K$^+$/2Cl$^-$ cotransport when transfected into NIH-3 T3 fibroblasts (Mayer and Sanders-Bush 1994). Recombinant 5-HT$_{2C}$ receptors expressed in 3 T3 cells have been shown to increase secretion of amyloid precursor proteins in a PKC- and PLA$_2$-dependent manner (Nitsch et al. 1996).

The 5-HT$_{2C}$ receptors in hypothalamic paraventricular nucleus are involved in the serotonergic stimulation of prolactin secretion (Rittenhouse et al. 1993; Bagdy 1996). In addition, 5-HT$_{2C}$ receptors have been shown to mediate vasopressin secretion (Bagdy et al. 1992) as well as oxytocin, corticosterone, and adrenocorticotropic hormone (ACTH) secretion (Bagdy 1996). At the same time, the 5-HT$_{2C}$ receptor markedly suppressed the neuronal release of dopamine (DA) and noradrenaline (NA) in the rat frontal cortex without affecting serotonergic transmission (Millan et al. 1998). Furthermore, it has been shown that inhibitory control of the nigrostriatal dopamine system occurs by 5-HT$_{2C}$ receptors localized within the terminal region, the dorsal striatum (Alex et al. 2005). In addition, another group using in vivo intracerebral microdialysis and CHO cells expressing 5-HT$_{2C}$ receptors has demonstrated that constitutive 5-HT$_{2C}$ receptor activity participates in the tonic inhibitory control of DA release in the rat striatum and nucleus accumbens in vivo (De Deurwaerdere et al. 2004). Thus, the 5-HT$_{2C}$ receptor has the unique ability to tonically regulate DA release by combined actions involving the effects of endogenous serotonin and constitutive activity of the receptor, and this ability may provide a functional basis for the fine tuning of central DA neuron function (Berg et al. 2008).

5.8 The 5-HT$_{2C}$ Receptor and Growth Regulation

5.8.1 5-HT$_{2C}$ Receptors Regulate Mitogenesis

Several studies have suggested that 5-HT$_{2C}$ receptor plays role in growth regulation (Westphal and Sanders-Bush 1996; Julius et al. 1989; Agarwal and Glasel 1997). Transfection of functional 5-HT$_{2C}$ receptors into NIH 3 T3 cells and continuous agonist activation of the receptor resulted in the development of a transformed phenotype (Julius et al. 1989). In addition, constitutively active 5-HT$_{2C}$ receptors were able to stimulate cell division in transfected NIH 3 T3 fibroblasts in the absence of an agonist (Westphal and Sanders-Bush 1996). Interestingly, agonist occupation of the 5-HT$_{2C}$ receptor activated PTX-sensitive G$_{\alpha i/o}$ proteins in addition to PTX-insensitive G$_{\alpha q/11}$ subtypes to control DNA synthesis and proliferation in

NIH-3 T3 cells (Westphal and Sanders-Bush 1996). On the other hand, the recombinant 5-HT$_{2C}$ receptor expressed in CCL36 hamster fibroblasts did not readily induce a transformed phenotype (Kahan et al. 1992); this suggests that 5-HT$_{2C}$ receptor effects on DNA synthesis and proliferation may be cell specific.

5.8.2   5-HT$_{2C}$ Receptors Activate Extracellular Signal-Regulated Kinases 1/2

Because the 5-HT$_{2C}$ receptor regulates cell proliferation (Sect. 5.8.1), it is not surprising that the receptor could activate mitogenic kinases. The first evidence that 5-HT$_{2C}$ receptor is able to activate extracellular signal-regulated kinases ½ (ERK1/2) when expressed in NIH 3 T3 cells was presented by Fitzgerald et al. (2000). Another group investigated the mechanism of coupling of 5-HT$_{2C}$ receptor to ERK1/2 using CHO cells that expressed the non-RNA-edited 5-HT$_{2C-INI}$ isoform of the human 5-HT$_{2C}$ receptor at levels comparable with those found in native neuronal preparations (Werry et al. 2005). The authors found that 5-HT$_{2C}$ receptor-stimulated ERK1/2 phosphorylation through a pathway that requires the involvement of the PTX-insensitive G proteins (most likely G$_{\alpha 12/13}$,), PLD, protein kinase C (PKC), and activation of the Raf/MEK/ERK module but that is independent of epidermal growth factor receptor (EGFR) tyrosine kinase and of nonreceptor tyrosine kinase Src, PLC, phosphoinositide 3-kinase (PI3-K), and receptor endocytosis (Werry et al. 2005). Later the same group expanded their studies of the mechanisms underlying 5-HT$_{2C}$ receptor-induced ERK1/2 activation using the non-RNA-edited 5-HT$_{2C-INI}$ isoform, fully edited 5-HT$_{2C-VGV}$ isoform, and partially edited but most abundant 5-HT$_{2C-VSV}$ isoform of the human 5-HT$_{2C}$ receptor expressed in CV1 kidney epithelial cells (Werry et al. 2008b). They found that 5-HT$_{2C}$ receptor–dependent activation of the ERK1/2 signaling cascade involves a variety of additional components, including activation of matrix metalloproteases, PKC, PI3-K, and transactivation of the EGFR, and that RNA editing of the 5-HT$_{2C}$ receptor has significant effects on its ability to signal to ERK1/2, changing both its temporal and pharmacological profiles, including the degree of dependence on specific effectors. Thus, while all three isoforms displayed a strong dependence on EGFR transactivation, the non-RNA-edited 5-HT$_{2C-INI}$ isoform did not utilize either PKC or PI3-K at early time points, the 5-HT$_{2C-VSV}$ isoform was connected with both PKC and PI3-K, and 5-HT$_{2C-VGV}$ isoform recruited PKC to a significant degree (Werry et al. 2008b).

These studies also support the notion that 5-HT$_{2C}$ receptor can generate different intracellular signals in a ligand-dependent manner (Berg et al. 1998) because agonist-directed trafficking of receptor stimuli was observed for some agonists when comparing ERK1/2 phosphorylation to IP$_3$ accumulation and intracellular Ca^{2+} elevations (Werry et al. 2005) and because the contribution of different signaling molecules utilized by 5-HT$_{2C}$ receptor to stimulate ERK1/2 was dependent on the agonist used to stimulate the receptor (Werry et al. 2008b). However, a recent study by Knauer et al. has shown that a partially edited isoform of human 5-HT$_{2C}$ receptor (5-HT$_{2C\text{-}ISV}$), expressed in low densities in CHO cells, efficiently couples to ERK1/2 activation via PTX-insensitive G proteins and sequential PLC-dependent hydrolysis of PIP$_2$, and Ca^{2+} mobilization, independent of both receptor and nonreceptor tyrosine kinases and PI3-K (Knauer et al. 2009). This study describes that 5-HT$_{2C\text{-}ISV}$ receptor-induced ERK1/2 activation resides mechanistically downstream of Ca^{2+} mobilization and PKC activation and does not support "agonist-directed trafficking" of receptor stimuli, and therefore is not consistent with the findings of Werry et al. (2005). One possible explanation of these discrepancies may be an effect of RNA editing, because Werry et al. studied the nonedited isoform of the human 5-HT$_{2C}$ receptor (Knauer et al. 2009).

On the other hand, Labasque et al. have demonstrated that 5-HT induced a rapid and sustained ERK1/2 activation in HEK-293 cells transiently transfected with 5-HT$_{2C}$ receptors, which was completely independent of G$_\alpha$-protein subunits known to couple to the receptor (Labasque et al. 2008). That 5-HT$_{2C}$ receptor-mediated ERK1/2 signaling required Ca^{2+}-dependent association of calmodulin (CaM) to the receptor C-terminus and subsequent recruitment of β-arrestin 2 by the receptor. Importantly, the Ca^{2+}/CaM-dependent mechanism of ERK1/2 activation by 5-HT was demonstrated not only in transfected HEK 293 cells but also in choroid plexus epithelial cells, which express the highest 5-HT$_{2C}$ receptor density, and in cortical neurons transfected with a GFP-5-HT$_{2C}$ receptor construct (Labasque et al. 2008). A role for β-arrestin in the activation of ERK1/2 also has been demonstrated in the dendrites of cultured PFC pyramidal neurons, which endogenously express 5-HT$_{2A}$ and 5-HT$_{2C}$ receptors. Stimulation of 5-HT$_{2A/C}$ receptors triggered the activation of ERK1/2 by a mechanism that is independent of PLC but requires β-arrestin, Src activation, and clathrin/dynamin-mediated endocytosis of the receptor (Yuen et al. 2008). One group speculated that 5-HT$_{2C}$ receptor–dependent ERK1/2 activation might provide the molecular substrate for increased neurogenesis induced by chronic treatment with 5-HT$_{2C}$ agonists, a phenomenon that possibly is involved in their antidepressant effects (Millan 2005).

5.9   Dimerization of the 5-HT$_{2C}$ Receptor

There is increasing evidence that many GPCRs exist as homodimers and heterodimers and that their oligomeric assembly plays functional roles affecting G-protein coupling, downstream signaling, and regulatory processes such as internalization (reviewed in Terrillon and Bouvier 2004). Using fluorescence resonance energy transfer (FRET) combined with confocal microscopy, Herrick-Davis et al. demonstrated that 5-HT$_{2C}$ receptors exist as constitutive homodimers on the plasma membrane of living HEK 293 cells expressing fluorescently tagged 5-HT$_{2C}$ receptors (Herrick-Davis et al. 2004). Further, the same group evaluated the effect of dimerization on receptor function by coexpressing wild-type 5-HT$_{2C}$ receptors with an inactive mutant (S138R) receptor in HEK 293 cells (Herrick-Davis et al. 2005).

It appears that inactive 5-HT$_{2C}$ receptors form nonfunctional heterodimers expressed on the plasma membrane, thus inhibiting wild-type receptor function. Studies of ligand/dimer/G-protein stoichiometry in living HEK 293 cells suggest that one 5-HT$_{2C}$ receptor dimer binds two molecules of ligand, resulting in the activation of one G protein (Herrick-Davis et al. 2005). The 5-HT$_{2C}$ receptor has two distinct dimerization interfaces, one between transmembrane (TM) helices IV and V, and the other between TM I helices (Mancia et al. 2008). The TM I interface is insensitive to the activation state of the receptor, whereas the IV/V interface is markedly sensitive to the type of ligands bound to the 5-HT$_{2C}$ receptors. Furthermore, the dimer is asymmetrical in its interaction with G$_{\alpha q}$ protein; activated 5-HT$_{2C}$ receptors also may be able to associate with other GPCRs (Mancia et al. 2008).

5.10 Proteins Interacting with the 5-HT$_{2C}$ Receptor

Although coupling to G proteins is a central event in the GPCR signaling, multiple recent studies indicate that the regulation and signaling of the 5-HT$_{2C}$ receptor also depends on interactions with a variety of additional protein partners. This section will describe some of these interactions.

5.10.1 PDZ Domain-Containing Proteins

To date, the best characterized proteins involved in the scaffolding of multiprotein complexes that are involved in the targeting, trafficking, and signaling of GPCRs are PSD-95/Disc large/Zonula occludens-1 (PDZ) domain-containing proteins (Nourry et al. 2003). 5-HT$_{2C}$ receptors contain a C-terminal sequence (SSV) corresponding to one of the class 1 PDZ recognition motifs (X-S/T-X-I/L/V). A multi-PDZ domain adaptor protein (MUPP1), which contains 13 PDZ domains, was the first 5-HT$_{2C}$ receptor–interacting protein identified using the yeast two-hybrid system (Ullmer et al. 1998). The C-terminus of the 5-HT$_{2C}$ receptor selectively interacts with the 10th PDZ domain of MUPP1, and this interaction leads to a conformational change in MUPP1 (Bécamel et al. 2001). Deletion of the 5-HT$_{2C}$ receptor PDZ recognition motif prevents phosphorylation of the receptor and delays resensitization of receptor responses in NIH 3 T3 fibroblasts, suggesting that interaction between MUPP1 and the 5-HT$_{2C}$ receptor is functionally significant (Backstrom et al. 2000). Thus, it was suggested that MUPP1 might function as a multivalent scaffold protein, which selectively assembles and targets 5-HT$_{2C}$ receptor–signaling complexes (Bécamel et al. 2001). Further, using a proteomic approach based on peptide affinity chromatography followed by mass spectrometry and/or immunoblotting, Bécamel et al. identified at least 15 additional proteins that interact with the C-terminal tail of 5-HT$_{2C}$ receptors, including scaffold proteins that contain one or several PDZ domains, such as the Veli3 (vertebrate homologue of the *Caenorabdis elegans* PDZ protein Lin7), postsynaptic density protein 95 (PSD95), and MPP3 protein of the membrane-associated guanylate kinases (MAGUK) p55 subfamily (Bécamel et al. 2002). These studies revealed the connection of 5-HT$_{2C}$ receptors to PDZ proteins exhibiting both presynaptic and postsynaptic localizations consistent with the differential distribution of the receptors at the synaptic junction. Thus, via association with Veli3, the 5-HT$_{2C}$ receptor interacts with the tripartite Veli3-CASK-Mint1 complex, which is located at the postsynaptic density, as well as with the presynaptic Veli3-CASK-Mint-Munc18 complex. Several studies using 5-HT$_{2C}$ receptors mutated on the PDZ recognition site show that interactions with PDZ proteins have a critical role in modulating signaling properties of 5-HT$_{2C}$ receptors and their desensitization (Backstrom et al. 2000; Gavarini et al. 2006). PSD95 and MPP3 proteins, the main PDZ-binding partners of the 5-HT$_{2C}$ receptor, appear to differentially modulate desensitization of receptor-associated Ca^{2+} responses in both heterologous cells and cultured cortical neurons, indicating that the functional activity of 5-HT$_{2C}$ receptors is modulated according to the repertoire of PDZ proteins coexpressed with the receptor (Gavarini et al. 2006).

Interestingly, the 5-HT$_{2A}$ and 5-HT$_{2C}$ receptors, two closely related GPCRs that share an identical canonical PDZ ligand, interact with distinct sets of PDZ proteins, which may contribute to their differences in signal transduction pathways (Bécamel et al. 2004).

5.10.2   Non-PDZ Proteins That Interact with 5-HT$_{2C}$ Receptors C Terminus

5-HT$_{2C}$ receptors also interact with non-PDZ proteins. These include CaM, which binds to the 5-HT$_{2C}$ receptor C-terminus (Bécamel et al. 2002). An additional putative CaM binding motif has been identified in the second intracellular loop of the 5-HT$_{2C}$ receptor (Turner and Raymond 2005). Labasque et al. have demonstrated that CaM binds to a prototypic Ca^{2+}-dependent "1-10" CaM-binding motif located in the proximal region of the 5-HT$_{2C}$ receptor C-terminus in an agonist-dependent manner (Labasque et al. 2008). Mutation of this motif inhibited both β-arrestin recruitment by the 5-HT$_{2C}$ receptor and 5-HT-induced ERK1/2 activation in HEK 293 cells. Therefore, CaM bound to the juxtamembrane region of the 5-HT$_{2C}$ receptor C-terminus might function to stabilize 5-HT$_{2C}$ receptor–β-arrestin interaction (Labasque et al. 2008).

Proteomic screens also have demonstrated interactions of the 5-HT$_{2C}$ receptor C-terminal domain with proteins of the cytoskeleton (β-actin, spectrin αII chain, α and β chains of CAPZ) and with two other proteins with unknown functions – PTD4, a putative GTP-binding protein and PICOT (PKC θ-interacting protein), which contains a thioredoxin homology domain that is involved in the regulation of the thioredoxin system (Bécamel et al. 2002).

5.10.3 Proteins That Interact with 5-HT$_{2C}$ Receptor Intracellular Loops

5-HT$_{2C}$ receptors interact with β-arrestin (mainly β-arrestin-2) via the second intracellular loop (Marion et al. 2004; 2006). β-Arrestin binding is tightly dependent on receptor phosphorylation on C-terminal serine and threonine residues and on RNA editing of 5-HT$_{2C}$ receptors (Marion et al. 2004). The unedited 5-HT$_{2C-INI}$ receptor is capable of interacting with β-arrestin-2 in the absence of agonist, leading to constitutive receptor internalization and its accumulation within endocytic vesicles. Application of inverse agonists results in the redistribution of 5- HT$_{2C-INI}$ receptors to the plasma membrane of HEK-293 cells (Marion et al. 2004; Chanrion et al. 2008). At the same time, fully edited 5-HT$_{2C-VGV}$ receptors, which display the lowest level of constitutive activity, do not associate with β-arrestin-2 in the absence of an agonist and are located primarily at the cell surface. However, upon agonist treatment, the fully edited 5-HT$_{2C-VGV}$ receptor associates with β-arrestin-2 and undergoes rapid internalization.

5-HT$_{2C}$ receptors physically interact via their third intracellular loop with the tumor suppressor phosphatase and tensin homolog (PTEN), an enzyme exhibiting both lipid and protein phosphatase activities (Ji et al. 2006). Interaction between 5-HT$_{2C}$ receptors and PTEN prevents agonist-induced receptor phosphorylation at C-terminal serine residues located in the receptor PDZ motif. Association of 5-HT$_{2C}$ receptors with PTEN occurs in dopaminergic neurons of the ventral tegmental area innervating the accumbens nucleus, which are tonically inhibited by activated 5-HT$_{2C}$ receptors (Ji et al. 2006). Therefore, regulation of 5-HT$_{2C}$ receptors functional activity mediated by the interaction with PTEN may play a role in mediating the reinforcing role of drugs (Bockaert et al. 2006).

5.11 Conclusion

The 5-HT$_{2C}$ receptor couples to a large array of signal transduction pathways. This pleiotropic coupling results from the activation of at least three distinct families of heterotrimeric G proteins, from physical interactions of numerous regulatory proteins with intracellular portions of the receptor, from receptor dimerization, and from cell- and ligand-specific effects. In that regard, the recently described phenomenon of "functional selectivity" implies that ligands induce unique, ligand-specific receptor conformations that frequently can result in differential activation of signal transduction pathways associated with that particular receptor. This differential activation may be expressed as differences in intrinsic activity and/or potency at one signaling pathway versus another that are not due to differences in affinity at the mediating receptor. This phenomenon has been observed for several families of GPCRs including serotonergic receptors (Urban et al. 2007). The 5HT$_{2C}$ receptor can generate differential intracellular signals in an agonist-dependent manner, as evidenced in both "classical" G-protein-mediated pathways and more complex pathways such as ERK1/2 activation. This pleiotropic coupling to multiple intracellular pathways suggests that the 5HT$_{2C}$ receptor is capable of agonist-directed trafficking of signals. Moreover, the complexity of the 5-HT$_{2C}$ receptor system, with its multiple receptor variants, activation of multiple G proteins, and modulation of both G-protein-dependent and G-protein-independent signaling pathways, suggests that ligand-specific effects can fine tune those pleiotropic coupling signals.

Acknowledgements The authors were supported by grants from the Department of Veterans Affairs (Merit Awards and a REAP Award to Maria N. Garnovskaya (MNG) and John R. Raymond (JRR)), the National Institutes of Health (DK52448 and GM63909 to JRR), AHA (GIA 0655445U to MNG), and a laboratory endowment jointly supported by the M.U.S.C. Division of Nephrology and Dialysis Clinics, Incorporated (JRR).

References

Abramowski D, Rigo M, Duc D, et al (1995) Localization of the 5-hydroxytryptamine2C receptor protein in human and rat brain using specific antisera. Neuropharmacology 34:1635–1645.

Agarwal D, Glasel JA (1997) Hormone-defined cell system for studying G-protein coupled receptor agonist-activated growth modulation: delta-opioid and serotonin-5HT2C receptor activation show opposite mitogenic effects. J Cell Physiol 171:61–74.

Aiyar J, Grissmer S, Chandy KG (1993) Full-length and truncated Kv1.3 K+channels are modulated by 5-HT1C receptor activation and independently by PKC. Am J Physiol Cell Physiol 265:C1571–C1578.

Alberts GL, Pregenzer JF, Im WB, et al (1999) Agonist-induced GTPgamma35S binding mediated by human 5-HT(2C) receptors expressed in human embryonic kidney 293 cells. Eur J Pharmacol 383(3):311–319.

Alex KD, Yavanian GJ, McFarlane HG, et al (2005) Modulation of dopamine release by striatal 5-HT2C receptors. Synapse 55:242–251.

Backstrom JR, Chang MS, Chu H, et al (1999) Agonist-directed signaling of serotonin 5-HT2C receptors: differences between serotonin and lysergic acid diethylamide (LSD). Neuropsychopharmacology 21:77 S–81 S.

Backstrom JR, Price RD, Reasoner DT, et al (2000) Deletion of the serotonin 5-HT2C receptor PDZ recognition motif prevents receptor phosphorylation and delays resensitization of receptor responses. J Biol Chem 275:23620–23626.

Bagdy G (1996) Role of the hypothalamic paraventricular nucleus in 5-HT1A, 5-HT2A and 5-HT2C receptor-mediated oxytocin, prolactin and ACTH/corticosterone responses. Behav Brain Res 73:277–280.

Bagdy G, Sved AF, Murphy DL, et al (1992) Pharmacological characterization of serotonin receptor subtypes involved in vasopressin and plasma renin activity responses to serotonin agonists. Eur J Pharmacol 210:285–289.

Bécamel C, Figge A, Poliak S, et al (2001) Interaction of serotonin 5-HT2C receptors with PDZ10 of the multi PDZ protein MUPP1. J Biol Chem 276:12974–12982.

Bécamel C, Alonso G, Galeotti N (2002) Synaptic multiprotein complexes associated with 5-HT2C receptors: a proteomic approach. EMBO J 21:2332–2342.

Bécamel C, Gavarini S, Chanrion B, et al (2004) The serotonin 5-HT2A and 5-HT2C receptors interact with specific sets of PDZ proteins. J Biol Chem 279:20257–20266.

Berg KA, Clarke WP, Sailstad C, et al (1994) Signal transduction differences between 5-hydroxytryptamine type 2A and type 2C receptor systems. Mol Pharmacol 46:477–484.

Berg KA, Maayani S, Clarke WP (1996) 5-Hydroxytryptamine2C receptor activation inhibits 5-hydroxytryptamine1B-like receptor function via arachidonic acid metabolism. Mol Pharmacol 50:1017–1023.

Berg KA, Maayani S, Goldfarb J, et al (1998) Effector pathway-dependent relative efficacy at serotonin type 2A and 2C receptors: evidence for agonist-directed trafficking of receptor stimulus. Mol Pharmacol 54:94–104.

Berg KA, Cropper JD, Niswender CM, et al (2001) RNA-editing of the 5-HT(2C) receptor alters agonist-receptor-effector coupling specificity. Br J Pharmacol 134:386–392.

Berg KA, Clarke WP, Cunningham KA, et al (2008) Fine-tuning serotonin 2C receptor function in the brain: molecular and functional implications. Neuropharmacology 55:969–976.

Bockaert J, Claeysen S, Bécamel C, et al (2006) Neuronal 5-HT metabotropic receptors: fine-tuning of their structure, signaling, and roles in synaptic modulation. Cell Tissue Res 326:553–572.

Burns CM, Chu H, Rueter SM, et al (1997) Regulation of serotonin-2C receptor G-protein coupling by RNA editing. Nature 387:303–308.

Canton H, Emeson RB, Barker EL, et al (1996) Identification, molecular cloning, and distribution of a short variant of the 5-hydroxytryptamine2C receptor produced by alternative splicing. Mol Pharmacol 50:799–807.

Carr DB, Cooper DC, Ulrich SL, et al (2002) Serotonin receptor activation inhibits sodium current and dendritic excitability in prefrontal cortex via a protein kinase C-dependent mechanism. J Neurosci 22:6846–6855.

Chang M, Zhang L, Tam JP, et al (2000) Dissecting G protein-coupled receptor signaling pathways with membrane-permeable blocking peptides. Endogenous 5-HT2C receptors in choroid plexus epithelial cells. J Biol Chem 275:7021–7029.

Chanrion B, Mannoury la Cour C, Gavarini S, et al (2008) Inverse agonist and neutral antagonist actions of antidepressants at recombinant and native 5-hydroxytryptamine2C receptors: differential modulation of cell surface expression and signal transduction. Mol Pharmacol 73:748–757.

Chen Y, Baez M, Yu L (1994) Functional coupling of the 5-HT2C serotonin receptor to G proteins in Xenopus oocytes. Neurosci Lett 179:100–102.

Chen CX, Cho DS, Wang Q, et al (2000) A third member of the RNA-specific adenosine deaminase gene family, ADAR3, contains both single- and double-stranded RNA binding domains. RNA 6:755–767.

Conn PJ, Sanders-Bush E, Hoffman BJ, et al (1986) A unique serotonin receptor in choroid plexus is linked to phosphatidylinositol turnover. Proc Natl Acad Sci USA 83:4086–4088.

Cussac D, Newman-Tancredi A, Duqueyroix D, et al (2002) Differential activation of Gq/11 and Gi(3) proteins at 5-hydroxytryptamine(2C) receptors revealed by antibody capture assays: influence of receptor reserve and relationship to agonist-directed trafficking. Mol Pharmacol 62:578–589.

Day M, Olson PA, Platzer J, et al (2002) Stimulation of 5-HT2 receptors in prefrontal pyramidal neurons inhibits Cav1.2 L-Type Ca2+ currents via a PLCbeta/IP3/Calcineurin signaling cascade. J Neurophysiol 87:2490–2504.

De Deurwaerdere P, Navailles S, Berg KA, et al (2004) Constitutive activity of the serotonin2C receptor inhibits in vivo dopamine release in the rat striatum and nucleus accumbens. J Neurosci 24:3235–3241.

DiMagno L, Dascal N, Davidson N, et al (1996) Serotonin and protein kinase C modulation of a rat brain inwardly rectifying K+channel expressed in Xenopus oocytes. Pflugers Arch/Eur J Physiol 431:335–340.

Eriksson KS, Stevens DR, Haas HL (2001) Serotonin excites tuberomammillary neurons by activation of Na+/Ca2+-exchange. Neuropharmacology 40:345–351.

Felder CC, Kanterman RY, Ma AL, et al (1990) Serotonin stimulates phospholipase A2 and the release of arachidonic acid in hippocampal neurons by a type 2 serotonin receptor that is independent of inositolphospholipid hydrolysis. Proc Natl Acad Sci USA 87:2187–2191.

Fitzgerald LW, Iyer G, Conklin, et al (1999) Messenger RNA editing of the human serotonin 5-HT2C receptor. Neuropsychopharmacology 21:82S–90S.

Fitzgerald LW, Burn TC, Brown BS, et al (2000) Possible role of valvular serotonin 5-HT2B receptors in the cardiopathy associated with fenfluramine. Mol Pharmacol 57:75–81.

Flomen R, Knight J, Sham P, et al (2004) Evidence that RNA editing modulates splice site selection in the 5-HT2C receptor gene. Nucleic Acids Res 32:2113–2122.

Foguet M, Nguyen H, Le H, et al (1992) Structure of the mouse 5-HT1C, 5-HT2 and stomach fundus serotonin receptor genes. Neuroreport 3:345–348.

Gavarini S, Bécamel C, Altier C (2006) Opposite effects of PSD-95 and MPP3 PDZ proteins on serotonin 5-hydroxytryptamine2C receptor desensitization and membrane stability. Mol Biol Cell 17:4619–4631.

Gohla A, Offermanns S, Wilkie TM, et al (1999) Differential involvement of Galpha 12 and Galpha 13 in receptor-mediated stress fiber formation. J Biol Chem 274:17901–17907.

Gurevich I, Tamir H, Arango V, et al (2002) Altered editing of serotonin 2C receptor premRNA in the prefrontal cortex of depressed suicide victims. Neuron 34:349–356.

Hartman JL IV, Northup JK (1996) Functional reconstitution in situ of 5-hydroxytryptamine2C (5HT2C) receptors with alpha q and inverse agonism of 5HT2C receptor antagonists. J Biol Chem 271:22591–22597.

Heisler LK, Chu HM, Tecott LH (1998) Epilepsy and obesity in serotonin 5-HT2C receptor mutant mice. Ann NY Acad Sci 861:74–78.

Heisler LH, Cowley MA, Tecott LH, et al (2002) Activation of central melanocortin pathways by fenfluramine. Science 297:609–611.

Herrick-Davis K, Grinde E, Niswender CM (1999) Serotonin 5-HT2C receptor RNA editing alters receptor basal activity: implications for serotonergic signal transduction. J Neurochem 73:1711–1717.

Herrick-Davis K, Grinde E, Mazurkiewicz JE (2004) Biochemical and biophysical characterization of serotonin 5-HT2C receptor homodimers on the plasma membrane of living cells. Biochemistry 43:13963–13971.

Herrick-Davis K, Grinde E, Harrigan TJ, et al (2005) Inhibition of serotonin 5-hydroxytryptamine2C receptor function through heterodimerization: receptor dimers bind two molecules of ligand and one G-protein. J Biol Chem 280:40144–40151.

Hoger JH, Walter AE, Vance D, et al (1991) Modulation of a cloned mouse brain potassium channel. Neuron 6:227–236.

Hoyer D, Clarke DE, Fozard JR, et al (1994) International Union of Pharmacology classification of receptors for 5-hydroxytryptamine (serotonin). Pharmacol Rev 46:157–203.

Huidobro-Toro JP, Valenzuela CF, Harris RA (1996) Modulation of GABA (A) receptor function by G protein-coupled 5-HT2C receptors. Neuropharmacology 35:1355–1363.

Hung BC, Loo DD, Wright EM (1993) Regulation of mouse choroid plexus apical Cl- and K+channels by serotonin. Brain Res 617:285–295.

Ji SP, Zhang Y, Van Cleemput J, et al (2006) Disruption of PTEN coupling with 5-HT2C receptors suppresses behavioral responses induced by drugs of abuse. Nat Med 12:324–329.

Julius D, MacDermott AB, Axel R, et al (1988) Molecular characterization of a functional cDNA encoding the serotonin 1c receptor. Science 241:558–564.

Julius D, Livelli TJ, Jessell TM, et al (1989) Ectopic expression of the serotonin 1c receptor and the triggering of malignant transformation. Science 244:1057–1062.

Kahan C, Julius D, Pouyssegur J, et al (1992) Effects of 5-HT1C-receptor expression on cell proliferation control in hamster fibroblasts: serotonin fails to induce a transformed phenotype. Exp Cell Res 200:523–527.

Kaufman MJ, Hirata F (1996) Cyclic GMP inhibits phosphoinositide turnover in choroid plexus: evidence for interactions between second messengers concurrently triggered by 5-HT2C receptors. Neurosci Lett 206:153–156.

Kaufman MJ, Hartig PR, Hoffman BJ (1995) Serotonin 5-HT2C receptor stimulates cyclic GMP formation in choroid plexus. J Neurochem 64:199–205.

Kenakin T (1995) Agonist-receptor efficacy II: agonist trafficking of receptor signals. Trends Pharmacol Sci 16:232–238.

Knauer CS, Campbell JE, Chio CL, et al (2009) Pharmacological characterization of mitogen-activated protein kinase activation by recombinant human 5-HT2C, 5-HT2A, and 5-HT2B receptors. Naunyn Schmiedeberg's Arch Pharmacol 379:461–471.

Labasque M, Reiter E, Becamel C, et al (2008) Physical interaction of calmodulin with the 5-hydroxytryptamine2C receptor C-terminus is essential for G protein-independent, arrestin-dependent receptor signaling. Mol Biol Cell 19:4640–4650.

Leysen JE (2004) 5-HT2 receptors. Curr Drug Targets: CNS Neurol Disord 3:11–26.

Lubbert H, Hoffman BJ, Snutch TP, et al (1987) cDNA cloning of a serotonin 5-HT1C receptor by electrophysiological assays of mRNA-injected Xenopus oocytes. Proc Natl Acad Sci USA 84:4332–4336.

Lucaites VL, Nelson DL, Wainscott DB, et al (1996) Receptor subtype and density determine the coupling repertoire of the 5-HT2 receptor subfamily. Life Sci 59:1081–1095.

Mancia F, Assur Z, Herman AG, et al (2008) Ligand sensitivity in dimeric associations of the serotonin 5HT2C receptor. EMBO Rep 9:363–369.

Marcoli M, Maura G, Tortarolo M, et al (1997) Serotonin inhibition of the NMDA receptor/nitric oxide/cyclic GMP pathway in rat cerebellum: involvement of 5-hydroxytryptamine2C receptors. J Neurochem 69:427–430.

Marion S, Weiner DM, Caron MG (2004) RNA editing induces variation in desensitization and trafficking of 5-hydroxytryptamine 2C receptor isoforms. J Biol Chem 279:2945–2954.

Marion S, Oakley RH, Kim KM, et al (2006) A beta-arrestin binding determinant common to the second intracellular loops of rhodopsin family G protein-coupled receptors. J Biol Chem 281:2932–2938.

Maura G, Marcoli M, Pepicelli O, et al (2000) Serotonin inhibition of the NMDA receptor/nitric oxide/cyclic GMP pathway in human neocortex slices: involvement of 5-HT2C and 5-HT1A receptors. Br J Pharmacol 130:1853–1858.

Mayer SE, Sanders-Bush E (1994) 5-Hydroxytryptamine type 2A and 2C receptors linked to Na+/K+/Cl- cotransport. Mol Pharmacol 45:991–996.

McGrew L, Chang MS, Sanders-Bush E (2002) Phospholipase D activation by endogenous 5-hydroxytryptamine 2C receptors is mediated by Galpha13 and pertussis toxin-insensitive G betagamma subunits. Mol Pharmacol 62:1339–1343.

McGrew L, Price RD, Hackler E, et al (2004) RNA editing of the human serotonin 5-HT2C receptor disrupts transactivation of the small G-protein RhoA. Mol Pharmacol 65:252–256.

Mengod G, Nguyen H, Le H, et al (1990) The distribution and cellular localization of the serotonin 1C receptor mRNA in the rodent brain examined by in situ hybridization histochemistry. Comparison with receptor binding distribution. Neuroscience 35:577–591.

Milatovich A, Hsieh CL, Bonaminio G, et al (1992) Serotonin receptor 1c gene assigned to X chromosome in human (band q24) and mouse (bands D-F4). Hum Mol Genet 1:681–684.

Millan MJ (2005) Serotonin 5-HT2C receptors as a target for the treatment of depressive and anxious states: focus on novel therapeutic strategies. Therapie 60:441–460.

Millan MJ, Dekeyne A, Gobert A (1998) Serotonin (5-HT)2C receptors tonically inhibit dopamine (DA) and noradrenaline (NA), but not 5-HT, release in the frontal cortex in vivo. Neuropharmacology 37:953–955.

Millan MJ, Marin P, Bockaert J, et al (2008) Signaling at G-protein-coupled serotonin receptors: recent advances and future research directions. Trends Pharmacol Sci 29:454–464.

Miller KJ (2005) Serotonin 5-HT2C receptor agonists: potential for the treatment of obesity. Mol Interv 5:282–291.

Niswender CM, Sanders-Bush E, Emeson RB (1998) Identification and characterization of RNA editing events within the 5-HT2C receptor. Ann NY Acad Sci 861:38–48.

Niswender CM, Copeland SC, Herrick-Davis K, et al (1999) RNA editing of the human serotonin 5-hydroxytryptamine2C receptor silences constitutive activity. J Biol Chem 274:9472–9478.

Niswender CM, Herrick-Davis K, Dilley GE, et al (2001) RNA editing of the human serotonin 5-HT2C receptor: alterations in suicide and implications for serotonergic pharmacotherapy. Neuropsychopharmacology 24:478–491.

Nitsch RM, Deng M, Growdon JH, et al (1996) Serotonin 5-HT2A and 5-HT2C receptors stimulate amyloid precursor protein ectodomain secretion. J Biol Chem 271:4188–4194.

Nourry C, Grant SG, Borg JP (2003) PDZ Domain Proteins: Plug and Play! Science's STKE 2003(179):re7.

Palacios JM, Markstein R, Pazos A (1986). Serotonin-1C sites in the choroid plexus are not linked in a stimulatory or inhibitory way to adenylate cyclase. Brain Res 380:151–154.

Panicker MM, Parker I, Miledi R (1991) Receptors of the serotonin 1C subtype expressed from cloned DNA mediate the closing of K+membrane channels encoded by brain mRNA. Proc Natl Acad Sci USA 88:2560–2562.

Pazos A, Cortes R, Palacios JM (1985) Quantitative autoradiographic mapping of serotonin receptors in the rat brain. II. Serotonin-2 receptors. Brain Res 346:231–249.

Pérez-León JA, Sarabia G, Miledi R, et al (2004) Distribution of 5-hydroxytriptamine2C receptor mRNA in rat retina. Mol Brain Res 125:140–142.

Qiu J, Xue C, Bosch MA, Murphy JG, et al (2007) Serotonin 5-hydroxytryptamine2C receptor signaling in hypothalamic proopiomelanocortin neurons:role in energy homeostasis in females. Mol Pharmacol 72:885–896.

Quick MW, Simon MI, Davidson N, et al (1994) Differential coupling of G protein alpha subunits to seven-helix receptors expressed in Xenopus oocytes. J Biol Chem 269:30164–30172.

Raymond JR, Mukhin YV, Gelasco A, et al (2001) Multiplicity of mechanisms of serotonin receptor signal transduction. Pharmacol Ther 92:179–212.

Raymond JR, Turner JH, Gelasco A, et al (2006) 5-HT receptor signal transduction pathways. In: Roth B, ed. The Serotonin Receptors: From Molecular Pharmacology to Human Therapeutics, Totowa, NJ: Humana.

Rittenhouse PA, Levy AD, Li Q, et al (1993) Neurons in the hypothalamic paraventricular nucleus mediate the serotonergic stimulation of prolactin secretion via 5-HT1C/2 receptors. Endocrinology 133:661–667.

Roberts MF (1996) Phospholipases: structural and functional motifs for working at an interface. FASEB J 10:1159–1172.

Roth BL, Willins DL, Kristiansen K, et al (1998) 5-Hydroxytryptamine 2-family receptors (5-hydroxytryptamine2A, 5-hydroxytryptamine2B, 5-hydroxytryptamine 2C): where structure meets function. Pharmacol Ther 79:231–257.

Saltzman AG, Morse B, Whitman MM, et al (1991) Cloning of the human serotonin 5-HT2 and 5-HT1C receptor subtypes. Biochem Biophys Res Commun 181:1469–1478.

Sanders-Bush E, Fentress H, Hazelwood L (2003) Serotonin 5-HT2 receptors: molecular and genomic diversity. Mol Interv 3:319–330.

Sodhi MS, Burnet PW, Makoff AJ, et al (2001) RNA editing of the 5-HT2C receptor is reduced in schizophrenia. Mol Psychiatry 6:373–379.

Speake T, Kibble JD, Brown PD (2004) Kv1.1 and Kv1.3 channels contribute to the delayed-rectifying K+conductance in rat choroid plexus epithelial cells. Am J Physiol Cell Physiol 286:C611–C620.

Stam NJ, Vanderheyden P, van Alebeek C, et al (1994) Genomic organisation and functional expression of the gene encoding the human serotonin 5-HT2C receptor. Eur J Pharmacol 269:339–348.

Tecott LH, Sun LM, Akana SF, et al (1995) Eating disorder and epilepsy in mice lacking 5-HT2C serotonin receptors. Nature 374:542–546.

Terrillon S, Bouvier M (2004) Roles of G-protein-coupled receptor dimerization. EMBO Rep 5:30–34.

Timpe LC, Fantl WJ (1994) Modulation of a voltage-activated potassium channel by peptide growth factor receptors. J Neurosci 14:1195–1201.

Tohda M, Tohda C, Oda H, et al (1995) Possible involvement of botulinum ADP-ribosyltransferase sensitive low molecular G-protein on 5-hydroxytryptamine (5-HT)-induced inositol phosphates formation in 5-HT2C cDNA transfected cells. Neurosci Lett 190:33–36.

Tohda M, Nomura M, Nomura Y (2006) Molecular pathopharmacology of 5-HT2C receptors and the RNA editing in the brain. J Pharmacol Sci 100:427–432.

Turner JH, Raymond JR (2005) Interaction of Calmodulin with the serotonin 5-hydroxytryptamine2A receptor. J Biol Chem 280:30741–30750.

Ullmer C, Schmuck K, Figge A, et al (1998) Cloning and characterization of MUPP1, a novel PDZ domain protein. FEBS Lett 424:63–68.

Urban JD, Clarke WP, von Zastrow M, et al (2007) Functional selectivity and classical concepts of quantitative pharmacology. J Pharmacol Exp Ther 320:1–13.

Werry TD, Gregory KJ, Sexton PM, et al (2005) Characterization of serotonin 5-HT2C receptor signaling to extracellular signal-regulated kinases 1 and 2. J Neurochem 93:1603–1615.

Werry TD, Loiacono R, Sexton PM, et al (2008) RNA editing of the serotonin 5HT2C receptor and its effects on cell signalling, pharmacology and brain function. Pharmacol Ther 119:7–23.

Werry TD, Stewart GD, Crouch MF, et al (2008) Pharmacology of 5HT2C receptor-mediated ERK1/2 phosphorylation: agonist-specific activation pathways and the impact of RNA editing. Biochem Pharmacol 76:1276–1287.

Westphal RS, Sanders-Bush E (1996) Differences in agonist-independent and -dependent 5-hydroxytryptamine2C receptor-mediated cell division. Mol Pharmacol 49:474–480.

Xie E, Zhu L, Zhao L, et al (1996) The human serotonin 5-HT2C receptor: complete cDNA, genomic structure, and alternatively spliced variant. Genomics 35:551–561.

Yuen EY, Jiang Q, Chen P, et al (2008) Activation of 5-HT2A/C receptors counteracts 5-HT1A Regulation of N-Methyl-D aspartate receptor channels in pyramidal neurons of prefrontal cortex. J Biol Chem 283:17194–17204.

Chapter 6
Homology Modeling of 5-HT$_{2C}$ Receptors

Nicolas Renault, Amaury Farce, and Philippe Chavatte

6.1 Introduction

Homology modeling is a computational approach very widely employed to recover theoretical structure of a macromolecule. Following the idea that structure is better conserved than sequence in a family of proteins sharing a common function, the main difficulty is to find a suitable template that is a protein of related activity and of properly similar sequence. The accepted limit in terms of sequence identity is about 30%, while even in cases of lower identity, the ratio of conserved residues is usually higher in the peptides involved in the function of the protein.

Despite targeting G-protein-coupled receptors (GPCRs) has become a challenge in drug design for many years, the structure of GPCRs is still hardly amenable to standard crystallographic or nuclear magnetic resonance (NMR) methods because their integral membrane protein nature renders them difficult to isolate and crystallize (White et al. 2001). However, homology modeling offers an alternative pathway through these difficulties by the construction of theoretical three-dimensional (3D) models based on the few crystallographic data yet available for GPCR structure. Even if the bovine rhodopsin has been the only structural template for 10 years (Palczewski et al. 2000), the homology modeling of monoamine GPCRs, particularly 5-HT$_{2C}$, should greatly benefit with the very recent discovery of more homologous crystal templates.

The homology modeling of the 5-HT$_{2C}$ receptor is specially challenging for rational drug design since few discovered ligands, from the cloning of this receptor 15 years ago (Xie et al. 1996; Stam et al. 1994; Saltzman et al. 1991) up to now, are found to be very selective for 5-HT$_{2C}$ among other 5-HT receptor subtypes but also G-protein-coupled neurotransmitter receptors like dopaminergic or adrenergic receptors (adrenoceptors). According to its investigated pharmacological profile, antagonist or agonist ligands must target, respectively, inactive or active states of the

P. Chavatte (✉)
Laboratoire de Chimie Thérapeutique, Université de Lille 2, EA 1043, 3 Rue du Professeur Laguesse, B.P. 83, F-59006, Lille, France
e-mail: philippe.chavatte@univ-lille2.fr

G. Di Giovanni et al. (eds.), *5-HT$_{2C}$ Receptors in the Pathophysiology of CNS Disease*,
The Receptors 22, DOI 10.1007/978-1-60761-941-3_6,
© Springer Science+Business Media, LLC 2011

receptor. Computational molecular dynamics of the homology models associated with data from site-directed mutagenesis are able to simulate these different boundary states of activation while bound to known agonist or antagonist ligands in order to provide insights on activation mechanisms or new relevant sites to be targeted.

On the basis of the several published works of 3D rhodopsin-templated homology models and previously evoked novel crystallographic structures, this chapter will describe the different ways to generate models able to answer the real challenges associated with the 5-HT$_{2C}$ receptor. These range from the investigation of strictly 5-HT$_{2C}$-dependent structural features in order to refine this actually detrimental ligand selectivity through an improved understanding of mechanisms leading to constitutive or ligand-induced activation of the receptor to the structural impact of genetic polymorphism or orthologous variations within mammalians.

6.2 Homology Modeling

6.2.1 Reference Structural Templates

Up to now the bovine rhodopsin (RHO) has been the only member of the eukaryotic GPCR family with an experimentally solved structure (Palczewski et al. 2000). As deduced from the first of the three classes of GPCRs, the rhodopsin-like family (class A), RHO is the representative protagonist of this largest protein family that includes hormone, neurotransmitter, and light receptors, all of which transducing extracellular signals through interaction with guanine nucleotide-binding (G) proteins. Although their activating ligands vary widely in structure and nature, the amino acid sequences of the receptors are very similar and are believed to adopt a common structural framework comprising seven transmembrane (TM) helices (Attwood and Findlay 1993; Birnbaumer 1990; Casey and Gilman 1988). However RHO remains unusual because it is highly abundant from natural sources and structurally stabilized by the covalently bound ligand 11-*cis*-retinal, which maintains the receptor in a dark-adapted, nonsignaling conformation. In contrast, all other class A GPCRs are activated by diffusible ligands and are expressed at relatively low levels in native tissues.

Considerable progress has arisen from the recent crystallization of GPCRs nearer to therapeutically relevant monoamine receptor, successively the human β2-adrenoceptor (ADRB2) (Cherezov et al. 2007; Rosenbaum et al. 2007), the meleagris β1-adrenoceptor (ADRB1) (Warne et al. 2008), and the human A$_{2A}$ adenosine receptor (A2AAR) (Jaakola et al. 2008). Their full-length sequence share 41%, 41%, and 36% homology, respectively, with the entire human 5-HT$_{2C}$ sequence versus 33% with the RHO sequence. Since they are also monoamine receptors, ADRB1 and ADRB2 get the highest score of sequence homology (41%) with the 5-HT$_{2C}$ receptor. Overall, these two crystal structures were reported in complex with the high-affinity antagonist cyanopindolol for ADRB1 and inverse agonist carazolol for ADRB2. This extensively improves the prediction of ligand binding sites and structural features of ligand accessibility in the other GPCR-type neurotransmitter receptors like the 5-HT$_{2C}$ receptor. Ligand-binding site accessibility is enabled by the second extracellular loop, which is held out of the binding cavity by two closely spaced disulfide bridges and a short helical segment within the loop. The ligand-binding pocket comprises 15 side chains from amino acid residues in four transmembrane alpha-helices and extracellular loop 2. Binding of either cyanopindolol to the β1-adrenergic receptor or carazolol to the β2-adrenergic receptor involves similar interactions mainly driven by one cluster of aromatic rings and one acidic side chain trapping the positive charge carried by the monoamine ligands like serotonin.

Additional insights are also gained from the examination of packing interactions and a network of hydrogen bonds, suggesting a conformational pathway from the ligand-, binding pocket to regions that interact with G proteins. Otherwise, a short well-defined helix in cytoplasmic loop 2 of ADRB1, not observed in either RHO or ADRB2, directly interacts by means of a tyrosine with the highly conserved DRY motif at the end of helix 3 that is essential for receptor activation. The A2AAR seems to be less interesting than the other structurally determined GPCRs with regard to modeling the structure of the 5-HT$_{2C}$ receptor. Indeed authors have found that the extracellular domain, combined with a subtle repacking of the transmembrane helices relative to the adrenergic and rhodopsin receptor structures, defines a pocket distinct from that of other relevant GPCR-type catecholamine receptors.

Superimposing the backbone of the TM regions (Fig. 6.1) shows that the four helix bundles of GPCR crystals are structurally very close, varying from a

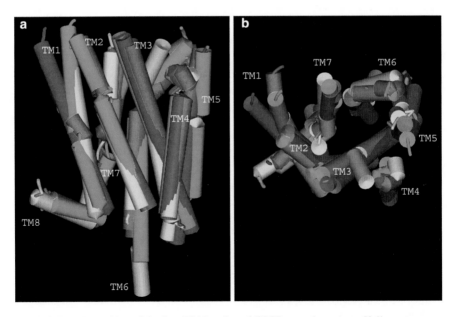

Fig. 6.1 Superimposition of the four TM bundles of GPCR crystal structure. Helices are represented as *cylinders* for RHO (*green*), ADRB1 (*red*), ADRB2 (*purple*), and A2AAR (*yellow*) in (**a**) a transversal view and (**b**) a longitudinal view from the extracellular side to the cytoplasmic side on the right (For interpretation of the colors in this figure, the reader is referred to the web version of this chapter)

2.44 Å root-mean-square deviation (RMSD) between the more homologous sequences, ADRB1 and ADRB2, to 5.39 Å between RHO and A2AAR (Table 6.1). Most structural variations occur in A2AAR, which shows a kink in the longitudinal axis of helices 2 and 3 in the top of the TM bundle at the extracellular side as well as a significantly longer helix 6 at the intracellular side. The second largest difference is a bend in the top of helix 1 of both adrenoceptors that is not found in RHO and A2AAR, which is comparatively straight. This structural difference may arise from the need for an accessible binding site in ADRB1 and ADRB2, which is provided in part by a lack of interactions between the N-terminus and extracellular loop segments. In contrast, the N-terminal region in RHO occludes the retinal-binding site through extensive interactions with the extracellular loops.

Consequently, ADRB1 and ADRB2 are preferential structural templates due to sequence homology and ligand-binding properties. However, modelers must take into account that these receptors have been modified in order to facilitate their crystallization, particularly cytoplasmic features. Thus, ADRB2 and A2AAR were engineered as fusion proteins in which T4 lysosyme replaces most of the third intracellular loop (IL3) of the GPCR. In the same manner, ADRB1 was modified in that six substitutions occur in TM 2, 5, and 7 and IL 1 and 3 as well as a 15-residues deletion in IL3, resulting in an extensive improvement of the protein thermostability. Although crystallized proteins retain near-native pharmacological properties and could consequently be relevant structural template for study of ligand-binding sites, the analysis of structural events in signal transducing should be weighted, particularly in cytoplasmic location.

6.2.2 Multiple Sequence Alignment

6.2.2.1 Prediction of Transmembrane Helices

Even though ADRB1 and ADRB2 seem to be the best structural templates to build the backbone of the 5-HT$_{2C}$ receptor from their 3D coordinates, the four crystal

Table 6.1 Calculated RMSD between the backbone of the four crystal structures of GPCRs

RMSD (Å)	Bovine rhodopsin	Human β1 adrenergic receptor	Meleagris β2 adrenergic receptor	Human A2A adenosine receptor
Bovine rhodopsin	0	4.80	4.21	5.39
Human β1 adrenergic receptor	–	0	2.44	4.74
Meleagris β2 adrenergic receptor	–	–	0	5.14
Human A2A adenosine receptor	–	–	–	0

structures are highly interesting in order to carry out a 3D prediction of TM segments in the 5-HT$_{2C}$ receptor. Indeed a profile alignment between a multiple sequence alignment, inherited from the structural superimposition of the four 7-TM bundles crystal structures, and the sequence of the human 5-HT$_{2C}$ receptor has been automatically produced (Thompson et al. 1994) and manually adjusted to delete gaps within the TM regions (Fig. 6.2). This permits one to emphasize the consensual delimiting positions of helices in the 5-HT$_{2C}$ sequence. In a parallel manner, bioinformatics methods have been used to predict the TM regions (Bray and Goddard 2008) on the assumptions that the outward facing sections of the TM helices must be hydrophobic because they are in contact with the hydrocarbon tails of the lipid bilayer and that the hydrophobic center of each helix should be at the center of the membrane (Donnelly et al. 1994).

The Table 6.2 summarizes the different TM predictions resulting of previous works based on RHO crystal structure as the only template or on the previously evoked bioinformatics method in comparison with the 3D-based prediction from the closest structural templates, ADRB1 and ADRB2. Sequence homology percentages within TM regions of RHO, ADRB1, ADRB2, and A2AAR, deduced from the TM prediction, are 44, 62, 62, and 53, respectively, and support the idea that TM regions of ADRB1 and ADRB2 are the best templates. In a general manner, the length of TM helices from the RHO-based prediction by Zuo is shorter than those of the other predictions. It appears that RHO-based prediction by Farce et al., Rashid et al., and the ADRB-based method apply the full-length 3D information of helices from crystal structures contrary to the other RHO-based method whereas all keep the same TM center. The most specific trait of ADRB-based prediction is the fourty percent longer C-terminal region of helix 5 compared with other TM5 of other predictions. Otherwise, even though 2D-based method shows a shift of the center in TM 4 and 6 toward their C-terminal region, all predictions seem to be homogeneous and independently provide crucial insights on ligand binding and activation mechanisms within the 5-HT$_{2C}$ receptor. If the seven TM helices are widely discussed for ligand binding and receptor activation, a small eighth helix exists in the C-terminal region of all crystal templates with a strictly conserved arginine residue, which is consequently labeled as Arg8.50. These regions present a good sequence alignment with the sequence of 5-HT$_{2C}$ and belong to segments further modeled as conserved regions.

6.2.2.2   Prediction of Common Structural Features

All reported discussions should be read while inspecting the multiple sequence alignment (Fig. 6.2). Important conserved features between the crystal structure and the 5-HT$_{2C}$ sequences are reported here using the Ballesteros and Weinstein numbering system to designate the more conserved residue in each helix as X.50, where X is the TM helix number, and the 5-HT$_{2C}$ numbering for loop regions.

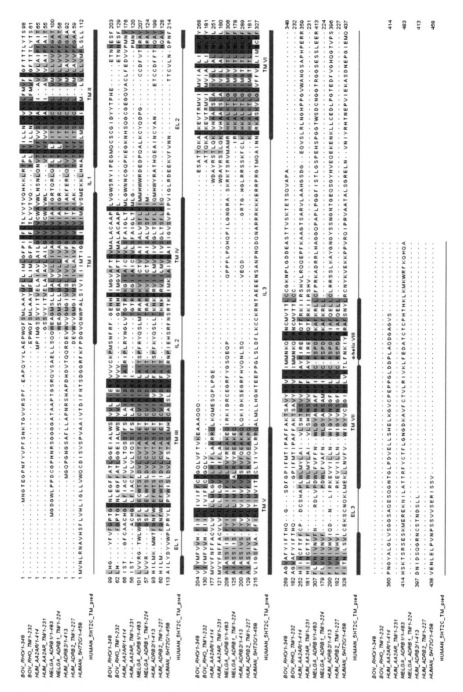

Fig. 6.2 Alignment of human 5-HT$_{2C}$ receptor and crystal template sequences with the Jalview sequence editor tools. The predicted TM helices of 5-HT$_{2C}$ (human_5-HT2C_TM_pred) are illustrated as *red bars* below the alignment and result from the delimiting positions of TM segments in crystal templates. Homology rates are rendered for residues overlined with graduated blue levels, whereas significant residues discussed in the text are labeled according to the Ballesteros and Weinstein numbering system to designate the most conserved residue in each helix as X.50, where X is the helix number (For interpretation of the colors in this figure, the reader is referred to the web version of this chapter)

Table 6.2 Predictions of TM segments (length within brackets) of the human 5-HT$_{2C}$ receptor

TM	Predicted TM region	Prediction method	Authors
1	DGVQNWPALSIVIIIMTIGGNILVIMAVSM (31)	RHO based	(Rashid et al. 2003)
	GVQNWPALSIVIIIMTIGGNILVIMAVSM (30)	RHO based	(Farce et al. 2006)
	NWPALSIVIIIMTIGGNILVIMAV (27)	RHO based	(Zuo et al. 2007)
	GVQNWPALSIVIIIMTIGGNILVIMAVSME (31)	Hydrophobic profile	(Bray and Goddard 2008)
	PDGVQNWPALSIVIIIMTIGGNILVIMAVSM (32)	ADRB based	(Renault et al. 2010)
2	ATNYFLMSLAIADMLVGLLVMPLSLLAI (28)	RHO based	(Rashid et al. 2003)
	ATNYFLMSLAIADMLVGLLVMPLSLLAILY (30)	RHO based	(Farce et al. 2006)
	FLMSLAIADMLVGLLVMPLS (20)	RHO based	(Zuo et al. 2007)
	TNYFLMSLAIADMLVGLLVMPLSLLAILYD (30)	Hydrophobic profile	(Bray and Goddard 2008)
	NATNYFLMSLAIADMLVGLLVMPLSLLAILY (31)	ADRB based	(Renault et al. 2010)
3	RYLCPVWISLDVLFSTASIMHLCAISLDRY (30)	RHO based	(Rashid et al. 2003)
	LPRYLCPVWISLDVLFSTASIMHLCAISLDRYVAI (35)	RHO based	(Farce et al. 2006)
	VWISLDVLFSTASIMHLCAISLDRY (25)	RHO based	(Zuo et al. 2007)
	RYLCPVWISLDVLFSTASIMHLCAISLDR (29)	Hydrophobic profile	(Bray and Goddard 2008)
	LPRYLCPVWISLDVLFSTASIMHLCAISLDRYVAIR (36)	ADRB based	(Renault et al. 2010)
4	AIMKIAIVWAISIGVSVPIP (20)	RHO based	(Rashid et al. 2003)
	RTKAIMKIAIVWAISIGVSVPIP (23)	RHO based	(Farce et al. 2006)
	TKAIMKIAIVWAISIGVSVPIPVI (24)	RHO based	(Zuo et al. 2007)
	AIMKIAIVWAISIGVSVPIPVIGL (24)	Hydrophobic profile	(Bray and Goddard 2008)
	SRTKAIMKIAIVWAISIGVSVPIPV (25)	ADRB based	(Renault et al. 2010)
5	DPNFVLIGSFVAFFIPLTIMVITYC (25)	RHO based	(Rashid et al. 2003)
	DPNFVLIGSFVAFFIPLTIMVITYC (25)	RHO based	(Farce et al. 2006)
	NFVLIGSFVAFFIPLTIMVIT (21)	RHO based	(Zuo et al. 2007)
	FVLIGSFVAFFIPLTIMVITYCLTIY (26)	Hydrophobic profile	(Bray and Goddard 2008)
	DPNFVLIGSFVAFFIPLTIMVITYCLTIYVLRRQ (34)	ADRB based	(Renault et al. 2010)
6	ERKASKVLGIVFFVFLIMWCPFFITNILSVL (32)	RHO based	(Rashid et al. 2003)
	INNERKASKVLGIVFFVFLIMWCPFFITNILSV (33)	RHO based	(Farce et al. 2006)
	NERKASKVLGIVFFVFLIMWCPFFITNI (28)	RHO based	(Zuo et al. 2007)
	LGIVFFVFLIMWCPFFITNILSVLCE (26)	Hydrophobic profile	(Bray and Goddard 2008)
	NERKASKVLGIVFFVFLIMWCPFFITNILSVL (32)	ADRB based	(Renault et al. 2010)

(continued)

Table 6.2 (continued)

TM	Predicted TM region	Prediction method	Authors
7	EKLLNVFVWIGYVCSGINPLVY (22)	RHO based	(Rashid et al. 2003)
	KLLNVFVWIGYVCSGINPLVYTLF (24)	RHO based	(Farce et al. 2006)
	KLLNVFVWIGYVCSGINPLVYTLPN (25)	RHO based	(Zuo et al. 2007)
	EKLLNVFVWIGYVCSGINPLVYT (23)	Hydrophobic profile	(Bray and Goddard 2008)
	MEKLLNVFVWIGYVCSGINPLVYT (24)	ADRB based	(Renault et al. 2010)

Overall Structure

Inspection of ADRB2 crystal (Cherezov et al. 2007; Rosenbaum et al. 2007) enables us to perceive the main interactions responsible for the maintaining its architecture and occurring in helical regions or some connecting loop regions conserved in the 5-HT$_{2C}$ sequence. Starting with helix 1 and 2, the side chain of the more conserved residue, Asn1.50, is included in a hydrogen bond with the carbonyl group of Ser7.46 whereas Asn2.40 hydrogen bonds with Tyr7.53. At the top of helix 3, Cys3.25, strictly conserved within the GPCR family, forms a disulfide bridge with a conserved cysteine residue (Cys127 in 5-HT$_{2C}$) of the extracellular loop 2 (EL2). Near this cysteine, Trp3.28, conserved within the GPCR-type monoamine receptors, 5-HT$_{2C}$ and ADRB, forms a π–π stacking interaction with the conserved tryptophan residue (respectively, numbered 99 and 120 in ADRB2 and 5-HT$_{2C}$) of the adjacent EL1. This interaction would restrain EL1 from pulling down toward the TM bundle and making the top of helices 2 and 3 closer. At the bottom of TM3, all of the mammalian monoamine GPCRs have a highly conserved DRY (ERY in rhodopsin) pattern that is involved in the receptor activation (Ballesteros et al. 2001; Jensen et al. 2001). Indeed, in RHO crystal, Arg3.50 is trapped by two salt bridges between Asp3.49 and Glu6.30, which would maintain the receptor in an inactive state. Previous studies on the G-coupled lutropin receptor (LHR) have shown that breaking of the charge-reinforced H-bonding interaction between Arg3.50 and Asp6.30 would increase the solvent accessibility of the cytosolic extensions of helices 3 and 6, which probably induce rotation of helices 3 and 6 to get drastic changes in the receptor–G protein interface (Angelova et al. 2002; Greasley et al. 2002; Zhang et al. 2005). In ADRB2 crystal, the salt bridges are disrupted by the presence of a sulfate ion, which is a stronger counterion than glutamate. The more conserved residue in helix 4, Trp4.50, also seems to be stabilizing for the receptor architecture since it is implied in an interhelical interaction by hydrogen bonding with Ser2.45 (Asn in RHO) of TM2, itself binding with the polar residue in position 3.42 (respectively, Ser, Thr, and His in RHO, ADRB2, and 5-HT$_{2C}$), whereas the reference residue in helix 5, Pro5.50, introduces a kink whose the angle could be decreased during the receptor activation (Crozier et al. 2007). Other conserved features in TM 3, 5, 6, and 7 are either hydrophobic clusters that maintain the stacking of the interior core of the protein and cover the outward section in order to interact with the lipid tails of the membrane or amino acids directly involved in the ligand binding, which will be discussed in Sect. 6.1.2.2.2.

The more recent rhodopsin crystal structures, available in the Protein Data Bank (Berman et al. 2000) under 1GZM (Li et al. 2004), 1L9H (Okada et al. 2002), and 1U19 (Okada et al. 2004) entries, have also revealed structural waters participating in an extensive hydrogen-bond network (Pardo et al. 2007) between conserved residues of helices 1, 2, 6, and 7, which are amenable to play a critical role in activation mechanism in a similar manner to that described above for the suggested impact of the DRY sequence in TM3 and resulting in a disrupted network of hydrogen bonds. This phenomenon was also analyzed in ADRB2 crystal structure taking care to decompose the examination of the extensive hydrogen bond network in order to reveal substantial differences with RHO-focused inspection (Fig. 6.3). The first such difference concerns the conformation of Tyr7.53, which is oriented toward one additional water included in the network of hydrogen bonds between Asn1.50 and Asn7.49 rather than toward the bottom of the helix bundle. The displacement of Tyr7.53 disrupts the π–π stacking interaction between side chains of Tyr7.53 and Phe7.60 (Tyr in 5-HT$_{2C}$) as well as a hydrogen bond with Asn2.40 and could suggest that the receptor is not in complete inactive state anymore, if we take into account that Tyr7.53 is strictly conserved in GPCRs and that ADRB2 is cocrystallized with carazolol, an inverse agonist. The second variation is a gain of one water-relayed hydrogen bond between helices 1 (Gly1.61) and 7 (Trp7.40). What is interesting in modeling GPCR-type monoamine receptors is the conserved network between 7.38, 6.47, and 6.51, which is strictly conserved within monoamine receptors and directly interacting with the ligand in ADRB2. On the assumption that rotation of TM6 could lead to the activation of the receptor, specific interactions of the ligand with this aromatic residue would be one of the starting events to switch off the hydrogen-bond signaling along the TM6.

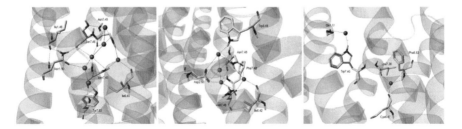

Fig. 6.3 Hydrogen bond network relayed by structural waters in ADRB2. The extensive hydrogen bond network is decomposed into three pictures. Interacting amino acids are represented in the transparent TM bundle as *sticks* depicted by atom type, whereas structural waters are illustrated as *spheres* and hydrogen bonds, as *dashed lines*

Ligand-Binding Site

In crystal structure of ADRB2, the cocrystallized ligand carazolol is caged in the top of helix bundle between TM 3, 5, 6, and 7 and EL2. The amino acids directly involved in the binding of ligand or the pocket plasticity and conserved in 5-HT$_{2C}$ receptor enable a prediction of a coarse fingerprint of interaction between serotonin and this receptor. This is supported by the neighbouring chemical structure of the two compounds since both include a benzyl group, itself in aromatic system (respectively, indol and carbazol groups in serotonin and carazolol), connected to a protonated nitrogen through a three-atom (serotonin) or four-atom (carazolol) spacer. The protonated amine of the ligand is bound by an electrostatic interaction with Asp3.32 and one hydrogen bond with Tyr7.43, two specific residues for ligand binding among all mammalian biogenic amine receptors (Bywater RP 2005). Moreover, Asp3.32 was found to anchor the terminal amine moiety of serotonin in 5-HT$_{2A}$ (Kristiansen et al. 2000). Conserved amino acids of ADRB2 TM5 in the vicinity of ligand position are Ser5.43 (first suggested to interact with the NH of the indole in serotonin (Ebersole et al. 2003) and then strongly suspected to bind to the 5-OH of serotonin (Braden and Nichols 2007)) and Phe5.47 (found to have significant interactions with serotonin in 5-HT$_{2A}$ (Shapiro et al. 2000) and more generally with agonist ligands (Salom et al. 2006)). However, inspection of ADRB2 crystal structure shows that there is no contact between Phe5.47 and the inverse agonist ligand but that it forms a π–π stacking interaction with Phe6.52, itself binding the ligand by an edge-to-face π–π interaction as well as Phe6.51, another strictly conserved residue within the monoamine receptors found to affect the binding of serotonin while mutated to a leucine residue (Choudhary et al. 1993). This suggests a different binding mode between the 5-HT receptor subtypes or between agonist and antagonist ligands.

Otherwise, the strictly conserved Trp6.48 in the GPCR family, previously described in the extended hydrogen-bond network going across the ADRB2 receptor, was found to cause an almost 1,000-fold decrease in serotonin binding in 5-HT$_{2A}$ (Roth et al. 1997). The ligand binding specificity associated with these conserved structural features within GPCR-type biogenic amine receptors must be distinguished from the ligand binding selectivity for receptor subtype of this GPCR subfamily. In this way the extracellular loop 2 (EL2) connecting helices 4 and 5 is suspected to play a crucial role from inspection of crystal structure and site-directed mutagenesis (Zhao et al. 1996; Wurch and Pauwels 2000). This is supported by the idea that EL2 has a very variable length and a low sequence homology, suggesting that it could partially determine the strict selectivity of a GPCR-type receptor. In addition to the idea that EL2 contributes to a hydrogen-bonding network that is thought to maintain rhodopsin in an inactive state (Klco et al. 2005), the examination of the two crystal structures of RHO and ADRB2 shows the propensity of EL2 to act as a gate for ligand accessibility. Moreover, chimeras of α1A and α1B adrenoceptors, built from the reciprocal exchange of three consecutive residues in EL2 (Zhao et al. 1996) as well as a single substitution in the canine 5-HT$_{1D}$ sequence shifting toward human 5-HT$_{1D}$ sequence (Wurch and Pauwels 2000), have induced the permutation of ligand selectivity within the two couples of engineered receptors. The lack of sequence homology and the variable length of EL2 render this loop very challenging for homology modeling of every GPCR.

6.2.3   Building of the Apo-5-HT$_{2C}$ 3D Models

Classically most of the 3D homology models are built using software packages like InsightII (Accelrys Inc, San Diego, CA), Jackal (Petrey et al. 2003), or Modeller (Sali and Blundell 1993), which are able to assign the coordinates of the structurally conserved regions from the template to the model on the basis of the alignment carried ou0074 from programs like ClustalW (Thompson et al. 1994) and manually optimized. According to the multiple sequence alignment (Fig. 6.2) over the eight helices, IL1 and IL2 connecting, respectively, TM1 and TM2 then TM3 and TM4, present no gap and enough sequence conservation to be considered as structurally conserved regions in the 5-HT$_{2C}$ receptor. If the coordinates of conserved amino acids are inherited from those of aligned residue in the template, the coordinates of nonconserved side chains are collected from rotamer libraries derived from high-resolution crystallographic structures. An ab initio method has also been employed to predict the structure of TM regions using only the amino acid sequence (Bray and Goddard 2008). As described in the preceding section, TM helical regions were predicted from hydrophobic analyses of the amino acid sequence using a program based on the Eisenberg hydrophobicity scale (Eisenberg et al. 1984). Each TM was then built into a canonical right-handed alpha helix with extended side-chain conformations and placed within a bundle according to geometric parameters of helix packing extracted from crystal structure of RHO and using a Newton–Euler inverse mass operator (NEIMO) torsional molecular dynamics (MD) method (Mathiowetz et al. 1994).

  In these two methods, the variable regions such as N-terminus, EL1, EL2, and EL3 are produced separately by various algorithms that are typically based on a systematic or empirical (Monte Carlo algorithm) conformational sampling of the loop backbone followed by steps of geometry optimization using molecular mechanics methods. In this case, iterative algorithms of energy minimization are employed by applying force fields including potential energy parameters of bonded atoms (torsional, bending, and stretching energies) and nonbonded energy parameters like electrostatic and van der Waals terms. These intrinsic values are extracted from experimental crystallographic, NMR, or infrared (IR) data. Typical force fields including parameters suitable for protein structures and commonly used for the refinement of 3D homology models are chemistry at Harvard macromolecular mechanics (CHARMM) (Brooks et al. 1983) and assisted model building and energy refinement (AMBER) (Cornell et al. 1995). Modeling EL2 of 5-HT$_{2C}$ takes advantage of the spatial restraint caused by the disulfide bridge (conserved within the GPCR family) between Cys207, the only cystein residue of this loop, and the conserved cystein of TM3, Cys127 (C3.25). As in most cases of loops longer than eight residues, N-terminus, IL3, and C-terminus are regions that are deliberately not entirely set up in the three published models of 5-HT$_{2C}$ (Bray and Goddard 2008; Zuo et al. 2007; Farce et al. 2006; Rashid et al. 2003) because their higher flexibility introduces too extensive structural bias in the overall structure and can not be modeled accurately.

The resultant structures of 3D modeling undergo a procedure of geometry optimization similar to that applied for the refinement of loops. The refinement of 3D models is assisted by a theoretical checking of the helix geometry in accordance with right, allowed, or forbidden regions of the Ramachandran map (Ramachandran 1963; Ramachandran et al. 1963). This diagram is a way to visualize dihedral angles ϕ against ψ of amino acid residues in protein structure. It shows the possible conformations of ϕ and ψ angles for a polypeptide. In addition to the large amount of data gathered by the comparison of the models, we have produced 3D homology models (Renault et al. 2010) on the basis of the alignment with the sequence of ADRB2 according to the protocol of homology modeling just described. Providing the vicinity of sequences relative to RHO, commonly used up to now but less homologous than monoamine receptors like adrenoceptors and because no such information has been published yet, this model of the 5-HT$_{2C}$ receptor will be compared with published 3D models in order to reveal new structural insights.

6.3   Modeling a Ligand-Bound State of the 5-HT$_{2C}$ Receptor

Most 3D homology models of therapeutically relevant GPCRs are conceived to design, from a training set of reference compounds, in situ fingerprints of interaction able to forecast the structure and efficiency of new potent drugs. Achieving the best prediction of structure-based qualitative 3D pharmacophores or quantitative structure-activity models requires a reliable correlation between in vitro and in silico amino acid residues found to be specifically bound by high-affinity ligands. Moreover, the binding mode of relevant ligands inside the predicted binding cavities needs a prior study of the structural features associated with the pharmacological profile of ligands. At a pharmacological point of view, receptors conform to the rules of thermodynamic equilibrium between the inactive (R) and active states (R*), which structurally means a gradient of conformational states between these two limit snapshot structures. Computational MD calculations enable the simulation of the collective motion of atoms in such macromolecular systems and result in the lowest-energy conformational states of receptor–ligand complexes theoretically corresponding to the ligand-induced activation state of the receptor.

6.3.1   Investigation of Antagonist Binding

Many strategies have been adopted by authors in order to study binding modes of 5-HT$_{2C}$ reference compounds whose associated structures and experimental data are represented in Table 6.3. Since templates used for homology modeling are either a ground state of RHO or ADRB2 cocrystallized with an inverse agonist ligand, both consequently not in an agonist-induced active state, resultant 3D models are presumed to be in an inactive state.

6.3.1.1   Identification of Ligand Binding Sites and Docking of Antagonists

In this way, Rashid et al. as well as Farce et al. have directly worked from the optimized and validated structure of the apo-5-HT$_{2C}$ to dock therein sarpogrelate (Rashid et al. 2003) and agomelatin (Millan et al. 2003) antagonist, respectively,

Table 6.3   Structure and affinity of reference 5-HT$_{2C}$ ligands

Agomelatin antagonist pKi 6.2	Sarpogrelate antagonist pKi 7.4
Methiothepin inverse agonist pKi 8.4	SB-228357 antagonist pKi 9.1

(continued)

Table 6.3 (continued)

Ritanserin antagonist pKi 9.6	Serotonin agonist pKi 7.8

Metergolin inverse agonist pKi 8.5   Azepinoindol agonist pKi 7.9

whereas modifications have been performed by Bray et al. before the docking of the methiothepin (Knight et al. 2004), metergolin (Knight et al. 2004), and ritanserin (Knight et al. 2004) reference antagonists in the ligand binding site. Indeed, in the later case, they were searching the best binding site for each studied ligand by virtually mutating all bulky nonpolar residues (leucine, isoleucine, valine, phelylalanine, and tryptophane) to alanine. Then the reshaped ligand binding site was divided into a set of overlapping regions in which were sampled different ligand conformations by docking calculations. For each of the 50 ligand conformations, the bulky non-polar residues were reversed to their original form then the geometry of ligands and closest residues was refined by energy minimization. Structures with the lowest total energy of the ligand-protein complex and with the best contacts between the respective functional groups of the ligands and the residues in the binding site were finally selected.

Once the ligand binding site has been identified in the apo-5-HT$_{2C}$ 3D models by virtually probing the solvent-accessible surface, the other authors have therein performed the docking calculations of the relevant ligands within a sphere radius. In these cases, flexible docking is commonly used and based on iterative genera-tions of sampled conformations of ligands for preferential electrostatic and hydro-phobic interactions with the amino acid residues of the protein. Bray et al. took on a systematic search of docking poses by an incremental building of ligand frag-ments (Kuntz 1992), whereas the others employed stochastic-oriented programs based on a genetic algorithm (Goodsell et al. 1996; Jones et al. 1995a; Jones et al. 1995b). In our ADRB-based 3D model of apo-5-HT$_{2C}$ we have manually preposi-tioned SB-228357, one potent member of the ((aryl)carbamoyl)indolin series (Bromidge et al. 1997; Bromidge et al. 1998; Bromidge et al. 2000; Bromidge et al. 1999), found to actually be the most selective 5-HT$_{2C}$ antagonists and voluminous enough to target the maximal spots of interactions in the vicinity of the binding site. A conformational sampling of the receptor–ligand complex has been then per-formed by simulated annealing. This molecular dynamics procedure, consists in ten heating cycles at 700 K, each followed by cooling steps down to 300 K, and permits to catch the lowest-energy conformations of the receptor–ligand complex. Annealing was simulated on all atoms of the E2 loop and the ligand as well as the side chains of TM amino acid residues involved in interaction with the ligand. Receptor–ligand complexes with the lowest total energy matching the two critical electrostatic (Asp3.32) and aromatic (Phe6.51 or Phe6.52) anchor sites were finally selected.

6.3.1.2   Analysis of 5-HT$_{2C}$-Antagonists Interactions

The receptor–ligand interactions brought out from all docking procedures in the four 3D models are summarized in the Table 6.4 and are distributed according to van der Waals (VdW) and aromatic (π–π stacking) as hydrophobic interactions or electro-static interactions gathering hydrogen bonds and salt bridges, respectively, H-bond and ionic in Table 6.4. The common fingerprint of interactions among all 3D models corresponds to the previously discussed patterns Asp3.32 and one member of the aromatic cluster (Phe5.47, Trp6.48, Phe6.51, Phe6.52), although not involved in metergolin and methiothepin interactions. Since these are conserved residues among GPCR-type monoamine receptors, specific 5-HT$_{2C}$ spots of interactions seem to be located over all the TM bundle with hydrophobic interaction from Trp3.28, Leu3.31, Val3.33, Phe3.35, Phe5.38, and Asn6.55 and polar interactions from Ser3.36 and Tyr7.43, all of which are described more than once in at least two different models. Other single interactions are coming from the relatively different chemical scaffolds of metergolin and methiothepin, which have additional hydrophobic interactions with TM4 hydrophobic residues (Ile4.60 and Pro4.61). Specific observations associ-ated to metergolin and methiothepin could be conferred by their particular pharma-cological profile of inverse agonist, which would induce a binding to another intermediate activation state of the receptor. Otherwise, a benzyl group of the more

Table 6.4 Interactions of 5-HT$_{2C}$ antagonists

Interactions of 5-HT$_{2C}$ antagonist		Sarpogrelate (Rashid's model)	Agomelatin (Farce's model)
TM2	Ala2.64(113)		
E1	Trp120		
TM3	Trp3.28 (130)	Aromatic	
	Leu3.31(133)	VdW	
	Asp3.32(134)	Ionic	H bond
	Val3.33(135)	VdW	VdW
	Phe3.35(137)		
	Ser3.36(138)	H bond	H bond
	Ser3.39(141)		
	Ile3.40(142)	VdW	
TM4	Ile4.60(189)		
	Pro4.61(190)		
E2	Arg195		
	Asn204		H bond
	Thr205		
	Asn210		
TM5	Phe5.38(214)		
	Val5.39(215)		
	Ser5.43(219)		
	Phe5.47(223)	Aromatic	
	Ile5.49(225)		
TM6	Trp6.48(324)	VdW	Aromatic
	Phe6.51(327)	VdW	Aromatic
	Phe6.52(328)	Aromatic	Aromatic
	Asn6.55(331)		
	Ile6.56(332)		
TM7	Asn7.36(351)		
	Val7.39(354)		
	Tyr7.43(358)	H bond	

	Bray's model		SB-228357
Ritanserin	Metergolin	Methiothepin	(Renault's model)
			VdW
			H bond
			Aromatic
Ionic	Ionic	Ionic	H bond
Aromatic	H bond		
	VdW		
	VdW		
	VdW	VdW	
			H bond
			H bond
			H bond
			H bond
	Aromatic		VdW
VdW	VdW		
		H bond	
		VdW	
H bond			
			Aromatic
VdW			
VdW			
			H bond
			H bond

weighty and voluminous sarpogrelate is implied in an aromatic interaction with the buried Phe5.47 residue. Only 3D models constructed in our lab highlighted amino acid residues of E2 loop that are directly involved in interactions with the ligands (Fig. 6.4). As previously reported, this loop was experimentally found to bring the receptor subtype selectivity within the class A GPCR family. In these models, it appears that only hydrogen bonds are involved. If such a bond establishes only once between agomelatin and Asn204, there are four hydrogen bonds between SB-228357 and E2 residues, Arg195, Asn204, Thr205, and Asn210. Moreover, E1 loop is also extensively represented in our ADRB-based model by interaction between the donor nitrogen of Trp120 indole and the acceptor nitrogen of the pyridyl group.

6.3.2 Investigation of Agonist Binding

Molecular dynamics simulations have been extensively used to study the dynamic properties of ground-state rhodopsin in the lipid bilayer (Crozier et al. 2003; Lau et al. 2007; Huber et al. 2004; Grossfield et al. 2006; Pitman et al. 2005; Schlegel et al. 2005; Cordomi and Perez 2007). The isomerization of retinal from 11-*cis* to all-*trans* was also simulated to shed light on the structural changes that lead to the lumi-rhodopsin (LUMI) state (Crozier et al. 2007; Lemaitre et al. 2005; Martinez-Mayorga et al. 2006; Saam et al. 2002; Kong and Karplus 2007), where LUMI is the first conformational

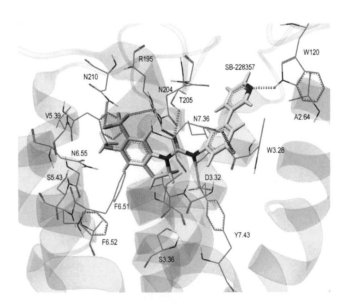

Fig. 6.4 Resultant docking of the selective antagonist SB-228357 in the ADRB-based three-dimensional (3D) model of 5-HT$_{2C}$. Amino acids residues are labeled according to the Ballesteros or 5-HT$_{2C}$ numbering as located in TM and E1&2 (W120, R195, N204, T205, N210) loops, respectively. Intermolecular hydrogen bonds are illustrated as *dashed cylinders*

state in which the retinal is found in the all-*trans* form. This computational method has also been used to generate a binding mode of 5-HT$_{2C}$ agonists in situ.

6.3.2.1   MD Simulations

Molecular dynamics consists in simulating the temperature increase in an atomic system. According to the thermodynamic rules, the provided kinetic energy induces motion of the atoms. After the system has reached a maximal total energy, it undergoes structural distortion as the conformational space is explored during the following equilibration stage. Numerical methods as the Verlet algorithm associated to molecular force fields (as previously described for energy minimization) permit the integration of Newton's equations of motion, which highlight collective atom motions in a macromolecular system. Nowadays, the commonly accepted method of carrying out MD simulations of membrane proteins is the explicit representation of the phospholipid bilayer solvated by water to provide the most optimal environment available with the current computational power (Ivanov et al. 2005; Elmore and Dougherty 2001; Xu et al. 2005a; Xu et al. 2005b). Works of Zuo et al. and Bray et al. consisted in hypothetically binding reference agonists in the presumed inactive starting structure and proceeding MD simulations during time periods long enough to equilibrate the total energy of the complex, 2 ns in a 241 POPC (palmitoyloleoyl-phosphatidylcholine) bilayer and 5 ns in a POPE (palmitoyloleoyl-phosphatidylethanolamine) bilayer, respectively, both with periodic boundary conditions. On one hand, apo-5-HT$_{2C}$ model of Zuo et al. was bound to an azepinoindol structure according to a prior 3D pharmacophore study from 18 derivative compounds of azepinoindol structures published as 5-HT$_{2C}$ agonists with high affinities (Ennis et al. 2003). The deduced 3D pharmacophore is a four-point fingerprint with three aromatic and one hydrogen bond donor/acceptor features. The possible binding site for this agonist was identified on the extracellular side of the TM domain and partially covered by EL2 loop through docking of the pharmacophoric elements and taking into account some published data about the position of the agonist binding site (Choudhary et al. 1993; Roth et al. 1997; Rashid et al. 2003; Kroeze et al. 2002; Kristiansen and Dahl 1996; Quirk et al. 2001; Roth et al. 1998; Wang et al. 1993). On the other hand, starting structure for the study of agonist binding by Bray et al. was a 5-HT$_{2C}$-serotonin complex derived from the same sampled docking as previously described for antagonist binding.

6.3.2.2   Analysis of 5-HT$_{2C}$-Agonists Interactions

As shown in Table 6.5, the main interactions observed in starting structures of 5-HT$_{2C}$-agonist complex come from the conserved patch of aromatic and acidic residues widely evoked in the preceding section. Nevertheless, some key interactions observed in the "active" conformation after MD simulations significantly change. Thus, the docking of all 18 agonists of the azepinoindol family in this

Table 6.5 Interactions of 5-HT$_{2C}$ agonists

		Azepinoindol derivative (Zuo's model)		Serotonin (Bray's model)	
	Interactions of 5-HT$_{2C}$ agonist	Starting complex	Final complex	Starting complex	Final complex
TM3	Ile3.29(131)		VdW		
	Asp3.32(134)	Ionic	Ionic	Ionic	Ionic
	Val3.33(135)		VdW		
	Ser3.36(138)		H bond	H bond	H bond
	Ser3.39(141)			H bond	
E2	Val202Asn210		VdW		H bond
TM5	Val5.39(215)		VdW		
	Ser5.43(219)		VdW	H bond	
	Phe5.47(223)	Aromatic	Aromatic	Aromatic	Aromatic
TM6	Trp6.48(324)			H bond	H bond
	Phe6.51(327)	Aromatic	Aromatic		
	Phe6.52(328)	Aromatic	Aromatic	Aromatic	Aromatic
	Asn6.55(331)		VdW		H bond
	Ile6.56(332)			VdW	
TM7	Val7.39(354)	VdW			
	Tyr7.48(358)	H bond	H bond		

"active" conformation results in a hydrogen-bond network more tightly packed between the protonated nitrogen and Asp3.32, Ser3.36, Asn6.55, the more repre-sented Tyr7.48 in all-azepinoindol docking, and one residue of E2 loop, namely, Asn210. In the same manner, both Asp3.32 and Ser3.36 maintain their polar inter-actions with the protonated amine group of serotonin as well as waters that enter the binding site and accumulate around the salt bridge between Asp3.32 and sero-tonin. However, the hydrogen bond with Ser5.43 and the VdW interactions with Ile6.56 are lost as the serotonin moves toward the intracellular end of the TM bundle. The hydrogen bond with Ser3.39 is lost as the hydroxyl of serotonin becomes involved in two other hydrogen bonds, as a strong hydrogen-bond accep-tor for the indole of Trp6.48 and a hydrogen-bond donor for the backbone oxygen of Ser3.36.

6.3.3   Selectivity of 5-HT$_{2C}$ Antagonist and Agonist Ligands over the Monoamine Receptor Family and the 5-HT Receptor Subfamily

In regards with a multiple sequence alignment (Fig. 6.5) of the human 5-HT recep-tor subtypes with several members of human GPCR-type monoamine receptors

Fig. 6.5 (continued) DRD1-5 (dopamine receptor ranging from type D1 to D5) and HRH1 (H1-histamine receptor). The critical amino acid residues for ligand binding are overlined and annotated as the Ballesteros and Weinstein numbering system, whereas the consensus sequence is shown at the *bottom* of each segment, along with a plot showing the degree of similarity at each residue position

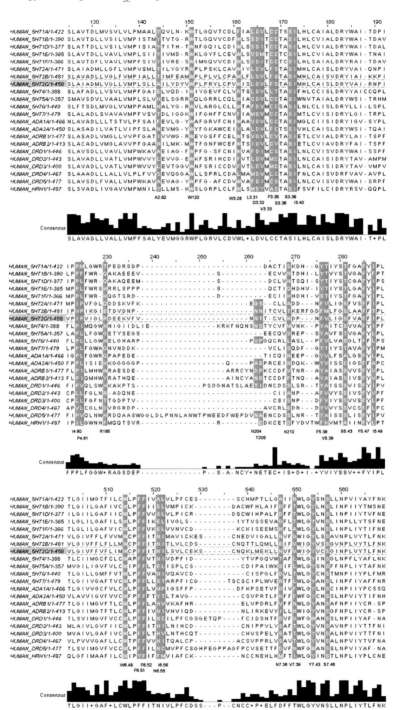

Fig. 6.5 Critical amino acids for monoamine receptors and the 5-HT$_{2C}$ selectivity. Three relevant regions of human 5-HT$_{2C}$ sequence (surrounded in a *dashed frame*) are aligned with human GPCR-type serotonin receptors and several members of human GPCR-type monoamine receptors, ADA1A (α1-adrenoceptor), ADA2A (α2-adrenoceptor), ADRB1 (β1-adrenoceptor), ADRB2 (β2-adrenoceptor),

(dopamine, histamine, or adrenalin receptors), we expect to provide some insights on single or groups of amino acid residues conferring a ligand selectivity for the human 5-HT$_{2C}$ receptor relatively to closest other subtypes. Focusing on the interactions previously picked in at least two models from TM2 to TM7, Trp3.28, Leu3.31, Val3.33, Phe5.38, Val5.39, Ser5.43, Ile5.49, Asn6.55, Asn7.36, and Tyr7.43 are not strictly conserved, but all variants are homologous according to their intrinsic properties of interaction, either aromatic (electronic π–π stacking contact), VdW (atomic contact) hydrophobic interactions, or ionic and hydrogen-bonding polar interactions. Phe3.35 and Ser3.36 appeal to us because they are homologous only among the 5-HT$_2$ subtypes and shift from aromatic and polar, respectively, toward VdW intrinsic properties. Other lonely interactions in different models like Ile4.60, Pro4.61, and Val7.39 are also homologous, whereas Ala2.64 is the only TM residue to divert from the polar amino acids of all the other receptors at this position.

Looking for all amino acids previously mentioned to bond either antagonist or agonist ligands shows that very few of these aligned residues are found solely in the human 5-HT$_{2C}$ receptor. Even Trp120 of extracellular loop EL1, which specifically hydrogen bonds with the more selective antagonists in our ADRB-based model, is strictly conserved among all the studied sequences. The only cluster to be clearly involved in selective interactions of 5-HT$_{2C}$ with ligands is the EL2 loop. Indeed, even though Asn204 and Thr205 are also found in EL2 of 5-HT$_{2B}$, the alignment match with ADRB2 and D5 can not be considered because of the consequent variation of the loop length that may render their 3D folding very different in this central region of the loop. On the contrary, the N-terminal or C-terminal residues of the loops get reliably less freedom of 3D folding due to their closeness to the spatially constrained α-helices. Thus, Asn210, the last C-terminal residue of EL2, is conserved in 5-HT$_{1B}$, 5-HT$_{1D,}$ and α-adrenoceptors and should be thus considered as a common structural feature, whereas Arg195, the fourth N-terminal residue of EL2, is not conserved and could therefore be labeled as a selective structural feature for the 5-HT$_{2C}$ subtype, enabling it to be targeted by selective ligands.

Consequently, drug design of 5-HT$_{2C}$ selective ligands should first respect the essential 3D pharmacophore common to GPCR-type monoamine receptors and integrate the discussed structural features to selectively target 5-HT$_{2C}$. As in most type-A GPCR receptors, antagonists are more voluminous than agonists are in that they should neutralize all interaction spots in a ligand binding cavity in order to prevent the unlocking and activation of the receptor. Thus, the design of 5-HT$_{2C}$ antagonists should integrate a central dipolar chemical group to consolidate specific interactions downward with D3.32, S3.36, and Y7.43 and especially earning hydrogen bonds upward with E2 loop (Arg195, Asn204, Thr205) and then two terminal aromatic ring assemblies targeted by aliphatic groups to earn specific VdW interactions in the hydrophobic cluster (Ala2.64, Trp120, Trp3.28). Agonists should be truncated from this aromatic pharmacophoric element to focus only on the dipolar group and one aromatic ring assembly near the TM5-TM6 aromatic cluster.

6.4 Structural Insights into Activation Mechanisms

Even though the biological response of the bovine rhodopsin is not induced by diffusible ligands, some characteristics of structural events leading to the activation of GPCRs have been highlighted only from this receptor up to now. Sequence homology should reveal structural insights into corresponding structural features of 5-HT$_{2C}$ since MD simulations performed on 5-HT$_{2C}$-agonists complexes (Bray and Goddard 2008; Zuo et al. 2007) have not shown particular and significant conformational changes through the receptor.

6.4.1 Agonist-Induced and Constitutive Activity

The study of the interactions between agonists or antagonists and the human 5-HT$_{2C}$ receptor tends to show that selectivity of the 5-HT$_{2C}$ subtype would be more sensitive to antagonist compounds. Moreover the high pharmacological potency of 5-HT$_{2C}$ antagonists has been shown. Indeed the maximal responses to highly potent 5-HT$_{2C}$ agonists were observed when the fractional occupancy of receptors was less than one (Hartman and Northup 1996) which suggests a pool of "spare receptors" with an intrinsic constitutive activity potentially sufficient to confer a transformed phenotype in cells transfected with the 5-HT$_{2C}$ gene (Julius et al. 1989). Thus, the specific pharmacological actions of certain receptor antagonists may reflect the direct consequences of binding the unoccupied constitutively active 5-HT$_{2C}$ receptors, rather than indirect actions mediated by a blockade of binding of endogenous agonists. Other type-A GPCRs, such as adrenoceptors, seemingly have little or no constitutive activity but constitutive activation is conferred by mutations in the third cytoplasmic loop (Cotecchia et al. 1992; Kjelsberg et al. 1992; Samama et al. 1993; Ren et al. 1993). Unfortunately, this region remains a challenge for 3D modeling because of its high variability. Besides, if GPCR-type monoamine receptors have relatively long IL3 segment in comparison with bovine rhodopsin, their structure cannot be studied in crystal structures because it is partially truncated (β1-adrenoceptor) or replaced with lysosome T4 (β2-adrenoceptor, A2A-adenosine receptor) in order to facilitate their crystallization.

Nevertheless, the recently published works about the signaling pathway in rhodopsin (Kong and Karplus 2007), reveal how the interaction network of coupled residues extends from the retinal-binding pocket to the cytoplasmic surface after 22 ns of MD runs. The interactions are mainly weakened or intensified while some charged residues of TM bundle (D2.50, E3.37, E3.49, R3.50, E6.31, K6.32, R8.50) but also in IL2 (R147) are involved and tend to be clustered in certain directions, instead of being randomly distributed. Compared with hydrophobic interactions, salt bridges and hydrogen bonds are most specific in direction, which could make them effective as molecular switches that respond to the distant perturbation of the ligand binding site. Inspecting the sequence alignment (Fig. 6.5) between RHO and 5-HT$_{2C}$ shows the conservation of most of these charged residues found except E3.37 in RHO mutated to a threonine in

5-HT$_{2C}$ and ADRB receptors. All other charged residues are either strictly conserved (D2.50, R3.50, R164, E6.31, R8.50) or homologous charged residues (D3.49, R6.32).

Thus the conductance of the signal toward the cytoplasmic surface could be the same as in bovine rhodospin. Moreover the starting structural event in bovine rhodopsin is the formation of a Schiff base between isomerized retinal and Lys296 inducing a continuous coupling pathway from retinal to the NPxxY TM7 motif through N1.50, S3.38 and D2.50. Since the amino acid residue corresponding to the RHO Lys296 (K7.43) is Tyr358 in 5-HT$_{2C}$, we suggest that polar interactions occurring between D3.32, S3.38 and Y7.43 could be perturbed by an intensified salt bridge between D3.32 and the protonated nitrogen of agonist ligands like previously discussed. This would disrupt hydrogen bond between D3.32 and Y7.43 whose the side chain conformation may switch downward to permit a hydrogen bond with the close D2.50 what would promote the signal pathway along the receptor toward the cytoplasmic region. Besides, D2.50 is also directly involved within the hydrogen bond network relayed by structural waters, as previously described in the section dedicated to the common features of monoamine receptors, and could be considered as a hinge residue to enhance and propagate the signal from the ligand binding pocket to the cytoplasmic end of the receptor. The below Fig. 6.6 summarizes the position of the key residues discussed to participate in the conductance of the signal pathway in 5-HT$_{2C}$. Implication of such pathway in monoamine receptors like 5-HT$_{2C}$ is supported by works (Jensen et al. 2001) which have shown agonist-induced spectral changes for preferred conformations of residues 6.33 and 6.34 in ADRB2, conserved in 5-HT$_{2C}$, and suggesting an agonist-promoted movement of the cytoplasmic part of TM6 away from the receptor core and upwards toward the membrane bilayer.

6.4.2   Structural Impact of RNA Editing

The 5-HT$_{2C}$ receptor is expressed in different isoforms as a result of mRNA editing (Niswender et al. 1998). Both INI (unedited) and VSV (a fully edited version) isoforms are the results of the variations I156V occurring in position 3.54 in TM3 from our ADBR-based TM prediction and N158S and I160V in the following IL2, well conserved in type-A GPCRs (Fig. 6.5). In regard with the multiple sequence alignments (Fig. 6.6), position 3.54 is dominantly occupied by an isoleucine residue but is a valine in bovine rhodopsin. In the same manner, the two other positions in IL2 are homologous with the outstanding serine variation instead of N158 and hydrophobic intrinsic properties conserved for the position 160. Investigation of the pharmacology for the agonist binding site of these two isoforms of the 5-HT$_{2C}$ receptor has shown that the INI isoform of the 5-HT$_{2C}$ receptor is pharmacologically similar to the VSV form but it couples more efficiently to G-proteins (Quirk et al. 2001).

It is possible to rationalize the structural impact of mutations on the thermodynamic stability of a system in buried regions or in the vicinity of ligand binding sites

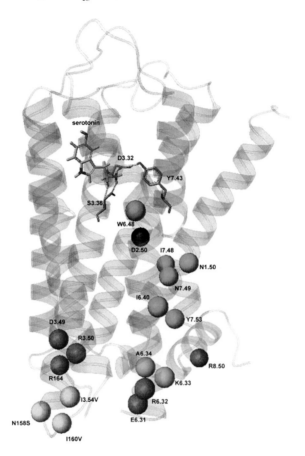

Fig. 6.6 Location of key residues for signal transducing in the ADRB-based 5-HT$_{2C}$ model. α-Carbon of key amino acid residues (represented as *spheres* and discussed in the text) to transduce the signal from the ligand binding pocket to the cytoplasmic regions are divided into five classes : (1) network of residues conserved among amino acids participating to the signal pathway revealed by Kong et al. (Kong and Karplus 2007) (*red*); (2) hydrogen bond network (*green*) assisted by structural waters (additional information in Fig. 6.3); (3) hinge residue (*black*) between the two previous networks; (4) variant amino acids from RNA editing isoforms (*yellow*); (5) amino acid residues involved in the agonist-promoted movement of the cytoplasmic part of TM6 (*cyan*). Docking of serotonin in the ligand-binding pocket shows polar interactions with the three amino acids residues that could be the starting event for the agonist-promoted signaling (For interpretation of the colors in this figure, the reader is referred to the web version of this chapter)

according to differences of hydrophobic, polar or volume intrinsic properties induced by the mutation. Thus it appears that such solvent-exposed variations would be well tolerated because there are no significant differences of these physical characteristics. However, this intracellular exposed region is in fine in the cytosol and hydrophobic positions 156 and 160 would be detrimental in this aqueous environment. Since intracellular IL2 and IL3 regions are suggested to be bound to G proteins, these positions could be a specific hydrophobic binding site for G-proteins.

Consequently variations would not change the overall thermodynamic stability of the 5-HT$_{2C}$ receptor and its ligand binding dynamics but would have incidence on the specific binding of G proteins and subsequent activity.

6.5   Conclusion

In the rational approach of drug design, computational tools have become increasingly useful. Numerous examples illustrate their impact in increasing the speed of drug discovery by helping in the generation of pertinent hypothesis. It should be still remembered that computational tools are mere tools, that is they can greatly help in the process of trials and errors to achieve a desired result but should not be substituted to an integrated research effort and should not confused with some kind of oracle always telling the truth without questioning their conclusions by trials. The more complex the problem to solve is, the harder it is to find a solution and the more integrated the various methods of investigation must be. In this context, the quest of selective 5-HT$_{2C}$ ligands is an example of the difficulties encountered in drug design and is therefore exemplary of the close imbrications of the various means to shed a new light on the same facts. We have described here the building of a series of static images of the not yet totally grasped dynamics of GPCRs activation, with the secondary but important problem of designing selective compounds. Although it is relatively easy to generate a model of a GPCR, given the right template, whose choice was fairly limited up to very recently and, most importantly, a good sequence alignment, it is another task to validate the model. Three steps must be adhered to in order to assure a good coherency of the final image of the receptor. First, the alignment must conserve as much as possible the characteristic sequences of the GPCR superfamily. Second, the structural elements of the putative model must be checked to verify they are physically plausible, in particular the relation between the helical domains and their lipidic environment. Third, the structurally valid model must comply with the experimental data gathered by in vitro experiments and observed in vivo responses. In the case of 5-HT$_{2C}$, the two first points are classical for the class A GPCR modeling. The third point is the clearest bottleneck in the building of the models, as the data are still relatively few. Due to the implication of the human cleverness and intuition in this process, as opposed to the two first steps, it should be coined tailoring of the model. In particular, rendering in silico the difference between the activated agonist-binding forms, constitutively active basal form and inactivated inverse agonist-binding forms require more facts on the residues implicated in the binding of the ligands and the atomic events giving rise to the signal transduction through the cellular membrane than currently available to build a model rather than to imagine a general scaffold of the receptor. The structural features of 5-HT$_{2C}$ are fairly well understood, even the length of the helical domains, give or take a few amino acids. The binding site has been approximated with a relative precision by computational methods for a fairly large selection of compounds of various degrees of affinity and different activity and the studies reported here are highly coherent despite their very different approaches. This shows the intrinsic power of computational tools for these common tasks of target-based knowledge generation. To the experience of the authors, in a short time span, the possibilities of developing a plausible model of 5-HT$_{2C}$ have greatly increased, partly due to the ever increasing sharpness of the available tools, but also, and for the most part, due to the ever growing body of data, collected from the compilation of the various investigations on this receptor, either by in vivo, in vitro, or in silico means.

It remains yet two equally interesting points. We could wonder if the models are even remotely close to a biologically sound reality. Here, the computational scientist should remain humble, propose some structural features that could be targeted with ease, by example by directed mutagenesis or design of a ligand with adequate substitution pattern, and hope that the experimental results will corroborate his own personal feeling, largely transcribed in the model he has put forth. It should be kept in mind that, by definition, a model is an oversimplification of the natural object in order to simulate its behaviour, and therefore should not be regarded as exact. It is the role of the modeler to find this delicate equilibrium between two goals. The one part of the equilibrium is a model that will be used to generate hypothesis of ligands interactions, usable to create new ligand structures. Such a model must be robust, that is not focused toward a precise set of ligand-receptor interactions. The other part of the equilibrium is a model that shows something of the dynamics of the ligand-receptor complex, that is a model tuned to represent as much of the known data about a ligand as possible, but that will have necessarily lost its capacity to be engaged in interaction with other ligands. This equilibrium is the key point of homology modeling, and the shift from hand-made models built around a ligand toward routine molecular dynamics simulations of the 5-HT$_{2C}$ complexes for larger series of structurally dissimilar compounds is a clear sign that it is still the hardest point to solve, in particular due to the unclear configurations of the different activation states of the GPCRs. The second interesting point that appears is that the current models, despite their possible flaws, have nonetheless provided some critical insights into the interactions formed by 5-HT$_{2C}$ ligands, among which some are selective. Comparing these insights with similar data from other GPCRs, or more directly comparing the residues in the structural elements of the GCPRs that interact with the ligands has generated a very large number of ideas. In turn, these ideas could be turned by experimental validation, including design of new ligands, into knowledge, which will contribute to a better characterization of the 5-HT$_{2C}$ subtype binding requirements. This will lead to some hard facts about its selectivity toward its congeners, given some more team works of biologists, modelers and chemists.

Acknowledgements We acknowledge the Institut de Recherche Servier for granting us the use of their 3-dimensional visualization tools, TriposBenchware (Tripos®) and DiscoveryStudio2.1 (Catalyst®), and specially thank A.Gohier for discussions, suggestions, and help with modeling and illustrations.

References

Angelova K, Fanelli F, Puett D (2002) A model for constitutive lutropin receptor activation based on molecular simulation and engineered mutations in transmembrane helices 6 and 7. J Biol Chem 277:32202–32213.

Attwood TK, Findlay JB (1993) Design of a discriminating fingerprint for G-protein-coupled receptors. Protein Eng 6:167–176.

Ballesteros JA, Jensen AD, Liapakis G, et al (2001) Activation of the beta 2-adrenergic receptor involves disruption of an ionic lock between the cytoplasmic ends of transmembrane segments 3 and 6. J Biol Chem 276:29171–29177.

Berman HM, Westbrook J, Feng Z, et al (2000) The protein data bank. Nucleic Acids Res 28:235–242.

Birnbaumer L (1990) G proteins in signal transduction. Annu Rev Pharmacol Toxicol 30:675–705.

Braden MR, Nichols DE (2007) Assessment of the roles of serines 5.43(239) and 5.46(242) for binding and potency of agonist ligands at the human serotonin 5-HT2A receptor. Mol Pharmacol 72:1200–1209.

Bray JK, Goddard WA 3 rd (2008) The structure of human serotonin 2c G-protein-coupled receptor bound to agonists and antagonists. J Mol Graph Model 27:66–81.

Bromidge SM, Duckworth M, Forbes IT, et al (1997) 6-Chloro-5-methyl-1-[[2-[(2-methyl-3-pyridyl)oxy]-5-pyridyl]carbamoyl]- indoline (SB-242084): the first selective and brain penetrant 5-HT2C receptor antagonist. J Med Chem 40:3494–3496.

Bromidge SM, Dabbs S, Davies DT, et al (1998) Novel and selective 5-HT2C/2B receptor antagonists as potential anxiolytic agents: synthesis, quantitative structure-activity relationships, and molecular modeling of substituted 1-(3-pyridylcarbamoyl)indolines. J Med Chem 41:1598–1612.

Bromidge SM, Dabbs S, Davies DT, et al (1999) Model studies on a synthetically facile series of N-substituted phenyl-N'-pyridin-3-yl ureas leading to 1-(3-pyridylcarbamoyl) indolines that are potent and selective 5-HT(2C/2B) receptor antagonists. Bioorg Med Chem 7:2767–2773.

Bromidge SM, Dabbs S, Davies DT, et al (2000) Biarylcarbamoylindolines are novel and selective 5-HT(2C) receptor inverse agonists: identification of 5-methyl-1-[[2-[(2-methyl-3-pyridyl) oxy]- 5-pyridyl]carbamoyl]-6-trifluoromethylindoline (SB-243213) as a potential antidepressant/anxiolytic agent. J Med Chem 43:1123–1134.

Brooks BR, Bruccoleri RE, Olafson BD, et al (1983) CHARMM: a program for macromolecular energy, minimization and dynamics calculations. J Comput Chem 4:187–217.

Bywater RP (2005) Location and nature of the residues important for ligand recognition in G-protein coupled receptors. J Mol Recognit 18:60–72.

Casey PJ, Gilman AG (1988) G protein involvement in receptor-effector coupling. J Biol Chem 263:2577–2580.

Cherezov V, Rosenbaum DM, Hanson MA, et al (2007) High-resolution crystal structure of an engineered human beta2-adrenergic G protein-coupled receptor. Science 318:1258–1265.

Choudhary MS, Craigo S, Roth BL (1993) A single point mutation (Phe340-->Leu340) of a conserved phenylalanine abolishes 4-[125I]iodo-(2,5-dimethoxy)phenylisopropylamine and [3H]mesulergine but not [3H]ketanserin binding to 5-hydroxytryptamine2 receptors. Mol Pharmacol 43:755–761.

Cordomi A, Perez JJ (2007) Molecular dynamics simulations of rhodopsin in different one-component lipid bilayers. J Phys Chem B 111:7052–7063.

Cornell WD, Cieplak P, Bayly CI, et al (1995) A second generation force field for the simulation of proteins, nucleic acids and organic molecules. J Am Chem Soc 117:5179–5197.

Cotecchia S, Ostrowski J, Kjelsberg MA, et al (1992) Discrete amino acid sequences of the alpha 1-adrenergic receptor determine the selectivity of coupling to phosphatidylinositol hydrolysis. J Biol Chem 267:1633–1639.

Crozier PS, Stevens MJ, Forrest LR, et al (2003) Molecular dynamics simulation of dark-adapted rhodopsin in an explicit membrane bilayer: coupling between local retinal and larger scale conformational change. J Mol Biol 333:493–514.

Crozier PS, Stevens MJ, Woolf TB (2007) How a small change in retinal leads to G-protein activation: initial events suggested by molecular dynamics calculations. Proteins 66:559–574.

Donnelly D, Overington JP, Blundell TL (1994) The prediction and orientation of alpha-helices from sequence alignments: the combined use of environment-dependent substitution tables, Fourier transform methods and helix capping rules. Protein Eng 7:645–653.

Ebersole BJ, Visiers I, Weinstein H, et al (2003) Molecular basis of partial agonism: orientation of indoleamine ligands in the binding pocket of the human serotonin 5-HT2A receptor determines relative efficacy. Mol Pharmacol 63:36–43.

Eisenberg D, Weiss RM, Terwilliger TC (1984) The hydrophobic moment detects periodicity in protein hydrophobicity. Proc Natl Acad Sci U S A 81:140–144.

Elmore DE, Dougherty DA (2001) Molecular dynamics simulations of wild-type and mutant forms of the Mycobacterium tuberculosis MscL channel. Biophys J 81:1345–1359.

Ennis MD, Hoffman RL, Ghazal NB, et al (2003) 2,3,4,5-Tetrahydro- and 2,3,4,5,11,11a-hexa-hydro-1H-[1,4]diazepino[1,7-a]indoles: new templates for 5-HT(2C) agonists. Bioorg Med Chem Lett 13:2369–2372.

Farce A, Dilly S, Yous S, et al (2006) Homology modelling of the serotoninergic 5-HT2c receptor. J Enzyme Inhib Med Chem 21:285–292.

Goodsell DS, Morris GM, Olson AJ (1996) Automated docking of flexible ligands: applications of AutoDock. J Mol Recognit 9:1–5.

Greasley PJ, Fanelli F, Rossier O, et al (2002) Mutagenesis and modelling of the alpha(1b)-adrenergic receptor highlight the role of the helix 3/helix 6 interface in receptor activation. Mol Pharmacol 61:1025–1032.

Grossfield A, Feller SE, Pitman MC (2006) A role for direct interactions in the modulation of rhodopsin by omega-3 polyunsaturated lipids. Proc Natl Acad Sci U S A 103:4888–4893.

Hartman JI, Northup JK (1996) Functional reconstitution in situ of 5-hydroxytryptamine2c (5HT2c) receptors with alphaq and inverse agonism of 5HT2c receptor antagonists. J Biol Chem 271:22591–22597.

Huber T, Botelho AV, Beyer K, et al (2004) Membrane model for the G-protein-coupled receptor rhodopsin: hydrophobic interface and dynamical structure. Biophys J 86:2078–2100.

Ivanov AA, Baskin, II, Palyulin VA, et al (2005) Molecular modeling and molecular dynamics simulation of the human A2B adenosine receptor. The study of the possible binding modes of the A2B receptor antagonists. J Med Chem 48:6813–6820.

Jaakola VP, Griffith MT, Hanson MA, et al (2008) The 2.6 angstrom crystal structure of a human A2A adenosine receptor bound to an antagonist. Science 322:1211–1217.

Jensen AD, Guarnieri F, Rasmussen SG, et al (2001) Agonist-induced conformational changes at the cytoplasmic side of transmembrane segment 6 in the beta 2 adrenergic receptor mapped by site-selective fluorescent labeling. J Biol Chem 276:9279–9290.

Jones G, Willett P, Glen RC (1995) A genetic algorithm for flexible molecular overlay and phar-macophore elucidation. J Comput Aided Mol Des 9:532–549.

Jones G, Willett P, Glen RC (1995) Molecular recognition of receptor sites using a genetic algorithm with a description of desolvation. J Mol Biol 245:43–53.

Julius D, Livelli TJ, Jessell TM, et al (1989) Ectopic expression of the serotonin 1c receptor and the triggering of malignant transformation. Science 244:1057–1062.

Kjelsberg MA, Cotecchia S, Ostrowski J, et al (1992) Constitutive activation of the alpha 1B-adrenergic receptor by all amino acid substitutions at a single site. Evidence for a region which constrains receptor activation. J Biol Chem 267:1430–1433.

Klco JM, Wiegand CB, Narzinski K, et al (2005) Essential role for the second extracellular loop in C5a receptor activation. Nat Struct Mol Biol 12:320–326.

Knight AR, Misra A, Quirk K, et al (2004) Pharmacological characterisation of the agonist radioligand binding site of 5-HT(2A), 5-HT(2B) and 5-HT(2C) receptors. Naunyn Schmiedebergs Arch Pharmacol 370:114–123.

Kong Y, Karplus M (2007) The signaling pathway of rhodopsin. Structure 15:611–623.

Kristiansen K, Dahl SG (1996) Molecular modeling of serotonin, ketanserin, ritanserin and their 5-HT2C receptor interactions. Eur J Pharmacol 306:195–210.

Kristiansen K, Kroeze WK, Willins DL, et al (2000) A highly conserved aspartic acid (Asp-155) anchors the terminal amine moiety of tryptamines and is involved in membrane targeting of the 5-HT(2A) serotonin receptor but does not participate in activation via a "salt-bridge disruption" mechanism. J Pharmacol Exp Ther 293:735–746.

Kroeze WK, Kristiansen K, Roth BL (2002) Molecular biology of serotonin receptors structure and function at the molecular level. Curr Top Med Chem 2:507–528.

Kuntz ID (1992) Structure-based strategies for drug design and discovery. Science 257:1078–1082.

Lau PW, Grossfield A, Feller SE, et al (2007) Dynamic structure of retinylidene ligand of rhodopsin probed by molecular simulations. J Mol Biol 372:906–917.

Lemaitre V, Yeagle P, Watts A (2005) Molecular dynamics simulations of retinal in rhodopsin: from the dark-adapted state towards lumirhodopsin. Biochemistry 44:12667–12680.

Li J, Edwards PC, Burghammer M, et al (2004) Structure of bovine rhodopsin in a trigonal crystal form. J Mol Biol 343:1409–1438.

Martinez-Mayorga K, Pitman MC, Grossfield A, et al (2006) Retinal counterion switch mechanism in vision evaluated by molecular simulations. J Am Chem Soc 128:16502–16503.

Mathiowetz AM, Jain A, Karasawa N, et al (1994) Protein simulations using techniques suitable for very large systems: the cell multipole method for nonbond interactions and the Newton-Euler inverse mass operator method for internal coordinate dynamics. Proteins 20:227–247.

Millan MJ, Gobert A, Lejeune F, et al (2003) The novel melatonin agonist agomelatine (S20098) is an antagonist at 5-hydroxytryptamine2C receptors, blockade of which enhances the activity of frontocortical dopaminergic and adrenergic pathways. J Pharmacol Exp Ther 306:954–964.

Niswender CM, Sanders-Bush E, Emeson RB (1998) Identification and characterization of RNA editing events within the 5-HT2C receptor. Ann N Y Acad Sci 861:38–48.

Okada T, Fujiyoshi Y, Silow M, et al (2002) Functional role of internal water molecules in rhodopsin revealed by x-ray crystallography. Proc Natl Acad Sci U S A 99:5982–5987.

Okada T, Sugihara M, Bondar AN, et al (2004) The retinal conformation and its environment in rhodopsin in light of a new 2.2 A crystal structure. J Mol Biol 342:571–583.

Palczewski K, Kumasaka T, Hori T, et al (2000) Crystal structure of rhodopsin: a G protein-coupled receptor. Science 289:739–745.

Pardo L, Deupi X, Dolker N, et al (2007) The role of internal water molecules in the structure and function of the rhodopsin family of G protein-coupled receptors. Chembiochem 8:19–24.

Petrey D, Xiang Z, Tang CL, et al (2003) Using multiple structure alignments, fast model building, and energetic analysis in fold recognition and homology modeling. Proteins 53 Suppl 6:430–435.

Pitman MC, Grossfield A, Suits F, et al (2005) Role of cholesterol and polyunsaturated chains in lipid-protein interactions: molecular dynamics simulation of rhodopsin in a realistic membrane environment. J Am Chem Soc 127:4576–4577.

Quirk K, Lawrence A, Jones J, et al (2001) Characterisation of agonist binding on human 5-HT2C receptor isoforms. Eur J Pharmacol 419:107–112.

Ramachandran GN (1963) Protein structure and crystallography. Science 141:288–291.

Ramachandran GN, Ramakrishnan C, Sasisekharan V (1963) Stereochemistry of polypeptide chain configurations. J Mol Biol 7:95–99.

Rashid M, Manivet P, Nishio H, et al (2003) Identification of the binding sites and selectivity of sarpogrelate, a novel 5-HT2 antagonist, to human 5-HT2A, 5-HT2B and 5-HT2C receptor subtypes by molecular modeling. Life Sci 73:193–207.

Ren Q, Kurose H, Lefkowitz RJ, et al (1993) Constitutively active mutants of the alpha 2-adrenergic receptor. J Biol Chem 268:16483–16487.

Renault N, Gohier A, Chavatte P, et al (2010) Novel structural insights for drug design of selective 5-HT(2C) inverse agonists from a ligand-biased receptor model. Eur J Med Chem 45: 5086–99.

Rosenbaum DM, Cherezov V, Hanson MA, et al (2007) GPCR engineering yields high-resolution structural insights into beta2-adrenergic receptor function. Science 318:1266–1273.

Roth BL, Shoham M, Choudhary MS, et al (1997) Identification of conserved aromatic residues essential for agonist binding and second messenger production at 5-hydroxytryptamine2A receptors. Mol Pharmacol 52:259–266.

Roth BL, Willins DL, Kristiansen K, et al (1998) 5-Hydroxytryptamine2-family receptors (5-hydroxytryptamine2A, 5-hydroxytryptamine2B, 5-hydroxytryptamine2C): where structure meets function. Pharmacol Ther 79:231–257.

Saam J, Tajkhorshid E, Hayashi S, et al (2002) Molecular dynamics investigation of primary photoinduced events in the activation of rhodopsin. Biophys J 83:3097–3112.

Sali A, Blundell TL (1993) Comparative protein modelling by satisfaction of spatial restraints. J Mol Biol 234:779–815.

Salom D, Lodowski DT, Stenkamp RE, et al (2006) Crystal structure of a photoactivated deprotonated intermediate of rhodopsin. Proc Natl Acad Sci U S A 103:16123–16128.

Saltzman AG, Morse B, Whitman MM, et al (1991) Cloning of the human serotonin 5-HT2 and 5-HT1C receptor subtypes. Biochem Biophys Res Commun 181:1469–1478.

Samama P, Cotecchia S, Costa T, et al (1993) A mutation-induced activated state of the beta 2-adrenergic receptor. Extending the ternary complex model. J Biol Chem 268:4625–4636.

Schlegel B, Sippl W, Holtje HD (2005) Molecular dynamics simulations of bovine rhodopsin: influence of protonation states and different membrane-mimicking environments. J Mol Model 12:49–64.

Shapiro DA, Kristiansen K, Kroeze WK, et al (2000) Differential modes of agonist binding to 5-hydroxytryptamine(2A) serotonin receptors revealed by mutation and molecular modeling of conserved residues in transmembrane region 5. Mol Pharmacol 58:877–886.

Stam NJ, Vanderheyden P, van Alebeek C, et al (1994) Genomic organisation and functional expression of the gene encoding the human serotonin 5-HT2C receptor. Eur J Pharmacol 269:339–348.

Thompson JD, Higgins DG, Gibson TJ (1994) CLUSTAL W: improving the sensitivity of progressive multiple sequence alignment through sequence weighting, position-specific gap penalties and weight matrix choice. Nucleic Acids Res 22:4673–4680.

Wang CD, Gallaher TK, Shih JC (1993) Site-directed mutagenesis of the serotonin 5-hydroxytrypamine2 receptor: identification of amino acids necessary for ligand binding and receptor activation. Mol Pharmacol 43:931–940.

Warne T, Serrano-Vega MJ, Baker JG, et al (2008) Structure of a beta1-adrenergic G-protein-coupled receptor. Nature 454:486–491.

White SH, Ladokhin AS, Jayasinghe S, et al (2001) How membranes shape protein structure. J Biol Chem 276:32395–32398.

Wurch T, Pauwels PJ (2000) Coupling of canine serotonin 5-HT(1B) and 5-HT(1D) receptor subtypes to the formation of inositol phosphates by dual interactions with endogenous G(i/o) and recombinant G(alpha15) proteins. J Neurochem 75:1180–1189.

Xie E, Zhu L, Zhao L, et al (1996) The human serotonin 5-HT2C receptor: complete cDNA, genomic structure, and alternatively spliced variant. Genomics 35:551–561.

Xu Y, Shen J, Luo X, et al (2005) Conformational transition of amyloid beta-peptide. Proc Natl Acad Sci U S A 102:5403–5407.

Xu Y, Barrantes FJ, Luo X, et al (2005) Conformational dynamics of the nicotinic acetylcholine receptor channel: a 35-ns molecular dynamics simulation study. J Am Chem Soc 127:1291–1299.

Zhang M, Mizrachi D, Fanelli F, et al (2005) The formation of a salt bridge between helices 3 and 6 is responsible for the constitutive activity and lack of hormone responsiveness of the naturally occurring L457R mutation of the human lutropin receptor. J Biol Chem 280:26169–26176.

Zhao MM, Hwa J, Perez DM (1996) Identification of critical extracellular loop residues involved in alpha 1-adrenergic receptor subtype-selective antagonist binding. Mol Pharmacol 50:1118–1126.

Zuo Z, Chen G, Luo X, et al (2007) Pharmacophore-directed homology modeling and molecular dynamics simulation of G protein-coupled receptor: study of possible binding modes of 5-HT2C receptor agonists. Acta Biochim Biophys Sin (Shanghai) 39:413–422.

Chapter 7
5-HT$_{2C}$ Receptor Dimerization

Katharine Herrick-Davis and Dinah T. Farrington

7.1 Introduction

G-protein-coupled receptor (GPCR) dimerization is a rapidly developing area of research investigating molecular mechanisms involved in receptor activation. Traditionally, GPCR were thought to function as monomeric units. However, in recent years dimer/oligomer formation has been reported for dozens of different GPCRs using a variety of different techniques including immunoprecipitation, resonance energy transfer, and bimolecular fluorescence complementation (reviewed in Angers et al. 2002; Park et al. 2004; Kerppola 2008). Homo- and hetero-dimerization between different families of GPCR have been reported to regulate ligand binding, second messenger activation, and receptor trafficking (George et al. 2002). GPCR dimerization has been reported to involve the N-terminal domain, transmembrane domains, and the C-terminal domain of interacting receptors, and agonists have been reported to increase, decrease, or have no effect on GPCR dimerization (Javitch 2004; Milligan 2004). While these studies suggest that dimerization may be a common property among GPCRs, the specific mechanisms and functional consequences of dimerization may be vastly different from one GPCR family to another. In the case of γ-aminobutyric acid B (GABA$_B$) receptors, the functional significance of dimerization is very clear. Heterodimerization between GABA$_B$R1 and GABA$_B$R2 receptors is essential for trafficking and expression of functional receptors on the plasma membrane (Jones et al. 1998; Kaupmann et al. 1998; White et al. 1998). However, for the majority of GPCRs studied to date the functional significance of dimerization remains unknown.

Twelve different serotonin receptors belonging to the GPCR superfamily have been identified in humans (Hoyer et al. 2002). Immunoprecipitation and resonance energy transfer techniques have been used as supportive evidence for 5-HT$_{1A}$,

K. Herrick-Davis (✉)
Center for Neuropharmacology & Neuroscience and Department of Psychiatry,
47 New Scotland Ave., MC-136 Albany Medical College, Albany, NY 12208, USA
e-mail: daviskh@mail.amc.edu

G. Di Giovanni et al. (eds.), *5-HT$_{2C}$ Receptors in the Pathophysiology of CNS Disease*,
The Receptors 22, DOI 10.1007/978-1-60761-941-3_7,
© Springer Science+Business Media, LLC 2011

5-HT$_{1B}$, 5-HT$_{1D}$, 5-HT$_{2C,}$ and 5-HT$_4$ receptor dimers (Xie et al. 1999; Salim et al. 2002; Herrick-Davis et al. 2004; Berthouze et al. 2007). This chapter focuses on the current literature related to 5-HT$_{2C}$ receptor dimerization. Specific topics to be addressed include the biochemical and biophysical properties of 5-HT$_{2C}$ homodimers, the homodimer biogenesis, the functional significance of homodimerization, the homodimer interface, and the heterodimerization with splice variants, isoforms, and other 5-HT receptor subtypes. Areas where there are gaps in our knowledge are identified and potential experiments to address these issues are discussed.

7.2   Homodimerization of 5-HT$_{2C}$ Receptors: Biochemical and Biophysical Properties

A combination of biochemical (immunoprecipitation and western blot) and biophysical (bioluminescence and fluorescence resonance energy transfer, BRET and FRET, respectively) techniques were used to determine if 5-HT$_{2C}$ receptors form homodimers, to determine the physical nature of the protein–protein interaction, and to determine the precise intracellular compartment in which dimer formation occurs. While each technique has its own limitations, all three techniques when used in a combined approach can provide supporting evidence for receptor dimerization. Coimmunoprecipitation experiments can provide evidence for protein–protein interactions, as long as proper controls are included to rule out false positive results arising during protein solubilization. BRET has the advantage that it can be performed in intact living cells, and it does not require the use of an external excitation source such as a laser. FRET, when combined with confocal microscopy, can be used to identify the precise cellular location of the FRET signal in an intact living cell. The 5-HT$_{2C}$ receptor homodimerization studies described in this chapter used fusion proteins, in which different probes were added to the C-terminus of the receptor. The presence of fluorescent proteins (CFP/YFP), hemagglutinin (HA), or Renilla luciferase (Rluc) tags on the C-terminal end of the receptor did not alter 5-HT binding affinity or potency (Herrick-Davis et al. 2004).

Previous studies have shown that detergent solubilized 5-HT$_{2C}$ receptors appear as multiple immunoreactive bands on Western blots, with multiple glycosylation states ranging from 40 to 60 kDa (Backstrom et al. 1995). We performed experiments to identify native 5-HT$_{2C}$ receptors in choroid plexus epithelial cells and to determine if Western blots of solubilized membrane proteins would provide supportive evidence for the existence of 5-HT$_{2C}$ homodimers. Freshly isolated choroid plexus tissue from E19 rat pups was immunostained with a 5-HT$_{2C}$ antibody to allow visualization of 5-HT$_{2C}$ receptors using fluorescence confocal microscopy (Fig. 7.1a). Freshly isolated choroid plexus tissue was sonication in Laemmli sample buffer and separated by polyacrylamide gel electrophoresis (PAGE) under nonreducing conditions (Fig. 7.1b). As expected, 5-HT$_{2C}$ immunoreactive bands were observed around 55 kDa. In addition, higher molecular weight bands the approximate size of multimers of 5-HT$_{2C}$ receptors were observed. Studies using

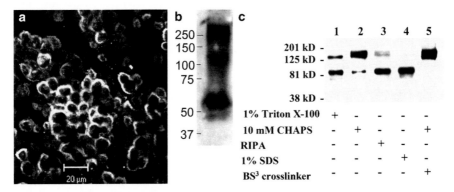

Fig. 7.1 5-HT$_{2C}$ receptors in choroid plexus. (**a**) Freshly isolated, intact choroid plexus tissue from an E19 rat pup, stained with 5-HT$_{2C}$ antibody (Santa Cruz, C-20) and Alexa 568. Scale bar = 20 μm. (**b**) Western blot: Choroid plexus tissue was briefly sonicated in Laemmli sample buffer and run under nonreducing conditions. Mature 5-HT$_{2C}$ receptor = 55 kDa. (**c**) Detergent and cross-linker sensitivity of 5-HT$_{2C}$ receptor homodimers. Cell membranes from a 5-HT$_{2C}$/YFP HEK 293 stable cell line were solubilized with different detergents as indicated (+), run on 10% PAGE, and probed with GFP antibody. Lanes 1–4: 5-HT$_{2C}$/YFP monomers (80–90 kDa) and dimers (160–180 kDa). Lane 5: Cells were pretreated with BS^3 cross-linker prior to solubilization in CHAPS (Reproduced from Herrick-Davis et al. 2004. With permission from American Chemical Society [ACS])

detergents, cross-linkers and reducing agents were performed to determine the chemical nature of the putative 5-HT$_{2C}$ receptor homodimers. Cell membranes were prepared from a 5-HT$_{2C}$/YFP HEK 293 stable cell line, solubilized using different detergents and analyzed by Western blot probed with GFP antibody (Herrick-Davis et al. 2004). As shown in Fig. 7.1c, immunoreactive bands the approximate size of 5-HT$_{2C}$/YFP fusion protein monomers are visible (80–90 kDa), along with higher molecular weight bands the approximate predicted size of 5-HT$_{2C}$/YFP dimers. As detergent stringency is gradually increased, by solubilization in 1% Triton X-100, 10 mM CHAPS, RIPA, or 1% SDS, the monomeric species becomes the predominant form identified on the blot (Fig. 7.1c). The addition of 10 mM DTT or 1% β-mercaptoethanol to membrane proteins solubilized in CHAPS did not reduce the higher molecular weight immunoreactive bands. These results indicate that the higher molecular weight bands observed on the western blot are detergent sensitive and do not contain disulfide bonds. These results are in contrast to studies of the 5-HT$_4$ receptor which clearly demonstrate the presence of a disulfide linkage required for homodimerization (Berthouze et al. 2007). When intact HEK 293 cells stably expressing 5-HT$_{2C}$/YFP were pretreated for 30 min with 5 mM BS^3 (a membrane-impermeable cross-linker), the higher molecular weight immunoreactive species was the predominant form observed on the blot (Fig. 7.1c). It is possible that the higher molecular weight bands observed on the western blots may represent 5-HT$_{2C}$ receptor homodimers or they may result from the copurification of accessory proteins along with 5-HT$_{2C}$ receptors, or both. Coimmunoprecipitation studies with differentially tagged 5-HT$_{2C}$ receptors (5-HT$_{2C}$/HA and 5-HT$_{2C}$/YFP) revealed the presence of immunoreactive bands the predicted size of 5-HT$_{2C}$ receptor homodimers only in membrane fractions prepared from cotransfected cells and not in samples prepared from singly transfected cells mixed posttransfection (Herrick-Davis et al. 2004).

Based on the results of the immunoprecipitation and Western blot studies, bioluminescence resonance energy transfer (BRET) experiments were performed to determine if similar results could be obtained in intact living cells. BRET is a proximity-based assay that determines if two proteins are close enough to associate with one another (Xu et al. 1999). When a bioluminescent donor protein and a fluorescent acceptor protein (with overlapping emission and excitation spectra) are within 1–10 nm of each other and their dipoles are aligned, resonance energy will be transferred from the donor to the acceptor, resulting in acceptor emission. BRET is measured as a ratio of donor and acceptor emissions (Angers et al. 2000). For BRET studies of the 5-HT$_{2C}$ receptor, Renilla luciferase (Rluc) expressed as a fusion protein on the receptor C-terminus was used as the donor and YFP served as the acceptor (Herrick-Davis et al. 2004). Human embryonic kidney (HEK) 293 cells transfected with the donor/acceptor pairs, as indicated in Fig. 7.2, were resuspended and placed in a cuvette along with 5-uM coelenterazine f. In the presence of Rluc and oxygen, coelenterazine f is converted to coelenteramide, resulting in the appearance of bioluminescence emission spectra with a λ_{max} of 475 nm. When HEK 293 cells are cotransfected with 5-HT$_{2C}$/Rluc and 5-HT$_{2C}$/YFP, and coelenterazine f is added, the characteristic emission spectra of YFP (λ_{max} 525 nm) appears, as an indicator of BRET between 5-HT$_{2C}$/Rluc and 5-HT$_{2C}$/YFP (Fig. 7.2a). Control experiments were performed using Rluc and YFP fusion proteins made with M$_4$-muscarinic and β$_2$-adrenergic receptors. Coexpression of 5-HT$_{2C}$/Rluc and 5-HT$_{2C}$/YFP produced a significant increase in the BRET ratio over cells transfected with 5-HT$_{2C}$/Rluc and pcDNA$_3$ vector, 5-HT$_{2C}$/Rluc and M$_4$-muscarinic/YFP, or β$_2$-adrenergic/Rluc and 5-HT$_{2C}$/YFP (Fig. 7.2b). BRET ratios obtained for cells cotransfected with 5-HT$_{2C}$/Rluc and 5-HT$_{2C}$/YFP were similar to BRET ratios measured in cells coexpressing β$_2$-adrenergic/Rluc and β$_2$-adrenergic/YFP. Pretreatment with 1 uM 5-HT did not produce a significant change in the BRET ratio.

Fluorescence resonance energy transfer (FRET) combined with confocal microscopy was performed to determine if 5-HT$_{2C}$ receptor homodimers could be visualized on the plasma membrane of living cells (Herrick-Davis et al. 2004, 2007). FRET operates on the same principle as BRET. FRET occurs when a fluorescent donor (CFP) is excited with an external laser and the resonance energy of the donor is transferred to an acceptor (YFP), resulting in excitation of the acceptor and a quenching of donor fluorescence. When the acceptor is removed by photobleaching (acceptor photobleaching), the donor is dequenched and an increase in donor fluorescence is observed (Bastiaens et al. 1996). A Zeiss LSM 510-Meta confocal imaging system was used to measure FRET between 5-HT$_{2C}$/CFP (donor) and 5-HT$_{2C}$/YFP (acceptor) on the plasma membrane of live HEK 293 cells (Fig. 7.3). Online fingerprinting and linear unmixing of CFP and YFP emission spectra were performed using the Zeiss LSM-510 Meta detector and AIM Software. Linear unmixing of CFP and YFP emission spectra eliminates the need for filters

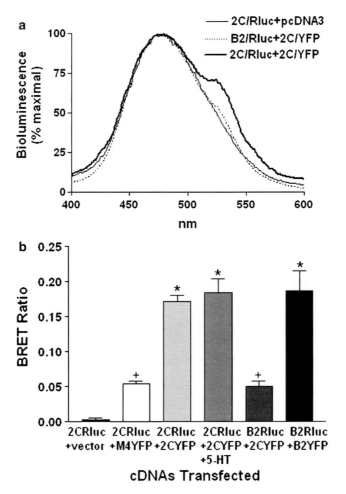

Fig. 7.2 5-HT$_{2C}$ receptor BRET. HEK 293 cells were transfected with cDNAs encoding 5-HT$_{2C}$ (2C), beta$_2$-adrenergic (β2), or M$_4$-muscarinic (M4) receptors tagged with renilla luciferase (Rluc) or YFP as indicated. (**a**) Emission spectra were collected using a Perkin Elmer LS-50B lumines- cence spectrophotometer following the addition of coelenterazine f. When BRET occurs, the YFP emission spectra (λ_{max} =525 nm) appear in addition to the Rluc emission spectra (λ_{max} =475 nm). (**b**) BRET ratios were calculated from the emission spectra as previously described (Angers et al. 2000). + 5-HT=cells pretreated with 10 uM 5-HT for 10 min. Data represent the mean±SEM of 4–6 experiments. + $p < 0.01$ versus 2C/Rluc+pcDNA3; *$p < 0.01$ versus 2C/Rluc+M4/YFP and B2/Rluc+2C/YFP (Reproduced from Herrick-Davis et al. 2004. With permission from ACS)

to separate CFP and YFP emission spectra and eliminates spectral cross-talk and bleed through that contribute to false FRET signals. Figure 7.3 illustrates how the acceptor photobleaching experiments were performed. The plasma membrane of live HEK 293 cells coexpressing 5-HT$_{2C}$/CFP and 5-HT$_{2C}$/YFP was visualized by confocal microscopy following excitation at 458 nm. The CFP and YFP emission spectra were separated, by linear unmixing, into two separate channels to generate

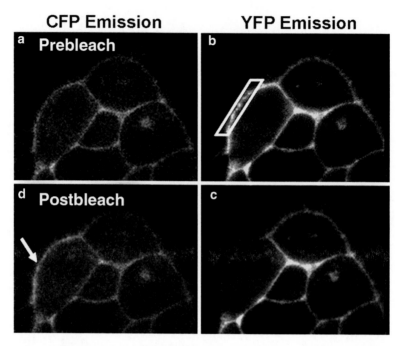

Fig. 7.3 Acceptor photobleaching FRET. Fluorescence confocal microscopy was used to visualize a 2-μm-thick optical cross-section of live HEK 293 cells coexpressing 5-HT$_{2C}$/CFP (donor) and 5-HT$_{2C}$/YFP (acceptor) on the plasma membrane. Prebleach CFP (**a**) and YFP (**b**) images were captured simultaneously following excitation at 458 nm and linear unmixing of their emission spectra. A region of plasma membrane (white rectangle) was photobleached at 514 nm for 10 s. Postbleach YFP (**c**) and CFP (**d**) images were captured simultaneously following excitation at 458 nm. Selective photobleaching of plasma membrane YFP fluorescence within the targeted region (**c**). Following YFP photobleaching, FRET is visualized as an increase in CFP fluorescence, as indicated by the arrow (**d**) (Adapted from Herrick-Davis et al. 2007. With permission from Elsevier)

the prebleach images shown in Fig. 7.3a and b. A region of plasma membrane selected for FRET measurement (indicated by the rectangle in Fig. 7.3b) was irradiated with the 514-nm laser for 10 s, and postbleach images were captured. Photobleaching of YFP fluorescence with the 514-nm laser resulted in a 95% decrease in YFP emission (Fig. 7.3c) and an increase in CFP emission (marked by the white arrow in Fig. 7.3d) corresponding to a FRET efficiency of 40%.

FRET measurements were made on select regions of plasma membrane in live cells coexpressing 5-HT$_{2C}$/CFP and 5-HT$_{2C}$/YFP using the acceptor photobleaching method depicted in Fig. 7.3. FRET efficiency was dependent on the donor-to-acceptor ratio (Fig. 7.4a) and independent of receptor expression level, as measured by YFP (acceptor) fluorescence (Fig. 7.4b). To determine if receptor movement within the plasma membrane was contributing to a false positive FRET signal, donor fluorescence was measured in cells transfected with 5-HT$_{2C}$/CFP prior to and after photobleaching. The average FRET efficiency in the singly transfected cells was

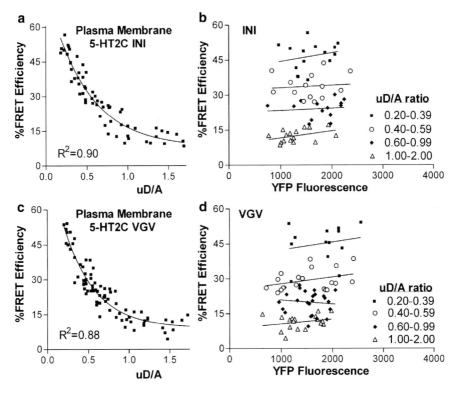

Fig. 7.4 FRET was measured on the plasma membrane of live HEK 293 cells expressing the INI or VGV isoform of CFP- and YFP-tagged 5-HT$_{2C}$ receptors. FRET efficiency was calculated as $100 \times$ [(CFP postbleach − CFP prebleach)/CFP postbleach] using the FRET macro in the Zeiss Aim Software. (**a**) and (**c**) FRET efficiency plotted versus uD/A ratio (postbleach CFP fluorescence/prebleach YFP fluorescence) for INI ($n = 60$ cells) and VGV ($N = 80$ cells). Nonlinear regression analyses (one-phase exponential decay) were performed using GraphPad Prism software. (**b**) and (**d**) FRET efficiency plotted versus YFP fluorescence by uD/A ratio for the same cells as shown in (**a**) and (**c**). Linear regression analyses performed using GraphPad Prism software yielded slopes that did not significantly differ from zero for all uD/A ratios examined (Images adapted from Herrick-Davis et al. 2007. With permission from Elsevier)

$0.3 \pm 1.8\%$. Similar results were obtained when FRET was measured in nonbleached regions of cell membrane. These results indicate that receptor migration within the region of interest (during the short photobleaching period) did not contribute significantly to the average FRET efficiency.

Many studies have used FRET as an indicator of GPCR dimerization (Rocheville et al. 2000; Cornea et al. 2001; McVey et al. 2001; Dinger et al. 2003; Canals et al. 2004). However, relatively few studies have combined FRET with live-cell confocal microscopy to demonstrate the cellular localization of the FRET signal. It is very important to demonstrate the specificity of the FRET signal. Two different patterns of receptor distribution on the plasma membrane can give rise to a positive FRET signal. A positive FRET signal can result from specific protein–protein interactions, such as dimer/oligomer formation (receptor clustering), or from protein overexpression resulting in high levels of donor and acceptor in close enough proximity to produce FRET because they are tightly packed in a small region of membrane (random proximity effect). The random proximity and clustering models have been tested experimentally in FRET studies using the membrane-anchored protein 5′ nucleotidase and the IgA receptor (Kenworthy and Edidin 1998; Wallrabe et al. 2003). These studies have suggested that FRET resulting from random proximity of donor and acceptor is dependent on the amount of acceptor expressed on the plasma membrane. However, FRET resulting from clustered proteins, such as dimers/oligomers, should be independent of acceptor expression levels and dependent on the ratio of donor to acceptor expressed on the cell membrane. In the 5-HT$_{2C}$ receptor FRET studies, FRET efficiency was independent of acceptor fluorescence and dependent on the donor-to-acceptor ratio (Fig. 7.4). These results indicate that the FRET signal results from the close proximity of receptor proteins in a clustered, nonrandom, distribution on the plasma membrane.

The results of the FRET studies investigating 5-HT$_{2C}$ receptor homodimerization demonstrate the dependence of FRET efficiency on the donor-to-acceptor ratio (Herrick-Davis et al. 2004, 2007). Therefore, the amount of FRET produced in a single cell will vary depending on the ratio of donor to acceptor expressed in that cell. This is an important consideration when interpreting the results of BRET and FRET experiments designed to determine the effect of ligand on GPCR dimer/oligomer formation. Experimental variables such as differences in donor-to-acceptor ratios between treatment groups, as well as differences in receptor expression levels, could contribute to the wide range of effects reported in the literature for agonist treatment on receptor dimerization measured by BRET and FRET. Our results highlight the importance of measuring the donor-to-acceptor ratio in the transfected cells and using cells with similar donor-to-acceptor ratios when BRET or FRET efficiencies are to be compared.

Many different classes of drugs have high affinity for 5-HT$_{2C}$ receptors, including hallucinogens, antipsychotics, antidepressants, anxiolytics and anorectic agents. Thus the 5-HT$_{2C}$ receptor has been identified as a potential target for drugs used to treat anxiety, depression, schizophrenia, and obesity(Hoyer et al. 2002; Herrick-Davis et al. 2000; Jones and Blackburn 2002; Roth et al. 2004; Li et al. 2005; Miller 2005). Therefore, experiments were designed to investigate the effect of drug treatment on the homodimer status of the 5-HT$_{2C}$ receptor. For these experiments, the homodimer status of two naturally occurring isoforms of the 5-HT$_{2C}$ receptor were compared: one that has no basal activity and the other which is constitutively active with respect to Gα_q signaling. RNA editing occurs in the second intracellular loop of the 5-HT$_{2C}$ receptor, changing amino acids 156, 158, and 160 from INI to VGV in the fully edited isoform (Burns et al. 1997) and results in the expression of 14 different 5-HT$_{2C}$ receptor isoforms in the human brain (Niswender et al. 1999; Fitzgerald et al. 1999) with varying levels of constitutive activity (Herrick-Davis et al. 1999). The unedited INI isoform has high basal activity, while the fully edited VGV isoform has very low basal activity (Herrick-Davis et al. 1999). Therefore, FRET experiments were performed in the absence and presence of agonist and inverse agonist to determine the relationship between the active and inactive conformations of the 5-HT$_{2C}$ receptor with respect to homodimerization.

Live-cell acceptor photobleaching FRET experiments were performed on cells coexpressing CFP- and YFP-tagged INI and VGV isoforms of the 5-HT$_{2C}$ receptor (Fig. 7.4a–d). FRET efficiencies measured within discrete regions of plasma membrane were plotted versus the uD/A ratio (postbleach CFP fluorescence divided by prebleach YFP fluorescence). Nonlinear regression analyses produced correlation coefficients of 0.90 and 0.88 for INI and VGV isoforms, respectively. When FRET efficiency was plotted versus YFP or acceptor fluorescence (by uD/A ratio) linear regression analyses revealed no correlation between the amount of FRET observed and the amount of YFP-tagged receptor expressed on the plasma membrane. In order to make a meaningful comparison of FRET efficiencies between the INI and VGV isoforms, FRET efficiencies were divided into four separate groups based on their uD/A ratios. The mean FRET efficiencies were different for each uD/A ratio, again demonstrating the dependence of FRET efficiency on the uD/A ratio. However, the mean FRET efficiencies for INI and VGV were the same when compared within a given uD/A range.

The effect of agonist and inverse agonist treatment on 5-HT$_{2C}$ receptor homodimerization was investigated. The CFP/YFP ratio, used as an indicator of FRET, was measured on the plasma membrane of cells coexpressing CFP- and YFP-tagged VGV or INI 5-HT$_{2C}$ receptors in the absence and presence of 5-HT or clozapine (Herrick-Davis et al. 2007). CFP/YFP ratios (CFP fluorescence devided by YFP fluorescence) were measured on the plasma membrane of cells coexpressing VGV/CFP and VGV/YFP at 30-s intervals over a 2-min period, followed by the application of 5-HT. There was no change in the CFP/YFP ratio over a 15-min period following the addition of 1-uM 5-HT (Herrick-Davis et al. 2007). In a similar experiment, the effect of the inverse agonist clozapine on homodimerzation of the INI isoform was monitored. Again, there was no change in the CFP/YFP ratio during a 15-min period following the addition of 1-uM clozapine (Herrick-Davis et al. 2007). These results are consistent with the hypothesis that 5-HT$_{2C}$ receptors are expressed on the plasma membrane as homodimers regardless of whether they are in an inactive or active conformation and do not dissociate upon agonist or inverse agonist binding. The results are consistent with a model in which 5-HT$_{2C}$ receptor homodimer formation occurs prior to receptor expression on the plasma membrane and remains unaltered following ligand binding.

7.3   Homodimer Biogenesis

There are several lines of evidence suggesting that GPCR dimerization occurs prior to receptor expression on the plasma membrane. The most notable example involves heterodimerization of class C GABA$_B$R1 and GABA$_B$R2 receptors, which has been demonstrated to be essential for receptor trafficking from the ER to the plasma membrane (Jones et al. 1998; Kaupmann et al. 1998; White et al. 1998;

Margeta-Mitrovic et al. 2000). Dimerization has been proposed as a general mechanism necessary for proper trafficking of class A GPCR to the plasma membrane (Terrillon and Bouvier 2004). Experimental evidence in favor of this model includes the identification of constitutive homodimers on the plasma membrane (Herrick-Davis et al. 2004; McVey et al. 2001), the identification of nontrafficking, mutant receptors that dimerize with and retain their wild-type counterparts within intracellular compartments (Benkirane et al. 1997; Zhu and Wess 1998; Karpa et al. 2000; Shioda et al. 2001; O'Dowd et al. 2005) and the detection of positive BRET signals in both plasma membrane and endomembrane-enriched sub-cellular fractions prepared from HEK 293 cells expressing vasopressin or β_2-adrenergic receptors (Terrillon et al. 2003; Salahpour et al. 2004). While these results suggest that GPCR dimerization may occur within intracellular compartments, direct evidence demonstrating the formation of class A GPCR homodimers within the ER and Golgi of intact living cells was not provided by these studies.

To address the issue of 5-HT$_{2C}$ receptor homodimer biogenesis, confocal microscopy combined with FRET was used to measure protein–protein proximity within the ER and Golgi of living cells (Herrick-Davis et al. 2006). The confocal microscopy-based FRET approach allows visualization of specific intracellular compartments, without compromising cellular integrity, and allows real-time monitoring of GPCR dimerization in the ER and Golgi apparatus. To begin this study, live-cell confocal fluorescence imaging was first performed on HEK 293 cells expressing an ER-YFP marker (a mutant YFP protein with an ER targeting and retention sequence) and Golgi-YFP marker (YFP containing a Golgi targeting sequence) to establish direct visualization of ER and Golgi membranes. ER membranes appear as a diffuse reticular network spreading outward from the nucleus to the plasma membrane (Fig. 7.5a). In contrast, Golgi membranes are visualized as dense punctate areas of fluorescence in the perinuclear region (Fig. 7.5b). A membrane impermeable dye, DiI, labels the plasma membrane (Fig. 7.5c).

Real-time imaging of CFP- and YFP-tagged 5-HT$_{2C}$ receptors was performed 12–24 h posttransfection to visualize 5-HT$_{2C}$ receptors following biosynthesis in the ER, trafficking through the Golgi, and subsequent expression on the plasma membrane (Herrick-Davis et al. 2006). In cells cotransfected with 5-HT$_{2C}$/CFP and the ER-YFP marker, CFP and YFP were colocalized within the ER 12 h posttransfection. In cells cotransfected with 5-HT$_{2C}$/CFP and the Golgi-YFP marker, CFP and YFP were colocalized within the Golgi 16 h posttransfection. By 20 h posttransfection, 5-HT$_{2C}$/CFP fluorescence was on the plasma membrane, while the Golgi-YFP marker remained in the Golgi (Herrick-Davis et al. 2006). Based on these results, a similar time course was performed in cells cotransfected with 5-HT$_{2C}$/CFP and 5-HT$_{2C}$/YFP. The cells were imaged live using confocal fluorescence imaging at various time points posttransfection. Ten hours after the addition of transfection reagents to HEK 293 cells, 5-HT$_{2C}$/CFP and 5-HT$_{2C}$/YFP receptor fluorescence began to emerge. Twelve hours posttransfection, confocal fluorescence imaging revealed many cells with a diffuse reticular pattern of fluorescence, similar to pattern of fluorescence observed for the ER-YFP marker (Fig. 7.5d). Sixteen hours posttransfection, most of the cells displayed a more clustered and

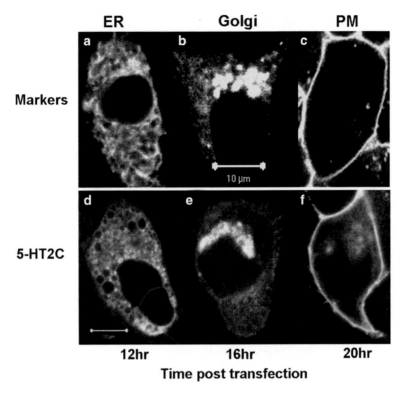

Fig. 7.5 Confocal microscopy of HEK 293 cells expressing markers for different cellular compartments (**a**)–(**c**) and at various time points posttransfection with 5-HT$_{2C}$ receptors (**d**)–(**f**). (**a**) HEK cell expressing the ER/YFP marker. (**b**) HEK cell expressing the Golgi/YFP marker. (**c**) DiI labeling of plasma membrane. (**d**) 5-HT$_{2C}$/YFP 12 h posttransfection. (**e**) 5-HT$_{2C}$/YFP 16 h posttransfection. (**f**) 5-HT$_{2C}$/YFP 20 h posttransfection. Scale bar = 10 μm

dense pattern of labeling with a distinct perinuclear distribution, in a manner very similar to the fluorescence from the Golgi-YFP marker (Fig. 7.5e). By 20 h posttransfection, most of the cells were expressing 5-HT$_{2C}$ receptors on the plasma membrane (Fig. 7.5f).

Confocal microscopy and acceptor photobleaching FRET were performed within discrete regions of the ER and Golgi, 12 and 16 h posttransfection, of HEK 293 cells coexpressing 5-HT$_{2C}$/CFP and 5-HT$_{2C}$/YFP (Herrick-Davis et al. 2006). Laser scanning confocal microscopy allows the photobleaching to be confined to very discrete intracellular regions. This minimizes the time required for irradiation of the acceptor fluorophore and makes the technique suitable for live-cell imaging. Acceptor photobleaching FRET experiments were performed in 50 living HEK 293 cells coexpressing 5-HT$_{2C}$/CFP and 5-HT$_{2C}$/YFP in the ER (12 h posttransfection) and in 50 cells coexpressing 5-HT$_{2C}$/CFP and 5-HT$_{2C}$/YFP in the Golgi (16 h posttransfection). The results are shown in Fig. 7.6a and b. A specific FRET signal resulting from proteins in a clustered distribution, such as dimers/oligomers, has

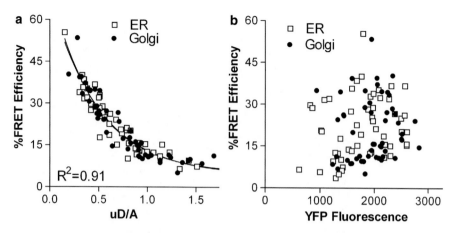

Fig. 7.6 Acceptor photobleaching FRET was measured in the ER and Golgi of live HEK 293 cells expressing CFP- and YFP-tagged 5-HT$_{2C}$ receptors. FRET efficiency was calculated as $100 \times [(CFP \text{ postbleach} - CFP \text{ prebleach})/CFP \text{ postbleach}]$ using Zeiss Aim Software. (**a**) FRET efficiency was plotted versus uD/A ratio (postbleach CFP fluorescence/prebleach YFP fluorescence) for cells coexpressing 5-HT$_{2C}$/CFP and 5-HT$_{2C}$/YFP in the ER ($n=50$) and in the Golgi ($n=50$). Nonlinear regression analysis (one-phase exponential decay, GraphPad Prism) yielded $R^2=0.91$. (**b**) FRET efficiency plotted versus YFP fluorescence for the same 100 cells as shown in Fig. 7.7a. Linear regression analysis (GraphPad Prism) revealed no correlation between FRET efficiency and YFP fluorescence levels ($r^2<0.1$).

been shown to be dependent on the ratio of donor to acceptor expressed in the cell, while FRET resulting from overexpression of randomly distributed proteins can be distinguished by a dependence on acceptor fluorescence or receptor expression level (Kenworthy and Edidin 1998; Wallrabe et al. 2003). Real-time FRET efficiencies measured for 5-HT$_{2C}$/CFP and 5-HT$_{2C}$/YFP fusion proteins in the ER and Golgi apparatus were dependent on the ratio of donor to acceptor (uD/A) expressed within a given cell (Fig. 7.6a) and independent of receptor expression level, measured as acceptor or YFP fluorescence (Fig. 7.6b). FRET efficiencies for a given uD/A ratio were similar for ER, Golgi, and plasma membrane. The mean FRET efficiency was 21.4% for ER, 20.3% for Golgi, and 21.5% for plasma membrane. These results suggest that 5-HT$_{2C}$ receptor homodimerization begins during receptor biosynthesis within the ER and is a naturally occurring step in 5-HT$_{2C}$ receptor maturation and processing (Herrick-Davis et al. 2006).

7.4 Functional Significance of 5-HT$_{2C}$ Receptor Dimerization: Determining the Ligand/Dimer/G-Protein Stoichiometry

Homodimerization and heterodimerization between different families of GPCR have been reported to regulate ligand binding, second messenger activation, and receptor trafficking (reviewed in Angers et al. 2002), (Javitch 2004), and (Milligan 2004).

While these studies suggest that dimerization may be a common property of GPCR, the specific mechanisms and functional consequences of dimerization may differ from one GPCR to another. For GABA$_B$ receptors the functional significance of dimerization is clear: heterodimerization of GABA$_B$R1 and GABA$_B$R2 is essential for proper plasma membrane targeting, ligand binding and signal transduction (Jones et al. 1998; Kaupmann et al. 1998; White et al. 1998; Margeta-Mitrovic et al. 2000). However, for many GPCR the functional significance of dimerization remains unknown. To address this issue, studies were performed to determine if dimerization plays a functional role in regulating the activity of 5-HT$_{2C}$ receptors. Radioligand binding, inositol phosphate (IP) signaling, confocal imaging and FRET were evaluated in HEK 293 cells coexpressing wild-type 5-HT$_{2C}$ receptors along with an inactive, mutant 5-HT$_{2C}$ receptor to determine the effect of dimerization on receptor function and to investigate the ligand/dimer/G-protein stoichiometry (Herrick-Davis et al. 2005). Mutation of serine 138 to arginine (S138R) in TMD III of the 5-HT$_{2C}$ receptor resulted in a loss of ligand binding and inositol phosphate (IP) production (Herrick-Davis et al. 2005). When HEK 293 cells were transfected with S138R, there was no detectable [^{3}H]mesulergine binding. In addition, there was no detectable basal or 5-HT-stimulated activation of IP production, even though wild-type receptors (VSV isoform) displayed moderate levels of both basal and 5-HT-stimulated IP production. These results indicate that S138R is devoid of both ligand binding and G-protein activation. GPCR activation has been reported to involve the coordinated movements of TMDs III and VI (Farrens et al. 1996; Gether et al. 1997). The S138R mutation may directly interfere with the ability of TMD III to adopt the proper conformation to achieve an active state of the receptor. This is supported by the observation that the S138R mutation eliminates receptor basal activity.

Fluorescence confocal microscopy was used to monitor cellular trafficking patterns of wild-type and mutant S138R 5-HT$_{2C}$ receptors in HEK 293 cells (Herrick-Davis et al. 2005). In cells expressing wild-type 5-HT$_{2C}$ receptors, significant constitutive receptor trafficking was observed between the plasma membrane and intracellular compartments. Within the first 3 min following the addition of 5-HT vesicular trafficking increased, followed by an overall shape change likely resulting from activation of 5-HT$_{2C}$ receptor activation of phospholipase D, Rho and Rac (McGrew et al. 2002). In contrast, in HEK 293 cells expressing S138R receptors there was no constitutive receptor trafficking, no receptor endocytosis in response to 5-HT, and no overall shape change during the 10-min observation period. In cells coexpressing wild-type and S138R receptors, constitutive and 5-HT-stimulated receptor trafficking was minimal. These results suggest that the ability of wild-type 5-HT$_{2C}$ receptors to respond to 5-HT stimulation is impaired when wild-type receptors are coexpressed with inactive S138R receptors (Herrick-Davis et al. 2005).

IP production was measured in cells coexpressing wild-type and S138R receptors to test the hypothesis that wild-type receptor function is impaired in the presence of S138R receptors. When wild-type 5-HT$_{2C}$ receptors were expressed in an S138R stable cell line, both basal and 5-HT-stimulated IP production were greatly reduced, with little change in 5-HT potency (Fig. 7.7a). Radioligand binding to

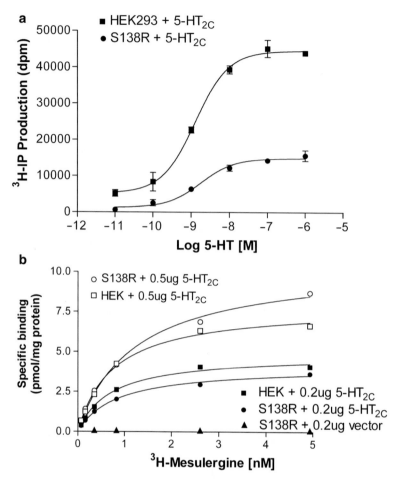

Fig. 7.7 (**a**) 5-HT dose-response curves for stimulation of ^{3}H-IP production were measured in HEK 293 cells and the S138R stable cell line following transfection with 0.2 ug of wild-type 5-HT$_{2C}$ plasmid DNA. Data represent the mean±SD of three experiments. (**b**) [^{3}H]Mesulergine saturation curves in HEK 293 cells and the S138R stable cell line transfected with 0.2 or 0.5 ug of wild-type 5-HT$_{2C}$ plasmid DNA or vector (pcDNA3) (Reprinted from Herrick-Davis et al. 2005. With permission from ASBMB)

wild-type 5-HT$_{2C}$ receptors was the same following expression in HEK 293 cells and in the S138R stable cell line with no change in kilodalton or Bmax (Fig. 7.7b), indicating that the decrease IP production was not due to decreased wild-type receptor expression or binding affinity. In addition, M$_1$-muscarinic receptors displayed normal IP signaling in the S138R cell line, demonstrating normal Gαq binding and signaling capabilities in the S138R stable cell line (Herrick-Davis et al. 2005).

Immunoprecipitation of S138R receptors following coexpression with wild-type HT$_{2C}$ receptors suggested that S138R may regulate the activity of wild-type receptors through a direct protein–protein interaction (Herrick-Davis et al. 2005). Therefore, live-cell acceptor photobleaching FRET experiments were performed on selected regions of plasma membrane in HEK 293 cells expressing CFP- and YFP-tagged receptors. Similar FRET efficiencies were obtained from cells expressing wild-type receptors and from cells coexpressing wild-type with S138R receptors (Herrick-Davis et al. 2005). FRET efficiency was dependent on the uD/A ratio and independent of receptor expression level. These results suggest that S138R receptors decrease the function of wild-type 5-HT$_{2C}$ receptors by forming inactive heterodimers that are expressed on the plasma membrane. As a result, the effective concentration of active, wild-type receptor homodimers is reduced, resulting in decreased IP signaling.

Taken together the results of these experiments indicate that the formation of a 5-HT$_{2C}$ heterodimer in which one protomer of the dimer is incapable of G-protein activation results in the silencing of G-protein signaling by the heterodimer. These results are consistent with a model of 5-HT$_{2C}$ receptor-mediated G-protein activation that requires the formation of a receptor dimer followed by the binding to and activation of a single G protein. The observation that ligand binding affinity and Bmax were unaltered following expression of wild-type receptors in the S138R cell line indicates that the formation of the wild-type/S138R heterodimer did not prevent the binding of ligand to the wild-type protomer of the heterodimer, clearly indicating that the heterodimer contains two separate ligand binding pockets. Thus the ligand/dimer/G-protein stoichiometry appears to be 2:1:1, consistent with a model in which one receptor dimer binds two molecules of ligand and one G protein (Herrick-Davis et al. 2005).

Dimer/oligomer formation may be a prerequisite for normal receptor trafficking and expression on the plasma membrane, as it may be necessary for passing ER quality control checkpoints that determine functionality. It is also possible that dimerization in the ER may be a prerequisite for trafficking to the plasma membrane as dimers may represent the basic metabotropic signaling unit. Studies involving the rhodopsin receptor support the hypothesis that the dimer may represent the basic signaling unit. Atomic force microscopy has been used to visualize rhodopsin receptors in native mammalian membranes as rows of dimeric complexes (Liang et al. 2003). In addition, the distance between the α- and γ-subunits of a single heterotrimeric G protein, which are the reported regions of contact with GPCR, is predicted to be too large to accommodate a single rhodopsin receptor (Hamm 2001). Studies using chemical cross-linking and purified leukotriene B4 receptors (LTB4) have demonstrated that an LTB4 homodimer forms a pentameric complex with a single heterotrimeric G protein (Baneres and Parello 2003). In addition, the D1–D2 dopamine receptor heterodimer has been reported to form a novel signaling complex in which ligand binding to both protomers results in Gα_q activation, but blockade of either protomer alone is sufficient to block signaling (O'Dowd et al. 2005). The results of these experiments are consistent with a model in which class A GPCR dimers interact with a single G protein and suggest that the dimer may represent the basic signaling unit.

7.5 Homodimer Interface

While GPCR are generally conceptualized as forming dimeric/oligomeric complexes, very little is understood about the molecular interactions between dimers and oligomers and how these interactions regulate receptor function. Several different techniques have been used to explore the GPCR dimer interface, including peptides derived from various transmembrane domains (TMDs), immunoprecipitation, Western blot, cysteine cross-linking, mutagenesis, FRET, and molecular modeling. For class A GPCR, potential dimer interfaces have been reported to include the N-terminus of δ-opiod and yeast α-factor receptors (Cvejic and Devi 1997; Overton and Blumer 2002), TMD 1 of V2-vasopressin, C5a, α_{1b}-adrenergic and rhodopsin receptors (Zhu and Wess 1998; Klco et al. 2003; Stanasila et al. 2003; Carrillo et al. 2004; Fotiadis et al. 2006), TMD 4 of C5a, rhodopsin, α_{1b}-adrenergic and D_2-dopamine receptors (Liang et al. 2003; Klco et al. 2003; Carrillo et al. 2004; Guo et al. 2005), TMD5 of rhodopsin receptors (Fotiadis et al. 2006), TMD 6 of β_2-adrenergic and D_2-dopamine receptors (Hebert et al. 1996; Ng et al. 1996), TMD 7 of α_{1b}-adrenergic and D_2-dopamine receptors (Stanasila et al. 2003; Ng et al. 1996), and the C-terminus of opiod receptors (Fan et al. 2005). The results of these experiments could be interpreted in several different ways: GPCR have multiple dimer interfaces to allow for oligomerization; members of the class A GPCR family use different mechanisms to form dimeric/oligomeric structures; or differences in the methods used to explore the dimer interface may yield different results. Clearly, additional studies designed to explore the GPCR dimer interface are warranted.

With respect to 5-HT$_{2C}$ receptors, Hendrickson and colleagues used a cysteine cross-linking approach in an attempt to elucidate potential dimer interfaces responsible for the formation of 5-HT$_{2C}$ homodimers (Mancia et al. 2008). In these studies, homology modeling based on the crystalline structure of rhodopsin provided a 5-HT$_{2C}$ receptor model that was used to identify residues with appropriate surface exposure for cysteine cross-linking. Candidate residues were mutated to cysteine, and transfected HEK 293 cells were treated with an oxidizing agent (or exposed to air); the formation of disulfide-linked dimers was evaluated by Western blot of detergent-solubilized cell lysates. These experiments yielded potential dimer interfaces at the extracellular end of TMD I (N55 and W56) and between TMD IV (I193) and TMD V (P213, N214). The putative TMD IV–V interface was sensitive to changes in receptor conformation following drug treatment. This observation is consistent with previous studies involving leukotriene B$_4$, glutamate, and D_2-dopamine receptors reporting conformational changes at the dimer interface during receptor activation and inactivation (Guo et al. 2005; Mesnier and Baneres 2004; Goudet et al. 2005). Previous studies have suggested that GPCR may use dimer interfaces at both TMD I and TMD IV–V to form oligomeric complexes (Milligan 2004; Fotiadis et al. 2006). However, higher-order oligomers of 5-HT$_{2C}$ receptors were not observed following cysteine cross-linking in cells coexpressing TMD I and TMD IV–V cysteine mutant receptors (Mancia et al. 2008). Based on these results and on experiments using receptor-Galpha fusion proteins, it was concluded that 5-HT$_{2C}$ receptor dimers are quasisymmetrical at the TMD IV–V interface and asymmetrical with respect to G-protein coupling (Mancia et al. 2008).

The results of the 5-HT$_{2C}$ receptor cysteine cross-linking experiments are consistent with prior reports of TMD I and TMD IV–V interfaces in dimer/oligomer models generated from the crystalline structure of rhodopsin. In contrast, the recently solved crystalline structure of the beta$_2$-adrenergic receptor (β_2-AR) reveals a very different type of dimeric interface (Cherezov et al. 2007): an inter-receptor link between TMD1 and helix 8 (H8) of two adjacent receptors. To date, there have been no studies examining the validity of this model in vivo, and thus the functional significance remains unknown. A review of the literature concerning GPCR dimer interface studies reveals that TMD I and TMD IV–V are the most frequently cited regions for dimer/oligomer formation. Thus, it remains controversial as to whether there are two dimer interfaces, one involving TMD I and another involving TMD IV–V, leading to the formation of higher-order oligomeric structures or there is just a single dimer interface.

Studies using cross-linking reagents may be useful in distinguishing between these two different models. For example, Western blot experiments showed that the 5-HT$_{2C}$ receptor is sensitive to cross-linking with the membrane impermeable cross-linker BS[3] (Herrick-Davis et al. 2004). Treatment with BS[3] resulted in the appearance of immunoreactive bands the predicted size of dimers, but oligomers were not detected. There are only four lysine residues in the 5-HT$_{2C}$ receptor that are exposed to the extracellular environment such that they could participate in cross-linking following treatment of intact cells with BS[3]. There is one lysine residue in the N-terminus near the beginning of TMD I, one lysine residue in the extracellular loop (EL) II, and two lysine residues in the EL III. If a symmetrical or quasisymetrical TMD I dimer interface is responsible for the formation of 5-HT$_{2C}$ homodimers, then the lysine residue located in the N-terminus near the top of TMD I would likely be a key player in cross-linking with BS[3]. Therefore, a loss in cross-linking following mutation of this lysine residue would support a TMD I dimer interface model. On the other hand, a loss of cross-linking following mutation of the lysine residue in EL II (between TMD IV and V) would favor a model in which TMD IV–V were in close proximity to each other in the homodimer. Experimental design to minimize the potential for dimer capture would need to be considered, such as using very low receptor expression levels and very short cross-linking times.

7.6   Heterodimerization: Splice Variants and Isoforms

Splice variants, lacking the full complement of TMDs, have been described for many different GPCR. These truncated receptors have been reported to have a dominant negative effect on wild-type receptor function through dimerization, resulting in trapping of wild-type receptors in the ER (Benkirane et al. 1997; Karpa et al. 2000; Grosse et al. 1997). Therefore, heterodimerization between wild-type receptors and splice variants may play an important role in normal physiological regulation of receptor function and in pathological conditions.

RNA splicing results in the production of a truncated 5-HT$_{2C}$ receptor (5-HT$_{2Ctr}$) containing the N-terminus and TMDs I–III followed by 96 unique amino acids resulting from a frame shift and premature stop codon (Canton et al. 1996). In human brain tissue, 5-HT$_{2Ctr}$ mRNA is found in all regions containing the full-length mRNA for the 5-HT$_{2C}$ receptor, including hippocampus, hypothalamus, frontal cortex, striatum, and olfactory tubercle (Canton et al. 1996). The mRNA for 5-HT$_{2Ctr}$ represents 60% of the total 5-HT$_{2C}$ mRNA found in choroid plexus and 20–30% of that found in other neuronal tissues. Western blots of membrane extracts from transfected NIH3T3 cells revealed immunoreactive bands the predicted size of the truncated protein, indicating that mRNA encoding 5-HT$_{2Ctr}$ is translated into protein. Radioligand binding and inositol phosphate production were not observed in the transfected cells, indicating that 5-HT$_{2Ctr}$ is nonfunctional (Canton et al. 1996).

Fluorescence confocal microscopy was used to compare the trafficking patterns of YFP-tagged full-length and splice variant 5-HT$_{2C}$ receptors. The full-length receptor was expressed on the plasma membrane, while the splice variant was retained inside the cell and did not traffic to the plasma membrane (Fig. 7.8). Cotransfection of HEK 293 cells with 5-HT$_{2Ctr}$/CFP and ER-YFP or Golgi-YFP markers revealed that 5-HT$_{2Ctr}$ was retained in the ER and did not traffic to the Golgi. When coexpressed with 5-HT$_{2Ctr}$ the full-length 5-HT$_{2C}$ receptors were retained inside the cell and a decrease in cell surface 5-HT$_{2C}$ receptor expression level was observed in an intact cell radioligand binding assay (Fig. 7.9). In cells coexpressing 5-HT$_{2C}$/CFP and 5-HT$_{2Ctr}$/YFP the mean FRET efficiency was 24.6 ± 1.3% ($n = 35$). Negligible FRET was observed in 10 cells expressing 5-HT$_{2Ctr}$/CFP alone (−2.7 ± 1.2%) or in combination with the ER-YFP marker (4.2 ± 2.3%). These results are consistent with the hypothesis that 5-HT$_{2Ctr}$ can associate with and inhibit plasma membrane targeting of 5-HT$_{2C}$ receptors.

Alternative splicing of 5-HT$_{2A}$ and 5-HT$_6$ RNA results in the synthesis of proteins that are truncated following TMD III (5-HT$_{2Atr}$ and 5-HT$_{6tr}$), and these truncated receptors are nonfunctional (Olsen et al. 1999; Guest et al. 2000). In contrast to our studies with 5-HT$_{2Ctr}$, splice variants of 5-HT$_{2A}$ and 5-HT$_6$ receptors have been reported to be expressed on cell membranes and not to interfere with wildtype receptor function (Olsen et al. 1999; Guest et al. 2000). However, these conclusions were based on the results of Western blots, and confocal microscopy was not performed to identify the trafficking pattern of these splice variants or to confirm plasma membrane expression. While coexpression studies indicate that 5-HT$_{2Ctr}$ forms heterodimers with full-length 5-HT$_{2C}$ receptors within the ER, thereby inhibiting normal trafficking of 5-HT$_{2C}$ receptors to the plasma membrane, the functional significance of this observation remains unknown. Additional studies demonstrating a change in 5-HT$_{2C}$ receptor expression or function in primary cultures coexpressing native splice variant and full-length receptors will be required to address this issue. At present, there are no specific antibodies for 5-HT$_{2Ctr}$ to verify expression levels in primary cultures or to demonstrate coimmunoprecipitation of endogenously expressed native and truncated 5-HT$_{2C}$ receptors.

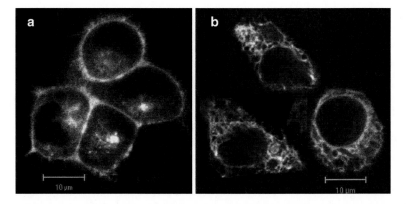

Fig. 7.8 Live-cell confocal imaging and FRET from HEK 293 cells expressing 5-HT$_{2C}$ and 5-HT$_{2Ctr}$. (**a**) 5-HT$_{2C}$/YFP expressed on the plasma membrane of transfected HEK 293 cells. (**b**) Intracellular 5-HT$_{2Ctr}$/YFP 24 h posttransfection. Note that the pattern of 5-HT$_{2Ctr}$ fluorescence is consistent with the pattern of fluorescence obtained with the ER/YFP marker (as shown in Fig. 7.5a)

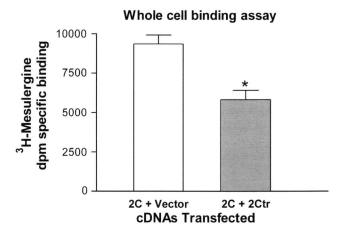

Fig. 7.9 Whole cell radioligand binding to intact HEK 293 cells following cotransfection with 5-HT$_{2C}$+pcDNA3 vector (control) or 5-HT$_{2C}$+5-HT$_{2Ctr}$. Intact cells were incubated with 2.5 nM [^{3}H]mesulergine for 30 min at 37 °C (±1 uM mianserin to define specific binding) and washed with PBS to remove unbound radioligand. Three percent TCA in PBS was added to the cells (20 min at 37°C) to release plasma membrane bound radioligand. The TCA solution was removed from the monolayer of cells, transferred to a scintillation vial containing Ecoscint Cocktail, and counted in a Beckman scintillation counter. Data represent the mean±SEM from five transfections. *$p < 0.01$

Another possible way that dimerization could regulate the functional properties of 5-HT$_{2C}$ receptors is through heterodimers of different 5-HT$_{2C}$ receptor isoforms with different functional characteristics. RNA editing occurs in the second intracellular loop of the 5-HT$_{2C}$ receptor, changing amino acids 156, 158, and 160 from INI to VGV in the fully edited isoform and results in the expression of 14 different

5-HT$_{2C}$ receptor isoforms in the human brain (Burns et al. 1997; Niswender et al. 1999; Fitzgerald et al. 1999) with varying levels of constitutive activity (Herrick-Davis et al. 1999). The unedited INI isoform has high basal activity, while the fully edited VGV isoform has little to no basal activity (Herrick-Davis et al. 1999). Furthermore, the VGV isoform differs in its binding affinity for several drugs including 5-HT and LDS (Burns et al. 1997; Fitzgerald et al. 1999). To determine if heterodimerization among 5-HT$_{2C}$ receptor isoforms is possible, a series of BRET measurements were made following coexpression of different receptor isoforms in HEK 293 cells (Fig. 7.10). Since it would not be possible to test all possible combinations of isoforms, BRET was measured between the most divergent and most similar isoforms. As expected from the results of our FRET experiments, the same BRET ratio was obtained for INI and VGV homodimers. Similar BRET ratios were produced following coexpression of INI and VGV, INI and VSV, and VSV with VGV (Fig. 7.10). These results suggest that there is no difference in the ability of the different isoforms to associate with one another.

Studies to address the functional significance of putative 5-HT$_{2C}$ isoform heterodimers have not been performed. Based on previous studies characterizing the ligand binding and constitutive signaling capabilities (functional properties) of the different isoforms, the two isoforms with the most divergent functional properties are the unedited (INI) and fully edited (VGV) isoforms (Burns et al. 1997; Niswender et al. 1999; Fitzgerald et al. 1999; Herrick-Davis et al. 1999). It is tempting to speculate that an INI-VGV heterodimer may have different functional

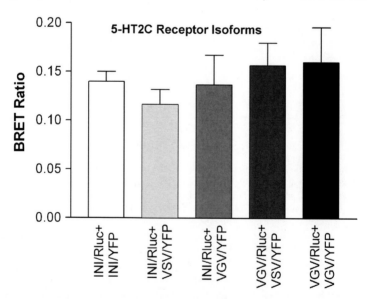

Fig. 7.10 5-HT$_{2C}$ receptor isoform BRET. HEK 293 cells were transfected with cDNAs encoding different 5-HT$_{2C}$ isoforms as indicated. Emission spectra were collected immediately following the addition of coelenterazine f. BRET ratios were calculated from the emission spectra as previously described (Herrick-Davis et al. 2004; Angers et al. 2000) (Data represent the mean±SEM from three experiments)

properties than either homodimeric species. However, it is not clear whether a single cell would have a significant portion of both unedited and fully edited isoforms to allow the formation of INI-VGV heterodimers. Studies from the original paper identifying 5-HT$_{2C}$ RNA editing reported the isoform distribution for several different brain regions (Burns et al. 1997). While multiple isoforms were identified in each brain region, it is noteworthy that the isoforms within an individual brain region tended to be similar in terms of their degree of RNA editing. For example, the majority of the isoforms identified in the choroid plexus were ones that were minimally edited and had the highest affinity for 5-HT and highest levels of constitutive activity, following expression in HEK 293 cells (Herrick-Davis et al. 1999). This observation suggests that it may be more likely to have heterodimers among isoforms with similar degrees of RNA editing. If this is the case, then heterodimers between VSV and VGV would be predicted to have the most dramatic change in functional characteristics since the VGV isoform has distinct differences in its ligand binding properties from the other isoforms, including VSV (Niswender et al. 1999; Fitzgerald et al. 1999). Of course, this remains speculative until it is demonstrated that native VSV and VGV isoforms exist as heterodimers in vivo.

7.7 Heterodimers: 5-HT$_{2C}$ and 5-HT$_{2A}$

To date, there are no published studies reporting the heterodimerization of 5-HT$_{2C}$ receptors with other members of the serotonin receptor family. Of the 13 different serotonin receptors identified in humans, the 5-HT$_{2A}$ receptor is structurally the most similar to the 5-HT$_{2C}$ receptor, and they are coexpressed in regions such as the hippocampus and cortex. Do 5-HT$_{2C}$ receptors form heterodimers with 5-HT$_{2A}$ receptors? Given the large body of evidence suggesting that heterodimers can have unique signaling properties, this is an appropriate question to ask. In preliminary studies, we have observed cotrafficking of CFP- and YFP-tagged 5-HT$_{2C}$ and 5-HT$_{2A}$ receptors in transfected HEK 293 cells. In addition, we have observed co-staining of 5-HT$_{2C}$ and 5-HT$_{2A}$ receptors within individual primary hippocampal neurons (unpublished observations). During these studies, we identified commercially available antibodies with good specificity for 5-HT$_{2C}$ (Santa Cruz, C-20) and 5-HT$_{2A}$ (Calbiochem, N22-41) receptors. When tested in 5-HT$_{2C}$ and 5-HT$_{2A}$ stable cells lines, no cross-reactivity was observed with either antibody. Coimmunoprecipitation of native 5-HT$_{2C}$ and 5-HT$_{2A}$ receptors from primary hippocampal cultures along with detailed ultra structural microscopy studies would provide a good starting point for exploring the possible existence of 5-HT$_{2C}$–5-HT$_{2A}$ heterodimers in vivo.

While many studies have reported the formation of GPCR heterodimers with unique ligand binding and signaling properties, very few studies have been able to demonstrate the existence of the putative heterodimers in vivo. Dimerization is believed to occur during receptor biosynthesis within the ER. Therefore, heterodimerization would require the coordinated synthesis of the two different receptor species. Thus, if heterodimerization of two different GPCR is to take place, then factors or signals that regulate or turn on the translation of each mRNA species must occur simultaneously, and protein synthesis of each protein species must occur within close enough proximity to allow the preferential formation of heterodimers over the formation of homodimers. A good example involves β_1- and β_2-adrenergic receptors that have been reported to form heterodimers in recombinant cells (Mercier et al. 2002). While the heterodimers display different functional properties in recombinant cells, near-field scanning microscopy studies reveal that native β_1- and β_2-adrenergic receptors in cardiomyocytes traffic in different membrane microdomains, inconsistent with the formation of heterodimers in vivo (Ianoul et al. 2005). Therefore, caution should be exercised in interpreting the physiological significance of heterodimerization studies performed in recombinant cells, even if there is evidence that both receptor species are coexpressed within a given cell type in vivo.

7.8 Conclusions

Studies using recombinant cell systems have provided evidence consistent with the hypothesis that 5-HT_{2C} receptors are expressed on the plasma membrane as homodimers (Herrick-Davis et al. 2004, 2005, 2006, 2007; Mancia et al. 2008). Homodimerization occurs regardless of whether the receptors are in an inactive or active conformation, and the homodimers do not dissociate upon agonist or inverse agonist binding (Herrick-Davis et al. 2007). 5-HT_{2C} receptor homodimerization begins during receptor biosynthesis within the ER and is a naturally occurring step in 5-HT_{2C} receptor maturation and processing (Herrick-Davis et al. 2006). Homodimerization may be a prerequisite for normal receptor trafficking and expression on the plasma membrane, as it may be necessary for passing ER quality-control checkpoints that determine functionality. It is also possible that dimerization in the ER may be a prerequisite for trafficking to the plasma membrane as dimers may represent the basic metabotropic signaling unit. The results or our experiments with the 5-HT_{2C} receptor suggest a ligand/dimer/G-protein stoichiometry of 2:1:1, consistent with a model in which one receptor dimer binds two molecules of ligand and one G protein (Herrick-Davis et al. 2005).

A review of the literature concerning the GPCR dimer interface indicates that TMD I and TMD IV–V are among the most frequently cited regions for class A GPCR dimer/oligomer formation. Consistent with these reports, cysteine cross-linking experiments involving the 5-HT_{2C} receptor reported the formation of disulfide bonds between residues in TMD I and between residues located in TMD IV and V (Mancia et al. 2008). Previous studies have suggested that GPCR may use dimer interfaces at both TMD I and TMD IV–V to form oligomeric complexes (Milligan 2004; Klco et al. 2003; Fotiadis et al. 2006). However, higher-order oligomers of 5-HT$_{2C}$ receptors were not observed following cysteine cross-linking (Mancia et al. 2008). Based on these results and on experiments using receptor-Galpha fusion proteins, it was concluded that 5-HT$_{2C}$ receptor dimers are quasisymmetrical at the TMD IV–V interface and asymmetrical with respect to G-protein coupling (Mancia et al. 2008). The cysteine cross-linking study revealed different receptor conformations at the putative TMD IV–V dimer interface for the agonist and antagonist bound states of the 5-HT$_{2C}$ receptor (Mancia et al. 2008). These results suggest that if 5-HT$_{2C}$ receptors do in fact use a TMD IV–V interface for homodimerization, then it is likely that conformational changes in one pro-tomer of the dimer could be translated across the dimer interface to the adjacent protomer.

Studies demonstrating that native GPCR form dimers/oligomers in vivo are lacking. Studies have reported dimeric/oligomeric D$_2$-dopamine receptors solubilized from brain membranes (Zawarynski et al. 1998), and coimmunoprecipi-tation of native mu- and delta-opiod receptors solubilized from spinal cord (Waldhoer et al. 2005). For 5-HT$_{2C}$ receptors, we have detected the presence of immunoreactive bands the predicted size of receptor homodimers on Western blots of solubilized membrane proteins prepared from freshly isolated choroid plexus tissue. To date, the most compelling evidence for native GPCR dimer/oligomer formation in intact tissue comes from atomic force microscopy studies of rhodopsin receptors in native disk membranes of mouse retina (Liang et al. 2003). While this study showed images of mouse retinal rhodopsin receptors organized in rows of dimmers, the structural organization of native GPCR in neuronal membranes may likely differ from that of rhodopsin receptors in the retina.

Future studies should aim toward studying dimer/oligomer formation of native receptors in live primary neuronal cultures. This will require the development of new tools and techniques specifically designed for labeling and monitoring native receptors in their natural cellular environment. For example, monoclonal antibodies (directed toward extracellular domains) that recognize the native receptor conformation would be valuable tools. If such antibodies were available, then fluorescent-labeled Fab fragments could be generated and used to directly label native receptors in live primary cultures for analysis by fluorescence correlation spectroscopy (FCS). FCS measures fluctuations in the fluorescence intensity of fluorescent molecules diffusing through a very small confocal volume, providing near single molecule detection sensitivity (Briddon and Hill 2008). FCS has been used to monitor ligand–receptor interactions and to identify receptor dimers/oligomers in recombinant cell systems transfected with cDNAs encoding fluorescent fusion proteins. The next step would be to perform FCS experiments using fluorescent Fab fragments to label native recep-tors. Photon-counting histogram analysis of the FCS data could be used to pro-vide information about the number of fluorescent molecules traveling together within a protein complex, such as dimers or tetramers/oligomers (Chen and Muller 2007). Novel techniques and approaches, as described above, will be essential for elucidating the true dimeric/oligomeric structure of native GPCR in their natural physiological environment.

References

Angers S, Salahpour A, Joly E, et al (2000) Detection of beta$_2$-adrenergic receptor dimerization in living cells using bioluminescence resonance energy transfer (BRET). Proc Natl Acad Sci 97:3684–3689.

Angers S, Salahpour A, Bouvier M (2002) Dimerization: an emerging concept for G protein-coupled receptor ontogeny and function. Annu Rev Pharmacol Toxicol 42:409–435.

Backstrom JR, Westphal RS, Canton H, et al (1995) Identification of rat serotonin 5-HT2C receptors as glycoproteins containing N-linked oligosaccharides. Mol Brain Res 33:311–318.

Baneres JL, Parello J (2003) Structure-based analysis of GPCR function: evidence for a novel pentameric assembly between the dimeric leukotriene B4 receptor BLT1 and the G-protein. J Mol Biol 329:815–829.

Bastiaens PI, Majoul IV, Verveer PJ, et al (1996) Imaging the intracellular trafficking and state of the AB5 quaternary structure of cholera toxin. EMBO J 15:4246–4253.

Benkirane M, Jin DY, Chun RF, et al (1997) Mechanism of transdominant inhibition of CCR5-mediated HIV-1 infection by ccr5delta32. J Biol Chem 272:30603–30606.

Berthouze M, Rivail L, Lucas A, et al (2007) Two transmembrane Cys residues are involved in 5-HT4 receptor dimerization. Biochem Biophys Res Commun 11:642–647.

Briddon SJ, Hill SJ (2008) Pharmacology under the microscope: the use of fluorescence correlation spectroscopy to determine properties of ligand-receptor complexes. Trends Pharmacol Sci 28:637–645.

Burns CM, Chu H, Rueter SM, et al (1997) Regulation of serotonin-2C receptor G-protein coupling by RNA editing. Nature 387:303–308.

Canals M, Burgueño J, Marcellino D, et al (2004) Homodimerization of adenosine A2A receptors: qualitative and quantitative assessment by fluorescence and bioluminescence energy transfer. J Neurochem 88:726–734.

Canton H, Emeson RB, Barker EL, et al (1996) Identification, molecular cloning, and distribution of a short variant of the 5-Hydroxytryptamine$_{2C}$ receptor produced by alternative splicing. Mol Pharmacol 50:799–807.

Carrillo JJ, Lopez-Gimenez JF, Milligan G (2004) Multiple interactions between transmembrane helices generate the oligomeric α_{1b}-adrenoceptor. Mol Pharmacol 66:1123–1137.

Chen J, Muller JD (2007) Determining the stoichiometry of protein heterocomplexes in living cells with fluorescence fluctuation spectroscopy. PNAS 104:3147–3152.

Cherezov V, Rosenbaum EM, Hanson MA (2007) High resolution crystal structure of an engineered human beta$_2$-adrenergic C protein-coupled receptor. Science 318:1258–1265.

Cornea A, Janovick JA, Maya-Núñez G, et al (2001) Gonadotropin-releasing hormone receptor microaggregation. Rate monitored by fluorescence resonance energy transfer. J Biol Chem 276:2153–2158.

Cvejic S, Devi LA (1997) Dimerization of the delta opioid receptor: implication for a role in receptor internalization. J Biol Chem 272:26959–26964.

Dinger MC, Bader JE, Kobor AD, et al (2003) Homodimerization of neuropeptide y receptors investigated by fluorescence resonance energy transfer in living cells. J Biol Chem 278:10562–10571.

Fan T, Varghese G, Nguyen T., et al (2005) A role for the distal carboxy tails in generating the novel pharmacology and G-protein activation profile of Mu and delta opiod receptor heterooligomers. J Biol Chem 280:38478–38488.

Farrens DL, Altenbach C, Yang K, et al (1996) Requirement of rigid-body motion of transmembrane helices for light activation of rhodopsin. Science 274:768–770.

Fitzgerald LW, Iyer G, Conklin DS, et al (1999) Messenger RNA editing of the human serotonin 5-HT2C receptor. Neuropsychopharmacology 21:82s–90s.

Fotiadis DM, Jastrzebska B, Philippsen A, et al (2006) Structure of the rhodopsin dimer: a working model for G-protein-coupled receptors. Curr Opin Struct Biol 16:252–259.

George SR, O'Dowd BF, Lee S (2002) G-protein-coupled receptor oligomerization and its potential for drug discovery. Nat Rev Drug Discov 1:808–820.

Gether U, Lin S, Ghanouni P, et al (1997) Agonists induce conformational changes in transmembrane domains III and VI of the beta2 adrenoceptor. EMBO J 16:6737–6747.

Goudet C, Kniazeff J, Hlavackova V, et al (2005) Asymmetric functioning of dimeric metabotropic glutamate receptors disclosed by positive allosteric modulators. J Biol Chem 280:24380–24385.

Grosse R, Schoneberg T, Schultz G, et al (1997) Inhibition of gonadotropin-releasing hormone receptor signaling by expression of a splice variant of the human receptor. Mol Endocrinol 11:1305–1318.

Guest PC, Salim K, Skynner HA, et al (2000) Identification and characterization of a truncated variant of the 5-hydroxytryptamine$_{2A}$ receptor produced by alternative splicing. Brain Res 876:238–244.

Guo W, Shi L, Filizola M, et al (2005) Crosstalk in G protein-coupled receptors: changes at the transmembrane homodimer interface determine activation. Proc Natl Acad Sci 102:17495–17500.

Hamm HE (2001) How activated receptors couple to G proteins. Proc Natl Acad Sci 98:4819–4821.

Hebert T E, Moffett S, Morello J P, et al (1996) A peptide derived from a beta2-adrenergic receptor transmembrane domain inhibits both receptor dimerization and activation, J Biol Chem 271:16384–16392.

Herrick-Davis K, Grinde E, Niswender CM (1999) Serotonin 5-HT$_{2C}$ receptor RNA editing alters receptor basal activity: implications for serotonergic signal transduction. J Neurochem 73:1711–1717.

Herrick-Davis K, Grinde E, Teitler M (2000) Inverse agonist activity of atypical antipsychotic drugs at human 5-hydroxytryptamine2C receptors. J Pharmacol Exp Ther 295:226–232.

Herrick-Davis K, Grinde E, Mazurkiewicz JE (2004) Biochemical and biophysical characterization of serotonin 5-HT$_{2C}$ receptor homodimers on the plasma membrane of living cells. Biochemistry 43:13963–13971.

Herrick-Davis K, Grinde E, Harrigan TJ, et al (2005) Inhibition of serotonin 5-hydroxytryptamine2C receptor function through heterodimerization: receptor dimers bind two molecules of ligand and one G-protein. J Biol Chem 280:40144–40151.

Herrick-Davis K, Weaver B, Grinde E, et al (2006) Serotonin 5-HT$_{2C}$ receptor homodimer biogenesis in the endoplasmic reticulum: Real-time visualization with confocal fluorescence resonance energy transfer. J Biol Chem 281:27109–27116.

Herrick-Davis K, Weaver BA, Grinde E (2007) Serotonin 5-HT(2C) receptor homodimerization is not regulated by agonist or inverse agonist treatment. Eur J Pharmacol 568:45–53.

Hoyer D, Hannon JP, Martin GR (2002) Molecular pharmacological and functional diversity of 5-HT receptors. Pharmacol Biochem Behav 71:533–554.

Ianoul A, Grant DD, Rouleau Y, et al (2005) Imaging nanometer domains of beta-adrenergic receptor complexes on the surface of cardiac myocytes. Nat Chem Biol 1:196–202.

Javitch JA (2004) The ants go marching two by two: oligomeric structure of G-protein-coupled receptors. Mol Pharmacol 66:1077–1082.

Jones BJ, Blackburn TP (2002) The medical benefit of 5-HT research. Pharmacol Biochem Behav 71:555–568.

Jones KA, Borowsky B, Tamm, JA, et al (1998) GABA(B) receptors function as a heteromeric assembly of the subunits GABA(B)R1 and GABA(B)R2. Nature 396:674–679.

Karpa KD, Lin R, Kabbani N, et al (2000) The dopamine D3 receptor interacts with itself and the truncated D3 splice variant d3nf: D3-D3nf interaction causes mislocalization of D3 receptors. Mol Pharmacol 58:677–683.

Kaupmann K, Malitschek B, Schuler V, et al (1998) GABA(B)-receptor subtypes assemble into functional heteromeric complexes. Nature 396:683–687.

Kenworthy AK, Edidin M (1998) Distribution of a glycosylphosphatidylinositol-anchored protein at the apical surface of MDCK cells examined at a resolution of <100 A using imaging fluorescence resonance energy transfer. J Cell Biol 142:69–84.

Kerppola TK (2008) Bimolecular fluorescence complementation (BiFC) analysis as a probe of protein interaction in living cells. Annu Rev Biophys 37:465–487.

Klco JM, Lassere TB, Baranski TJ (2003) C5a receptor oligomerization. Disulfide trapping reveals oligomers and potential contact surfaces in a G-protein-coupled receptor. J Biol Chem 278:35345–35353.

Li Z, Ichikawa J, Huang M, et al (2005) ACP–103, a 5-HT2A/2C inverse agonist, potentiates haloperidol-induced dopamine release in rat medial prefrontal cortex and nucleus accumbens. Psychopharmacology 183:144–153.

Liang Y, Fotiadis D, Filipek S, et al (2003) Organization of the G protein-coupled receptors rhodopsin and opsin in native membranes. J Biol Chem 278:21655–21662.

Mancia F, Assur Z, Herman AG, et al (2008) Ligand sensitivity in dimeric associations of the serotonin 5HT2c receptor. EMBO Rep 9:363–369.

Margeta-Mitrovic M, Jan YN, Jan LY (2000) A trafficking checkpoint controls GABA(B) receptor heterodimerization. Neuron 27:97–106.

McGrew L, Chang MS, Sanders-Bush E (2002) Phospholipase D activation by endogenous 5-hydroxytryptamine 2C receptors is mediated by Galpha13 and pertussis toxin-insensitive Gbetagamma subunits. Mol Pharmacol 62:1339–1343.

McVey M, Ramsay D, Kellett E, et al (2001) Monitoring receptor oligomerization using time-resolved fluorescence resonance energy transfer and bioluminescence resonance energy transfer. The human delta -opioid receptor displays constitutive oligomerization at the cell surface, which is not regulated by receptor occupancy. J Biol Chem 276:14092–14099.

Mercier JF, Salahpur A, Angers S, et al (2002) Quantitative assessment of B1- and B2-adrenergic receptor homo- and heterodimerization by bioluminescence resonance energy transfer. J Biol Chem 277:44925–44931.

Mesnier D, Baneres JL (2004) Cooperative conformational changes in a G-protein-coupled receptor dimer, the leukotriene B(4) receptor BLT1. J Biol Chem 279:49664–49670.

Miller KJ (2005) Serotonin 5-ht2c receptor agonists: potential for the treatment of obesity. Mol Interv 5:282–291.

Milligan G (2004) G protein-coupled receptor dimerization: function and ligand pharmacology. Mol Pharmacol 66:1–7.

Ng GY, O'Dowd BF, Lee SP, et al (1996) Dopamine D2 receptor dimers and receptor-blocking peptides. Biochem Biophys Res Commun 227:200–204.

Niswender CM, Copeland SC, Herrick-Davis K, et al (1999) RNA editing of the human serotonin 5-hydroxytryptamine 2C receptor silences constitutive activity. J Biol Chem 274:9472–9478.

O'Dowd BF, Ji X, Alijaniaram M, et al (2005) Dopamine receptor oligomerization visualized in living cells. J Biol Chem 280:37225–37235.

Olsen MA, Nawoschik SP, Schurman BR, et al (1999) Identification of a human 5-HT6 receptor variant produced by alternative splicing. Brain Res Mol Brain Res 64:255–263.

Overton MC, Blumer KJ (2002) The extracellular N-terminal domain and transmembrane domains 1 and 2 mediate oliogmerization of a yeast G-protein-coupled receptor. J Biol Chem 277:41463–41472.

Park PS, Filipek S, Wells JW, et al (2004) Oligomerization of G protein-coupled receptors: past, present, and future. Biochemistry 43:15643–15656.

Rocheville M, Lange DC, Kumar U, et al (2000) Receptors for dopamine and somatostatin: formation of hetero-oligomers with enhanced functional activity. Science 288:154–157.

Roth BL, Hanizavareh SM, Blum AE (2004) Serotonin receptors represent highly favorable molecular targets for cognitive enhancement in schizophrenia and other disorders. Psychopharmacology 174:17–24.

Salahpour A, Angers S, Mercier J F, et al (2004) Homodimerization of the beta 2-adrenergic receptor as a pre-requisite for cell surface targeting. J.Biol.Chem 279:33390–33397.

Salim K, Fenton T, Bacha J, et al (2002) Oligomerization of G-protein-coupled receptors shown by selective co-immunoprecipitation. J Biol Chem 277:15482–15485.

Shioda T, Nakayama EE, Tanaka Y, et al (2001) Naturally occurring deletional mutation in the C-terminal cytoplasmic tail of CCR5 affects surface trafficking of CCR5. J Virol 75:3462–3468.

Stanasila L, Perez JB, Vogel H, et al (2003) Oligomerization of the α_{1a}- and α_{1b}-adrenergic receptor subtypes. J Biol Chem 278:40239–40251.

Terrillon A, Bouvier M (2004) Roles of G-protein-coupled receptor dimerization. EMBO Rep 5:30–34.

Terrillon S, Durroux T, Mouillac B, et al (2003) Oxytocin and vasopressin V1a and V2 receptors form constitutive homo- and heterodimers during biosynthesis. Mol Endocrinol 17:677–691.

Waldhoer M, Fong J, Jones RM, et al (2005) A heterodimer-selective agonist shows in vivo relevance of G protein-coupled receptor dimers. Proc Natl Acad Sci U S A 102:9050–9055.

Wallrabe H, Elangovan M, Burchard A, et al (2003) Confocal FRET microscopy to measure clustering of ligand-receptor complexes in endocytic membranes. Biophys J 85:559–571.

White JH, Wise A, Main MJ, et al (1998) Heterodimerization is required for the formation of a functional GABA(B) receptor. Nature 396:679–682.

Xie Z, Lee S P, O'Dowd B F, et al (1999) Serotonin 5-HT1B and 5-HT1D receptors form homodimers when expressed alone and heterodimers when co-expressed. FEBS Letters 456:63–67.

Xu Y, Piston D, Johnson CH (1999) A bioluminescence resonance energy transfer (BRET) system: application to interacting circadian clock proteins. Proc Natl Acad Sci 96:151–156.

Zawarynski P, Tallerico T, Seeman P, et al (1998) Dopamine D2 receptor dimers in human and rat brain. FEBS Lett 441:383–386.

Zhu X, Wess J (1998) Truncated V2 vasopressin receptors as negative regulators of wild-type V2 receptor function. Biochemistry 37:15773–15784.

Chapter 8
RNA Editing of 5-HT$_{2C}$ Receptor and Neuropsychiatric Diseases

Kazuya Iwamoto, Miki Bundo, and Tadafumi Kato

8.1 Introduction

Serotonin receptor 2C (HTR2C) encodes a seven-transmembrane-spanning, G-protein-coupled receptor that activates phospholipase C. Mice deficient for this gene showed a range of behavioral alterations related to neuropsychiatric disorders, including seizure, enhanced exploration of a novel environment, and dysregulation of anxiety-related behaviors (Tecott et al. 1995; Rocha et al. 2002; Heisler et al. 2007). In addition, several studies reported the significant associations between genetic variations of HTR2C and mental disorders such as major depression (MD) and bipolar disorder (BD) (Lerer et al. 2001; Massat et al. 2007). Downregulation of this gene was also reported in postmortem brains of patients with BD and schizophrenia (SZ) (Castensson et al. 2003; Iwamoto et al. 2004). These results implicated the potential importance of this gene in the pathophysiology of neuropsychiatric disorders. Furthermore, what makes this gene a unique candidate is the fact that pre-mRNA of this gene undergoes adenosine-to-inosine (A-to-I) RNA editing by adenosine deaminases (Burns et al. 1997).

8.1.1 A-to-I RNA Editing and ADARs

RNA editing is one of posttranscriptional events, and this modifies pre-mRNA sequences, resulted in the change of genomically encoded information (Bass 2002;

K. Iwamoto (✉)
Laboratory for Molecular Dynamics of Mental Disorders, RIKEN Brain Science Institute, 2-1 Hirosawa, Wako, Saitama Japan, 351-0198
and
Department of Molecular Psychiatry, Graduate School of Medicine, The University of Tokyo, 7-3-1 Hongo, Bunkyo-ku, Tokyo Japan, 113-8655
e-mail: kaziwamoto-tky@umin.ac.jp

G. Di Giovanni et al. (eds.), *5-HT$_{2C}$ Receptors in the Pathophysiology of CNS Disease*, 157
The Receptors 22, DOI 10.1007/978-1-60761-941-3_8,
© Springer Science+Business Media, LLC 2011

Keegan et al. 2001; Maas et al. 2003). Consequence of RNA editing is dependent on the context of target site, and affects splice pattern, intracellular localization, and protein function. In mammals, there are two dominant types of RNA editing: C-to-U and A-to-I. The C-to-U RNA editing includes conversion of the cytidine (C) to uridine (U) in apolipoprotein B pre-mRNA by APOBEC1 cytidine deaminases (Navaratnam and Sarwar 2006). The latter, A-to-I RNA editing includes conversion of adenosine (A) to inosine (I), and is mediated by adenosine deaminases acting on RNA (ADARs). Because translation machinery recognizes inosine in pre-mRNA as guanosine, change of amino acid sequences can occur. The target RNA for A-to-I RNA editing included neurotransmitter receptors such as glutamate and serotonin receptors, although recent studies revealed its more frequent targets would be repetitive sequence regions in the mRNAs and noncoding RNAs (Nishikura 2006).

In mammals, three ADARs (ADAR1, ADAR2, and ADAR3) have been identified (Fig. 8.1) (Keegan et al. 2001; Nishikura 2006). They all have conserved double-stranded RNA-binding domain and deaminase domain in the C-terminal of the protein. In the N-terminal, there are specific domains of each ADAR such as presence or absence of Z-DNA-binding domain and single-stranded RNA-binding domain. ADAR1 and ADAR2 express ubiquitously, while ADAR3 showed restricted expression of the subregions of brain. Some reports showed that ADAR3 have no deaminase activity (Chen et al. 2000; Melcher et al. 1996) and biological function of ADAR3 remains to be defined.

The locus of ADAR2 (22q22.3) has been suggested to be associated with familial BD by the linkage studies (Aita et al. 1999; Detera-Wadleigh et al. 1996). Amore et al. examined cDNA sequences of ADAR2 in seven patients with familial BDs (Amore et al. 2004), and Kostyrko et al. performed mutation screening in 60 patients with BD (Kostyrko et al. 2006). Neither study identified polymorphisms or mutations specific to BD patients. On the other hand, ADAR1 (1q21.3) has been reported to be involved in dyschromatosis symmetrica hereditaria, which is characterized by hyperpigmented and hypopigmented macules on extremities (Miyamura et al. 2003). However, there is no attempt to examine the relationship between genetic variations of ADAR1 and mental disorders further.

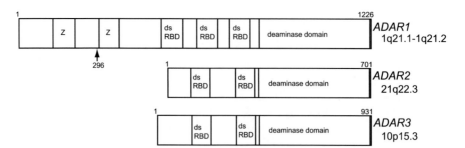

Fig. 8.1 Structure of human adenosine deaminases acting on RNA (*ADAR*) proteins. Domain structure and their genomic locations were shown. Alternative translation site of ADAR1 is denoted by arrow (*Z* Z DNA-binding domain, dsRBD double-stranded RNA-binding domain)

8.2   RNA Editing of HTR2C

In HTR2C, A-to-I RNA editing occurs at five positions (termed *sites A* to *E*), and resulted in the amino acid change at three positions (Fig. 8.2). Despite the high sequence homologies across 14 identified serotonin receptors, including other members of 5-HT$_2$ subfamily (HTR2A and HTR2B), HTR2C is the only receptor that undergoes RNA editing. The nonedited HTR2C isoform, which contains a genomically encoded sequence, showed the amino acids isoleucine at position 156 (I156), asparagine at position 158 (N158), and isoleucine at position 160 (I160, abbreviate as INI). If A-to-I RNA editing occurs at A and D sites, the resultant pre-mRNA would encode the VNV isoform. Theoretically, there are 32 transcript combinations in total, and they could generate up to 24 distinctive iso-forms (Table 8.1). The distribution of HTR2C isoforms was differed across brain regions (Burns et al. 1997) and has been believed to have physiological and pathophysiological roles in vivo. The editing site is located within the second intracellular loop of the receptor, which is important for the association with G proteins. In addition, to activate phospholipase C via coupling with Gq protein (Berg et al. 1994), the nonedited INI isoform also activates phospholipase D (McGrew et al. 2002) and ERK1/2 (Werry et al. 2005) via coupling with Gα12/13. It has been established that RNA editing generally results in the reduc-tion of receptor activity by modulating serotonin potency and G-protein-coupling

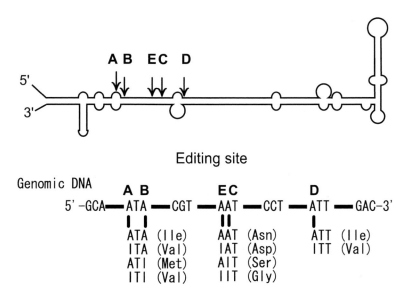

Fig. 8.2 RNA editing of HTR2C. *Top*: Predicted secondary structure of the human HTR2C pre-mRNA. Double-stranded RNA structure is recognized by adenosine deaminases acting on RNAs (ADARs). In addition to the five sites, it has been reported that the sixth editing site (site F) is in the intoronic sequence (Flomen et al. 2004). Functional significance remains to be eluci-dated, though. *Bottom*: Pattern of amino acid change

Table 8.1 *HTR2C* isoform distribution in the human and rat prefrontal cortex[a]

Isoform	Human (%)	Rat (%)
VSV	25.2	14.8
VNV	18.0	41.8
VSI	12.6	11.7
INI	9.0	6.3
ISV	6.3	1.1
VDV	5.4	1.6
VGV	5.4	0.5
VNI	5.4	17
INV	3.6	1.4
VGI	3.6	0.8
MNI	1.8	1.2
IDI	0.9	nd
IGV	0.9	0.5
ISI	0.9	nd
MNV	0.9	nd
IDV	nd	nd
IGI	nd	nd
MDI	nd	nd
MDV	nd	nd
MGI	nd	nd
MGV	nd	nd
MSI	nd	nd
MSV	nd	nd
VDI	nd	1.2

nd not detected

[a]Data were obtained by sequencing of individual bacterial clone (N > 100) that contained HTR2C cDNA sequence (Iwamoto, unpublished data)

activity (Berg et al. 2008; Marion et al. 2004; Niswender et al. 1999). In addition, RNA editing may affect the cellular localization of HTR2C (Marion et al. 2004). While nonedited INI isoform showed agonist-independent internalization, fully edited VGV resides in membrane.

Molecular mechanism that contributes to differential RNA editing level across different brain regions remains elucidated. Expression level of ADAR1 and ADAR2 in a given sample will be one of important factors, because in vivo and in vitro studies revealed that ADARs has differential editing activities in the editing sites. For example, A and B sites were preferentially edited by ADAR1, whereas D site was exclusively edited by ADAR2 (Hartner et al. 2004; Higuchi et al. 2000; Liu et al. 1999). In addition, it has been suggested that splice event of HTR2C was closely linked to RNA editing (Flomen et al. 2004). It has been also reported that HBII-52, one of small nucleolar RNAs (snoRNAs), regulated alternative splicing of

HTR2C (Cavaille et al. 2000; Kishore and Stamm 2006) and influenced RNA editing level (Vitali et al. 2005). Therefore, regulation of splicing and related factors may also have important role in the RNA editing level of HTR2C.

8.3   RNA Editing of HTR2C in Mental Disorders: Postmortem Brain Studies

Several groups examined RNA editing status of HTR2C in the brains of patients with mental disorders (Table 8.2). Due to the difference of technologies used and limited number of brain samples, the results are inconsistent across studies. In the early studies, Niswender et al. reported no significant alterations of RNA editing in patients with MD or SZ (Niswender et al. 2001). They examined RNA editing status of A, C, and D sites by a primer-extension-based method, yielding quantitative editing values. In a post hoc analysis, they observed significantly increased RNA editing at site A in suicide victims. In contrast, Sodhi et al. reported global decrease of RNA editing for patients with SZ (Sodhi et al. 2001). They examined RNA editing status by cloning-sequencing method. This yielded both editing levels of five sites and resultant isoform patterns, but it required exhaustive analysis for the precise estimation. Through the analysis, they reported significant decrease of RNA editing at site B and tendency of decrease at other four sites. At the isoform level, they found significant increase of the nonedited INI isoform in SZ compared with controls. Since the cause of death of patients has not been reported, the effect of suicide was not clear. Iwamoto et al. examined RNA editing level of sites A and D in the prefrontal

Table 8.2 RNA editing studies using postmortem prefrontal cortex samples

Disease	n	Findings	Isoform Pattern	Reference
Schizophrenia	15	Not altered	Not altered	Dracheva et al. (2003)
Bipolar disorder	34	Not altered	Decrease of VGV	Dracheva et al. (2008)
Schizophrenia	35	Not altered	Not altered	Dracheva et al. (2008a, b)
Major depression	6	Increase of E and C, decrease of D	Decrease of VGI, increase of VNI	Gurevich et al. (2002a, b)
Bipolar disorder	12	Not altered	nd	Iwamoto and Kato (2003)
Major depression	11	Increase of D	nd	Iwamoto and Kato (2003)
Schizophrenia	13	Not altered	nd	Iwamoto and Kato (2003)
Major depression	13	Not altered	nd	Niswender et al. (2001)
Schizophrenia	13	Not altered	nd	Niswender et al. (2001)
Schizophrenia	5	Decrease of B	Increase of INI	Sodhi et al. (2001)

nd not determined

cortex of patients with mental disorders by a primer-extension-based method (Iwamoto and Kato 2002, 2003). Although they did not detect significant alteration in the RNA editing levels, they reported a trend for increase of RNA editing at D site in MD. In addition, they also reported an increased editing at site A in suicide victims in a post hoc analysis (Iwamoto and Kato 2003). Gurevich et al. examined RNA editing status in suicide victims with MD by a cloning-sequencing method. They reported altered RNA editing of increased editing at site E, decreased at site D, and a trend for increase at site C for patients (Gurevich et al. 2002a).

Among the studies, currently, Dracheva et al. provided the most comprehensive results, as they used relatively large number of subjects and analyzed by extensive cloning-sequencing methods. In their first report using prefrontal cortex of SZ, they reported no alterations of the RNA editing level at each of five sites or isoform distribution (Dracheva et al. 2003). Using the different cohort of SZ and BD, they again did not find significant editing alterations in SZ (Dracheva et al. 2008a). However, they detected significant decrease of VGV isoform in the prefrontal cortex of BD. Moreover, they also found significant increase of RNA editing at sites A, C, and D and the increase of VSV isoform in suicide victims, supporting the association of editing alteration and suicide. They also reported altered splicing pattern of HTR2C in the brains of suicide victims, which suggests the altered regulation of splicing was involved in the altered RNA editing in the suicide victims (Dracheva et al. 2008b).

8.4   RNA Editing of HTR2C in Mental Disorders: Cellular and Animal Studies

RNA editing status of HTR2C could be changed in response to stress and drugs in cellular and animal models (Iwamoto et al. 2009). The adverse effects of interferons, which are clinically used for treatment of hepatitis, cirrhosis, and cancer, include depression (Schaefer et al. 2002). Among the three ADARs, expression of ADAR1 is induced by interferon. This resulted in the generation of the ADAR1 protein with higher molecular weight (150 kDa) than that produced from a constitutive promoter (110 kDa) (Fig. 8.1) (Patterson and Samuel 1995). Using glioblastoma cell lines, Yang et al. reported that interferon induced expression of 150 kDa ADAR1 and resulted in increased RNA editing at sites A, B, and C and decrease at site D (Yang et al. 2004). They suggested the involvement of RNA editing in the pathophysiology of depression caused by interferon treatment.

In mice, RNA editing status of HTR2C was differed among different inbred strains (Englander et al. 2005; Hackler et al. 2006; Du et al. 2006). In the forebrain, both 129Sv and C57BL/6 mice strains showed modest to high level of RNA editing. In contrast, BALB/c showed less editing at all five sites, resulting in the drastic increase of nonedited INI isoform (Englander et al. 2005). In amygdala, C57BL/6J showed higher editing level at site C compared with BALB/cJ or DBA/2J. This difference resulted in increase of the VSV isoform (Hackler et al. 2006). These differences at basal editing level are considered involved in the different behavioral characteristics among mice strains, such as level of anxiety and stress reactivity. Recently, functional consequence of altered RNA editing in vivo was directly tested. Kawahara et al. created the knock-in mice of fully edited (VGV) and nonedited (INI) HTR2C (Kawahara et al. 2008). While the INI mice were reported to grow normally, the VGV mice showed severe growth retardation by the activation of the sympathetic nervous system and increased energy expenditure.

Treatment of para-chlorophenylalanine (pCPA), an inhibitor of tryptophan hydroxylase, resulted in the reduction of serotonin level in brain in animal models. This reduction is associated with altered RNA editing including decreased editing at site E (Gurevich et al. 2002b). Conversely, DOI [(±)-1-(-4-iodo-2,5-dimethoxyphenyl)-2-aminopropane], a 5-HT2 receptor agonist, treated mice showed increased RNA editing at site E, suggesting the receptor activity-dependent regulation of RNA editing. Results of treatment and other drugs on normal rodents (Iwamoto and Kato 2002; Gurevich et al. 2002a; Englander et al. 2005; Sodhi et al. 2005) are summarized in Table 8.3. Although the direction of change was complex, some of the drugs affected the RNA editing status in rodents and provided an important resource to interpret the results obtained from postmortem brains.

The learned helplessness (LH) is one of the most validated animal models of depression (Overmier and Seligman 1967; Telner and Singhal 1984). In this model, after pretreatment with repeated inescapable shocks such as foot electroshock, animals with LH show decreased ability to escape unfavorable situations. Since this phenotype can be improved by antidepressants, the LH model has been widely used for studying the depression and the actions of antidepressants. In the LH rats, RNA editing levels of sites A, B, C, and D showed nonsignificant increase and that of site E showed significant increase compared with normal rats (Iwamoto et al. 2005). Treatment with antidepressants such as fluoxetine and imipramine altered RNA editing level. Fluoxetine significantly decreased editing level at sites A, B, and E in the LH rats. Imipramine also significantly decreased editing level at site E. At the isoform level, both antidepressants significantly decreased level of nonedited INI isoform. Importantly, these changes were reported to be associated with improvement of altered behavior in the LH rats (Iwamoto et al. 2005; Nakatani et al. 2004). In the other models of depression, forced swimming (Englander et al. 2005), water maze (Du et al. 2007), and maternal separation (Bhansali et al. 2007) have been reported to alter RNA editing status. Although direction of change in animal models was not consistent across studies, drug treatments such as fluoxetine (Englander et al. 2005; Iwamoto et al. 2005; Bhansali et al. 2007) altered or reversed RNA editing pattern, which associated with improvement of behavioral abnormalities in each animal models.

8.5   Conclusion

Postmortem brain studies, cellular and animal studies, and pharmacological studies suggest the involvement of RNA editing of HTR2C in major mental disorders. However, reported results were tended to be discordant, and observed differences

Table 8.3 Effect of drugs on RNA editing of brains of control animals

Drug	Animal	Dose and duration	Findings	Isoform pattern	Reference
Haloperidol	Rat (SD)	1 mg/kg, 14 days	Decrease of D	Increase of VNI, decrease of VNV	Sodhi et al. (2005)
Risperidone	Rat (SD)	0.5 mg/kg, 14 days	Increase of B, decrease of C and D	Increase of VNV	Sodhi et al. (2005)
Clozapine	Rat (SD)	25 mg/kg, 14 days	Not altered	Not altered	Sodhi et al. (2005)
Chlorpromazine	Rat (SD)	15 mg/kg, 14 days	Not altered	Not altered	Sodhi et al. (2005)
Cocaine	Rat (SD)	15 mg/kg, 7 days	Not altered	nd	Iwamoto and Kato (2002)
Reserpine	Rat (SD)	1 mg/kg, 2 days	Not altered	nd	Iwamoto and Kato (2002)
Fluoxetine	Mice (129Sv)	5 mg/kg, 3 days	Decrease of E	Not altered	Gurevich et al. (2002)
Fluoxetine	Mice (129Sv)	5–10 mg/kg, 28 days	Increase of D, decrease of C and E	Increase of VNV	Gurevich et al. (2002)
Fluoxetine	Mice (C57Bl/6)	18 mg/kg, 24 days	Not altered	Not altered	Englander et al. (2005)
Fluoxetine	Mice (Balb/c)	18 mg/kg, 24 days	Increase of A, B, C and D	Various changes such as decrease of INI	Englander et al. (2005)

nd not determined.

were very subtle. Further studies using larger and independent samples are needed.

Acknowledgments This study was supported by Grant-in-Aid from the Japanese Ministry of Health, Welfare, and Labor. We thank Junko Ueda for help in preparation of the manuscript.

References

Aita VM, Liu J, Knowles JA, et al (1999) A comprehensive linkage analysis of chromosome 21q22 supports prior evidence for a putative bipolar affective disorder locus. Am J Hum Genet 64:210–217.

Amore M, Strippoli P, Laterza C, et al (2004) Sequence analysis of ADARB1 gene in patients with familial bipolar disorder. J Affect Disord 81:79–85.

Bass BL (2002) RNA editing by adenosine deaminases that act on RNA. Annu Rev Biochem 71:817–846.

Berg KA, Clarke WP, Sailstad C, et al (1994) Signal transduction differences between 5-hydroxytryptamine type 2A and type 2C receptor systems. Mol Pharmacol 46:477–484.

Berg KA, Dunlop J, Sanchez T, et al (2008) A conservative, single-amino acid substitution in the second cytoplasmic domain of the human Serotonin2C receptor alters both ligand-dependent and -independent receptor signaling. J Pharmacol Exp Ther 324:1084–1092.

Bhansali P, Dunning J, Singer SE, et al (2007) Early life stress alters adult serotonin 2C receptor pre-mRNA editing and expression of the alpha subunit of the heterotrimeric G-protein G q. J Neurosci 27:1467–1473.

Burns CM, Chu H, Rueter SM, et al (1997) Regulation of serotonin-2C receptor G-protein coupling by RNA editing. Nature 387:303–308.

Castensson A, Emilsson L, Sundberg R, et al (2003) Decrease of serotonin receptor 2C in schizophrenia brains identified by high-resolution mRNA expression analysis. Biol Psychiatry 54:1212–1221.

Cavaille J, Buiting K, Kiefmann M, et al (2000) Identification of brain-specific and imprinted small nucleolar RNA genes exhibiting an unusual genomic organization. Proc Natl Acad Sci U S A 97:14311–14316.

Chen CX, Cho DS, Wang Q, et al (2000) A third member of the RNA-specific adenosine deaminase gene family, ADAR3, contains both single- and double-stranded RNA binding domains. RNA 6:755–767.

Detera-Wadleigh SD, Badner JA, Goldin LR, et al (1996) Affected-sib-pair analyses reveal support of prior evidence for a susceptibility locus for bipolar disorder, on 21q. Am J Hum Genet 58:1279–1285.

Dracheva S, Elhakem SL, Marcus SM, et al (2003) RNA editing and alternative splicing of human serotonin 2C receptor in schizophrenia. J Neurochem 87:1402–1412.

Dracheva S, Patel N, Woo DA, et al (2008a) Increased serotonin 2C receptor mRNA editing: a possible risk factor for suicide. Mol Psychiatry 13:1001–1010.

Dracheva S, Chin B, and Haroutunian V (2008b) Altered serotonin 2C receptor RNA splicing in suicide: association with editing. Neuroreport 19:379–382.

Du Y, Davisson MT, Kafadar K, et al (2006) A-to-I pre-mRNA editing of the serotonin 2C receptor: comparisons among inbred mouse strains. Gene 382:39–46.

Du Y, Stasko M, Costa AC, et al (2007) Editing of the serotonin 2C receptor pre-mRNA: Effects of the Morris Water Maze. Gene 391:186–197.

Englander MT, Dulawa SC, Bhansali P, et al (2005) How stress and fluoxetine modulate serotonin 2C receptor pre-mRNA editing. J Neurosci 25:648–651.

Flomen R, Knight J, Sham P, et al (2004) Evidence that RNA editing modulates splice site selection in the 5-HT2C receptor gene. Nucleic Acids Res 32:2113–2122.

Gurevich I, Tamir H, Arango V, et al (2002a) Altered editing of serotonin 2C receptor pre-mRNA in the prefrontal cortex of depressed suicide victims. Neuron 34:349–356.

Gurevich I, Englander MT, Adlersberg M, et al (2002b) Modulation of serotonin 2C receptor editing by sustained changes in serotonergic neurotransmission. J Neurosci 22:10529–10532.

Hackler EA, Airey DC, Shannon CC, et al (2006) 5-HT(2C) receptor RNA editing in the amygdala of C57BL/6J, DBA/2J, and BALB/cJ mice. Neurosci Res 55:96–104.

Hartner JC, Schmittwolf C, Kispert A, et al (2004) Liver disintegration in the mouse embryo caused by deficiency in the RNA-editing enzyme ADAR1. J Biol Chem 279:4894–4902.

Heisler LK, Zhou L, Bajwa P, et al (2007) Serotonin 5-HT(2C) receptors regulate anxiety-like behavior. Genes Brain Behav 6:491–496.

Higuchi M, Maas S, Single FN, et al (2000) Point mutation in an AMPA receptor gene rescues lethality in mice deficient in the RNA-editing enzyme ADAR2. Nature 406:78–81.

Iwamoto K and Kato T (2002) Effects of cocaine and reserpine administration on RNA editing of rat 5-HT2C receptor estimated by primer extension combined with denaturing high-performance liquid chromatography. Pharmacogenomics J 2:335–340.

Iwamoto K and Kato T (2003) RNA editing of serotonin 2C receptor in human postmortem brains of major mental disorders. Neurosci Lett 346:169–172.

Iwamoto K, Kakiuchi C, Bundo M, et al (2004) Molecular characterization of bipolar disorder by comparing gene expression profiles of postmortem brains of major mental disorders. Mol Psychiatry 9:406–416.

Iwamoto K, Nakatani N, Bundo M, et al (2005) Altered RNA editing of serotonin 2C receptor in a rat model of depression. Neurosci Res 53:69–76.

Iwamoto K, Bundo M, and Kato T (2009) Serotonin receptor 2C and mental disorders: genetic, expression and RNA editing studies. RNA Biol 6:248–253.

Kawahara Y, Grimberg A, Teegarden S, et al (2008) Dysregulated editing of serotonin 2C receptor mRNAs results in energy dissipation and loss of fat mass. J Neurosci 28:12834–12844.

Keegan LP, Gallo A, and O'Connell MA (2001) The many roles of an RNA editor. Nat Rev Genet 2:869–878.

Kishore S and Stamm S (2006) The snoRNA HBII-52 regulates alternative splicing of the serotonin receptor 2C. Science 311:230–232.

Kostyrko A, Hauser J, Rybakowski JK, et al (2006) Screening of chromosomal region 21q22.3 for mutations in genes associated with neuronal Ca2+ signalling in bipolar affective disorder. Acta Biochim Pol 53:317–320.

Lerer B, Macciardi F, Segman RH, et al (2001) Variability of 5-HT2C receptor cys23ser polymorphism among European populations and vulnerability to affective disorder. Mol Psychiatry 6:579–585.

Liu Y, Emeson RB, and Samuel CE (1999) Serotonin-2C receptor pre-mRNA editing in rat brain and in vitro by splice site variants of the interferon-inducible double-stranded RNA-specific adenosine deaminase ADAR1. J Biol Chem 274:18351–18358.

Maas S, Rich A, and Nishikura K (2003) A-to-I RNA editing: recent news and residual mysteries. J Biol Chem 278:1391–1394.

Marion S, Weiner DM, and Caron MG (2004) RNA editing induces variation in desensitization and trafficking of 5-hydroxytryptamine 2c receptor isoforms. J Biol Chem 279:2945–2954.

Massat I, Lerer B, Souery D, et al (2007) HTR2C (cys23ser) polymorphism influences early onset in bipolar patients in a large European multicenter association study. Mol Psychiatry 12:797–798.

McGrew L, Chang MS, and Sanders-Bush E (2002) Phospholipase D activation by endogenous 5-hydroxytryptamine 2C receptors is mediated by Galpha13 and pertussis toxin-insensitive Gbetagamma subunits. Mol Pharmacol 62:1339–1343.

Melcher T, Maas S, Herb A, et al (1996) A mammalian RNA editing enzyme. Nature 379:460–464.

Miyamura Y, Suzuki T, Kono M, et al (2003) Mutations of the RNA-specific adenosine deaminase gene (DSRAD) are involved in dyschromatosis symmetrica hereditaria. Am J Hum Genet 73:693–699.

Nakatani N, Aburatani H, Nishimura K, et al (2004) Comprehensive expression analysis of a rat depression model. Pharmacogenomics J 4:114–126.

Navaratnam N and Sarwar R (2006) An overview of cytidine deaminases. Int J Hematol 83:195–200.

Nishikura K (2006) Editor meets silencer: crosstalk between RNA editing and RNA interference. Nat Rev Mol Cell Biol 7:919–931.

Niswender CM, Copeland SC, Herrick-Davis K, et al (1999) RNA editing of the human serotonin 5-hydroxytryptamine 2C receptor silences constitutive activity. J Biol Chem 274:9472–9478.

Niswender CM, Herrick-Davis K, Dilley GE, et al (2001) RNA editing of the human serotonin 5-HT2C receptor. alterations in suicide and implications for serotonergic pharmacotherapy. Neuropsychopharmacology 24:478–491.

Overmier JB and Seligman ME (1967) Effects of inescapable shock upon subsequent escape and avoidance responding. J Comp Physiol Psychol 63:28–33.

Patterson JB and Samuel CE (1995) Expression and regulation by interferon of a double-stranded-RNA-specific adenosine deaminase from human cells: evidence for two forms of the deaminase. Mol Cell Biol 15:5376–5388.

Rocha BA, Goulding EH, O'Dell LE, et al (2002) Enhanced locomotor, reinforcing, and neuro-chemical effects of cocaine in serotonin 5-hydroxytryptamine 2C receptor mutant mice. J Neurosci 22:10039–10045.

Schaefer M, Engelbrecht MA, Gut O, et al (2002) Interferon alpha (IFNalpha) and psychiatric syndromes: a review. Prog Neuropsychopharmacol Biol Psychiatry 26:731–746.

Sodhi MS, Burnet PW, Makoff AJ, et al (2001) RNA editing of the 5-HT(2C) receptor is reduced in schizophrenia. Mol Psychiatry 6:373–379.

Sodhi MS, Airey DC, Lambert W, et al (2005) A rapid new assay to detect RNA editing reveals antipsychotic-induced changes in serotonin-2C transcripts. Mol Pharmacol 68:711–719.

Tecott LH, Sun LM, Akana SF, et al (1995) Eating disorder and epilepsy in mice lacking 5-HT2c serotonin receptors. Nature 374:542–546.

Telner JI and Singhal RL (1984) Psychiatric progress. The learned helplessness model of depression. J Psychiatr Res 18:207–215.

Vitali P, Basyuk E, Le Meur E, et al (2005) ADAR2-mediated editing of RNA substrates in the nucleolus is inhibited by C/D small nucleolar RNAs. J Cell Biol 169:745–753.

Werry TD, Gregory KJ, Sexton PM, et al (2005) Characterization of serotonin 5-HT2C receptor signaling to extracellular signal-regulated kinases 1 and 2. J Neurochem 93:1603–1615.

Yang W, Wang Q, Kanes SJ, et al (2004) Altered RNA editing of serotonin 5-HT2C receptor induced by interferon: implications for depression associated with cytokine therapy. Brain Res Mol Brain Res 124:70–78.

Chapter 9
Serotonergic Control of Adult Neurogenesis: Focus on 5-HT$_{2C}$ Receptors

Annie Daszuta

9.1 Historical Perspective

It took several decades to agree that new neurons are generated in adult brain but less time to establish that they are functionally integrated in two discrete regions: the olfactory bulb (OB) and the hippocampus (Ming and Song 2005; Lledo et al. 2006; Zhao et al. 2008). While our current knowledge on adult neurogenesis in mammals is mainly based on rodent studies, recent in vivo data have confirmed the occurrence of neural stem cells in the human brain (Manganas et al. 2007), already detected in postmortem tissues (Eriksson et al. 1998). Originally described by Altman (Altman 1969) using [^{3}H]thymidine autoradiography, the presence of new neural cells in adult brain was rediscovered by Gould et al. in the early 1990s (Gould et al. 1992; Cameron et al. 1993) in studies based on the administration of 5-bromo-2′-deoxyuridine (BrdU), a thymidine analogue taken up by cells synthesizing DNA in preparation for division. BrdU-labeled neurons are visualized using a method, immunocytochemistry, which can be combined with or used in parallel with several other techniques or models, including novel transgenic mouse lines (Yamaguchi et al. 2000; Encinas and Enikolopov 2008). Together, these tools allow the visualization of discrete stages of neurogenesis and ascertain the neuronal identity of the new cells. Indeed, the term *neurogenesis* refers to the combined processes of cell proliferation, survival, differentiation, maturation, and synaptic integration in the preexisting circuitry (Duan et al. 2008). Adult neurogenesis thus appears to be an extreme form of structural remodeling, compared with the subtle modification in synaptic morphology mediating functional plasticity of neural circuitry. Furthermore, numerous factors can modulate the number of newborn cells, suggesting that adult neurogenesis represents an adaptive response to various challenges imposed by an external and/or internal environment, under physiological or pathological conditions.

A. Daszuta (✉)
IBDML, UMR 6216, CNRS, Marseille, France
e-mail: daszuta@ibdml.univ-mrs.fr

G. Di Giovanni et al. (eds.), *5-HT$_{2C}$ Receptors in the Pathophysiology of CNS Disease*,   169
The Receptors 22, DOI 10.1007/978-1-60761-941-3_9,
© Springer Science+Business Media, LLC 2011

9.2 Neurogenic Brain Regions

New neurons are continuously generated in two neurogenic regions in the adult brain: the subgranular zone (SGZ) of dentate gyrus (DG) in the hippocampal formation and the subventricular zone (SVZ) of lateral ventricles, as illustrated in Fig. 9.1 (Ming and Song 2005; Lledo et al. 2006; Abrous et al. 2005; Christie and Cameron 2006; Ehninger and Kempermann 2008). In both regions, the glial origin of adult neural stem cells and their multipotentiality, giving rise to neurons, astrocytes, and oligodendrocytes, have been well established (Doetsch et al. 1999; Gage 2000; Alvarez-Buylla et al. 2001). In the SVZ, newborn cells migrate over a long distance via the rostral migratory stream (RMS) and differentiate mainly into inhibitory γ-aminobutyric acid (GABA)-ergic interneurons in the granular layer of

Neurogenic zones and Serotonergic innervation

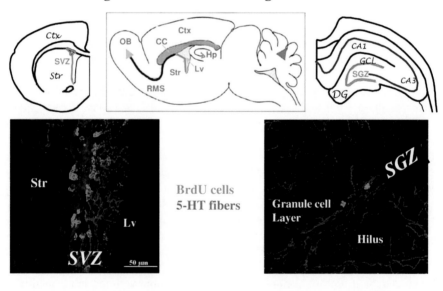

Fig. 9.1 Schematic representation of the neurogenic zones in adult rat brain and visualization of the corresponding serotonergic innervation. The parasagittal view shows the location of the hippocampus (Hp), the lateral ventricle (Lv), and the striatum (Str) limiting the subventricular zone (SVZ) that lines the wall of the lateral ventricle. From there, neuroblasts migrate along the rostral migratory stream (RMS) towards the olfactory bulb (OB). Frontal sections illustrate the precise location of the SVZ and the subgranular zone (SGZ) of the granule cell layer (GCL) in the dentate gyrus (DG) of the hippocampal formation. Immunostaining for serotonergic fibers (5-HT fibers) shows them in close proximity to dividing cells labeled with bromodeoxyuridine (BrdU), a marker of cell division, in the SVZ and the SGZ. Notably, a dense plexus of 5-HT fibers is seen crossing the SVZ and the ependymal layer (Ctx cortex, CC corpus callosum)

the OB and into a small proportion of peri-glomerular cells expressing tyrosine hydroxylase (Alvarez-Buylla and Garcia-Verdugo 2002). By contrast, in the DG, neural stem cells and progenitors are aligned on the hilar border and can easily integrate the granule cell layer. These new glutamatergic cells develop dendritic arborization into the molecular layer and axonal projections (mossy fibers) to area CA3 and finally integrate the hippocampal three-synaptic circuitry (Kempermann et al. 2004a). Another difference between the two neurogenic zones is the level of cell production, which is much higher in the SVZ (about 30,000 new cells per day) compared with the SGZ (about 9,000 new cells per day), as quantified in young rodent (Alvarez-Buylla and Garcia-Verdugo 2002; Cameron and McKay 2001). Although a large number of progenitors are created through proliferation, only about 50% of newborn cells are selected for long-term survival (Winner et al. 2002; Kempermann et al. 2003). These cells are continuously added to the dentate gyrus (Cameron and McKay 2001), while they replace old cells in the olfactory bulb in phase with the turnover of sensory inputs (Lledo et al. 2006).

In spite of the low rate of adult neuronal production, the constant integration of newborn cells into the functional circuitry and their unique properties compared with more mature neurons suggest that these neurons may have a very specific function (Kempermann et al. 2004b; Doetsch and Hen 2005; Lledo and Saghatelyan 2005; Aimone et al. 2006). This function has not been fully elucidated and remains a topic of intense investigation and discussion (Scharfman and Hen 2007; Kuhn et al. 2007; Ortega-Perez et al. 2007; Kempermann 2008; Aimone et al. 2009). However, a number of studies have suggested the implication of adult neurogenesis in learning and memory process (Abrous et al. 2005). More specifically, newborn bulbar neurons are presumed to contribute to essential aspects of olfactory function, notably in odor discrimination (Lledo et al. 2006), whereas new granule cells in the DG have been involved in hippocampus-dependent behaviors (Leuner et al. 2006; Bruel-Jungerman et al. 2007; Shors 2008) and are presumed to be particularly important in episodic memory (Aimone et al. 2006). Until recently, a major difficulty in testing whether bulbar or hippocampal newborn neurons are required for specific tasks was the lack of selective method to block neurogenesis, without the use of irradiation or neuro-toxic approaches. Transgenic strategies aimed to ablate neural precursors provided new information indicating that hippocampal newborn neurons contribute to contextual and spatial memory formation and that bulbar neurogenesis is required not for acquisition but for short-term retention of odor-associated memory (Dupret et al. 2008; Imayoshi et al. 2008; Jessberger et al. 2009).

9.3   Regulation of Adult Neurogenesis

Adult neurogenesis is controlled by a wide array of intrinsic and extrinsic factors that have been extensively studied in the case of the hippocampus (Ming and Song 2005; Zhao et al. 2008; Abrous et al. 2005). One of the first characteristics of adult neurogenesis is its exponential decrease over life span (Seki and Arai 1995; Kuhn et al. 1996). Hippocampal neurogenesis is present at much lower level in aged animals but still can be reactivated by general stimuli or appropriate signals (Kempermann et al. 2002; Drapeau and Abrous Nora 2008). Adult neurogenesis is also largely influenced by the genetic background, as demonstrated by comparing basal levels of hippocampal cell proliferation in various mice strains (Kempermann et al. 1997, 2006), although these differences may also depend on the phase of neurogenesis examined (Schauwecker 2006). Finally, gender and estrous cycle are considered as potent intrinsic modulators of hippocampal neurogenesis in rats (Galea et al. 2006) but not in mice (Lagace et al. 2007), emphasizing the genetic dependency of these processes. On the other hand, at a systemic level, physical exercise and enriched environment versus stressful experiences, including sleep deprivation, increase and decrease neurogenesis respectively, via interactions between multiple cellular pathways and molecular factors (Olson et al. 2006; Mirescu and Gould 2006; Fabel and Kempermann 2008; Meerlo et al. 2008).

9.3.1 Neurogenic Niches

An increasing number of "potential" neurogenic zones have been discovered during the last decade, in accordance with the fact that neural stem cells can be isolated from several areas in adult brain (Gould 2007). However, the only zones where neurogenesis has been consistently found are the SGZ and the SVZ. Indeed, the microenvironments of the SGZ and SVZ contain specific factors that are permissive for adult neurogenesis, whereas the lack of these factors as well as inhibitory molecules restricts differentiation to neural lineages in other brain regions. The existence of specific neurogenic niches where a number of mechanisms direct the differentiation, proliferation, and survival of newborn cells has underlined the role of the glial environment, astrocytes, and oligodendrocytes and other cell types such as ependymal or endothelial cells and neuroblasts (Doetsch 2003; Ma et al. 2005; Ninkovic and Gotz 2007; Riquelme et al. 2008). For instance, a number of trophic factors such as vascular endothelial growth factor (VEGF), brain derived neurotrophic factor (BDNF), and insulin-like growth factor-1 (IGF-1) have been shown to modulate one or several phases of adult neurogenesis after intracerebral or peripheral infusions (Schmidt and Duman 2007). Interestingly, neurogenesis appears to occur in intimate association with angiogenesis, and both processes can be activated by similar stimuli (Palmer et al. 2000).

More diffusible factors such as hormones, peptides, and neurotransmitters may act and interact to affect adult neurogenesis (Ming and Song 2005; Hagg 2005; Vaidya et al. 2007). Indeed, the neurogenic regions are richly innervated by neurotransmitter systems of local and distant origins. For instance, glucocorticoids, which mediate some of the deleterious effects of stress, have inhibitory effects on hippocampal neurogenesis both directly and indirectly by activating the glutamatergic transmission (Mirescu and Gould 2006). In contrast to its role as an inhibitory neurotransmitter on mature cells, GABA released locally depolarizes neural progenitors and early newborn neurons. Thus, GABA was shown to inhibit proliferation and promote maturation and differentiation of these cells (Ge et al. 2007). In the SVZ, GABA serves as a feedback regulator of neural production and migration (Bordey 2007), and in the SGZ, GABA-ergic mechanisms regulate differentiation and the timing of synaptic integration of new neurons (Markwardt and Overstreet-Wadiche 2008). As already mentioned, glutamatergic signaling, mainly through N-methyl-D-aspartate (NMDA) receptors, regulates adult neurogenesis. However, the role of NMDA receptor activation is complex since the progenitor cells do not express functional NMDA receptors (Nacher et al. 2007). Increases in cell proliferation have been observed after both NMDA (Joo et al. 2007) and NMDA antagonist administration (Nacher and McEwen 2006), depending on the delay posttreatment. One hypothesis is that this effect could be mediated by GABA, whose release by hippocampal interneurons requires NMDA activation (Matsuyama et al. 1997). Indeed, interactions between various neurotransmitters or between neurotransmitters and hormones or trophic factors released near or into the neurogenic niches may account for complex in vivo regulation of adult neurogenesis.

Every monoaminergic input (dopaminergic [DA], noradrenergic [NA], and serotonergic [5-HT] afferents) affects proliferation and neurogenesis in both neurogenic region (Hagg 2005; Vaidya et al. 2007). The strong DA innervation of the striatum has triggered a number of studies investigating the DA control of cell proliferation and neurogenesis in the SVZ-OB system (Borta and Hoglinger 2007), with the hope of using the pool of progenitors to replace the DA content lost in patients with Parkinson disease (Geraerts et al. 2007). However, conflicting data have been reported since DA depletion leads to an increase (Liu et al. 2006; Aponso et al. 2008), or a decrease in cell proliferation in the SVZ (Baker et al. 2004; Hoglinger et al. 2004; Freundlieb et al. 2006), followed by a shift in differentiation of dopaminergic cells in the OB (Winner et al. 2006). Activation of DA-D2 receptors has been shown to inhibit (Kippin et al. 2005) or to rescue and stimulate cell proliferation in the SVZ, an effect mediated by CNTF (ciliary neurotrophic factor) (Hoglinger et al. 2004; Yang et al. 2008). Further, DA-D3 receptors strongly expressed at the SVZ level are also positively involved in the regulation of cell proliferation in rats (Coronas et al. 2004; Van Kampen et al. 2004) but not in mice (Baker et al. 2005), suggesting the influence of genetic background.

By contrast, data concerning NA control of adult neurogenesis are rather scarce, but all suggest a positive implication of this catecholamine. Chronic increase in NA level produces significant enhancement in cell proliferation in the SGZ (Malberg et al. 2000), while NA depletion induces the opposite effect without affecting survival and differentiation (Kulkarni et al. 2002). However, by using an α_2-adrenergic receptor antagonist leading to a general increase in NA transmission and a blockade of postsynaptic α_2-noradrenergic receptors, a selective increase in cell survival and differentiation was also shown in the hippocampus (Rizk et al. 2006), and in the OB (Bauer et al. 2003). These results suggest that various adrenergic receptor subtypes have specific involvement in NA regulation of neurogenesis phases.

9.3.2 Serotonergic Regulation of Adult Neurogenesis

Among monoamines, the 5-HT control of adult neurogenesis has been the most largely investigated, notably when studying the consequences of serotonergic antidepressants. Indeed, chronic increases in extracellular level of 5-HT by selective serotonin reuptake inhibitors (SSRIs) fluoxetine, citalopram, or paroxetine have repeatedly been shown to increase hippocampal neurogenesis in basal conditions and reverse the suppressive effects of stress or treatments associated with stressful events (Malberg et al. 2000; Lee et al. 2001; Malberg and Duman 2003; Encinas et al. 2006; Chen et al. 2006; Jayatissa et al. 2006; Qiu et al. 2007; Marcussen et al. 2008; Wang et al. 2008). Recent studies also indicated that fluoxetine-induced increases in neurogenesis can depend on age (Cowen et al. 2008; Couillard-Despres et al. 2009), on genetic background (Miller et al. 2008), and on interactions with the hypothalamic–pituitary–adrenal axis activity (Huang and Herbert 2006).

Consistent with a stimulatory role of 5-HT, we provided the first evidence that long-lasting 5-HT depletion induced by neurotoxic lesion of 5-HT-containing cells in the raphé nuclei produced significant decreases in cell proliferation and neurogenesis (Brezun and Daszuta 1999). This effect was reversed by grafting embryonic raphé tissue, containing 5-HT neurons, in the hippocampus (Brezun and Daszuta 2000a), suggesting a local 5-HT control. Other kinds of 5-HT depletion may reproduce these effect (Ueda et al. 2005; Rosenbrock et al. 2005; Zhang et al. 2006), or not (Huang and Herbert 2005; Jha et al. 2008). Reasons might be in genetic and gender differences, as well as the kind of neurotoxic lesion (intra cerebro ventricular (ICV) vs intra-raphé injections of 5,7-DHT) and duration of 5-HT depletion. Indeed, injured 5-HT neurons have remarkable capacities for regrowth (Azmitia et al. 1978), as shown by the progressive reinnervation of the hippocampus observed after a partial lesion, which was also associated to a recovery of cell proliferation (Brezun and Daszuta 2000b). Although the mechanism of action of neurotransmitters and the receptor subtypes that are expressed by neural precursors and surrounding cells in neurogenic niches remain to be precisely determined, several 5-HT receptor subtypes, such as 5-HT_{1A} and 5-HT_{1B}, 5-HT_{2A}, and 5-HT_{2C}, and 5-HT_4, are involved in the regulation of adult neurogenesis (Jha et al. 2008; Banasr et al. 2004; Lucas et al. 2007).

9.3.2.1 Implication of 5-HT_{1A} and 5-HT_{1B} Receptors

A number of studies have also demonstrated the positive implication of 5-HT_{1A} receptors in the regulation of hippocampal neurogenesis (Huang and Herbert 2005; Banasr et al. 2004; Radley and Jacobs 2002; Santarelli et al. 2003). Further, both acute and chronic activation of 5-HT1A receptors by 8-OHDPAT administration increase cell proliferation in the SVZ and, later, on neurogenesis in the OB (Banasr et al. 2004). As observed with fluoxetine, this mediation depends on the genetic background (Holick et al. 2008) and complex interactions with corticosterone level (Huang and Herbert 2006). Postsynaptic 5-HT$_{1A}$ receptors, not somatic 5-HT$_{1A}$ autoreceptors that desensitize over time, are involved in this effect in the SGZ (Huang and Herbert 2005; Banasr et al. 2004). Indeed, the increase in cell proliferation induced by chronic administration of 8-OHDPAT persists in 5-HT depleted rats, indicating that this stimulatory effect is not dependent on the integrity of 5-HT transmission. Interestingly, findings that are more recent suggest that chronic activation of 5-HT$_{1A}$ receptors has also a positive effect on the survival of hippocampal granule cells (Soumier et al. 2010). This result is consistent with the expression of 5-HT$_{1A}$ receptors by mature granule cells (Pompeiano et al. 1992; Kia et al. 1996) and is further supported by the neuroprotective action observed after stimulation of 5-HT$_{1A}$ receptors in various models of brain damage (Schaper et al. 2000; Mauler and Horvath 2005; Bezard et al. 2006).

As for 5-HT$_{1A}$, the 5-HT$_{1B}$ agonist sumatriptan stimulates cell proliferation in the SGZ through activation of 5-HT$_{1B}$ heteroreceptors (Banasr et al. 2004). However, in the SVZ, the decrease and increase in cell proliferation observed after administration of 5-HT1B agonist and antagonist (GR 127935), respectively, are more likely due to the decrease and increase in 5-HT transmission modulated by 5-HT1B terminal autoreceptors, since sumatriptan does not affect cell proliferation in 5-HT depleted rats (Banasr et al. 2004). These data suggest that multiple receptor subtypes may be involved in the 5-HT control of neurogenesis in forebrain regions.

9.3.2.2   Implication of 5-HT$_{2C}$ Receptors

Forebrain Regions

In the SVZ, cell proliferation is enhanced by administration of the 5-HT$_{2A/2C}$ agonist DOI, and this effect is reproduced by the selective 5-HT$_{2C}$ agonist, RO 600175 (Banasr et al. 2004). The lack of effect of the 5-HT$_{2C}$ antagonist SB206553 is consistent with a phasic 5-HT stimulation of cell proliferation in this region. Further, increases in neurogenesis in the OB following either acute or chronic administration of RO600175 indicate that the 5-HT$_{2C}$ receptors involved in this modulation do not desensitize over time (Banasr et al. 2004). By contrast, this treatment does not affect the survival of new neurons in the OB (Soumier et al. 2010), supporting the view that differential mechanisms are involved in the regulation of proliferation and survival of newly formed cells. Since choroid plexus expresses one of the highest concentrations of 5-HT$_{2C}$ receptors in the brain (Palacios et al. 1990), secretion of trophic factors by this structure (Chodobski and Szmydynger-Chodobska 2001), such as VEGF, may indirectly contribute to the effects of 5-HT$_{2C}$ agonists on cell proliferation in the SVZ.

More surprisingly, RO 600175 treatment also produces an increase in gliogenesis in the striatum and prefrontal cortex (Soumier et al. 2010). These two regions have been considered as showing clear adult neurogenic activity in response to brain injury (Magavi et al. 2000; Kokaia and Lindvall 2003). However, a low neurogenic activity is also present in basal conditions, which has been attributed to progenitors derived from the SVZ and/or local parenchymal progenitors (Cameron and Dayer 2008; Luzzati et al. 2006). We showed that in both regions, RO 600175 increases the total number of newly formed cells mainly expressing a glial marker (NG2, a marker of oligodendrocyte progenitors). This treatment did not induce a neuronal phenotype or affect the survival of these new cells. The regional analysis in the upper versus deeper cortical layers and medial versus lateral striatal subregions showed similar increases in every region. No gradient in the number of new cells could be detected, suggesting that these changes probably target local progenitors and do not involve a migration of progenitors from the SVZ. However, the lack of overlap with the distribution of the 5-HT2C receptors in these regions also favors the hypothesis that these effects are indirect.

Hippocampus

In the DG, acute or chronic activation of 5-HT$_{2A/2C}$ receptors does not affect cell proliferation or neurogenesis, as also observed with the selective 5-HT$_{2C}$ agonist (Banasr et al. 2004; Jha et al. 2006). By contrast, chronic blockade of 5-HT$_{2A/2C}$ (Jha et al. 2006; Soumier et al. 2009) or 5-HT$_{2C}$ receptors (Soumier et al. 2010), but not acute (Banasr et al. 2004), stimulates this process. These results underpin the need for long-term inactivation of 5-HT$_{2C}$ receptors. Consistent with the constitutive activity of this receptor subtype (Berg et al. 2005), only the selective 5-HT$_{2C}$ receptor inverse agonist (SB 243,213) induced an increase in cell proliferation, while the neutral antagonist (SB 242,084) had no effect. Notably, it has been demonstrated that 5-HT$_{2C}$ receptor inverse agonists are more efficacious in enhancing DA release (De Deurwaerdere and Spampinato 2001), probably via GABA-ergic inhibitory interneurons (Invernizzi et al. 2007). As previously mentioned, GABA has a negative influence on cell proliferation and GABA-ergic interneurons are known to express 5-HT$_{2C}$ receptors in various brain regions (Invernizzi et al. 2007; Serrats et al. 2005; Liu et al. 2007). Thus, besides the "more direct" 5-HT$_{2C}$-mediated GABA-ergic influence on precursor cells (Ge et al. 2007), a 5-HT$_{2C}$-mediated activation of NA (Millan et al. 2000) or DA projections can modulate hippocampal neurogenesis (Hoglinger et al. 2004; Kulkarni et al. 2002).

Regarding cell survival, none of the selective 5-HT$_{2C}$ agonist or antagonists affected this phase of neurogenesis in every brain region examined (Soumier et al. 2010). Administration of 5-HT$_{2A/2C}$ agonist has been shown to modulate hippocampal and cortical levels of BDNF (Vaidya et al. 1997), a trophic factor known to promote particularly cell differentiation and survival (Sairanen et al. 2007). However, this effect on BDNF levels involves 5-HT$_{2A}$ and not 5-HT$_{2C}$ receptors. In line with this result, we did not detect changes in hippocampal BDNF levels following administration of SB 242,084 or SB 243,213. Altogether, these data reinforce the lack of implication of 5-HT$_{2C}$ in cell survival.

9.3.2.3   5-HT$_{2C}$ Receptors, Adult Neurogenesis, and Depression–Anxiety Disorders

Both hippocampal neurogenesis and trophic factor levels show opposite changes in stressful situations and following antidepressant treatments. These observations led, a few years ago, to "the neurogenic/neurotrophic theory of depression and antidepressant action" (Duman and Monteggia 2006), based on the hypothesis that changes in neuroplasticity may compromise or favor neuronal function (Duman et al. 1997; Jacobs et al. 2000). A large number of studies and reviews have contributed to the debate (Sahay and Hen 2007; Paizanis et al. 2007; Pittenger and Duman 2008; Czeh and Lucassen 2007; Feldmann et al. 2007; Vollmayr et al. 2007; Fuchs 2007). Most of them agree with the fact that reducing neurogenesis by using different methods does not result in depressive-like behaviors, while induction of neurogenesis is required for the behavioral action of antidepressants (Santarelli et al. 2003; Vollmayr et al. 2003; Airan et al. 2007).

Interestingly, 5-HT$_{2C}$ receptors are thought to play a complex role in the regulation of mood, since both 5-HT$_{2C}$ *agonists* and *antagonists* have antidepressant-like effects (Millan 2005; Rosenzweig-Lipson et al. 2007). Likewise, both 5-HT$_{2C}$ agonist and antagonist increase neurogenesis, although in a region-dependent manner: Activation of 5-HT$_{2C}$ receptors leads to a stimulation of cell proliferation in the SVZ, while the blockade has the same effect in the SGZ. One possible explanation is that this 5-HT$_{2C}$ region-dependent regulation may be related to mRNA editing of this receptor subtype. Indeed, the constitutive activity at 5-HT$_{2C}$ sites is modulated by mRNA editing that generates unique isoforms of proteins in a cell- and/or tissue-specific manner, the unedited isoform displaying the greatest basal activity (Niswender et al. 1999). Thus, it would be of interest to compare 5-HT$_{2C}$ mRNA editing in different limbic structures, including hippocampus, following various stressful events and antidepressant treatments. In line with this, it has to be noted that increased 5-HT$_{2C}$ mRNA editing in cortical regions has been associated with depression and suicide (Gurevich et al. 2002).

Interestingly, we also demonstrated that 5-HT$_{2C}$ agonist increases cortical gliogenesis. Growing evidence indicates that glial elements are involved in the pathophysiology of depression (Rajkowska and Miguel-Hidalgo 2007) and that cortical gliogenesis is modulated by stress and antidepressants (Banasr and Duman 2007). Thus, the question of a possible functional implication of increased gliogenesis by activation of 5-HT$_{2C}$ receptors constitutes a promising target of investigation.

Finally, regarding the brain circuits involved in depression and anxiety, we observed a selective implication of the ventral hippocampus (VH) in cell proliferation following blockade of 5-HT$_{2C}$ receptors. These data may be related to anatomical and functional differences reported when comparing the ventral (temporal pole) and the dorsal (septal pole) hippocampus (Moser and Moser 1998; Bannerman et al. 2004). The projections of the VH to the prefrontal cortex and its strong connection with the amygdala support the view that this subregion is particularly involved in "emotional circuitry" and is more specialized for the control of anxiety and depression-related functions, whereas the DH is more implicated in cognitive functions (Bannerman et al. 2004; Engin and Treit 2007). This regional dissociation regarding the implication of new neurons in cognitive processes may not be so rigorous (Snyder et al. 2008). However, recent studies demonstrated selective decreases in neurogenesis in the VH following various stress exposures that elicit depressive-like behavior in the adult (Jayatissa et al. 2006; Kim et al. 2005; Lagace et al. 2006). Prenatal stress exposure also induced a reduction in neurogenesis in the VH selectively, which is associated with anxious behavior (Zuena et al. 2008). Conversely, the selective increase in cell proliferation in the VH following 5-HT_{2C} antagonist administration is consistent with the preferential implication of these receptors in emotional response. In contrast to their sparse expression in DH, 5-HT_{2C} receptors are, to a major extent, expressed in VH (Holmes et al. 1995), and their activation by injecting 5-HT_{2C} agonists in this subregion produced anxiogenic effects (Alves et al. 2004). In line with these data, agomelatine, a new antidepressant with anxiolytic property, acting partly as a 5-HT_{2C} antagonist, also increases, selectively, cell proliferation and neurogenesis in this subregion (Banasr et al. 2006). Altogether, these results suggest that the control of adult neurogenesis in the VH may be related to anxiolytic-antidepressant effects.

9.4 Conclusions

In conclusion, the adult neurogenesis theory of major depression has given an entirely new perspective on this mood disorder. Up to now, the only study available in humans did not show evidence that depression is associated with decreases in hippocampal cell proliferation, while this could be the case for schizophrenia (Reif et al. 2006). Various methodological problems may limit the interpretations resulting from this study (comments in References Sahay and Hen (2007) and Kempermann et al. (2008)), and more experimental and clinical work is needed to go further in this topic. However, a novel theory around the concept relating hippocampal neurogenesis to cognitive disturbances found in various psychiatric disorders, including depression and schizophrenia, may help to explain the functional contribution of newly formed neurons (Aimone et al. 2006; Kempermann et al. 2008; Becker and Wojtowicz 2007). More cellular and molecular studies aimed to dissect the mechanisms of action used by different modulators, particularly neurotransmitters in neurogenesis, will also result in a better understanding of the function of these new neurons in the adult brain.

Acknowledgments I wish to thank my PhD students: J.M. Brezun, M. Banasr, and A. Soumier, who have performed these studies on serotonin regulation of adult neurogenesis and largely contributed to their development, as well as the IC2N group of IBDML and L. Kerkerian Le Goff, S. Lorte, and F. Masmejean, particularly, and with special thanks to M. Hery. I also wish to thank our colleagues from S. Maccari group in Lille University. Finally, I am also very grateful to E. Mocaer, C. Gabriel-Gracia, and M. Millan from Servier Company for their help and support over the years.

References

Abrous DN, Koehl M, Le Moal M (2005) Adult neurogenesis: from precursors to network and physiology. Physiol Rev 85:523–569.

Aimone JB, Wiles J, Gage FH (2006) Potential role for adult neurogenesis in the encoding of time in new memories. Nat Neurosci 9:723–727.

Aimone JB, Wiles J, Gage FH (2009) Computational influence of adult neurogenesis on memory encoding. Neuron 61:187–202.

Airan RD, Meltzer LA, Roy M, et al (2007) High-speed imaging reveals neurophysiological links to behavior in an animal model of depression. Science 317:819–823.

Altman J (1969) Autoradiographic and histological studies of postnatal neurogenesis. IV. Cell proliferation and migration in the anterior forebrain, with special reference to persisting neurogenesis in the olfactory bulb. J Comp Neurol 137:433–457.

Alvarez-Buylla A, Garcia-Verdugo JM (2002) Neurogenesis in adult subventricular zone. J Neurosci 22:629–634.

Alvarez-Buylla A, Garcia-Verdugo JM, Tramontin AD (2001) A unified hypothesis on the lineage of neural stem cells. Nat Rev Neurosci 2:287–293.

Alves SH, Pinheiro G, Motta V, et al (2004) Anxiogenic effects in the rat elevated plus-maze of 5-HT(2C) agonists into ventral but not dorsal hippocampus. Behav Pharmacol 15:37–43.

Aponso PM, Faull RL, Connor B (2008) Increased progenitor cell proliferation and astrogenesis in the partial progressive 6-hydroxydopamine model of Parkinson's disease. Neuroscience 151:1142–1153.

Azmitia EC, Buchan AM, Williams JH (1978) Structural and functional restoration by collateral sprouting of hippocampal 5-HT axons. Nature 274:374–376.

Baker SA, Baker KA, Hagg T (2004) Dopaminergic nigrostriatal projections regulate neural precursor proliferation in the adult mouse subventricular zone. Eur J Neurosci 20:575–579.

Baker SA, Baker KA, Hagg T (2005) D3 dopamine receptors do not regulate neurogenesis in the subventricular zone of adult mice. Neurobiol Dis 18:523–527.

Banasr M, Duman RS (2007) Regulation of neurogenesis and gliogenesis by stress and antidepressant treatment. CNS Neurol Disord Drug Targets 6:311–320.

Banasr M, Hery M, Printemps R, et al (2004) Serotonin-induced increases in adult cell proliferation and neurogenesis are mediated through different and common 5-HT receptor subtypes in the dentate gyrus and the subventricular zone. Neuropsychopharmacology 29:450–460.

Banasr M, Soumier A, Hery M, et al (2006) Agomelatine, a new antidepressant, induces regional changes in hippocampal neurogenesis. Biol Psychiatry 59:1087–1096.

Bannerman DM, Rawlins JN, McHugh SB, et al (2004) Regional dissociations within the hippocampus—memory and anxiety. Neurosci Biobehav Rev 28:273–283.

Bauer S, Moyse E, Jourdan F, et al (2003) Effects of the alpha 2-adrenoreceptor antagonist dexefaroxan on neurogenesis in the olfactory bulb of the adult rat in vivo: selective protection against neuronal death. Neuroscience 117:281–291.

Becker S, Wojtowicz JM (2007) A model of hippocampal neurogenesis in memory and mood disorders. Trends Cogn Sci 11:70–76.

Berg KA, Harvey JA, Spampinato U, et al (2005) Physiological relevance of constitutive activity of 5-HT2A and 5-HT2C receptors. Trends Pharmacol Sci 26:625–630.

Bezard E, Gerlach I, Moratalla R, et al (2006) 5-HT1A receptor agonist-mediated protection from MPTP toxicity in mouse and macaque models of Parkinson's disease. Neurobiol Dis 23:77–86.

Bordey A (2007) Enigmatic GABAergic networks in adult neurogenic zones. Brain Res Rev 53:124–134.

Borta A, Hoglinger GU (2007) Dopamine and adult neurogenesis. J Neurochem 100:587–595.

Brezun JM, Daszuta A (1999) Depletion in serotonin decreases neurogenesis in the dentate gyrus and the subventricular zone of adult rats. Neuroscience 89:999–1002.

Brezun JM, Daszuta A (2000a) Serotonin may stimulate granule cell proliferation in the adult hippocampus, as observed in rats grafted with foetal raphe neurons. Eur J Neurosci 12:391–396.

Brezun JM, Daszuta A (2000b) Serotonergic reinnervation reverses lesion-induced decreases in PSA-NCAM labeling and proliferation of hippocampal cells in adult rats. Hippocampus 10:37–46.

Bruel-Jungerman E, Rampon C, Laroche S (2007) Adult hippocampal neurogenesis, synaptic plasticity and memory: facts and hypotheses. Rev Neurosci 18:93–114.

Cameron HA, Dayer AG (2008) New interneurons in the adult neocortex:small, sparse, but significant? Biol Psychiatry 63:650–655.

Cameron HA, McKay RD (2001) Adult neurogenesis produces a large pool of new granule cells in the dentate gyrus. J Comp Neurol 435:406–417.

Cameron HA, Woolley CS, McEwen BS, et al (1993) Differentiation of newly born neurons and glia in the dentate gyrus of the adult rat. Neuroscience 56:337–344.

Chen H, Pandey GN, Dwivedi Y (2006) Hippocampal cell proliferation regulation by repeated stress and antidepressants. Neuroreport 17:863–867.

Chodobski A, Szmydynger-Chodobska J (2001) Choroid plexus:target for polypeptides and site of their synthesis. Microsc Res Tech 52:65–82.

Christie BR, Cameron HA (2006) Neurogenesis in the adult hippocampus. Hippocampus 16:199–207.

Coronas V, Bantubungi K, Fombonne J, et al (2004) Dopamine D3 receptor stimulation promotes the proliferation of cells derived from the post-natal subventricular zone. J Neurochem 91:1292–1301.

Couillard-Despres S, Wuertinger C, Kandasamy M, et al (2009) Ageing abolishes the effects of fluoxetine on neurogenesis. Mol Psychiatry 14:856–864.

Cowen DS, Takase LF, Fornal CA, et al (2008) Age-dependent decline in hippocampal neurogenesis is not altered by chronic treatment with fluoxetine. Brain Res 1228:14–19.

Czeh B, Lucassen PJ (2007) What causes the hippocampal volume decrease in depression? Are neurogenesis, glial changes and apoptosis implicated? Eur Arch Psychiatry Clin Neurosci 257:250–260.

De Deurwaerdere P, Spampinato U (2001) The nigrostriatal dopamine system: a neglected target for 5-HT2C receptors. Trends Pharmacol Sci 22:502–504.

Doetsch F (2003) A niche for adult neural stem cells. Curr Opin Genet Dev 13:543–550.

Doetsch F, Hen R (2005) Young and excitable: the function of new neurons in the adult mammalian brain. Curr Opin Neurobiol 15:121–128.

Doetsch F, Caille I, Lim DA, Garcia-Verdugo JM, Alvarez-Buylla A (1999) Subventricular zone astrocytes are neural stem cells in the adult mammalian brain. Cell 97:703–716.

Drapeau E, Abrous Nora D (2008) Stem cell review series: role of neurogenesis in age-related memory disorders. Aging Cell 7:569–589.

Duan X, Kang E, Liu CY, Ming GL, Song H (2008) Development of neural stem cell in the adult brain. Curr Opin Neurobiol 18:108–115.

Duman RS, Monteggia LM (2006) A neurotrophic model for stress-related mood disorders. Biol Psychiatry 59:1116–1127.

Duman RS, Heninger GR, Nestler EJ (1997) A molecular and cellular theory of depression. Arch Gen Psychiatry 54:597–606.

Dupret D, Revest JM, Koehl M, Ichas F, De Giorgi F, Costet P, Abrous DN, Piazza PV (2008) Spatial relational memory requires hippocampal adult neurogenesis. PLoS ONE 3:e1959.

Ehninger D, Kempermann G (2008) Neurogenesis in the adult hippocampus. Cell Tissue Res 331:243–250.

Encinas JM, Enikolopov G (2008) Identifying and quantitating neural stem and progenitor cells in the adult brain. Methods Cell Biol 85:243–72.

Encinas JM, Vaahtokari A, Enikolopov G (2006) Fluoxetine targets early progenitor cells in the adult brain. Proc Natl Acad Sci USA 103:8233–8238.

Engin E, Treit D (2007) The role of hippocampus in anxiety: intracerebral infusion studies. Behav Pharmacol 18:365–374.

Eriksson PS, Perfilieva E, Bjork-Eriksson T, et al (1998) Neurogenesis in the adult human hippocampus. Nat Med 4:1313–1317.

Fabel K, Kempermann G (2008) Physical activity and the regulation of neurogenesis in the adult and aging brain. Neuromolecular Med 10:59–66.

Feldmann RE Jr, Sawa A, Seidler GH (2007) Causality of stem cell based neurogenesis and depression—to be or not to be, is that the question? J Psychiatr Res 41:713–723.

Freundlieb N, Francois C, Tande D, et al (2006) Dopaminergic substantia nigra neurons project topographically organized to the subventricular zone and stimulate precursor cell proliferation in aged primates. J Neurosci 26:2321–2325.

Fuchs E (2007) Neurogenesis in the adult brain: is there an association with mental disorders? Eur Arch Psychiatry Clin Neurosci 257:247–249.

Gage FH (2000) Mammalian neural stem cells. Science 287:1433–1438.

Galea LA, Spritzer MD, Barker JM, Pawluski JL (2006) Gonadal hormone modulation of hippocampal neurogenesis in the adult. Hippocampus 16:225–232.

Ge S, Pradhan DA, Ming GL, Song H (2007) GABA sets the tempo for activity-dependent adult neurogenesis. Trends Neurosci 30:1–8.

Geraerts M, Krylyshkina O, Debyser Z, et al (2007) Concise review:therapeutic strategies for Parkinson disease based on the modulation of adult neurogenesis. Stem Cells 25:263–270.

Gould E (2007) How widespread is adult neurogenesis in mammals? Nat Rev Neurosci 8:481–488

Gould E, Cameron HA, Daniels DC, et al (1992) Adrenal hormones suppress cell division in the adult rat dentate gyrus. J Neurosci 12:3642–3650.

Gurevich I, Tamir H, Arango V, et al (2002) Altered editing of serotonin 2C receptor pre-mRNA in the prefrontal cortex of depressed suicide victims. Neuron 34:349–356.

Hagg T (2005) Molecular regulation of adult CNS neurogenesis: an integrated view. Trends Neurosci 28:589–595.

Hoglinger GU, Rizk P, Muriel MP, et al (2004) Dopamine depletion impairs precursor cell proliferation in Parkinson disease. Nat Neurosci 7:726–735.

Holick KA, Lee DC, Hen R, et al (2008) Behavioral effects of chronic fluoxetine in BALB/cJ mice do not require adult hippocampal neurogenesis or the serotonin 1A receptor. Neuropsychopharmacology 33:406–417.

Holmes MC, French KL, Seckl JR (1995) Modulation of serotonin and corticosteroid receptor gene expression in the rat hippocampus with circadian rhythm and stress. Brain Res Mol Brain Res 28:186–192.

Huang GJ, Herbert J (2005) The role of 5-HT1A receptors in the proliferation and survival of progenitor cells in the dentate gyrus of the adult hippocampus and their regulation by corticoids. Neuroscience 135:803–813.

Huang GJ, Herbert J (2006) Stimulation of neurogenesis in the hippocampus of the adult rat by fluoxetine requires rhythmic change in corticosterone. Biol Psychiatry 59:619–624.

Imayoshi I, Sakamoto M, Ohtsuka T, Takao K, Miyakawa T, Yamaguchi M, Mori K, Ikeda T, Itohara S, Kageyama R (2008) Roles of continuous neurogenesis in the structural and functional integrity of the adult forebrain. Nat Neurosci 11:1153–1161.

Invernizzi RW, Pierucci M, Calcagno E, et al (2007) Selective activation of 5-HT(2C) receptors stimulates GABA-ergic function in the rat substantia nigra pars reticulata: a combined in vivo electrophysiological and neurochemical study. Neuroscience 144:1523–1535.

Jacobs BL, Praag H, Gage FH (2000) Adult brain neurogenesis and psychiatry: a novel theory of depression. Mol Psychiatry 5:262–269.

Jayatissa MN, Bisgaard C, Tingstrom A, et al (2006) Hippocampal cytogenesis correlates to escitalopram-mediated recovery in a chronic mild stress rat model of depression. Neuropsychopharmacology 31:2395–2404.

Jessberger S, Clark RE, Broadbent NJ, Clemenson GD Jr, Consiglio A, Lie DC, Squire LR, Gage FH (2009) Dentate gyrus-specific knockdown of adult neurogenesis impairs spatial and object recognition memory in adult rats. Learn Mem 16:147–154.

Jha S, Rajendran R, Davda J, et al (2006) Selective serotonin depletion does not regulate hippocampal neurogenesis in the adult rat brain: differential effects of p-chlorophenylalanine and 5,7-dihydroxytryptamine. Brain Res 1075:48–59.

Jha S, Rajendran R, Fernandes KA, et al (2008) 5-HT2A/2C receptor blockade regulates progenitor cell proliferation in the adult rat hippocampus. Neurosci Lett 441:210–214.

Joo JY, Kim BW, Lee JS, et al (2007) Activation of NMDA receptors increases proliferation and differentiation of hippocampal neural progenitor cells. J Cell Sci 120:1358–1370.

Kempermann G (2008) The neurogenic reserve hypothesis:what is adult hippocampal neurogenesis good for? Trends Neurosci 31:163–169.

Kempermann G, Kuhn HG, Gage FH (1997) Genetic influence on neurogenesis in the dentate gyrus of adult mice. Proc Natl Acad Sci USA 94:10409–10414.

Kempermann G, Gast D, Gage FH (2002) Neuroplasticity in old age: sustained fivefold induction of hippocampal neurogenesis by long-term environmental enrichment. Ann Neurol 52:135–143.

Kempermann G, Gast D, Kronenberg G, Yamaguchi M, Gage FH (2003) Early determination and long-term persistence of adult-generated new neurons in the hippocampus of mice. Development 130:391–399.

Kempermann G, Jessberger S, Steiner B, Kronenberg G (2004a) Milestones of neuronal development in the adult hippocampus. Trends Neurosci 27:447–452.

Kempermann G, Wiskott L, Gage FH (2004b) Functional significance of adult neurogenesis. Curr Opin Neurobiol 14:186–191.

Kempermann G, Chesler EJ, Lu L, Williams RW, Gage FH (2006) Natural variation and genetic covariance in adult hippocampal neurogenesis. Proc Natl Acad Sci USA 103:780–785.

Kempermann G, Krebs J, Fabel K (2008) The contribution of failing adult hippocampal neurogenesis to psychiatric disorders. Curr Opin Psychiatry 21:290–295.

Kia HK, Miquel MC, Brisorgueil MJ, et al (1996) Immunocytochemical localization of serotonin1A receptors in the rat central nervous system. J Comp Neurol 365:289–305.

Kim SJ, Lee KJ, Shin YC, et al (2005) Stress-induced decrease of granule cell proliferation in adult rat hippocampus: assessment of granule cell proliferation using high doses of bromodeoxyuridine before and after restraint stress. Mol Cells 19:74–80.

Kippin TE, Kapur S, van der Kooy D (2005) Dopamine specifically inhibits forebrain neural stem cell proliferation, suggesting a novel effect of antipsychotic drugs. J Neurosci 25:5815–5823.

Kokaia Z, Lindvall O (2003) Neurogenesis after ischaemic brain insults. Curr Opin Neurobiol 13:127–132.

Kuhn HG, Dickinson-Anson H, Gage FH (1996) Neurogenesis in the dentate gyrus of the adult rat: age-related decrease of neuronal progenitor proliferation. J Neurosci 16:2027–2033.

Kuhn HG, Cooper-Kuhn CM, Boekhoorn K, Lucassen PJ (2007) Changes in neurogenesis in dementia and Alzheimer mouse models: are they functionally relevant? Eur Arch Psychiatry Clin Neurosci 257:281–289.

Kulkarni VA, Jha S, Vaidya VA (2002) Depletion of norepinephrine decreases the proliferation, but does not influence the survival and differentiation, of granule cell progenitors in the adult rat hippocampus. Eur J Neurosci 16:2008–2012.

Lagace DC, Yee JK, Bolanos CA, et al (2006) Juvenile administration of methylphenidate attenuates adult hippocampal neurogenesis. Biol Psychiatry 60:1121–1130.

Lagace DC, Fischer SJ, Eisch AJ (2007) Gender and endogenous levels of estradiol do not influence adult hippocampal neurogenesis in mice. Hippocampus 17:175–180.

Lee HJ, Kim JW, Yim SV, et al (2001) Fluoxetine enhances cell proliferation and prevents apoptosis in dentate gyrus of maternally separated rats. Mol Psychiatry 6:610, 725–728.

Leuner B, Gould E, Shors TJ (2006) Is there a link between adult neurogenesis and learning? Hippocampus 16:216–224.

Liu BF, Gao EJ, Zeng XZ, Ji M, Cai Q, Lu Q, Yang H, Xu QY (2006) Proliferation of neural precursors in the subventricular zone after chemical lesions of the nigrostriatal pathway in rat brain. Brain Res 1106:30–39.

Liu S, Bubar MJ, Lanfranco MF, et al (2007) Serotonin2C receptor localization in GABA neurons of the rat medial prefrontal cortex:implications for understanding the neurobiology of addiction. Neuroscience 146:1677–1688.

Lledo PM, Saghatelyan A (2005) Integrating new neurons into the adult olfactory bulb: joining the network, life-death decisions, and the effects of sensory experience. Trends Neurosci 28:248–254.

Lledo PM, Alonso M, Grubb MS (2006) Adult neurogenesis and functional plasticity in neuronal circuits. Nat Rev Neurosci 7:179–193.

Lucas G, Rymar VV, Du J, et al (2007) Serotonin(4) (5-HT(4)) receptor agonists are putative antidepressants with a rapid onset of action. Neuron 55:712–725.

Luzzati F, De Marchis S, Fasolo A, et al (2006) Neurogenesis in the caudate nucleus of the adult rabbit. J Neurosci 26:609–621.

Ma DK, Ming GL, Song H (2005) Glial influences on neural stem cell development: cellular niches for adult neurogenesis. Curr Opin Neurobiol 15:514–520.

Magavi SS, Leavitt BR, Macklis JD (2000) Induction of neurogenesis in the neocortex of adult mice. Nature 405:951–955.

Malberg JE, Duman RS (2003) Cell proliferation in adult hippocampus is decreased by inescapable stress: reversal by fluoxetine treatment. Neuropsychopharmacology 28:1562–1571.

Malberg JE, Eisch AJ, Nestler EJ, et al (2000) Chronic antidepressant treatment increases neurogenesis in adult rat hippocampus. J Neurosci 20:9104–9110.

Manganas LN, Zhang X, Li Y, et al (2007) Magnetic resonance spectroscopy identifies neural progenitor cells in the live human brain. Science 318:980–985.

Marcussen AB, Flagstad P, Kristjansen PE, et al (2008) Increase in neurogenesis and behavioural benefit after chronic fluoxetine treatment in Wistar rats. Acta Neurol Scand 117:94–100.

Markwardt S, Overstreet-Wadiche L (2008) GABAergic signalling to adult-generated neurons. J Physiol 586:3745–3749.

Matsuyama S, Nei K, Tanaka C (1997) Regulation of GABA release via NMDA and 5-HT1A receptors in guinea pig dentate gyrus. Brain Res 761:105–112.

Mauler F, Horvath E (2005) Neuroprotective efficacy of repinotan HCl, a 5-HT1A receptor agonist, in animal models of stroke and traumatic brain injury. J Cereb Blood Flow Metab 25:451–459.

Meerlo P, Mistlberger RE, Jacobs BL, Craig Heller H, McGinty D (2008) New neurons in the adult brain: The role of sleep and consequences of sleep loss. Sleep Med Rev 13:187–194.

Millan MJ (2005) Serotonin 5-HT2C receptors as a target for the treatment of depressive and anxious states: focus on novel therapeutic strategies. Therapie 60:441–460

Millan MJ, Gobert A, Rivet JM, et al (2000) Mirtazapine enhances frontocortical dopaminergic and corticolimbic adrenergic, but not serotonergic, transmission by blockade of alpha2-adrenergic and serotonin2C receptors:a comparison with citalopram. Eur J Neurosci 12:1079–1095.

Miller BH, Schultz LE, Gulati A, et al (2008) Genetic regulation of behavioral and neuronal responses to fluoxetine. Neuropsychopharmacology 33:1312–1322.

Ming GL, Song H (2005) Adult neurogenesis in the mammalian central nervous system. Annu Rev Neurosci 28:223–250.

Mirescu C, Gould E (2006) Stress and adult neurogenesis. Hippocampus 16:233–238.

Moser MB, Moser EI (1998) Functional differentiation in the hippocampus. Hippocampus 8:608–619.

Nacher J, McEwen BS (2006) The role of N-methyl-D-asparate receptors in neurogenesis. Hippocampus 16:267–270.

Nacher J, Varea E, Miguel Blasco-Ibanez J, et al (2007) N-methyl-d-aspartate receptor expression during adult neurogenesis in the rat dentate gyrus. Neuroscience 144:855–864.

Ninkovic J, Gotz M (2007) Signaling in adult neurogenesis:from stem cell niche to neuronal networks. Curr Opin Neurobiol 17:338–344.

Niswender CM, Copeland SC, Herrick-Davis K, et al (1999) RNA editing of the human serotonin 5-hydroxytryptamine 2C receptor silences constitutive activity. J Biol Chem 274:9472–9478.

Olson AK, Eadie BD, Ernst C, Christie BR (2006) Environmental enrichment and voluntary exercise massively increase neurogenesis in the adult hippocampus via dissociable pathways. Hippocampus 16:250–260.

Ortega-Perez I, Murray K, Lledo PM (2007) The how and why of adult neurogenesis. J Mol Histol 38:555–562.

Paizanis E, Hamon M, Lanfumey L (2007) Hippocampal neurogenesis, depressive disorders, and antidepressant therapy. Neural Plast 2007:73754.

Palacios JM, Waeber C, Hoyer D, et al (1990) Distribution of serotonin receptors. Ann N Y Acad Sci 600:36–52.

Palmer TD, Willhoite AR, Gage FH (2000) Vascular niche for adult hippocampal neurogenesis. J Comp Neurol 425:479–494.

Pittenger C, Duman RS (2008) Stress, depression, and neuroplasticity: a convergence of mechanisms. Neuropsychopharmacology 33:88–109.

Pompeiano M, Palacios JM, Mengod G (1992) Distribution and cellular localization of mRNA coding for 5-HT1A receptor in the rat brain: correlation with receptor binding. J Neurosci 12:440–453.

Qiu G, Helmeste DM, Samaranayake AN, et al (2007) Modulation of the suppressive effect of corticosterone on adult rat hippocampal cell proliferation by paroxetine. Neurosci Bull 23:131–136.

Radley JJ, Jacobs BL (2002) 5-HT1A receptor antagonist administration decreases cell proliferation in the dentate gyrus. Brain Res 955:264–267.

Rajkowska G, Miguel-Hidalgo JJ (2007) Gliogenesis and glial pathology in depression. CNS Neurol Disord Drug Targets 6:219–233.

Reif A, Fritzen S, Finger M, et al (2006) Neural stem cell proliferation is decreased in schizophrenia, but not in depression. Mol Psychiatry 11:514–522.

Riquelme PA, Drapeau E, Doetsch F (2008) Brain micro-ecologies: neural stem cell niches in the adult mammalian brain. Philos Trans R Soc Lond B Biol Sci 363:123–137.

Rizk P, Salazar J, Raisman-Vozari R, et al (2006) The alpha2-adrenoceptor antagonist dexefaroxan enhances hippocampal neurogenesis by increasing the survival and differentiation of new granule cells. Neuropsychopharmacology 31:1146–1157.

Rosenbrock H, Bloching A, Weiss C, et al (2005) Partial serotonergic denervation decreases progenitor cell proliferation in the adult rat hippocampus, but has no effect on rat behavior in the forced swimming test. Pharmacol Biochem Behav 80:549–556.

Rosenzweig-Lipson S, Dunlop J, Marquis KL (2007) 5-HT2C receptor agonists as an innovative approach for psychiatric disorders. Drug News Perspect 20:565–571.

Sahay A, Hen R (2007) Adult hippocampal neurogenesis in depression. Nat Neurosci 10:1110–1115.

Sairanen M, O'Leary OF, Knuuttila JE, et al (2007) Chronic antidepressant treatment selectively increases expression of plasticity-related proteins in the hippocampus and medial prefrontal cortex of the rat. Neuroscience 144:368–374.

Santarelli L, Saxe M, Gross C, et al (2003) Requirement of hippocampal neurogenesis for the behavioral effects of antidepressants. Science 301:805–809.

Schaper C, Zhu Y, Kouklei M, et al (2000) Stimulation of 5-HT(1A) receptors reduces apoptosis after transient forebrain ischemia in the rat. Brain Res 883:41–50.

Scharfman HE, Hen R (2007) Neuroscience. Is more neurogenesis always better? Science 315:336–338.

Schauwecker PE (2006) Genetic influence on neurogenesis in the dentate gyrus of two strains of adult mice. Brain Res 1120:83–92.

Schmidt HD, Duman RS (2007) The role of neurotrophic factors in adult hippocampal neurogenesis, antidepressant treatments and animal models of depressive-like behavior. Behav Pharmacol 18:391–418.

Seki T, Arai Y (1995) Age-related production of new granule cells in the adult dentate gyrus. Neuroreport 6:2479–2482.

Serrats J, Mengod G, Cortes R (2005) Expression of serotonin 5-HT2C receptors in GABAergic cells of the anterior raphe nuclei. J Chem Neuroanat 29:83–91.

Shors TJ (2008) From stem cells to grandmother cells: how neurogenesis relates to learning and memory. Cell Stem Cell 3:253–258.

Snyder JS, Radik R, Wojtowicz JM, et al (2008) Anatomical gradients of adult neurogenesis and activity: young neurons in the ventral dentate gyrus are activated by water maze training. Hippocampus 19:360–370.

Soumier A, Lortet S, Masmejean F, et al (2009) Cellular and molecular mechanisms underlying increased adult hippocampal neurogenesis induced by agomelatine. Neuropsychopharmacology 34:2390–2403.

Soumier A, Banasr M, Kerkerian-Le Gogg L, et al (2010) Region and phase-dependent effects of 5-HT1A and 5-HT2C receptor activation on adult neurogenesis. Eur Neuropsychopharmacol 20:336–345.

Ueda S, Sakakibara S, Yoshimoto K (2005) Effect of long-lasting serotonin depletion on environmental enrichment-induced neurogenesis in adult rat hippocampus and spatial learning. Neuroscience 135:395–402.

Vaidya VA, Marek GJ, Aghajanian GK, et al (1997) 5-HT2A receptor-mediated regulation of brain-derived neurotrophic factor mRNA in the hippocampus and the neocortex. J Neurosci 17:2785–2795.

Vaidya VA, Vadodaria KC, Jha S (2007) Neurotransmitter regulation of adult neurogenesis: putative therapeutic targets. CNS Neurol Disord Drug Targets 6:358–374.

Van Kampen JM, Hagg T, Robertson HA (2004) Induction of neurogenesis in the adult rat subventricular zone and neostriatum following dopamine D3 receptor stimulation. Eur J Neurosci 19:2377–2387.

Vollmayr B, Simonis C, Weber S, et al (2003) Reduced cell proliferation in the dentate gyrus is not correlated with the development of learned helplessness. Biol Psychiatry 54:1035–1040.

Vollmayr B, Mahlstedt MM, Henn FA (2007) Neurogenesis and depression: what animal models tell us about the link. Eur Arch Psychiatry Clin Neurosci 257:300–303.

Wang JW, David DJ, Monckton JE, et al (2008) Chronic fluoxetine stimulates maturation and synaptic plasticity of adult-born hippocampal granule cells. J Neurosci 28:1374–1384.

Winner B, Cooper-Kuhn CM, Aigner R, Winkler J, Kuhn HG (2002) Long-term survival and cell death of newly generated neurons in the adult rat olfactory bulb. Eur J Neurosci 16:1681–1689.

Winner B, Geyer M, Couillard-Despres S, et al (2006) Striatal deafferentation increases dopaminergic neurogenesis in the adult olfactory bulb. Exp Neurol 197:113–121.

Yamaguchi M, Saito H, Suzuki M, et al (2000) Visualization of neurogenesis in the central nervous system using nestin promoter-GFP transgenic mice. Neuroreport 11:1991–1996.

Yang P, Arnold SA, Habas A, et al (2008) Ciliary neurotrophic factor mediates dopamine D2 receptor-induced CNS neurogenesis in adult mice. J Neurosci 28:2231–2241.

Zhang L, Guadarrama L, Corona-Morales AA, et al (2006) Rats subjected to extended L-tryptophan restriction during early postnatal stage exhibit anxious-depressive features and structural changes. J Neuropathol Exp Neurol 65:562–570.

Zhao C, Deng W, Gage FH (2008) Mechanisms and functional implications of adult neurogenesis. Cell 132:645–660.

Zuena AR, Mairesse J, Casolini P, et al (2008) Prenatal restraint stress generates two distinct behavioral and neurochemical profiles in male and female rats. PLoS ONE 3:e2170.

Chapter 10
The Constitutive Activity of 5-HT$_{2C}$ Receptors as an Additional Modality of Interaction of the Serotonergic System

Sylvia Navailles and Philippe De Deurwaerdère

10.1 Introduction

It is generally acknowledged that neurotransmitters influence their targets in a tonic and/or phasic manner via their receptors. Additionally, numerous G-protein-coupled receptors (GPCR) may spontaneously activate intracellular pathways independently from the presence of the endogenous neurotransmitter. The so-called constitutive activity of GPCR is quite easy to study on heterologous cell expression system in vitro in which the level of endogenous ligand can be controlled and the pharmacological ligands can be used and/or characterized adequately. In vivo, none of the two later requirements can be properly controlled so that the postulate that the constitutive activity of a GPCR can physiologically influence cellular and system activity should be viewed with caution. In addition, it can not be a universal property for a given receptor, as the influence of one receptor varies among tissues and regions in the body. Yet, the existence of physiological constitutive activity has been proposed for numerous GPCR including the histaminergic H$_3$ autoreceptor (Morisset et al. 2000), the serotonin-2A receptor (Harvey et al. 1999), and the serotonin-2C (5-HT$_{2C}$) receptor (De Deurwaerdère et al. 2004), implying that a given receptor could, in theory, cumulate phasic, tonic, and constitutive influences.

The 5-HT$_{2C}$ receptor is one of the main examples available to illustrate the diversity of interaction of a receptor toward its various cerebral targets, putatively mixing the three above-mentioned modalities of interaction. This peculiarity would be a physiological property of 5-HT$_{2C}$ receptors in the basal ganglia, a group of subcortical brain regions involved in motor control. A closer look at the available data in vivo, however, raises concerns about the transposition of the concept of constitutive activity from in vitro settings to the living animals. This chapter presents briefly how 5-HT$_{2C}$ receptors share their constitutive activity with the phasic and tonic controls in theory. Then, we will present the experimental arguments

S. Navailles (✉)
Laboratoire Mouvement Adaptation Cognition - UMR CNRS 5227 Université Victor Segalen Bordeaux 2146 rue Léo Saignât, Zone Nord, Bât. 2A 33076, Bordeaux, France
e-mail: sylvia.navailles@u-bordeaux2.fr

G. Di Giovanni et al. (eds.), *5-HT$_{2C}$ Receptors in the Pathophysiology of CNS Disease*,   187
The Receptors 22, DOI 10.1007/978-1-60761-941-3_10,
© Springer Science+Business Media, LLC 2011

obtained from the study of various 5-HT$_{2C}$ drugs (numerous antagonists, inverse agonists) in the control of dopamine release in vivo. This will raise the idea that the control exerted by the constitutive activity of 5-HT$_{2C}$ receptors is dependent on the use of anesthetics, the level of activity of dopamine neurons, and the region of basal ganglia considered. Then, we will bring up data on Fos expression in basal ganglia and oral activity in rats favoring the existence of the three modalities of interaction of 5-HT$_{2C}$ receptors in freely behaving animals. These new data will directly and indirectly address whether the three modalities, namely, phasic, tonic, and constitutive, are related to the same 5-HT$_{2C}$ receptors in basal ganglia.

10.1.1 Constitutive Activity of 5-HT$_{2C}$ Receptors: In vitro Consideration and Pharmacology

The constitutive activity of seven-transmembrane receptor is a concept now widely accepted and well characterized in heterologous recombinant systems in vitro (Costa and Herz 1989; Milligan et al. 1995; Berg et al. 2001; Kenakin 2001, 2004). The capacity of this receptor to regulate cellular signaling systems in the absence of occupancy by a ligand could be considered as artificial property as it depends on the density of the receptor at cell surface and the total absence of the endogenous ligand, two conditions that can be easily controlled in vitro (Milligan et al. 1995; Milligan and Bond 1997). Nevertheless, a large number of seven-transmembrane receptors exhibit constitutive activity in vitro (Seifert and Wenzel-Seifert 2002), which suggests that inverse agonism might have broad pharmacological relevance.

5-HT$_{2C}$ receptors couple to multiple effectors systems (Fig. 10.1) (Berg et al. 2008) and multistate receptor models predict that constitutive activity should depend on the responses measured (De Deurwaerdère et al. 2004; Berg et al. 2005). It has been shown that 5-HT$_{2C}$ receptors display a higher degree of constitutive activity towards phospholipase C (PLC) compared with phospholipase A$_2$ (PLA$_2$) (Barker et al. 1994; Berg et al. 1999; Herrick-Davis et al. 1999; Niswender et al. 1999). Numerous 5-HT$_{2C}$ compounds previously thought to act as antagonists, such as mesulergine, ritanserin, mianserin, and the prototypical atypical antipsychotic drugs clozapine and olanzapine, have been characterized as inverse agonists in vitro on PLC signaling pathway (Fig. 10.1b) (Rauser et al. 2001). Indeed, these compounds decrease the high basal response induced by constitutively active 5-HT$_{2C}$ receptors on myo-[^{3}H]inositol phosphate (IP) accumulation (Berg et al. 1998, 1999, 2005; Barker et al. 1994). Their effect is opposite to the effects of 5-HT$_{2C}$ agonists in these heterologous systems. Further, 5-HT$_{2C}$ antagonists are able to block, in a concentration-dependent manner, the effects induced by both agonists and inverse agonists (Fig. 10.1a) (Berg et al. 1998; Kennett et al. 1996, 1997; Price et al. 2001). In Chinese hamster ovary (CHO) cells expressing human 5-HT$_{2C}$ receptors at a density to optimize its constitutive activity (Berg et al. 1998), SB 206553 behaves as a strong inverse agonist at PLC-dependent, PLA$_2$-dependent, and activation of Gα$_i$-dependent responses (De Deurwaerdère et al. 2004). SB 242084 is equally

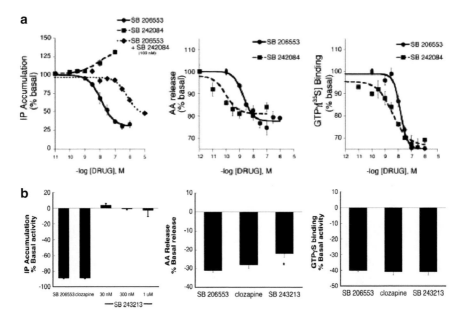

Fig. 10.1 Pharmacological properties of 5-HT$_{2C}$ inverse agonists (SB 206553 and clozapine) and 5-HT$_{2C}$ antagonists (SB 242084 and SB 243213) in CHO-1C7 cells. The effects of each compound were measured as the percentage of basal activity of three effectors systems: PLC, PLA$_2$ and activation of Gαi. Respectively, the amount of myo-[^{3}H]inositol phosphate (*IP*) accumulated, [^{3}H] arachidonic acid (*AA*) released, and [^{35}S]GTPγS bound was determined after 25 min of incubation with the indicated drugs. (**a**) Basal IP accumulation, AA release, and [^{35}S]GTPγS binding levels were as follows: 5,378±789 dpm, 1,544±167 dpm, and 72±13 pmol/mg protein, respectively. Data are expressed as the mean±SEM of four to six independent experiments. (**b**) Basal IP accumulation, AA release, and [^{35}S]GTPγS binding levels were as follows: 6,577±407 dpm, 1,222±42 dpm, and 89±7 pmol/mg protein, respectively. Data are expressed as the mean±SEM of three to five independent experiments (*$p<0.05$ compared with the effect of SB 206553, Student's paired *t* test) (**a**: From De Deurwaerdère et al. 2004. Copyright permission from the Society for Neuroscience. **b**: Adapted from Berg et al. 2006. Copyright permission from the American Society for Pharmacology and Experimental Therapeutics)

effective as an inverse agonist towards PLA$_2$ and Gα_i activation but displays low-efficacy agonism toward PLC (De Deurwaerdère et al. 2004). SB 243213 behaves as a partial inverse agonist at PLA$_2$ and as a full inverse agonist at Gα_i activation, while it displays neutral antagonism at PLC (Burns et al. 1997). SB 243213 and SB 242084 correspond pharmacologically to protean ligands (Kenakin 2001; Berg et al. 2005). According to their pharmacological properties at PLC-dependent responses, SB 242084 and SB 243213 induce a rightward shift of the inhibition of IP accumulation induced by SB 206553 (Fig. 10.1a) (De Deurwaerdère et al. 2004; Berg et al. 2006).

Other cellular effectors such as desensitization mechanisms occurring via G-protein-coupled receptor kinase (GRK) and receptor internalization (down-regulation) are targeted by the constitutive activity of 5-HT$_{2C}$ receptors (Berg et al. 1999, 2006). These mechanisms triggered by a prolonged exposure of the receptor to inverse agonists lead to enhanced responsiveness of both ligand-dependent and ligand-independent 5-HT$_{2C}$ receptor activation through the G$_{q/11}$–PLC signaling pathway without change in receptor density (Berg et al. 1999). The molecular plasticity of the 5-HT$_{2C}$ receptor is further supported by the posttranscriptional process of editing. Adenosine-to-inosine editing events of the 5-HT$_{2C}$ mRNA leads to changes in amino acids in the second intracellular loop of the receptor, a domain known to play an important role in G-protein coupling, desensitization mechanisms, and isomerization. All these mechanisms are able to modulate the level of constitutive receptor activity (Burstein et al. 1995; Marion et al. 2004; Werry et al. 2008). RNA editing generates different mRNA transcripts that encode for numerous receptor isoforms (Burns et al. 1997; Fitzgerald et al. 1999). In general, the constitutive activity of RNA-edited isoforms of the 5-HT$_{2C}$ receptor towards PLC is reduced compared with the nonedited receptor (Berg et al. 2005, 2008; Herrick-Davis et al. 1999; Niswender et al. 1999). Along with a specific distribution of RNA-edited 5-HT$_{2C}$ isoforms throughout the brain (Burns et al. 1997) and a regulation by 5-HT (Gurevich et al. 2002; Schmauss 2005), it has been suggested that this process may allow for a fine-tuning of 5-HT transmission via the 5-HT$_{2C}$ receptor (Berg et al. 2008; Herrick-Davis et al. 1999; Niswender et al. 1999).

10.2   A Puzzling Control of In vivo DA Release Exerted by 5-HT$_{2C}$ Receptors

The collection of microdialysis data in the literature brings an overview of the different effects induced by numerous 5-HT$_{2C}$ agonists and antagonists/inverse agonists on dopamine (DA) release (Tables 10.1 and 10.2). The phasic inhibitory control exerted by 5-HT$_{2C}$ receptors is illustrated by the overall decrease in DA release induced by the systemic administration of selective and nonselective 5-HT$_{2C}$ agonists (see Table 10.2 for references). The maximal inhibition in DA release reaches up to 35–40% below basal levels in the striatum (STR), the nucleus accumbens (NAc), and the medial prefrontal cortex (mPFC), as shown by the commonly used and potent 5-HT$_{2C}$ agonist Ro 60-0175 (Gobert et al. 2000; Millan et al. 1998; Navailles et al. 2004; De Deurwaerdère and Spampinato 2001; Di Matteo et al. 2001). On the other hand, the tonic inhibitory control exerted by 5-HT$_{2C}$ receptors is revealed by the overall increase in DA release induced by the systemic administration of selective and nonselective 5-HT$_{2C}$ antagonists/inverse agonists (see Table 10.1 for references). Compared with 5-HT$_{2C}$ agonists, 5-HT$_{2C}$ antagonists induce a larger range of effects on basal DA release.

Numerous studies report that 5-HT$_{2C}$ receptors also exert an inhibitory control on stimulated DA release, although in restricted conditions. 5-HT$_{2C}$ antagonists facilitate DA release enhanced by morphine (Porras et al. 2002), haloperidol (Lucas et al. 2000), MK-801 and phencyclidine (Hutson et al. 2000), and cocaine (Navailles et al. 2004) but not by amphetamine (Porras et al. 2002). 5-HT$_{2C}$ agonists

Table 10.1 Effect of the systemic administration of 5-HT2C antagonists on DA release in the striatum, the nucleus accumbens (NAc) and the frontal cortex of freely-moving (FM) and anesthetized rats (Anesthetic: halothane and chloral hydrate).

	Striatum		NAc		Frontal cortex		References
	Anesthesia	FM	Anesthesia	FM	Anesthesia	FM	
5-HT$_{2C}$ antagonists							
SB 242084							
0.5	0		0				De Deurwaerdère et al. (2004); Di Matteo et al. (1999, 2000a, 2004); Gobert et al. (2000); Hutson et al. (2000); Li et al. (2005a); Millan et al. (1998); Navailles et al. (2006a); Pozzi et al. (2002)
0.63						40	
1	20, 19		25, 25	0, 0		0	
2.5			0				
3	25		30				
5	5		16				
10	24, 10	0	28, 35	0		75, 75, 86	
SB 243213							
0.5	0		0				Berg et al. (2006); Di Matteo et al. (2004); Navailles et al. (2006a); Shilliam and Dawson (2005)
1	15		25, 25				
3			30				
10			35	0			
S32006							
10		0		0		70	Dekeyne et al. (2008)
Non selective 5-HT$_{2C}$ antagonists							
SB 206553: 5-HT$_{2B/2C}$ antagonist/inverse agonist							
0.63	20		50, 15			50	De Deurwaerdère et al. (2004); Di Giovanni et al. (1999); Di Matteo et al. (1998); Gobert and Millan 1999; Gobert et al. (2000); Lucas et al. (2000); Navailles et al. (2004, 2006b); Porras et al. (2002)
1			75			50	
2.5							
5	40, 33, 46, 45	26	41, 42, 42, 50				
10	90	30	80	45		130	

(continued)

Table 10.1 (continued)

	Striatum		NAc		Frontal cortex		References
	Anesthesia	FM	Anesthesia	FM	Anesthesia	FM	
Mesulergine: DA-D$_2$ & 5-HT$_{2A/2C}$ antagonist							
0.1	27		32, 20				see Fig. 10.3; Di Matteo et al. (1998)
0.2	45		70, 20				
0.5	55		85				
Clozapine: non selective aminergic & 5-HT$_{2C}$ inverse agonist							
0.3	13		15				Adachi et al. (2005, 2008); Ichikawa et al. (2001, 2005); Jaskiw et al. (2005); Kuroki et al. (1999a); Li et al. (2005b); Navailles et al. (2006b); Rowley et al. (2000)
1	25		27				
1.25						0	
3	35		48				
5				0, 25		70	
10	60	20, 40		40, 30		223, 209	
20						150-450, 125	
Mirtazapine: a2 adrenergic & 5-HT$_{2A/2B/2C3}$ antagonist							
0.63						80	Devoto et al. (2004); Millan et al. (2000)
5						140	
10		15		0		150, 160	
Mianserin: a2 adrenergic & 5-HT$_{2A/2B/2C}$ antagonist							
1				0		370	Di Matteo et al. (2000b); Tanda et al. (1996)
2.5			0				
5			34	0		600	
10				0		600	
Ritanserin: 5-HT$_{2A/2B/2C}$ antagonist							
0.63		0, 0	0, 0	144			Andersson et al. (1995); Devaud et al. (1992); Di Giovanni et al. (1999); Di Matteo et al. (1998); Kuroki et al. (1999); Lucas et al. (1997, 2000); Marcus et al. (1996); Nomikos et al. (1994); Tanda et al. (1996); Pehek (1996); Pehek and Bi (1997)
1			190			0, 0, 0	
1.25		0, 0	0				
1.5	0		0	19		0, 25	
3				21		47	
5						0, 45	

Table 10.2 Effect of the systemic administration of 5-HT2C agonists on DA release in the striatum, the nucleus accumbens (NAc) and the frontal cortex of freely-moving (FM) and anesthetized rats (Anesthetic: halothane and chloral hydrate).

	Striatum		NAc		Frontal cortex		References
	Anesthesia	FM	Anesthesia	FM	Anesthesia	FM	
Selective 5-HT$_{2C}$ agonists							
WAY 163909							
3	0	0		0		0	Marquis et al. (2007)
10	0	0		-48		25	
Non selective 5-HT$_{2C}$ agonists							
Ro 60-0175: 5-HT$_{2A/2B/2C}$ agonist							De Deurwaerdere et al. (2004; Di Matteo et al. (1999, 2000a, 2004); Gobert et al. (2000); Millan et al. (1998); Navailles et al. (2004, 2006a, 2006b); Pozzi et al. (2002)
0.3	0		0				
1	0		0, -26, -26, -27				
1.25						-25	
2.5		-35		-30		-35, 0	
3	-35, 0		-40, 42				
MK 212: 5-HT$_{1B/2B/2C/3}$ agonist							Di Giovanni et al. (2000); Willins and Meltzer (1998)
1	0		-25				
5				0			
mCPP: 5-HT$_{1B/2B/2C}$ agonist							Di Giovanni et al. (2000); Eriksson et al. (1999)
1	0			-27			
2.5		24					

reduce DA release enhanced by morphine (Willins and Meltzer 1998), haloperidol (Navailles et al. 2004), and nicotine (Di Matteo et al. 2004) but not that induced by cocaine (Navailles et al. 2004; Willins and Meltzer 1998). These data suggest that 5-HT$_{2C}$ receptors modulate DA release via an action at DA neuron firing rate and therefore control DA exocytosis that is sensitive to changes in DA neuron impulse activity. Furthermore, this control is dependent on the degree of DA neuron activity, given that the stimulatory effect of 5-HT$_{2C}$ antagonists occurs only on DA release induced by low dose of haloperidol (0.01 mg/kg), whereas the inhibitory effect of 5-HT$_{2C}$ agonists is triggered by a higher dose of haloperidol (0.1 mg/kg) (Lucas et al. 2000; Navailles et al. 2004; Navailles et al. 2006b). Strikingly, the effects induced by the blockade of 5-HT$_{2C}$ receptors on activated DA release substantially differ depending upon the 5-HT$_{2C}$ antagonist used. SB 206553, in contrast to SB 242084 and SB 243213, stimulates further the increase in DA release induced by 0.01 mg/kg haloperidol (Lucas et al. 2000; Navailles et al. 2006b). Moreover, SB 206553 facilitated cocaine-induced DA release with a sharper increase and a rapid onset of potentiation compared with SB 242084 (Navailles et al. 2004). Finally, the differences in the magnitude of effects across various 5-HT$_{2C}$ antagonists/inverse agonists bring up a puzzling picture of the tonic inhibitory control exerted by 5-HT$_{2C}$ receptors on DA release.

The experimental conditions (presence or absence of anesthesia, type of anesthesia), the selectivity of compounds towards 5-HT$_{2C}$ receptors, and the pharmacological profile of these ligands may account for the distinct effects of 5-HT$_{2C}$ antagonists (Gobert et al. 2000; Di Giovanni et al. 1999; Di Matteo et al. 1998). Nonetheless, it has been also proposed that the effects of 5-HT$_{2C}$ drugs may not be paralleled by changes of extracellular levels of 5-HT in feeding behavior (Curzon et al. 1997) as in the control of DA release (Gobert et al. 2000; Willins and Meltzer 1998). This raises the possibility that the way the drug interacts with the receptor, i.e., its pharmacological profile, is a decisive factor contributing to its ability to affect basal DA release. In this chapter, we try to provide answers to the following question: Is there a role for the constitutive activity of 5-HT$_{2C}$ receptors in the regulation of DA release?

10.3 Pharmacological Evidence That the Constitutive Activity of 5-HT$_{2c}$ Receptors Inhibits Basal DA Release In vivo

10.3.1 Experimental Arguments Using the Prototypical Inverse Agonist SB 206553

Along with the fact that the PLC pathway is a prominent target of 5-HT$_{2C}$ receptors in basal ganglia (Wolf and Schultz 1997), the pharmacological characterization of 5-HT$_{2C}$ compounds at PLC-dependent responses has been taken into account to determine their in vivo profile on DA release. First, their effects on DA release have been assessed through dose–response studies and reveal large differences in their magnitude of effects (De Deurwaerdère et al. 2004; Berg et al. 2005; De Deurwaerdère and Spampinato 2001). While SB 206553 dose dependently enhances DA release, SB 242084 and/or SB 243213 induce an increase in DA release that reaches their maximal effect at low doses (Fig. 10.2). Second, the 5-HT_{2C} antagonists SB 242084 and SB 243213 block the excitatory effect of SB 206553 (De Deurwaerdère et al. 2004; Berg et al. 2006). Altogether, these results suggest that SB 206553 behaves as a 5-HT_{2C} inverse agonist and increases DA release by silencing the constitutive activity of 5-HT_{2C} receptors.

In contrast to heterologous recombinant systems in vitro, the continuous presence of endogenous 5-HT in vivo (Sharp et al. 1989) may confound this interpretation. However, the effect of SB 206553 on DA release remains insensitive to the decrease

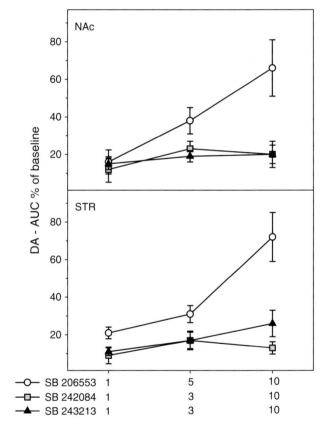

Fig. 10.2 Dose-dependent differences in the magnitude of the increase in basal accumbal (NAc) and striatal (STR) dopamine release induced by the 5-HT_{2C} inverse agonist SB 206553 and the 5-HT_{2C} antagonists SB 242084 and SB 243213. Data are averaged over 2.5 h monitoring and expressed as the mean ± Standard Error Mean (SEM) percentages of baseline corresponding to the area under the curve (AUC). The 5-HT_{2C} compounds were administered intraperitoneally (doses in mg/kg) ($n = 5$–7 animals) (Adapted from De Deurwaerdère et al. 2004. Copyright permission from The Society for Neuroscience)

in 5-HT terminal activity induced by either peripheral administration of the 5-HT$_{1A}$ autoreceptor agonist 8-OH-DPAT or by intra-raphé injections of the neurotoxin 5,7-dihydroxytryptamine (De Deurwaerdère et al. 2004). Therefore, the difference between SB 206553 and SB 242084 or SB 243213 may be a consequence of their distinct intrinsic pharmacological properties and suggest that the constitutive activity of the 5-HT$_{2C}$ receptor participates in the regulation of DA release in vivo. The parallel pharmacological effects of 5-HT$_{2C}$ compounds on IP accumulation and DA release suggest that the marked increase in DA release produced by SB 206553 may involve its inverse agonist property toward the PLC effector pathway (De Deurwaerdère et al. 2004).

The small increase in basal DA release induced by SB 242084 and SB 243213 could be consequent to either their full inverse agonist activity at the PLA$_2$ and Gα_i pathways or the selective blockade of a small endogenous inhibitory tone exerted by 5-HT itself at 5-HT$_{2C}$ receptors (De Deurwaerdère et al. 2004; Berg et al. 2006). Recently, it has been proposed that SB 243213 behaves as a partial inverse agonist with a weaker efficacy compared with SB 206553 on DA release (Berg et al. 2006). Indeed, SB 243213 produces a small increase in DA release by itself and antagonizes both the agonist- (Ro 60-0175) and inverse agonist- (SB 206553) induced changes in DA release in the NAc (Berg et al. 2006). The small excitatory effect of SB 242084 was also insensitive to a pretreatment with 8-OH-DPAT (Navailles et al. unpublished data) suggesting that its effect is not related to the level of 5-HT extracellular levels. Like SB 243213, the effect of SB 242084 on basal DA release could result from a partial inverse agonism with a weaker efficacy than SB 206553. These results may also illustrate the protean agonism property of SB 242084 and SB 243213 (Arrang et al. 2008).

10.3.2   Indirect Evidence Using the Nonselective 5-HT$_{2C}$ Compounds Clozapine and Mesulergine

To date, only SB 206553 has been characterized as a 5-HT$_{2C}$ inverse agonist in vivo. Several other 5-HT$_{2C}$ compounds displaying 5-HT$_{2C}$ inverse agonist properties at PLC-dependent responses in vitro, such as mesulergine, clozapine, ritanserin, mianserin, or mirtazapine, could potentially join the class of 5-HT$_{2C}$ inverse agonist in vivo. Of note, all these compounds increase basal DA release in vivo (see Table 10.1), but their weak selectivity toward 5-HT$_{2C}$ receptors indicates the need for caution in interpreting these data. The pharmacological approach described above has been used to show that the increase in subcortical DA release induced by an optimal dose of clozapine involves the participation of its inverse agonist property at 5-HT$_{2C}$ receptors in vivo (Navailles et al. 2006b).

Indeed, clozapine significantly increases striatal and accumbal DA release at 1 mg/kg, a dose that induces only 5–10% of DA-D$_2$ receptor occupancy (Schotte et al. 1993). Clozapine reverses the inhibitory effect of the 5-HT$_{2C}$ agonist Ro

60-0175, suggesting that clozapine already recruits its 5-HT$_{2C}$ component at this dose. Additionally, clozapine does not affect the facilitatory effect of the 5-HT$_{2C}$ inverse agonist SB 206553 on DA release (Navailles et al. 2006b), arguing against the idea that clozapine acts as a simple 5-HT$_{2C}$ antagonist in vivo (Prinssen et al. 2000; Di Matteo et al. 2002). Finally, the antagonists SB 243213 and SB 242084 block the effect of clozapine on DA release (Navailles et al. 2006b). These findings demonstrate that clozapine, in line with its inverse agonist profile in vitro at PLC-dependent responses (Fig. 10.1) (Rauser et al. 2001; Lefkowitz 1993; Herrick-Davis et al. 2000), enhances subcortical DA release by acting as an inverse agonist at 5-HT$_{2C}$ receptors to silence their constitutive activity in vivo.

Similarly, mesulergine displays inverse agonist property at 5-HT$_{2C}$ receptors in vitro (Barker et al. 1994). However, mesulergine has been classically defined and used as a 5-HT$_{2C}$ antagonist in vivo (Pazos et al. 1984; Kennett and Curzon 1988; Prisco et al. 1994). In line with this, mesulergine, at the optimal dose of 0.1 mg/kg (Fig. 10.3), blocks the decrease in subcortical DA release induced by Ro 60-0175. Interestingly, the selective 5-HT$_{2C}$ antagonist SB 242084 prevents a further increase in subcortical DA release induced by mesulergine (Navailles et al. unpublished data), confirming that mesulergine may affect DA neuron function independently of its DA-D$_2$ antagonist properties (Di Matteo et al. 1998; Prisco et al. 1994; Van Wijngaarden et al. 1990). Although requiring further investigation, these data suggest that mesulergine may also behave as a 5-HT$_{2C}$ inverse agonist in vivo.

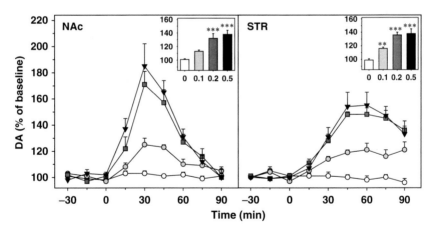

Fig. 10.3 Dose–response effect of mesulergine (0.1, 0.2, and 0.5 mg/kg) on extracellular levels of dopamine (DA) in the nucleus accumbens (NAc) and the striatum (STR) of halothane-anesthe-tized rats. Mesulergine or its corresponding vehicle was subcutaneously administered at time zero. Baseline is calculated from the three fractions preceding the pharmacological treatment. Data represent mean ± SEM percentages of baseline in each sample (time courses) or averaged over 90 min monitoring (insets). Doses are shown in milligrams per kilogram ($n = 5$–7 animals per group). $**p < 0.01$, $***p < 0.001$ versus the vehicle (0) group (Fisher PLSD test). Mesulergine at 0.1 mg/kg, a dose that does not involve the blockade of D$_2$ receptors (Di Matteo et al. 1998), increases DA release in the STR (+16%) and the NAc (+13%)

10.4   Role of Terminal Regions of Dopaminergic Neurons in the Effects Evoked by Inverse Agonists and Agonists

One emerging idea in the will to fully understand how 5-HT$_{2C}$ receptors control DA function is that this control may engage regionally distinct 5-HT$_{2C}$ populations (Filip and Cunningham 2002, 2003; Fletcher et al. 2004). Compelling data assessing the effect of locally administered 5-HT$_{2C}$ compounds (Table 10.3) begin to provide an interesting picture in this matter and suggest that the constitutive activity of 5-HT$_{2C}$ receptors participates in a region-dependent manner in the control of DA function.

The terminal regions of DA neurons in the NAc and the STR – but not the mPFC – may be a major site for the endogenous constitutive activity of 5-HT$_{2C}$ receptors on DA release. Indeed, the 5-HT$_{2C}$ inverse agonist SB 206553 increases accumbal DA release only when it is locally applied into the NAc and the STR but not into the VTA or the cortex (Navailles et al. 2006a; Alex et al. 2005). Moreover, SB 242084 applied into the NAc but not into the VTA is able to prevent the increase in accumbal DA release induced by the systemic administration of SB 206553. The intra-NAc infusion of clozapine also increases accumbal DA release (Ferré and Artigas 1995), but at the high concentration used (10 μmol/L), mechanisms other than the blockade of 5-HT$_{2C}$ receptors also have to be considered. Although a deeper pharmacological investigation would be required, these data suggest that the control exerted by the constitutive activity of 5-HT$_{2C}$ receptors may be restricted to subcortical DA terminal regions.

If the constitutive activity corresponds to that described theoretically from in vitro data, then we should expect that the intra-NAc infusion of a 5-HT$_{2C}$ agonist inhibits DA release. However, data have reported either no effect (Navailles et al. 2008) or even an increase in DA release (Parsons and Justice 1993). The lack of effect could not be related to a strong occupation of 5-HT$_{2C}$ receptors by endogenous 5-HT because the local application of the protean ligands SB 243213 and SB 242084 does not affect NAc DA release. An inhibitory action arising from 5-HT$_{2C}$ receptors located into the NAc can be observed only when agonists are systemically administered (Table 10.2). Indeed, the intra-VTA and/or intra-NAc application of 5-HT$_{2C}$ antagonists is effective in preventing Ro 60-0175-induced inhibition of accumbal DA release (Navailles et al. 2006a). This indicates that distinct populations of 5-HT$_{2C}$ receptors participate in the control of DA release. Some of them, outside the NAc, are likely to trigger the phasic inhibitory control exerted by 5-HT$_{2C}$ receptors that are not able to inhibit accumbal DA release per se. Consequently, it is possible that NAc 5-HT$_{2C}$ receptors respond globally in an opposite manner to the inverse agonist and the agonist, but in different situations. These findings suggest that the 5-HT$_{2C}$ receptor per se is not directly responsible for promoting its own constitutive activity. The constitutive activity exerted by 5-HT$_{2C}$ receptors in vivo would not correspond to the concept raised in vitro.

Compared with a preferential constitutive influence of 5-HT$_{2C}$ receptors on striatal and accumbal DA release (Tables 10.1 and 10.3), the tonic 5-HT$_{2C}$ receptor control may be predominant on the mesocortical DA pathway (Gobert et al. 2000; Millan et al. 1998; Pozzi et al. 2002; Dekeyne et al. 2008). Although no study has assessed the local effect of selective 5-HT$_{2C}$ antagonists on cortical DA release (Table 10.3), data obtained from their systemic administration in freely moving animals show that the overall blockade of endogenous 5-HT increases cortical, but not subcortical, DA release (Table 10.1). Despite the lack of data on the influence of the systemic administration of 5-HT$_{2C}$ antagonists on cortical DA release in anesthetized rats, the decrease in endogenous 5-HT levels induced by halothane (Kalén et al. 1988) suggests that the tonic influence of 5-HT$_{2C}$ receptors on cortical DA release might be blunted in the presence of anesthesia (see Section 10.5).

10.5   The Constitutive Activity of 5-HT$_{2C}$ Receptors in the Control of Subcortical DA Release Is Magnified by Halothane Anesthesia

As reported in Table 10.1, the effects on in vivo DA release induced by 5-HT$_{2C}$ antagonists versus 5-HT$_{2C}$ inverse agonists vary depending upon the experimental condition (presence or absence of anesthesia). All the studies showing the participation of the constitutive activity of 5-HT$_{2C}$ receptors have been performed in halothane-anesthetized animals (De Deurwaerdère et al. 2004; Berg et al. 2006; Navailles et al. 2006b). It appears evident that halothane anesthesia may have facilitated the occurrence of the constitutive activity of 5-HT$_{2C}$ receptors in vivo.

The first consideration to take into account is the potential modification of 5-HT levels by halothane. In freely moving rats, 5-HT levels are about 60% higher than in the presence of halothane (Kalén et al. 1988). Despite these high 5-HT levels, the 5-HT$_{2C}$ antagonists SB 242084 and SB 243213 do not affect subcortical DA release (Gobert et al. 2000; Li et al. 2005a; Hutson et al. 2000; Millan et al. 1998; Shilliam and Dawson 2005), while the 5-HT$_{2C}$ inverse agonist SB 206553 still increases DA release, although to a lesser extent than under anesthesia (Gobert et al. 2000; Lucas et al. 2000; see Table 10.1). These data further support the fact that the efficacy of SB 206553 is independent from endogenous 5-HT tone (De Deurwaerdère et al. 2004). Basal DA extracellular levels are also insensitive to changes in 5-HT levels in the presence of halothane. Indeed, the decrease in endogenous 5-HT tone induced by either the administration of the 5-HT$_{1A}$ agonist 8-OH-DPAT or the selective lesion of 5-HT neurons in the raphé nuclei do not affect basal DA release in the NAc and the STR of halothane-anesthetized rats (De Deurwaerdère et al. 2004;

Table 10.3 Effect of the local application of 5-HT$_{2c}$ compounds into the ventral tegmental area (*VTA*), the nucleus accumbens (*NAc*), the striatum (*STR*) or the frontal cortex (*FCX*) on DA release in the striatum, the NAc and the frontal cortex of freely moving (*FM*) and anesthetized rats

	Striatum		NAc		Frontal cortex		References
	Anesthetic	FM	Anesthetic	FM	Anesthetic	FM	
Selective 5-HT$_{2c}$ antagonists							
SB 242084							
VTA			0				Navailles et al. (2006a, 2008); Leggio et al. (2008)
N accumbens			0				
FCX			0				
SB 243213							
VTA			0				Navailles et al. (2006a); Leggio et al. (2008)
FCX			0				
Nonselective 5-HT$_{2c}$ antagonists							
SB 206553: 5-HT$_{2B/2C}$ antagonist/inverse agonist							
VTA			0				Navailles et al. (2006a); Alex et al. (2005)
N accumbens			25				
STR		42–112					
FCX							
Clozapine: Nonselective aminergic and 5-HT$_{2c}$ inverse agonist							
N accumbens				75		0	Ferré and Artigas (1995)
Ritanserin: 5-HT$_{2A/2B/2C}$ antagonist							
FCX						70	Pehek (1996); Benloucif et al. (1993)
STR	0						

De Deurwaerdère and Spampinato 1999; De Deurwaerdère et al. 1998). Thus, the larger increase in DA release induced by SB 206553 in halothane-anesthetized compared with freely moving rats suggests that halothane may favor the expression of the constitutive activity of 5-HT$_{2C}$ receptors. Similarly, the paradoxical loss of effect of selective 5-HT$_{2C}$ antagonists on DA release in freely moving rats suggests that the increase observed in halothane-anesthetized rats may reflect the direct influence of halothane on DA release as well.

Most anesthetics, and particularly halothane, exert an excitatory influence on DA neuron activity (Bunney et al. 1973; Mereu et al. 1984). Basal DA extracellular levels in the STR and the NAc are almost twice in halothane-anesthetized compared with awake animals (Table 10.4) (Lucas et al. 2000). The excitation of the nigrostriatal DA pathway seems to be specific to halothane anesthesia because chloral hydrate, though doubling accumbal DA levels, does not alter basal DA levels in the STR compared with waking conditions (Table 10.4). Strikingly, 5-HT$_{2C}$ antagonists barely alter DA release in the STR of chloral hydrate-anesthetized rats, an effect that has lead some authors to propose a preferential or selective control of 5-HT$_{2C}$ receptors on the mesoaccumbal DA pathway (Di Matteo et al. 1999; Di Giovanni et al. 1999; Di Matteo et al. 2001). In addition to its influence at DA cell bodies, halothane alters DA function at DA terminal regions since it reduces the activity of striatal efferent neurons of type I (Kelland et al. 1991), neurons known to express 5-HT$_{2C}$ receptors mRNA and proteins (Pompeiano et al. 1994; Ward and Dorsa 1996). These simultaneous actions of halothane at different levels of DA transmission may have participated to the maximal increase in DA release induced by SB 206553 compared with waking conditions. Halothane has been shown to increase clozapine-induced subcortical DA release (Adachi et al. 2005) and, as for SB 206553, highlights the ability of clozapine to silence the constitutive activity of 5-HT$_{2C}$ receptors. The combined evidence of these data draw attention to the fact that halothane may trigger the constitutive activity of 5-HT$_{2C}$ receptors in the control of subcortical DA release in vivo.

Table 10.4 Basal DA levels (pg/sample) assessed by in vivo microdialysis in the striatum (*STR*), the nucleus accumbens (*NAc*) and the frontal Cortex (*FCX*) of freely moving (*FM*) and anesthetized rats (Halothane and chloral hydrate)

	FM	Anesthesia	
		Halothane	Chloral hydrate
Dopamine: basal levels			
STR	9.2±0.4	17.3±1.6	8.3±1.1
NAc	3.6±0.9	5.4±0.7	5.15±0.5
FCX	1.0±0.1	nd	nd

nd not determined

10.6 Influence of 5-HT$_{2C}$ Receptors in Behaving Animals

10.6.1 Behavioral Effects Induced by 5-HT$_{2C}$ Receptors: General Consideration

It has been reported for many years that the stimulation of 5-HT$_{2C}$ receptors alters behaviors per se in rodents including penile erection, grooming, stereotypes, oral dyskinesia, decrease in impulsivity, locomotor activity and feeding behavior, and increase in anxiety (Jones and Blackburn 2002; Giorgetti and Tecott 2004; Millan et al. 2005). Interestingly, some authors, using new agonists, have reported that 5-HT$_{2C}$ receptors may also display anxiolytic/antidepressant properties (Rosenzweig-Lipson et al. 2007), underscoring a possible existence of multiple and opposite behavioral outputs induced by 5-HT$_{2C}$ drugs of a similar pharmacological class. The stimulation of 5-HT$_{2C}$ receptors has also been shown to alter behaviors associated with perturbations of central DA transmission. In particular, 5-HT$_{2C}$ receptor stimulation decreases the effects on locomotor hyperactivity and self-administration associated with drugs of abuse (Grottick et al. 2000; Tomkins et al. 2002). However, 5-HT$_{2C}$ receptors have a widespread distribution in the brain, and as for the control exerted on DA neuron activity, several loci in the brain may intervene in the regulation of behaviors controlled by 5-HT$_{2C}$ receptors. This has been clearly shown by K. Cunningham and colleagues (Filip and Cunningham 2002, 2003) highlighting various loci expressing 5-HT$_{2C}$ receptors able to interfere on basal locomotor activity as well as on the acute behavioral consequences of cocaine intoxication. These loci may exert opposite roles on locomotor activity, the mPFC being excitatory and the NAc being inhibitory on basal and cocaine behavioral effects. These data are fundamental points to postulate the existence of a functional organization of 5-HT$_{2C}$ receptors in the brain. Nevertheless, it becomes difficult to imagine, in case of a putative influence of a constitutive activity on behavior, the effect of 5-HT$_{2C}$ inverse agonists compared with agonists, if such agonists already recruit opposite influences in their behavioral effects.

The data obtained with various 5-HT$_{2C}$ antagonists are often consistent with a blockade of the noxious effects involving the phasic influence of endogenous 5-HT on 5-HT$_{2C}$ receptors. Thus, blockade of 5-HT$_{2C}$ receptors has been shown to elicit anxiolytic properties in various rodent models (Kennett et al. 1996, 1997; Dekeyne et al. 2008; Millan et al. 2005; Olsen et al. 2002; Wood et al. 2001). Most of these effects are reported with both antagonists and inverse agonists using very large range of doses. These findings are in agreement with data obtained in 5-HT$_{2C}$ knockout mice (Heisler et al. 2007). Previous and recent studies have reported an increase in body weight in 5-HT$_{2C}$ receptor knockout mice compared with their littermates (Tecott et al. 1995; Nonogaki et al. 2003). However, data reporting the effect of distinct 5-HT$_{2C}$ antagonists on food intake and weight gain remain inconsistent (Wood et al. 2001; Bonhaus et al. 1997; Thomsen et al. 2008). It would be interesting to determine whether a chronic treatment with the full inverse agonist

SB 206553 induces an increase in body weight, as the ability of 5-HT drugs to affect feeding behavior is not directly paralleled by changes in 5-HT availability (Curzon et al. 1997). Finally, the 5-HT$_{2C}$ antagonist SB 242084 increases the expression of impulsive behavior, whereas agonists decrease it (Fletcher et al. 2007). Lesion of 5-HT neurons tends to increase impulsive behavior in line with the effects elicited by SB 242084 (Fletcher et al. 2007; Winstanley et al. 2004). The fact that the 5-HT$_{2B/2C}$ antagonist SDZ SER082 systematically differed from the effect elicited by SB 242084 has been interpreted as a possible inverse agonism profile of this compound (Talpos et al. 2006). Interestingly, neurobiological bases of 5-HT action on impulsive behavior include the NAc (Robinson et al. 2008), likely housing a constitutive form of 5-HT$_{2C}$ receptors in the control of DA release (Navailles et al. 2006a). However, without a full pharmacological characterization, it is impossible to support that the differences of the effects elicited by SDZ SER082 and SB 242084 are related to blockade of endogenous constitutive activity of 5-HT$_{2C}$ receptors, and several reasons proposed by Fletcher et al. (2007) may account for this difference.

Regarding the interaction with DA transmission, the available data indicate that in most cases, both neutral antagonists (SB 242084, SB 243213) and inverse agonists (SB 206553) are facilitating the behavioral effects elicited by drugs of abuse or antiparkinsonian drugs. In one case, however, SB 206553 behaved differentially with respect to its regimen, facilitating locomotor hyperactivity induced by cocaine when administered at low doses (1 or 2 mg/kg) and inhibiting it when administered at a higher dose (4 mg/kg) (McCreary and Cunningham 1999). In the 6-hydroxy-dopamine rat model of Parkinson's disease, SB 206553 elicited by itself contralateral rotations when administered directly into the altered substantia nigra (Fox et al. 1998). These data suggest that SB 206553 differs sometimes from other antagonists in behavioral responses. This difference could account for its inverse agonism property (De Deurwaerdère et al. 2004) although a proper pharmacological characterization has never been provided.

10.6.2   Possible Involvement of Constitutive Activity of 5-HT$_{2C}$ Receptors in the Control of Oral Function

It has been shown for several years that purposeless oral movement is also a common and extremely sensitive behavioral output consequent to the administration of nonselective 5-HT$_{2C}$ agonist including m-CPP and TFMPP (Stewart et al. 1989; Gong et al. 1992; Wolf et al. 2005). The effect is dependent on 5-HT$_{2C}$ receptors as numerous 5-HT$_{2C}$ antagonists abolish the abnormal oral response induced by the agonists. This abnormal behavior has been related to an alteration of basal ganglia function involving subthalamic nucleus and striatal 5-HT$_{2C}$ receptors (Plech et al. 1995; Eberle-Wang et al. 1996; De Deurwaerdère and Chesselet 2000).

Nonselective and selective 5-HT$_{2C}$ antagonists do not affect oral activity by themselves when administered at 1 or 2 mg/kg (Wolf et al. 2005). As in microdialysis studies, a higher dose might be required to unmask tonic or constitutive controls. We have recently obtained behavioral data showing that higher doses of SB 206553 (3–20 mg/kg), but not SB 243213 (1–10 mg/kg), induced a dose-dependent enhancement of purposeless oral movements (De Deurwaerdère et al. 2008). The effect elicited by SB 206553 is not associated with the other classical above-mentioned behaviors elicited by agonists. Most importantly, SB 243213 pretreatment abolishes purposeless oral movements induced by SB 206553 or the agonist m-CPP (Fig. 10.4) (De Deurwaerdère et al. 2008). These pharmacological data are in line with the possibility that SB 206553 behaves on oral movements as a 5-HT$_{2C}$ inverse agonist, suggesting that the constitutive activity of 5-HT$_{2C}$ receptors could play an endogenous role in maintaining optimal oral responses. It is difficult to determine if the same 5-HT$_{2C}$ receptors underline purposeless oral movements triggered by the opposite class of pharmacological agents. Indeed, several brain regions and circuits in basal ganglia may interfere with this behavior. These results suggest again that basal ganglia are housing a constitutive control, exerted by 5-HT$_{2C}$ receptors, that is operating in freely moving animals.

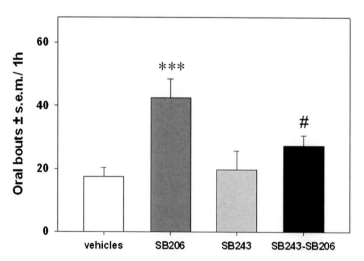

Fig. 10.4 The 5-HT$_{2C}$ antagonist SB 243213 blocks the effect of the 5-HT$_{2C}$ inverse agonist SB 206553 on purposeless oral movements in naïve rats. Histograms represent the number of oral bouts (±SEM) observed during 1 h for each experimental group ($n=6$–7 rats per group). Measurements have been performed according to references Eberle-Wang et al. (1996) and De Deurwaerdère and Chesselet (2000). Rats were pretreated with SB 243213 (SB243: 1 mg/kg intraperitoneally [ip]) or its vehicle 1 h before the ip administration of SB 206553 (SB206: 10 mg/kg) or its vehicle. ***$p<0.001$ versus the vehicles group; #$p<0.05$, versus the SB 206553 group (Fisher PLSD test)

10.7   Effect of Various 5-HT$_{2C}$ Agents in the Control of Fos Expression in Basal Ganglia

The use of anatomo-functional markers, without precisely addressing cellular mechanisms of action of pharmacological compounds, gives a picture of the effects of drugs in a tissue. It is interesting to look at the data concerning 5-HT$_{2C}$ receptors in the basal ganglia where all modalities of interaction of 5-HT$_{2C}$ receptors may occur.

Several studies have reported that 5-HT$_{2C}$ receptor stimulation may affect the expression of the proto-oncogene c-Fos in the brain. In basal ganglia, the effect is noticeable in the medial STR, the subthalamic nucleus, and with lower magnitudes of effects, in the lateral striatum, the entopeduncular nucleus and the substantia nigra pars reticulata (De Deurwaerdère and Chesselet 2000; Cook and Whirtshafter 1995; Stark et al. 2006). Numerous studies have failed to report substantial effect of 5-HT$_{2C}$ antagonists on indirect cellular markers of neuronal activity (Moorman and Leslie 1996; Rouillard et al. 1996; Javed et al. 1998; Gardier et al. 2000; Stark et al. 2008). Again, this might be due to the lack of selectivity of the antagonists for 5-HT$_{2C}$ receptors (mianserin, ritanserin, metergoline), to the low tone of endogenous 5-HT in subcortical regions or to the pharmacological profile of these antagonists. In a series of experiments, we have observed with (1 mg/kg) the mixed 5-HT$_{2B/2C}$ antagonist SDZ SER082 and the 5-HT$_{2C}$ antagonist SB 243213 a Fos labeling in striatal and subthalamic regions (De Deurwaerdère et al. 2010). A high dose of SB 206553 (10 mg/kg) also enhanced Fos in the entopeduncular nucleus and the substantia nigra pars reticulata (De Deurwaerdère et al. 2008). Increase in Fos expression or other neuronal markers have not been observed in other brain areas suggesting that the link between 5-HT$_{2C}$ receptors and subcortical DA transmission is strong in basal ganglia.

The expression of Fos in basal ganglia after 5-HT$_{2C}$ agonists or antagonists suggests that all these class of drugs affect cellular activity in basal ganglia. The distribution of the effects elicited by these drugs is similar in medial sectors of basal ganglia but seems to be higher with antagonists in lateral sectors. The effects associated with antagonists can not be clearly associated with a specific pharmacological interaction of the drugs with the receptor. Nevertheless, the distinct pattern of their effects suggests mechanisms more complex than a simple blockade of endogenous 5-HT at 5-HT$_{2C}$ receptors. Interestingly, if part of the effects of the antagonists is related to the removal of a tonic inhibitory influence exerted by 5-HT$_{2C}$ receptors on gene expression, this would suggest that 5-HT$_{2C}$ receptors are able to exert an opposite role in the control of gene expression in the striatum and the subthalamic nucleus.

10.8   Synthesis and Remarks

The concept of constitutive activity implying that an agonist increases and an inverse agonist decreases the coupling efficiency of a receptor toward its intracellular effector pathway is a pharmacological outcome of in vitro preparations. Its transposition to cerebral function is probably not as clear. Indeed, even if this concept fits with the opposite effects of agonist versus inverse agonist in the control of subcortical DA release in halothane-anesthetized rats regarding 5-HT$_{2C}$ receptors (De Deurwaerdère et al. 2004; Berg et al. 2005), the neurobiological basis of these opposite effects could imply the participation of 5-HT$_{2C}$ receptors at different loci in the brain. Consequently, there is no in vivo evidence to date that a single population of 5-HT$_{2C}$ receptor is able to share its activity from constitutive to tonic and phasic responses.

The existence of a constitutive activity in vitro is associated with the number of copies of the receptor expressed by the heterologous systems, and it is likely that this high expression for numerous GPCR is barely reached endogenously in vivo. There would be two other possibilities for a receptor to display a constitutive activity in vivo. First, the constitutive activity would be related to the very high capacity of the receptor to interact with G proteins. This property has been confirmed in vitro for native forms of 5-HT$_{2C}$ receptors (unedited form of the mRNA) (Berg et al. 1999; Niswender et al. 1999) but not for 5-HT$_{2A}$ receptors (Berg et al. 1999; Grotewiel and Sanders-Bush 1999) although they have been proposed to be constitutively active in conditioned behavioral responses (Harvey et al. 1999). In addition, the constitutive activity for 5-HT$_{2C}$ receptors has been associated with poorly edited forms for which the coupling efficiency and affinity for ligands are high compared with the edited forms (Fitzgerald et al. 1999; Quirk et al. 2001). In vivo, most 5-HT$_{2C}$ receptors are expressed in edited forms (Burns et al. 1997; Schmauss 2003). The reduced functional properties of these edited receptors may require higher doses of inverse agonists and agonists to unmask the constitutive and phasic controls on DA release. This suggests that the nature of the 5-HT$_{2C}$ receptor control may not be only related to the receptor itself. Another possibility would be the own metabolic and electrical activity of the cell expressing the receptor. Regarding 5-HT$_{2C}$ receptors, it is indeed evident that the activity of the cells expressing the receptor dramatically varies in the basal ganglia, ranging from hyperpolarized cells in the striatum to a tonic and sustained activity of discharge of neurons in the substantia nigra pars reticulata and the entopeduncular nucleus. Consequently, the cell's own activity would narrow the impact of 5-HT$_{2C}$ receptors from constitutive to tonic/phasic responses, depending on the host cell considered.

10.9   Conclusions

In conclusion, 5-HT$_{2C}$ receptors can exert tonic, phasic and constitutive controls in the brain of mammals. It seems that these activities, when relevant, are not mediated by the same population of 5-HT$_{2C}$ receptors. It is likely that the receptor per se may not fully account for this property and that the activity of the cell carrying the receptor imposes the modality of interaction. From a pathophysiological point of view, changes of cellular activity in a system might favor or displace one modality of interaction to another independently from changes in 5-HT extracellular levels, a hypothesis already proposed (Curzon et al. 1997) with regard to the effects of 5-HT drugs in the control of feeding behavior.

References

Adachi YU, Satomoto M, Higuchi H, et al (2005) Halothane enhances dopamine metabolism at presynaptic sites in a calcium-independent manner in rat striatum. Br J Anaesth 95:485–494.

Adachi YU, Aramaki Y, Satomoto M, et al (2008) Halothane attenuated haloperidol and enhanced clozapine-induced dopamine release in the rat striatum. Neurochem Int 43:113–119.

Alex KD, Yavanian GJ, McFarlane HG, et al (2005) Modulation of dopamine release by striatal 5-HT2C receptors. Synapse 55:242–251.

Andersson JJ, Nomikos GG, Marcus M, et al (1995) Ritanserin potentiates the stimulatory effects of raclopride on neuronal activity and dopamine release selectively in the mesolimbic dopaminergic system. Naunyn Schmiedeberg's Arch Pharmacol 352:374–385.

Arrang JM, Morisset S, Gbahou F (2008) Constitutive activity of the histamine H3 receptor. Trends Pharmacol Sci 28:350–357.

Barker EL, Westphal RS, Schimdt D, et al (1994) Constitutively active 5-hydroxytryptamine2C receptors reveal novel inverse agonist activity of receptor ligands. J Biol Chem 269:11687–11690.

Benloucif S, Keegan MJ, Galloway MP (1993) Serotonin-facilitated dopamine release in vivo: pharmacological characterization. J Pharmacol Exp Ther 265:373–377.

Berg KA, Maayani S, Goldfarb J, et al (1998) Effector pathway-dependent relative efficacy at serotonin type 2A and 2C receptors: evidence for agonist-directed trafficking of receptor stimulus. Mol Pharmacol 54:94–104.

Berg KA, Stout BD, Cropper JD, et al (1999) Novel actions of inverse agonists on 5-HT2C receptor systems. Mol Pharmacol 55:863–872.

Berg KA, Cropper JD, Niswender CM, et al (2001) RNA-editing of the 5-HT2C receptor alters agonist-receptor-effector coupling specificity. Br J Pharmacol 134:386–392.

Berg KA, Harvey JA, Spampinato U, et al (2005) Physiological relevance of constitutive activity of 5-HT2A and 5-HT2C receptors. Trends Pharmacol Sci 26:625–630.

Berg KA, Navailles S, Sanchez TA, et al (2006) Differential effects of 5-methyl-1-[[2-[(2-methyl-3-pyridyl)oxy]-5-pyridyl]carbamoyl]-6-trifluoromethylindone   (SB   243213)   on 5-hydroxytryptamine(2C) receptor-mediated responses. J Pharmacol Exp Ther 319:260–268.

Berg KA, Clarke WP, Cunningham KA, et al (2008) Fine-tuning serotonin2C receptor function in the brain: Molecular and functional implications. Neuropharmacology 55:969–976.

Bonhaus DW, Weinhardt KK, Taylor M, et al (1997) RS 02221: a novel high affinity and selective, 5-HT2C receptor antagonist. Neuropharmacology 36:621–629.

Bunney BS, Walters JR, Roth RH, et al (1973) Dopaminergic neurons: effect of antipsychotic drugs and amphetamine on single cell activity. J Pharmacol Exp Ther 185:560–571.

Burns CM, Chu H, Rueter SM, et al (1997) Regulation of serotonin2C receptor G-protein coupling by RNA editing. Nature 387:303–307.

Burstein ES, Spalding TA, Brauner-Osborne H, et al (1995) Constitutive activation of muscarinic receptors by the G-protein Gq. FEBS Lett 363:261–3.

Cook DF, Whirtshafter D (1995) Serotonin agonist-induced c-fos expression in the rat striatum. Soc Neurosci Abstr 21:1424.

Costa T, Herz A (1989) Antagonists with negative intrinsic activity at delta opiod receptors coupled to GTP-binding proteins. Proc Natl Acad Sci U S A 86:7321–7325.

Curzon G, Gibson EL, Oluyomi AO (1997) Appetite suppression by commonly used drugs depends on 5-HT receptors but not on 5-HT availability. Trends Pharmacol Sci 18:21–25.

De Deurwaerdère P, Chesselet MF (2000) Nigrostriatal lesions alter oral dyskinesia and c-fos expression induced by the serotonin agonist 1-(m-chlorophenyl)piperazine in adult rats. J Neurosci 20:5170–5178.

De Deurwaerdère P, Spampinato U (1999) Role of serotonin2A and serotonin2B/2C receptor subtypes in the control of accumbal and striatal dopamine release elicited in vivo by dorsal raphe nucleus electrical stimulation. J Neurochem 73:1033–1042.

De Deurwaerdère P, Spampinato U (2001) The nigrostriatal dopamine system: a neglected target for 5-HT2C receptors. Trends Pharmacol Sci 22:502–503.

De Deurwaerdère P, Stinus L, Spampinato U (1998) Opposite change of in vivo dopamine release in the rat nucleus accumbens and striatum that follows electrical stimulation of dorsal raphe nucleus: role of 5-HT3 receptors. J Neurosci 18:6528–6538.

De Deurwaerdère P, Navailles S, Berg KA, et al (2004) Constitutive activity of the serotonin2C receptor inhibits in vivo dopamine release in the rat striatum and nucleus accumbens. J Neurosci 24:3235–3241.

De Deurwaerdère P, Kadiri N, Trannois A, et al (2008) Continuous controls exerted by Serotonin2C receptor in rat basal ganglia involve both serotonergic tone and its constitutive activity: regional distribution on c-Fos expression and behavioral aspects. EPHAR, Serotonin Club Meeting, Oxford, UK. July 16-19. In Fundamental and Clinical Pharmacology, 22 (suppl 2):SCP036.

De Deurwaerdère P, Le Moine C and Chesselet M-F (2010) Selective blockade of Serotonin2C receptor enhances Fos expression specifically in the striatum and the subthalamic nucleus within the basal ganglia. Neurosci Lett 469:251–255.

Dekeyne A, Mannoury la Cour C, Gobert A, et al (2008) S32006, a novel 5-HT2C receptor antagonist displaying broad-based antidepressant and anxiolytic properties in rodent models Psychopharmacology (Berl) 199:549–568.

Devaud LL, Hollingsworth EB, Cooper BR (1992) Alterations in extracellular and tissue levels of biogenic amines in rat brain induced by the serotonin2 receptor antagonist, ritanserin. J Neurochem 59:1459–1466.

Devoto P, Flore G, Pira L, et al (2004) Mirtazapine-induced corelease of dopamine and noradrenaline from noradrenergic neurons in the medial prefrontal and occipital cortex. Eur J Pharmacol 487:105–111.

Di Giovanni G, De Deurwaerdère P, Di Mascio M, et al (1999) Selective blockade of serotonin2C/2B receptors enhances mesolimbic and mesostriatal dopaminergic function: a combined in vivo electrophysiological and microdialysis study. Neuroscience 91:587–597.

Di Giovanni G, Di Matteo V, Di Mascio M, et al (2000) Preferential modulation of mesolimbic vs. nigrostriatal dopaminergic function by serotonin(2C/2B) receptor agonists: a combined in vivo electrophysiological and microdialysis study. Synapse 35:53–61.

Di Matteo V, Di Giovanni G, Di Mascio M, et al (1998) Selective blockade of serotonin2C/2B receptors enhances dopamine release in the rat nucleus accumbens. Neuropharmacology 37:265–272.

Di Matteo V, Di Giovanni G, Di Mascio M, et al (1999) SB 242084, a selective serotonin2C receptor antagonist, increases dopaminergic transmission in the mesolimbic system. Neuropharmacology 38:1195–1205.

Di Matteo V, Di Giovanni G, Di Mascio M, et al (2000a) Biochemical and electrophysiological evidence that RO 60-0175 inhibits mesolimbic dopaminergic function through serotonin2C receptors. Brain Res 865:85–90.

Di Matteo V, Di Mascio M, Di Giovanni G, et al (2000b) Acute administration of amitriptyline and mianserin increases dopamine release in the rat nucleus accumbens: possible involvement of serotonin2C receptors. Psychopharmacology 150:45–51.

Di Matteo V, De Blasi A, Di Giulio C, et al (2001) Role of 5-HT2C receptors in the control of central dopamine function. Trends Pharmacol Sci 22:229–232.

Di Matteo V, Cacchio M, Di Giulio C, et al (2002) Biochemical evidence that the atypical antipsychotic drugs clozapine and risperidone block 5-HT2C receptors in vivo. Pharmacol Biochem Behav 71:607–613.

Di Matteo V, Pierucci M, Esposito E (2004) Selective stimulation of serotonin2c receptors blocks the enhancement of striatal and accumbal dopamine release induced by nicotine administration. J Neurochem 89:418–429.

Eberle-Wang K, Lucki I, Chesselet MF (1996) A role for the subthalamic nucleus in 5-HT2C-induced oral dyskinesia. Neuroscience 72:117–28.

Eriksson E, Engberg G, Bing O, et al (1999) Effects of mCPP on the extracellular concentrations of serotonin and dopamine in rat brain Neuropsychopharmacology 20:287–296.

Ferré S, Artigas F (1995). Clozapine decreases serotonin extracellular levels in the nucleus accumbens by a dopamine receptor-independent mechanism. Neurosci Lett 187:61–64.

Filip M, Cunningham KA (2002) Serotonin 5-HT(2C) receptors in the nucleus accumbens regulate expression of the hyperlocomotive and discriminative stimulus effects of cocaine. Pharmacol Biochem Behav 71:745–56.

Filip M, Cunningham KA (2003) Hyperlocomotive and discriminative stimulus effects of cocaine are under the control of serotonin(2C) (5-HT(2C)) receptors in rat prefrontal cortex. J Pharmacol Exp Ther 306:734–743.

Fitzgerald LW, Iyer G, Conklin DS, et al (1999) Messenger RNA editing of the human serotonin 5-HT2C receptor. Neuropsychopharmacology 21:82S–90S.

Fletcher PJ, Chintoh AF, Sinyard J, et al (2004) Injection of the 5-HT2C receptor agonist Ro60-0175 into the ventral tegmental area reduces cocaine-induced locomotor activity and cocaine self-administration. Neuropsychopharmacology 29:308–318.

Fletcher PJ, Tampakeras M, Sinyard J, et al (2007) Opposing effects of 5-HT2A and 5-HT2C receptor antagonists in the rat and mouse on premature responding in the five-choice serial reaction time test. Psychopharmacology (Berl) 195:223–234.

Fox SH, Moser B, Brotchie JM (1998) Behavioral effects of 5-HT2C receptor antagonism in the substantia nigra zona reticulata of the 6-hydroxydopamine-lesioned rat model of Parkinson's disease. Exp Neurol 151:35–49.

Gardier AM, Moratalla R, Cuéllar B, et al (2000) Interaction between the serotoninergic and dopaminergic systems in d-fenfluramine-induced activation of c-fos and jun B genes in rat striatal neurons. J Neurochem 74:1363–1373.

Giorgetti M, Tecott LH (2004) Contributions of 5-HT2C receptors to multiple actions of central serotonin systems. Eur J Pharmacol 488:1–9.

Gobert A, Millan MJ (1999) Serotonin (5-HT)2A receptor activation enhances dialysate levels of dopamine and noradrenaline, but not 5-HT, in the frontal cortex of freely-moving rats. Neuropharmacology 38:315–317.

Gobert A, Rivet JM, Lejeune F, et al (2000) Serotonin2C receptors tonically suppress the activity of mesocortical dopaminergic and adrenergic, but not serotonergic, pathways: A combined dialysis and electrophysiological analysis in the rat. Synapse 36:205–221.

Gong L, Kostrzewa RM, Fuller RW, et al (1992) Supersensitization of the oral response to SKF 38393 in neonatal 6-OHDA-lesioned rats is mediated through a serotonin system. J Pharmacol Exp Ther 261:1000–1007.

Grotewiel MS, Sanders-Bush E (1999) Differences in agonist-independent activity of 5-HT2A and 5-HT2C receptors revealed by heterologous expression. Naunyn-Schmiedeberg's Arch Pharmacol 359:21–27.

Grottick AJ, Fletcher PJ, Higgins GA (2000) Studies to investigate the role of 5-HT2C receptors on cocaine-and food-maintained behavior. J Pharmacol Exp Ther 295:1183–1191.

Gurevich I, Englander MT, Adlesberg M, et al (2002) Modulation of serotonin2C receptor editing by sustained changes in serotonergic neurotransmission. J Neurosci 22:10529–10532.

Harvey JA, Welsh SE, Hood H, et al (1999). Effect of 5-HT2 antagonists on cranial nerve reflex in the rabbit: evidence for inverse agonism. Psychopharmacology 141:163–168.

Heisler LK, Zhou L, Bajwa P, et al (2007) Serotonin 5-HT2C receptros regulate anxiety-like behavior. Genes Brain Behav 6:491–496.

Herrick-Davis K, Grinde E, Niswender CM (1999) Serotonin 5-HT2C receptor RNA editing alters receptor basal activity: implications for serotonergic signal transduction. J Neurochem 73:1711–1717.

Herrick-Davis K, Grinde E, Teitler M (2000) Inverse agonist activity of atypical antipsychotic drugs at human 5-hydroxytryptamine2C receptors. J Pharmacol Exp Ther 295:226–232.

Hutson PH, Barton CL, Jay M, et al (2000) Activation of mesolimbic dopamine function by phencyclidine is enhanced by 5-HT2C/2B receptor antagonists: neurochemical and behavioural studies. Neuropharmacology 39:2318–2328.

Ichikawa J, Dai J, Meltzer HY (2001) DOI, a 5-HT2A/2C receptor agonist, attenuates clozapine-induced cortical dopamine release. Brain Res 907:151–155.

Ichikawa J, Chung Y-C, Dai J, et al (2005). Valproic acid potentiates both typical and atypical antipsychotic-induced prefrontal cortical dopamine release. Brain Res 1052:56–62.

Jaskiw GE, Kirkbride B, Newbould E, et al (2005) Clozapine-induced dopamine release in the medial prefrontal cortex is augmented by a moderate concentration of locally administered tyrosine but attenuated by high tyrosine concentrations or by tyrosine depletion. Psychopharmacology (Berl) 179:713–724.

Javed A, Van de Kar LD, Gray TS (1998) The 5-HT1A and 5-HT2A/2C receptor antagonists WAY-100635 and ritanserin do not attenuate D-fenfluramine-induced fos expression in the brain. Brain Res 791 :67–74.

Jones BJ, Blackburn TP (2002) The medical benefit of 5-HT research. Pharmacol Biochem Behav 71:555–568.

Kalén P, Strecker RE, Rosengren E, et al (1988). Endogenous release of neuronal serotonin and 5-hydroxyindoleacetic acid in the caudate-putamen of the rat as revealed by intracerebral dialysis coupled to high-performance liquid chromatography with fluorimetric detection. J Neurochem 51:1422–1435.

Kelland MD, Chiodo LA, Freeman AS (1991) Dissociative anesthesia and striatal neuronal electrophysiology. Synapse 9:75–78.

Kenakin T (2001) Inverse, protean, and ligand-selective agonism: matters of receptor conformation FASEB J 15:598–611.

Kenakin T (2004) Efficacy as a vector: the relative prevalence and paucity of inverse agonism. Mol Pharmacol 65:2–11.

Kennett GA, Curzon G (1988) Evidence that mCPP may have behavioural effects mediated by central 5-HT1C receptors. Br J Pharmacol 94:137–147.

Kennett GA, Wood MD, Bright F, Cilia J, Piper DC, Gager T, Thomas DR, Baxter GS, Forbes IT, Ham P, Blackburn TP (1996) In vitro and in vivo profile of SB 206553, a potent 5-HT2C/5-HT2B receptor antagonist with anxiolytic-like properties. Br J Pharmacol 117:427–434.

Kennett GA, Wood MD, Bright F, et al (1997) SB 242084, a selective and brain penetrant 5-HT2C receptor antagonist. Neuropharmacology 36:609–620.

Kuroki T, Meltzer HY, Ichikawa J (1999) Effects of antipsychotic drugs on extracellular dopamine levels in rat medial prefrontal cortex and nucleus accumbens. J Pharmacol Exp Ther 288:774–781.

Lefkowitz RJ (1993) G-protein-coupled receptors. Turned on to ill effect. Nature 365:603–604.

Leggio GM, Cathala A, Moison D, et al (2008) Serotonin(2C) receptors in the medial prefrontal cortex facilitate cocaine-induced dopamine release in the rat nucleus accumbens. Neuropharmacology 56:507–513.

Li Z, Ichikawa J, Huang M, et al (2005a) ACP-103, a 5-HT2A/2C inverse agonist, potentiates haloperidol-induced dopamine release in rat medial prefrontal cortex and nucleus accumbens. Psychopharmacology (Berl) 183:144–153.

Li Z, Huang M, Ichikawa J, et al (2005b) N-desmethylclozapine, a major metabolite of clozapine, increases cortical acetylcholine and dopamine release in vivo via stimulation of M1 muscarinic receptors. Neuropsychopharmacology 30:1986–1995.

Lucas G, Bonhomme N, De Deurwaerdère P, et al (1997) 8-OH-DPAT, a 5-HT1A agonist and ritanserin, a 5-HT2A/C antagonist, reverse haloperidol-induced catalepsy in rats independently of striatal dopamine release. Psychopharmacology 131:57–63.

Lucas G, De Deurwaerdère P, Caccia S, et al (2000) The effect of serotonergic agents on haloperidol-induced striatal dopamine release in vivo: opposite role of 5-HT2A and 5-HT2C receptor subtypes and significance of the haloperidol dose used. Neuropharmacology 39:1053–1063.

Marcus MM, Nomikos GG, Svensson TH (1996) Differential actions of typical and atypical antipsychotic drugs on dopamine release in the core and shell of the nucleus accumbens. Eur Neuropsychopharmacol 6:29–38.

Marion S, Weiner DM, Caron MG (2004) RNA editing induces variation in desensitization and trafficking of 5-hydroxytryptamine2C receptor isoforms. J Biol Chem 279:2945–54.

Marquis KL, Sabb AL, Logue SF, et al (2007) WAY-163909 [(7bR,10aR)-1,2,3,4,8,9,10,10a-octahydro-7bH-cyclopenta-[b][1,4]diazepino[6,7,1hi]indole]: A novel 5-hydroxytryptamine2C receptor-selective agonist with preclinical antipsychotic-like activity. J Pharmacol Exp Ther 320:486–496.

McCreary AC, Cunningham KA (1999) Effects of the 5-HT2C/2B antagonist SB 206553 on hyperactivity induced by cocaine. Neuropsychopharmacology 20:556–564.

Mereu G, Fanni B, Gessa GL (1984) General anesthetics prevent dopaminergic neuron stimulation by neuroleptics. In: Usdin E, Carlsson A, Dahlstrom A, Engel J (eds.). Catecholamines: Neuropharmacology and Central Nervous System – Theoretical Aspects. Alan R Liss Inc.: New York, pp. 353–358.

Millan MJ, Dekeyne A, Gobert A (1998) Serotonin (5-HT)2C receptors tonically inhibit dopamine (DA) and noradrenaline (NA), but no 5-HT, release in the frontal cortex in vivo. Neuropharmacology 37:953–955.

Millan MJ, Gobert A, Rivet JM, et al (2000) Mirtazapine enhances frontocortical dopaminergic and corticolimbic adrenergic, but not serotonergic, transmission by blockade of alpha2-adrenergic and serotonin2C receptors: a comparison with citalopram. Eur J Neurosci 12:1079–1095.

Millan MJ, Brocco M, Gobert A, et al (2005) Anxiolytic properties of agomelatine, an antidepressant with melatoninergic and serotonergic properties: role of 5-HT2C receptor blockade. Psychopharmacology (Berl) 177:448–458.

Milligan G, Bond RA (1997) Inverse agonism and the regulation of receptor number. Trends Pharmacol Sci 18: 468–474.

Milligan G, Bond RA, Lee M (1995) Inverse agonism: pharmacological curiosity or potential therapeutic strategy? Trends Pharmacol Sci 16:10–13.

Moorman JM, Leslie RA (1996) P-chloroamphetamine induces c-fos in rat brain: a study of serotonin2A/2C receptor function. Neuroscience 72:129–139.

Morisset S, Rouleau A, Ligneau X, et al (2000) High constitutive activity of native H3 receptors regulates histamine neurons in brain. Nature 408:860–862.

Navailles S, De Deurwaerdère P, Porras G, et al (2004) In vivo evidence that 5-HT2C receptor antagonist but not agonist modulates cocaine-induced dopamine outflow in the rat nucleus accumbens and striatum. Neuropsychopharmacology 29:319–326.

Navailles S, Moison D, Ryczko D, et al (2006a) Region-dependent regulation of mesoaccumbens dopamine neurons in vivo by the constitutive activity of central serotonin2C receptors. J Neurochem 99:1311–1319.

Navailles S, De Deurwaerdère P, Spampinato U (2006b) Clozapine and haloperidol differentially alter the constitutive activity of central serotonin2C receptors in vivo. Biol Psychiatry 59:568–575.

Navailles S, Moison D, Cunningham KA, et al (2008) Differential regulation of the mesoaccumbens dopamine circuit by serotonin2C receptors in the ventral tegmental area and the nucleus accumbens: an in vivo microdialysis study with cocaine. Neuropsychopharmacology 33:237–246.

Niswender CM, Copeland SC, Herrick-Davis K, et al (1999) RNA editing of the human serotonin 5-hydroxytryptamine2C receptor silences constitutive activity. J Biol Chem 274:9472–9478.

Nomikos GG, Iurlo M, Andersson JL, et al (1994) Systemic administration of amperozide, a new atypical antipsychotic drug, preferentially increases dopamine release in the rat medial prefrontal cortex. Psychopharmacology (Berl) 115:147–156.

Nonogaki K, Abdallah L, Goulding EH, et al (2003) Hyperactivity and reduced energy cost of physical activity in serotonin 5-HT(2C) receptor mutant mice. Diabetes 52:315–320.

Olsen CK, Hogg S, Lapiz MD (2002) Tonic immobility in guinea pigs: a behavioural response for detecting an anxiolytic-like effect? Behav Pharmacol 13:261–269.

Parsons LH, Justice Jr JB (1993) Perfusate serotonin increases extracellular dopamine in the nucleus accumbens as measured by in vivo microdialysis. Brain Res 606:195–199.

Pazos A, Hoyer D, Palacios JM (1984) The binding of serotonergic ligands to the porcine choroid plexus: characterization of a new type of serotonin recognition site. Eur J Pharmacol 106:539–546.

Pehek EA (1996) Local infusion of the serotonin antagonists ritanserin or ICS 205,930 increases in vivo dopamine release in the rat medial prefrontal cortex. Synapse 24:12–18.

Pehek EA, Bi Y (1997) Ritanserin administration potentiates amphetamine-stimulated dopamine release in the rat prefrontal cortex. Prog Neuropsychopharmacol Biol Psychiatry 21:671–682.

Plech A, Brus R, Kalbfleisch J H, et al (1995) Enhanced oral activity responses to intrastriatal SKF 38393 and m-CCP are attenuated by intrastriatal mianserin in neonatal 6-OHDA-lesionned rats. Psychopharmacology (Berl.) 119:466–473.

Pompeiano M, Palacios JM, Mengod G (1994) Distribution of the serotonin 5-HT2 receptor family mRNAs: comparison between 5-HT2A and 5-HT2C receptors. Brain Res Mol Brain Res 23:163–178.

Porras G, Di Matteo V, Fracass C, et al (2002) 5-HT2A and 5-HT2C/2B receptor subtypes modulate dopamine release induced in vivo by amphetamine and morphine in both the nucleus accumbens and striatum. Neuropsychopharmacology 26:311–324.

Pozzi L, Acconcia S, Ceglia I, et al (2002) Stimulation of 5-hydroxytryptamine (5-HT(2C)) receptors in the ventrotegmental area inhibits stress-induced but not basal dopamine release in the rat prefrontal cortex. J Neurochem 82:93–100.

Price RD, Weiner DM, Chang MS, et al (2001) RNA editing of the human serotonin 5-HT2C receptor alters receptor-mediated activation of G13 protein. J Biol Chem 276:55663–55668.

Prinssen EP, Koek W, Kleven MS (2000) The effects of antipsychotics with 5-HT(2C) receptor affinity in behavioural assays selective for 5-HT(2C) receptor antagonist properties compounds. Eur J Pharmacol 388:57–67.

Prisco S, Pagannone S, Esposito E (1994) Serotonin-dopamine interaction in the rat ventral tegmental area: an electrophysiological study in vivo. J Pharmacol Exp Ther. 271:83–90.

Quirk K, Lawrence A, Jones J, et al (2001) Characterisation of agonist binding on human 5-HT2C receptor isoforms. Eur J Pharmacol 419:107–112.

Rauser L, Savage JE, Meltzer HY, et al (2001) Inverse agonist actions of typical and atypical antipsychotic drugs at the human 5-hydroxytryptamine(2C) receptor. J Pharmacol Exp Ther 299:83–89.

Robinson ES, Dalley JW, Theobald DE, et al (2008) Opposing roles for 5-HT2A and 5-HT2C receptors in the nucleus accumbens on inhibitory response control in the 5-choice serial reaction time task. Neuropsychopharmacology 33:2398–2406.

Rosenzweig-Lipson S, Sabb A, Stack G, et al (2007) Antidepressant-like effects of the novel, selective, 5-HT2C receptor agonist WAY-163909 in rodents. Psychopharmacology (Berl) 192:159–170.

Rouillard C, Bovetto S, Gervais J, et al (1996) Fenfluramine-induced activation of the immediate-early gene c-fos in the striatum: possible interaction between serotonin and dopamine. Brain Res Mol Brain Res 37:105–115.

Rowley HL, Needham PL, Kilpatrick IC, et al (2000) A comparison of the acute effects of zotepine and other antipsychotics on rat cortical dopamine release in vivo. Naunyn Schmiedeberg's Arch Pharmacol 361:187–192.

Schmauss C (2003) Serotonin 2C receptors: suicide, serotonin, and runaway RNA editing. Neuroscientist 9:237–242.

Schmauss C (2005) Regulation of serotonin2C receptor pre-mRNA editing by serotonin. Int Rev Neurobiol 63:83–100.

Schotte A, Janssen PFM, Megens AA, et al (1993) Occupancy of central neurotransmitter receptors by risperidone, clozapine and haloperidol, measured ex vivo by quantitative autoradiography. Brain Res 631:191–202.

Seifert R, Wenzel-Seifert K (2002) Constitutive activity of G-protein coupled receptors: cause of disease and common property of wild-type receptors. Naunyn Schmiedebergs Arch Pharmacol 366:381–416.

Sharp T, Bramwell SR, Grahame-Smith DG (1989) 5-HT1 agonists reduce 5-hydroxytryptamine release in rat hippocampus in vivo as determined by brain microdialysis. Br J Pharmacol 96:283–290.

Shilliam CS, Dawson LE (2005) The effect of clozapine on extracellular dopamine levels in the shell subregion of the rat nucleus accumbens is reversed following chronic administration: comparison with a selective 5-HT2C receptor antagonist. Neuropsychopharmacology 30:372–80.

Stark JA, Davies KE, Williams SR, et al (2006) Functional magnetic resonance imaging and c-Fos mapping in rats following an anoretic dose of m-chlorophenylpiperazine. Neuroimage 31:1228–1237.

Stark JA, McKie S, Davies KE, et al (2008) 5-HT2C antagonism blocks blood oxygen level-dependent pharmacological-challenge magnetic resonance imaging signal in rat brain areas related to feeding. Eur J Neurosci 27:457–465.

Stewart BR, Jenner P, Marsden CD (1989) Induction of purposeless chewing behaviour in rats by 5-HT agonist drugs. Eur J Pharmacol 162:101–107.

Talpos JC, Wilkinson LS, Robbins TW (2006) A comparison of multiple 5-HT receptors in two tasks measuring impulsivity. J Psychopharmacol 20:47–58.

Tanda G, Bassareo V, DI Chiara G (1996) Mianserin markedly and selectively increases extracellular dopamine in the prefrontal cortex as compared to the nucleus accumbens of the rat. Psychopharmacology (Berl) 123:127–130.

Tecott LH, Sun LM, Akana SF, et al (1995) Eating disorder and epilepsy in mice lacking 5-HT2C serotonin receptors. Nature 74:542–546.

Thomsen WJ, Grottick AJ, Menzaghi F, et al (2008) Lorcaserin, a novel selective human 5-hydroxytryptamine2C agonist: in vitro and in vivo pharmacological characterization. J Pharmacol Exp Ther 325:577–587.

Tomkins DM, Joharchi N, Tampakeras M, et al (2002) An investigation of the role of 5-HT2C receptors in modifying ethanol self-administration behaviour. Pharmacol Biochem Behav 71:735–744.

Van Wijngaarden I, Tulp MTM, Soudijn W (1990) The concept of selectivity in 5-HT receptor research. Eur J Pharmacol 188:301–312.

Ward RP, Dorsa DM (1996) Colocalization of serotonin receptor subtypes 5-HT2A, 5-HT2C, and 5-HT6 with neuropeptides in rat striatum. J Comp Neurol 370:405–414.

Werry TD, Stewart GD, Crouch MF, et al (2008) Pharmacology of 5HT(2C) receptor-mediated ERK1/2 phosphorylation: agonist-specific activation pathways and the impact of RNA editing. Biochem Pharmacol 76:1276–87.

Willins DL, Meltzer HY (1998) Serotonin 5-HT2C agonists selectively inhibit morphine-induced dopamine efflux in the nucleus accumbens. Brain Res 781:291–299.

Winstanley CA, Theobald DE, Dalley JW, et al (2004) 5-HT2A and 5-HT2C receptor antagonists have opposing effects on a measure of impulsivity: interactions with global 5-HT depletion. Psychopharmacology (Berl) 176:376–385.

Wolf WA, Schultz L (1997) The serotonin 5-HT2C receptor is a prominent receptor in basal ganglia: evidence from functional studies on serotonin-mediated phosphoinositide hydrolysis. J Neurochem 69:1449–1458.

Wolf AW, Bieganski GJ, Guillen V, et al (2005) Enhanced 5-HT2C receptor signalling is associated with haloperidol-induced "early onset" vacuous chewing in rats : implications for antipsychotic drug therapy. Psychopharmacology 182:84–94.

Wood MD, Reavill C, Trail B, et al (2001) SB-243213, a selective 5-HT2C receptor inverse agonist with improved anxiolytic profile: lack of tolerance and withdrawal anxiety. Neuropharmacology 41:186–199.

Chapter 11
The 5-HT$_{2C}$ Receptor Subtype Controls Central Dopaminergic Systems: Evidence from Electrophysiological and Neurochemical Studies

Giuseppe Di Giovanni, Ennio Esposito, and Vincenzo Di Matteo

11.1 Introduction

There is now an extensive scientific literature regarding the functional interaction between serotonin (5-HT) and dopamine (DA)-containing neurons in the brain (Di Giovanni et al. 2008). In recent years, research on this matter has been spurred by the acquisition of important new insights on the molecular biology of 5-HT receptor subtypes and by the availability of 5-HT receptor knockout mice (Bonasera and Tecott 2000; Hoyer et al. 2002).

Central serotonergic and dopaminergic systems play an important role in regulating normal and abnormal behaviors (Koob 1992; Roth et al. 1992; Fibiger 1995). Moreover, dysfunctions of 5-HT and DA neurotransmission are involved in the pathophysiology of various neuropsychiatric disorders including schizophrenia, depression, and drug abuse (Di Giovanni et al. 2008; Koob 1992; Roth et al. 1992; Fibiger 1995; Brown and Gershon 1993). Thus, the development of a number of relatively selective pharmacological agents with agonist or antagonist activity at 5-HT$_{2C}$ receptor subtype has allowed investigators to better understand the functional role of this receptor in the control of central dopaminergic function, as it widely contributes to the serotonergic regulation of a number of behavioral and physiological processes involving both limbic and striatal DA pathways (Jenck et al. 1998; Di Matteo et al. 2001; Higgins and Fletcher 2003; Giorgetti and Tecott 2004). Therefore, the physiology, pharmacology, and anatomical distribution of the 5-HT$_{2C}$ receptors in the central nervous system (CNS) as well as experimental data regarding the effect of 5-HT$_{2C}$ selective agents on the neuronal activity of DA neurons of the ventral tegmental area

V. Di Matteo (✉) and E. Esposito
Istituto di Ricerche Farmacologiche "Mario Negri," Consorzio Mario Negri Sud,
66030, Santa Maria Imbaro, Chieti, Italy
e-mail: vdimatteo@negrisud.it

G. Di Giovanni (✉)
Department of Physiology and Biochemistry, University of Malta, Msida,
MSD 2080, Msida, Malta
e-mail: giuseppe.digiovanni@um.edu.mt

G. Di Giovanni et al. (eds.), *5-HT$_{2C}$ Receptors in the Pathophysiology of CNS Disease*,
The Receptors 22, DOI 10.1007/978-1-60761-941-3_11,
© Springer Science+Business Media, LLC 2011

(VTA) and substantia nigra pars compacta (SNc) and the changes of basal DA release in the striatum, nucleus accumbens, and cortex are reviewed in this chapter, introduced by a brief description of the functional neuroanatomy of dopaminergic and serotonergic systems. Finally, the potential use of 5-HT$_{2C}$ agents in the treatment of depression, schizophrenia, Parkinson disease (PD), and drug abuse will be also examined, given its prominence as the receptor by which the serotonergic system affects both mesolimbic and nigrostriatal DA function, and its consequent involvement in the regulation of a number of behavioral and physiological processes.

11.2 Dopamine Systems

Dopamine-containing neurons of the ventral mesencephalon have been designated as A8, A9, and A10 cell groups: These neurons can be collectively designated as the mesotelencephalic DA system (Dahlström and Fuxe 1964). Historically, the mesolimbic DA system was defined as originating in the A10 cells of the VTA and projecting to structures closely associated with the limbic system. This system was considered separated from the nigrostriatal DA system, which originates from the more lateral substantia nigra (A9 cell group) (Dahlström and Fuxe 1964; Bannon and Roth 1983; Roth et al. 1987; Kalivas 1993; White 1996).

The mesolimbic and mesocortical DA system appear critically involved in modulation of the functions subserved by cortical and limbic regions such as motivation, emotional control, and cognition (Le Moal and Simon 1991). Substantial evidence indicates that the mesolimbic pathway, particularly the DA cells innervating accumbal areas, is implicated in the reward value of both natural and drug reinforcers, such as sexual behavior or psychostimulants, respectively (Koob 1992; Di Chiara and Imperato 1988; Salamone et al. 2007). Furthermore, animal studies have shown that lesion of DA terminals in the nucleus accumbens induces hypoexploration, enhanced latency in the initiation of motor responses, disturbances in organizing complex behaviors, and inability to switch from one to another behavioral activity (Le Moal and Simon 1991). Hence the mesolimbic DA system seems important for acquisition and regulation of goal-directed behaviors, established and maintained by natural or drug reinforcers (Le Moal and Simon 1991; Kiyatkin 1995).

The medial prefrontal cortex (mPFC) is generally associated with cognitive functions including working memory, planning and execution of behavior, inhibitory response control and maintenance of focused attention (Le Moal and Simon 1991). In addition, the mesolimbic DA pathway is sensitive to a variety of physical and psychological stressors (Roth and Elsworth 1995). Indeed, recent studies have indicated that stress-induced activation of the mesocortical DA neurons may be obligatory for the behavioral expression of such stimuli (Morrow et al. 1999).

The nigrostriatal DA system, which originates from the substantia nigra (A9 cell group), is one of the best studied because of its involvement in the pathogenesis of Parkinson disease (Grace and Bunney 1985). In mammals, the substantia nigra (SN) is a heterogeneous structure that includes two distinct compartments: the substantia nigra pars compacta (SNc) and the substantia nigra pars reticulata (SNr).

The SNc represents the major source of striatal DA and, as already mentioned, its degeneration causes Parkinson disease. On the other hand, the SNr mainly contains γ-aminobutyric acid (GABA)-ergic neurons that constitute one of the major efferences of the basal ganglia (Grace and Bunney 1985).

11.3   Serotonin Systems

Virtually all parts of the CNS receive innervation from serotonergic fibers arising from cell bodies of the two main subdivisions of the midbrain serotonergic nuclei, the dorsal raphé (DR) and the median raphé (MR) (Azmitia and Segal 1978; Van der Kooy and Attori 1980; Steinbush 1984; Hervé et al. 1987; Van Bockstaele et al. 1993, 1994; Moukhles et al. 1997). Serotonin-containing cell bodies of the raphé nuclei send projections to dopaminergic cells both in the VTA and the SN and to their terminal fields in the nucleus accumbens, prefrontal cortex, and striatum (Van der Kooy and Attori 1980; Steinbush 1984; Hervé et al. 1987; Van Bockstaele et al. 1993, 1994; Moukhles et al. 1997). Electron microscopy demonstrates the presence of synaptic contacts of [^{3}H]5-HT labeled terminals with both dopaminergic and nondopaminergic dendrites in all subnuclei of the VTA and of the SN pars compacta and reticulata (Kalivas 1993; Hervé et al. 1987; Moukhles et al. 1997).

11.4   Serotonin Modulation of DA Neuronal Activity

The first electrophysiological study aimed at investigating the effect of 5-HT on DA electrical activity was performed on neurorochemically identified DA neurons in the SNc (Aghajanian and Bunney 1974). Microiontophoretic application of 5-HT was found to exert only weak inhibitory effects on the firing rate of DA neurons (Aghajanian and Bunney 1974). However, most of the early electrophysiological studies on this subject were carried out recording the electrical activity of neurons indiscriminately, in both the SNc and the SNr, and without using the criteria for the neurochemical identification of the neurons recorded (Dray et al. 1976, 1978; Fibiger and Miller 1977).

11.4.1   Serotonergic Lesion on DA Neuronal Activity

The intracerebroventricular (i.c.v.) injection of the toxin 5,7-dihydroxytryptamine (5,7-DHT) produces a robust and selective decrease in brain 5-HT levels (Prisco et al. 1994; Prisco and Esposito 1995; Di Mascio et al. 1999; Guiard et al. 2008). In fact, the toxin treatment resulted in significant depletions of 5-HT in the corpus striatum (−66.5%) and hippocampus (−90%) (Di Mascio et al. 1999) and in the frontal cortex (−87%) (Guiard et al. 2008). Conversely, 5,7-DHT lesions did not cause any change in the basal interspike interval (ISI) standard characteristics (firing rate, bursting activity) either in the VTA (Prisco et al. 1994; Prisco and Esposito 1995) or in the

SNc (Kelland et al. 1990). The basal firing rate of VTA dopaminergic neurons in
5-HT-depleted rats was higher than in control animals, but this difference was not
statistically significant (Prisco et al. 1994; Prisco and Esposito 1995). In addition,
depletion of brain 5-HT had little effect on the basal activity of nigrostriatal DA neurons. However, both 5-HT depleters *para*-chlorophenylalanine (PCPA) and 5,7-DHT
produced small but significant reductions in the conduction velocity of these cells
(Kelland et al. 1990). Minabe et al. (1996) showed that depletion of brain 5-HT by
administration of PCPA produced a significant decrease in the number of spontaneously active DA cells in both the SNc (52%) and VTA (63%) areas, compared with
controls. The burst firing analysis indicated that there was a significant increase in the
mean interspike interval, with a decrease in both the burst firing pattern and the
number of bursts of SNc and VTA DA neurons in PCPA treated animals. The intravenous (IV) administration of 5-hydroxytryptophan and the peripheral aromatic acid
decarboxylase inhibitor benserazide, which restores 5-HT content, reversed both the
decrease in the number of spontaneously active SNc and VTA DA neurons, as well
as the decrease in the percentage of VTA DA neurons exhibiting a bursting pattern.

It was noticeable that although we did not reveal any modification of DA conventional firing characteristics on 5-HT depleted animals (Prisco et al. 1994), in successive experiments, using the nonlinear prediction method combined with Gaussian-scaled
surrogate, we showed a decrease in chaos of the electrical activity of VTA DA neurons, extracellularly recorded in vivo lesioned with 5,7-DHT (Di Mascio et al. 1999).
The term *chaos* is here intended to describe a highly erratic yet deterministic behavior.
Moreover, in the control (unlesioned) group a positive correlation was found between
the functional operator (Ψ), equivalent to the density power spectrum of the signals
and the interspike intervals, and the chaos content measure by nonlinear prediction
S score; a relation that was lost in the lesioned group (Di Mascio et al. 1999). We
suggested that the decreased chaotic behavior of dopaminergic neurons in lesioned
rats, which have a decreased serotonergic tone, might represent a preclinical phenomenon of the onset of psychiatric disorders in humans. In fact, chaotic behavior, more
than periodic or stochastic behavior, makes biological systems more capable of
responding to different stimuli without causing damage.

Recently, Guiard et al. (2008) showed contrasting results with the above findings, supporting the hypothetical inhibitory 5-HT tone upon DA neurons. In their
experimental conditions, 5,7-DHT lesions led to a 36% increase in the discharge
rate of VTA DA neurons in rats. This enhancement of DA neuronal activity resulted
from a higher number of bursts and spikes per burst (Guiard et al. 2008).

11.4.2   Electrical Stimulation of the Dorsal and Medial Raphé
           Nuclei on DA Neuron Activity

Microiontophoretic application of 5-HT caused mainly inhibition of neuronal activity,
but excitation and biphasic effects were also observed (Dray et al. 1976). These
early findings led the authors of the studies to conclude "…the substantia nigra

receives a direct monosynaptic inhibitory input from the DR and MR nuclei and those pathways use 5-HT as a neurotransmitter serving to tonically inhibit DA-ergic neurons..." (Dray et al. 1978). Dray et al. (1978) also studied the effect of the stimulation of the MR in the rat, revealing a predominant reduction of the activity of cells in the SN. This evidence was not confirmed by Fibiger and Miller (1977); in fact, single pulse stimulation of the MR was relatively ineffective in their experimental conditions. Instead, they showed that a selective single stimuli delivered to the DR was capable of causing complete cessation of spontaneously active nigral cells for periods ranging from 25 to 180 ms. The discrepancy between these two studies is probably due to the fact (Dray et al. 1976), by stimulating the MR, could have activated serotonergic axons from the DR, which descend in the vicinity of the MR. Since PCPA blocked the inhibitory effects of DR stimulation and did not significantly influence the spontaneous activity of cells in the SNc, it follows that the serotoninergic projection from the DR influences the dopaminergic cells in a phasic rather than a tonic way. Hence, it is only under certain circumstances that this inhibition is manifest.

Following these studies, Kelland et al. (1990, 1993) found that electrical stimulation at low to moderate frequencies of DR inhibits all VTA DA neurons, as compared with inhibiting only SNc DA neurons with a basal activity below 4 Hz, which they defined as "slowly firing." The inhibition of VTA neurons was significant at stimulation rates of 0.1 Hz and higher for slow cells and at rates of 0.5 Hz and higher for fast cells. Maximal inhibition at 10.0 Hz was $85 \pm 10\%$ and $60 \pm 16\%$, respectively, and there was no significant difference between the slow and fast cell groups (Kelland et al. 1993). The inhibition in the SNc was seen only on slow cells, was significant at stimulation rates of 0.5 Hz and higher, and reached a maximal inhibition of $66 \pm 7\%$ at 10.0 Hz (Kelland et al. 1990, 1993). This preferential influence of 5-HT on a subset of nigrostriatal DA neurons raises significant issues. This data suggests that slow nigrostriatal DA neurons may fire more slowly because they are under a tonic inhibitory influence by 5-HT. A likely correlate is that fast cells may also receive some preferential excitatory input, such as the excitatory input from the pedunculopontine tegmental nucleus, subthalamus, and cortex. Thus, slow cells and fast cells, under basal conditions, may represent distinct populations of DA neurons with differential afferent regulation. In addition, these data help to explain why roughly half of the DA cells (the fast cells) might not respond to administration of 5-HT and related compounds.

Furthermore, Trent and Tepper (1991) reported that electrical stimulation of DR decreased the somatodendritic excitability of DA neurons in the SNc, expressed a reduction of the proportion of neostriatal-evoked antidromic responses. Conversely, the DR stimulation did not significantly affect the basal firing rate and pattern of the SNc recorded, consistent with dissociation between the effects of raphé input on somatodendritic excitability and neuronal firing. The reduction in excitability is due, at least in part, to 5-HT release in the SNc, as indicated by the reversal of the effect by the 5-HT antagonist metergoline and by the abolition of the effect in rats depleted of serotonin by PCPA treatment and to the subsequent reinstatement of the effect by 5-HTP administration. Stimulation of the raphé (DR) nucleus activates a serotonergic raphé-nigral pathway synapting on dopaminergic neurons. Synaptically released 5-HT activates a 5-HT receptor on the nigral somadendrite, which triggers a local dendritic release of DA, either by dendritic depolarization or by a direct action on a somatodendritic calcium conductance. The dendritically released DA in turn activates somatodendritic D_2-dopamine autoreceptors which hyperpolarize the somatodendritic membrane locally and results in conduction failure of the SD component of the antidromic response due to either the dendritic hyperpolarization or the underlying increase in potassium conductance or both (Trent and Tepper 1991). The authors further supported the thesis that, in vivo, the serotonergic input from the DRN acts in a phasic rather than tonic manner to regulate the presynaptic (DA-releasing) functions of nigral dopaminergic dendrites locally, without affecting neuronal excitability as a whole.

The latest evidence available is given by Gervais and Rouillard (2000). In vivo stimulation of 5-HT neurons in the DR inhibited most of the recorded DA neurons in A9, while the majority of DA neurons in A10 were excited. Some cells exhibited an inhibition–excitation response, while in other DA neurons the initial response was an excitation followed by an inhibition. In SNc, 56% of the DA cells recorded were initially inhibited and 31% of the DA cells were initially excited. In contrast, 63% of VTA DA cells were initially excited and 34% were initially inhibited. Depletion of endogenous 5-HT by 5,7-DHT and PCPA almost completely eliminated the inhibition–excitation response in both SNc and VTA DA cells, without changing the percentage of DA cells initially excited. Consequently, the proportion of DA neurons that were not affected by DRN stimulation increased after 5-HT depletion (from 13% to 60% in the SNc and from 6% to 31% in the VTA). In several DA cells, DRN stimulation caused important changes in firing rate and firing pattern. These data strongly confirm that the 5-HT input from the DRN is mainly inhibitory. It also suggests that 5-HT afferences modulate SNc and VTA DA neurons in an opposite manner. These results also suggest that non-5-HT inputs from DRN can also modulate mesencephalic DA neurons and have an excitatory role on the activity of these neurons (Gervais and Rouillard 2000).

The precise nature of the interaction between 5-HT and DA has been difficult to elucidate, in that both inhibitory and excitatory roles for 5-HT have been suggested. However, these discrepancies may be attributable to the differential distribution and to the diverse functional roles of 5-HT receptors subtypes within the dopaminergic systems (Hoyer et al. 1994; Barnes and Sharp 1999). Thus, much attention has been devoted to the role of 5-HT_2 receptor family in the control of central DA activity, because of the moderate to dense localization of both transcript and protein for the 5-HT_{2A} and 5-HT_{2C} receptors in the SN and VTA as well as in DA terminal regions of the rat forebrain (Pompeiano et al. 1994; Abramowski et al. 1995; Doherty and Pickel 2000; Nocjar et al. 2002). It is therefore of interest to review briefly the principal characteristics of the 5-HT_2 receptor family.

11.5 The 5-HT$_2$ Receptor Family

5-HT$_2$ receptors form a closely related subgroup of G-protein-coupled receptors, are functionally linked to the phosphatidylinositol hydrolysis pathway, and are currently classified as 5-HT$_{2A}$, 5-HT$_{2B}$, and 5-HT$_{2C}$ subtypes, based on their close structural homology and pharmacology (Hoyer et al. 1994; Barnes and Sharp 1999; Boess and Martin 1994). There is a high sequence homology (>80% in the transmembrane regions) between the mouse, rat, and human 5-HT$_{2C}$ receptors (Barnes and Sharp 1999), and it is not surprising that many compounds bind with high affinity to all three receptor subtypes. 5-HT$_{2C}$ receptors are widely distributed throughout the brain and have been proposed as the main mediators of the different actions of 5-HT in the CNS (Hoyer et al. 1994; Barnes and Sharp 1999; Boess and Martin 1994). High levels of 5-HT$_{2C}$ mRNA or protein expression have been found in the choroid plexus, the frontal cortex, in limbic structures such as hippocampus, septum, and hypothalamus and in the striatum, nucleus accumbens, rhombencephalon, and spinal cord. The presence of these receptors has also been demonstrated on DA and non-DA cells in the VTA, the SNc, and the SNr (Grace and Bunney 1985; Molineaux et al. 1989; Wright et al. 1995; Sharma et al. 1997; Clemett et al. 2000; Bubar and Cunningham 2007). The regional and cellular distribution of 5-HT$_{2C}$ receptors was also investigated in the human brain. The main sites of mRNA 5-HT$_{2C}$ receptors or protein expression were the choroid plexus, cerebral cortex, hippocampus, amygdala, some components of the basal ganglia, and other limbic structures (Abramowski et al. 1995; Pasqualetti et al. 1999), suggesting that this receptor might be involved in the regulation of different human brain function and might play a role in the pathophysiology of several mental disorders (Jenck et al. 1998; Di Matteo et al. 2001; Higgins and Fletcher 2003; Giorgetti and Tecott 2004; Kennett 1993; Baxter et al. 1995; Alex and Pehek 2007).

There is now evidence that the 5-HT$_{2C}$ receptor is mainly located postsynaptically within dopaminergic, GABA-ergic, cholinergic, substance P, dynorphin, and other systems (Barnes and Sharp 1999; Bubar and Cunningham 2007; Ward and Dorsa 1996; Eberle-Wang et al. 1997). Interestingly, the studies by Eberle-Wang et al. (1997) showed the presence of 5-HT$_{2C}$ mRNA within inhibitory GABA-ergic interneurons making direct synaptic contact with SNc and VTA dopaminergic cell bodies. Other immunohistochemical and electrophysiological studies demonstrated an important role of 5-HT$_{2C}$ receptors, localized on non-DA neurons, presumably GABA-ergic, in the regulation of DA cells in the VTA (Bubar and Cunningham 2007; Van Bockstaele and Pickel 1995; Steffensen et al. 1998), in the medial prefrontal cortex (Liu et al. 2007), and in the SNc (Di Giovanni et al. 2001; Invernizzi et al. 2007).

Recent studies found a somatodentritic localization of 5-HT$_{2A}$ receptors on DA neurons in both the parabrachial and paranigral subdivisions of the VTA (Doherty and Pickel 2000; Nocjar et al. 2002), which project mainly to the prefrontal cortex and nucleus accumbens, respectively. In addition, 5-HT$_{2A}$ immunoreactivity was also expressed on non-DA cells in the VTA, providing a potential anatomical basis for the modulation of DA neurons in the VTA either directly by 5-HT$_{2A}$ receptors localized on DA cell or indirectly through receptors present on non-DA (presumably GABA-ergic) neurons (Doherty and Pickel 2000; Nocjar et al. 2002). These receptors were also found at high concentrations in various cortical regions (Doherty and Pickel 2000; Wright et al. 1995). It is likely that 5-HT$_{2A}$ receptors could affect DA function by acting at the level of dopaminergic nerve terminals, although no direct evidence for the presence of 5-HT$_{2A}$ receptors on such terminals has been provided so far.

11.6   5-HT$_{2C}$ Receptors and Dopamine Function

Ugedo et al. (1989) showed that systemic administration of ritanserin, a 5-HT$_2$ antagonist, was capable of increasing both the firing rate and the bursting activity of DA neurons in the SNc and the VTA. These effects were prevented by 5-HT depletion induced by PCPA (Ugedo et al. 1989). These authors concluded that "… these results suggest that 5-HT exerts an inhibitory control of midbrain DA cell activity mediated by 5-HT$_2$ receptors …" (Ugedo et al. 1989), obviously referring to 5-HT$_2$ as the receptors that were subsequently defined 5-HT$_{2A}$ (Hoyer et al. 2002). However, it is important to point out that the doses of ritanserin (0.5–2.0 mg/kg i.v.) used in the study of (Ugedo et al. 1989) were too high to selectively block 5-HT$_{2A}$ receptors, in that it has been shown that ritanserin is a potent 5-HT antagonist, which also binds with high affinity to 5-HT$_{2C}$ receptors (Boess and Martin 1994). It is therefore impossible to discriminate the relative contribution of 5-HT$_{2A}$ and 5-HT$_{2C}$ in the disinhibitory effect of ritanserin reported in the study by Ugedo et al. (1989). In a subsequent study, the same research group (Andersson et al. 1995) found that ritanserin (1.0 mg/kg i.v.) increased the firing rate, the burst firing, and the variation coefficient of DA neurons in the VTA but not in the SNc. Moreover, ritanserin pretreatment significantly enhanced the stimulatory effects of low doses of raclopride (10–20 μg/kg i.v.) on the burst firing of VTA DA neurons (Andersson et al. 1995). These data indicated that unselective blockade of 5-HT$_2$ receptors by ritanserin can preferentially increase the activity of DA neurons in the VTA, but they did not clearly establish which receptor subtype (5-HT$_{2A}$ or 5-HT$_{2C}$ or both?) is involved in this effect. This picture was further complicated by the evidence that ritanserin (0.1–6.4 mg/kg i.v.) had no consistent effects on the basal firing rate of SNc DA neurons but significantly reversed the inhibition induced by both direct and indirect DA agonists (Shi et al. 1995). However, the effect of ritanserin was apparently mediated by a mechanism independent of 5-HT, but it was due to its ability to selectively block DA autoreceptors (Shi et al. 1995). Although 5-HT$_{2A}$ receptors are localized on a subset of DA cells in the VTA (Nocjar et al. 2002), the use of selective ligands has not revealed a clear role for these receptors in modulating dopaminergic neuronal activity within the nuclei. Blockade of 5-HT$_{2A}$ receptors by the potent and selective 5-HT$_{2A}$ antagonist MDL 100907 {R(+)-alpha-(2,3-dimethoxyphenyl)-1-[2-(4-fluorophenylethyl)]-4-piperidine-methanol} (Shi et al. 1995; Minabe et al. 2001) and SR 46349B {but-2-enedioic acid; 4-[(E)-3-(2-dimethylaminoethoxyamino)-3-(2-fluorophenyl)prop-2-enylidene]cyclohexa-2,5-dien-1-one} (Di Giovanni et al. 1999), or their activation with the agonist (±)-DOI (±)-1-(2,5-dimethoxy-4-iodophenyl)-2-aminopropane hydrochloride (Di Matteo et al. 2000a), had no significant effect on the basal activity of SNc and VTA DA neurons and on their inhibition of direct and indirect DA agonists (Shi et al. 1995). A cell-per-track study showed that MDL 100907 behaves as an atypical APD (Sorensen et al. 1993). MDL 100907 (1.0 mg/kg, i.p.) produced only small increases in the number of active SNc and VTA DA neurons after acute administration (Sorensen et al. 1993) but at 0.1 mg/kg, i.p. significantly increased the number of spontaneously active SNc and VTA DA neurons. When administered chronically, MDL 100907 (1.0 mg/kg, i.p.) selectively reduced the number of spontaneously active VTA neurons (Sorensen et al. 1993), whereas at lower doses (0.03-0.1 mg/kg, i.p.) it was active in reducing the number of the cells in both the DA nuclei.

In spite of these findings, a functional role for 5-HT$_{2A}$ receptors localized on VTA DA neurons has yet to be determined. However, in another study 5-HT$_{2A}$ antagonists were found to block the inhibitory effect of amphetamine on the basal firing rate of DA neurons, in both the SNc and the VTA (Sorensen et al. 1993) or the amphetamine-induced DA release in the nucleus accumbens and striatum (Porras et al. 2002). Conversely, the selective blockade of 5-HT$_{2A}$ does not modify the effect of morphine (Porras et al. 2002). Thus, this receptor subtype might modulate the activity of both the nigro-striatal and mesocorticolimbic dopaminergic systems only when specific neuronal circuitry mechanisms are activated. Further investigations into the circuitry of this regulation indicated that 5-HT$_{2A}$ receptors on cortical projections regulate DA cellular activity. 5-HT$_{2A}$ receptors seem to be unable to modulate DA function under resting conditions (Porras et al. 2002; Di Giovanni et al. 2000).

Compelling evidence has been given about the lack of influence on DA cell function by the selective blockade of central 5-HT$_{2B}$ receptors (Di Giovanni et al. 1999; Di Matteo et al. 2000a; Gobert et al. 2000).

Sixteen years ago, the issue regarding the role of 5-HT on the control of the electrical activity of DA neurons in the SNc and the VTA was quite confused and controversial. At that time, a study undertaken in our laboratory shed some light on the subject (Prisco et al. 1994). Systemic administration of mesulergine produced a significant increase in the basal firing rate of VTA DA neurons, whereas ritanserin, used at doses that selectively block 5-HT$_{2A}$ receptors, caused a slight, statistically significant decrease in the basal activity of these neurons (Prisco et al. 1994). These data, although obtained by using partially selective antagonists, represent the first evidence of a differential effect of 5-HT$_{2A}$ and 5-HT$_{2C}$ receptors upon DA-containing neurons in the VTA (Prisco et al. 1994). Thus, the data obtained by Prisco et al. (1994) supported the conclusion that 5-HT exerts an inhibitory action on DA neurons in the VTA through the 5-HT$_{2C}$ receptor subtype (which at that time was still named 5-HT$_{1C}$). Subsequent studies confirmed a selective involvement of 5-HT$_{2C}$ receptors on the basis of the evidence that the inhibitory effect of the mixed 5HT$_{2A/2B/2C}$ receptor agonists 1-(meta-chlorophenyl)piperazine (mCPP) and 6-chloro-2-(1-piperazinyl) piperazine (MK 212) on the activity of VTA DA-containing neurons and on accumbal

DA release was completely prevented by SB 242084 {6-chloro-5-methyl-1-[2-(2-methylpyridiyl-3-oxy)-pyrid-5-yl carbamoyl] indoline}, a selective 5-HT$_{2C}$ receptor antagonist (Di Giovanni et al. 2000). Moreover, SB 242084 blocked the inhibitory action of RO 60-0175 [(S)-2-(chloro-5-fluoro-indol-1-yl)-1-methylethylamine 1:1 C$_4$H$_4$O$_4$], a selective 5-HT$_{2C}$ receptor agonist (Di Matteo et al. 2000a; Martin et al. 1998). Another series of studies clearly indicated a selective involvement of 5-HT$_{2C}$ receptors for the suppressive influence of 5-HT on the activity of dopaminergic pathways. In fact, a series of in vivo electrophysiological and neurochemical studies showed that 5-methyl-1-(3-pyridylcarbamoyl)-1,2,3,5-tetrahydropyrrolo[2,3-f]indole (SB 206553), a selective 5-HT$_{2C/2B}$ receptor inverse agonist (Kennett et al. 1996; De Deurwaerdère et al. 2004), and 6-chloro-5-methyl-l-[2-(2-methylpyridiyl-3-oxy)-pyrid-5-yl carbamoyl] indoline (SB 242084), the most potent and selective 5-HT$_{2C}$ receptor antagonist available (Kennett et al. 1996), increased the basal firing rate and the bursting activity of VTA DA neurons (Fig. 11.1) and enhanced DA release in both rat nucleus accumbens (Fig. 11.2) and prefrontal cortex (Di Giovanni et al. 1999; Gobert et al. 2000; De Deurwaerdère et al. 2004; De Deurwaerdère and Spampinato 1999; Di Matteo et al. 1999; Gobert and Millan 1999). Conversely, systemic administration of RO 60-0175 had opposite effects (Figs. 11.3 and 11.5) (Kennett et al. 1996; Di Matteo et al. 1999; Millan et al. 1998). SB 206553 and SB 242084 were also found to potentiate pharmacological-induced accumbal DA release (Porras et al. 2002; Hutson et al. 2000; Navailles et al. 2004) and stress-stimulated DA outflow in

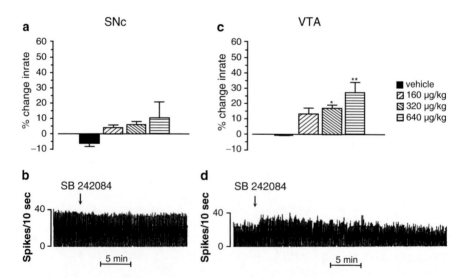

Fig. 11.1 Effect of SB242084 on the firing rate of SNc and VTA DA neurons. (**a**) and (**c**) Histograms showing mean percentage of change (±SEM) in firing rate of DA neurons in the SNc (**a**) and the VTA (**c**) after i.v. SB 242084 (*n*=6–8) (**b**) and (**d**) Representative rate histograms showing the lack of effect of i.v. SB 242084 (640 µg/kg) in the SNc (**b**) and the typical excitatory response in the VTA (**d**). *$P<0.05$, **$P<0.01$ compared with control group; one-way ANOVA followed by Tukey test (Modified from Di Matteo et al. 1999)

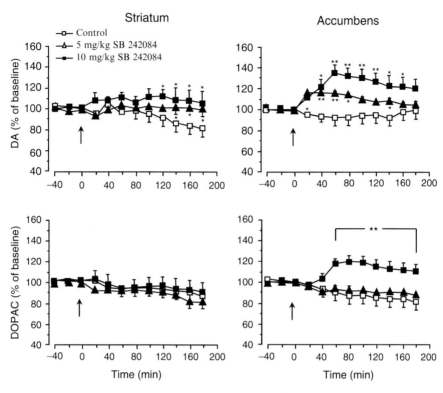

Fig. 11.2 Time course of the effect of i.p. administration of 5 (▲) and 10 mg/kg SB 242084 (■) on extracellular DA and DOPAC levels in the striatum (*left column*) and the nucleus accumbens (*right column*). (□) Control group treated with vehicle. SB 242084 was administered at the time indicated by vertical *arrows*. Each data point represents mean percentage ± SEM of the baseline value calculated from three samples before SB 242084 injection. Each experiment was carried out on five to six animals per group. *$P<0.05$, **$P<0.01$ compared with control group; two-way ANOVA followed by Tukey test (Modified from Di Matteo et al. 1999)

the rat prefrontal cortex (Pozzi et al. 2002), while stimulation of 5-HT$_{2C}$ receptors by RO 60-0175 in the VTA suppressed it (Pozzi et al. 2002), suggesting a role of these receptors on evoked accumbal DA release also. On the other hand, 5-HT$_{2C}$ receptor agonists such as mCPP, MK 212, and RO 60-0175 did not significantly affect the activity of SNc DA neurons and the in vivo DA release in the striatum (Figs. 11.4 and 11.5) (Gobert et al. 2000; Di Matteo et al. 1999), and the mixed 5HT$_{2B/2C}$ antagonist SB 206553 caused only a slight increase in the basal activity of DA neurons in the nigrostriatal pathway (Di Giovanni et al. 1999), suggesting that the serotonergic system controls both basal and stimulated impulse flow-dependent release of DA preferentially in the mesocorticolimbic system by acting through 5-HT$_{2C}$ receptors. Consistently, it has been reported that SB 243213 {5-methyl-1-[[-2-[(2-methyl-3-pyridyl)oxy]-5-pyridyl]carbamoyl]-6-trifluoromethylindoline hydrochloride}, a new selective 5-HT$_{2C}$ receptor inverse agonist (Wood et al. 2001; Berg et al. 2006), at a dose of 3 mg/kg, i.p.

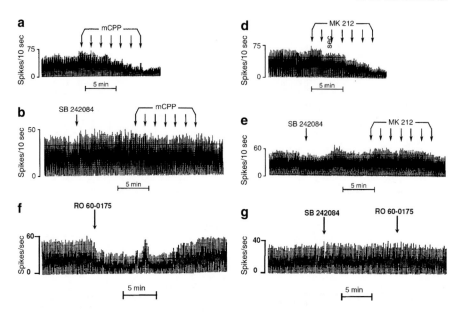

Fig. 11.3 Effect of mCPP, MK 212, and RO 60-0175 on the firing rate of VTA DA neurons and the blockade by SB 242084 of their inhibitory actions. Representative rate histograms showing the typical inhibitory effects produced by i.v. administration of mCPP and MK 212 (5, 5, 10, 20, 40, 80, 160 µg/kg at *arrows*) and RO 60-0175 (320 µg/kg at *arrows*) in control rats (**a**), (**d**), and (**f**) Representative rate histograms showing that i.v. SB 242084 (200 µg/kg) prevents the inhibitory effects of mCPP and MK 212 (5, 5, 10, 20, 40, 80, 160 µg/kg at *arrows*) (**b**), (**e**), and (**g**) (Modified from references Di Giovanni et al. 2000; Di Matteo et al. 1999, respectively)

significantly decreased only the number of spontaneously active VTA DA neurons and modified the pattern discharge but did not affect the number of spontaneously active SNc DA cells, whereas the 10 mg/kg, i.p. dose altered the firing pattern of DA neurons in both the SNc and the VTA (Blackburn et al. 2002).

Moreover, a study carried out in our laboratory has shown that mCPP excites non-DA (presumably GABA-containing) neurons in both the SNr and the VTA by activating 5-HT$_{2C}$ receptors (Di Giovanni et al. 2001). One interesting finding of that study was the differential effect exerted by mCPP on subpopulations of SNr neurons. Thus, mCPP caused a marked excitation of presumed GABA-ergic SNr projection neurons, whereas it did not affect SNr GABA-containing interneurons that exert a direct inhibitory influence on DA neurons in the substantia nigra (Di Giovanni et al. 2001). On the other hand, all non-DA neurons in the VTA were equally excited by mCPP. It is tempting to speculate that this differential response to mCPP might be the basis of the preferential inhibitory effect of 5-HT$_{2C}$ agonists on the mesocorticolimbic versus the nigrostriatal DA function. Other in vivo electrophysiological and neurochemical studies have confirmed and extended the above mentioned data, namely, that 5-HT exerts a direct excitatory effect on GABA-ergic neurons in the substantia nigra pars reticulata and VTA by acting on 5-HT$_{2C}$ receptors (Invernizzi et al. 2007; Bankson and Yamamoto 2004). In fact, about 50% of SNr neurons are excited by the selective

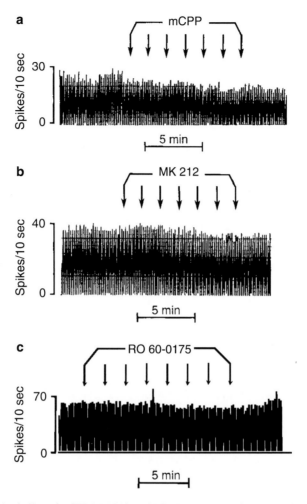

Fig. 11.4 Lack of effect of mCPP, MK 212, and RO 60-0175 on the firing rate of SNc DA neurons. Representative rate histograms showing the typical effects produced by i.v. administration of mCPP and MK 212 (5, 5, 10, 20, 40, 80, 160 μg/kg at *arrows*) and RO 60-0175 (5, 10, 20, 40, 80, 160, 320, 640 μg/kg at *arrows*) (Modified from Di Giovanni et al. 2000; Di Matteo et al. 1999, respectively)

5-HT$_{2C}$ receptors agonist RO 60-0175, and this effect is counteracted by SB 243213; moreover, microiontophoretic application of RO 60-0175 clearly showed a direct effect of the 5-HT$_{2C}$ receptors on the SNr neurons, antagonized by SB 243213 (Invernizzi et al. 2007). Infusion of RO 60-0175 and mCPP by reverse dialysis significantly increased extracellular levels of GABA in the SNr (Invernizzi et al. 2007). Nevertheless, intra-VTA infusion of SB 206553 has been shown to attenuate 3,4-methylenedioxymethamphetamine (MDMA)-induced increase GABA levels in the VTA and to potentiate the concurrent increase in accumbal DA release (Bankson and Yamamoto 2004).

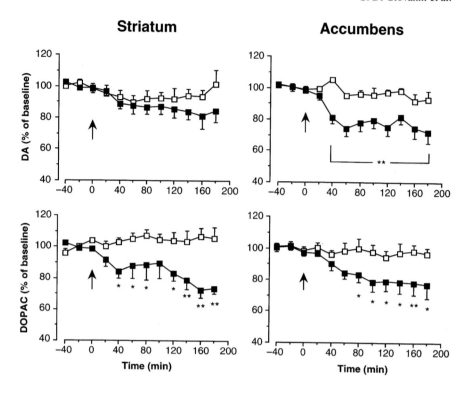

Fig. 11.5 Time course of the effect of i.p. administration of 1 mg/kg RO 60-0175 (■) on extracellular DA and DOPAC levels in the striatum (*left column*) and in the nucleus accumbens (*right column*). (□) Control group treated with vehicle. RO 60-0175 was administered at the time indicated by vertical *arrows*. Each data point represents mean percentage±SEM of the baseline value calculated from three samples before RO 60-0175 injection. Each experiment was carried out on five animals per group. *$P<0.05$, **$P<0.01$ compared with control group; two-way ANOVA followed by Tukey test (Modified from reference Di Matteo et al. 1999)

Although recent studies showed that systemic administration of 5-HT$_{2C}$ receptor agonists, including RO 60-0175, do not significantly decrease the activity of nigrostriatal dopaminergic neurons (Di Giovanni et al. 2000; Di Matteo et al. 1999), such treatment decreases DA efflux in the striatum (Gobert and Millan 1999; Navailles et al. 2004; Alex et al. 2005), while systemic administration of SB 206553 and SB 242084 enhance it (Porras et al. 2002; Di Giovanni et al. 2000; De Deurwaerdère and Spampinato 1999; Navailles et al. 2004). A recent study has shown that the 5-HT$_{2C}$ receptor inverse agonist-induced increase in accumbal and striatal DA release is insensitive to the depletion of extracellular 5-HT, suggesting that constitutive activity of the 5-HT$_{2C}$ receptors participates in the tonic inhibitory control that they exert upon DA release in both the nucleus accumbens and striatum (De Deurwaerdère et al. 2004). Furthermore, biochemical evidence indicates that both VTA and accumbal 5-HT$_{2C}$ receptors participate in the phasic inhibitory control exerted by central 5-HT$_{2C}$ receptors on mesoaccumbens DA neurons (Navailles et al. 2006a, 2008) and that the nucleus accumbens shell region constitutes the major site for the expression of the tonic inhibitory control involving the constitutive activity of 5-HT$_{2C}$ receptors (Navailles et al. 2006a). There is also evidence that 5-HT$_{2C}$ receptors can modulate the phasic activity of the dopaminergic nigrostriatal system. Indeed, SB 206553 has been shown to potentiate cocaine-, morphine-, and haloperidol-induced increases in DA outflow in the rat striatum (Porras et al. 2002; Navailles et al. 2004, 2006b), and systemic administration of RO 60-0175 was found to attenuate haloperidol-induced DA release in the same area (Navailles et al. 2004), as well as nicotine-induced increase in DA activity in the nigrostriatal system (Di Matteo et al. 2004; Pierucci et al. 2004).

11.7 5-HT$_{2C}$–Dopamine Interactions in Psychiatric Disorders

11.7.1 Depression

Although dopamine has received little attention in biological research on depression, as compared with other monoamines such as serotonin and noradrenaline, current research on the dopaminergic system is about to change this situation. It is now well established that disturbances of mesolimbic and nigrostriatal DA function are involved in the pathophysiology of depression (Fibiger 1995; Brown and Gershon 1993). Moreover, stress promotes profound and complex alterations involving DA release, metabolism, and receptor densities in the mesolimbic system (Puglisi-Allegra et al. 1991; Cabib and Puglisi-Allegra 1996). It seems that exposure to unavoidable/uncontrollable aversive experiences leads to inhibition of DA release in the mesoaccumbens DA system as well as impaired responding to rewarding and aversive stimuli. These alterations could elicit stress-induced expression and exacerbation of some depressive symptoms in humans (Cabib and Puglisi-Allegra 1996). Thus, in view of the hypothesis that disinhibition of the mesocorticolimbic DA system underlies the mechanism of action of several antidepressant drugs (Cervo and Samanin 1987; Cervo and Samanin 1988; Cervo et al. 1990; D'Aquila et al. 2000; Di Matteo et al. 2000b, c), the disinhibitory effect of SB 206553 and SB 242084 on the mesolimbic DA system might open new possibilities for the employment of 5-HT$_{2C}$ receptor antagonists as antidepressants (Di Matteo et al. 1998, 1999, 2000b, c, 2001). This hypothesis is consistent with the suggestion that 5-HT$_{2C}$ receptor blockers might exert antidepressant activity (Jenck et al. 1998; Di Matteo et al. 2000b, 2001; Giorgetti and Tecott 2004; Baxter et al. 1995). In this respect, it is interesting to note that several antidepressant drugs have been shown to bind with submicromolar affinity to 5-HT$_{2C}$ receptors in the pig brain and to antagonize mCPP-induced penile erections in rats, an effect mediated through the stimulation of central 5-HT$_{2C}$ receptors (Jenck et al. 1993, 1994, 1998). Based on those findings, Di Matteo et al. (2000c) have carried out experiments showing that acute administration of amitriptyline and mianserin, two antidepressants with high affinity for 5-HT$_{2C}$ receptors, enhances DA release in the rat nucleus accumbens by blocking these receptor subtypes, in addition to their other pharmacological properties. Interestingly, amitriptyline and mianserin have been tested in the chronic mild stress-induced anhedonia model of depression and were found to be effective in reversing the stress effects (Sampson et al. 1991; Moreau et al. 1994). The antianhedonic effects of tricyclic antidepressants, mianserin, and fluoxetine were abolished by pretreatment with D_2/D_3 receptor antagonists, thus indicating an involvement of DA in the antidepressant effect of various drugs in this model (Sampson et al. 1991; Willner 1995). The ability of antidepressants, such as tricyclics, selective serotonin reuptake inhibitors (SSRIs), and mianserin, to affect DA systems via indirect mechanisms was also reported by studies of Tanda et al. (1994, 1996), suggesting that potentiation of DA release in the rat cortex may indicate its role in the therapeutic action of antidepressants. The chronic mild stress procedure, which induces a depression-like state in animals, was shown to enhance 5-HT_{2C} receptor-mediated function, as measured in vivo by mCPP-induced penile erections. In contrast, two different antidepressant treatments [72-h rapid eye movement (REM) sleep deprivation and 10-day administration of moclobemide, a reversible inhibitor of monoamine oxidase type A] resulted in a reduction of this 5-HT_{2C} receptor-mediated function (Moreau et al. 1993). This was interpreted as an indication that the 5-HT_{2C} receptor may be altered and presumably may exist in a dysregulated (hypersensitive) state in depressive illness. Thus, adaptive processes resulting from chronic antidepressant treatment (i.e., desensitization and/or downregulation of 5-HT_{2C} receptors) may play an important role in reversing the 5-HT_{2C} receptor system supersensitivity resulting from a depressive state (Jenck et al. 1998; Moreau et al. 1996).

In contrast to most other receptors, 5-HT_{2C} is not classically regulated. Indeed, 5-HT_{2C} receptors appear to decrease their responsiveness not only upon chronic agonist stimulation but also, paradoxically, after chronic treatment with antagonists (Van Oekelen et al. 2003; Serretti et al. 2004). This mechanism appears to be related to an internalization process that removes activated cell surface receptors from the plasma membrane involving a phosphorylation step and possible degradation in lysosomes (Van Oekelen et al. 2003). As a large number of psychotropic drugs, including atypical antipsychotics, antidepressants, and anxiolytics, can all induce downregulation of 5-HT_{2C} receptors, it has been suggested that this receptor adaptation plays a role in the therapeutic action of these drugs (Van Oekelen et al. 2003; Serretti et al. 2004). In this respect, it is interesting to note that chronic treatment with 5-HT_2 agonists or antagonists resulted in a paradoxical downregulation at the 5-HT_{2A} and 5-HT_{2C} receptors (Moreau et al. 1996; Van Oekelen et al. 2003; Serretti et al. 2004; Barker and Sanders-Bush 1993; Pranzatelli et al. 1993; Newton and Elliott 1997), and it seems that the downregulation state occurring after chronic exposure to mianserin in isolated systems as well as in cell cultures is a direct receptor-mediated mechanism of this drug at these receptors (Newton and Elliott 1997). Therefore, the downregulating capacity of 5-HT_{2C} agonists and antagonists may play a particularly important role in treating the supersensitivity of 5-HT_{2C} receptors resulting from a depressive state (Jenck et al. 1998; Moreau et al. 1996; Serretti et al. 2004).

The possible involvement of 5-HT$_{2C}$ receptors in the pathogenesis of depressive disorders and in the mode of action of antidepressants is further substantiated by several other observations. For example, acute administration of fluoxetine caused a dose-dependent inhibition of the firing rate of VTA DA neurons (Prisco and Esposito 1995) and a decreased DA release in both the nucleus accumbens and the striatum (Ichikawa and Meltzer 1995), but it did not affect the activity of DA cells in the SNc (Prisco and Esposito 1995). A similar effect, though less pronounced, has been observed with citalopram (Prisco and Esposito 1995). Furthermore, mesulergine, an unselective 5-HT$_{2C}$ receptor antagonist (Boess and Martin 1994), as well as the lesion of 5-HT neurons by the neurotoxin 5,7-DHT, prevented fluoxetine-induced inhibition of VTA DA cells (Prisco and Esposito 1995). These results indicate that fluoxetine inhibits the mesolimbic DA pathway by enhancing the extracellular level of 5-HT, which would act through 5-HT$_{2C}$ receptors (Prisco and Esposito 1995). This study also demonstrated that fluoxetine-induced inhibition of DA neurons in the VTA was no longer observed after chronic treatment (21 days) with this drug. Interestingly, mCPP inhibited the firing activity of VTA DA neurons in control animals but not in those chronically treated with fluoxetine (Prisco and Esposito 1995). The authors suggested that 5-HT$_{2C}$ receptors might be downregulated after repeated fluoxetine administration. Consistent with this hypothesis is the evidence that chronic treatment with sertraline and citalopram, two SSRIs, induces tolerance to the hypolocomotor effect of mCPP (Maj and Moryl 1992).

This hyposensitivity of 5-HT$_{2C}$ receptors might be a key step for the achievement of an antidepressant effect. Indeed, it is possible to argue that the acute inhibitory effect of fluoxetine on mesolimbic DA system would mask its clinical efficacy in the early stage of treatment. This masking effect disappears when the hyposensitivity of 5-HT$_{2C}$ receptors occurs. A series of studies carried out in our laboratory has shown that acute administration of SSRIs such as paroxetine, sertraline, and fluvoxamine causes a slight but significant decrease in the basal firing rate of VTA DA neurons (Di Mascio et al. 1998). Therefore, it is conceivable that, similar to fluoxetine, these SSRIs could reduce mesocorticolimbic DA transmission by activating 5-HT$_{2C}$ receptors. Furthermore, employing complementary electrophysiological and neurochemical approaches and both acute and chronic administration routes, it was found that mirtazapine, nefazodone, and agomelatine, three effective and innovative antidepressants, elicit a robust and pronounced enhancement in the activity of mesocorticolimbic DA pathways. These actions were ascribed to their antagonistic properties at inhibitory, tonically active 5-HT$_{2C}$ receptors that desensitize after repeated drug administration (Millan et al. 2000, 2003; Dremencov et al. 2005). Interestingly, agomelatine has shown antidepressant efficacy in clinical trials (Lôo et al. 2002; Pandi-Perumal et al. 2006; Zupancic and Guilleminault 2006), and indeed, it was found to be effective in treating severe depression associated with anxiety symptoms, with a better tolerability and lower adverse effects than other antidepressants such as paroxetine (Lôo et al. 2002). Likewise, the novel benzourea derivative S32006 showed potent 5-HT$_{2C}$ receptor antagonistic properties, enhanced the activity of mesocortical dopaminergic and adrenergic projections, displaying a broad-based profile of antidepressant properties upon acute and/or repeated administration, and exhibited both rapid and sustained anxiolytic actions (Dekeyne et al. 2008).

11.7.2   Schizophrenia

Both hypo- and hyperfunction of dopaminergic systems may occur in schizophrenic patients, perhaps even simultaneously, albeit in a region-specific manner (Davis et al. 1991; Svensson et al. 1993, 1995). Thus, whereas a dopaminergic hyperfunction of the mesolimbic system may underlie the development of positive symptoms, a dopaminergic hypofunction of the cortical projections may well be related to the negative symptomatology in schizophrenia. Given the critical role of cortical DA in cognitive functioning (Arnsten et al. 1994; Sawaguchi and Goldman-Rakic 1994), the hypothesized cortical DA hypofunction may therefore also be implicated in the cognitive disturbances frequently experienced by schizophrenic patients. Hence, it appears likely that both the negative symptoms and cognitive disturbances of schizophrenia may be associated with a hypofunction of the mesocortical DA system.

Currently used antipsychotic drugs are usually divided into two main classes on the basis of their liability to induce neurological side effects after long-term treatment. Drugs defined as typical antipsychotics (e.g., chlorpromazine, haloperidol, trifluopromazine) are known to induce, following repeated administration, various extrapyramidal side effects (EPS) including Parkinson-like syndrome and tardive dyskinesia (Meltzer and Nash 1991). On the other hand, chronic treatment with atypical antipsychotic drugs (e.g., clozapine, risperidone, sertindole, zotepine) is associated with a low incidence of neurological side effects (Meltzer and Nash 1991). Moreover, atypical antipsychotic drugs do not increase plasma prolactin levels in humans (Meltzer and Nash 1991). The hypothesis that typical antipsychotics produce their clinical effects, as well as EPS, by blocking DA D_2 receptors in the mesolimbic and nigrostriatal systems, respectively (Meltzer and Nash 1991), is now generally accepted. In contrast, the mechanisms responsible for the clinical effects of atypical antipsychotic drugs are still not clear. The most relevant hypothesis on the mode of action of the atypical antipsychotics is that their action depends on their interaction with central 5-HT_{2A} or 5-HT_{2C} receptor subtypes, more than with D_2 receptors (Roth et al. 1992; Meltzer and Nash 1991; Meltzer et al. 1989). Numerous studies show that several antipsychotic drugs exhibit appreciable affinity for central 5-HT_2 receptors (Meltzer and Nash 1991; Schotte et al. 1996) and induce significant blockade of these receptors in human brain (Farde et al. 1995). Early clinical studies indicated that the selective $5\text{-HT}_{2A/2C}$ receptor antagonist ritanserin (Leysen et al. 1985; Schotte et al. 1989) could ameliorate negative symptoms as well as attenuate exciting EPS in schizophrenics treated with classical antipsychotic drugs (Bersani et al. 1990; Miller et al. 1990). The importance of 5-HT_2 receptor antagonism in the pharmacology of schizophrenia is further underlined by the fact that clozapine is indeed a potent 5-HT_{2A} receptor antagonist and exhibits a high ratio of 5-HT$_{2A}$ to D$_2$ receptor affinities (Schmidt et al. 1995; Ashby and Wang 1996). In fact, by examining in vitro receptor binding data, Meltzer et al. (1989) found that typical and atypical antipsychotics could be distinguished on the basis of their 5-HT$_{2A}$ to D$_2$ receptor binding ratios. Accordingly, they suggested that the mechanism of action of atypical antipsychotic drugs is based on their ability to achieve a balanced 5-HT$_{2A}$ to D$_2$ receptor antagonistic action and not on their absolute affinity for these receptors per se.

Such hypotheses have encouraged the development of novel antipsychotic drugs with combined antiserotonergic and antidopaminergic properties. Indeed, agents acting at multireceptor sites appear to be more promising as antipsychotic drugs, and recent data show that blockade of DA receptors and combined antagonism at 5-HT$_{2A}$ as well as at 5-HT$_{2C}$ receptors may be involved in the therapeutic effects of novel antipsychotics (Meltzer 1999; Bonaccorso et al. 2002; Jones and Blackburn 2002). Earlier studies demonstrated that administration of ritanserin, a mixed 5-HT$_{2A/2C}$ receptor antagonist, increased nigrostriatal and mesocorticolimbic DA efflux (Devaud et al. 1992; Pehek 1996; Pehek and Bi 1997). Interestingly, ritanserin, has been reported to potentiate the D$_{2/3}$ receptor antagonist raclopride-induced DA release in the mPFC and nucleus accumbens but not in the striatum (Andersson et al. 1995). Another putative atypical antipsychotic drug SR46349B, which shares both 5-HT$_{2A}$ and 5-HT$_{2C}$ receptor antagonism, increased cortical DA release and potentiated haloperidol-induced DA release in both mPFC and nucleus accumbens, suggesting that 5-HT$_{2C}$ receptor antagonism may also contribute to the potentiation of DA release produced by haloperidol (Bonaccorso et al. 2002). A novel putative atypical antipsychotic ACP-103, inverse agonist at both 5-HT$_{2A}$ and 5-HT$_{2C}$ receptors, increased DA release in the mPFC but not in the nucleus accumbens and potentiated low dose of haloperidol-induced DA release in the mPFC while inhibiting that in the nucleus accumbens (Li et al. 2005). Taken together, these data suggest that combined 5-HT$_{2A/2C}$ receptor antagonism may be more advantageous than selective 5-HT$_{2A}$ antagonism alone as an adjunct to D$_2$ antagonism to improve cognition end negative symptoms in schizophrenia.

Recent data show that atypical antipsychotic drugs (clozapine, sertindole, olanzapine, ziprasidone, risperidone, zotepine, tiospirone, fluperlapine, tenilapine), which produce little or no EPS while improving negative symptoms of schizophrenia, exert substantial inverse agonist activity at 5HT$_{2C}$ receptors (Herrick-Davis et al. 2000; Rauser et al. 2001). Thus, 5-HT$_{2C}$ receptor inverse agonism might underlie the unique clinical properties of atypical antipsychotic drugs (Herrick-Davis et al. 2000).

Antagonism at 5-HT$_{2C}$ receptors by several antipsychotics was also observed in vivo. Indeed, clozapine produces an increase in extracellular levels of DA in the nucleus accumbens (Di Matteo et al. 2002; Shilliam and Dawson 2005), reverses the inhibition of accumbal DA release induced by the 5-HT$_{2C}$ agonist RO 60-0175 (Di Matteo et al. 2002) and blocks the hypolocomotion induced by the 5-HT$_{2C}$ agonist mCPP (Prinssen et al. 2000). It is worth noting that clozapine, like several atypical APDs, behaves as a 5HT$_{2C}$ inverse agonist in heterologus expression systems in vitro and in vivo (Navailles et al. 2006a, b; Herrick-Davis et al. 2000;

Rauser et al. 2001). Thus, the 5-HT$_{2C}$ receptor inverse agonism might underlie the unique clinical properties of atypical APDs (Navailles et al. 2006a; Herrick-Davis et al. 2000). The modification of 5-HT$_{2C}$ receptors constitutive activity may also participate in the effects of the typical APD haloperidol. Indeed, it has been reported that the increase in striatal DA release induced by haloperidol is dramatically potentiated by the 5-HT$_{2C}$ inverse agonist SB 206553 (Navailles et al. 2006a). Therefore, bearing in mind that haloperidol does not bind to 5-HT$_{2C}$ receptors, it was suggested that it could act at the level of a common effector pathway (Navailles et al. 2006b).

A preferential increase of DA release in medial prefrontal cortex seems to be a common mechanism of action of atypical antipsychotic drugs, an effect that might be relevant for their therapeutic action on negative symptoms of schizophrenia (Kuroki et al. 1999). In this respect, it is important to note that the selective 5-HT$_{2C}$ receptor antagonist SB 242084 (Kennett et al. 1997) markedly increases DA release in the frontal cortex of awake rats (Gobert et al. 2000; Millan et al. 1998). Thus, it is possible to argue that blockade of 5-HT$_{2C}$ receptors might contribute to the preferential effect of atypical antipsychotics on DA release in the prefrontal cortex. Interestingly, there is preclinical evidence indicating that 5-HT$_{2C}$-receptor blockade is responsible for reducing EPS: 5-HT$_{2C}$ but not 5-HT$_{2A}$ receptor antagonists were capable of inhibiting haloperidol-induced catalepsy in rats (Reavill et al. 1999). Moreover, the blockade of dopaminergic neurotransmission in the nucleus accumbens via D$_2$ receptor antagonism or partial agonism is considered the primary mechanism underlying antipsychotic efficacy for the positive symptoms (i.e., hallucinations, delusions, and thought disorder) of schizophrenia. Thus, an alternative approach to blocking dopamine D$_2$ receptors may be to reduce the activity of the mesolimbic pathway without affecting that of the nigrostriatal system, thus avoiding potential extrapyramidal side-effect liabilities.

The selective effects shown by the 5-HT$_{2C}$ receptor agonists on the mesolimbic DA pathway suggest that 5-HT$_{2C}$ receptor agonists should have antipsychotic efficacy without the EPS associated with typical antipsychotics. To this end, recently, the antipsychotic efficacy of the selective 5-HT$_{2C}$ receptor agonist WAY-163909 was preclinically evaluated by in vivo microdialysis, electrophysiology, and various animal models of schizophrenia (Marquis et al. 2007; Grauer et al. 2009), showing selectivity for the mesolimbic system and an interesting profile similar to that of an atypical antipsychotic when given acutely or chronically in mice and rats, facilitating cortical dopaminergic neurotransmission and reducing that of the nucleus accumbens without affecting the nigrostriatal DA activity.

11.7.3   Parkinson Disease

Another interesting application of the data regarding the functional role of 5-HT$_{2C}$ receptors in the basal ganglia is the possible use of 5-HT$_{2C}$ receptor antagonists in the treatment of Parkinson disease (Fox and Brotchie 1999;

Nicholson and Brotchie 2002; Di Giovanni et al. 2006). The neural mechanisms underlying the generation of Parkinsonian symptoms are thought to involve reduced activation of primary motor and premotor cortex and supplementary motor areas, secondary to overactivation of the output regions of the basal ganglia, i.e., SNr and globus pallidus internus (GPi) (Albin et al. 1989), largely because of excessive excitatory drive from the subthalamic nucleus (STN), consequent to DA loss in the striatum (Nicholson and Brotchie 2002; Utter and Basso 2008). Hence, it is theoretically possible that antagonists at the 5-HT$_{2C}$, which act directly to reduce STN neural activity, may have positive therapeutic benefits in acronymus Parkinson disease (PD).

Therapy of Parkinson disease consists mainly of amelioration of the symptoms with classical dopaminomimetics (Hagan et al. 1997). This treatment, however, is characterized by declining efficacy and occurrence of disabling side effects (Agid 1998). Functional inhibition of GPi or STN has provided an alternative to lesioning, by deep brain stimulation associated with modest side effects (Rodriguez et al. 1998). As already mentioned, 5-HT$_{2C}$ receptors are located in the SNr and medial segment of the pallidal complex in the rat and human brain (Azmitia and Segal 1978; Pasqualetti et al. 1999), and enhanced 5-HT$_{2C}$ receptor-mediated transmission within the output regions of the basal ganglia in parkinsonism appears to contribute to their overactivity (Fox and Brotchie 1999). In addition, 5-HT$_{2C}$-like receptor binding is increased in a rat model of Parkinsonism (Radja et al. 1993) and in human Parkinsonian patients (Fox and Brotchie 2000a). Interestingly, systemic administration of SB 206553 enhanced the anti-Parkinsonian action of the DA D$_1$ and D$_2$ agonists in the 6-hydroxydopamine-lesioned rats (Fox et al. 1998; Fox and Brotchie 2000b), suggesting that the use of a 5-HT$_{2C}$ receptor antagonist in combination with a DA receptor agonist may reduce the reliance upon dopamine replacement therapies and may thus reduce the problems associated with long term use of currently available antiparkinsonian agents (Fox and Brotchie 1999).

On the other hand, there is also evidence that 5-HT increases the firing rate of STN neurons by acting, in part, through 5-HT$_{2C}$ receptors (Stanford et al. 2005; Xiang et al. 2005). In this respect, a recent in vivo study showed an increase in the percentage of subthalamic neurons exhibiting burst-firing pattern with no change in firing rate after unilateral lesion of the nigrostriatal pathway compared with normal rats. Moreover, the systemic and local administration of m-CPP increased the firing rate of subthalamic neurons in the lesioned and sham-operated rats, reversed by the subsequent administration of SB242084 (Zhang et al. 2009). Recently, De Deur waerdère et al. (2010) showed that from a functional point of view, 5-HT$_{2C}$ receptors exert a tonic control in basal ganglia that impacts cellular activity only in the two brain areas receiving cortical inputs, i.e., striatum and nucleus subthalamic. This suggests that the 5-HT$_{2C}$ receptors may functionally contribute to the regional organization of the basal ganglia network and its main afferents. Therefore, these data further support the view that 5-HT$_{2C}$ receptor antagonists may be useful as an adjuvant treatment to dopamine agonists to treat motor complications of PD (Di Giovanni et al. 2006).

11.7.4 Drugs of Abuse

Substantial evidence indicates that the mesolimbic pathway, particularly the dopaminergic system innervating accumbal areas, is implicated in the reward value of both natural and drug reinforcers, such as sexual behavior or psychostimulants, respectively (Koob 1992; Di Chiara and Imperato 1988; Salamone et al. 2007; Spanagel and Weiss 1999). The fact that drugs of abuse act through different cellular mechanisms leads to the possibility that their effects on DA release could be modulated differentially by each of the $5\text{-}HT_2$ receptor subtypes. As an example, it has been reported that the increased locomotor activity, as well as the accumbal DA release, elicited by phencyclidine is further enhanced by the blockade of $5\text{-}HT_{2C}$ receptors (Gobert and Millan 1999), while antagonism at $5\text{-}HT_{2A}$ receptors has opposite effects (Maurel-Remy et al. 1995). A similar picture emerges when considering the influence of these receptors on MDMA (aka *ecstasy*)-induced effects on DA neuron activity. Thus, the selective $5\text{-}HT_{2A}$ antagonist MDL 100,907 significantly reduced the hyper-locomotion and stimulated DA release produced by MDMA, while the selective $5\text{-}HT_{2C}$ antagonists SB 242084 and SB 206553 potentiated it (Schmidt et al. 1992; Kehne et al. 1996; Bankson and Cunningham 2002; Fletcher et al. 2002a).

It was recently found that SB 206553 administration potentiates both the enhancement of DA release in the nucleus accumbens and striatum and the increased DA neuron firing rate induced by morphine in both the VTA and the SNc (Porras et al. 2002). Consistent with these findings, stimulation of central $5\text{-}HT_{2C}$ receptors has been shown to inhibit morphine-induced increase in DA release in the nucleus accumbens of freely moving rats (Willins and Meltzer 1998). A series of studies showed that blockade of $5\text{-}HT_{2A}$ or $5\text{-}HT_{2C}$ receptors had opposite effects on cocaine-induced locomotor activity. Thus, $5\text{-}HT_{2A}$-receptor blockade with M100,907 attenuated cocaine-induced locomotion, whereas $5\text{-}HT_{2C}$ blockade with SB 242084 or SB 206553 enhanced cocaine-induced activity (McCreary and Cunningham 1999; McMahon and Cunningham 2001; O'Neill et al. 1999; Fletcher et al. 2002b). Consistent with these data obtained in rats, $5\text{-}HT_{2C}$ receptor null mutant mice showed enhanced cocaine-induced elevations of DA levels in the nucleus accumbens and marked increase in locomotor response to cocaine as compared with wild-type mice, suggesting that selective $5\text{-}HT_{2C}$ receptor agonist treatments may represent a promising novel approach for treating cocaine abuse and dependence (Rocha et al. 2002). In line with this hypothesis, it was found that RO 60-0175 reduced cocaine-reinforced behavior by stimulating $5\text{-}HT_{2C}$ receptors (Grottick et al. 2000) and that intra-VTA injection of the same $5\text{-}HT_{2C}$ receptor agonist reduced the enhancement of DA outflow in the nucleus accumbens induced by a systemic injection of cocaine, while intra-VTA administration of SB 242084 had no effect (Alex et al. 2005). Moreover, it was also shown that RO 60-0175 reduced ethanol- and nicotine-induced self-administration and hyperactivity (Grottick et al. 2000; Tomkins et al. 2002).

Consistent with this evidence, we showed that the selective activation of $5\text{-}HT_{2C}$ receptors by RO 60-0175 blocks the stimulatory action of nicotine on SNc DA neuronal activity and DA release in the corpus striatum (Di Matteo et al. 2004; Pierucci et al. 2004). The mesolimbic DA system appeared to be less sensitive to the inhibitory effect of 5-HT$_{2C}$ receptors activation on nicotine-induced stimulation; indeed a higher dose of RO 60-0175 was necessary to prevent the enhancement of VTA DA neuronal firing elicited by acute nicotine. Furthermore, pretreatment with the 5-HT$_{2C}$ agonist did not affect nicotine-induced DA release in the nucleus accumbens (Di Matteo et al. 2004; Pierucci et al. 2004). Interestingly, in animals treated repeatedly with nicotine, pretreatment with RO 60-0175 reproduced the same pattern of effects on the enhancement in DA neuronal firing caused by challenge with nicotine, resulting effective only at a higher dose in preventing nicotine excitation in the VTA compared with the SNc. Furthermore, the 5-HT$_{2C}$ receptors agonist counteracted nicotine-induced DA release both in the striatum and in the nucleus accumbens in rats chronically treated with this alkaloid, even if this effect was observed only with the highest dose of RO 60-0175 (Di Matteo et al. 2004; Pierucci et al. 2004). Therefore, we hypothesized that after repeated nicotine exposure an upregulation of 5-HT$_{2C}$ receptors occurs only in the DA mesolimbic system and the blocking of its hyperfunction by 5-HT$_{2C}$ receptor activation might be a useful approach in reducing nicotine reward and eventually helping in smoking cessation. Further, it was also found that nicotine-induced increase in VTA DA neuronal activity can be prevented by 5-HT$_{2C}$ agonism and that 5-HT$_{2C}$ receptor agonist effects can be induced intracellularly using the protein peptide Tat-3L4F, which prevents 5-HT$_{2C}$ receptor dephosphorilation induced by the phosphatase and tensin homologue deleted on chromosome 10 (PTEN) (Ji et al. 2006; Müller and Carey 2006). Thus, systemic administration of Tat-3L4F or the 5-HT$_{2C}$ receptor agonist RO 60-0175 suppressed the increased firing rate of VTA dopaminergic neurons induced by tetrahydrocannabinol (THC), the psychoactive ingredient of marijuana (Ji et al. 2006; Müller and Carey 2006). Using behavioral tests, it was found that Tat-3L4F or RO 60-0175 blocks conditioned place preference of THC or nicotine and that RO 60-0175 but not Tat-3L4F produces anxiogenic effects, thus providing a preclinical basis for treating drug addiction with the Tat-3L4F peptide (Ji et al. 2006; Müller and Carey 2006; Maillet et al. 2008). Thus, these and other preclinical studies indicate that 5-HT$_{2A}$ receptor antagonists and/or 5-HT$_{2C}$ receptor agonists may effectively reduce craving and/or relapse and, likewise, enhance abstinence from several drugs of abuse and that 5-HT$_{2C}$ receptor agonists may also effectively reduce cocaine intake in active cocaine users.

11.8   Conclusion

Serotonergic and dopaminergic systems are closely related in the CNS, and the involvement of 5-HT$_2$ receptor family in the control of central DA activity is now well established. Twenty-five years of 5-HT$_{2C}$ receptors research have generated detailed information on the molecular biology and regional and cellular localization of these receptors. The main effect of 5-HT upon DA neurons is an inhibition, i.e., a control mediated generally by 5-HT$_{2C}$ receptors, and this applies especially to the VTA. It appears that the inhibitory effect of 5-HT$_{2C}$ receptors is indirectly mediated through excitation of GABA-ergic VTA interneurons impinging on DA-containing neurons, although 5-HT$_{2C}$ receptors might be present on VTA DA neurons. A series of studies have shown that the serotonergic system exerts phasic and tonic control on DA function in the mesocorticolimbic system by acting through 5-HT$_{2C}$ receptors.

Based on these findings, it has been suggested that 5-HT$_{2C}$ receptor antagonists might be useful in the treatment of depression. This hypothesis has been confirmed by preliminary clinical trials showing antidepressant activity of drugs acting as 5-HT$_{2C}$ receptor antagonists. Several other studies indicate that selective 5-HT$_{2C}$ ligands may serve for the treatment of other neuropsychiatric illness such as schizophrenia, Parkinson disease, and drug abuse. In addition, many atypical antipsychotic drugs display antagonism at both 5-HT$_{2C}$ and 5-HT$_{2A}$ receptors, which might be the basis of their capability to ameliorate negative symptoms, as well as to attenuate EPS in schizophrenic patients treated with classical antipsychotic drugs. A combination of 5-HT$_{2C}$ antagonists and dopamine agonists to reduce the problems associated with the long-term use of currently available anti-Parkinsonian agents has also been proposed. However, the possible use of 5-HT$_{2C}$ agonists for the treatment of drug addiction is still under investigation.

Acknowledgments The authors wish to thank Dr. Clare Austen for the English revision and Ms Barbara Mariani for her help in preparing the manuscript.

References

Abramowski D, Rigo M, Due D, et al (1995) Localization of 5-hydroxytryptamine$_{2C}$ receptor protein in human and rat brain using specific antisera. Neuropharmacology 35:1635–1645.
Aghajanian GK, Bunney BS (1974) DA-ergic and nonDA-ergic neurons of the substantia nigra: differential responses to putative transmitters. In: Boissier JR, Hippius H, Pichot P, eds. Proceedings of the IX congress of the college of international neuropsychopharmacology, Amstrerdam, Excerpta Medica. pp 444–452.
Agid Y (1998) Levodopa: is toxicity a myth? Neurology 50:858–863.
Albin R, Young AB, Penney JB (1989) The functional anatomy of basal ganglia disorders. Trends Neurosci 12:366–375.
Alex KD, Pehek EA (2007) Pharmacologic mechanisms of serotonergic regulation of dopamine neurotransmission. Pharmacol Ther 113:296–320.
Alex KD, Yavanian GJ, McFarlane HG, et al (2005) Modulation of dopamine release by striatal 5-HT$_{2C}$ receptors. Synapse 55:242–251.
Andersson JL, Nomikos GG, Marcus M, et al (1995) Ritanserin potentiates the stimulatory effects of raclopride on neuronal activity and dopamine release selectively in the mesolimbic DA-ergic system. Naunyn-Schmiedeberg's Arch Pharmacol 352:374–385.
Arnsten AF, Cai JX, Murphy BL, et al (1994) Dopamine D$_1$ receptor mechanisms in the cognitive performance of young adult and aged monkeys. Psychopharmacology 116:143–151.
Ashby CR, Wang RY (1996) Pharmacological actions of the atypical antipsychotic drug clozapine. A review. Synapse 24:349–394.

Azmitia EC, Segal M (1978) An autoradiographic analysis of the differential ascending projections of the dorsal and median raphé nuclei in the rat. J Comp Neurol 179:641–668.

Bankson GM, Cunningham KA (2002) Pharmacological studies of the acute effects of (+)-3,4-Methylenedioxymethamphetamine on locomotor activity: role of 5-HT$_{1B/1D}$ and 5-HT$_2$ receptors. Neuropsychopharmacology 26:40–52.

Bankson MG, Yamamoto BK (2004) Serotonin-GABA interactions modulate MDMA-induced mesolimbic dopamine release. J Neurochem 91:852–859.

Bannon MJ, Roth RH (1983) Pharmacology of mesocortical dopamine neurons. Pharmacol Rev 35:53–68.

Barker EL, Sanders-Bush E (1993) 5-Hydroxytryptamine$_{1C}$ receptor density and mRNA levels in choroid plexus epithelial cells after treatment with mianserin and (–)-1-(4-bromo-2, 5-dimethoxyphenyl)-2-aminopropane. Mol Pharmacol 44:725–730.

Barnes NM, Sharp T (1999) A review of central 5-HT receptors and their function. Neuropharmacology 38:1083–1152.

Baxter GS, Kennett GA, Blaney F, et al (1995) 5-HT2 receptor subtypes: a family reunited? Trends Pharmacol Sci 16:105–110.

Berg KA, Navailles S, Sanchez TA, et al (2006) Differential effects of 5-methyl-1-[[2-[(2-methyl-3-pyridyl)oxy]-5-pyridyl]carbamoyl]-6-trifluoromethylindone   (SB   243213)   on 5-Hydroxytryptamine$_{2C}$ receptor-mediated responses. J Pharmacol Exp Ther 319:260–268.

Bersani G, Grispini A, Marini S, et al (1990) 5-HT2 antagonist ritanserin in neuroleptic-induced parkinsonism: a double-blind comparison with orphenadrine and placebo. Clin Neuropharmacol 13:500–506.

Blackburn TP, Minabe Y, Middlemiss DN, et al (2002) Effect of acute and chronic administration of the selective 5-HT$_{2C}$ receptor antagonist SB-243213 on midbrain dopamine neurons in the rat: an in vivo extracellular single cell study. Synapse 46:129–139.

Boess FG, Martin IL (1994) Molecular biology of 5-HT receptors. Neuropharmacology 33:275–317.

Bonaccorso S, Meltzer HY, Li Z, et al (2002) SR46349-B, a 5-HT$_{2A/2C}$ receptor antagonist, potentiates haloperidol-induced dopamine release in rat medial prefrontal cortex and nucleus accumbens. Neuropsychopharmacology 27:430–441.

Bonasera SJ, Tecott LH (2000) Mouse models of serotonin receptor function: toward a genetic dissection of serotonin systems. Pharmacol Ther 88:133–142.

Brown AS, Gershon S (1993) Dopamine and depression. J Neural Transm 91:75–109.

Bubar MJ, Cunningham KA (2007) Distribution of serotonin 5-HT$_{2C}$ receptors in the ventral tegmental area. Neuroscience 146:286–297.

Cabib S, Puglisi-Allegra S (1996) Stress, depression and the mesolimbic dopamine system. Psychopharmacology 128:331–342.

Cervo L, Samanin R (1987) Evidence that dopamine mechanisms in the nucleus accumbens are selectively involved in the effect of desipramine in the forced swimming test. Neuropharmacology 26:1469–1472.

Cervo L, Samanin R (1988) Repeated treatment with imipramine and amitriptyline reduced the immobility of rats in the swimming test by enhancing dopamine mechanisms in the nucleus accumbens. J Pharm Pharmacol 40:155–156.

Cervo L, Grignaschi G, Samanin R (1990) The role of the mesolimbic dopaminergic system in the desipramine effect in the forced swimming test. Eur J Pharmacol 178:129–133.

Clemett DA, Punhani T, Duxon MS, et al (2000) Immunohistochemical localisation of the 5-HT$_{2C}$ receptor protein in the rat CNS. Neuropharmacology 39:123–132.

D'Aquila PS, Collu M, Gessa GL, et al (2000) The role of dopamine in the mechanism of action of antidepressant drugs. Eur J Pharmacol 405:365–373.

Dahlström A, Fuxe K (1964) Evidence for the existence of monoamine-containing neurons in the central nervous system. I. Demonstration of monoamines in the cell bodies of brain stem neurons. Acta Physiol Scand 62:1–55.

Davis KL, Kahn RS, Ko G, et al (1991) Dopamine in schizophrenia: a review and reconceptualization. Am J Psychiatry 148:1474–1486.

De Deurwaerdère P, Spampinato U (1999) Role of serotonin$_{2A}$ and serotonin$_{2B/2C}$ receptor subtypes in the control of accumbal and striatal dopamine release elicited in vivo by dorsal raphe nucleus electrical stimulation. J Neurochem 73:1033–1042.

De Deurwaerdère P, Navailles S, Berg KA, et al (2004) Constitutive activity of the serotonin$_{2C}$ receptor inhibits *in vivo* dopamine release in the rat striatum and nucleus accumbens. J. Neurosci 24:3235–3241.

De Deurwaerdère P, Le Moine C, Chesselet MF (2010) Selective blockade of serotonin 2C receptor enhances Fos expression specifically in the striatum and the subthalamic nucleus within the basal ganglia. Neurosci Lett 469:251–255.

Dekeyne A, Mannoury la Cour C, Gobert A, et al (2008) S32006, a novel 5-HT$_{2C}$ receptor antagonist displaying broad-based antidepressant and anxiolytic properties in rodent models. Psychopharmacology 19:549–568.

Devaud LL, Hollingsworth EB, Cooper BR (1992) Alterations in extracellular and tissue levels of biogenic amines in rat brain induced by the serotonin$_2$ receptor antagonist, ritanserin. J. Neurochem 59:1459–1466.

Di Chiara G, Imperato A (1988) Drugs abused by humans preferentially increase synaptic dopamine concentrations in the mesolimbic system of freely moving rats. Proc Natl Acad Sci USA 85:5274–5278.

Di Giovanni G, De Deurwaerdère P, Di Mascio M, et al (1999) Selective blockade of serotonin$_{2C/2B}$ receptors enhances mesolimbic and mesostriatal dopaminergic function: a combined in vivo electrophysiological and microdialysis study. Neuroscience 91:587–597.

Di Giovanni G, Di Matteo V, Di Mascio M, et al (2000) Preferential modulation of mesolimbic versus nigrostriatal dopaminergic function by serotonin$_{2C/2B}$ receptor agonists: a combined in vivo electrophysiological and microdialysis study. Synapse 35:53–61.

Di Giovanni G, Di Matteo V, La Grutta V, et al (2001) m-Chlorophenylpiperazine excites non-dopaminergic neurons in the rat substantia nigra and ventral tegmental area by activating serotonin-2C receptors. Neuroscience 103:111–116.

Di Giovanni G, Di Matteo V, Pierucci M, et al (2006) Serotonin involvement in the basal ganglia pathophysiology: could the 5-HT$_{2C}$ receptor be a new target for therapeutic strategies? Curr Med Chem 13:3069–3081.

Di Giovanni G, Di Matteo V, Esposito E, eds. (2008) Serotonin-dopamine interaction: experimental evidence and therapeutic relevance. Prog Brain Res 172.

Di Mascio M, Di Giovanni G, Di Matteo V, et al (1998) Selective serotonin reuptake inhibitors reduce the spontaneous activity of dopaminergic neurons in the ventral tegmental area. Brain Res Bull 46:547–554.

Di Mascio M, Di Giovanni G, Di Matteo V, et al (1999) Decreased chaos of midbrain DA-ergic neurons after serotonin denervation. Neuroscience 91:587–597.

Di Matteo V, Di Giovanni G, Di Mascio M, et al (1998) Selective blockade of serotonin$_{2C/2B}$ receptors enhances dopamine release in the rat nucleus accumbens. Neuropharmacology 37:265–272.

Di Matteo V, Di Giovanni G, Di Mascio M, et al (1999) SB 242084, a selective serotonin$_{2C}$ receptor antagonist, increases dopaminergic transmission in the mesolimbic system. Neuropharmacology 38:1195–1205.

Di Matteo V, Di Giovanni G, Di Mascio M, et al (2000a) Biochemical and electrophysiological evidence that RO 60-0175 inhibits mesolimbic dopaminergic function through serotonin$_{2C}$ receptors. Brain Res 865:85–90.

Di Matteo V, Di Giovanni G, Esposito E (2000) SB 242084: a selective 5-HT$_{2C}$ receptor antagonist. CNS Drug Reviews 6:195–205.

Di Matteo V, Di Mascio M, Di Giovanni G, et al (2000) Acute administration of amitriptyline and mian[39]serin increases dopamine release in the rat nucleus accumbens: possible involvement of serotonin$_{2C}$ receptors. Psychopharmacology 150:45–51.

Di Matteo V, De Blasi A, Di Giulio C, et al (2001) Role of 5-HT$_{2C}$ receptors in the control of central dopamine function. Trends Pharmacol Sci 22:229–232.

Di Matteo V, Cacchio M, Di Giulio C, et al (2002) Biochemical evidence that the atypical antipsychotic drugs clozapine and risperidone block 5-HT$_{2C}$ receptors in vivo. Pharmacol Biochem Behav 71:607–613.

Di Matteo V, Pierucci M, Esposito E (2004) Selective stimulation of serotonin$_{2C}$ receptors blocks the enhancement of striatal and accumbal dopamine release induced by nicotine administration. J Neurochem 89:418–429.

Doherty MD, Pickel V (2000) Ultrastructural localization of serotonin 2A receptor in dopaminergic neurons in the ventral tegmental area. Brain Res 864:176–185.

Dray A, Gonye N, Oakley NR, et al (1976) Evidence for the existence of a raphe projection to the substantia nigra in the rat. Brain Res 113:45–57.

Dray A, Davies J, Oakley NR, et al (1978) The dorsal and medial raphe projections to the substantia nigra in the rat: electrophysiological, biochemical and behavioural observations. Brain Res 151:431–442.

Dremencov E, Newman ME, Kinor N, et al (2005) Hyperfunctionality of serotonin-2C receptor-mediated inhibition of accumbal dopamine release in an animal model of depression is reversed by antidepressant treatment. Neuropharmacology 48:34–42.

Eberle-Wang K, Mikeladze Z, Uryu K, et al (1997) Pattern of expression of the serotonin$_{2C}$ receptor messenger RNA in the basal ganglia of adult rats. J Comp Neurol 384:233–247.

Farde L, Nyberg S, Oxenstierna G, et al (1995) Positron emission tomography studies on D2 and 5-HT2 receptor binding in risperidone-treated schizophrenic patients. J Clin Psychopharmacol 15:19S–23S.

Fibiger HC (1995) Neurobiology of depression: focus on dopamine. In: Gessa G, Fratta W, Pani L, Serra G, eds. Depression and mania: from neurobiology to treatment. New York, Raven Press. pp 1–17.

Fibiger HC, Miller JJ (1977) An anatomical and electrophysiological investigation of the serotonergic projection from the dorsal raphe nucleus to the substantia nigra in the rat. Neuroscience 2:975–987.

Fletcher PJ, Phil D, Grottick AJ, et al (2002) Differential effects of the 5-HT$_{2A}$ receptor antagonist M100,907 and the 5-HT$_{2C}$ receptor antagonist SB242,084 on cocaine-induced locomotor activity, cocaine self-administration and cocaine-induced reinstatement of responding. Neuropsychopharmacology 27:576–586.

Fletcher PJ, Korth KM, Robinson SR, et al (2002) Multiple 5-HT receptors are involved in the effects of acute MDMA treatment: studies on locomotor activity and responding for conditioned reinforcement. Psychopharmacology 162:282–291.

Fox SH, Brotchie JM (1999) A role for 5-HT$_{2C}$ receptor antagonists in the treatment of Parkinson's disease? Drug News Perspect 12:477–483.

Fox SH, Brotchie JM (2000) 5-HT$_{2C}$ receptor binding is increased in the substantia nigra pars reticulata in Parkinson's disease. Mov Disord 15:1064–1069.

Fox SH, Brotchie JM (2000) 5-HT$_{2C}$ receptor antagonists enhance the behavioural response to dopamine D1 receptor agonists in the 6-hydroxydopamine-lesioned rat. Eur J Pharmacol 398:59–64.

Fox SH, Moser B, Brotchie JM (1998) Behavioural effects of 5-HT$_{2C}$ receptor antagonism in the substantia nigra zona reticulata of the 6-hydroxydopamine-lesioned rat model of Parkinson's disease. Exp Neurol 151:35–49.

Gervais J, Rouillard C (2000) Dorsal raphe stimulation differentially modulates DA-ergic neurons in the ventral tegmental area and substantia nigra. Synapse 35:281–291.

Giorgetti M, Tecott L (2004) Contribution of 5-HT$_{2C}$ receptors to multiple action of central serotonin systems. Eur J Pharmacol 488:1–9.

Gobert A, Millan MJ (1999) Serotonin (5-HT)$_{2A}$ receptor activation enhances dialysate levels of dopamine and noradrenaline, but not 5-HT, in the frontal cortex of freely-moving rats. Neuropharmacology 38:315–317.

Gobert A, Rivet J-M, Lejeune F, et al (2000) Serotonin$_{2C}$ receptors tonically suppress the activity of mesocortical dopaminergic and adrenergic, but not serotonergic, pathways: a combined dialysis and electrophysiological analysis in the rat. Synapse 36:205–221.

Grace A, Bunney B (1985) Dopamine. In: Rogawski MA, Barker JL Eds., Neurotransmitter action in the vertebrate nervous system. New York, Plenum Press. pp 285–319.

Grauer SM, Graf R, Navarra R, et al (2009) WAY-163909, a 5-HT$_{2C}$ agonist, enhances the preclinical potency of current antipsychotics. Psychopharmacology 204:37–48.

Grottick AJ, Fletcher PJ, Higgins GA (2000) Studies to investigate the role of 5-HT$_{2C}$ receptors on cocaine- and food-maintained behavior. J Pharmacol Exp Ther 295:1183–1191.

Guiard BP, El Mansari M, Merali Z, et al (2008) Functional interactions between dopamine, serotonin and norepinephrine neurons: an in-vivo electrophysiological study in rats with monoaminergic lesions. Int J Neuropsychopharmacol 11:625–639.

Hagan JJ, Middlemiss DN, Sharp PC, et al (1997) Parkinson's disease: prospects for improved drug therapy. Trends Pharmacol Sci 18:156–63.

Herrick-Davis K, Grinde E, Teitler M (2000) Inverse agonist activity of atypical antipsychotic drugs at human 5-hydroxytryptamine2C receptors. J Pharmacol Exp Ther 295:226–232.

Hervé D, Pickel VM, Joh TH, et al (1987) Serotonin axon terminals in the ventral tegmental area of the rat: fine structure and synaptic input to dopaminergic neurons. Brain Res 435:71–83.

Higgins GA, Fletcher PJ (2003) Serotonin and drug reward: focus on 5-HT$_{2C}$ receptors. Eur J Pharmacol 480:151–162.

Hoyer D, Clarke DE, Fozard JR, et al (1994) VII. International union of pharmacology classification of receptors for 5-hydroxytryptamine (serotonin). Pharmacol Rev 46:157–203.

Hoyer D, Hannon JP, Martin GR (2002) Molecular, pharmacological and functional diversity of 5-HT receptors. Pharmacol Biochem Behav 71:533–554.

Hutson PH, Barton CL, Jay M, et al (2000) Activation of mesolimbic dopamine function by phencyclidine is enhanced by 5-HT$_{2C/2B}$ receptor antagonists: neurochemical and behavioural studies. Neuropharmacology 39:2318–2328.

Ichikawa J, Meltzer HY (1995) Effect of antidepressants on striatal and accumbens extracellular dopamine levels. Eur J Pharmacol 281:255–261.

Invernizzi RW, Pierucci M, Calcagno E, et al (2007) Selective activation of 5-HT$_{2C}$ receptors stimulates GABA-ergic function in the rat substantia nigra pars reticulata: a combined in vivo electrophysiological and neurochemical study. Neuroscience 144:1523–1535.

Jenck F, Moreau J-L, Mutel V, et al (1993) Evidence for a role of 5-HT$_{1C}$ receptors in the antiserotonergic properties of some antidepressant drugs. Eur J Pharmacol 231:223–229.

Jenck F, Moreau J-L, Mutel V, et al (1994) Brain 5-HT$_{1C}$ receptors and antidepressants. Prog neuropsychopharmacol & Biol Psychiatry 18:563–574.

Jenck F, Bös J, Wichmann J, Stadler H, Martin JR, Moreau JL (1998) The role of 5-HT$_{2C}$ receptors in affective disorders. Expert Opin Investig Drugs 7:1587–1599.

Ji S-P, Zhang Y, Van Cleemput J, et al (2006) Disruption of PTEN coupling with 5-HT$_{2C}$ receptors suppresses behavioral responses induced by drugs of abuse. Nat Med 12:324–329.

Jones BJ, Blackburn TP (2002) The medical benefit of 5-HT research. Pharmacol Biochem Behav 71:555–568.

Kalivas PW (1993) Neurotransmitter regulation of dopamine neurons in the ventral tegmental area. Brain Res Rev 18:75–113.

Kehne JH, Ketteler HJ, McCloskey TC, et al (1996) Effects of the selective 5-HT$_{2A}$ receptor antagonist MDL 100,907 on MDMA-induced locomotor stimulation in rats. Neuropsychopharmacology 15:116–124.

Kelland MD, Freeman AS, Chiodo LA (1990) Serotonergic afferent regulation of the basic physiology and pharmacological responsiveness of nigrostriatal dopamine neurons. J Pharmacol Exp Ther 253:803–811.

Kelland MD, Freeman AS, Rubin J, et al (1993) Ascending afferent regulation of rat midbrain dopamine neurons. Brain Res Bull 31:539–546.

Kennett GA (1993) 5-HT$_{1C}$ receptors and their therapeutic relevance. Curr Opin Investig Drugs 2:317–362.

Kennett GA, Wood MD, Bright F, et al (1996) In vitro and in vivo profile of SB 206553, a potent 5-HT$_{2C}$/5HT$_{2B}$ receptor antagonist with anxiolytic-like properties. Br J Pharmacol 117:427–434.

Kennett GA, Wood MD, Bright F, et al (1997) SB 242084, a selective and brain penetrant 5-HT$_{2C}$ receptor antagonist. Neuropharmacology 36:609–620.

Kiyatkin EA (1995) Functional significance of mesolimbic dopamine. Neurosci Biobehav Rev 19:573–598.

Koob GF (1992) Drugs of abuse: anatomy, pharmacology and function of reward pathways. Trends Pharmacol Sci 13:177–184.

Kuroki T, Meltzer HY, Ichikawa J (1999) Effects of antipsychotic drugs on extracellular dopamine levels in rat medial prefrontal cortex and nucleus accumbens. J Pharmacol Exp Ther 288:774–781.

Le Moal M, Simon H (1991) Mesocorticolimbic dopaminergic network: functional and regulatory roles. Physiol Rev 71:155–234.

Leysen JE, Gommeren W, Van Gompel P, et al (1985) Receptor-binding properties in vitro and in vivo of ritanserin: a very potent and long acting serotonin-S2 antagonist. Mol Pharmacol 27:600–611.

Li Z, Ichikawa J, Huang M, et al (2005) ACP-103, a 5-HT$_{2A/2C}$ inverse agonist, potentiates haloperidol-induced dopamine release in rat medial prefrontal cortex and nucleus accumbens. Psychopharmacology 183:144–153.

Liu S, Bubar MJ, Lanfranco MF, et al (2007) Serotonin$_{2C}$ receptor localization in GABA neurons of the rat medial prefrontal cortex: implications for understanding the neurobiology of addiction. Neuroscience 146:1677–1688.

Lôo H, Hale A, D'Haenen H (2002) Determination of the dose of agomelatine, a melatonergic agonist and selective 5-HT$_{2C}$ antagonist, in the treatment of major depressive disorder: a placebo-controlled dose range study. Int Clin Psychopharmacol 17:239–247.

Maillet JC, Zhang Y, Li X, et al (2008) PTEN-5-HT$_{2C}$ coupling: a new target for treating drug addiction. Prog Brain Res 172:407–420.

Maj J, Moryl E (1992) Effects of sertraline and citalopram given repeatedly on the responsiveness of 5-HT receptor subpopulations. J Neural Transm Gen Sect 88:143–156.

Marquis KL, Sabb AL, Logue SF, et al (2007) WAY-163909 [(7bR,10aR)-1,2,3,4,8,9,10,10a-Octahydro-7bH-cyclopenta-[b][1,4]diazepino[6,7,1hi]indole]a novel 5-hydroxytryptamine 2C receptor-selective agonist with preclinical antipsychotic-like activity. J Pharmacol Exp Ther 320:486–496.

Martin JR, Bös M, Jenck F, et al (1998) 5-HT$_{2C}$ agonists: pharmacological characteristics and therapeutical potential, J Pharmacol Exp Ther 286:913–924.

Maurel-Remy S, Bervoets K, Millan MJ (1995) Blockade of phencyclidine-induced hyperlocomotion by clozapine and MDL 100,907 in rats reflects antagonism of 5-HT$_{2A}$ receptors. Eur J Pharmacol 280:R9–R11.

McCreary AC, Cunningham KA (1999) Effects of the 5-HT$_{2C/2B}$ antagonist SB 206553 on hyperactivity induced by cocaine. Neuropsychopharmacology 20:556–564.

McMahon LR, Cunningham KA (2001) Antagonism of 5-Hydroxytryptamine$_{2A}$ receptors attenuates the behavioral effects of cocaine in rats. J Pharmacol Exp Ther 297:357–363.

Meltzer HY (1999) The role of serotonin in antipsychotic drug action. Neuropsychopharmacology 21:106S–115S.

Meltzer HY, Nash JF (1991) VII. Effects of antipsychotic drugs on serotonin receptors. Pharmacol Rev 43:587–604.

Meltzer HY, Matsubara S, Lee JC (1989) Classification of typical and atypical antipsychotic drugs on the basis of dopamine D1, D2 and serotonin2 pKi values. J Pharmacol Exp Ther 251:238–246.

Millan MJ, Dekene A, Gobert A (1998) Serotonin (5-HT)$_{2C}$ receptors tonically inhibit dopamine (DA) and noradrenaline (NA), but not 5-HT release in the frontal cortex in vivo. Neuropharmacology 37:953–955.

Millan MJ, Gobert A, Rivet J-M, et al (2000) Mirtazapine enhances frontocortical dopaminergic and corticolimbic adrenergic, but not serotonergic, transmission by blockade of α_2-adrenergic and serotonin$_{2C}$ receptors: a comparison with citalopram. Eur J Neurosci 12:1079–1095.

Millan MJ, Gobert A, Lejeune F, et al (2003) The novel melatonin agonist agomelatine (S20098) is an antagonist at 5-hydroxytryptamine$_{2C}$ receptors, blockade of which enhances the activity of frontocortical dopaminergic and adrenergic pathways. J Pharmacol Exp Ther 306:954–964.

Miller CH, Fleischhacker WW, Ehrmann H, et al (1990) Treatment of neuroleptic induced akathisia with the 5-HT2 antagonist ritanserin. Psychopharmacol Bull 26:373–376.

Minabe Y, Emori K, Ashby CR Jr (1996) The depletion of brain serotonin levels by para-chlorophenylalanine administration significantly alters the activity of midbrain dopamine cells in rats: an extracellular single cell recording study. Synapse 22:46–53.

Minabe Y, Hashimoto K, Watanabe KI, et al (2001) Acute and repeated administration of the selective 5-HT(2A) receptor antagonist M100907 significantly alters the activity of midbrain dopamine neurons: an in vivo electrophysiological study. Synapse 40:102–112.

Molineaux SM, Jessell TM, Axel R, et al (1989) 5-HT$_{1C}$ receptor is a prominent serotonin receptor subtype in the central nervous system. Proc Natl Acad Sci USA 86:6793–6797.

Moreau J-L, Jenck F, Martin JR, et al (1993) Effect of repeated mild stress and two antidepressant treatments on the behavioral response to 5-HT$_{1C}$ receptor activation in rats. Psychopharmacology 110:140–144.

Moreau J-L, Bourson A, Jenck F, et al (1994) Curative effects of the atypical antidepressant mianserin in the chronic mild stress-induced anhedonia model of depression. J Psychiatry Neurosci 19:51–56.

Moreau J-L, Bös M, Jenck F, et al (1996) 5-HT$_{2C}$ receptor agonists exhibit antidepressant - like properties in the anhedonia model of depression in rats. Eur Neuropsychopharmacol 6:169–175.

Morrow BA, Elsworth JD, Zito C, et al (1999) Biochemical and Behavioral anxiolytic-like effects of R(+) HA-966 at the level of the ventral tegmental area in rats. Psychopharmacology 143:227–234.

Moukhles H, Bosler O, Bolam JP, et al (1997) Quantitative and morphometric data indicate precise cellular interactions between serotonin terminals and postsynaptic targets in rat substantia nigra. Neuroscience 76:1159–1171.

Müller CP, Carey RJ (2006) Intracellular 5-HT$_{2C}$-receptor dephosphorylation: a new target for treating drug addiction. Trends Pharmacol Sci 27:455–458.

Navailles S, De Deurwaerdère PD, Porras G, et al (2004) In vivo evidence that 5-HT$_{2C}$ receptor antagonist but not agonist modulates cocaine-induced dopamine outflow in the rat nucleus accumbens and striatum. Neuropsychopharmacology 29:319–326.

Navailles S, Moison D, Ryczko D, et al (2006) Region-dependent regulation of mesoaccumbens dopamine neurons in vivo by the constitutive activity of central serotonin$_{2C}$ receptors. J Neurochem 99:1311–1319.

Navailles S, De Deurwaerdère PD, Spampinato U (2006) Clozapine and Haloperidol differentially alter the constitutive activity of central serotonin$_{2C}$ receptors in vivo. Biol Psychiatry 59:568–575.

Navailles S, Moison D, Cunningham KA, et al (2008) Differential regulation of the mesoaccumbens dopamine circuit by serotonin$_{2C}$ receptors in the ventral tegmental area and the nucleus accumbens: an in vivo microdialysis study with cocaine. Neuropsychopharmacology 33:237–246.

Newton RA, Elliott JM (1997) Mianserin-induced down-regulation of human 5-hydroxytryptamine$_{2A}$ and 5-Hydroxytryptamine$_{2C}$ receptors stably expressed in the human neuroblastoma cell line SH-SY5Y. J Neurochem 69:1031–1038.

Nicholson SL, Brotchie JM (2002) 5-hydroxytryptamine (5-HT, serotonin) and Parkinson's disease-opportunities for novel therapeutics to reduce the problems of levodopa therapy. Eur J Neurol 9:1–6.

Nocjar C, Roth BL, Pehek EA (2002) Localization of 5-HT$_{2A}$ receptors on dopamine cells in subnuclei of the midbrain A10 cell group. Neuroscience 111:163–176.

O'Neill MF, Heron-Maxwell CL, Shaw G (1999) 5-HT$_2$ receptor antagonism reduces hyperactivity induced by amphetamine, cocaine, and MK-801 but not D$_1$ agonist C-APB. Pharmacol, Biochem Behav 63:237–243.

Pandi-Perumal SR, Srinivasan V, Cardinali DP, et al (2006) Could agomelatine be the ideal antidepressant? Expert Rev Neurother 6:1595–1608.

Pasqualetti M, Ori M, Castagna M, et al (1999) Distribution and cellular localization of the serotonin type 2C receptor messenger RNA in human brain. Neuroscience 92:601–611.

Pehek EA (1996) Local infusion of the serotonin antagonists ritanserin or ICS 205,930 increases in vivo dopamine release in the rat medial prefrontal cortex. Synapse 24:12–18.

Pehek EA, Bi Y (1997) Ritanserin administration potentiates amphetamine-stimulated dopamine release in the rat prefrontal cortex. Prog Neuropsychopharmacol & Biol Psychiatry 21:671–682.

Pierucci M, Di Matteo V, Esposito E (2004) Stimulation of serotonin$_{2C}$ receptors blocks the hyperactivation of midbrain dopamine neurons induced by nicotine administration. J Pharmacol Exp Ther 309:109–118.

Pompeiano M, Palacios JM, Mengod G (1994) Distribution of the serotonin 5-HT$_2$ receptor family mRNAs: comparison between 5-HT$_{2A}$ and 5-HT$_{2C}$ receptors. Brain Res Mol Brain Res 23:163–178.

Porras G, Di Matteo V, Fracasso C, et al (2002) 5-HT$_{2A}$ and 5-HT$_{2C/2B}$ receptor subtypes modulate dopamine release induced in vivo by amphetamine and morphine in both the rat nucleus accumbens and striatum. Neuropsychopharmacology 26:311–324.

Pozzi L, Acconcia S, Ceglia I, et al (2002) Stimulation of 5-hydroxytryptamine (5-HT$_{2C}$) receptors in the ventrotegmental area inhibits stress-induced but not basal dopamine release in the rat prefrontal cortex. J Neurochem 82:93–100.

Pranzatelli MR, Murthy JN, Tailor PT (1993) Novel regulation of 5-HT$_{1C}$ receptors: down-regulation induced both by 5-HT$_{1C/2}$ receptor agonists and antagonists. Eur J Pharmacol 244:1–5.

Prinssen EPM, Koek W, Kleven MS (2000) The effects of antipsychotics with 5-HT$_{2C}$ receptor affinity in behavioral assays selective for 5-HT$_{2C}$ receptor antagonist properties of compounds. Eur J Pharmacol 388:57–67.

Prisco S, Esposito E (1995) Differential effects of acute and chronic fluoxetine administration on the spontaneous activity of dopaminergic neurones in the ventral tegmental area. Br J Pharmacol 116:1923–1931.

Prisco S, Pagannone S, Esposito E (1994) Serotonin–dopamine interaction in the rat ventral tegmental area: an electrophysiological study in vivo. J Pharmacol Exp Ther 271:83–90.

Puglisi-Allegra S, Imperato A, Angelucci L, et al (1991) Acute stress induces time-dependent responses in dopamine mesolimbic system. Brain Res 554:217–222.

Radja F, Descarrier L, Dewar KM, et al (1993) Serotonin 5-HT$_1$ and 5-HT$_2$ receptors in adult rat brain after destruction of nigrostriatal dopamine neurons: a quantitative autoradiographic study. Brain Res 606:273–285.

Rauser L, Savage JE, Meltzer HY, et al (2001) Inverse agonist actions of typical and atypical antipsychotic drugs at the human 5-hydroxytryptamine$_{2C}$ receptor. J Pharmacol Exp Ther 299:83–89.

Reavill C, Kettle A, Holland V, et al (1999) Attenuation of haloperidol-induced catalepsy by a 5-HT$_{2C}$ receptor antagonist. Br J Pharmacol 126:572–574.

Rocha BA, Goulding EH, O'Dell LE, et al (2002) Enhanced locomotor, reinforcing, and neurochemical effects of cocaine in serotonin 5-hydroxytryptamine 2C receptor mutant mice. J Neurosci 22:10039–10045.

Rodriguez MC, Obeso JA, Olanow CW (1998) Subthalamic nucleus-mediated excitoxicity in Parkinson's disease: a target for neuroprotection. Ann Neurol 44(Suppl):S175–S188.

Roth RH, Elsworth JD (1995) Biochemical pharmacology of midbrain dopamine neurons. In: Bloom FE, Kupfer DJ Eds Psychopharmacology: the fourth generation of progress. New York, Raven Press. pp 227–243.

Roth RH, Wolf ME, Deutch AY (1987) Neurochemistry of midbrain dopamine systems. In: Meltzer, HY Ed Psychopharmacology: the third generation of progress. New York, Raven Press. pp 81–94.

Roth BL, Roland D, Ciaranello D, et al (1992) Binding of typical and atypical antipsychotic agents to transiently expressed 5-HT$_{1C}$ receptors. J Pharmacol Exp Ther 260:1361–1365.

Salamone JD, Correa M, Farrar A, et al (2007) Effort-related functions of nucleus accumbens dopamine and associated forebrain circuits. Psychopharmacology 191:461–482.

Sampson D, Muscat R, Willner P (1991) Reversal of antidepressant action by dopamine antagonists in an animal model of depression. Psychopharmacology 104:491–495.

Sawaguchi T, Goldman-Rakic PS (1994) The role of D_1 dopamine receptor in working memory: local injections of dopamine antagonists into the prefrontal cortex of rhesus monkeys performing an oculomotor delayed-response task. J Neurophysiol 71:515–528.

Schmidt CJ, Fadayel GM, Sullivan CK, et al (1992) $5-HT_2$ receptors exert a state-dependent regulation of dopaminergic function: studies with MDL 100,907 and the amphetamine analogue, 3,4-methylenedioxymethamphetamine. Eur J Pharmacol 223:65–74.

Schmidt CJ, Sorensen SM, Kehne JH, et al (1995) The role of $5-HT_{2A}$ receptors in antipsychotic activity. Life Sci 25:2209–2222.

Schotte A, de Bruyckere K, Janssen PF, et al (1989) Receptor occupancy by ritanserin and risperidone measured using ex vivo autoradiography. Brain Res 500:295–301.

Schotte A, Janssen PFM, Gommeren W, et al (1996) Risperidone compared with new and reference antipsychotic drugs: in vitro and in vivo receptor binding. Psychopharmacology 124:57–73.

Serretti A, Artioli P, De Ronchi D (2004) The $5-HT_{2C}$ receptor as a target for mood disorders. Expert Opin Ther Targets 8:1–9.

Sharma A, Punhani T, Fone KCF (1997) Distribution of the 5-hydroxytryptamine$_{2C}$ receptor protein in adult rat brain and spinal cord determined using a receptor-directed antibody: effect of 5,7,-dihydroxytryptamine. Synapse 27:45–56.

Shi W-X, Nathaniel P, Bunney BS (1995) Ritanserin, a $5-HT_{2A/2C}$ antagonist, reverses direct dopamine agonist-induced inhibition of midbrain dopamine neurons. J Pharmacol Exp Ther 274:735–740.

Shilliam CS, Dawson LA (2005) The effect of clozapine on extracellular dopamine levels in the shell subregion of the rat nucleus accumbens is reversed following chronic administration: comparison with a selective $5-HT_{2C}$ receptor antagonist. Neuropsychopharmacology 30:372–380.

Sorensen SM, Kehne JH, Fayadel GM, et al (1993) Characterization of the $5-HT_2$ receptor antagonist MDL 100907 as a putative atypical antipsychotic: behavioural, electrophysiological and neurochemical studies. J Pharmacol Exp Ther 266:684–691.

Spanagel R, Weiss F (1999) The dopamine hypothesis of reward: past and current status. Trends Neurosci 22:521–527.

Stanford IM, Kantaria MA, Chahal HS, et al (2005) 5 Hydroxytryptamine induced excitation and inhibition in the subthalamic nucleus: action at $5-HT_{2C}$, $5-HT_4$ and $5-HT_{1A}$ receptors. Neuropharmacology 49:1228–1234.

Steffensen SC, Svingos AL, Pickel VM, et al (1998) Electrophysiological characterization of GABAergic neurons in the ventral tegmental area. J Neurosci 18:8003–8015.

Steinbush HWM (1984) Serotonin-immunoreactive neurons and their projections in the CNS. In: Björklund A, Hökfelt T Kuhar MJ, eds. Handbook of chemical neuroanatomy: classical transmitter receptors in the CNS, Part II. Amsterdam, Elsevier. pp 68–125.

Svensson TH, Nomikos GG, Andersson JL (1993) Modulation of dopaminergic neurotransmission by 5-HT2 antagonist. In: Vanhouette PM, Saxena PR, Paoletti R, Brunello N, Jackson AS Eds Serotonin: from cell biology to pharmacology and therapeutics. Dordrecht, Kluwer Academic Publishers. pp 263–270.

Svensson TH, Mathe JM, Andersson JL, et al (1995) Mode of action of atypical neuroleptics in relation to the phencyclidine model of schizophrenia: role of 5-HT2 receptor and alpha 1-adrenoceptor antagonism. J Clin Psychopharmacol 15:11S–18S.

Tanda G, Carboni E, Frau R, et al (1994) Increase of extracellular dopamine in the prefrontal cortex: a trait of drugs with antidepressant potential? Psychopharmacology 155:285–288.

Tanda G, Bassareo V, Di Chiara G (1996) Mianserin markedly and selectively increases extracellular dopamine in the prefrontal cortex as compared to the nucleus accumbens of the rat. Psychopharmacology 123:127–130.

Tomkins DM, Joharchi N, Tampakeras M, et al (2002) An investigation of the role of 5-HT$_{2C}$ receptors in modifying ethanol self-administration behaviour. Pharmacol Biochem Behav 71:735–744.

Trent F, Tepper JM (1991) Dorsal raphé stimulation modifies striata-evoked antidromic invasion of nigral dopaminergic neurons in vivo. Exp Brain Res 84:620–630.

Ugedo L, Grenhoff J, Svensson TH (1989) Ritanserin, a 5-HT$_2$ receptor antagonist, activates midbrain dopamine neurons by blocking serotonin inhibition. Psychopharmacology 98:45–50.

Utter AA, Basso MA (2008) The basal ganglia: an overview of circuits and function. Neurosci Biobehav Rev 32:333–342.

Van Bockstaele EJ, Pickel VM (1995) GABA-containing neurons in the ventral tegmental area project to the nucleus accumbens in rat brain. Brain Res 682:215–221.

Van Bockstaele EJ, Biswas A, Pickel VM (1993) Topography of serotonin neurons in the dorsal raphé nucleus that send axon collaterals to the rat prefrontal cortex and nucleus accumbens. Brain Res 624:188–198.

Van Bockstaele EJ, Cestari DM, Pickel VM (1994) Synaptic structure and connectivity of serotonin terminals in the ventral tegmental area: potential sites for modulation of mesolimbic dopamine neurons. Brain Res 647:307–322.

Van der Kooy D, Attori T(1980) Dorsal raphé cells with collateral projections to the caudate-putamen and substantia nigra: a fluorescent retrograde double labeling study in the rat. Brain Res 186:1–7.

Van Oekelen D, Luyten WH, Leysen J E (2003) 5-HT$_{2A}$ and 5-HT$_{2C}$ receptors and their atypical regulation properties. Life Sci 72:2429–49.

Ward RP, Dorsa DM (1996) Colocalization of serotonin receptor subtypes 5-HT$_{2A}$, 5-HT$_{2C}$ and 5-HT$_6$ with neuropeptides in rat striatum. J Comp Neurol 370:405–414.

White FJ (1996) Synaptic regulation of mesocorticolimbic dopamine neurons. Annu Rev Neurosci 19:405–436.

Willins DL, Meltzer HY (1998) Serotonin 5-HT$_{2C}$ agonists selectively inhibit morphine-induced dopamine efflux in the nucleus accumbens. Brain Res 781:291–299.

Willner P (1995) Animal models of depression: validity and applications. In: Gessa G, Fratta W, Pani L, Serra G, eds. Depression and mania: from neurobiology to treatment. New York, Raven Press. pp 19–41.

Wood MD, Reavill C, Trail B, et al (2001) SB-243213; a selective 5-HT$_{2C}$ receptor inverse agonist with improved anxiolytic profile: lack of tolerance and withdrawal anxiety. Neuropharmacology 41:186–199.

Wright DE, Seroogy KB, Lundgren KH, et al (1995) Comparative localization of serotonin$_{1A,1C,}$ and $_2$ receptor subtype mRNAs in rat brain. J Comp Neurol 351:357–373.

Xiang Z, Wang L, Kitai ST (2005) Modulation of spontaneous firing in rat subthalamic neurons by 5-HT receptor subtypes. J Neurophysiol 93:1145–1157.

Zhang QJ, Liu X, Liu J, et al (2009) Subthalamic neurons show increased firing to 5-HT$_{2C}$ receptor activation in 6-hydroxydopamine-lesioned rats. Brain Res 1256:180–189.

Zupancic M, Guilleminault C (2006) Agomelatine: a preliminary review of a new antidepressant. CNS Drugs 20:981–992.

Chapter 12
The Role of 5-HT$_{2C}$ Receptors in the Pathophysiology and Treatment of Depression

Eliyahu Dremencov, Joost H.A. Folgering, Sandra Hogg, Laurence Tecott, and Thomas I.F.H. Cremers

12.1 Introduction

Major depression is a severe disorder that affects, at least once in lifetime, 10–20% of world population (Kaplan and Sadock 1995; World Health Organization 2002). The keys of the disease are depressed mood and anhedonia. However, depression is also characterized also by various and significant affective, cognitive, and physical disturbances, such as anxiety, fatigue, impaired memory and concentration difficulties, sleep disturbances, chronic pain, and sexual dysfunctions (Kaplan and Sadock 1995; DSM-IV 2000). Depression is currently considered the fourth major cause of disability worldwide and estimated to become the second cause of disability by 2020 (World Health Organization report 2002). Several different neurotransmitters and their receptors are involved in the pathophysiology of this disease, which is still poorly understood. The current chapter focuses on the role of serotonin-2C (5-HT$_{2C}$) receptors in the pathophysiology and treatment of depression.

12.2 Role of 5-HT$_{2C}$ Receptors in the Regulation of Monoamine Neurotransmission

It is well established that monoamine systems play a central role in pathophysiology of affective disorders and in antidepressant response: Monoamine neurons densely innervate the limbic areas of the brain, modulate behavioral and emotional

E. Dremencov (✉)
Brains On-Line BV, Groningen, The Netherlands
e-mail: e.dremencov@brainsonline.org

T.I.F.H. Cremers (✉)
Brains On-Line US LLC, San-Francisco, CA, USA
e-mail: thomas.cremers@brainsonline.org

G. Di Giovanni et al. (eds.), *5-HT$_{2C}$ Receptors in the Pathophysiology of CNS Disease*,
The Receptors 22, DOI 10.1007/978-1-60761-941-3_12,
© Springer Science+Business Media, LLC 2011

functions, and serve as a primary target for all known antidepressant medication (Dremencov et al. 2002, 2003; Dremencov 2009; Tremblay and Blier 2006). Therefore, receptors regulating monoamine transmission might play an important role in the response to antidepressant drugs. 5-HT_{2C} receptors have been demonstrated to modulate 5-HT, norepinephrine (NE) and dopamine (DA) transmission in the brain.

The effects of agonists and antagonists of 5-HT_{2C} receptors on the firing activity of 5-HT, NE, and DA neurons are summarized in Table 12.1. It can be concluded that 5-HT_{2C} receptors negatively regulate the firing activity of 5-HT and DA neurons in the dorsal raphé nucleus (DRN) and ventral tegmental area (VTA), respectively (Boothman et al. 2003, 2006; Di Giovanni et al. 2000, 2001; Di Matteo et al. 2000; Sotty et al. 2009; Millan 2006; Dremencov et al. 2009). 5-HT_{2C} receptor agonists decrease the firing activity of 5-HT and DA neurons. This inhibition of firing of 5-HT and DA neurons and that produced by selective serotonin reuptake inhibitors (SSRIs) citalopram and escitalopram are reversed by 5-HT_{2C} receptor antagonists. One study demonstrated that SB242084, a selective antagonist of 5-HT_{2C} receptors, also increases DA neuronal firing activity on its own (Chenu et al. 2009).

A study by Millan et al. (2003) demonstrated that the mixed melatonin $\text{MT}_{1/2}$ receptor agonist and 5-HT_{2C} receptor antagonist agomelatine increases the firing activity of NE neurons and NE release in the prefrontal cortex (PFC). It was also shown that this effect of agomelatine was not reversed by selective melatonin antagonist, indicating that NE effect of agomelatine is mediated via 5-HT rather than the melatonin agonistic properties of this drug. However, a more recent study by Dremencov et al. (2007a) has shown that SB242084 has no effect on NE neuronal firing activity. It may therefore be suggested that 5-HT_{1A} and/or $5\text{-HT}_{2A/2B}$ antagonistic properties of agomelatine contribute to the stimulatory effect of this drug on NE neuronal firing activity (Millan et al. 2003).

The effects of agonists and antagonists of 5-HT_{2C} receptors on the release of 5-HT, NE, and DA are summarized in Table 12.2. The antagonists of 5-HT_{2C} receptors, administered individually, increase extracellular NE and DA levels in PFC but not in striatum and 5-HT levels in the nucleus accumbens (NAc), but not in hippocampus or PFC (Millan et al. 2003; Dremencov et al. 2005; Cremers et al. 2004, 2007; Alex et al. 2005). However, 5-HT_{2C} receptor antagonists enhanced selective serotonin reuptake inhibitor (SSRI)-induced increase in extracellular 5-HT levels (Cremers et al. 2004, 2007). Since this effect was observed after both systemic (Cremers et al. 2004) and local (Cremers et al. 2007) administration of 5-HT_{2C} receptor antagonists, it is possible that their potentiating effect on SSRI-induced increases in 5-HT levels is mediated via both the reversal of inhibition of 5-HT neurons in the DRN (Sotty et al. 2009) and local stimulation of 5-HT release from the neuronal terminals (Dremencov et al. 2005).

In general, it can be stated that 5-HT_{2C} receptors negatively regulate monoamine transmission in the brain via the suppression of firing of monoamine neurons and/or via local inhibition of transmitter release from the terminals of monoamine neurons. The mechanism of 5-HT_{2C}-mediated inhibition of 5-HT and DA neuronal firing

Table 12.1 Effects of agonists and antagonists of 5-HT$_{2C}$ receptors on 5-HT, NE, and DA neuronal firing activity

Compound	Effect	References	Note
5-HT neuronal firing activity in the DRN			
DOI	Decrease	Boothman et al. (2003)	Mixed 5-HT$_{2A/2C}$ agonist
WAY161503	Decrease	Boothman et al. (2006)	Selective 5-HT$_{2C}$ agonist
SB242084	No effect by its own	Sotty et al. (2009)	Selective 5-HT$_{2C}$ antagonist
	Reverses WAY161,503-induced decrease of 5-HT neuronal activity	Boothman et al. (2006)	
	Diminishes citalopram-induced decrease of 5-HT neuronal activity	Sotty et al. (2009)	
NE neuronal firing activity in the LC			
SB242084	No effect	Dremencov et al. (2007b)	
Agomelatine	Increase	Millan et al. (2003)	Mixed melatonin and 5-HT$_{2C/2B/2A/1A}$ agonist
DA neuronal firing activity in the VTA			
SB242084	Increase	Chenu et al. 2009; Di Mascio et al. (1999)	
	No effect by its own; Reverses escitalopram-induced decrease of DA neuronal activity	Dremencov et al. (2009)	
RO600175	Decrease	Di Matteo et al. 2000; Millan et al. (2003)	Selective 5-HT$_{2C}$ antagonist
Agomelatine	Reverses RO600175-induced inhibition	Millan et al. (2003)	
mCPP	Decrease	Di Giovanni et al. (2000)	Mixed 5-HT$_{2C/2B}$ agonist
MK212	Decrease	Di Giovanni et al. (2000)	Mixed 5-HT$_{2C/2B}$ agonist

Table 12.2 Effects of agonists and antagonists of 5-HT$_{2C}$ receptors on 5-HT, NE, and DA release in the rat brain

Compound	Brain area	Effect	References
	5-HT release		
Ketanserin (mixed 5-HT$_{2A/2C}$ antagonist, systemic administration)	Hippocampus	No effect by its own Potentiate citalopram, fluoxetine and sertraline-induced increase in 5-HT levels	Cremers et al. (2004)
	Prefrontal cortex	No effect on its own Potentiate citalopram-induced increase in 5-HT levels	
SB242084 (local and systemic)	Hippocampus	Potentiate citalopram-induced increase in 5-HT levels	Cremers et al. (2007)
RS 102221 (local)	Nucleus accumbens	Increase	Dremencov et al. (2005)
	NE release		
Agomelatine (systemic)	Prefrontal cortex	Increase	Millan et al. (2003)
	Striatum	No effect	
	DA release		
SB 206553 (mixed 5-HT$_{2C/2B}$ reverse agonist, local)	Prefrontal cortex	Increase	Alex et al. (2005)
mCPP (mixed 5-HT$_{2C/2B}$ agonist, systemic)	Prefrontal cortex	Decrease Reverse SB 206553-induced increase in DA levels	
Agomelatine (systemic)	Prefrontal cortex	Increase	Millan et al. (2003)
	Striatum	No effect	
RS 102221 (local)	Nucleus accumbens (Sprague-Dawley rats)	Slight increase	Dremencov et al. (2005)
	Nucleus accumbens (FSL rats)	Robust Increase	

activity is not yet completely understood. 5-HT$_{2C}$ receptors are coupled to G_{Q11} and G_{13} proteins (Cussac et al. 2002; McGrew et al. 2002; Theile et al. 2009; Westphal et al. 1995). The G_{Q11}-mediated signal transduction pathway of 5-HT$_{2C}$ receptors is relatively well characterized. This pathway includes the activation of phospholipase C (PLC) and neuronal excitation due to the inositol triphosphate (IP$_3$)-induced Ca^{2+} influx into the neuronal cytoplasm (Bockaert et al. 2006). However, it was shown that 5-HT$_{2C}$ receptors stimulate the firing activity of γ-aminobutyric acid (GABA) interneurons in the Vental tegmental area VTA (Di Giovanni et al. 2001). It is thus possible that the inhibitory effect of 5-HT$_{2C}$ receptors on DA neuronal firing activity is mediated, at least in part, via GABA neurons and GABA$_A$ and/or GABA$_B$ receptors (Alex and Pehek 2007; Dremencov et al. 2006). The study by Cremers et al. (2007) demonstrated that phaclofen, a GABA$_B$-receptor agonist, enhanced the citalopram-induced increase in hippocampal 5-HT levels in the same way as SB242084, a 5-HT$_{2C}$ receptor antagonist. It suggests that the inhibitory effect of 5-HT$_{2C}$ receptors on 5-HT neuronal firing activity is also mediated, at least in part, via GABA interneurons and GABA$_B$ receptors.

12.3   5-HT$_{2C}$ Receptors in Depressive Patients and in Animal Models of Depression

In depressed patients and in animal models of depression, some abnormalities in the molecular biology, expression, and functioning of 5-HT$_{2C}$ receptors have been observed. Two studies have reported an increase in the density of 5-HT$_2$ receptors in the brains of depressed suicide victims (Hrdina et al. 1993; McKeith et al. 1987). Conversely, tricyclic antidepressants and SSRIs were found to decrease the expression of 5-HT$_2$ receptors in the rat brain (Hrdina and Vu 1993; Green et al. 1986; Mikuni and Meltzer 1984). However, these early studies did not distinguish between 5-HT$_{2C}$ receptors and other subtypes of the 5-HT$_2$ family.

It was previously reported that mRNA encoding for 5-HT$_{2C}$ receptors undergoes posttranscriptional modification, or editing. There are five sites within the 5-HT$_{2C}$ pre-mRNA (A, B, C, D, and E), where adenosine can be converted into inosine (Burns et al. 1997). The editing at C and E sites results in substitution of asparagine by serine and in generation of an additional aspartate within the sequence of the intracellular loop of 5-HT$_{2C}$ receptor, responsible for G-protein coupling. It was suggested that this modification resulted in decreased activity of 5-HT$_{2C}$ receptors (Niswender et al. 1999). It is interesting that 5-HT$_{2C}$ is the only G-protein-coupled receptor known to undergo RNA editing (Gurevich et al. 2002).

It was reported that editing at the D site is significantly decreased and at the E site is significantly increased in the brains of depressed suicide victims (Gurevich et al. 2002). Interestingly, the SSRI fluoxetine increases the D-site and decreases E-site 5-HT$_{2C}$ pre-mRNA editing in the mouse brain (Gurevich et al. 2002). Another study demonstrated that exposure of mice to stress or chronic antidepressant administration differentially affects the editing of 5-HT$_{2C}$ pre-mRNA. Thus, stress increase enhanced the editing of 5-HT$_{2C}$ pre-mRNA at site C, and chronic fluoxetine administration enhanced editing at sites A, B, C, and D (Englander et al. 2005). It can be summarized that the posttranscriptional modifications of 5-HT$_{2C}$ pre-mRNA might play a role in pathophysiology of affective disorders. However, the functional outcome of the observed receptor alterations remains to be determined.

A functional abnormality of 5-HT$_{2C}$ receptors was reported in Flinder sensitive line (FSL) rats, a putative animal model of depression. It was found that the local perfusion with RS 102221, a selective antagonist of 5-HT$_{2C}$ receptors, produce a greater increase in DA levels in the NAc of FSL rats than in control Sprague–Dawley animals (Dremencov et al. 2005). This leads to the suggestion that FSL rats are characterized by hyperfunctionality of 5-HT$_{2C}$ receptors, resulting in decreased DA levels in the NAc, absence of 5-HT-induced accumbal DA release, and depressive-like behavior, such as immobility during the swim test (Dremencov et al. 2004). Interestingly, chronic administration of various antidepressant drugs, such as desipramine, paroxetine, nefazodone, mirtazapine, and venlafaxine, decreases the RS 102221-induced DA release, increases accumbal DA levels, and normalizes immobility time of FSL rats (Dremencov et al. 2006) (Table 12.3).

The chronic mild stress assay is commonly considered to be among the most predictive models of antidepressant action in the rat. The regimen of chronic unpredictable stress, which is essential to produce the anhedonic behavior on which the paradigm is based, also results in a 5-HT$_{2C}$-mediated increase in penile erection (Berendsen and Broekkamp 1997). Isolation rearing in the rat has also been demonstrated to produce changes that can be likened to symptoms of depression; specifically the model is characterized by increased behavioral and hormonal responses to stress. An increased responsiveness to metachlorophenylpiperazine (mCPP) on anxiety-like behavior and plasma corticosterone levels following isolation rearing suggests an increase in sensitivity of 5-HT$_{2C}$ receptors (Fone et al. 1996). These observations implicate 5-HT$_{2C}$ receptors in the mediation of the depression-like symptoms exhibited by these models, just as they are implicated in the changes brought about by the selective breeding used to generate the FSL rats described above.

The pharmacological evidence for a role of 5-HT$_{2C}$ receptors in the mediation and modulation of mood or depressive symptoms is somewhat mixed. The mCPP is the most extensively used tool to study the function of 5-HT$_{2C}$ receptors in the clinic; however, it is only tenfold more selective for 5-HT$_{2C}$ receptors over 5-HT$_{2A}$

Table 12.3 The effect of antidepressant treatment on depressive-like behavior of FSL rats, on basal accumbal dopamine levels, and on RS 102221-induced dopamine release in the nucleus accumbens

	Time course	DMI 7 days	DMI 14 days	PRX 7 days	PRX 14 days	NFZ 7 days	MRT 7 days	VFX 7 days
Immobility during swim test	↑	0	↓	0	↓	↓	↓	↓
Basal dopamine in the NAC	↓	↑	0/↑	↑	↓	↑	↑	NE
RS 102221-induced dopamine release	↑	0	↓	NE	NE	↓	NE	NE

Behavioral and neurochemical parameters of untreated FSL rats are provided in comparison with untreated control Sprague-Dawley rats. Behavioral and neurochemical parameters of FSL rats treated with desipramine (*DMI*) for 7 and 14 days, with paroxetine (*PRX*) for 7 and 14 days, and with nefazodone (*NFZ*), mirtazapine (*MRT*) and venlafaxine (*VFX*) for 7 days are provided in comparison with FSL rats treated with corresponding vehicle for corresponding time course: (↑) increase; (↓) decrease, (0) no effect, (NE) not examined

and 5-HT$_{2B}$ receptors, on which it also has a stimulatory effect (Rajkumar et al. 2009). There appears to be a consensus that mCPP precipitates anxiogenic responses in man, most notably those with panic disorder or obsessive compulsive disorder diagnoses (Kennett et al. 1994; Stefanski et al. 1999), and in behavioral models in rodents (Blackburn et al. 1993; Griebel et al. 1991) and pigeons (Gleeson et al. 1989).

Despite the selectivity issues with mCPP, there are numerous examples supporting an anxiolytic effect of both selective and nonselective 5HT$_{2C}$ receptor antagonists. Ritanserin, mianserin, and mesulergine, but not ketanserin, have anxiolytic effects on punishment responses in the rat (Cervo and Samanin 1995). Mianserin, as well as a number of other 5-HT$_{2A/2B/2C}$ receptor antagonists (I-NP, ICI 169 and 369, and LY 53857) also produced an anxiolytic effect in the rat social interaction test (Cervo and Samanin 1995). The selective 5-HT$_{2B/2C}$ antagonist SB200646 produces an anxiolytic effect in both neophobia and conflict models of anxiety (Kennett et al. 1995), and the most selective 5-HT$_{2C}$ antagonist extensively studied to date, SB 242084, has also been demonstrated to be anxiolytic in a multitude of animal models (Martin et al. 2002).

Whilst the support for the use of 5-HT$_{2C}$ antagonists in the treatment of anxiety is quite clear, their effects on anhedonia, another of the key diagnostic criteria for depression, are less so. A forced swim test experiment with mice demonstrated that the antidepressant-like effects of imipramine and desipramine are mediated, at least in part, via the blockade of 5-HT$_{2C}$ receptors (Redrobe and Bourin 1997). However, a putative antidepressant effect of 5-HT$_{2C}$-receptor blockade is not in accord with effects of 5-HT$_{2C}$ receptor agonists in the chronic mild stress paradigm. The 5-HT$_{2C}$ agonists Ro 60-0175 and Ro 60-0332 were observed to reduce stress-induced intracranial self-stimulation on the VTA (Moreau et al. 1996), which, based on the pharmacological validation of the model is suggestive of an antidepressant effect. However, the direct effect of 5-HT$_{2C}$ receptor stimulation on spontaneous VTA firing activity (Di Giovanni et al. 2000) may confound the finding. More recently the 5-HT$_{2C}$ receptor agonist, WAY-163909, has been shown to be active across a range of models for the detection of antidepressant effect, with both acute and chronic administration (Rosenzweig-Lipson et al. 2007).

The possible link between antidepressant and or anxiolytic effect was also demonstrated in a study by Rocha et al. (1994), who evaluated the expression of 5HT$_{2C}$ receptor and levels of neophobia following chronic administration of mianserin or eltoprazine. Mianserin, but not eltoprazine, produced an anxiolytic effect that was associated with a reduction in 5-HT$_{2C}$ receptor binding in the amygdala.

There is clear evidence for a role of 5-HT$_{2C}$ receptors in the treatment of anxiety and/or depression. On the whole, 5-HT$_{2C}$-receptor blockade appears to have anxiolytic potential, whilst findings are mixed, possibly because of limited studies with optimal pharmacological tools on the role of agonists and antagonists in antidepressant models. The mechanism of action of 5-HT$_{2C}$ receptors in depression may be based on the inhibition of firing activity of 5-HT and DA neurons and of monoamine release from the nerve terminals. The SSRI-induced inhibition of 5-HT and DA neurons, which may play a role in the lack of adequate response or in slow onset of action, is mediated, at least in part, via 5-HT$_{2C}$ receptors. The antagonists of 5-HT$_{2C}$ receptors might be beneficial in the treatment of depression because of their ability to reverse SSRI-induced inhibition of 5-HT and DA neurons and to potentiate the increase in 5-HT levels.

12.4 Efficacy of 5-HT$_{2C}$ Antagonists Alone and in Combination with SSRIs in Clinical Treatment of Depression

Several compounds that are clinically effective antidepressants are potent 5-HT$_{2C}$ antagonists (Millan 2006). Nefazodone, trazodone, and remeron all block 5-HT$_{2C}$ receptors; however, their pharmacological profile also encompasses blockade of 5-HT$_{2A}$ receptors as well as histamine 1 and alpha 2 receptors (Hyttel 1982). Agomelatonine is a potent 5-HT$_{2C}$ antagonist that has been shown to be effective in treatment of major depression. However, this compound also activates melatonin receptors, which prevents concluding that the clinical effectiveness is merely due to blockade of 5-HT$_{2C}$ receptors (Goodwin et al. 2009).

Depression is characterized by a high degree of comorbidity with closely related dysfunctions like anxiety and insomnia (van Mill et al. 2010). Although the efficacy of treatment of depression with selective 5-HT$_{2C}$ antagonists remains to be elucidated, it has been shown that blockade of 5-HT$_{2C}$ receptors alleviates anxiety and improves quality of sleep (Stahl et al. 2010). Reversing these comorbid dysfunctions adds to the overall treatment of depression.

The current first choice in the treatment of depression is SSRIs. Although safe in their clinical use, their effectiveness and side-effect profile urges the development of newer powerful antidepressants (Dremencov 2009). The onset and sustained efficacy of SSRIs has been a matter of debate for decades. The combination of SSRIs with autoreceptor antagonists like 5-HT$_{1B}$ and HT$_{1A}$ would relieve the autorestraining properties of the serotonergic system in response to SSRI administration and push the clinical efficacy (Mongeau et al. 1997; Blier et al. 1998). In addition, 5-HT$_{2C}$ antagonists have been shown to be able to augment the biochemical effectiveness of SSRIs in preclinical studies (Cremers et al. 2004; Cremers et al. 2007). Thus, systemic administration of ketanserin potentiated fluoxetine-induced increase in cortical and hippocampal 5-HT levels (Cremers et al. 2004). Either systemic or local (into the hippocampus) administration of SB 242084 enhanced citalopram-induced increase in hippocampal 5-HT levels (Cremers et al. 2007).

Given the possible clinical effectiveness of 5-HT$_{2C}$ antagonists and their activity in treating comorbid dysfunctions, this approach is appealing. In addition, 5-HT$_{2C}$ antagonists have also been shown to reverse several SSRI-induced side effects, e.g., anxiogenesis and sexual dysfunction, further adding to the attractiveness of synthesizing combined serotonin uptake inhibitors/5-HT$_{2C}$ antagonists. Lu 24530 is a compound with combined 5-HT reuptake inhibition and 5-HT$_{2C}$ antagonistic properties that is currently in phase II of clinical evaluations (see http://clinicaltrials.gov/ct2/show/NCT00599911?term=trial+NCT00599911&rank=1).

12.5   Conclusions

An abundant amount of evidence implicates 5HT$_{2C}$ receptor functionality in depression and anxiety. Multiple lines of research have shown the effectiveness of 5HT$_{2C}$ antagonists in the treatment of depression when studied preclinically. Interestingly, several clinically effective antidepressants are already potent 5HT$_{2C}$ antagonists. Future studies will show the clinical effectiveness of 5HT$_{2C}$ antagonists alone or in combination with SSRI.

References

Alex KD, Pehek EA (2007) Pharmacologic mechanisms of serotonergic regulation of dopamine neurotransmission. Pharmacol Ther 113:296–320.

Alex KD, Yavanian GJ, McFarlane HG, et al (2005) Modulation of dopamine release by striatal 5-HT2C receptors. Synapse 55:242–251.

American Psychiatric Association (2000). Diagnostic and statistical manual (DSM) IV.

Berendsen HH, Broekkamp CL (1997) Indirect in vivo 5-HT1A-agonistic effects of the new anti-depressant mirtazapine. Psychopharmacology 133:275–282.

Blackburn TP, Baxter GS, Kennett GA, et al (1993) BRL 46470A: a highly potent, selective and long acting 5-HT3 receptor antagonist with anxiolytic-like properties. Psychopharmacology 110:257–264.

Blier P, Pineyro G, el Mansari M, et al (1998) Role of somatodendritic 5-HT autoreceptors in modulating 5-HT neurotransmission. Ann N Y Acad Sci 861:204–216.

Bockaert J, Claeysen S, Becamel C, et al (2006) Neuronal 5-HT metabotropic receptors: fine-tuning of their structure, signaling, and roles in synaptic modulation. Cell Tissue Res 326: 553–572.

Boothman LJ, Allers KA, Rasmussen K, et al (2003) Evidence that central 5-HT2A and 5-HT2B/C receptors regulate 5-HT cell firing in the dorsal raphe nucleus of the anaesthetised rat. Br J Pharmacol 139:998–1004.

Boothman L, Raley J, Denk F, et al (2006) In vivo evidence that 5-HT(2C) receptors inhibit 5-HT neuronal activity via a GABAergic mechanism. Br J Pharmacol 149:861–869.

Burns CM, Chu H, Rueter SM, et al (1997) Regulation of serotonin-2C receptor G-protein coupling by RNA editing. Nature 387:303–308.

Cervo L, Samanin R (1995) 5-HT1A receptor full and partial agonists and 5-HT2C (but not 5-HT3) receptor antagonists increase rates of punished responding in rats. Pharmacol Biochem Behav 52:671–676.

Chenu F, El Mansari M, Blier P (2009) Long-term administration of monoamine oxidase inhibitors alters the firing rate and pattern of dopamine neurons in the ventral tegmental area. Int J Neuropsychopharmacol 12:475–485.

Cremers TI, Giorgetti M, Bosker FJ, et al (2004) Inactivation of 5-HT(2C) receptors potentiates consequences of serotonin reuptake blockade. Neuropsychopharmacology 29:1782–1789.

Cremers TI, Rea K, Bosker FJ, et al (2007) Augmentation of SSRI effects on serotonin by 5-HT2C antagonists: mechanistic studies. Neuropsychopharmacology 32:1550–1557.

Cussac D, Newman-Tancredi A, Duqueyroix D, et al (2002) Differential activation of Gq/11 and Gi(3) proteins at 5-hydroxytryptamine(2C) receptors revealed by antibody capture assays: influence of receptor reserve and relationship to agonist-directed trafficking. Mol Pharmacol 62:578–589.

Di Giovanni G, Di Matteo V, Di Mascio M, et al (2000) Preferential modulation of mesolimbic vs. nigrostriatal dopaminergic function by serotonin (2C/2B) receptor agonists: a combined in vivo electrophysiological and microdialysis study. Synapse 35:53–61.

Di Giovanni G, Di Matteo V, La Grutta V, et al (2001) m-Chlorophenylpiperazine excites non-dopaminergic neurons in the rat substantia nigra and ventral tegmental area by activating serotonin-2C receptors. Neuroscience 103:111–116.

Di Mascio M, Di Giovanni G, Di Matteo V, et al (1999) Decreased chaos of midbrain dopaminergic neurons after serotonin denervation. Neuroscience 92:237–243.

Di Matteo V, Di Giovanni G, Di Mascio M, et al (2000) Biochemical and electrophysiological evidence that RO 60-0175 inhibits mesolimbic dopaminergic function through serotonin(2C) receptors. Brain Res 865:85–90.

Dremencov E (2009) Aiming at new targets for the treatment of affective disorders. Curr Drug Targets 10:1049.

Dremencov E, Gur E, Lerer B, et al (2002) Effects of chronic antidepressants and electroconvulsive shock on serotonergic neurotransmission in the rat hypothalamus. Prog Neuropsychopharmacol Biol Psychiatry 26:1029–1034.

Dremencov E, Gur E, Lerer B, et al (2003) Effects of chronic antidepressants and electroconvulsive shock on serotonergic neurotransmission in the rat hippocampus. Prog Neuropsychopharmacol Biol Psychiatry 27:729–739.

Dremencov E, Gispan-Herman I, Rosenstein M, et al (2004) The serotonin-dopamine interaction is critical for fast-onset action of antidepressant treatment: in vivo studies in an animal model of depression. Prog Neuropsychopharmacol Biol Psychiatry 28:141–147.

Dremencov E, Newman ME, Kinor N, et al (2005) Hyperfunctionality of serotonin-2C receptor-mediated inhibition of accumbal dopamine release in an animal model of depression is reversed by antidepressant treatment. Neuropharmacology 48:34–42.

Dremencov E, Weizmann Y, Kinor N, et al (2006) Modulation of dopamine transmission by 5HT2C and 5HT3 receptors: a role in the antidepressant response. Curr Drug Targets 7:165–175.

Dremencov E, El Mansari M, Blier P (2007a) Distinct electrophysiological effect of paliperidone and risperidone on the firing activity of rat serotonin and norepinephrine neurons. Psychopharmacology 194:63–72.

Dremencov E, El Mansari M, Blier P (2007b) Noradrenergic augmentation of escitalopram response by risperidone: electrophysiologic studies in the rat brain. Biol Psychiatry 61: 671–678.

Dremencov E, El Mansari M, Blier P (2009) Effects of sustained serotonin reuptake inhibition on the firing of dopamine neurons in the rat ventral tegmental area. J Psychiatry Neurosci 34:223–229.

Englander MT, Dulawa SC, Bhansali P, et al (2005) How stress and fluoxetine modulate serotonin 2C receptor pre-mRNA editing. J Neurosci 25:648–651.

Fone KC, Shalders K, Fox ZD, et al (1996) Increased 5-HT2C receptor responsiveness occurs on rearing rats in social isolation. Psychopharmacology 123:346–352.

Gleeson S, Ahlers ST, Mansbach RS, et al (1989) Behavioral studies with anxiolytic drugs. VI. Effects on punished responding of drugs interacting with serotonin receptor subtypes. J Pharmacol Exp Ther 250:809–817.

Goodwin GM, Emsley R, Rembry S, et al (2009) Agomelatine prevents relapse in patients with major depressive disorder without evidence of a discontinuation syndrome: a 24-week randomized, double-blind, placebo-controlled trial. J Clin Psychiatry 70:1128–1137.

Green AR, Heal DJ, Goodwin GM (1986) The effects of electroconvulsive therapy and antidepressant drugs on monoamine receptors in rodent brain--similarities and differences. Ciba Found Symp 123:246–267.

Griebel G, Misslin R, Pawlowski M, et al (1991) m-Chlorophenylpiperazine enhances neophobic and anxious behaviour in mice. Neuroreport 2:627–629.

Gurevich I, Tamir H, Arango V, et al (2002) Altered editing of serotonin 2C receptor pre-mRNA in the prefrontal cortex of depressed suicide victims. Neuron 34:349–356.

Hrdina PD, Vu TB (1993) Chronic fluoxetine treatment upregulates 5-HT uptake sites and 5-HT2 receptors in rat brain: an autoradiographic study. Synapse 14:324–331.

Hrdina PD, Demeter E, Vu TB, et al (1993) 5-HT uptake sites and 5-HT2 receptors in brain of antidepressant-free suicide victims/depressives: increase in 5-HT2 sites in cortex and amygdala. Brain Res 614:37–44.

Hyttel J (1982) Citalopram--pharmacological profile of a specific serotonin uptake inhibitor with antidepressant activity. Prog Neuropsychopharmacol Biol Psychiatry 6:277–295.

Kaplan HI, Sadock BJ (1995) Comprehensive textbook of psychiatry. VI, 2804, Baltimore, MD: Williams & Wilkins, p I-122.

Kennett GA, Wood MD, Glen A, et al (1994) In vivo properties of SB 200646A, a 5-HT2C/2B receptor antagonist. Br J Pharmacol 111: 797–802.

Kennett GA, Bailey F, Piper DC, et al (1995) Effect of SB 200646A, a 5-HT2C/5-HT2B receptor antagonist, in two conflict models of anxiety. Psychopharmacology 118: 178–182.

Martin JR, Ballard TM, Higgins GA (2002) Influence of the 5-HT2C receptor antagonist, SB-242084, in tests of anxiety. Pharmacol Biochem Behav 71:615–625.

McGrew L, Chang MS, Sanders-Bush E (2002) Phospholipase D activation by endogenous 5-hydroxytryptamine 2C receptors is mediated by Galpha13 and pertussis toxin-insensitive Gbetagamma subunits. Mol Pharmacol 62:1339–1343.

McKeith IG, Marshall EF, Ferrier IN, et al (1987) 5-HT receptor binding in post-mortem brain from patients with affective disorder. J Affect Disord 13:67–74.

Mikuni M, Meltzer HY (1984) Reduction of serotonin-2 receptors in rat cerebral cortex after sub-chronic administration of imipramine, chlorpromazine, and the combination thereof. Life Sci 34:87–92.

Millan MJ (2006) Multi-target strategies for the improved treatment of depressive states: Conceptual foundations and neuronal substrates, drug discovery and therapeutic application. Pharmacol Ther 110:135–370.

Millan MJ, Gobert A, Lejeune F, et al (2003) The novel melatonin agonist agomelatine (S20098) is an antagonist at 5-hydroxytryptamine2C receptors, blockade of which enhances the activity of frontocortical dopaminergic and adrenergic pathways. J Pharmacol Exp Ther 306:954–964.

Mongeau R, Blier P, de Montigny C (1997) The serotonergic and noradrenergic systems of the hippocampus: their interactions and the effects of antidepressant treatments. Brain Res Brain Res Rev 23:145–195.

Moreau JL, Bos M, Jenck F, et al (1996) 5HT2C receptor agonists exhibit antidepressant-like properties in the anhedonia model of depression in rats. Eur Neuropsychopharmacol 6:169–175.

Niswender CM, Copeland SC, Herrick-Davis K, et al (1999) RNA editing of the human serotonin 5-hydroxytryptamine 2C receptor silences constitutive activity. J Biol Chem 274:9472–9478.

Rajkumar R, Pandey DK, Mahesh R, et al (2009) 1-(m-Chlorophenyl)piperazine induces depressogenic-like behaviour in rodents by stimulating the neuronal 5-HT(2A) receptors: proposal of a modified rodent antidepressant assay. Eur J Pharmacol 608:32–41.

Redrobe JP, Bourin M (1997) Partial role of 5-HT2 and 5-HT3 receptors in the activity of antidepressants in the mouse forced swimming test. Eur J Pharmacol 325:129–135.

Rocha B, Rigo M, Di Scala G, et al (1994) Chronic mianserin or eltoprazine treatment in rats: effects on the elevated plus-maze test and on limbic 5-HT2C receptor levels. Eur J Pharmacol 262:125–131.

Rosenzweig-Lipson S, Sabb A, Stack G, et al (2007) Antidepressant-like effects of the novel, selective, 5-HT2C receptor agonist WAY-163909 in rodents. Psychopharmacology 192:159–170.

Sotty F, Folgering JH, Brennum LT, et al (2009) Relevance of dorsal raphe nucleus firing in serotonin 5-HT(2C) receptor blockade-induced augmentation of SSRIs effects. Neuropharmacology 57:18–24.

Stahl SM, Fava M, Trivedi MH, et al (2010) Agomelatine in the treatment of major depressive disorder: an 8-week, multicenter, randomized, placebo-controlled trial. J Clin Psychiatry 71:616–626.

Stefanski R, Ladenheim B, Lee SH, et al (1999) Neuroadaptations in the dopaminergic system after active self-administration but not after passive administration of methamphetamine. Eur J Pharmacol 371:123–135.

Theile JW, Morikawa H, Gonzales RA, et al (2009) Role of 5-hydroxytryptamine2C receptors in Ca2 + -dependent, ethanol potentiation of GABA release onto ventral tegmental area dopamine neurons. J Pharmacol Exp Ther 329:625–633.

Tremblay P, Blier P (2006) Catecholaminergic strategies for the treatment of major depression. Curr Drug Targets 7:149–158.

van Mill JG, Hoogendijk WJ, Vogelzangs N, et al (2010) Insomnia and sleep duration in a large cohort of patients with major depressive disorder and anxiety disorders. J Clin Psychiatry 71:239–246.

Westphal RS, Backstrom JR, Sanders-Bush E (1995) Increased basal phosphorylation of the constitutively active serotonin 2C receptor accompanies agonist-mediated desensitization. Mol Pharmacol 48:200–205.

World Health Organization (2002) The world health report 2002.

Chapter 13
5-HT$_{2C}$ Receptors and Suicidal Behavior

Fabio Panariello, Naima Javaid, and Vincenzo De Luca

13.1 Introduction

Suicide is defined as the act of intentionally ending one's own life. Nonfatal suicidal thoughts and behaviors (hereafter called *suicidal behaviors*) are classified more specifically into three categories. The first of these is *suicide ideation*, which refers to thoughts of engaging in behavior intended to end one's life. A *suicide plan* refers to the formulation of a specific means by which one intends to die. Finally, a *suicide attempt* is the engagement in potentially self-injurious behavior where the intention to die is present to some extent.

Suicide is a serious public health problem in the USA and around the world. Over 30,000 people in the USA and approximately 1 million people worldwide die by suicide each year, making it one of the leading causes of death (World Health Organization 1996; U.S. Public Health Service 1999; U.S. Department of Health and Human Services 2000). Reports from the World Health Organization (WHO) indicate that suicide accounts for the largest share of the intentional injury burden in developed countries (Mathers et al. 2003) and that suicide is projected to become an even greater contributor to the global burden of disease over the coming decades (Murray and Lopez 1996; Mathers and Loncar 2006). Moreover, in most Western countries, a gender paradox of suicidal behavior is observed. Rates of suicidal ideation and behavior are higher in females than males. However, mortality from suicide is typically lower for females.

Suicidal behavior (death and attempts) is usually a complication associated with a broad spectrum of psychiatric conditions, most commonly with mood disorders (Mann 2003). Suicidal behavior is also present in patients with schizophrenia, substance and alcohol abuse, personality disorders, and anxiety disorders among others (Mann 2003).

V. De Luca (✉)
Centre for Addiction and Mental Health, Department of Psychiatry, University
of Toronto, Toronto, Ontario, Canada
e-mail: vincenzo_deluca@camh.net

G. Di Giovanni et al. (eds.), *5-HT$_{2C}$ Receptors in the Pathophysiology of CNS Disease*,
The Receptors 22, DOI 10.1007/978-1-60761-941-3_13,
© Springer Science+Business Media, LLC 2011

13.2   Clinical Relevance

Several lines of evidence, coming from studies of family history of suicide and more definitively from twin and adoption studies, suggest that suicidal behavior has a genetic component partly independent from the familial transmission of major psychiatric disorders (Garfinkel et al. 1982; Pfeffer et al. 1994; Roy et al. 1997). This is supported by clinical data. In fact, about 10% of those who commit or attempt suicide have no identifiable psychiatric illness (Oquendo et al. 2008). It is proposed that suicidal behavior could be considered a separate diagnostic category documented on a sixth axis as part of the newer classification of mental diseases within the Diagnostic and Statistical Manual of Mental Disorders, 5th Edition (DSM-V) (Oquendo et al. 2008). This is because suicidal behavior meets the criteria for diagnostic validity set forth by Robins and Guze (1970), and it does so to the same extent as most conditions treated by clinicians. In fact, such behavior is described so well clinically that research has pointed towards postmortem and in vivo laboratory markers (Mann 2003; Posner et al. 2007).

It is now possible to perform a strict differential diagnosis for suicidal behavior, and recent data from follow-up studies confirm higher rates in those with a previous diagnosis of psychiatric illness and a family history of suicidal behavior (Posner et al. 2007; Oquendo et al. 2006; Brent et al. 2002). Voracek and Loibl (2007) showed that the heritability of suicidal behavior as a broader phenotype (attempts, thoughts, and plans) is between 30% and 55%. Data from twin studies (Roy 1993; Statham et al. 1998) strongly supports the role of a genetic component in suicidal behavior, specifically, suicide attempts and completion.

Independently transmitted factors may partly explain the genetic risks for major depression (Gershon 1990) and for suicide (Roy 1993; Roy et al. 1995). Adoption studies rule out effects of a shared environment that may influence results of twin studies and indicate that genetic transmission of suicide risk is independent from transmission of major psychiatric disorders such as mood disorders or schizophrenia.

13.3   Neurobiology of Suicide and the Serotonergic System

Regarding the neurobiological background of suicidal behavior, there is a compelling body of evidence pointing towards variations in the monoaminergic, neurotrophin and hypothalamic–pituitary–adrenal (HPA) axis systems (Mann 2003; Westrin 2000; Strohle and Holsboer 2003; Pfennig et al. 2005; Duman and Monteggia 2006). Although approximately 60% of suicides occur in the context of depressive disorders, major depression and suicidal behavior appear independently related to altered indices of the neurotransmitter serotonin (5-HT). The first wave of genetic studies sought to identify genes involved in suicide or attempted suicide through linkage studies or specific single nucleotide polymorphisms (SNPs) in association studies. Candidate genes for association studies have been generally selected based on evidence from neurobiological studies in suicide. Neurobiological factors are not generally used by clinicians to determine the risk of suicidal behavior. It is not known whether understanding these underlying factors can improve the prediction of suicidal behavior; however, an understanding of neurobiology may assist in the development of new treatment approaches.

Although knowledge regarding the neurobiology of suicide is still limited, the 5-HT system has drawn attention and been studied more extensively than other neurotransmitter systems. Studies have shown that traits like impulsivity and aggression, which are related to lower 5-HT activity, are more closely associated with suicide than are objective severity of depression or psychosis (Mann et al. 1999).

Neurobiological evidence for the "serotonergic hypothesis" for major depressive illness and suicide comes from studies on the brain, cerebrospinal fluid (CSF), and even platelets (Mann et al. 1989, 1997, 2000; Arango and Mann 1992; Arango and Underwood 1997). A reduction in the concentration of the 5-HT metabolite 5-*Hydroxy*-3-Indole Acetic Acid (5-HIAA) in the CSF of suicide attempters was first reported by Asberg in 1976 (Asberg et al. 1976). Subsequently, many researchers have examined the concentration of 5-HIAA in order to point the turnover of the neurotransmitter in the brain rather than in the spinal cord (Lester 1995). Since Asberg's study in 1976, several, but not all, findings have confirmed the original observation. Furthermore, the altered 5-HIAA concentrations appear to be more strongly associated with suicidal behavior than with depression. This reduction in 5-HIAA occurs in patients who make a lethal suicide attempt, independent of the psychiatric diagnosis.

Recent studies have examined the relationship between suicide, impulsivity and aggression, and lower levels of CSF 5-HIAA. Mann and colleagues (Mann et al. 1997) found lower levels of CSF 5-HIAA in depressed patients with a history of a high lethality or well-planned suicide attempt compared with depressed patients with a history of only low lethality suicide attempts. Lower levels of CSF 5-HIAA are also reported in suicide attempters with schizophrenia (Ninan et al. 1984) and may predict suicidal behavior in this population as well (Cooper et al. 1992). Moreover, the degree of metabolite reduction seems to be correlated with the lethality of the attempt.

Although several studies have found the association between CSF 5-HT or 5-HIAA and suicidal behavior, the possible causes of this reduction are still not completely understood. A few hypotheses have been proposed: reduced release of neurotransmitter, fewer serotonergic neurons or poor innervation of target regions, impaired 5-HT synthesis, or augmented autoinhibitory actions of 5-HT itself.

To investigate the role of the 5-HT system, platelets are also commonly studied as they share with 5-HT neuron terminals a common embryonic origin and the presence of 5-HT transporter, 5-HTR2A, and monoamino oxidase B (MAOB). Overall, these studies indicate that platelet 5-HT$_{2A}$ receptors are increased in depression. The increase in platelet 5-HT$_{2A}$ receptors reported in suicidal patients may be related to suicidal behavior and may be state related, independent of diagnosis (Pandey 1997).

Postmortem findings in suicide completers have then provided further supporting evidence for abnormalities in the serotonergic system. Two important aspects of this research need to be considered. First, studies of postmortem tissue permit direct examination of the brain with high anatomical resolution and the ability to perform quantitative measures. Second, because the studies are performed postmortem, there are several clinical and experimental limitations. The clinical limitations include difficulties in obtaining accurate information regarding the patient's psychiatric and medication histories, particularly with socially isolated individuals. The experimental limitations involve postmortem delay artifacts, the potential confounding effects of drugs and medication, and the limited availability of the tissue.

Presynaptic and postsynaptic 5-HT receptor alterations have been observed in the prefrontal cortex (PFC) of people who commit suicide and are consistent with reduced serotonergic function. Some but not all investigators have found the binding to the 5-HT transporter to be reduced, particularly in ventral and lateral prefrontal cortical regions. Moreover, changes in the binding to postsynaptic 5-HT_{1A} and 5-HT_{2A} receptors in the PFC have been reported in most studies.

The data from postmortem studies and neuroimaging findings suggest that reduced serotonergic function in the ventral PFC constitutes a critical element in the vulnerability to suicidal behavior. The importance of the PFC in suicide is thought to involve its role in mediating behavioral inhibition. Impaired PFC function may underlie a reduced capacity to resist the impulse to act on suicidal thoughts.

The 5-HT system has been most extensively investigated and heavily scrutinized for genetic variants potentially contributing to serotonergic system dysfunction and, thereby, suicidal behavior.

5-HT synthesizing neurons innervating the forebrain in animals, and presumably in humans, are located primarily in the dorsal raphé nucleus (DRN) and the median raphé nucleus (MRN) in the brainstem. The ascending dual serotonergic projections previously described in the rat have also been demonstrated in nonhuman primates as arising from both the DRN and the MRN.

The finding of reduced 5-HT transporter binding in the PFC of completed suicides, in particular, raises the possibility that reduced serotonergic innervation is associated with suicidal behaviors. Alternatively, a normal or increased number of serotonergic neurons may be deficient in the extent of innervation and the number of 5-HT transporter sites synthesized. Instead, the finding of reduced 5-HT in the DRN and the MRN and of receptor changes in the PFC of suicide victims led us to hypothesize that the reduced 5-HT indices may be due to fewer 5-HT synthesizing neurons in the DRN in suicide victims. An alteration in the number of 5-HT neurons in suicide victims may reflect a neurodevelopmental effect and, perhaps, a genetic role in the neurobiology of suicide.

Emerging approaches attempting to better understand the genetics of suicide have the goal of investigating functional genomics using microarray technologies to profile expression of thousands of genes simultaneously or performing a genome-wide array for hundreds of thousands of SNPs. Although association studies have been the most common design, replication of findings has proven difficult for individual SNP association studies for several reasons. These include the effects of ethnicity/race-related stratification, the effect of interactions with environmental factors, differences in study sample size and composition with respect to diagnosis, and the complexity of the suicidal behavior phenotype. In that regard, different definitions and phenotypes of suicidality, including suicide, nonfatal attempts, or suicidal ideation, are all partly heritable and may involve different genes. The environment is also an element of complexity, particularly during developmental periods in childhood, as it can influence the effect of genetic variants on neurobiological function.

Genetic association studies that investigated the serotonergic system in suicidal behavior have focused mainly on the 5-HT transporter, 5-HT$_{1A}$, 5-HT$_{1B}$, 5-HT$_{2A}$, TPH1 and TPH2. In addition, the involvement of monoamine oxidase A (MAOA), which is involved in the breakdown of monoamines including 5-HT, has also been found. With regards to 5-HT receptors, the most studied has been the 5-HT$_{2A}$. In the PFC of suicide victims compared with that of nonsuicides, higher levels of post-mortem 5-HT$_{2A}$ receptor binding were discovered. Moreover, suicide victims with and without depressive diagnoses show evidence that 5-HT$_{2A}$ receptors are upregulated in the dorsal PFC.

There are now seven known families of serotonin receptors, 5-HT$_{1-7}$, comprising 14 structurally and pharmacologically distinct mammalian 5-HT receptor subtypes in total (Barnes and Sharp 1999). At the molecular level, it has been established that the 5-HT receptor family is mostly transmembrane G-protein-coupled metabotropic receptors; however, the 5-HT$_3$ receptor is a ligand-gated ion channel. Studies of the 5-HT$_{2A}$ receptor gene and suicidal behavior have largely focused on the T102C SNP. Published data are not consistent in the association between this SNP and suicide behavior. Metaanalysis of nine studies found no association between the T102C polymorphism and suicide attempt or completed suicide, and a recent expanded metaanalysis of 25 studies further confirmed this lack of association (Li and He 2007). Several sets of data from clinical and preclinical experimental paradigms have suggested that the 5-HT$_{1B}$ receptor gene may also be involved in the aggressive/impulsive endophenotype of suicidal behavior. The G861C SNP was the most studied SNP, with increased expression of the C allele in aggressive and impulsive phenotypes. Other 5-HT receptors have not been studied to similar extents with respect to genetic involvement in suicidal behavior. An overrepresentation of 5-HT$_{1A}$ 1018G allele in suicides compared with controls has been reported by some authors, but others have found no association. Studies of the 5-HT$_{2C}$ and 5-HT$_6$ receptors and a study of seven other 5-HT receptor genes seem to indicate no association with suicidality.

13.4 Support for the Serotonin 2C Receptor

The 5-HT$_{2C}$ receptor has been implicated as a potential therapeutic target in a wide variety of conditions including obesity, anxiety, depression, obsessive–compulsive disorder, schizophrenia, migraine, and erectile dysfunction. The 5-HT$_2$ receptor family currently accommodates three receptor subtypes, 5-HT$_{2A}$, 5-HT$_{2B,}$ and 5-HT$_{2C}$ receptors, which are similar in terms of their molecular structure, pharmacology, and signal transduction pathways. Because the binding properties of 5-HT toward the

5-HT$_{2C}$ subtype appeared to resemble 5-HT$_{1A}$ and 5-HT$_{1B}$ sites more than the 5-HT$_{2A}$ sites, its original designation was 5-HT$_{1C}$. Subsequently, the molecular cloning of the cDNA encoding this receptor revealed it to be a member of the superfamily of G-protein-coupled receptors highly related to the 5-HT$_{2A}$ receptor subtype (Lübbert et al. 1987; Julius et al. 1988), with which it shares substantial amino acid identity (Julius et al. 1990). The 5-HT$_{2C}$ appellation has replaced 5-HT$_{1C}$ to carry the receptor from the 5-HT$_1$ to the 5-HT$_2$ receptor family. The two characteristics of all genes in the 5-HT$_2$ receptor family are that they have either two (in the case of both the 5-HT$_{2A}$ and 5-HT$_{2B}$ receptors) or three introns (5-HT$_{2C}$ receptors) in their coding sequence (Yu et al. 1991; Chen et al. 1992; Stam et al. 1992) and that all are coupled positively with the phospholipase C signaling pathway and mobilize intracellular calcium.

Interestingly, the 5-HT$_{2C}$ receptor is the only known G-protein-coupled receptor whose mRNA undergoes posttranscriptional editing to yield different receptor isoforms. The 5-HT$_{2C}$ receptor gene is a large gene spanning at least 230 kb on chromosome Xq24 (Xie et al. 1996) that produces a G-protein-coupled receptor. The 5-HT$_{2C}$ receptor increases phospholipase C activity in choroid plexus where, it has been suggested, it may regulate CSF formation as a result of its ability to mediate cGMP formation (Kaufman et al. 1995). Unlike the 5-HT$_{2A}$ and 5-HT$_{2B}$ receptors, there is little evidence for expression of 5-HT$_{2C}$ receptors outside the central nervous system (CNS) (Julius et al. 1988). In addition to the very high levels detected in the choroid plexus, 5-HT$_{2C}$ binding sites are widely distributed and present in areas of cortex (olfactory nucleus, pyriform, cingulate, and retrosplenial), limbic system (nucleus accumbens, hippocampus, amygdala), and the basal ganglia (caudate nucleus, substantia nigra).

The involvement of 5-HT$_{2C}$ receptors in psychiatric disorders has been hypothesized based largely on pharmacological and clinical studies. A number of atypical and typical antipsychotic agents, including clozapine, loxapine, and chlorpromazine, have a relatively high affinity for 5-HT$_{2C}$ binding sites as well as 5-HT$_{2A}$. Some conventional and atypical antidepressants (e.g., tricyclics, doxepin, mianserin, and trazodone) also exhibit a similar binding affinity (Canton et al. 1990, 1996; Roth and Ciaranello 1991; Jenck et al. 1993). In addition, hallucinogenic drugs such as lysergic acid diethylamide (LSD) and psilocybin, which produce behavioral and cerebral metabolic disturbances that resemble symptoms of acute schizophrenia (Vollenweider et al. 1997; Vollenweider et al. 1998), are agonists at 5-HT$_{2A}$ and 5-HT$_{2C}$ receptors (Burris et al. 1991; Egan et al. 1998). The pharmacological properties of the 5-HT$_{2C}$ receptor have also suggested a role in depression (Moreau et al. 1996; Martin et al. 1998) as well as anxiety (Kennett et al. 1997), obesity, obsessive–compulsive disorder, schizophrenia, migraine, and erectile dysfunction. Intriguing evidence also exists for a potential role of cortical 5-HT$_{2C}$ receptors in the pathophysiology of depression. In a study published in 2002, Gurevich and colleagues (Gurevich et al. 2002) found that altered patterns of 5-HT$_{2C}$ receptor editing were detected in post-mortem brains from suicide victims with a history of major depression. In fact, in addition to variations of genomic sequence, posttranscriptional modification may also have some pathophysiological significance, and among these modifications, editing appears the most relevant in the case of 5-HT$_{2C}$ receptor. Burns and colleagues (Burns et al. 1997) discovered that the 5-HT$_{2C}$ receptor is modified by RNA editing, a posttranscriptional event that generates 5-HT$_{2C}$ isoforms with distinct functional properties.

RNA editing events affect five positions within the human 5-HT$_{2C}$ receptor. These are located in a region encoding the second intracellular loop of the receptor that converts specific mRNA adenosine residues into inosine by adenosine deaminases. Since inosine pairs with cytosine and is read as guanosine during translation, this modification can lead to amino acid substitution and altered function. Sequence analysis of complementary DNA isolates from dissected brain regions have indicated the tissue-specific expression of seven major 5-HT$_{2C}$ receptor isoforms encoded by 11 distinct RNA species. The editing of 5-HT$_{2C}$ receptor messenger RNAs alters the amino acid coding potential of the predicted second intracellular loop of the receptor and can lead to a 10- to 15-fold reduction in the efficacy of the interaction between receptors and their G proteins (Lappalainen et al. 1995). Two main functional consequences of editing events have been characterized. The first one is an isoform of the 5-HT$_{2C}$ receptor that couples less efficiently to the phospholipase C signaling cascade, resulting in a decreased potency for 5-HT transmission (Burns et al. 1997; Fitzgerald et al. 1999; Niswender et al. 1999). The second consequence leads to a production of 5-HT$_{2C}$ receptor isoforms that differ in their ability to interact with G proteins in the absence of agonist.

In summary, studies about the editing of the 5-HT$_{2C}$ receptor mRNA gene suggest that regardless of the final editing combination, the most common consequence is the downregulation in the activity of 5-HT$_{2C}$ receptors. Given that lower serotonergic activity is thought to be a major contributor to suicide risk, there has been additional scrutiny on the role of gene alteration, such as SNPs or modification due to the editing process, in suicide behavior. The results found by Gurevich et al. (2002) indicate significantly different 5-HT$_{2C}$ pre-mRNA editing site preferences in brains of suicide victims. In addition, the changes in editing site preferences detected in suicide victims with a history of major depression are opposite from changes in corresponding editing site preferences detected in mice treated chronically with the antidepressant drug fluoxetine.

These findings provide one of the earliest pieces of evidence that functional modification in the 5-HT$_{2C}$ gene may be involved in depression and suicide behavior. Another interesting study focused on RNA editing of the 5-HT$_{2C}$ receptor was performed by Niswender et al. (2001) showing that the fully edited human 5-HT$_{2C}$ receptor isoform exhibits a marked reduction in sensitivity to LSD and atypical antipsychotic drugs, suggesting a possible role for editing of the 5-HT$_{2C}$ receptor mRNA in the etiology and pharmacotherapy of schizophrenia. They also examined the efficiency of editing of the 5-HT$_{2C}$ receptor mRNA in the PFC of subjects diagnosed with schizophrenia against matched controls. A second group of psychiatric patients diagnosed with major depression was analyzed to test the specificity of any observed changes. Although this examination revealed no significant differences in RNA editing, the subjects who had committed suicide exhibited a statistically significant elevation of editing at the A site regardless of their diagnosis. The A site is predicted to change the amino acid sequence in the second intracellular loop of the 5-HT_{2C} receptor, resulting in an increase in 5-HT_{2C} receptors containing a valine at amino acid 156. These findings have further suggested that alterations in RNA editing may contribute to suicidal behavior.

Besides the editing process that leads to the existence of functionally distinct isoforms of $5\ \text{HT}_{2C}$ receptors, Lappalainen et al. (1995) have identified a nonsynonymous Cys23Ser polymorphism in the first hydrophobic region of the human 5-HT_{2C} receptor. In a recent study this polymorphism has been associated with psychotic symptoms in Alzheimer disease (Holmes et al. 1998) and with an increased risk for hospitalization in schizophrenic patients (Segman et al. 1997). The Csy23Ser polymorphism in the 5-HT_{2C} receptor gene (alias 68G/C, rs6318) is also associated with mood disorder phenotypes (Kõks et al. 2006). Thus, genes that control the brain 5-HT pathways, such as the 5-HT_{2C} receptor gene, seem to consistently be good candidates for mediating genetic susceptibility to mood disorders.

In one of the first studies that addressed the 5-HT_{2C} receptor gene polymorphism variant as risk factor to suicide-related behavior, Stefulj et al. (2004) tested the association of the polymorphism in the 5-HT_{2C} receptor coding region (Cys23Ser) with suicide commitment. The study was based on two independent samples, one of German and the other of Slavic/Croatian ethnicity. No significant differences in allele or genotype frequencies between victims and controls were demonstrated. Previously, Pooley et al. (2003) studied six serotonergic gene polymorphisms in a well-characterized sample of 129 deliberate self-harm subjects and 329 comparison subjects. Among the addressed polymorphisms, they also scrutinized the 5-HT_{2C} receptor (HTR2C Cys23Ser). None of the other polymorphisms, with the exception of the tryptophan hydroxylase (TPH) A779 allele, was associated with deliberate self-harm. Within the deliberate self-harm group, there were no associations with impulsivity, suicide risk, lifetime history of depression, or family history of deliberate self-harm.

More recently, Serretti and colleagues (Serretti et al. 2007) have studied gene variants of both HTR2C and HTR1A receptors in suicide attempters and completers with different psychiatric diagnosis. The sample was composed of 167 German suicide attempters, 92 White Caucasian individuals who committed suicide, 312 German healthy subjects, 152 Italian suicide attempters, and 131 Italian healthy volunteers. They did not find any association of 5-HT_{2C} receptor gene variants with suicide-, anger-, or aggression-related behavior in their sample. Only a marginal association between HTR2C and impulsive suicide was found, and this could be in line with reports of impulsivity modulation operated by HTR2C. However, this finding must be considered very cautiously given the low significance value. In the same paper, they have also investigated anger- and aggression-related traits as these represent intermediate phenotypes of suicidal behavior and observed mild associations. Brain 5-HT more than any other neurotransmitters has been implicated in the neural control of aggression-related behavior, and more specifically, several findings are consistent with the fact that reduced 5-HT_{2C} receptor-mediated function may play a pivotal role in the elevation of aggressive behavior. Serretti et al. (2007) have consistently observed a marginal association between HTR2C and aggression.

In an intriguing study, Baxter et al. (2001) showed this in an animal model where the administration of a 5-HT$_{2C}$ receptor agonist evokes some ritualistic behavior of aggressive intent in nondominant lizard males.

In a recent paper, Serretti et al. (2009) also investigated possible associations between personality traits and a panel of markers in both HTR1A and HTR2C receptors in a German sample of suicide attempters and controls and in an Italian sample of mood disorder patients. Analyzing HTR2C haplotypes in the suicide attempter sample, they have found a trend towards significance for reward dependence (RD) score. This link was a replication of the study performed by Ebstein et al. (1997). These findings make it possible to hypothesize a link between personality and HTR2C SNPs, since the presence of similar findings are present in the literature. A recent study by De Luca et al. (2008) analyzed HTR2C variants (Cys23Ser and a common STR in the promoter) in suicide attempters. Their sample was composed of 306 families with at least one member affected by bipolar disorder, and they have studied HTR2C haplotypes with respect to attempter status and the severity of suicidal behavior. The goals of their study were to investigate the association between the HTR2C gene with suicide attempt and severe suicidal behavior in their bipolar sample and to explore the hypothesis that the HTR2C and MAOA genes interact to influence the above-mentioned outcomes in that sample. The main finding of this study was that HTR2C markers do not confer risk for suicide attempt in bipolar disorder. The genetic analysis that they have performed incorporating the clinical-demographics factors was very valuable since many of the genetic reports published on 5-HT genes and suicidality did not take these factors into account; however, after the correction they have found just a trend with one of studied allele protecting against suicide attempt. Their haplotype analysis did not find significant associations between the HTR2C haplotype and suicide attempts; however, very few studies have explored the relationship between suicidality and the HTR2C haplotype, as it is located on the X chromosome.

Other findings on 5-HT$_{2C}$ receptor isoforms from the editing process with relation to suicidal behavior have been investigated by the Dracheva group. In their first study (Dracheva et al. 2008a), they compared the frequencies of 5-HT$_{2C}$ receptor mRNA editing variants in postmortem prefrontal cortices of subjects who committed suicide and in which the suicides occurred in the context of either schizophrenia or bipolar disorder. The control subjects were represented by subjects who died from other causes with either schizophrenia, bipolar disorder, or no psychiatric diagnosis. They have identified 5-HT$_{2C}$ receptor mRNA editing variations that were associated with suicide but not with the comorbid psychiatric diagnoses. These variations consisted of a significant increase in the pool of mRNA variants that encode one of the most prevalent and highly edited isoforms of 5-HT$_{2C}$ receptor that exhibit low functional activity. This increased expression may significantly influence the serotonergic function in the brain. In another study (Dracheva et al. 2008b), this group has pointed out that the ratio between the two 5-HT$_{2C}$ receptor splice variants (RNAsp2/RNAsp1), was significantly increased in patients who committed suicide relative to patients who died of other causes.

Regarding the different expression of mRNA of 5-HT_{2C} receptor among subjects who committed suicide, Anisman et al. (2008) have described differences that exist in mRNA expression of several 5-HT receptor subtypes (5-HT_{1A}, 5-HT_{1B}, 5-HT_{2A}, 5-HT_{2C}) in brain regions that have been implicated in depression (the frontopolar cortex, orbitofrontal cortex, and hippocampus), stressor reactivity and anxiety (the amygdala), and HPA functioning (the paraventricular nucleus of the hypothalamus). The analysis of 5-HT_{2C} receptor expression has yielded a significant interaction between the variables of sex and cause of death. The follow-up tests have confirmed that, among men, 5-HT_{2C} receptor mRNA expression did not differ as a function of the cause of death, whereas among women, 5-HT_{2C} receptor expression was lower in suicide subjects than in control subjects. The expression of 5-HT_{2C} receptor was likewise found not to differ as a function of the cause of either death or sex, although there was a modest trend for lower 5-HT_{2C} receptor expression in those who committed suicide when compared with control subjects. 5-HT_{2C} receptor mRNA expression in the amygdala and frontotemporal cortex was lower in suicide samples than in control samples. In conclusion, 5-HT_{2C} receptor, much like several other 5-HT receptor subtypes, has been associated with depression and suicide, but these receptor differences vary across brain regions and are moderated by other variables such as the patient's sex.

13.5 Conclusions

Summarizing studies of 5-HT_{2C} receptor expression and variations in suicide matters, the most complex alterations in 5-HT_{2C} receptor were in the pre-mRNA editing found in brains of depressed suicide victims. In these brains, 5-HT_{2C} receptor isoforms with reduced function are expressed at significantly increased levels, suggesting that the regulation of editing by synaptic 5-HT is defective. Moreover, postmortem studies on brain tissues of patients with schizophrenia and major depression found distinct site-specific alterations of this editing in the PFC, a brain region expressing a large number of differently edited 5-HT_{2C} receptor mRNA isoforms.

Acknowledgments Dr. De Luca has been supported by a NARSAD Young Investigator Award and an American Foundation for Suicide Prevention Pilot Grant.

References

Anisman H, Du L, Palkovits M (2008) Serotonin receptor subtype and p11 mRNA expression in stress-relevant brain regions of suicide and control subjects. J Psychiatry Neurosci 33:131–141.

Arango V, Mann JJ (1992) Relevance of serotonergic postmortem studies to suicidal behaviour. Int Rev Psychiatry 4:131–140.

Arango V, Underwood MD (1997) Serotonin chemistry in the brain of suicide victims. In: Mans R, Silverman M, Canetto S, eds. Review of Suicidology. New York: Guilford Press.

Asberg M, Traskman L, Thoren P (1976) 5-HIAA in the cerebrospinal fluid: A biochemical suicide predictor? Arch Gen Psychiatry 33:1193–1197.

Barnes NM, Sharp T (1999) A review of central 5-HT receptors and their function. Neuropharmacology 38:1083–1152.

Baxter LR Jr., Clark EC, Ackermann RF, et al (2001) Brain mediation of Anolis social dominance displays. II. Differential forebrain serotonin turnover, and effects of specific 5-HT receptor agonists. Brain Behav Evol 57:184–201.

Brent DA, Oquendo M, Birmaher B, et al (2002) Familial pathways to early-onset suicide attempt: Risk for suicidal behaviour in offspring of mood-disordered suicide attempters. Arch Gen Psychiatry 59:801–807.

Burns CM, Chu H, Rueter SM, et al (1997) Regulation of serotonin-2C receptor G-protein coupling by RNA editing. Nature 387:303–308.

Burris KD, Breeding M, Sanders-Bush E (1991) (+)Lysergic acid diethylamide, but not its nonhallucinogenic congeners, is a potent serotonin 5HT1C receptor agonist. J Pharmacol Exp Ther 258:891–896.

Canton H, Verriele L, Colpaert FC (1990) Binding of typical and atypical antipsychotics to 5-HT1C and 5-HT2 sites: Clozapine potently interacts with 5-HT1C sites. Eur J Pharmacol 191:93–96.

Canton H, Emeson RB, Barker EL, et al (1996) Identification, molecular cloning, and distribution of a short variant of the 5–hydroxytryptamine2C receptor produced by alternative splicing. Mol Pharmacol 50:799–807.

Chen K, Yang W, Grimsby J, et al (1992) The human 5-HT2 receptor is encoded by a multiple intron-exon gene. Mol Brain Res 14:20–26.

Cooper SJ, Kelly CB, King DJ (1992) 5-Hydroxyindoleacetic acid in cerebrospinal fluid and prediction of suicidal behaviour in schizophrenia. Lancet 340:940–941.

De Luca V, Tharmaligam S, Strauss J, et al (2008) 5-HT2C receptor and MAO-A interaction analysis: No association with suicidal behaviour in bipolar patients. Eur Arch Psychiatry Clin Neurosci 258:428–433.

Dracheva S, Patel N, Woo DA, et al (2008a) Increased serotonin 2C receptor mRNA editing: A possible risk factor for suicide. Mol Psychiatry 13:1001–1010.

Dracheva S, Chin B, Haroutunian V (2008b) Altered serotonin 2C receptor RNA splicing in suicide: association with editing. Neuroreport 19:379–382.

Duman RS, Monteggia LM (2006) A neurotrophic model for stress-related mood disorders. Biol Psychiatry 59:1116–1127.

Ebstein RP, Segman R, Benjamin J, et al (1997) 5-HT2C (HTR2C) serotonin receptor gene polymorphism associated with the human personality trait of reward dependence. Interaction with dopamine D4 receptor (D4DR) and dopamine D3 receptor (D3DR) polymorphisms. Am J Med Genet 74:65–72.

Egan C, Herrick-Davis K, Teitler M (1998) Creation of a constitutively activated state of the 5-HT2A receptor by site-directed mutagenesis: revelation of inverse agonist activity of antagonists. Ann N Y Acad Sci 861:136–139.

Fitzgerald LW, Iyer G, Conklin DS, et al (1999) Messenger RNA editing of the human serotonin 5-HT2C receptor. Neuropsychopharmacology 21(2 Suppl):82S–90S.

Garfinkel BD, Froese A, Hood J (1982) Suicide attempts in children and adolescents. Am J Psychiatry 139:1257–1261.

Gershon ES (1990) Genetics. In: Goodwin FK, Jamison KR (eds). Maniac Depressive Illness. New York: Oxford University Press.

Gurevich I, Tamir H, Arango V, et al (2002) Altered editing of serotonin 2C receptor pre-mRNA in the prefrontal cortex of depressed suicide victims. Neuron 34:349–356.

Holmes C, Arranz MJ, Powell JF, et al (1998) 5-HT2A and 5-HT2C receptor polymorphisms and psychopathology in late onset Alzheimer's disease. Hum Mol Genet 7:1507–1509.

Jenck F, Moreau JL, Mutel V, et al (1993) Evidence for a role of 5-HT1C receptors in the antiserotonergic properties of some antidepressant drugs. Eur J Pharmacol 231:223–229.

Julius D, MacDermott AB, Axel R, et al (1988) Molecular characterization of a functional cDNA encoding the serotonin 1c receptor. Science 241:558–564.

Julius D, Huang KN, Livelli TJ, et al (1990) The 5HT2 receptor defines a family of structurally distinct but functionally conserved serotonin receptors. Proc Natl Acad Sci USA 87:928–932.

Kaufman MJ, Hartig PR, Hoffman BJ (1995) Serotonin 5-HT2C receptor stimulates cyclic GMP formation in choroid plexus. J Neurochem 64:199–205.

Kennett GA, Wood MD, Bright F, et al (1997) SB 242084, a selective and brain penetrant 5-HT2C receptor antagonist. Neuropharmacology 36:609–620.

Kõks S, Nikopensius T, Koido K, et al (2006) Analysis of SNP profiles in patients with major depressive disorder. Int J Neuropsychopharmacol 9:167–174.

Lappalainen J, Zhang L, Dean M, et al (1995) Identification, expression, and pharmacology of a Cys23-Ser23 substitution in the human 5-HT2c receptor gene (HTR2C). Genomics 27:274–279.

Lester D (1995) The concentration of neurotransmitter metabolites in the cerebrospinal fluid of suicidal individuals: a meta-analysis. Pharmacopsychiatry 28:45–50.

Li D, He L (2007) Meta-analysis supports association between serotonin transporter (5-HTT) and suicidal behaviour. Mol Psychiatry 12:47–54.

Lübbert H, Hoffman BJ, Snutch TP, et al (1987) cDNA cloning of a serotonin 5-HT1C receptor by electrophysiological assays of mRNA-injected Xenopus oocytes. Proc Natl Acad Sci USA 84:4332–4336.

Mann JJ (2003) Neurobiology of suicidal behaviour. Nat Rev Neurosci 4:819–828.

Mann JJ, Arango V, Marzuk PM, et al (1989) Evidence for the 5-HT hypothesis of suicide. A review of post-mortem studies. Br J Psychiatry Suppl (8):7–14.

Mann JJ, Malone KM, Nielsen DA, et al (1997) Possible association of a polymorphism of the tryptophan hydroxytase gene with suicidal behaviour in depressed patients. Am J Psychiatry 154:1451–1453.

Mann JJ, Waternaux C, Haas GL, et al (1999) Towards a clinical model of suicidal behaviour in psychiatric patients. Am J Psychiatry 156:181–189.

Mann JJ, Huang Y, Underwood MD, et al (2000) A serotonin transporter gene promoter polymorphism (5-HTTLPR) and prefrontal cortical binding in major depression and suicide. Arch Gen Psychiatry 57:729–738.

Martin P, Waters N, Schmidt CJ, et al (1998) Rodent data and general hypothesis: antipsychotic action exerted through 5-Ht2A receptor antagonism is dependent on increased serotonergic tone. J Neural Transm 105:365–396.

Mathers CD, Loncar D (2006) Projections of global mortality and burden of disease from 2002 to 2030. PLoS Med 3:e442.

Mathers CD, Bernard C, Iburg KM, et al (2003) Global burden of disease in 2002: Data sources, methods and results. Paper presented at the Global Programme on Evidence for Health Policy. Geneva, Switzerland, World Health Organization.

Moreau JL, Bös M, Jenck F, et al (1996) 5HT2C receptor agonists exhibit antidepressant-like properties in the anhedonia model of depression in rats. Eur Neuropsychopharmacol 6:169–175.

Murray CL, Lopez AD, eds. (1996) The global burden of disease: a comprehensive assessment of mortality and disability from diseases, injuries, and risk factors in 1990 and projected to 2020. Cambridge, MA, Harvard University Press.

Ninan PT, van Kammen DP, Scheinin M, et al (1984) CSF 5-hydroxyindoleacetic acid levels in suicidal schizophrenic patients. Am J Psychiatry 141:566–569.

Niswender CM, Copeland SC, Herrick-Davis K, et al (1999) RNA editing of the human serotonin 5-hydroxytryptamine 2C receptor silences constitutive activity. J Biol Chem 274:9472–9478.

Niswender CM, Herrick-Davis K, Dilley GE, et al (2001) RNA editing of the human serotonin HTR2C. Alterations in suicide and implications for serotonergic pharmacotherapy. Neuropsychopharmacology 24:478–491.

Oquendo MA, Currier D, Mann JJ (2006) Prospective studies of suicidal behaviour in major depressive and bipolar disorders: what is the evidence for predictive risk factors? Acta Psychiatr Scand 114:151–158.

Oquendo MA, Baca-García E, Mann JJ, et al (2008) Issues for DSM-V: Suicidal behaviour as a separate diagnosis on a separate axis. Am J Psychiatry 165:1383–1384.

Pandey GN (1997) Altered serotonin function in suicide. Evidence from platelet and neuroendocrine studies. Ann N Y Acad Sci 836:182–200.

Pfeffer CR, Normandin L, Kakuma T (1994) Suicidal children grow up: Suicidal behaviour and psychiatric disorders among relatives. J Am Acad Child Adolesc Psychiatry 33:1087–1097.

Pfennig A, Kunzel HE, Kern N, et al (2005) Hypothalamus-pituitary-adrenal system regulation and suicidal behaviour in depression. Biol Psychiatry 57:336–342.

Pooley EC, Houston K, Hawton K, et al (2003) Deliberate self-harm is associated with allelic variation in the tryptophan hydroxylase gene (TPH A779C), but not with polymorphisms in five other serotonergic genes. Psychol Med 33:775–783.

Posner K, Oquendo MA, Gould M, et al (2007) Columbia Classification Algorithm of Suicide Assessment (C-CASA): classification of suicidal events in the FDA's pediatric suicidal risk analysis of antidepressants. Am J Psychiatry 164:1035–1043.

Robins E, Guze SB (1970) Establishment of diagnostic validity in psychiatric illness: its application to schizophrenia. Am J Psychiatry 126:983–987.

Roth BL, Ciaranello RD (1991) Chronic mianserin treatment decreases 5-HT2 receptor binding without altering 5-HT2 receptor mRNA levels. Eur J Pharmacol 207:169–172.

Roy A (1993) Genetic and biologic risk factors for suicide in depressive disorders. Psychiatric Q 64:345–358.

Roy A, Segal NL, Sarchiapone M (1995) Attempted suicide among living co twins of twin suicide victims. Am J Psychiatry 152:1075–1076.

Roy A, Rytander G, Sarchiapone M (1997) Genetics of suicide: Family studies and molecular genetics. Ann N Y Acad Sci 836:135–157.

Segman RH, Ebstein RP, Heresco-Levy U, et al (1997) Schizophrenia, chronic hospitalization and the 5-HT2C receptor gene. Psychiatr Genet 7:75–78.

Serretti A, Mandelli L, Giegling I, et al (2007) HTR2C and HTR1A gene variants in German and Italian suicide attempters and completers. Am J Med Genet B Neuropsychiatr Genet 144:291–299.

Serretti A, Calati R, Giegling I, et al (2009) Serotonin receptor HTR1A and HTR2C variants and personality traits in suicide attempters and controls J Psychiatr Res 43:519–525.

Stam NJ, Vanhuizen F, Vanalebeek C, et al (1992) Genomic organization, coding sequence and functional expression of human 5-HT2 and 5-HT1A receptor genes. Eur J Pharmacol 227:153–162.

Statham DJ, Heath AC, Madden PAF, et al (1998) Suicidal behaviour: an epidemiological and genetic study. Psychol Med 28:839–855.

Stefulj J, Buttner A, Kubat M, et al (2004) 5HT-2C receptor polymorphism in suicide victims. Association studies in German and Slavic populations. Eur Arch Psychiatry Clin Neurosci 254:224–227.

Strohle A, Holsboer F (2003) Stress responsive neurohormones in depression and anxiety. Pharmacopsychiatry 36:207–214.

U.S. Department of Health and Human Services (2000) With understanding and improving health and objectives for improving health, 2 vols. Healthy People 2010. 2nd.ed. Washington, DC: U.S. Department of Health and Human Services.

U.S. Public Health Service (1999) The Surgeon General's call to action to prevent suicide. Washington, DC: U.S. Public Health Service.

Vollenweider FX, Leenders KL, Scharfetter C, et al (1997) Positron emission tomography and fluorodeoxyglucose studies of metabolic hyperfrontality and psychopathology in the psilocybin model of psychosis. Neuropsychopharmacology 16:357–372.

Vollenweider FX, Vollenweider-Scherpenhuyzen MF, Bäbler A, et al (1998) Psilocybin induces schizophrenia-like psychosis in humans via a serotonin-2 agonist action. Neuroreport 9:3897–3902.

Voracek M, Loibl LM (2007) Genetics of suicide: a systematic review of twin studies. Wien Klin Wochenschr 119:463–475.

Westrin A (2000) Stress system alterations and mood disorders in suicidal patients. A review. Biomed Pharmacother 54:142–145.

World Health Organization (1996) Prevention of suicide: guidelines for the formulation and implementation of national strategies. Geneva, World Health Organization.

Xie E, Zhao L, Levine AJ, et al (1996) Human serotonin 5-HT2C receptor gene: Complete cDNA, genomic structure, and alternatively spliced variant. EMBL accession number U49516.

Yu L, Nguyen H, Le H, et al (1991) The mouse 5-HT1C receptor contains eight hydrophobic domains and is X-linked. Mol Brain Res 11:143–149.

Chapter 14
The 5-HT$_{2C}$ Receptor as a Target for Schizophrenia

Herbert Y. Meltzer, Liwen Sun, and Hitoshi Hashimoto

14.1 The Role of the 5-HT$_{2C}$ Receptor in Schizophrenia and Antipsychotic Drug

14.1.1 The 5-HT$_{2C}$ Receptor in Schizophrenia and Its Relation to Other 5-HT Receptors and 5-HT Mechanism

The serotonin-2C (5-HT$_{2C}$) receptor is one of several of the 14 distinct 5-HT receptors that has been implicated in schizophrenia as a target for antipsychotic drugs to diminish psychosis, to diminish side effects of some antipsychotic drugs (e.g., catalepsy), or to improve cognition, a major need in the treatment of schizophrenia. The 5-HT$_{2C}$ receptor has also been of interest as the basis for adverse side effects of antipsychotic drugs, particularly weight gain (Kroeze et al. 2003). While this is an impressive array of potential relevance to schizophrenia, it pales in comparison with that for the 5-HT$_{2A}$ receptor, i.e., the 5-HT receptor that has received the most attention of any 5-HT receptor in this regard (Meltzer and Huang 2008). There is considerable evidence that 5-HT$_{1A}$ and 5-HT$_7$ receptors are also important for schizophrenia, especially for cognition, or that they modulate antipsychotic drug action (Meltzer and Huang 2008). To a lesser extent, the 5-HT$_3$, 5-HT$_4$, 5-HT$_5$, and 5-HT$_6$ receptors have also been considered in various contexts as targets for schizophrenia. This chapter will, of course, focus mainly on the role of the 5-HT$_{2C}$ receptor and mention the others mainly to clarify interactions with the 5-HT$_{2C}$ receptor. To avoid repetition with other chapters in this volume that provide in-depth consideration of specific aspects of 5-HT$_{2C}$ receptors relevant to its potential role in the pathophysiology and treatment of schizophrenia, the consideration given to some of these subjects in this chapter will mainly be to discuss how the 5-HT$_{2C}$ receptor

H.Y. Meltzer (✉)
Department of Psychiatry and Pharmacology, Vanderbilt University School of Medicine, Nashville, TN, 37212, USA
herbert.meltzer@vanderbilt.edu

G. Di Giovanni et al. (eds.), *5-HT$_{2C}$ Receptors in the Pathophysiology of CNS Disease*, The Receptors 22, DOI 10.1007/978-1-60761-941-3_14, © Springer Science+Business Media, LLC 2011

relates to those subjects as they pertain to schizophrenia. Thus, the reader is referred for in-depth consideration to the following chapters that discuss issues that are of definite or potential importance to schizophrenia (and bipolar disorder) and its treatment. These include

1. Basic science issues such as 5-HT$_{2C}$ receptor neuroanatomy (Chap. 2), signal transduction (Chap. 5), neurogenesis (Chap. 9), control of central dopamine function (Chap. 11), PTEN-5-HT$_{the}$ coupling (Chap. 16), the basal ganglia (Chap. 18), RNA editing (Chap. 8), attentional processes, learning, and motor control.
2. The role of the 5-HT$_{2C}$ receptor in the regulation of specific behaviors, e.g., suicide (Chap. 13), reward (Chap. 15), eating behavior (Chap. 17), and sleep (Chap. 20)
3. The role of the 5-HT$_{2C}$ receptor in the regulation of specific clinical syndromes or antipsychotic drug-related issues, e.g., depression and anxiety (Chaps. 8 and 12), tardive dyskinesia (Chap. 19), and Alzheimer disease (Chap. 25).

14.2   5-HT$_{2C}$ Receptors Are Relevant to Both Schizophrenia and Bipolar Disorder

14.2.1   Overlap Between Schizophrenia and Bipolar Disorder

Schizophrenia is best thought of as a syndrome rather than a discrete disease. Its boundaries with a number of other psychiatric disorders are not clearly delineated, and there is no pathognomonic symptom or biomarker that defines it. It is characterized by a variety of disturbances in reality testing, cognition, affect, volition, and mood, which vary in intensity from one patient to another and over time, with or without treatment. It is usually first diagnosed in the middle of the second to the middle of the third decade of life, earlier in males than females, but may appear even earlier or later in some individuals.

It is highly heritable although not entirely so, with environmental factors contributing to onset and modifying its course. These factors are broadly conceptualized as stress, with maternal influenza in the second trimester and secular famine contributing as well. Strikingly, only half of monozygotic twins are concordant for schizophrenia. Multiple genes of small effects are generally believed to be the basis for the genetic vulnerability to schizophrenia, but there is increasing evidence that highly penetrant, single, large deletions or insertions (copy number variations) may be causal. There is much overlap between schizophrenia and bipolar disorder with regard to susceptibility genes, phenomenology, and response to treatment; therefore, the ensuing discussion of the 5-HT$_{2C}$ receptor as a target for schizophrenia and its treatment is likely to be highly relevant to bipolar disorder as well. Schizophrenia affects about 1% of the population worldwide, and bipolar disorder affects a somewhat greater number, but not all bipolar patients have periods of psychosis, which is one of the hallmarks of schizophrenia. However, not all patients with schizophrenia, e.g., those with catatonia or simple schizophrenia, have hallucinations or delusions. Although the first delineation of schizophrenia as a separate disorder of brain function by Emil Kraepelin, at the turn of the twentiethh century, emphasized cognitive dysfunction as the central feature of the syndrome (he called it *dementia praecox*), its reconceptualization by Bleuler in the early twentieth century emphasized symptoms, particularly delusions and hallucinations.

14.2.2   The Role of DA and 5-HT in Psychosis

The discovery of chlorpromazine, the first antipsychotic drug, in 1952, and, shortly thereafter, the central role of dopamine (DA) receptor blockade in its action and the role of DA in the etiology of delusions following chronic amphetamine administration led to neglect of the critical importance of cognitive dysfunction for schizophrenia, a serious error, which has been corrected over the last 20 years. Since delusions and hallucination both responded well to chlorpromazine, and amphetamine is known to enhance synaptic DA levels, the etiology of these two symptoms was linked to enhanced dopaminergic function (Angrist and Gershon 1974).

We have reviewed the history of the development of the theory of the role of 5-HT in the etiology of schizophrenia elsewhere (Meltzer et al. 2003). The pivotal discovery that lysergic acid diethylamide (LSD) ingestion produces visual hallucinations, followed by the findings of similar effects of mescaline and 4-iodo-2,5-dimethoxyamphetamine (DOI), all three of which are agonists at 5-HT$_{2A}$ and 5-HT$_{2C}$ receptors, stimulated a major effort to determine which of these two receptors was more involved in the causation of hallucinations. Furthermore, as result of the LSD finding, great interest arose in the possibility that endogenous production of 5-HT agonists with psychotomimetic properties could be a principal cause of schizophrenia. However, this theory could not be confirmed, clearing the way for a prolonged period of extreme interest in the DA hypothesis of schizophrenia. The limitations in this hypothesis became apparent with the limited efficacy of typical antipsychotic drugs.

14.3   Serotonin Dopamine Receptor Antagonist Hypothesis of Schizophrenia

14.3.1   Affinities for 5-HT$_{2A}$, D$_2$ Receptors and 5-HT$_{2C}$ Receptors

Interest in the role of 5-HT in schizophrenia was revived by the hypothesis that clozapine and other antipsychotics with minimal potential to impair motor function (so-called atypical antipsychotic drug) were more potent 5-HT$_{2A}$ than DA D$_2$ antagonists (Meltzer 1989). This hypothesis, now generally accepted, led to the development of a group of antipsychotic drugs that are among the most widely used of the antipsychotic drugs. The drugs currently approved or close to approval are listed in Table 14.1, along with their affinities for human 5-HT_{2A}, D_2, and 5-HT_{2C} receptors, as well as their $5\text{-HT}_{2A}/D_2$, $5\text{-HT}_{2C}/D_2$, and $5\text{-HT}_{2A}/_{2C}$ ratios. These data are obtained from the National Institute of Mental Health (NIMH) Psychopharmacology Drug Screening Program Website. For most drugs, the ligand used to determine the 5-HT_{2C} affinity was [^{3}H]mesulergine; for 5-HT_{2A} receptors, [^{3}H]ketanserin; and for DA, [^{3}H]spiperone. Where available, the tissue source was human cortex; for some, it was cloned receptors expressed in cell lines. It should be noted that aripiprazole is a partial agonist at the D_2 receptor and achieves much weaker blockade of the 5-HT_{2A} receptor than would be expected based on its Ki value. It is evident that the atypical antipsychotic drugs, with the exception of aripiprazole, have higher affinities for 5-HT_{2A} than D_2 receptors have. Only 6 of 12 atypicals have a higher affinity for the 5-HT_{2C} than the D_2 receptor has. Iloperidone, lurasidone, melperone, risperidone, sertindole, and aripiprazole do not. All atypicals with the exception of asenapine have a higher affinity for the 5-HT_{2A} than 5-HT_{2C} receptor has. None of the typical antipsychotic drugs has a higher affinity for the 5-HT_{2C} than the D_2 receptor has. Thus, from this perspective, it is unlikely that there is a systematic effect of the 5-HT_{2C} receptor for the action of all antipsychotic drugs as there appears to be for the D_2 receptor or for all atypical antipsychotic drugs, as there appears to be for the 5-HT_{2A} receptor. Nevertheless, it is quite possible that for some atypical antipsychotic drugs, at clinically relevant doses, the 5-HT_{2C} receptor is important for some actions. This will be discussed subsequently.

14.3.2 5-HT_{2C} Receptor and Dopamine Function in Schizophrenia

Abnormalities in dopaminergic function in the limbic regions (nucleus accumbens and the extended amygdala), frontal cortex, cingulate cortex, and hippocampus have been considered important causes of psychosis (limbic) and cognitive dysfunction (cortical–hippocampal) in patients with schizophrenia. The data supporting the role of DA in these components of the illness suggest enhanced dopaminergic activity in the limbic region as the basis for psychosis and hypodopaminergic function in the cortico-hippocampal regions as the basis for cognitive impairment and negative symptoms. As psychosis, negative symptoms, and cognitive impairment in schizophrenia represent separate dimensions of the illness, i.e., they are virtually completely uncorrelated in severity, course, and response to available treatments, they more than likely result from independent processes, even if some of the pathways involved are the same (Meltzer 1992). Regardless of the ultimate cause(s) of the hyperdopaminergic activity that leads to psychosis, of which there could be many, blockade of D_2 receptors by typical antipsychotic drugs is effective to reduce psychosis in about 70% of patients (Meltzer and Stahl 1976). The other 30% of schizophrenia patients have no or minimal response to

Table 14.1 Affinities of antipsychotic drugs for 5-HT$_{2A}$, D$_2$ and 5-HT$_{2C}$ receptors and their ratios

Drug	5-HT$_{2A}$	D$_2$	5-HT$_{2C}$	5-HT$_{2A}$/D2	5-HT$_{2C}$/D2	5-HT$_{2A}$/5-HT$_{2C}$
Atypical antipsychotic drugs						
Clozapine	2.59	210	8.0	0.012	0.038	0.32
Asenapine	0.77	2.0	0.2	0.39	0.26	3.85
Iloperidone	0.20	3.0	14.0	0.067	4.67	0.014
Lurasidone	0.47	0.99	415	0.47	419	0.0011
Olanzapine	1.48	72	11.0	0.021	0.15	0.13
Melperone	102	180	2,100	0.57	11.7	0.049
Quetiapine	221	770	615	0.29	0.79	0.36
Risperidone	0.15	4.9	26.0	0.031	5.31	0.0058
Sertindole	0.2	3.3	14.0	0.061	4.24	0.01
Ziprasidone	0.12	6.9	0.9	0.017	0.13	0.13
Zotepine	2.6	8.0	3.2	0.33	0.40	0.81
Aripiprazole[a]	8.7	0.95	22.4	9.16	23.6	0.39
Typical antipsychotic drugs						
Chlorpromazine	1.4	0.66	25	2.12	3.12	0.056
Fluphenazine	189	0.54	1,386	350	2,567	0.14
Haloperidol	61	2.6	3,085	23.5	1,187	0.020
Perphenazine	5.6	1.4	132	4.0	94.3	0.042
Pimozide	19	29	874	0.67	30.1	0.22
Thioridazine	10	10	60	1.0	6.0	0.17

Data obtained from NIMH PDP Website: Where possible, data are from the same laboratory, usually from the PDP laborator (Bymaster et al. 2003; Richelson and Souder 2000). Human cortex or cloned cells were the tissue source for all data

[a]Partial DA agonist: ratios should not be interpreted in the same way as for the DA antagonists

typical antipsychotic drugs, but about two thirds will respond to higher doses of clozapine or olanzapine, when given for up to 6 months (Meltzer 1995). The failure of typical antipsychotic drugs to improve psychosis in these patients might be related to abnormalities of the postsynaptic D_2 receptor, i.e., failure to block D_2 receptor-stimulated signaling pathways. For such patients, inhibition of the release of DA in the nucleus accumbens might be expected to be beneficial. Some, but not all, 5-HT_{2C} agonists inhibit DA efflux in the nucleus accumbens (Willins and Meltzer 1998; Di Matteo et al. 1999). This would be predicted to decrease psychotic symptoms. Some 5-HT_{2C} receptor agonists, e.g., mCPP, also modestly decrease DA efflux in the dorsal striatum, while others, e.g., Ro60-0175 and WAY 163909, have no effect (Alex et al. 2005). This would be predicted to inhibit motor function and cause extrapyramidal side effects. As will be discussed, WAY 163909, is a recently discovered 5-HT_{2C} agonist reported to be effective in multiple models of antipsychotic activity (Marquis et al. 2007). However, 5-HT_{2C} receptor agonists also suppress the efflux of DA in the medial prefrontal cortex (Pozzi et al. 2002), which would be expected to modulate cognitive function. The functional effect of this modulation would depend on whether specific cognitive functions, which are DA dependent, would benefit or be impaired by a decrease in dopaminergic activity.

14.4   Postmortem Studies of 5-HT_{2C} Function

Decreased messenger ribonucleic acid (mRNA) expression of 5-HT_{2C} receptors has been reported in postmortem prefrontal cortex tissues in 55 patients with schizophrenia and 55 controls (Castensson et al. 2003). Castensson et al. (2005) reported that the decrease in expression of the 5-HT_{2C} receptor was not related to drug action and might be a core feature of the illness. Decreased expression of the 5-HT_{2C} receptor has also been reported in bipolar disorder patients (Iwamoto and Kato 2003). Diminished 5-HT_{2C} receptor expression would be expected to facilitate limbic DA release.

14.5   Genetic and Pharmacogenetics of 5-HT_{2C} Receptors in Schizophrenia and Antipsychotic Drug Action

Three single nucleotide polymorphisms of the 5-HT_{2C} receptor have been extensively studied in relation to risk for schizophrenia and prediction of response to or side effects of antipsychotic drugs. The most studied of these has been Cys23Ser, a coding nonsynonymous single nucleotide polymorphism (SNP) that results in an amino acid substitution of cysteine for serine at position 23, followed by rs3813929 (−759C/T) and rs518147 (−697G/C), which are SNPs located within the promoter and 5′ untranslated region of the X-chromosome 5-HT_{2C} receptor gene, respectively (Drago and Serretti 2009). These 5-HT_{2C} receptor SNPs have attracted substantial interest in genetic association studies in relation to psychotropic drug effects (including antipsychotic drug-induced weight gain (Ellingrod et al. 2005; Miller et al. 2005; Reynolds et al. 2002; Ryu et al. 2007)), extrapyramidal side effects of antipsychotic drugs (Gunes et al. 2007) (including tardive dyskinesia; Segman et al. 2000), symptomatic improvement following clozapine (Sodhi et al. 1995), and association with major affective disorder (Gutierrez et al. 1996; Lerer et al. 2001) and eating disorders (Hu et al. 2003; Ribases et al. 2008). The Cys23Ser SNP was examined for linkage to schizophrenia in 207 nuclear families using the transmission disequilibrium test. No preferential transmission of the Cys23Ser 5-HT$_{2C}$ receptor alleles in either German or Palestinian-Arab groups alone or combined was found (Murad et al. 2001). A study of 5-HT$_{2C}$ and 5-HT$_{2A}$ receptor SNPs and psychotic symptomatology in late-onset Alzheimer disease showed an association between the 5-HT$_{2C}$ receptor SNPs rs6318 (Cys23Ser) and visual hallucinations and depression, as well as between the 5-HT$_{2A}$ receptor SNP 102T/C and visual and auditory hallucinations (Holmes et al. 1998).

Based on the association of the Cys23ser with psychopathology of Alzheimer disease and response to clozapine, we examined the association between all three 5-HT$_{2C}$ receptor SNPs and (1) the severity of psychopathology in patients with schizophrenia or schizoaffective disorder at baseline and (2) as predictors of response to treatment with antipsychotic drugs in patients with schizophrenia or schizoaffective disorder, in a sample of 186 Caucasian and Afro-American patients with schizophrenia or schizoaffective disorder. The Ser23 allele carriage was strongly associated with higher baseline scores of the Brief Psychiatric Rating Scale (BPRS) (total scores [$p=0.01$] and psychosis cluster [$p=0.007$]), but not with BPRS negative symptom or anxiety/depression subscale scores. The -759C/T SNP but not the -697G/C SNP was significantly associated with psychosis ($p=0.04$). However, the haplotype (-759C)–(-697C)–(23Ser), $p=0.0009$, and (-759C)–(23Ser), $p=0.0006$, were both strongly associated with the BPRS psychosis subscale. The number of prior hospitalizations was significantly greater in -759C and -679G carriers compared with the -759T and -679G carriers, respectively. In addition, the Ser23 allele carriage was significantly associated with improvement in negative symptoms and at trend level, psychosis, following treatment with antipsychotic drugs at 6 weeks ($p=0.05$) and 6 months ($p=0.0005$). These results suggest that these three 5-HT$_{2C}$ receptor polymorphisms may be related to the genetic influence on severity of positive symptoms and response to antipsychotic drugs for patients with schizophrenia or schizoaffective disorder (Meltzer et al., submitted).

14.6   Editing and Alternative Splicing of the 5-HT$_{2C}$ Receptor in Schizophrenia

One of the distinctive features of the 5-HT$_{2C}$ receptor is that its mRNA undergoes site-specific conversion of adenosine to inosine (A-to-I) by posttranscriptional RNA editing (see Chap. 8 for a discussion of editing). This leads to multiple receptor isoforms with dramatically different constitutive activity, expression patterns, and efficacy of antipsychotic drugs (Burns et al. 1997; Morabito and Emeson 2009; Niswender et al. 2001). These observations suggest that human 5-HT_{2C} receptor gene (HTR2C) RNA editing may be involved in the maintenance of appropriate serotonergic neurotransmission and influence treatment response in psychiatric disorders.

There has been some investigation of the relevance of editing of the 5-HT_{2C} receptor for schizophrenia. In a study of brain tissue from frontal cortex of only five controls and five subjects with schizophrenia, reduced RNA editing, increased expression of the unedited 5-HT(2C-INI) isoform in schizophrenia ($p=0.001$), and decreased expression of the 5-HT(2C-VSV) and 5-HT(2C-VNV) isoforms were found in the schizophrenia group. Since the unedited 5-HT(2C-INI) couples more efficiently to G proteins than the other isoforms do, the authors suggested enhanced $5\text{-HT}_{2C}\text{R}$-mediated effects might be present in schizophrenia (Sodhi et al. 2001). However, Dracheva et al. (2003) found no differences in editing or alternative splicing of the 5-HT_{2C} receptor in a comparison of 15 elderly patients with schizophrenia and 15 matched controls. Sodhi et al. (2005) have also suggested that chronic treatment with antipsychotic drugs might alter editing of the 5-HT_{2C} receptor (Sodhi et al. 2005).

14.7 Antipsychotic Drugs: Importance of 5-HT_{2C} Receptor Antagonism in the Action of $5\text{-HT}_{2A}/D_2$ Antagonist Antipsychotic

After the discovery of the importance of 5-HT_{2A} receptor blockade (inverse agonism), initiated by research to uncover the mechanism of action of clozapine, which has multiple receptor affinities, interest focused on other clozapine high affinity binding sites, including the 5-HT_{2C} receptor. The 5-HT_{2C} receptor was proposed to be an important site of action of atypical antipsychotic drugs by a number of investigators (Canton et al. 1990; Cussac et al. 2000; Duinkerke et al. 1993a; Herrick-Davis et al. 1998, 2000; Kuoppamaki et al. 1995), particularly for improvement in negative symptoms (Duinkerke et al. 1993b). However, this was based on research with ritanserin and ketanserin, which are not specific for the 5-HT_{2C} receptor. Roth et al. (Roth et al. 1992), after examining the 5-HT_{2C} receptor affinities of a large number of typical and atypical antipsychotic drugs (see Table 14.1), concluded that 5-HT_{2C} receptor affinity alone or in relation to D_2 receptor affinity did not differentiate typical and atypical antipsychotic drugs.

Herrick-Davis et al. (Herrick-Davis et al. 2000) next suggested that the atypical antipsychotic drugs, but not the typicals, were inverse agonists at a naturally occurring constitutively active isoforms of 5-HT_{2C} receptor, However, this difference was not confirmed by Rauser et al. (Rauser et al. 2001), who found that many typical and atypical antipsychotic drugs were 5-HT_{2C} inverse agonists, although the affinity of the typical drugs was much weaker than that of most of the atypical antipsychotics. Nevertheless, other investigators have also suggested specific differences among antipsychotic drugs with regards to activity as inverse agonists at variously edited 5-HT$_{2C}$ receptors or that the ability to downregulate 5-HT$_{2C}$ receptors may be important to individual differences in antipsychotic drug action. For example, sertindole, which has nanomolar affinity for the 5-HT$_{2C}$ receptor (Table 14.1), after 21-day treatment, did not induce significant changes in the density of 5-HT$_{2C}$ receptors, whereas clozapine treatment equally decreased 5-HT$_{2C}$ receptor binding for agonist and antagonist sites by about one third. However, only sertindole induced a highly significant decrease in agonist binding (Hietala et al. 2001). These authors suggested that the latter effect might be relevant to effects of sertindole on anxiety, cognition, and weight gain. However, no convincing clinical or preclinical correlates of such differences between clozapine and sertindole in favor of sertindole have been reported. It is conceivable, however, that the downregulation of 5-HT$_{2C}$ receptors by clozapine could contribute to its advantages for treatment resistant patients and for suicide risk reduction.

Rueter et al. (2000) studied the effect of clozapine in wild-type and 5-HT$_{2C}$ receptor null mice littermates in an effort to clarify the interaction of 5-HT$_{2A}$ and 5-HT$_{2C}$ receptors and their importance to the action of clozapine. The predominantly 5-HT$_{2A}$ agonist DOI, which is hallucinogenic, mCPP, a predominantly 5-HT$_{2C}$ agonist with weak 5-HT$_{2A}$ properties, and 5-HT were applied microiontophoretically to the orbitofrontal cortex (OFC) and the head of the caudate nucleus. All three induced current-dependent inhibition of neuronal firing activity in both brain regions for both the wild-type and 5-HT$_{2C}$ null mice. Clozapine antagonized only mCPP in the wild-type mice. It was not effective to prevent the inhibitory effect of DOI. This suggests that clozapine may require the contribution of 5-HT$_{2C}$ receptor blockade to prevent the hallucinogenic properties of DOI. Dalton et al. (2004) used 5-HT$_{2C}$ null mice to show that in the absence of protective inhibitory effects of 5-HT$_{2C}$ receptor stimulation, enhanced response to 5-HT$_{2A}$ and 5-HT$_{1B}$ stimulation may occur. The implications of this for schizophrenia are unknown.

5-HT$_{2C}$ and 5-HT$_{2A}$ receptor signalings have been shown to depend upon the postsynaptic density protein (PSD-95), a scaffolding protein that has been shown to be necessary for the regulation of ionotropic glutamatergic signaling, to be a target for antipsychotic drugs, and to be necessary for some behavioral effects of clozapine (Abbas et al. 2009). Studies in PSD-95 knockout mice indicated that PSD-95 is crucial for normal 5-HT$_{2A}$ and 5-HT$_{2C}$ expression in vivo and downstream signaling by promoting apical dendritic targeting and stabilizing 5-HT$_{2C}$ and 5-HT$_{2A}$ receptor turnover in vivo. The apical dendrites are targets for clozapine action (Willins et al. 1999).

14.8   Effects of 5-HT$_{2C}$ Inverse Agonists in Animal Models of Psychosis

There are three commonly used models for antipsychotic activity: blockade of conditioned avoidance response, antagonism of locomotor activity produced by amphetamine or the *N*-methyl-D-asparate (NMDA)-receptor antagonists, phencyclidine

(PCP) or MK-801, and blockade of the disruption of prepulse inhibition by indirect or direct DA agonists or NMDA receptor antagonists. Atypical and typical antipsychotic drugs are active in all these models, with atypical antipsychotic drugs, apparently because of their 5-HT$_{2A}$ antagonist properties, preferentially effective in blocking NMDA-receptor blockade-mediated hyperlocomotion. Pretreatment of rats with the 5-HT$_{2C/2B}$ receptor antagonist SB 221284 (0.1–1 mg/kg, intraperitoneal (IP)]) or with the selective 5-HT$_{2C}$ receptor antagonist SB 242084 (1 mg/kg, IP), at doses shown to block mCPP-\induced hypolocomotion, significantly *enhanced* the hyperactivity induced by PCP or MK-801. Neither compound altered locomotor activity when administered alone. Both 5-HT$_{2C}$ antagonists enhanced the ability of PCP to increase the efflux of DA from the nucleus accumbens while having no effect on their own in that regard (Hutson et al. 2000). These results demonstrate that blockade of 5-HT$_{2C}$ receptors enhance the activation of mesolimbic DA neuronal function in hypoglutamatergic rodents, a state consistent with a deficit thought to be important to the pathophysiology of schizophrenia.

SB 243213 {5-methyl-1-[[-2-[(2-methyl-3-pyridyl)oxy]-5-pyridyl]carbamoyl]-6-trifluoromethylindoline hydrochloride} is a highly selective 5-HT$_{2C}$ receptor inverse agonist (Wood et al. 2001; Berg et al. 2006). In vivo studies indicated anxiolytic-like activity in both the social interaction and Geller–Seifter conflict tests, and it attenuated haloperidol-induced catalepsy. However, it did not affect amphetamine-, MK-801-, or PCP-induced hyperactivity, suggesting it would not be useful as an antipsychotic drug. However, it or other 5-HT$_{2C}$ inverse agonists might be of benefit in reducing anxiety or extrapyramidal symptoms (EPS) in some patients. There are no studies of the effect of 5-HT$_{2C}$ antagonists on prepulse inhibition in rodents. 5-HT$_{2C}$ null mice have some deficits in prepulse inhibition (Sun and Meltzer, in preparation). There have been, to our knowledge, no clinical trials of specific 5-HT$_{2C}$ inverse agonists as monotherapy for schizophrenia or as adjunctive agents for those antipsychotic drugs that lack affinity for 5-HT$_{2C}$ receptors. These agents might be useful to reduce anxiety or EPS, but it appears unlikely that they will improve the ability of antipsychotic agents to reduce psychotic symptoms. Indeed, the results reported by Hutson et al. (2000) suggest the possibility of exacerbating psychotic symptoms following selective blockade of 5-HT$_{2C}$ receptors.

14.9   5-HT$_{2C}$ Receptors and Dopamine Efflux in Relation to Schizophrenia and Antipsychotic Drug Action

An interesting difference in the effect of haloperidol and clozapine on DA efflux in the nucleus accumbens and striatum was reported by Navailles et al. (2006). Using microdialysis to measure DA efflux, these authors showed differential effects of these two prototypical antipsychotic drugs on DA release in relation to the constitutive effect of 5-HT$_{2C}$ receptors to inhibit DA release in these two regions. The ability of haloperidol 0.01 mg/kg to increase DA efflux in both regions was potentiated by the 5-HT$_{2C}$ inverse agonist SB 206553 (5 mg/kg IP) {5-methyl-1-(3-pyridylcarbamoyl)-1,2,3,5-tetrahydropyrrolo[2,3-f]indole} but not by the competitive antagonist SB 242084 or the inverse agonist SB 243213 (1 mg/kg IP). However, the effect of clozapine at a dose in which it shows 5-HT$_{2C}$ inverse agonist activity (1 mg/kg) was blocked by SB243213 and SB242084. Since haloperidol itself has no activity at the 5-HT$_{2C}$ receptor, the authors suggest that haloperidol promoted the constitutive activity of the 5-HT$_{2C}$ receptor. This is an intriguing suggestion, which, if valid, would suggest an important role of the 5-HT$_{2C}$ receptor in the action of haloperidol and the possible basis for interactions between haloperidol and clozapine. Clozapine, like SB206553, might inhibit the effect of haloperidol on the constitutive activity of the 5-HT$_{2C}$ receptor, thereby altering effects on striatal and accumbal DA efflux. Whether this holds true for the cortex as well remains to be studied.

14.10   WAY 163969: A 5-HT$_{2C}$ Agonist Treatment for Schizophrenia

The 5-HT$_{2C}$-selective receptor agonist, WAY-163909 [12][1,4]diazepino[12] indole],was reported to be active in various animal models of schizophrenia (Marquis et al. 2007). At doses of 1.7–30 mg/kg IP, it decreased apomorphine-induced climbing and behaved like an atypical in that it did not cause catalepsy. Similar to 5-HT$_{2A}$ antagonists, e.g., M100907, WAY 163909 (0.3–3 mg/kg subcutaneous [SC]) was reported to be more effective in inhibiting PCP-induced locomotor activity than d-amphetamine with no effect on spontaneous activity. WAY 163909 (1.7 to 17 mg/kg IP) reversed MK-801 {5H-dibenzo[a,d]cyclohepten-5,10-imine (dizocilpine maleate}- and DOI [1-(2,5-dimethoxy-4-iodophenyl)-2-aminopropane]-disrupted prepulse inhibition of startle (PPI) and improved PPI in DBA/2N mice. Like all known antipsychotic drugs, WAY 163909 reduced conditioned-avoidance responding, an effect blocked by the 5-HT$_{2B/2C}$ receptor antagonist SB 206553. At very high doses, WAY 163909 (10 mg/kg SC) selectively decreased extracellular levels of DA in the nucleus accumbens without affecting the striatum. It showed some selectivity in targeting DA neurons in the ventral tegmental area, sparing the substantia nigra with both acute and chronic (21-day) administration. WAY 163909 was recently reported to augment the ability of haloperidol or clozapine to block apomorphine-induced climbing behavior in mice without enhancing catalepsy (Grauer et al. 2009a). It also potentiated the ability of both antipsychotic drugs to block avoidance responses, suggesting that WAY 163909 might be useful as adjunctive treatment for schizophrenia, similar to what has been postulated for the 5-HT$_{2A}$ inverse agonist pimavanserin (Grauer et al. 2009b). These effects were attributed to the ability of 5-HT$_{2C}$ agonism to selectively suppress DA release in the nucleus accumbens and inhibit the firing of DA neurons in the ventral tegmentum, sparing the striatum and the substantia nigra. These unique results are very suggestive of efficacy in schizophrenia, but only clinical testing can verify the potential of what would truly be a new generation of antipsychotic agents.

14.11   Effect of Meta-Chlorophenylpiperazine, a 5-HT$_{2C}$ Agonist as a Challenge Agent in Schizophrenia

There are conflicting data on the effect of the acute administration of mCPP, a mixed 5-HT$_{2A/2C}$ agonist, to patients with schizophrenia, with one study reporting a decrease in psychotic symptoms (Kahn et al. 1992); three others, a mild exacerbation of positive symptoms (Iqbal et al. 1991; Krystal et al. 1993); and two, no significant effect (Maes and Meltzer 1996; Koreen et al. 1997). The reasons for these discrepancies are unclear. It is not related to route of administration, severity of symptoms, or gender. Maes and Meltzer (1996) also reported no differences in the mCPP-induced prolactin, cortisol, or temperature responses in patients with schizophrenia who were neuroleptic free at the time of study. Clozapine has been reported to antagonize the cortisol response to mCPP in patients with schizophrenia, consistent with its 5-HT$_{2C}$ antagonist properties (Kahn et al. 1994; Owen et al. 1993). Two studies reported no difference in the cortisol or adrenocorticotropic hormone (ACTH) response between patients with schizophrenia and controls but both found that the greater the cortisol or ACTH response to mCPP prior to subsequent treatment with clozapine, the better the response to clozapine (Owen et al. 1993; Kahn et al. 1993). A possible explanation of the relationship between the cortisol or ACTH response to mCPP and the subsequent response to clozapine is that subjects with relatively enhanced 5-HT$_{2C}$ receptor activation throughout the brain, compared with other patients with schizophrenia, might benefit from inhibition of such activity by clozapine. However, there is no evidence as yet to suggest that patients with schizophrenia as a group have enhanced 5-HT$_{2C}$ receptor activity relative to controls. The opposite may be the case, as suggested by the mRNA expression studies previous mentioned (Castensson et al. 2003, 2005). It should also be noted that the cortisol response to mCPP in rats was blocked by the selective 5-HT$_{2A}$ inverse agonist MDL 100907 but not by the selective 5-HT$_{2C}$ inverse agonist SB 242084 (Hemrick-Luecke and Evans 2002). It will be of interest to determine the endocrine, temperature, and behavioral responses to WAY 100639 in patients with schizophrenia and normal controls and to follow this up with a clinical trial with WAY 163909 or other antipsychotic drugs to determine if the cortisol or ACTH response in the unmedicated state is predictive of response. It would also be of interest to couple these to data from genotyping of the three common SNPs of the 5-HT$_{2C}$ receptor previously discussed.

14.12   5-HT$_{2C}$ Receptors: Cognitive Impairment in Schizophrenia

The role of 5-HT$_{2C}$ receptors in attention and learning will be discussed in detail in Chaps. 23 and 24. As discussed previously, the ability of 5-HT$_{2C}$ receptor stimulation to tonically inhibit DA release in the frontal cortex and hippocampus, one of a number of mechanisms that fine-tune DA release in these two brain regions crucial for cognition, has been linked to the cognitive deficits in schizophrenia that are the principal cause of enduring disability in schizophrenia (Green 2006). The dentate gyrus of the hippocampus has been identified as an important locus of deficits in working, long-term verbal, and spatial memory in schizophrenia (Takao and Miyakawa 2009). The dentate gyrus is also a key region for neurogenesis, a process that may be disturbed in schizophrenia, along with plasticity at the synaptic level (Reif et al. 2007). In 5-HT$_{2C}$ null mice, Tecott et al. (1998) found that long-term potentiation was impaired in the medial perforant pathway of the dentate gyrus. This was suggested to be the basis for impairment of performance by these mice in the Morris water maze, a measure of spatial learning and reduced aversion to a novel environment. No general deficit in learning or spatial discrimination was noted.

The PCP model of cognitive impairment is widely utilized to model some of the disturbances in cognition in schizophrenia. When given acutely or chronically, it causes a hypoglutamatergic deficit by blocking the NMDA receptor. 5-HT$_{2C}$ receptors were recently suggested to be important to reversal learning in female rats impaired in this task by subchronic PCP treatment (McLean et al. 2009). Reversal learning, a test of cognitive flexibility has been found to be impaired in patients with schizophrenia, possibly the result of abnormalities in frontal lobe functioning as well as other subcortical systems which are interconnected with the prefrontal cortex, e.g., the striatum, thalamus and dopaminergic systems. Adult female hooded Lister rats were trained to perform an operant reversal-learning task and then received subchronic PCP, which impaired reversal phase performance ($p<0.01$ to 0.001), with no effect in the initial phase. Acute treatment with the selective 5-HT$_{2C}$ receptor antagonist SB 243213A (1.0, 3.0, and 10.0 mg/kg IP) significantly attenuated the deficit in reversal learning but only at the dose of 10 mg/kg ($p<0.05$; Floresco et al. 2009). Thus, 5-HT$_{2C}$ receptor agonism may correct the hypoglutamatergic deficit postulated to be present in patients with schizophrenia (Bymaster et al. 2003; Richelson and Souder 2000).

14.13   Conclusions

The studies reviewed here provide some evidence for the conclusion that the 5-HT$_{2C}$ receptor plays a significant role in key elements of the pathophysiology of schizophrenia, including psychosis, negative symptoms, cognitive impairment, as well as the mechanism of action and side effects of antipsychotic drugs. Since the 5-HT$_{2A}$ receptor is so clearly involved in the action of atypical antipsychotic drugs and aspects of the pathophysiology of schizophrenia and since there is clear evidence for interactions between the 5-HT$_{2A}$ and 5-HT$_{2C}$ receptors, by default, the 5-HT$_{2C}$ receptor is important for schizophrenia. The strongest direct evidence may be the genetic studies that have shown that one or more of three common polymorphisms of the 5-HT$_{2C}$ receptor is associated with positive or negative symptoms, weight gain, and extrapyramidal side effects secondary to treatment on some antipsychotic drugs. In addition, the expression of the 5-HT$_{2C}$ receptor in postmortem specimens from patients with schizophrenia has been reported to be reduced, but this finding needs to be replicated.

The two reports that the cortisol or ACTH response to the 5-HT$_{2A/2C}$ receptor agonist mCPP in patients with schizophrenia predicts subsequent response to treatment with clozapine are inconclusive because this might be related to the 5-HT$_{2A}$ rather than the 5-HT$_{2C}$ component of the action of mCPP. Further studies with 5-HT$_{2C}$ null versus wild-type mice are needed to determine if diminished 5-HT$_{2C}$ receptor function in a mouse model is suggestive of a schizophrenia-like phenotype. Although 5-HT$_{2A}$ receptor stimulation is thought to be the basis for hallucinogenic action of LSD and related drugs, a role for 5-HT$_{2C}$ receptors with regard to hallucinations or delusions in man remains a viable possibility. 5-HT$_{2C}$ antagonism does not seem to be a necessary feature of atypical antipsychotic drug action, but it could have a secondary influence.

Finally, WAY 163909, a 5-HT$_{2C}$ receptor agonist, has shown activity as a cognitive enhancing antipsychotic agent in a variety of animal models. It will be of interest to determine if 5-HT$_{2C}$ agonism has a role as an adjunctive agent in the treatment of schizophrenia.

In summary, there is reason to intensively investigate the role of the 5-HT$_{2C}$ receptor in the pathophysiology of schizophrenia and the action of antipsychotic drugs and to further test 5-HT$_{2C}$ agonists for their ability to improve cognition and psychosis in patients with schizophrenia and related disorders.

Acknowledgments Supported, in part, by a grant from the Weisman Family Foundation and a Distinguished Investigator Award from NARSAD. Ki determinations were generously provided by the National Institute of Mental Health's Psychoactive Drug Screening Program, Contract # HHSN-271-2008-00025-C (NIMH PDSP). The NIMH PDSP is Directed by Bryan L. Roth, M.D., Ph.D. at the University of North Carolina at Chapel Hill and Project Officer Jamie Driscol at NIMH, Bethesda, MD.

References

Abbas AI, Yadav PN, Yao WD, et al (2009) PSD-95 is essential for hallucinogen and atypical antipsychotic drug actions at serotonin receptors. J Neurosci 29:7124–7136.

Alex KD, Yavanian GJ, McFarlane HG, et al (2005) Modulation of dopamine release by striatal 5-HT2C receptors. Synapse 55:242–251.

Angrist B, Gershon S (1974) Dopamine and psychotic states: preliminary remarks. Adv Biochem Psychopharmacol 12:211–219.

Berg KA, Navailles S, Sanchez TA, et al (2006) Differential effects of 5-methyl-1-[[2-[(2-methyl-3-pyridyl)oxy]-5-pyridyl]carbamoyl]-6-trifluoro methylindone (SB 243213) on 5-hydroxytryptamine(2C) receptor-mediated responses. J Pharmacol Exp Ther 319:260–268.

Burns CM, Chu H, Rueter SM, et al (1997) Regulation of serotonin-2C receptor G-protein coupling by RNA editing. Nature 387:303–308.

Bymaster FP, Felder CC, Tzavara E, et al (2003) Muscarinic mechanisms of antipsychotic atypicality. Prog Neuropsychopharmacol Biol Psychiatry 27:1125–1143.

Canton H, Verriele L, Colpaert FC (1990) Binding of typical and atypical antipsychotics to 5-HT1C and 5-HT2 sites: clozapine potently interacts with 5-HT1C sites. Eur J Pharmacol 191:93–96.

Castensson A, Emilsson L, Sundberg R, et al (2003) Decrease of serotonin receptor 2C in schizophrenia brains identified by high-resolution mRNA expression analysis. Biol Psychiatry 54:1212–1221.

Castensson A, Aberg K, McCarthy S, et al (2005) Serotonin receptor 2C (HTR2C) and schizophrenia: examination of possible medication and genetic influences on expression levels. A J Med Genet B Neuropsychiatr Genet 134B:84–89.

Cussac D, Newman-Tancredi A, Nicolas JP, et al (2000) Antagonist properties of the novel antipsychotic, S16924, at cloned, human serotonin 5-HT2C receptors: a parallel phosphatidylinositol and calcium accumulation comparison with clozapine and haloperidol. Naunyn Schmiedebergs Arch Pharmacol 361:549–554.

Dalton GL, Lee MD, Kennett GA, et al (2004) mCPP-induced hyperactivity in 5-HT2C receptor mutant mice is mediated by activation of multiple 5-HT receptor subtypes. Neuropharmacology 46:663–671.

Di Matteo, V, Di Giovanni G, Di Mascio M, et al (1999): SB 242084, a selective serotonin2C receptor antagonist, increases dopaminergic transmission in the mesolimbic system. Neuropharmacology 38:1195–1205.

Dracheva S, Elhakem SL, Marcus SM, et al (2003) RNA editing and alternative splicing of human serotonin 2C receptor in schizophrenia. J Neurochem 87:1402–1412.

Drago A, Serretti A (2009) Focus on HTR2C A possible suggestion for genetic studies of complex disorders. Am J Med Genet B Neuropsychiatr Genet 150B:601–637.

Duinkerke SJ, Botter PA, Jansen AA, et al (1993) Ritanserin, a selective 5-HT2/1C antagonist, and negative symptoms in schizophrenia. A placebo-controlled double-blind trial. Br J Psychiatry 163:451–455.

Ellingrod VL, Perry PJ, Ringold JC, et al (2005) Weight gain associated with the -759C/T polymorphism of the 5HT2C receptor and olanzapine. Am J Med Genet B Neuropsychiatr Genet 134B:76–78.

Floresco SB, Zhang Y, Enomoto T(2009) Neural circuits subserving behavioral flexibility and their relevance to schizophrenia. Behav Brain Res 204:396–409.

Grauer SM, Graf R, Navarra R, et al (2009) WAY-163909, a 5-HT2C agonist, enhances the preclinical potency of current antipsychotics. Psychopharmacology (Berlin) 204:37–48.

Green MF (2006) Cognitive impairment and functional outcome in schizophrenia and bipolar disorder. J Clin Psychiatry 67:e12.

Gunes A, Scordo MG, Jaanson P, et al (2007) Serotonin and dopamine receptor gene polymorphisms and the risk of extrapyramidal side effects in perphenazine-treated schizophrenic patients. Psychopharmacology (Berlin) 190:479–484.

Gutierrez B, Fananas L, Arranz MJ, et al (1996) Allelic association analysis of the 5-HT2C receptor gene in bipolar affective disorder. Neurosci Lett 212:65–67.

Hemrick-Luecke SK, Evans DC (2002) Comparison of the potency of MDL 100,907 and SB 242084 in blocking the serotonin (5-HT)(2) receptor agonist-induced increases in rat serum corticosterone concentrations: evidence for 5-HT(2A) receptor mediation of the HPA axis. Neuropharmacology 42:162–169.

Herrick-Davis K, Grinde E, Gauthier C, et al (1998) Pharmacological characterization of the constitutively activated state of the serotonin 5-HT2C receptor. Ann N Y Acad Sci 861:140–145.

Herrick-Davis K, Grinde E, Teitler M (2000) Inverse agonist activity of atypical antipsychotic drugs at human 5-hydroxytryptamine2C receptors. J Pharmacol Exp Ther 295:226–232.

Hietala J, Kuonnamaki M, Palvimaki EP, et al (2001) Sertindole is a serotonin 5-HT2c inverse agonist and decreases agonist but not antagonist binding to 5-HT2c receptors after chronic treatment. Psychopharmacology (Berlin) 157:180–187.

Holmes C, Arranz MJ, Powell JF, et al (1998) 5-HT2A and 5-HT2C receptor polymorphisms and psychopathology in late onset Alzheimer's disease. Hum Mol Genet 7:1507–1509.

Hu X, Giotakis O, Li T, et al (2003) Association of the 5-HT2c gene with susceptibility and minimum body mass index in anorexia nervosa. Neuroreport 14:781–783.

Hutson PH, Barton CL, Jay M, et al (2000) Activation of mesolimbic dopamine function by phencyclidine is enhanced by 5-HT(2C/2B) receptor antagonists: neurochemical and behavioral studies. Neuropharmacology 39:2318–2328.

Iqbal N, Asnis GM, Wetzler S, et al (1991) The MCPP challenge test in schizophrenia: hormonal and behavioral responses. Biol Psychiatry 30:770–778.

Iwamoto K, Kato T (2003) RNA editing of serotonin 2C receptor in human postmortem brains of major mental disorders. Neurosci Lett 346:169–172.

Kahn RS, Siever LJ, Gabriel S, et al (1992) Serotonin function in schizophrenia: effects of metachlorophenylpiperazine in schizophrenic patients and healthy subjects. Psychiatry Res 43:1–12.

Kahn RS, Davidson M, Siever L, et al (1993) Serotonin function and treatment response to clozapine in schizophrenic patients. Am J Psychiatry 150:1337–1342.

Kahn RS, Davidson M, Siever LJ, et al (1994) Clozapine treatment and its effect on neuroendocrine responses induced by the serotonin agonist, m-chlorophenylpiperazine. Biol Psychiatry 35:909–912.

Koreen AR, Lieberman JA, Alvir J, et al (1997) The behavioral effect of m-chlorophenylpiperazine (mCPP) and methylphenidate in first-episode schizophrenia and normal controls. Neuropsychopharmacology 16:61–68.

Kroeze WK, Hufeisen SJ, Popadak BA, et al (2003) H1-histamine receptor affinity predicts shortterm weight gain for typical and atypical antipsychotic drugs. Neuropsychopharmacology 28:519–526.

Krystal JH, Seibyl JP, Price LH, et al (1993) m-Chlorophenylpiperazine effects in neuroleptic-free schizophrenic patients. Evidence implicating serotonergic systems in the positive symptoms of schizophrenia. Arch Gen Psychiatry 50:624–635.

Kuoppamaki M, Palvimaki EP, Hietala J, et al (1995) Differential regulation of rat 5-HT2A and 5-HT2C receptors after chronic treatment with clozapine, chlorpromazine and three putative atypical antipsychotic drugs. Neuropsychopharmacology 13:139–150.

Lerer B, Macciardi F, Segman RH, et al (2001) Variability of 5-HT2C receptor cys23ser polymorphism among European populations and vulnerability to affective disorder. Mol Psychiatry 6:579–585.

Maes M, Meltzer HY (1996) Effects of meta-chlorophenylpiperazine on neuroendocrine and behavioral responses in male schizophrenic patients and normal volunteers. Psychiatry Res 64:147–159.

Marquis KL, Sabb AL, Logue SF, et al (2007) WAY-163909 [(7bR,10aR)-1,2,3,4,8,9,10,10a-octahydro-7bH-cyclopenta-[b][1, 4]diazepino[6,7,1hi]indole]: a novel 5-hydroxytryptamine 2C receptor-selective agonist with preclinical antipsychotic-like activity. J Pharmacol Exp Ther 320:486–496.

McLean SL, Woolley ML, Thomas D, et al (2009) Role of 5-HT receptor mechanisms in sub-chronic PCP-induced reversal learning deficits in the rat. Psychopharmacology (Berlin) 206:403–414.

Meltzer H (1989) Serotonergic dysfunction in depression. Br J Psychiatry (suppl):25–31.

Meltzer HY (1992) Dimensions of outcome with clozapine. Br J Psychiatry (suppl):46–53.

Meltzer HY (1995) Clozapine: is another view valid? Am J Psychiatry 152:821–825.

Meltzer HY, Huang M (2008) In vivo actions of atypical antipsychotic drug on serotonergic and dopaminergic systems. Prog Brain Res 172:177–197.

Meltzer HY, Stahl SM (1976) The dopamine hypothesis of schizophrenia: a review. Schizophr Bull 2:19–76.

Meltzer HY, Li Z, Kaneda Y, et al (2003) Serotonin receptors: their key role in drugs to treat schizophrenia. Prog Neuropsychopharmacol Biol Psychiatry 27:1159–1172.

Miller DD, Ellingrod VL, Holman TL, et al (2005) Clozapine-induced weight gain associated with the 5HT2C receptor -759C/T polymorphism. Am J Med Genet B Neuropsychiatr Genet 133B:97–100.

Morabito MV, Emeson RB (2009) RNA editing as a therapeutic target for CNS disorders. Neuropsychopharmacology 34:246.

Murad I, Kremer I, Dobrusin M, et al (2001) A family-based study of the Cys23Ser 5HT2C serotonin receptor polymorphism in schizophrenia. Am J Med Genet 105:236–238.

Navailles S, De DP, Spampinato U (2006) Clozapine and haloperidol differentially alter the constitutive activity of central serotonin2C receptors in vivo. Biol Psychiatry 59:568–575.

Niswender CM, Herrick-Davis K, Dilley GE, et al (2001) RNA editing of the human serotonin 5-HT2C receptor. alterations in suicide and implications for serotonergic pharmacotherapy. Neuropsychopharmacology 24:478–491.

Owen RR, Jr., Gutierrez-Esteinou R, Hsiao J, et al (1993) Effects of clozapine and fluphenazine treatment on responses to m-chlorophenylpiperazine infusions in schizophrenia. Arch Gen Psychiatry 50:636–644.

Pozzi L, Acconcia S, Ceglia I, et al (2002) Stimulation of 5-hydroxytryptamine (5-HT(2C)) receptors in the ventrotegmental area inhibits stress-induced but not basal dopamine release in the rat prefrontal cortex. J Neurochem 82:93–100.

Rauser L, Savage JE, Meltzer HY, et al (2001) Inverse agonist actions of typical and atypical antipsychotic drugs at the human 5-hydroxytryptamine(2C) receptor. J Pharmacol Exp Ther 299:83–89.

Reif A, Schmitt A, Fritzen S, et al (2007) Neurogenesis and schizophrenia: dividing neurons in a divided mind? Eur Arch Psychiatry Clin Neurosci 257:290–299.

Reynolds GP, Zhang ZJ, Zhang XB (2002) Association of antipsychotic drug-induced weight gain with a 5-HT2C receptor gene polymorphism. Lancet 359:2086–2087.

Ribases M, Fernandez-Aranda F, Gratacos M, et al (2008) Contribution of the serotoninergic system to anxious and depressive traits that may be partially responsible for the phenotypical variability of bulimia nervosa. J Psychiatr Res 42:50–57.

Richelson E, Souder T (2000) Binding of antipsychotic drugs to human brain receptors focus on newer generation compounds. Life Sci 68:29–39.

Roth BL, Ciaranello RD, Meltzer HY (1992) Binding of typical and atypical antipsychotic agents to transiently expressed 5-HT1C receptors. J Pharmacol Exp Ther 260:1361–1365.

Rueter LE, Tecott LH, Blier P (2000): In vivo electrophysiological examination of 5-HT2 responses in 5-HT2C receptor mutant mice. Naunyn Schmiedebergs Arch Pharmacol 361:484–491.

Ryu S, Cho EY, Park Ty, et al (2007) -759 C/T polymorphism of 5-HT2C receptor gene and early phase weight gain associated with antipsychotic drug treatment. Prog Neuropsychopharmacol Biol Psychiatry 31:673–677.

Segman RH, Heresco-Levy U, Finkel B, et al (2000) Association between the serotonin 2C receptor gene and tardive dyskinesia in chronic schizophrenia: additive contribution of 5-HT2Cser and DRD3gly alleles to susceptibility. Psychopharmacology (Berlin) 152:408–413.

Sodhi MS, Arranz MJ, Curtis D, et al (1995) Association between clozapine response and allelic variation in the 5-HT2C receptor gene. Neuroreport 7:169–172.

Sodhi MS, Burnet PW, Makoff AJ, et al (2001) RNA editing of the 5-HT(2C) receptor is reduced in schizophrenia. Mol Psychiatry 6:373–379.

Sodhi MS, Airey DC, Lambert W, et al (2005) A rapid new assay to detect RNA editing reveals antipsychotic-induced changes in serotonin-2C transcripts. Mol Pharmacol 68:711–719.

Takao K, Miyakawa T (2009) Intrauterine environment-genome interaction and children's development (4): Brain-behavior phenotyping of genetically-engineered mice using a comprehensive behavioral test battery on research of neuropsychiatric disorders. J Toxicol Sci 34 (suppl 2):SP293–SP305

Tecott LH, Logue SF, Wehner JM, et al (1998) Perturbed dentate gyrus function in serotonin 5-HT2C receptor mutant mice. Proc Natl Acad Sci USA 95:15026–15031.

Willins DL, Meltzer HY (1998) Serotonin 5-HT2C agonists selectively inhibit morphine-induced dopamine efflux in the nucleus accumbens. Brain Res 781:291–299.

Willins DL, Berry SA, Alsayegh L, et al (1999) Clozapine and other 5-hydroxytryptamine-2A receptor antagonists alter the subcellular distribution of 5-hydroxytryptamine-2A receptors in vitro and in vivo. Neuroscience 91:599–606.

Wood MD, Reavill C, Trail B, et al (2001) SB-243213; a selective 5-HT2C receptor inverse agonist with improved anxiolytic profile: lack of tolerance and withdrawal anxiety. Neuropharmacology 41:186–199.

Chapter 15
Serotonin and Reward-Related Behavior: Focus on 5-HT$_{2C}$ Receptors

Paul J. Fletcher and Guy A. Higgins

15.1 Introduction

Abnormal functioning of brain 5-hydroxytryptamine (5-HT; serotonin) has been linked to a variety of psychiatric disorders including mood disorders, schizophrenia, anxiety, addiction, and impulse control disorders (Dalley et al. 2008; Oades 2008; Deakin 1998; Remington 2008; Bubar and Cunningham 2006; Higgins and Fletcher 2003a). The 5-HT system is also an important target for different classes of psychopharmacological agents including antidepressants, anxiolytics, and antipsychotics (Jones and Blackburn 2002). The pervasiveness of serotonergic involvement in a variety of disorders is perhaps not surprising given the widespread innervation of the central nervous system (CNS) by 5-HT neurons, the involvement of 5-HT in multiple physiological and behavioral functions, and the fact that 5-HT interacts with and modulates the activity of many other neurotransmitter systems. One such interaction is with brain dopamine (DA) systems, and there is a considerable literature on this interaction (Di Matteo et al. 2008; Di Giovanni et al. 2008; Alex and Pehek 2007). Dopamine systems play a major role in mediating reward-related processes and particularly the behavioral effects of many drugs of abuse. Not surprisingly then a number of studies have investigated the effects of manipulating 5-HT function on DA-dependent behaviors and on the effects of drugs of abuse.

Several studies have shown that elevations in brain 5-HT levels resulting from treatments such as fluoxetine or L-tryptophan reduce amphetamine and cocaine self-administration (Carroll et al. 1990; Fletcher et al. 1999; Howell and Byrd 1995; Peltier and Schenk 1993; Porrino et al. 1989; Richardson and Roberts 1991).

P.J. Fletcher (✉)
Section of Biopsychology, Centre for Addiction and Mental Health, 250 College Street, Toronto, ON, Canada, M5T 1R8
and
Departments of Psychiatry and Psychology, University of Toronto, Toronto, ON, Canada
e-mail: Paul_Fletcher@camh.net

G.A. Higgins
CanCog Technologies, 120 Carlton Street, Toronto, Ontario, M5A 4K2, Canada

G. Di Giovanni et al. (eds.), *5-HT$_{2C}$ Receptors in the Pathophysiology of CNS Disease*,
The Receptors 22, DOI 10.1007/978-1-60761-941-3_15,
© Springer Science+Business Media, LLC 2011

In contrast, widespread depletion of 5-HT enhances responding for amphetamine and cocaine in some (Lyness et al. 1980; Loh and Roberts 1990) though not all studies (Fletcher et al. 1999). These types of manipulations also alter behaviors such as brain stimulation reward, conditioned place preference, aspects of feeding behavior, and impulsive behavior.

In the last decade, our understanding of 5-HT systems in terms of its receptor diversity, pharmacology, distribution, and function has progressed considerably. To date, 14 distinct 5-HT receptors have been cloned (Hoyer et al. 2002; Barnes and Sharp 1999). During this period a number of studies have appeared investigating the roles of these different 5-HT receptor subtypes on reward-related behaviors. The 5-HT$_{2C}$ receptor has attracted considerable interest in this regard, and as shown by the review presented here; activation of this target appears to exert an inhibitory influence over reward-related behaviors, especially those induced by drugs of abuse.

15.2   Distribution, Electrophysiology, and Neurochemistry of 5-HT$_{2C}$ Receptors

The 5-HT$_{2C}$ receptor subtype was first identified using radioligand binding of [^{3}H]5-HT and [^{3}H]mesulergine to pig choroid plexus tissue (Pazos et al. 1984). This novel binding site was originally termed the 5-HT$_{1C}$ receptor by virtue of its high affinity for [^{3}H]5-HT, although it soon became apparent that this receptor belonged to the 5-HT$_2$ subclass based on sequence homology, pharmacology and signal transduction mechanisms (Hoyer 1988; Hoyer et al. 1994).

Initial mapping studies of the 5-HT$_{2C}$ receptor utilized an autoradiography technique using [^{3}H]5-HT and [^{3}H]mesulergine, and relied on competition studies to characterize the site as 5-HT$_{2C}$ (or 5-HT$_{1C}$ at the time) (Hoyer et al. 1985). Limitations of this technique prevented a detailed understanding of receptor distribution in terms of cellular location, although a relatively broad distribution of 5-HT$_{2C}$ receptors within regions of cortex (pyriform, cingulate, prefrontal), limbic areas (nucleus accumbens, amygdala, hippocampus), and midbrain (ventral tegmental area [VTA], substantia nigra [SNR]) was identified (Hoyer et al. 1985; Mengod et al. 1990; Pazos and Palacios 1985). The reasonable overlap between 5-HT$_{2C}$ receptor messenger ribonucleic acid (mRNA) expression and receptor binding sites supports a predominantly postsynaptic rather than presynaptic localization (Mengod et al. 1990). Compared with the choroid plexus, the density of the 5-HT$_{2C}$ binding site is at least an order of magnitude lower in these forebrain regions in both human and rat (Marazziti et al. 1999). More recently, using a validated antibody raised to the 5-HT$_{2C}$ receptor (Bubar et al. 2005), Bubar and Cunningham (2007) and Liu et al. (2007) have begun to produce a more detailed map of 5-HT$_{2C}$ receptor distribution within specific regions of the rat brain, notably the VTA and the medial prefrontal cortex (mPFC). In the VTA, double labeling immunofluorescence demonstrates the existence of 5-HT$_{2C}$ receptors on γ-aminobutyric acid (GABA) interneurons as well as on DA cell bodies, in approximately equal proportions

(Bubar and Cunningham 2007). Although the significance of this differential expression of 5-HT$_{2C}$ receptors across cell types is not yet known, it may have functional implications for how 5-HT$_{2C}$ receptors control the activity of VTA neurons.

The majority of electrophysiological studies designed to study the functions of 5-HT$_{2C}$ receptors have focused on how this receptor modifies the activity of DA neurons. Stimulation of 5-HT afferents originating from the midbrain dorsal and median raphe reduces the firing of DA cell bodies of the VTA and SNR (Dray et al. 1978; Prisco et al. 1994) suggesting an inhibitory influence of 5-HT on dopaminergic neurons. An extensive series of experiments conducted by Esposito and colleagues, have characterized this inhibitory effect as 5-HT$_{2C}$ receptor mediated (Di Matteo et al. 2004). For example, the moderately selective 5-HT$_{2C}$ receptor agonists Ro 60-0175 and MK212 reduced basal firing rate of VTA DA neurons (Di Matteo et al. 2000; Di Giovanni et al. 2000). A downstream consequence of these electrophysiological changes is reduced DA efflux in terminal regions such as nucleus accumbens (Di Matteo et al. 2000; Di Giovanni et al. 2000) and frontal cortex (Gobert et al. 2000). Thus, activation of 5-HT$_{2C}$ receptors reduces the functional activity of mesocorticolimbic DA neurons. The raphe inputs to the VTA and SNR appear to exert a degree of tonic inhibitory control, since the selective 5-HT$_{2C}$ antagonist SB 242084 increases VTA and SNR cell firing rates, with concomitant increases in DA release in terminal zones (Gobert et al. 2000; Di Giovanni et al. 1999; Di Matteo et al. 1998, 1999).

5-HT$_{2C}$ receptors couple preferentially to G$_{q/11}$ to increase inositol phosphates, cytosolic [Ca^{2+}], and increase cell excitability (Millan et al. 2008). Thus, the inhibition of dopaminergic cells is presumably through excitation of the inhibitory GABA interneurons that synapse onto the DA cell bodies (Di Giovanni et al. 2001; Invernizzi et al. 2007). The consequence of activation of the 5-HT$_{2C}$ receptors localized directly on the DA VTA neurons is presently unclear. Immunohistochemical studies show a rostrocaudal gradient of 5-HT$_{2C}$ receptor expression within the VTA, and also a differential expression across VTA subnuclei (Bubar and Cunningham 2007). This may suggest a complex regulatory mechanism through which 5-HT$_{2C}$ receptors modulate VTA cell firing.

Within the prefrontal cortex, a significant proportion of 5-HT$_{2C}$ receptors are localized to GABA interneurons within layers V/VI (Liu et al. 2007). Thus 5-HT$_{2C}$ receptor activation would be predicted to increase GABA-ergic activity, and consequently inhibit the output of the glutamatergic pyramidal cells, which send excitatory afferents to components of the mesolimbic pathway, including both the VTA and nucleus accumbens (Sesack et al. 1989, 2003).

In addition to the emerging complexity posed by differential expression of 5-HT$_{2C}$ receptors across distinct cell types, this receptor is also subject to RNA editing (Burns et al. 1997), which may produce up to 24 distinct receptor isoforms (Fitzgerald et al. 1999; Niswender et al. 2001). The edited region is restricted to the second intracellular loop important for receptor coupling to G proteins Indeed, the functional consequence of these various edited isoforms is a tenfold difference in sensitivity (EC$_{50}$) to 5-HT (Burns et al. 1997). The fact that isoform expression may be modified by disease state and that chronic treatment with a 5-HT$_2$ agonist can shift the 5-HT$_{2C}$ receptor to isoforms that activate G protein less efficiently suggest that this may be an important means by which 5-HT$_{2C}$ function may be regulated (Gurevich et al. 2002). Furthermore, the fact that this receptor has evolved to be the only known G-protein-coupled receptor (GPCR) that undergoes RNA editing argues further for its functional importance. The first report of mice expressing two distinct 5-HT$_{2C}$ isoforms, 5-HT$_{2CINI}$ (unedited) and 5-HT$_{2CVGV}$ (fully edited), highlights a striking phenotypic difference with the VGV mice being hyperphagic and having reduced fat mass compared with INI counterparts (Kawahara et al. 2008).

15.3   Effects of 5-HT$_{2C}$ Receptor Ligands on Responses to Drugs of Abuse

Most of the work on the effects of 5-HT$_{2C}$ receptor ligands on the stimulant, discriminative stimulus properties, and primary rewarding effects of drugs of abuse has been done using cocaine, nicotine, and to a lesser extent, ethanol.

15.3.1   Cocaine and Other Psychomotor Stimulants

The pharmacological effects of cocaine include blockade of the transporters for 5-HT, norepinephrine, and DA. The action of cocaine to elevate extracellular levels of DA appears to be the primary neurochemical event mediating a number of the behavioral effects of cocaine. Locomotor activity induced by cocaine is attenuated by 6-hydroxydopamine (6-OHDA) lesions of the mesolimbic DA system and treatment with DA receptor antagonists (Kelly and Iversen 1976; Spyraki et al. 1982). Disrupting DA function also blocks the discriminative stimulus properties of cocaine in conventional operant conditioning-based drug-discrimination tasks (Callahan et al. 1991). The effectiveness of cocaine to serve as a positive reinforcer in tests of intravenous cocaine self-administration is also diminished by 6-OHDA lesions of the mesolimbic DA system (Roberts et al. 1980), by DA receptor antagonists (Caine and Koob 1994), and by deletion of the DA transporter gene (Thomsen et al. 2009). The modulation of the functioning of mesolimbic DA function by 5-HT$_{2C}$ receptors provided the impetus for examining the effects of 5-HT$_{2C}$ receptor ligands on behavioral responses to cocaine.

15.3.1.1   Locomotor Activity

A consistent finding regarding the effects of 5-HT$_{2C}$ receptor ligands on the expression of cocaine-induced behavior is that 5-HT$_{2C}$ receptor agonists reduce cocaine-induced locomotor activity. This has been demonstrated for MK212 and Ro60-0175 in several studies (Grottick et al. 2000; Filip et al. 2004). In contrast, cocaine-induced locomotion was increased by the 5-HT$_{2B/2C}$ receptor antagonist SB 206553 (McCreary and Cunningham 1999) and by the more selective 5-HT$_{2C}$ receptor antagonist SB 242084 (Fletcher et al. 2002, 2006). The locomotor stimulant effect of cocaine was also enhanced in 5-HT$_{2C}$ receptor knockout mice, compared with their wild-type controls (Rocha et al. 2002). Thus, the results of experiments using pharmacological and gene-deletion approaches converge to show that impaired transmission through 5-HT$_{2C}$ receptors can enhance the locomotor stimulant effect of cocaine. Collectively, these data show that the unconditioned locomotor stimulant effect of cocaine is altered in a bidirectional fashion by 5-HT$_{2C}$ receptor agonists and antagonists.

Limited evidence suggests that the effects of other psychomotor stimulants are modified by 5-HT$_{2C}$ receptor ligands. The 5-HT$_{2B/2C}$ receptor antagonist SB206553 modestly potentiated the locomotor stimulant effect of the DA releaser amphetamine, the combined DA and 5-HT releaser 3,4-methylenedioxy-N-methylamphetamine (MDMA) (Bankson and Cunningham 2002), and a cocktail of the DA reuptake blocker mazindol plus the 5-HT reuptake blocker fluvoxamine (McMahon and Cunningham 2001). The more selective 5-HT$_{2C}$ receptor antagonist SB 242084 enhanced the locomotor stimulant effects of MDMA, amphetamine, the DA reuptake blocker methylphenidate (Fletcher et al. 2006) and the N-methyl-D-aspartate (NMDA) antagonists phencyclidine and dizocilpine (Hutson et al. 2000). Thus, enhancement of drug-induced locomotor activity by 5-HT$_{2C}$ receptor blockade is seen across a number of psychomotor stimulants with differing pharmacological actions.

Pairing a specific environment with certain drugs can elicit conditioned responses when animals are subsequently exposed to that environment in a drug-free state. In the case of psychomotor stimulants such as cocaine, this can be seen as a conditioned hyperactivity response. As with the unconditioned locomotor stimulant effects of cocaine, the expression of cocaine-induced conditioned locomotor activity was attenuated by MK212 and enhanced by SB 242084 (Liu and Cunningham 2006). A notable feature of the responses of these two 5-HT$_{2C}$ ligands is that they were obtained in the absence of cocaine, indicating that their effects on activity are not due to pharmacokinetic interactions with cocaine.

15.3.1.2 Drug Discrimination

Drug discrimination assays in which rats learn to produce one response under the influence of a drug, and a different response under the influence of a vehicle injection, measure the interoceptive or stimulus properties of that drug. This assay has been used extensively to model the subjective effects of drugs. Rats readily learn to discriminate cocaine from vehicle, and this discrimination is DA-dependent since it is blocked by DA receptor antagonists, and generalises to other drugs that elevate DA activity (Callahan et al. 1991). The mixed 5-HT$_{1B/2C}$ agonist *meta*-chlorophenylpiperazine (mCPP) and the more selective 5-HT$_{2C}$ receptor agonist MK 212 partially antagonized the discriminative stimulus properties of cocaine (Callahan and Cunningham 1995). mCPP also produced a rightward shift in the cocaine dose-response curve. Given that 5-HT_{1B} agonists induce a leftward shift in the cocaine dose response curve, it seems likely that the effects of mCPP results from 5-HT_{2C} rather than 5-HT_{1B}, receptor stimulation (Callahan and Cunningham 1995). These data show that activation of 5-HT_{2C} receptors diminishes the discriminative state induced by cocaine.

15.3.1.3   Cocaine Self-administration

Intravenous drug self-administration procedures are the most widely used method to study the direct reinforcing effects of drugs of abuse. Our group was the first to examine the impact of selective 5-HT_{2C} receptor ligands on cocaine self-administration. In our first study we trained rats to self-administer cocaine under a fixed ratio of 5, 60-s timeout schedule (FR5TO60s) of reinforcement (Grottick et al. 2000). The effect of Ro60-0175 on cocaine self-administration was measured and compared with the effect seen in a different group of rats responding for food under the identical FR5TO60s schedule. The results demonstrated a dose-related reduction in the rate of cocaine self-administration over a similar dose range to that which reduced food maintained responding, even though response rates varied markedly between the two reinforcers (Grottick et al. 2000). This effect to reduce the rate of cocaine self-administration was subsequently replicated in a later study, which also showed that the effects of Ro60-0175 were stable over eight daily injections (Fletcher et al. 2008a).

In a progressive ratio schedule of reinforcement the number of responses required to earn successive infusions of the drug escalate, usually according to an exponential function. The number of ratios completed (or infusions earned) increases in a linear fashion with increasing dose of the self-administered drug. The ratio at which responding ceases is termed the *breaking point* and provides a good measure of the direct reinforcing effects of the drug. Consistent with results from the FR5 schedule, Ro60-0175 also reduced breaking points for cocaine (Grottick et al. 2000; Fletcher et al. 2008a) (see Fig. 15.1a).

While these observations might suggest a generalized suppression of motivated behavior, control experiments demonstrated little or no motor or neurological impairments, at least over the dose range of Ro60-0175 (Grottick et al. 2000, 2001; Fletcher et al. 2008a) that reduced responding for cocaine. In the cocaine self-administration studies, initial rates of responding and mean interinfusion intervals were similar under drug and control conditions, suggesting an earlier termination of responding following Ro60-0175 treatment. Finally, the number of responses emitted on the progressive ratio schedule by rats treated with the highest dose of Ro60-0175 were considerably higher than response rates emitted under vehicle treatment on the FR5TO60s schedule (Grottick et al. 2000). Consequently, these data suggest that 5-HT_{2C} receptor stimulation has a generalized effect to reduce motivation for cocaine and food that is not the product of some indiscriminate neurological or motor disturbance.

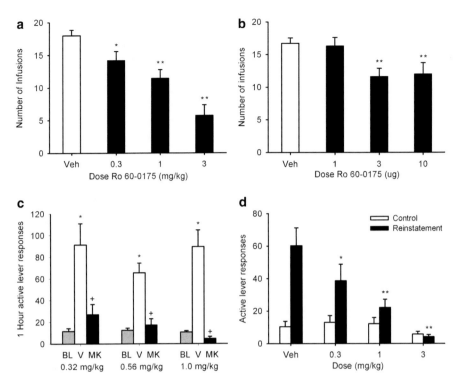

Fig. 15.1 Effects of 5-HT$_{2C}$ receptor agonists on several behavioral effects of cocaine. (**a**) Systemic injection of Ro60-0175 subcutaneous (*SC*) reduced the number of cocaine infusions earned on a progressive ratio schedule. * and ** indicate significant differences from Veh (Redrawn from Grottick et al. 2000). (**b**) Microinjection of Ro60-0175 into the ventral tegmental area reduced the number of cocaine infusions earned on a progressive ratio schedule. ** indicates significant difference from Veh (Redrawn from Fletcher et al. 2004). (**c**) MK212 reduced the ability of a priming injection of cocaine (10 mg/kg) to reinstate bar pressing in rats that had undergone extinction of cocaine self-administration. At each dose of MK 212 response rates are shown for a baseline test, without cocaine (BL), a test with vehicle and cocaine (V), and a test with MK212 and cocaine (MK). Graph element * indicates significantly different from BL; +, significantly different from V (Redrawn from (**b**) in Neisewander and Acosta 2007) (**d**) Ro60-0175 reduced the reinstatement of bar pressing induced by reexposure to the context in which drug self-administration occurred (Reinstatement). In a control group, responding remained low in a context that differed from the self-administration context and was not altered by Ro60-0175 (Responses on an inactive lever were uniformly low and not altered by Ro60-0175; data not shown) (Data are redrawn from Fletcher et al. 2008a)

In contrast to the effects of Ro60-0175, SB 242084 enhanced responding for cocaine on a PR schedule of reinforcement (Fletcher et al. 2002). This effect did appear to be dependent upon the infusion dose of cocaine. Thus, responding was enhanced only at low to moderate doses of cocaine. Consistent with this finding, and with effects on cocaine-stimulated locomotor activity, the 5-HT$_{2C}$ receptor knockout mouse also displayed an enhanced motivation to seek cocaine on the PR schedule (Rocha et al. 2002).

15.3.1.4 Reinstatement of Cocaine Seeking

Three classes of stimuli elicit drug craving and relapse in humans. These stimuli can also trigger reinstatement of drug-seeking behavior in animals, and this has prompted the development of preclinical models to study factors relating to relapse to drug-seeking behavior (Epstein et al. 2006; Shaham et al. 2003). The typical experimental design involves a period of drug self-administration followed by a period in which the drug-seeking response is extinguished by ensuring that responses are no longer reinforced by drug delivery. Reinstatement tests can then be conducted using a noncontingent injection of the previously self-administered drug, exposure to a stressor, or reexposure to the drug associated cues. Reinstatement is then measured as an increased emission of the previously reinforced response relative to a baseline control. The mechanisms involved in reinstatement of drug-seeking behavior elicited by these three types of stimuli are separable but overlap to some degree in terms of circuitry and neurochemical substrates. The mesocorticolimbic DA pathways including projections to the mPFC and amygdala are especially important for mediating reinstatement induced by priming and drug-paired cues (Epstein et al. 2006; Shaham et al. 2003; Crombag et al. 2008; Schmidt et al. 2005; See 2005).

A generally consistent picture is emerging regarding the effects of 5-HT$_{2C}$ receptor agonists in reinstatement models. In rats previously trained to self-administer cocaine, Ro60-0175 reduced responding reinstated by priming injections of cocaine (Grottick et al. 2000). In contrast the 5-HT$_{2C}$ receptor antagonist SB 242084 enhanced the ability of cocaine to reinstate responding (Fletcher et al. 2002). A more recent study found that MK212 also attenuated the priming effect of cocaine (see Fig. 15.1c) and that this action was prevented by SB 242084 (Neisewander and Acosta 2007). However, in comparison with results from the earlier study (Fletcher et al. 2002), SB 242084 did not enhance the effects of cocaine to reinstate responding. Although the reason for the discrepancy between these studies is not known, there is quite a marked difference in doses of SB 242084 used between the two experiments: 0.5 mg/kg (Fletcher et al. 2002) versus 3 mg/kg (Neisewander and Acosta 2007), and possible loss of receptor selectivity could have occurred in the latter study.

One commonly used behavioral procedure for measuring cue-induced reinstatement involves training rats to self-administer cocaine, extinguishing the operant response in the absence of cue presentations, and then delivering cues in a response contingent manner during reinstatement tests. In this procedure the 5-HT releaser d-fenfluramine reduced cue reinstatement. The effect of d-fenfluramine was blocked by SB 242084 indicating a major role for 5-HT$_{2C}$ receptors in the expression of fenfluramine's action (Burmeister et al. 2004). Using a pharmacologically more selective approach, it was then shown that MK212 reduced cue-induced reinstatement and that this effect was prevented by SB 242084 (Neisewander and Acosta 2007). Finally, using a slightly modified procedure, Ro60-1075 dose-dependently attenuated cue-induced reinstatement over the dose range 0.1 to 1 mg/kg (Burbassi and Cervo 2008). Again, this was a 5-HT$_{2C}$ receptor-mediated effect since it was blocked by SB 242084. The results of these studies converge to show that 5-HT$_{2C}$ receptor stimulation attenuates cue induced responding based on cocaine self-administration.

Cue-induced reinstatement can also be elicited by contextual cues, as well as discrete cues; Shaham and colleagues have devised a procedure in which the entire context, including discrete drug-paired cues, in which the drug is experienced serves as the stimulus complex for reinstatement (Crombag et al. 2002). This procedure has good face validity in that a common trigger for relapse in addicts may be a return to the home environment (Childress et al. 1993). Basically this test involves letting rats self-administer drug in one environment, extinguishing responding in another environment, reinstatement is elicited by then reexposing animals to the original context. In keeping with the results of previous studies outlined above, this contextual reinstatement was attenuated by Ro60-1075 (Fletcher et al. 2008a) (see Fig. 15.1d). Again, this effect of Ro60-0175 was blocked by SB 242084, which did not alter reinstatement in its own right.

Stress induced reinstatement has typically been studied using footshock stress. However, a number of reports have shown that extinguished responding for a variety of drugs can be reinstated by the pharmacological stressor yohimbine (Shepard et al. 2004; Lee et al. 2004a). In keeping with results from reinstatement procedures using cocaine priming and cue reexposure yohimbine induced reinstatement was attenuated by Ro60-0175 (Fletcher et al. 2008a). This effect was reversed by SB 242084. By itself, SB 242084 did not alter yohimbine-induced reinstatement.

15.3.1.5 Interaction Between 5-HT$_{2C}$ Receptor Ligands and Cocaine's Neurochemical Effects

As described above, 5-HT$_{2C}$ receptor agonists reduce the basal activity of VTA neurons, and extracellular levels of DA in terminal areas whereas 5-HT$_{2C}$ receptor antagonists exert the opposite effects (Di Matteo et al. 2008; Di Giovanni et al. 2008). Behaviorally, 5-HT$_{2C}$ receptor agonists and antagonists inhibit and enhance respectively some behavioral effects of cocaine, and a logical prediction is that cocaine's impact on DA neurotransmission would be altered by 5-HT$_{2C}$ receptor ligands. This prediction is true in the case of a 5-HT$_{2C}$ antagonist; SB 242084 enhanced the ability of cocaine to elevate levels of DA in the nucleus accumbens (Navailles et al. 2004). However, Ro60-0175 failed to alter cocaine-induced DA overflow in nucleus accumbens or dorsal striatum (Navailles et al. 2004). This finding is difficult to reconcile with the reports that systemically injected 5-HT$_{2C}$ agonists reduced several cocaine-induced behaviors, that Ro60-0175 injected into the VTA attenuated the effects of cocaine on DA efflux (Navailles et al. 2008) and behavior (Fletcher et al. 2008a), and that systemically injected Ro60-0175 attenuated the ability of nicotine to elevate accumbens levels of DA (Di Matteo et al. 2004). A potential limitation of the study by Navailles et al. (Navailles et al. 2004) is that it was conducted in anesthetised and therefore nonbehaving rats challenged with acute injections of cocaine and Ro60-0175. It will be important to extend this work into drug-experienced, behaving animals to more closely resemble conditions under which behavioral interactions between 5-HT$_{2C}$ agonists and cocaine have been found.

15.3.2   Nicotine

The pharmacological actions and physiological effects of nicotine are complex. It is widely accepted that the reinforcing effects of nicotine are mediated in part by interactions between nicotinic acetylcholine receptors (nAChRs), GABA, glutamate, and DA in the VTA (for reviews, see Markou 2008; Di Matteo et al. 2007). A major neurochemical event that is involved in the behavioral effects of nicotine, including locomotor activity and reinforcement, is indirect activation of mesolimbic DA neurons (Clarke et al. 1988; Di Chiara 2000; Corrigall et al. 1994).

15.3.2.1   Locomotor Activity

The critical importance of the VTA as a substrate for the acute reinforcing effect of nicotine (as well as other drugs of abuse) provided us with a rationale to examine the effects of Ro60-0175 on the motor stimulant and reinforcing effects of nicotine and cocaine (Grottick et al. 2000, 2001; Higgins and Fletcher 2003b). In rats previously exposed to nicotine, Ro60-0175 blocked the hyperactivity induced by an acute challenge injection of nicotine (Fig. 15.2a); this effect was reversed by SB 242084 (Grottick et al. 2001). Further experiments revealed that Ro60-0175 given concomitantly with nicotine blocked the development of sensitization to the locomotor stimulant effect of nicotine. Two recent studies found that WAY 161503 (Hayes et al. 2009) and WAY 163909 (Zaniewska et al. 2009) also blocked the effects of nicotine to stimulate locomotion in rats with prior nicotine exposure and that these effect were reversed by SB 242084.

15.3.2.2   Drug Discrimination

As with cocaine, rats readily discriminate nicotine from saline. Again this nicotine-induced discriminative state is partially DA dependent since it is blocked by DA receptor antagonists. Recently it has been shown that the 5-HT$_{2C}$ receptor agonists Ro60-0175 and WAY 163909 attenuated the discriminative stimulus properties of nicotine (Zaniewska et al. 2007; Quarta et al. 2007). For both drugs the reduction in the expression of the nicotine cue was shown to be dependent on 5-HT$_{2C}$ receptor stimulation since the effects of the agonist were blocked by the 5-HT$_{2C}$ receptor antagonist SB 242084 (Zaniewska et al. 2007). By itself, SB 242084 did not alter the nicotine cue, although it engendered some nicotine appropriate responding (~30%). Taken together, the results show that while tonic activation of 5-HT$_{2C}$ receptors does not play a major role in the subjective effects of nicotine, pharmacological activation of 5-HT$_{2C}$ receptors diminishes the discriminative stimulus properties of nicotine.

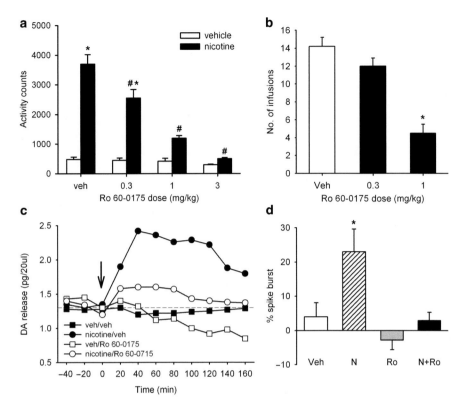

Fig. 15.2 5-HT$_{2C}$ receptor agonists alter behavioral, electrophysiological, and neurochemical effects of nicotine. (**a**) Effect of Ro 60-0175 on hyperactivity induced by nicotine (0.4 mg/kg) in rats previously exposed to nicotine (ten daily injections of nicotine 0.4 mg/kg). * indicates significantly different from vehicle/vehicle; #, indicates significantly different from vehicle/nicotine (Redrawn from Fig. 1a and b in Grottick et al. 2001). (**b**) Ro 60-0175 reduced self-administration of nicotine (0.03 mg/infusion) on an FR5TO1min schedule of reinforcement (Redrawn from Fig. 2b in (Grottick et al. 2001). (**c**) Nicotine (1 mg/kg i.p) increased extracellular dopamine levels in the nucleus accumbens. This effect was reduced by 1 mg/kg Ro60-0175 (*arrow*) (Adapted from Fig. 4 in Di Matteo et al. 2004). (**d**) Effect of nicotine (IV) and Ro 60-0175 (0.1 mg/kg IV) on the firing pattern of VTA dopamine neurons. The data represent the mean ± SEM difference between the percentage of spikes occurring in bursts during baseline versus postdrug periods. The data show that nicotine increases burst firing of VTA dopamine neurons and that this effect is blocked by Ro 60-0175 (Data adapted from Table 2 in Pierucci et al. 2004)

15.3.2.3   Self-administration and Conditioned Place Preference

In the only published report of the effects of a 5-HT$_{2C}$ receptor agonist on nicotine self-administration, Ro60-0175 reduced rates of responding for nicotine on a FR5TO60s schedule (Grottick et al. 2001) (see Fig. 15.2b). This occurred over a similar dose range for reductions in cocaine self-administration (Grottick et al. 2000). To the best of our knowledge, no further studies of 5-HT$_{2C}$ receptor ligands on nicotine self-administration or reinstatement of nicotine seeking behavior have been conducted.

Another technique that has been used to study the rewarding effects of drugs is the conditioned place preference technique. In this procedure noncontingent drug injections are paired with a distinct environmental context, whereas noncontingent injections of vehicle are paired with a different context. Many drugs of abuse elicit a preference for the drug-paired context when the animal is allowed a choice between the drug and vehicle-associated contexts. The effects of 5-HT$_{2C}$ receptor agonists on the acquisition of a nicotine-induced conditioned place preference have been examined with mixed results. In rats, doses of WAY 161503 that blocked nicotine-induced locomotion failed to alter nicotine-induced place preference (Hayes et al. 2009). In contrast, Ro60-0175 blocked the ability of nicotine to induce a place preference in mice (Ji et al. 2006). In this latter study Ro60-0175 also prevented place conditioning with tetrahydrocannabinol (THC). Whether these discrepancies relate to the use of different species or result from the use of different 5-HT$_{2C}$ receptor agonists is not known.

15.3.2.4 Interaction Between 5-HT$_{2C}$ Receptor Ligands and Nicotine's Electrophysiological and Neurochemical Effects

Electrophysiological work showed that in animals previously exposed to nicotine, the basal level of activity of midbrain DA cells in the VTA and the SNR was not altered. However, a challenge injection of nicotine increased the burst firing activity of these neurons, and this effect was attenuated by Ro60-0175 (Pierucci et al. 2004). In animals previously exposed to nicotine, an acute challenge with nicotine increased DA efflux in the nucleus accumbens and the dorsal striatum. Consistent with the electrophysiological findings these effects were blocked with Ro60-0175 (Di Matteo et al. 2004). At the present time, it seems that the effect of 5-HT$_{2C}$ receptor agonists to reduce nicotine-induced or nicotine-maintained behaviors most likely involves an alteration of the ability of nicotine to alter mesolimbic DA function (Di Matteo et al. 2004; Pierucci et al. 2004; Fletcher et al. 2008b) (see Fig. 15.2c and d).

15.3.3 Ethanol

A small number of studies have examined the effects of 5-HT$_{2C}$ receptor ligands on ethanol maintained behavior. In rats trained to discriminate ethanol from saline, mCPP produced ethanol-appropriate responding. However, the fact that this effect was blocked by a selective 5-HT$_{1B}$ receptor antagonist, GR 127935, as well as the 5-HT$_{2B/2C}$ receptor antagonist SB 206553 makes it difficult to draw definitive conclusions about the role of 5-HT$_{2C}$ receptors in this response (Maurel et al. 1998). *Meta*-chlorophenylpiperazine also reduced operant responding for ethanol, but since no antagonist studies were performed, the contribution of 5-HT$_{2C}$ receptors to this response is not clear (Maurel et al. 1999a). Likewise, although a high dose of mCPP reduced ethanol drinking, it is also not clear to what extent this response involves 5-HT$_{2C}$ receptor stimulation (Maurel et al. 1999b).

Fenfluramine reduced operant responding for ethanol, and this effect was reversed by SB 242084 suggesting a primary role for 5-HT$_{2C}$ receptors in the action of fenfluramine (Tomkins et al. 2002). SB 242084 alone increased responding for ethanol, making it difficult to determine whether the reversal of fenfluramine's effect represents a functional rather than a pharmacological antagonism at 5-HT$_{2C}$ receptors. The latter possibility is supported by the finding that ethanol maintained responding was reduced by a specific 5-HT$_{2C}$ receptor agonist, Ro60-0175 (Tomkins et al. 2002). These bidirectional effects of Ro60-1075 and SB 242084 on responding for ethanol suggest a tonic modulation of this behavior by 5-HT$_{2C}$ receptor-mediated neurotransmission.

15.3.4 Opioids

Opioid agonists such as heroin and morphine activate midbrain dopaminergic neurons and elevate extracellular levels of DA in striatal regions. The rewarding effects of opioids may depend in part on dopaminergic systems, although there is clear evidence of DA-independent mechanisms as well (Laviolette et al. 2002; Pettit et al. 1984; van Ree et al. 1999). The indirect dopaminergic effect of opioids can be altered by 5-HT$_{2C}$ receptor ligands. The agonist MK212 inhibited the increase in extracellular levels of DA in the nucleus accumbens induced by morphine (Willins and Meltzer 1998). The 5-HT$_{2B/2C}$ antagonist SB 206553 enhanced the ability of morphine to increase the firing rate of midbrain DA neurons and to increase levels of DA in the nucleus accumbens (Porras et al. 2002). Consistent with this latter result, SB 242084 produced a modest potentiation of the ability of morphine to stimulate locomotor activity in rats (Fletcher et al. 2006). We are not aware of any published studies of the effects of 5-HT$_{2C}$ receptor ligands on reward-related behavioral effects of opioids.

In summary, these results show that multiple behavioral effects of cocaine, nicotine, and ethanol are reduced or attenuated by 5-HT$_{2C}$ receptor agonists. Under certain experimental conditions the effects of these drugs can be enhanced by blocking 5-HT$_{2C}$ receptors, suggesting some endogenous tonic control over these behaviors by 5-HT$_{2C}$ receptor-mediated neurotransmission.

15.4 Effects of 5-HT$_{2C}$ Receptor Ligands on Brain Stimulation Reward

Rats will readily self-administer brief trains of electrical stimulation to specific brain regions such as the lateral hypothalamus or VTA. Measuring changes in brain stimulation reward (BSR) can be used as a technique for investigating the circuitry of the brain's reward systems and for determining the effects of drugs on that circuitry (Carlezon and Chartoff 2007; Wise 2002). Manipulations of brain 5-HT function have long been known to modulate BSR (Redgrave 1978; van der Kooy et al. 1977). Greenshaw and colleagues have recently examined the effects of 5-HT$_{2C}$ receptor ligands on BSR via electrodes placed in the VTA (Hayes et al. 2008). Using a rate-frequency threshold analysis, it was found that systemic administration of the 5-HT$_{1A/1B/2C}$ agonist trifluromethylphenylpiperazine (TFMPP) and the selective 5-HT$_{2C}$ agonist WAY 161503 increased rate-frequency thresholds without altering maximal response rates. The effect of TFMPP was blocked by SB 242084 indicating that 5-HT$_{2C}$ receptor activation is the critical pharmacological action of this drug. By itself, SB 242084 did not alter rate–frequency thresholds. The profile of behavioral change induced by TFMPP and WAY 161503 suggests that 5-HT$_{2C}$ receptor activation reduces the rewarding effects of electrical stimulation of the VTA without affecting the capacity to respond. These results show that while the rewarding effects of brain stimulation can be reduced by pharmacological activation of 5-HT$_{2C}$ receptors, 5-HT acting via 5-HT$_{2C}$ receptors likely does not play a tonic role in the expression of brain stimulation reward.

15.5   Effects of 5-HT$_{2C}$ Receptor Ligands on Impulsive Behavior

Impulsivity is a characteristic of normal everyday behavior, but excessive impulsivity may take on a pathological nature that is exhibited in a diversity of psychiatric disorders including aggression, attention deficit hyperactivity disorder (ADHD), drug abuse, and eating disorders (Hollander and Rosen 2000; Moeller et al. 2001). Impulsivity is not a unitary construct and encompasses a variety of different types of behavior, which may well be independent of each other (Evenden 1999). Two forms of impulsivity have been especially well studied in animal models. Impulsive action, which refers to the tendency to make premature responses in anticipation of an expected event, reflects a loss of inhibitory control over behavior. Impulsive choice is a description of the situation in which individuals prefer a small reward that is available immediately to a larger reward that is available after a delay. Operationally, this is studied in animals tests by allowing subjects to choose between one food pellet that is immediately contingent on making an operant response, or three or four pellets that are made available some time after the operant response. It has been argued that the type of delay discounting that underlies choice behavior in this type of situation is reduced in drug abuse (Perry and Carroll 2008; Perry et al. 2005). In other words, the immediate effects of a taking a drug (e.g., euphoria) are valued over the long-term benefits of drug abstinence (e.g., good health, successful interpersonal relationships, or steady employment).

Impulsiveness and a propensity to drug abuse are related to each other in several ways. Evidence suggests that impulsiveness may lead to drug abuse; that drug abuse itself may lead to impulsivity; and that impulsivity and drug abuse are linked through a variety of common other factors such as reactivity to rewarding stimuli, sex, and early experience (Perry and Carroll 2008). Using nicotine as an example, studies in animals have shown that impulsive choice and impulsive action differentially predict acquisition and maintenance of nicotine self-administration, as well as reinstatement of nicotine (Diergaarde et al. 2008) and perhaps alcohol seeking (Le et al. 2008). A recent study indicates that adolescent exposure to nicotine enhances impulsive action in adulthood (Counotte et al. 2009). In humans, high impulsivity may be associated with increased sensitivity to the reinforcing effects of nicotine and may predict a more rapid relapse to smoking following a period of abstinence (Doran et al. 2004; Mitchell 2004; VanderVeen et al. 2008; Perkins et al. 2008). Individual differences in impulsive choice in rats may predict subsequent acquisition of cocaine self-administration (Perry et al. 2005). Finally, a history of drug self-administration in rats may enhance impulsive action, although the magnitude and duration of this effect differs across different drugs (Dalley et al. 2005a, b).

Serotonin has long been linked to impulsivity. Early work in this area, typically involving nonselective generalized manipulation of brain 5-HT function, suggests that low 5-HT activity is linked to impulsivity (Soubrie 1986). It is now apparent that the relationship between 5-HT and impulsive behavior is much more complex depending on the type of impulsivity, the specific 5-HT receptor, and the brain region involved. A number of studies have now shown that altering 5-HT$_{2C}$ receptor-mediated function has an inhibitory effect, particularly on impulsive action.

On a differential reinforcement of low rate of responding (DRL) schedule animals are reinforced for spacing their responses a specified minimum time period apart. Responses occurring earlier than the schedule value are not reinforced, and the schedule clock is reset. For example, on a DRL24s schedule responses are reinforced only if they occur at least 24 s after the previous response. Blocking 5-HT$_{2C}$ receptors with SB 242084 disrupted responding on a DRL24s schedule by increasing the likelihood that responses were made earlier than the 24-s value (Higgins et al. 2003). The exact behavioral processes affected by SB 242084 are not clear, but this profile of behavior is consistent with impaired behavioral inhibition. More direct evidence that impulsive action is increased by SB 242084 comes from studies using the five-choice serial reaction time (5-CSRT) test. This test is primarily used to measure visual attention and requires animals to detect and respond to brief light stimuli (Robbins 2002). Poor response inhibition in this test is observed as an increase in premature responses, occurring before the onset of stimulus presentation. SB 242084 produced a significant increase in premature responding without consistently affecting accuracy of performance (Higgins et al. 2003; Fletcher et al. 2007; Winstanley et al. 2004). This effect was seen under standard baseline conditions when levels of premature responding were quite low, as well as under conditions where premature response rate was elevated by manipulating the rate and predictability of trials. The increase in impulsive action was also seen in both rats and mice (Fletcher et al. 2007). Selective destruction of brain 5-HT neurons achieved through intracerebral infusions of the neurotoxin 5,7-dihydoxytryptamine produce the same effect (Fletcher 1995; Harrison et al. 1997; Wogar et al. 1992). The results with SB 242084 imply that loss of 5-HT$_{2C}$ receptor-mediated neurotransmission is the main mediator of this effect. In contrast to the effects of SB

242084, activating 5-HT$_{2C}$ receptors using Ro60-0175 (Quarta et al. 2007; Fletcher et al. 2007) and WAY 163909 (Navarra et al. 2008) reduced premature responding under basal conditions and when premature responding was elevated through unpredictable or long ITIs. In general, the reductions in premature responding induced by 5-HT$_{2C}$ receptor agonists were not accompanied by consistent changes in other behaviors such as accuracy of responding. Indeed, Quarta et al. (2007) explicitly concluded that the ability of Ro60-1075 to reduce premature responding could not be explained by a sedative action of this drug.

The results from these tests show consistently, across different laboratories, that 5-HT$_{2C}$ receptor agonists reduce while 5-HT$_{2C}$ receptor antagonists enhance impulsive responding. This bidirectional modulation of impulsive action by 5-HT$_{2C}$ receptor ligands is similar to that observed for aspects of the behavioral effects of cocaine, including cocaine reinforcement and reinstatement for responding. Thus, manipulation of 5-HT$_{2C}$ receptor function produces a consistent spectrum of behavioral changes that encompasses aspects of both drug-abuse-related behavior and impulsivity.

15.6   Effects of 5-HT$_{2C}$ Receptor Ligands on Feeding Behavior

It has long been known that manipulation of 5-HT function alters feeding behavior and body weight gain in both rodents and humans. For example, many drugs that indirectly elevate 5-HT function, including the 5-HT releaser fenfluramine, reduce food intake and body weight gain (Halford et al. 2007). Indeed, fenfluramine was used clinically for many years as an appetite suppressant and in the treatment of obesity, before its withdrawal in 1997 because of risks of pulmonary hypertension and cardiac valvulopathy. In contrast, under some conditions, lowering 5-HT activity led to increased food intake (Dourish et al. 1985; Fletcher 1988). A number of studies showed that the ability of fenfluramine to reduce food intake is dependent upon 5-HT$_{1B}$ and 5-HT$_{2C}$ receptors (Lee et al. 2004b; Neill and Cooper 1989; Vickers et al. 2001).

The importance of the 5-HT$_{2C}$ receptor in the control of food intake is further shown by the fact that 5-HT$_{2C}$ receptor agonists reliably reduce food intake and body weight gain. This has been demonstrated numerous times for the nonselective agents mCPP and TFMPP, as well as for more selective compounds including Ro60-0175, Org 12962, VER 23779, BTV-933, YM348, WAY 161503, and lorcaserin (APD356) (Bickerdike 2003; Clifton et al. 2000a; Dalton et al. 2006; Hayashi et al. 2005; Kennett and Curzon 1988; Rowland et al. 2001; Schreiber and De Vry 2002; Somerville et al. 2007; Lam et al. 2008; Rosenzweig-Lipson et al. 2006). Indeed this latter compound has entered phase III clinical trials for the treatment of obesity.

Some evidence suggests that serotonergic drugs act specifically on the process of satiation. As satiety develops, animals show a characteristic behavioral profile that is termed the *behavioral satiety sequence* (Blundell et al. 1985). In this sequence, feeding gives way to a period of activity, then grooming, and finally a period of rest or inactivity. In the cases of fenfluramine and some 5-HT$_{2C}$ receptor agonists food intake reduction is characterized by an earlier onset of the behavioral sequence of satiety (Somerville et al. 2007; Halford et al. 1998; Vickers et al. 1999; Hewitt et al. 2002). The finding that the temporal sequencing of behavior is intact under the influence of these drugs implies that their ability to reduce feeding is through a specific alteration of motivational processes rather than some nonspecific sedative or disruptive action.

A dominant view in this field is that serotonergic systems influence feeding through part of the normal regulatory physiological mechanisms related to energy expenditure (Currie and Coscina 1997; Leibowitz and Alexander 1998). It has been specifically suggested that 5-HT$_{2C}$ receptor agonists influence feeding via an interaction with hypothalamic melanocortin mechanisms (Lam et al. 2008). Several earlier papers raised the possibility that drugs such as fenfluramine and fluoxetine may also act to alter rewarding properties of food (Gray and Cooper 1996; Hoebel et al. 1988; Leander 1987; Schwartz et al. 1989). Given the similarity in behavioral effects between fenfluramine and 5-HT$_{2C}$ receptor agonists and the involvement of 5-HT$_{2C}$ receptors in mediating some of the effects of fenfluramine, it is possible that 5-HT$_{2C}$ receptor agonists may reduce feeding by influencing food reward. To the best of our knowledge, this has not yet been tested. However, one study has found that the 5-HT$_{2C}$ receptor agonist VER23779 reduced responding for food on a second-order schedule of reinforcement (Somerville et al. 2007). In mice, mCPP dose-dependently reduced responding for palatable food on a progressive ratio schedule; interestingly, mCPP acted synergistically with the CB1 receptor antagonist rimonabant to reduce responding (Ward et al. 2008). All of these findings suggest that 5-HT$_{2C}$ receptor stimulation can reduce appetitive as well as consummatory aspects of feeding.

The finding that 5-HT$_{2C}$ receptor agonists reduce food intake has been confirmed many times, but the situation with 5-HT$_{2C}$ receptor antagonists is much less clear. At first sight the fact that 5-HT$_{2C}$ receptor knockout mice show an increase in daily food intake and become overweight in comparison to their wild-type littermate controls (Tecott et al. 1995) is consistent with the effects of 5-HT$_{2C}$ receptor agonists to reduce feeding. However, these mice do not show reliably enhanced food intake in short palatability-induced feeding tests, which are sensitive to the effects of 5-HT$_{2C}$ agonists (Dalton et al. 2006). These mice also do not show an enhanced motivation to seek food, as shown by the fact that they respond no differently than controls on a food reinforced progressive ratio schedule (Rocha et al. 2002). Similarly, the 5-HT$_{2C}$ receptor antagonist SB 242084 does not increase food intake in short feeding tests (Vickers et al. 1999; Hewitt et al. 2002; Clifton et al. 2000b). Although, another 5-HT$_{2C}$ receptor antagonist, RS102221, increased food intake (Bonhaus et al. 1997a), the fact that this compound failed to reverse motor effects of mCPP and appears to have limited brain penetration (Bonhaus et al. 1997b) casts doubt on this effect being mediated by central 5-HT$_{2C}$ receptors.

15.7 Neuroanatomical Locus of Effects

The work reviewed in previous sections clearly demonstrates that manipulating 5-HT function via 5-HT$_{2C}$ receptors modulates in a consistent fashion a number of reward-related behaviors, including the behavioral responses to psychoactive drugs such as cocaine and nicotine. All of the work reviewed to date has involved the systemic injection of 5-HT$_{2C}$ receptor ligands, and so the possible sites in the brain where these effects are mediated are not known. Identification of critical brain sites can be achieved by examining the behavioral effects of microinjecting 5-HT$_{2C}$ receptor ligands into discrete brain regions. A small number of studies have begun to do this, focussing on areas of the mesocorticolimbic DA system.

15.7.1 Ventral Tegmental Area

Given that systemically injected 5-HT$_{2C}$ receptor ligands alter the firing rate of dopaminergic VTA neurons (Di Giovanni et al. 2000; Di Matteo et al. 1999, 2000), the population of 5-HT$_{2C}$ receptors in this region is a logical candidate for a site mediating the actions of these drugs. The first study to examine the effects of manipulating 5-HT$_{2C}$ receptor function in the VTA found that blocking these receptors with RS102221 did not modify cocaine-stimulated activity (McMahon et al. 2001). A subsequent study showed that intra-VTA injections of SB 242084 did not modify cocaine-induced overflow of DA in the nucleus accumbens (Navailles et al. 2008). In contrast, local infusion of Ro60-0175 in the VTA dose-dependently reduced cocaine stimulated locomotor activity and responding for cocaine on a progressive ratio schedule (Fletcher et al. 2004) (see Fig. 15.1b). Analysis of cumulative response records showed that the temporal structure of responding was unaffected by Ro60-0175 except for an earlier cessation of responding. This pattern of activity is consistent with a reduction in the reinforcing efficacy of cocaine induced by local activation of VTA 5-HT$_{2C}$ receptors. Experiments involving in vivo microdialysis further showed that the enhancement of extracellular DA levels in the nucleus accumbens that is induced by systemically injected cocaine was also blocked by intra-VTA injections of Ro60-1075 (Navailles et al. 2008). A further study found that intra-VTA injections of Ro60-0175 prevented the rise in extracellular DA in the medial prefrontal cortex (mPFC) induced by immobilization stress (Pozzi et al. 2002).

Collectively these data show that selective activation of 5-HT$_{2C}$ receptors in the VTA is sufficient to alter cocaine-induced locomotion and reinforcement, and that these effects likely result from attenuation of cocaine's ability to elevate DA levels in terminal regions. However, selective blockade of these VTA receptors seemingly does not alter either the behavioral or the neurochemical effects of cocaine. This implies that a separate population of 5-HT$_{2C}$ receptors is responsible for mediating the effects of 5-HT$_{2C}$ receptor blockade on the behavioral effects of cocaine.

15.7.2   Prefrontal Cortex

One study has reported that local infusions of MK212 into the mPFC attenuates both the locomotor stimulant, and the discriminative stimulus effects, of cocaine (Filip and Cunningham 2003). Blockade of 5-HT$_{2C}$ receptors with RS102221 enhanced both of these effects. Recent data suggests that both effects appear to be independent of any modulation of cocaine's effects on DA function (Leggio et al. 2009). Whether 5-HT$_{2C}$ receptors in the mPFC play a role in mediating effects of 5-HT$_{2C}$ receptor ligands on cocaine or other drug self-administration or reinstatement remains to be determined.

15.7.3   Nucleus Accumbens

Systemically injected 5-HT$_{2C}$ receptor agonists reduce cocaine stimulated locomotor activity and attenuate the discriminative stimulus properties of cocaine (see Sects. 15.3.1.1 and 15.3.1.2). In contrast, infusions of MK212 or Ro60-0175 into the shell region of the nucleus accumbens enhanced the locomotor activating effect of cocaine (Filip and Cunningham 2002). In the drug discrimination procedure both drugs elicited some cocaine-like responding and enhanced the discriminative stimulus properties of a submaximal dose of cocaine (Filip and Cunningham 2002). The same study also reported that blocking 5-HT$_{2C}$ receptors with RS102221 dose-dependently attenuated the locomotor activating and stimulus effects of cocaine. These results clearly demonstrate some functional importance for 5-HT$_{2C}$ receptors in the nucleus accumbens in modulating the behavioral effects of cocaine. However, in all cases the effects are opposite to those resulting from systemically injected 5-HT$_{2C}$ receptor agonists and antagonists. Using reverse microdialysis to locally perfuse the nucleus accumbens with Ro60-10175 produced biphasic effects on cocaine-induced DA overflow in the nucleus accumbens (Navailles et al. 2008). A low concentration enhanced and a higher concentration reduced DA overflow. However, it is not clear exactly how these effects map onto the finding that MK212 exerted a unidirectional behavioral effect, namely, enhancing cocaine stimulated locomotion. In contrast to these effects, one further study showed that responding for brain stimulation reward via electrodes implanted into the VTA was attenuated by systemically injected WAY 161503 but not by infusions into the nucleus accumbens (Hayes et al. 2008). A general conclusion to be drawn from this work is that local manipulation of 5-HT$_{2C}$ receptor activity in the nucleus accumbens does not reproduce the effects of systemically injected ligands on cocaine-mediated effects or BSR. Thus, the nucleus accumbens is not the primary site of action mediating the behavioral effects of these drugs when given systemically. The lack of effect of intraaccumbens WAY 161503 further suggests that 5-HT$_{2C}$ receptors in this site are not involved in reward-related behavior.

Regarding impulsive action, infusion of SB 242084 into the nucleus accumbens increased premature responding in the 5-CSRT test (Robinson et al. 2008), but the effect was very small (increase from approximately three to seven responses) when compared with effects of this drug injected systemically (increase from ~5 to 20 responses). Whether 5-HT$_{2C}$ receptor agonists reduce impulsive action (Fletcher et al. 2007) via nucleus accumbens 5-HT$_{2C}$ receptors is not known.

15.7.4   Hypothalamic and Other Areas

Work beginning in the 1980s identified hypothalamic areas such as the paraventricular nucleus (PVN) as a primary site where 5-HT acts to suppress feeding (Leibowitz and Alexander 1998). Some evidence indicates that both 5-HT$_{1B}$ and 5-HT$_{2A}$ receptor activation in this site reduces food intake (Currie and Coscina 1997). Recently, Ro60-0175 infused into the PVN induced a moderate reduction in food intake (Lopez-Alonso et al. 2007). Other hypothalamic sites such as the arcuate nucleus have been implicated in mediating the effects of systemically injected 5-HT$_{2C}$ receptor agonists on food intake. A c-Fos mapping study of the effects of VER23799 showed activation not only in expected feeding-related sites such as the PVN but also in the basolateral amygdala (Somerville et al. 2007). Given the suggestion that 5-HT$_{2C}$ receptor agonist-induced reductions in feeding could arise in part through altered functioning of food reward systems, the effects of infusing 5-HT$_{2C}$ receptor agonists into extrahypothalamic sites such as the basolateral amygdala, the VTA, or mPFC may be worth examining. The fact that lesions of the PVN do not alter the anorectic effects of TFMPP and d-fenfluramine, which act partly via 5-HT$_{2C}$ receptors, is further evidence that perhaps extrahypothalamic sites are in involved in mediating the effects of these types of drugs (Fletcher et al. 1993).

15.7.5   Summary: Localization Studies

In summary, a detailed picture of how and where 5-HT$_{2C}$ receptor agonists and antagonists act in the brain to produce their effects on reward-related behavior is lacking. The available data suggest that the VTA and the mPFC are possible sites of action for the well-described inhibitory effects of systemically injected 5-HT$_{2C}$ receptor agonists on aspects of drug abuse-related behaviors and impulsivity. To fully understand how 5-HT, acting via 5-HT$_{2C}$ receptors, controls these behaviors future experiments need to systematically compare the effects of 5-HT$_{2C}$ receptor ligands microinjected into regions such as the VTA, the mPFC and the nucleus accumbens on behavioral measures such as drug self-administration, reinstatement, feeding and impulse control.

15.8   Therapeutic Potential of 5-HT$_{2C}$ Receptor Drugs for Reward-Related Behavioral Problems

Pharmaceutical research into the identification of subtype selective 5-HT$_{2C}$ agonists has been an ongoing effort since the mid-1990s, driven largely by the potential of such drugs to treat obesity (Nilsson 2006). This has not been an immediately successful enterprise, largely because of the high sequence homology between the 5-HT$_{2A}$, 5-HT$_{2B}$, and 5-HT$_{2C}$ receptor subtypes (Hoyer et al. 2002; Barnes and Sharp 1999) and the negative impact of direct 5-HT$_{2A}$ and 5-HT$_{2B}$ receptor activation. The hallucinogenic properties of the indoleamine and phenylalkylamine drug classes are believed to be attributable to 5-HT$_{2A}$ receptor agonism (Geyer and Vollenweider 2008) and the cardiac valvulopathies associated with the antiparkinsonian drug pergolide (Antonini and Poewe 2007), and the 5-HT releaser dex fenfluramine are likely 5-HT$_{2B}$ receptor mediated (Fitzgerald et al. 2000; Rothman et al. 2000). Therefore, from a safety perspective alone, 5-HT$_{2C}$ receptor agonists with a high degree of subtype selectivity are necessary.

A number of subtype selective 5-HT$_{2C}$ agonists have now appeared in the patent literature (Nilsson 2006). The most advanced of these, lorcaserin (formerly ADP356), has recently completed a phase III trial for obesity. As the first selective 5-HT$_{2C}$ agonist to undergo extensive clinical evaluation, lorcaserin could serve as a useful guide when considering the therapeutic potential of this drug class. Lorcaserin was discovered by Arena Pharmaceuticals (Smith et al. 2008) and based on in vitro screening across clonal cell lines expressing h5-HT$_2$ receptors it has approximately 15- and 100-fold selectivity for h5-HT$_{2C}$ against h5-HT$_{2A}$ and h5-HT$_{2B}$ receptors (Thomsen et al. 2008). In animal models of obesity, lorcaserin reduces body weight gain (Thomsen et al. 2008). The inhibitory effects of this drug on feeding and motor behavior are absent in 5-HT$_{2C}$ receptor knockout mice (Fletcher et al. 2009) confirming a 5-HT$_{2C}$ receptor-mediated action. In a randomized, double-blind, parallel arm phase II study in 469 men and women with BMI 30–45 kg/m^2, lorcaserin (10 mg q.d., 15 mg q.d., 10 mg b.i.d.) was well tolerated over a 12-week treatment period, with dose-related and progressive weight loss compared with placebo (Smith et al. 2009). The magnitude of this change relative to placebo reached 3.6 kg at the 10 mg b.i.d dose, which is comparable to weight reductions produced by sibutramine and rimonabant over equivalent or longer-term trials (see Smith et al. 2009). Importantly, regular echocardiographic measurement failed to identify any treatment-related effects on cardiac function, and no adverse effects on cognition, perception, or mood were reported. Most recently, these promising results in terms of efficacy, tolerability, and cardiac safety, have been extended in a phase III placebo controlled trial involving over 3,000 subjects (Arena press release, March 30, 2009, http://invest.arenapharm.com/releasedetail.cfm?ReleaseID =373684). Taken together, these data suggest that as a drug class 5-HT$_{2C}$ receptor agonists have a therapeutic potential for the treatment of obesity, and providing that suitable selectivity for the 5-HT$_{2C}$ receptor is met, such drugs at least on current evidence, appear to be well tolerated following sustained dosing.

Because 5-HT$_{2C}$ receptors modulate reward-related behaviors, a logical question about the therapeutic potential of drugs like lorcaserin is whether this can be extended to the treatment of drug abuse. Several neurochemical systems, including DA, endogenous opioids, and cannabinoids, that are involved in drug abuse and dependence are also involved in the control of different aspects of feeding behavior (Kirkham 2009; Barbano and Cador 2007; Barbano et al. 2009). Similarly, the neuropeptides leptin and orexin that are involved in mediating aspects of feeding behavior also modulate reward-related behaviors (Borgland et al. 2006; Fulton et al. 2000). Thus, there is convergence and overlap of brain mechanisms and systems involved in reward-related behaviors relevant to obesity and substance abuse (Trinko et al. 2007; Volkow and Wise 2005). It is feasible then that a serotonergic drug that is effective in treating obesity may have potential in treating addiction.

There are some reports of serotonergic drugs that may affect weight gain in obese individuals, acting to reduce addictive behaviors, although in each case effect size is modest. For example, some reports document a short-term benefit of fluoxetine and sertraline during the first few weeks of a smoking cessation trial and similar trends in alcohol dependence trials (Covey et al. 2002; Hitsman et al. 1999; Niaura et al. 2002; Cornelius et al. 1999; Naranjo and Knoke 2001; Naranjo et al. 1992). In general however, medications such as SSRIs have not proven to be consistently effective in treating drug abuse and dependence (Walsh and Cunningham 1997). These nonselective serotonergic drugs enhance brain 5-HT function in an indiscriminate fashion, such that 5-HT neurotransmission is increased through all of the various receptor subtypes. This may not always be desirable since different 5-HT receptor subtypes may act in an opposing fashion to modulate neuronal activity and behavioral output. For example, the 5-HT$_{2A}$ receptor appears to modulate aspects of cocaine-induced behaviors and impulsivity in a fashion opposite to the 5-HT$_{2C}$ receptor (Fletcher et al. 2007, 2002; Higgins et al. 2003). As reviewed in this chapter, animal-based work clearly points to the 5-HT$_{2C}$ receptor as a modulator of reward-related behavior. Therefore, compared with nonselective serotonergic agents, 5-HT$_{2C}$ receptor agonists represent a rational and evidence-based (at least from preclinical studies) treatment strategy. Finally, by virtue of subtype selectivity and orthosteric agonist properties, 5-HT$_{2C}$ receptor agonists may activate the 5-HT$_{2C}$ receptor pathway to a greater magnitude than nonselective and indirect 5-HT agonists such as dexfenfluramine and fluoxetine. This may translate into improved efficacy.

With the emergence of positive phase III data for lorcaserin in obesity, and a new drug application (NDA) filing in 2010, it is possible that opportunities to evaluate the therapeutic potential of this compound for drug dependence (e.g. for nicotine or alcohol) could emerge. If lorcaserin is approved for the treatment of obesity, it will likely be followed by several other selective 5-HT$_{2C}$ agonists currently in Phase I/II development, providing further opportunity to evaluate this drug class for treating drug abuse and dependence. Agonists with a superior 5-HT$_{2C}$ versus 5-HT$_{2A/2B}$ selectivity to lorcaserin may allow an even greater magnitude of 5-HT$_{2C}$ receptor activation and potentially efficacy, due to lesser concerns for off-target activity. An even longer term goal may be to find drugs that can modulate RNA editing patterns of 5-HT$_{2C}$ receptors in a manner that may beneficially treat drug dependency states. Whether RNA editing of the 5-HT$_{2C}$ receptor can be therapeutically exploited is an intriguing possibility for future drug discovery (Werry et al. 2008).

15.9 Conclusions

A number of different laboratories have now reported that 5-HT$_{2C}$ receptor agonists consistently reduce a variety of what can be generally described as reward-related behaviors. These include feeding behavior, the unconditioned stimulant and discriminative stimulus properties of drugs of abuse (notably, cocaine and nicotine), reinstatement of cocaine seeking, and responding for BSR. Additionally 5-HT$_{2C}$ receptor agonists reduce impulsive action, a behavior that is linked to several clinical problems of dysregulated motivation. None of these effects can be accounted for in terms of nonselective sedative effects, cognitive deficits, or motor impairments. In general, it appears that 5-HT$_{2C}$ receptor agonists produce a general dampening effect on a number of motivated behaviors. This type of behavioral profile may indicate that 5-HT$_{2C}$ receptor agonists have the potential for use as medications to treat substance abuse and dependence. At least one 5-HT$_{2C}$ receptor agonist, lorcaserin, has reached advanced stage of clinical testing for obesity, displaying clinical efficacy with reasonable tolerability. This means that the opportunity of testing 5-HT$_{2C}$ receptor agonists for their ability to treat drug abuse, drug dependence, and ancillary behaviors such as impulsivity is potentially close at hand.

Acknowledgements Work described in this chapter conduced by P.J. Fletcher has been supported by the Canadian Institutes of Health Research and the Natural Sciences and Engineering Research Council of Canada.

References

Alex KD, Pehek EA (2007) Pharmacologic mechanisms of serotonergic regulation of dopamine neurotransmission. Pharmacol Ther 113:296–320.

Antonini A, Poewe W (2007) Fibrotic heart-valve reactions to dopamine-agonist treatment in Parkinson's disease. Lancet Neurol 6:826–829.

Bankson MG, Cunningham KA (2002) Pharmacological studies of the acute effects of (+)-3,4-methylenedioxymethamphetamine on locomotor activity: role of 5-HT$_{1B/1D}$ and 5-HT$_2$ receptors. Neuropsychopharmacology 26:40–52.

Barbano MF, Cador M (2007) Opioids for hedonic experience and dopamine to get ready for it. Psychopharmacology (Berl) 191:497–506.

Barbano MF, Le Saux M, Cador M (2009) Involvement of dopamine and opioids in the motivation to eat: influence of palatability, homeostatic state, and behavioral paradigms. Psychopharmacology (Berl) 203:475–487.

Barnes NM, Sharp T (1999) A review of central 5-HT receptors and their function. Neuropharmacology 38:1083–1152.

Bickerdike MJ (2003) 5-HT$_{2C}$ receptor agonists as potential drugs for the treatment of obesity. Curr Top Med Chem 3:885–897.

Blundell JE, Rogers PJ, Hill AJ (1985) Behavioral structure and mechanisms of anorexia: calibration of natural and abnormal inhibition of eating. Brain Res Bull 15:371–376.

Bonhaus DW, Weinhardt KK, Taylor M, et al (1997) RS-102221: a novel high affinity and selective, 5-HT$_{2C}$ receptor antagonist. Neuropharmacology 36:621–629.

Borgland SL, Taha SA, Sarti F, et al (2006) Orexin A in the VTA is critical for the induction of synaptic plasticity and behavioral sensitization to cocaine. Neuron 49:589–601.

Bubar MJ, Cunningham KA (2006) Serotonin 5-HT$_{2A}$ and 5-HT$_{2C}$ receptors as potential targets for modulation of psychostimulant use and dependence. Curr Top Med Chem 6:1971–1985.

Bubar MJ, Cunningham KA (2007) Distribution of serotonin 5-HT$_{2C}$ receptors in the ventral tegmental area. Neuroscience 146:286–297.

Bubar MJ, Seitz PK, Thomas ML, et al (2005) Validation of a selective serotonin 5-HT$_{2C}$ receptor antibody for utilization in fluorescence immunohistochemistry studies. Brain Res 1063:105–113.

Burbassi S, Cervo L (2008) Stimulation of serotonin$_{2C}$ receptors influences cocaine-seeking behavior in response to drug-associated stimuli in rats. Psychopharmacology (Berl) 196:15–27.

Burmeister JJ, Lungren EM, Kirschner KF, et al (2004) Differential roles of 5-HT receptor subtypes in cue and cocaine reinstatement of cocaine-seeking behavior in rats. Neuropsychopharmacology 29:660–668.

Burns CM, Chu H, Rueter SM, et al (1997) Regulation of serotonin-$_{2C}$ receptor G-protein coupling by RNA editing. Nature 387:303–308.

Caine SB, Koob GF (1994) Effects of dopamine D-1 and D-2 antagonists on cocaine self-administration under different schedules of reinforcement in the rat. J Pharmacol Exp Ther 270:209–218.

Callahan PM, Cunningham KA (1995) Modulation of the discriminative stimulus properties of cocaine by 5-HT$_{1B}$ and 5-HT$_{2C}$ receptors. J Pharmacol Exp Ther 274:1414–1424.

Callahan PM, Appel JB, Cunningham KA (1991) Dopamine D1 and D2 mediation of the discriminative stimulus properties of d-amphetamine and cocaine. Psychopharmacology (Berl) 103:50–55.

Carlezon WA, Jr., Chartoff EH (2007) Intracranial self-stimulation (ICSS) in rodents to study the neurobiology of motivation. Nat Protoc 2:2987–2995.

Carroll ME, Lac ST, Asencio M, et al (1990) Fluoxetine reduces intravenous cocaine self-administration in rats. Pharmacol Biochem Behav 35:237–244.

Childress AR, Hole AV, Ehrman RN, et al (1993) Cue reactivity and cue reactivity interventions in drug dependence. NIDA Res Monogr 137:73–95.

Clarke PB, Fu DS, Jakubovic A, et al (1988) Evidence that mesolimbic dopaminergic activation underlies the locomotor stimulant action of nicotine in rats. J Pharmacol Exp Ther 246:701–708.

Clifton PG, Lee MD, Dourish CT (2000) Similarities in the action of Ro 60-0175, a 5-HT$_{2C}$ receptor agonist and d-fenfluramine on feeding patterns in the rat. Psychopharmacology (Berl) 152:256–267.

Cornelius JR, Perkins KA, Salloum IM, et al (1999) Fluoxetine versus placebo to decrease the smoking of depressed alcoholic patients. J Clin Psychopharmacol 19:183–184.

Corrigall WA, Coen KM, Adamson KL (1994) Self-administered nicotine activates the mesolimbic dopamine system through the ventral tegmental area. Brain Res 653:278–284.

Counotte DS, Spijker S, Van de Burgwal LH, et al (2009) Long-lasting cognitive deficits resulting from adolescent nicotine exposure in rats. Neuropsychopharmacology 34:299–306.

Covey LS, Glassman AH, Stetner F, et al (2002) A randomized trial of sertraline as a cessation aid for smokers with a history of major depression. Am J Psychiatry 159:1731–1737.

Crombag HS, Grimm JW, Shaham Y (2002) Effect of dopamine receptor antagonists on renewal of cocaine seeking by reexposure to drug-associated contextual cues. Neuropsychopharmacology 27:1006–1015.

Crombag HS, Bossert JM, Koya E, et al (2008) Review. Context-induced relapse to drug seeking: a review. Philos Trans R Soc Lond B Biol Sci 363:3233–3243.

Currie PJ, Coscina DV (1997) Stimulation of 5-HT$_{2A}$/$_{2C}$ receptors within specific hypothalamic nuclei differentially antagonizes NPY-induced feeding. Neuroreport 8:3759–3762.

Dalley JW, Laane K, Pena Y, et al (2005) Attentional and motivational deficits in rats withdrawn from intravenous self-administration of cocaine or heroin. Psychopharmacology (Berl) 182:579–587.

Dalley JW, Theobald DE, Berry D, et al (2005) Cognitive sequelae of intravenous amphetamine self-administration in rats: evidence for selective effects on attentional performance. Neuropsychopharmacology 30:525–537.

Dalley JW, Mar AC, Economidou D, et al (2008) Neurobehavioral mechanisms of impulsivity: fronto-striatal systems and functional neurochemistry. Pharmacol Biochem Behav 90:250–260.

Dalton GL, Lee MD, Kennett GA, et al (2006) Serotonin $_{1B}$ and $_{2C}$ receptor interactions in the modulation of feeding behavior in the mouse. Psychopharmacology (Berl) 185:45–57.

Deakin JF (1998) The role of serotonin in panic, anxiety and depression. Int Clin Psychopharmacol 13(suppl 4):S1–S5.

Di Chiara G (2000) Role of dopamine in the behavioral actions of nicotine related to addiction. Eur J Pharmacol 393:295–314.

Di Giovanni G, De Deurwaerdere P, Di Mascio M, et al (1999) Selective blockade of serotonin-$_{2C/2B}$ receptors enhances mesolimbic and mesostriatal dopaminergic function: a combined in vivo electrophysiological and microdialysis study. Neuroscience 91:587–597.

Di Giovanni G, Di Matteo V, Di Mascio M, et al (2000) Preferential modulation of mesolimbic vs. nigrostriatal dopaminergic function by serotonin($_{2C/2B}$) receptor agonists: a combined in vivo electrophysiological and microdialysis study. Synapse 35:53–61.

Di Giovanni G, Di Matteo V, La Grutta V, et al (2001) m-Chlorophenylpiperazine excites non-dopaminergic neurons in the rat substantia nigra and ventral tegmental area by activating serotonin-$_{2C}$ receptors. Neuroscience 103:111–116.

Di Giovanni G, Di Matteo V, Pierucci M, et al (2008) Serotonin-dopamine interaction: electro-physiological evidence. Prog Brain Res 172:45–71.

Di Matteo V, Di Giovanni G, Di Mascio M, et al (1998) Selective blockade of serotonin$_{2C/2B}$ receptors enhances dopamine release in the rat nucleus accumbens. Neuropharmacology 37:265–272.

Di Matteo V, Di Giovanni G, Di Mascio M, et al (1999) SB 242084, a selective serotonin$_{2C}$ receptor antagonist, increases dopaminergic transmission in the mesolimbic system. Neuropharmacology 38:1195–1205.

Di Matteo V, Di Giovanni G, Di Mascio M, et al (2000) Biochemical and electrophysiological evidence that Ro 60-0175 inhibits mesolimbic dopaminergic function through serotonin$_{2C}$ receptors. Brain Res 865:85–90.

Di Matteo V, Pierucci M, Esposito E (2004) Selective stimulation of serotonin$_{2C}$ receptors blocks the enhancement of striatal and accumbal dopamine release induced by nicotine administration. J Neurochem 89:418–429.

Di Matteo V, Pierucci M, Di Giovanni G, et al (2007) The neurobiological bases for the pharmacotherapy of nicotine addiction. Curr Pharm Des 13:1269–1284.

Di Matteo V, Di Giovanni G, Pierucci M, et al (2008) Serotonin control of central dopaminergic function: focus on in vivo microdialysis studies. Prog Brain Res 172:7–44.

Diergaarde L, Pattij T, Poortvliet I, et al (2008) Impulsive choice and impulsive action predict vulnerability to distinct stages of nicotine seeking in rats. Biol Psychiatry 63:301–308.

Doran N, Spring B, McChargue D, et al (2004) Impulsivity and smoking relapse. Nicotine Tob Res 6:641–647.

Dourish CT, Hutson PH, Curzon G (1985) Low doses of the putative serotonin agonist 8-hydroxy-2-(di-n-propylamino) tetralin (8-OH-DPAT) elicit feeding in the rat. Psychopharmacology (Berl) 86:197–204.

Dray A, Davies J, Oakley NR, et al (1978) The dorsal and medial raphe projections to the substantia nigra in the rat: electrophysiological, biochemical and behavioral observations. Brain Res 151:431–442.

Epstein DH, Preston KL, Stewart J, et al (2006) Toward a model of drug relapse: an assessment of the validity of the reinstatement procedure. Psychopharmacology (Berl) 189:1–16.

Evenden JL (1999) Varieties of impulsivity. Psychopharmacology (Berl) 146:348–361.

Filip M, Cunningham KA (2002) Serotonin 5-HT$_{2C}$ receptors in nucleus accumbens regulate expression of the hyperlocomotive and discriminative stimulus effects of cocaine. Pharmacol Biochem Behav 71:745–756.

Filip M, Cunningham KA (2003) Hyperlocomotive and discriminative stimulus effects of cocaine are under the control of serotonin$_{2C}$ (5-HT$_{2C}$) receptors in rat prefrontal cortex. J Pharmacol Exp Ther 306:734–743.

Filip M, Bubar MJ, Cunningham KA (2004) Contribution of serotonin (5-hydroxytryptamine; 5-HT) 5-HT$_2$ receptor subtypes to the hyperlocomotor effects of cocaine: acute and chronic pharmacological analyses. J Pharmacol Exp Ther 310:1246–1254.

Fitzgerald LW, Iyer G, Conklin DS, et al (1999) Messenger RNA editing of the human serotonin 5-HT$_{2C}$ receptor. Neuropsychopharmacology 21:82S–90S.

Fitzgerald LW, Burn TC, Brown BS, et al (2000) Possible role of valvular serotonin 5-HT(2B) receptors in the cardiopathy associated with fenfluramine. Mol Pharmacol 57:75–81.

Fletcher PJ (1988) Increased food intake in satiated rats induced by the 5-HT antagonists methysergide, metergoline and ritanserin. Psychopharmacology (Berl) 96:237–242.

Fletcher PJ (1995) Effects of combined or separate 5,7-dihydroxytryptamine lesions of the dorsal and median raphe nuclei on responding maintained by a DRL 20s schedule of food reinforcement. Brain Res 675:45–54.

Fletcher PJ, Currie PJ, Chambers JW, et al (1993) Radiofrequency lesions of the PVN fail to modify the effects of serotonergic drugs on food intake. Brain Res 630:1–9.

Fletcher PJ, Korth KM, Chambers JW (1999) Depletion of brain serotonin does not alter d-amphetamine self-administration under a variety of schedule and access conditions. Psychopharmacology (Berl) 146:185–193.

Fletcher PJ, Grottick AJ, Higgins GA (2002) Differential effects of the 5-HT$_{2A}$ receptor antagonist M100907 and the 5-HT$_{2C}$ receptor antagonist SB242084 on cocaine-induced locomotor activity, cocaine self-administration and cocaine-induced reinstatement of responding. Neuropsychopharmacology 27:576–586.

Fletcher PJ, Chintoh AF, Sinyard J, et al (2004) Injection of the 5-HT$_{2C}$ receptor agonist Ro60-0175 into the ventral tegmental area reduces cocaine-induced locomotor activity and cocaine self-administration. Neuropsychopharmacology 29:308–318.

Fletcher PJ, Sinyard J, Higgins GA (2006) The effects of the 5-HT$_{2C}$ receptor antagonist SB242084 on locomotor activity induced by selective, or mixed, indirect serotonergic and dopaminergic agonists. Psychopharmacology (Berl) 187:515–525.

Fletcher PJ, Tampakeras M, Sinyard J, et al (2007) Opposing effects of 5-HT$_{2A}$ and 5-HT$_{2C}$ receptor antagonists in the rat and mouse on premature responding in the five-choice serial reaction time test. Psychopharmacology (Berl) 195:223–234.

Fletcher PJ, Rizos Z, Sinyard J, et al (2008) The 5-HT$_{2C}$ receptor agonist Ro60-0175 reduces cocaine self-administration and reinstatement induced by the stressor yohimbine, and contextual cues. Neuropsychopharmacology 33:1402–1412.

Fletcher PJ, Le AD, Higgins GA (2008) Serotonin receptors as potential targets for modulation of nicotine use and dependence. Prog Brain Res 172:361–383.

Fletcher PJ, Tampakeras M, Sinyard J, et al (2009) Characterizing the effects of 5-HT$_{2C}$ receptor ligands on motor activity and feeding behavior in 5-HT$_{2C}$ receptor knockout mice. Neuropharmacology 57:259–267.

Fulton S, Woodside B, Shizgal P (2000) Modulation of brain reward circuitry by leptin. Science 287:125–128.

Geyer MA, Vollenweider FX (2008) Serotonin research: contributions to understanding psychoses. Trends Pharmacol Sci 29:445–453.

Gobert A, Rivet JM, Lejeune F, et al (2000) Serotonin$_{2C}$ receptors tonically suppress the activity of mesocortical dopaminergic and adrenergic, but not serotonergic, pathways: a combined dialysis and electrophysiological analysis in the rat. Synapse 36:205–221.

Gray RW, Cooper SJ (1996) D-fenfluramine's effects on normal ingestion assessed with taste reactivity measures. Physiol Behav 59:1129–1135.

Grottick AJ, Fletcher PJ, Higgins GA (2000) Studies to investigate the role of 5-HT$_{2C}$ receptors on cocaine- and food-maintained behavior. J Pharmacol Exp Ther 295:1183–1191.

Grottick AJ, Corrigall WA, Higgins GA (2001) Activation of 5-HT$_{2C}$ receptors reduces the locomotor and rewarding effects of nicotine. Psychopharmacology (Berl) 157:292–298.

Gurevich I, Englander MT, Adlersberg M, et al (2002) Modulation of serotonin $_{2C}$ receptor editing by sustained changes in serotonergic neurotransmission. J Neurosci 22:10529–10532.

Halford JC, Wanninayake SC, Blundell JE (1998) Behavioral satiety sequence (BSS) for the diagnosis of drug action on food intake. Pharmacol Biochem Behav 61:159–168.

Halford JC, Harrold JA, Boyland EJ, et al (2007) Serotonergic drugs: effects on appetite expression and use for the treatment of obesity. Drugs 67:27–55.

Harrison AA, Everitt BJ, Robbins TW (1997) Central 5-HT depletion enhances impulsive responding without affecting the accuracy of attentional performance: interactions with dopaminergic mechanisms. Psychopharmacology (Berl) 133:329–342.

Hayashi A, Suzuki M, Sasamata M, et al (2005) Agonist diversity in 5-HT$_{2C}$ receptor-mediated weight control in rats. Psychopharmacology (Berl) 178:241–249.

Hayes DJ, Clements R, Greenshaw AJ (2008) Effects of systemic and intra-nucleus accumbens 5-HT$_{2C}$ receptor compounds on ventral tegmental area self-stimulation thresholds in rats. Psychopharmacology (Berl) 203:579–588.

Hayes DJ, Mosher TM, Greenshaw AJ (2009) Differential effects of 5-HT$_{2C}$ receptor activation by WAY 161503 on nicotine-induced place conditioning and locomotor activity in rats. Behav Brain Res 197:323–330.

Hewitt KN, Lee MD, Dourish CT, et al (2002) Serotonin $_{2C}$ receptor agonists and the behavioral satiety sequence in mice. Pharmacol Biochem Behav 71:691–700.

Higgins GA, Fletcher PJ (2003) Serotonin and drug reward:focus on 5-HT$_{2C}$ receptors. Eur J Pharmacol 480:151–162.

Higgins GA, Enderlin M, Haman M, et al (2003) The 5-HT$_{2A}$ receptor antagonist M100,907 attenuates motor and "impulsive-type" behaviors produced by NMDA receptor antagonism. Psychopharmacology (Berl) 170:309–319.

Hitsman B, Pingitore R, Spring B, et al (1999) Antidepressant pharmacotherapy helps some cigarette smokers more than others. J Consult Clin Psychol 67:547–554.

Hoebel BG, Hernandez L, McClelland RC, et al (1988) Dexfenfluramine and feeding reward. Clin Neuropharmacol 11(suppl 1):S72–S85.

Hollander E, Rosen J (2000) Impulsivity. J Psychopharmacol 14:S39–S44.

Howell LL, Byrd LD (1995) Serotonergic modulation of the behavioral effects of cocaine in the squirrel monkey. J Pharmacol Exp Ther 275:1551–1559.

Hoyer D (1988) Molecular pharmacology and biology of 5-HT$_{1C}$ receptors. Trends Pharmacol Sci 9:89–94.

Hoyer D, Engel G, Kalkman HO (1985) Molecular pharmacology of 5-HT1 and 5-HT2 recognition sites in rat and pig brain membranes: radioligand binding studies with [3H]5-HT, [3H]8-OH-DPAT (-)[125I]iodocyanopindolol, [3H]mesulergine and [3H]ketanserin. Eur J Pharmacol 118:13–23.

Hoyer D, Clarke DE, Fozard JR, et al (1994) International Union of Pharmacology classification of receptors for 5-hydroxytryptamine (Serotonin). Pharmacol Rev 46:157–203.

Hoyer D, Hannon JP, Martin GR (2002) Molecular, pharmacological and functional diversity of 5-HT receptors. Pharmacol Biochem Behav 71:533–554.

Hutson PH, Barton CL, Jay M, et al (2000) Activation of mesolimbic dopamine function by phencyclidine is enhanced by 5-HT$_{2C/2B}$ receptor antagonists: neurochemical and behavioral studies. Neuropharmacology 39:2318–2328.

Invernizzi RW, Pierucci M, Calcagno E, et al (2007) Selective activation of 5-HT$_{2C}$ receptors stimulates GABA-ergic function in the rat substantia nigra pars reticulata: a combined in vivo electrophysiological and neurochemical study. Neuroscience 144:1523–1535.

Ji SP, Zhang Y, Van Cleemput J, et al (2006) Disruption of PTEN coupling with 5-HT$_{2C}$ receptors suppresses behavioral responses induced by drugs of abuse. Nat Med 12:324–329.

Jones BJ, Blackburn TP (2002) The medical benefit of 5-HT research. Pharmacol Biochem Behav 71:555–568.

Kawahara Y, Grimberg A, Teegarden S, et al (2008) Dysregulated editing of serotonin $_{2C}$ receptor mRNAs results in energy dissipation and loss of fat mass. J Neurosci 28:12834–12844.

Kelly PH, Iversen SD (1976) Selective 6OHDA-induced destruction of mesolimbic dopamine neurons: abolition of psychostimulant-induced locomotor activity in rats. Eur J Pharmacol 40:45–56.

Kennett GA, Curzon G (1988) Evidence that hypophagia induced by mCPP and TFMPP requires 5-HT$_{1C}$ and 5-HT$_{1B}$ receptors; hypophagia induced by RU 24969 only requires 5-HT1B receptors. Psychopharmacology (Berl) 96:93–100.

Kirkham TC (2009) Cannabinoids and appetite: food craving and food pleasure. Int Rev Psychiatry 21:163–171.

Lam DD, Przydzial MJ, Ridley SH, et al (2008) Serotonin 5-HT$_{2C}$ receptor agonist promotes hypophagia via downstream activation of melanocortin 4 receptors. Endocrinology 149:1323–1328.

Laviolette SR, Nader K, van der Kooy D (2002) Motivational state determines the functional role of the mesolimbic dopamine system in the mediation of opiate reward processes. Behav Brain Res 129:17–29.

Le AD, Funk D, Harding S, et al (2008) Intra-median raphe nucleus (MRN) infusions of muscimol, a GABA-A receptor agonist, reinstate alcohol seeking in rats: role of impulsivity and reward. Psychopharmacology (Berl) 195:605–615.

Leander JD (1987) Fluoxetine suppresses palatability-induced ingestion. Psychopharmacology (Berl) 91:285–287.

Lee B, Tiefenbacher S, Platt DM, et al (2004) Pharmacological blockade of alpha2-adrenoceptors induces reinstatement of cocaine-seeking behavior in squirrel monkeys. Neuropsychopharmacology 29:686–693.

Lee MD, Somerville EM, Kennett GA, et al (2004) Reduced hypophagic effects of d-fenfluramine and the 5-HT$_{2C}$ receptor agonist mCPP in 5-HT1B receptor knockout mice. Psychopharmacology (Berl) 176:39–49.

Leggio GM, Cathala A, Moison D, et al (2009) Serotonin$_{2C}$ receptors in the medial prefrontal cortex facilitate cocaine-induced dopamine release in the rat nucleus accumbens. Neuropharmacology 56:507–513.

Leibowitz SF, Alexander JT (1998) Hypothalamic serotonin in control of eating behavior, meal size, and body weight. Biol Psychiatry 44:851–864.

Liu S, Cunningham KA (2006) Serotonin$_{2C}$ receptors (5-HT$_{2C}$ R) control expression of cocaine-induced conditioned hyperactivity. Drug Alcohol Depend 81:275–282.

Liu S, Bubar MJ, Lanfranco MF, et al (2007) Serotonin$_{2C}$ receptor localization in GABA neurons of the rat medial prefrontal cortex: implications for understanding the neurobiology of addiction. Neuroscience 146:1677–1688.

Loh EA, Roberts DC (1990) Break-points on a progressive ratio schedule reinforced by intravenous cocaine increase following depletion of forebrain serotonin. Psychopharmacology (Berl) 101:262–266.

Lopez-Alonso VE, Mancilla-Diaz JM, Rito-Domingo M, et al (2007) The effects of 5-HT1A and 5-HT$_{2C}$ receptor agonists on behavioral satiety sequence in rats. Neurosci Lett 416:285–288.

Lyness WH, Friedle NM, Moore KE (1980) Increased self-administration of d-amphetamine after destruction of 5-hydroxytryptaminergic neurons. Pharmacol Biochem Behav 12:937–941.

Marazziti D, Rossi A, Giannaccini G, et al (1999) Distribution and characterization of [3H]mesulergine binding in human brain postmortem. Eur Neuropsychopharmacol 10:21–26.

Markou A (2008) Review. Neurobiology of nicotine dependence. Philos Trans R Soc Lond B Biol Sci 363:3159–3168.

Maurel S, Schreiber R, De Vry J (1998) Role of 5-HT$_{1B}$, 5-HT$_{2A}$ and 5-HT$_{2C}$ receptors in the generalization of 5-HT receptor agonists to the ethanol cue in the rat. Behav Pharmacol 9:337–343.

Maurel S, De Vry J, Schreiber R (1999) 5-HT receptor ligands differentially affect operant oral self- administration of ethanol in the rat. Eur J Pharmacol 370:217–223.

Maurel S, De Vry J, De Beun R, et al (1999) 5-HT$_{2A}$ and 5-HT$_{2C}$/5-HT$_{1B}$ receptors are differentially involved in alcohol preference and consummatory behavior in cAA rats. Pharmacol Biochem Behav 62:89–96.

McCreary AC, Cunningham KA (1999) Effects of the 5-HT$_{2C/2B}$ antagonist SB 206553 on hyperactivity induced by cocaine. Neuropsychopharmacology 20:556–564.

McMahon LR, Cunningham KA (2001) Role of 5-HT$_{2A}$ and 5-HT$_{2B/2C}$ receptors in the behavioral interactions between serotonin and catecholamine reuptake inhibitors. Neuropsychopharmacology 24:319–329.

McMahon LR, Filip M, Cunningham KA (2001) Differential regulation of the mesoaccumbens circuit by serotonin 5-hydroxytryptamine (5-HT)$_{2A}$ and 5-HT$_{2C}$ receptors. J Neurosci 21:7781–7787.

Mengod G, Pompeiano M, Martinez-Mir MI, et al (1990) Localization of the mRNA for the 5-HT2 receptor by in situ hybridization histochemistry. Correlation with the distribution of receptor sites. Brain Res 524:139–143.

Millan MJ, Marin P, Bockaert J, et al (2008) Signaling at G-protein-coupled serotonin receptors: recent advances and future research directions. Trends Pharmacol Sci 29:454–464.

Mitchell SH (2004) Measuring impulsivity and modeling its association with cigarette smoking. Behav Cogn Neurosci Rev 3:261–275.

Moeller FG, Barratt ES, Dougherty DM, et al (2001) Psychiatric aspects of impulsivity. Am J Psychiatry 158:1783–1793.

Naranjo CA, Knoke DM (2001) The role of selective serotonin reuptake inhibitors in reducing alcohol consumption. J Clin Psychiatry 62(suppl 20):18–25.

Naranjo CA, Poulos CX, Bremner KE, et al (1992) Citalopram decreases desirability, liking, and consumption of alcohol in alcohol-dependent drinkers. Clin Pharmacol Ther 51:729–739.

Navailles S, De Deurwaerdere P, Porras G, et al (2004) In vivo evidence that 5-HT$_{2C}$ receptor antagonist but not agonist modulates cocaine-induced dopamine outflow in the rat nucleus accumbens and striatum. Neuropsychopharmacology 29:319–326.

Navailles S, Moison D, Cunningham KA, et al (2008) Differential regulation of the mesoaccumbens dopamine circuit by serotonin$_{2C}$ receptors in the ventral tegmental area and the nucleus accumbens: an in vivo microdialysis study with cocaine. Neuropsychopharmacology 33:237–246.

Navarra R, Comery TA, Graf R, et al (2008) The 5-HT$_{2C}$ receptor agonist WAY 163909 decreases impulsivity in the 5-choice serial reaction time test. Behav Brain Res 188:412–415.

Neill JC, Cooper SJ (1989) Evidence that d-fenfluramine anorexia is mediated by 5-HT1 receptors. Psychopharmacology (Berl) 97:213–218.

Neisewander JL, Acosta JI (2007) Stimulation of 5-HT$_{2C}$ receptors attenuates cue and cocaine-primed reinstatement of cocaine-seeking behavior in rats. Behav Pharmacol 18:791–800.

Niaura R, Spring B, Borrelli B, et al (2002) Multicenter trial of fluoxetine as an adjunct to behavioral smoking cessation treatment. J Consult Clin Psychol 70:887–896.

Nilsson BM (2006) 5-Hydroxytryptamine $_{2C}$ (5-HT$_{2C}$) receptor agonists as potential antiobesity agents. J Med Chem 49:4023–4034.

Niswender CM, Herrick-Davis K, Dilley GE, et al (2001) RNA editing of the human serotonin 5-HT$_{2C}$ receptor. alterations in suicide and implications for serotonergic pharmacotherapy. Neuropsychopharmacology 24:478–491.

Oades RD (2008) Dopamine-serotonin interactions in attention-deficit hyperactivity disorder (ADHD). Prog Brain Res 172:543–565.

Pazos A, Palacios JM (1985) Quantitative autoradiographic mapping of serotonin receptors in the rat brain. I. Serotonin-1 receptors. Brain Res 346:205–230.

Pazos A, Hoyer D, Palacios JM (1984) The binding of serotonergic ligands to the porcine choroid plexus: characterization of a new type of serotonin recognition site. Eur J Pharmacol 106:539–546.

Peltier R, Schenk S (1993) Effects of serotonergic manipulations on cocaine self-administration in rats. Psychopharmacology (Berl) 110:390–394.

Perkins KA, Lerman C, Coddington SB, et al (2008) Initial nicotine sensitivity in humans as a function of impulsivity. Psychopharmacology (Berl) 200:529–544.

Perry JL, Carroll ME (2008) The role of impulsive behavior in drug abuse. Psychopharmacology (Berl) 200:1–26.

Perry JL, Larson EB, German JP, et al (2005) Impulsivity (delay discounting) as a predictor of acquisition of IV cocaine self-administration in female rats. Psychopharmacology (Berl) 178:193–201.

Pettit HO, Ettenberg A, Bloom FE, et al (1984) Destruction of dopamine in the nucleus accumbens selectively attenuates cocaine but not heroin self-administration in rats. Psychopharmacology (Berl) 84:167–173.

Pierucci M, Di Matteo V, Esposito E (2004) Stimulation of serotonin$_{2C}$ receptors blocks the hyperactivation of midbrain dopamine neurons induced by nicotine administration. J Pharmacol Exp Ther 309:109–118.

Porras G, Di Matteo V, Fracasso C, et al (2002) 5-HT$_{2A}$ and 5-HT$_{2C/2B}$ receptor subtypes modulate dopamine release induced in vivo by amphetamine and morphine in both the rat nucleus accumbens and striatum. Neuropsychopharmacology 26:311–324.

Porrino LJ, Ritz MC, Goodman NL, et al (1989) Differential effects of the pharmacological manipulation of serotonin systems on cocaine and amphetamine self-administration in rats. Life Sci 45:1529–1535.

Pozzi L, Acconcia S, Ceglia I, et al (2002) Stimulation of 5-hydroxytryptamine (5-HT$_{2C}$) receptors in the ventrotegmental area inhibits stress-induced but not basal dopamine release in the rat prefrontal cortex. J Neurochem 82:93–100.

Prisco S, Pagannone S, Esposito E (1994) Serotonin-dopamine interaction in the rat ventral tegmental area: an electrophysiological study in vivo. J Pharmacol Exp Ther 271:83–90.

Quarta D, Naylor CG, Stolerman IP (2007) The serotonin $_{2C}$ receptor agonist Ro-60-0175 attenuates effects of nicotine in the five-choice serial reaction time task and in drug discrimination. Psychopharmacology (Berl) 193:391–402.

Redgrave P (1978) Modulation of intracranial self-stimulation behavior by local perfusions of dopamine, noradrenaline and serotonin within the caudate nucleus and nucleus accumbens. Brain Res 155:277–295.

Remington G (2008) Alterations of dopamine and serotonin transmission in schizophrenia. Prog Brain Res 172:117–140.

Richardson NR, Roberts DC (1991) Fluoxetine pretreatment reduces breaking points on a progressive ratio schedule reinforced by intravenous cocaine self-administration in the rat. Life Sci 49:833–840.

Robbins TW (2002) The 5-choice serial reaction time task: behavioral pharmacology and functional neurochemistry. Psychopharmacology (Berl) 163:362–380.

Roberts DC, Koob GF, Klonoff P, et al (1980) Extinction and recovery of cocaine self-administration following 6-hydroxydopamine lesions of the nucleus accumbens. Pharmacol Biochem Behav 12:781–787.

Robinson ES, Dalley JW, Theobald DE, et al (2008) Opposing roles for 5-HT$_{2A}$ and 5-HT$_{2C}$ receptors in the nucleus accumbens on inhibitory response control in the 5-choice serial reaction time task. Neuropsychopharmacology 33:2398–2406.

Rocha BA, Goulding EH, O'Dell LE, et al (2002) Enhanced locomotor, reinforcing, and neurochemical effects of cocaine in serotonin 5-hydroxytryptamine $_{2C}$ receptor mutant mice. J Neurosci 22:10039–10045.

Rosenzweig-Lipson S, Zhang J, Mazandarani H, et al (2006) Antiobesity-like effects of the 5-HT$_{2C}$ receptor agonist WAY 161503. Brain Res 1073-1074:240–251.

Rothman RB, Baumann MH, Savage JE, et al (2000) Evidence for possible involvement of 5-HT(2B) receptors in the cardiac valvulopathy associated with fenfluramine and other serotonergic medications. Circulation 102:2836–2841.

Rowland NE, Robertson K, Lo J, et al (2001) Cross tolerance between anorectic action and induction of Fos-ir with dexfenfluramine and 5HT$_{1B/2C}$ agonists in rats. Psychopharmacology (Berl) 156:108–114.

Schmidt HD, Anderson SM, Famous KR, et al (2005) Anatomy and pharmacology of cocaine priming-induced reinstatement of drug seeking. Eur J Pharmacol 526:65–76.

Schreiber R, De Vry J (2002) Role of 5-HT$_{2C}$ receptors in the hypophagic effect of m-CPP, ORG 37684 and CP-94,253 in the rat. Prog Neuropsychopharmacol Biol Psychiatry 26:441–449.

Schwartz DH, McClane S, Hernandez L, et al (1989) Feeding increases extracellular serotonin in the lateral hypothalamus of the rat as measured by microdialysis. Brain Res 479:349–354.

See RE (2005) Neural substrates of cocaine-cue associations that trigger relapse. Eur J Pharmacol 526:140–146.

Sesack SR, Deutch AY, Roth RH, et al (1989) Topographical organization of the efferent projections of the medial prefrontal cortex in the rat: an anterograde tract-tracing study with Phaseolus vulgaris leucoagglutinin. J Comp Neurol 290:213–242.

Sesack SR, Carr DB, Omelchenko N, et al (2003) Anatomical substrates for glutamate-dopamine interactions: evidence for specificity of connections and extrasynaptic actions. Ann N Y Acad Sci 1003:36–52.

Shaham Y, Shalev U, Lu L, et al (2003) The reinstatement model of drug relapse: history, methodology and major findings. Psychopharmacology (Berl) 168:3–20.

Shepard JD, Bossert JM, Liu SY, et al (2004) The anxiogenic drug yohimbine reinstates methamphetamine seeking in a rat model of drug relapse. Biol Psychiatry 55:1082–1089.

Smith BM, Smith JM, Tsai JH, et al (2008) Discovery and structure-activity relationship of (1R)-8-chloro-2,3,4,5-tetrahydro-1-methyl-1H-3-benzazepine (Lorcaserin), a selective serotonin 5-HT$_{2C}$ receptor agonist for the treatment of obesity. J Med Chem 51:305–313.

Smith SR, Prosser WA, Donahue DJ, et al (2009) Lorcaserin (APD356), a selective 5-HT$_{2C}$ agonist, reduces body weight in obese men and women. Obesity (Silver Spring) 17:494–503.

Somerville EM, Horwood JM, Lee MD, et al (2007) 5-HT$_{2C}$ receptor activation inhibits appetitive and consummatory components of feeding and increases brain c-fos immunoreactivity in mice. Eur J Neurosci 25:3115–3124.

Soubrie P (1986) Reconciling the role of central serotonin neurons in human and animal behavior. Behavioral and Brain Sciences 9:319–364.

Spyraki C, Fibiger HC, Phillips AG (1982) Cocaine-induced place preference conditioning: lack of effects of neuroleptics and 6-hydroxydopamine lesions. Brain Res 253:195–203.

Tecott LH, Sun LM, Akana SF, et al (1995) Eating disorder and epilepsy in mice lacking 5-HT$_{2C}$ serotonin receptors. Nature 374:542–546.

Thomsen WJ, Grottick AJ, Menzaghi F, et al (2008) Lorcaserin, a novel selective human 5-hydroxytryptamine$_{2C}$ agonist: in vitro and in vivo pharmacological characterization. J Pharmacol Exp Ther 325:577–587.

Thomsen M, Hall FS, Uhl GR, et al (2009) Dramatically decreased cocaine self-administration in dopamine but not serotonin transporter knock-out mice. J Neurosci 29:1087–1092.

Tomkins DM, Joharchi N, Tampakeras M, et al (2002) An investigation of the role of 5-HT$_{2C}$ receptors in modifying ethanol self-administration behavior. Pharmacol Biochem Behav 71:735–744.

Trinko R, Sears RM, Guarnieri DJ, et al (2007) Neural mechanisms underlying obesity and drug addiction. Physiol Behav 91:499–505.

van der Kooy D, Fibiger HC, Phillips AG (1977) Monoamine involvement in hippocampal self-stimulation. Brain Res 136:119–130.

van Ree JM, Gerrits MA, Vanderschuren LJ (1999) Opioids, reward and addiction: an encounter of biology, psychology, and medicine. Pharmacol Rev 51:341–396.

VanderVeen JW, Cohen LM, Cukrowicz KC, et al (2008) The role of impulsivity on smoking maintenance. Nicotine Tob Res 10:1397–1404.

Vickers SP, Clifton PG, Dourish CT, et al (1999) Reduced satiating effect of d-fenfluramine in serotonin 5-HT$_{2C}$ receptor mutant mice. Psychopharmacology (Berl) 143:309–314.

Vickers SP, Dourish CT, Kennett GA (2001) Evidence that hypophagia induced by d-fenfluramine and d-norfenfluramine in the rat is mediated by 5-HT$_{2C}$ receptors. Neuropharmacology 41:200–209.

Volkow ND, Wise RA (2005) How can drug addiction help us understand obesity? Nat Neurosci 8:555–560.

Walsh SL, Cunningham KA (1997) Serotonergic mechanisms involved in the discriminative stimulus, reinforcing and subjective effects of cocaine. Psychopharmacology (Berl) 130:41–58.

Ward SJ, Lefever TW, Jackson C, et al (2008) Effects of a Cannabinoid1 receptor antagonist and Serotonin$_{2C}$ receptor agonist alone and in combination on motivation for palatable food: a dose-addition analysis study in mice. J Pharmacol Exp Ther 325:567–576.

Werry TD, Loiacono R, Sexton PM, et al (2008) RNA editing of the serotonin 5HT$_{2C}$ receptor and its effects on cell signalling, pharmacology and brain function. Pharmacol Ther 119:7–23.

Willins DL, Meltzer HY (1998) Serotonin 5-HT$_{2C}$ agonists selectively inhibit morphine-induced dopamine efflux in the nucleus accumbens. Brain Res 781:291–299.

Winstanley CA, Theobald DE, Dalley JW, et al (2004) 5-HT$_{2A}$ and 5-HT$_{2C}$ receptor antagonists have opposing effects on a measure of impulsivity:interactions with global 5-HT depletion. Psychopharmacology (Berl) 176:376–385.

Wise RA (2002) Brain reward circuitry: insights from unsensed incentives. Neuron 36:229–240.

Wogar MA, Bradshaw CM, Szabadi E (1992) Impaired acquisition of temporal differentiation performance following lesions of the ascending 5-hydroxytryptaminergic pathways. Psychopharmacology (Berl) 107:373–378.

Zaniewska M, McCreary AC, Przegalinski E, et al (2007) Effects of the serotonin 5-HT$_{2A}$ and 5-HT$_{2C}$ receptor ligands on the discriminative stimulus effects of nicotine in rats. Eur J Pharmacol 571:156–165.

Zaniewska M, McCreary AC, Filip M (2009) Interactions of serotonin (5-HT)(2) receptor-targeting ligands and nicotine: Locomotor activity studies in rats. Synapse 63:653–661.

Chapter 16
Tat-3L4F: A Novel Peptide for Treating Drug Addiction by Disrupting Interaction Between PTEN and 5-HT$_{2C}$ Receptor

Amy Hu, Lintao Jia, Jean-Christian Maillet, and Xia Zhang

16.1 Introduction

From an evolutionary perspective, reinforcement of natural stimulation that enhances survival or the perpetuation of genes (such as food and sex) would be beneficial to the continuation of the species. The system of conditioning through reward and gratification remains one of the most fundamental examples of such reinforcement. However, there is a small fraction of pharmacological substances that possess the ability to abnormally intensify this reward function in the brain, known as *drugs of abuse*, which include the psychostimulants (cocaine and amphetamine), the opiates (heroin and opioids), nicotine, ethanol, and marijuana (i.e., cannabis or cannabinoids) (Hyman et al. 2006). These abused drugs act on specific receptors in the brain to mediate their actions, each resulting in distinct behavioral and physiological responses. While reasons for individual self-administration may differ, a major motivational factor may be to experience the euphoric sensations resulting from stimulation of the reward circuit of the brain. However, drugs of abuse differ from natural rewards as they have little survival value and may cause a state known as drug addiction, which is a complex affliction of the brain, where initial voluntary drug consumption leads to compulsive drug seeking, loss of self-control, and eventually habituation (Everitt and Robbins 2005).

Drug addiction had been traditionally viewed as the motivation of an addict to take drug results from the desire to experience the rewarding (e.g., hedonic) effects of the drug as well as from the desire to avoid the punishing (e.g., anhedonic or aversive) consequences of drug withdrawal. Intensive studies during the past decades have added new knowledge about the mechanisms of drug addiction.

X. Zhang (✉)
Departments of Psychiatry and Cellular and Molecular Medicine,
Institute of Mental Health Research, University of Ottawa,
1145 Carling Ave, Ottawa, ON, Canada, K1Z 7K4
e-mail: Xia.Zhang@rohcg.on.ca.

G. Di Giovanni et al. (eds.), *5-HT$_{2C}$ Receptors in the Pathophysiology of CNS Disease*,
The Receptors 22, DOI 10.1007/978-1-60761-941-3_16,
© Springer Science+Business Media, LLC 2011

The current understanding of drug addiction is that repeated drug use abnormally stimulates neurons responding to natural reinforcers such as food and sex, leading to long-lasting adaptation changes of brain reward pathways (Koob and Moal 1997; Nestler 2004) and aberrant learning processes (Robbins and Everitt 1999, 2002). Drug addiction has many phases, including initiation or euphoria phase (i.e., drug-induced rewarding or reinforcing effects), maintenance or dependence phase (i.e., compulsive drug taking), tolerance, withdrawal episodes, protracted abstinence, and craving and relapse (or reinstatement). The latter stages are characterized by lasting neural adaptations in the brain, promoting obsessive drug-seeking behavior via decreased value of natural rewards and impaired cognitive control (Kalivas and Volkow 2005). Although initial mechanisms for various drugs may differ, the underlying reward mechanism is of integral importance since it serves as a common pathway for virtually all abused drugs. Hence, numerous studies have focused on understanding drug-induced rewarding effects and the brain structures associated with them.

The main system in the brain that mediates natural and drug-induced reward effects is known as the *mesolimbic dopamine pathway*, which consists of dopamine neurons in the ventral tegmental area (VTA), a mammalian midbrain region, projecting nerve fibers (i.e., axons) to innervate the nucleus accumbens (NAc), a neuronal cluster in the forebrain (Alex and Pehek 2007; Laviolette and van der Kooy 2004). The VTA is populated by dopamine, serotonin (5-hydroxytryptamine, 5-HT), gamma-aminobutyric acid (GABA), and glutamatergic neurons (Bubar and Cunningham 2006; Yamaguchi et al. 2007). Stimulation of the VTA dopamine neurons leads to increased release of the neurotransmitter dopamine in the NAc, resulting in the acute sensations of euphoria or reward known to initiate drug addiction.

Since dopamine is crucial for the onset of addiction, the regulation of dopamine neural processes is of particular interest. The 5-HT neurons from dorsal raphe (DR) nucleus densely innervate the VTA and NAc, thereby making them well situated to exert influence on the mesolimbic dopamine pathways (Bubar and Cunningham 2006; Halliday and Tork 1989). Although it is well known that 5-HT has the ability to mediate dopamine neurotransmission (Fletcher et al. 2004), the direction of modulation differs according to different subtypes of the 5-HT receptor population that are stimulated (Fletcher et al. 2004). The 5-HT_2 receptor family in particular is linked to the control of central dopaminergic activity (Di Matteo et al. 2000). It has been established that 5-HT_{2C}, one of the three subtypes of the 5-HT_2 receptor, plays a prominent role in inhibiting dopamine neurotransmission in the midbrain (Di Matteo et al. 2002; Fletcher et al. 2004). Studies using the 5-HT_{2C} agonist RO600175 displayed a reduction in the firing rate of VTA dopamine neurons, resulting in decreased levels of dopamine release in the NAc and frontal cortex. The 5-HT_{2C} antagonist, SB 242084, reversed this effect (Di Matteo et al. 2000; Fletcher et al. 2004). When administered alone, SB 242084 resulted in an increase of the basal firing rate of VTA dopamine neurons (Di Matteo et al. 2000; Fletcher et al. 2004). These results support that 5-HT_{2C}

exerts tonic inhibitory control over the mesolimbic dopamine system, the reward center of the brain. Thus, it is plausible that 5-HT$_{2C}$ influences drug addiction. Research exploring this relationship discovered that treatment with RO600175 attenuated drug-induced activity and self-administration of both cocaine and nicotine, effects that were reversed by SB 242084 (Grottick et al. 2000, 2001). Further evidence demonstrated that administration of RO600175 could also dampen self-administration of food, supporting the generality of 5-HT$_{2C}$ in regulating the reward pathway of the brain (Grottick et al. 2001). The 5-HT$_{2C}$ receptor belongs to a superfamily of a G-protein-coupled receptor (GPCR) consisting of seven transmembrane domains, one extracellular N-terminal, one intracellular C-terminal, three extracellular loops, and three intracellular loops. The 5-HT$_{2C}$ receptor is involved in phosphatidylinositol hydrolysis (Muller and Carey 2006). The three intracellular loops of the 5-HT$_{2C}$ receptor contain phosphorylation sites that may be activated by serotonin or agonists such as RO600175 (Fletcher et al. 2004). Phosphorylation of 5-HT$_{2C}$ receptor prevents receptor desensitization, enhances resensitization, and is necessary for efficient signaling (Backstrom et al. 2000; Muller and Carey 2006). Therefore, by regulating the phosphorylation of 5-HT$_{2C}$ receptors, it may be possible to exert control over the mesolimbic pathway (Fig. 16.1).

The phosphorylation state of receptors can be modulated by phosphatases, enzymes that hydrolyze phosphate groups from their substrates. Phosphatase and tensin homologue deleted on chromosome 10 (PTEN) is a dual lipid–protein phosphatase that functions as a tumor suppressor involved in cell cycling, translation, and apoptosis (Simpson and Parsons 2001). PTEN utilizes its lipid phosphatase activity to dephosphorylate PtdIns-3,4,5-P3, opposing the phosphatidylinositol 3-kinase/Akt pathway involved in cellular survival and proliferation (Leslie and Downes 2004; Simpson and Parsons 2001). Moreover, the protein phosphatase activity of PTEN allows it to modulate the mitogen-activated kinase pathway, which plays a similarly vital role to the Akt pathway in regulating proliferation and differentiation (Waite and Eng 2002). It was discovered that PTEN is widely expressed in the brain where it is involved in neuronal development and adult neuronal function (Lachyankar et al. 2000). In addition, we recently found that PTEN regulates hippocampal extrasynaptic signaling through direct protein–protein interaction with the N-methyl-D-aspartate (NMDA) subtype of glutamate receptors (Ning et al. 2004). The significance of interaction of PTEN with other proteins in the brain, the phosphatase ability of PTEN, and the wide distribution of PTEN in most brain neurons, if not all, suggest that PTEN may play prominent roles in various brain functions through its ability to interact with other proteins and protein phosphorylation. Thus, considering that the phosphorylation state of 5-HT$_{2C}$ receptor may impact the activity of the mesolimbic dopamine system as described above, we initially hypothesized that PTEN may directly interact with 5-HT$_{2C}$ receptor in VTA dopamine pathway so as to govern the phosphorylation of 5-HT$_{2C}$ receptor, indirectly controlling the reward pathway of the brain (Fig. 16.1).

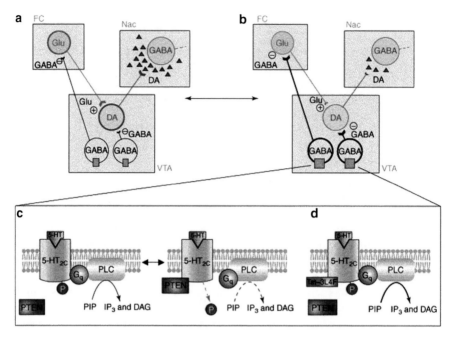

Fig. 16.1 PTEN modifies 5-HT$_{2C}$-receptor function in the ventral tegmental area (*VTA*), and disruption of PTEN-5-HT$_{2C}$ receptor coupling reduces drug-of-addiction-induced DA activity increases in the nucleus accumbens (*NAc*). 5-HT$_{2C}$ receptors are found in the VTA in gamma-aminobutyric acid (*GABA*) and, possibly, DA neurons. (**a**) Low-level activation of 5-HT$_{2C}$ receptors in VTA GABA-containing neurons causes only a small inhibition of frontal cortex (*FC*) glutamate-containing (*Glu*) neurons and of VTA DA-containing neurons (thin lines), resulting in a high level of DA activity in the NAc. (**b**) High-level activation of 5-HT$_{2C}$ receptors in VTA GABA-containing neurons causes an increase in GABA activity in the VTA and FC (thick lines) that suppresses the firing of DA neurons in the VTA directly or by reduced activation of FC projections. This results in reduced NAc DA activity. (**c**) 5-HT$_{2C}$ receptors associate with Gq and activate PLC activity. PLC catalyses PIP breakdown, resulting in formation of the second messengers inositol trisphosphate (*IP$_3$*) and diacylglycerol (*DAG*). Phosphorylation (*P*) of the 5-HT$_{2C}$ receptor prevents receptor desensitization after 5-HT or agonist binding. The protein–protein association of PTEN-5-HT$_{2C}$ receptor causes receptor dephosphorylation and, thereby, inactivation. (**d**) Prevention of PTEN-5-HT$_{2C}$ receptor coupling by the peptide Tat-3L4F increases 5-HT$_{2C}$ receptor (Reprinted from Muller and Carey 2006. With permission from Elsevier Science)

16.2 PTEN and 5-HT$_{2C}$ Receptor Interaction

16.2.1 Involvement of the Third Intracellular Loop of 5-HT$_{2C}$ Receptor in the Physical Interaction Between 5-HT$_{2C}$ Receptor and PTEN

To test the hypothesis, the first question we have to answer is whether PTEN interacts physically with 5-HT$_{2C}$ receptors. Because PC12 cells express both PTEN (Lachyankar et al. 2000; Musatov et al. 2004) and 5-HT$_{2C}$ receptors

(Flomen et al. 2004), we tried to identify the possible existence of PTEN:5-HT$_{2C}$ receptor complexes in cultured PC12 cells using coimmunoprecipitation (Co-IP) assay. A complex was immunoprecipitated from PC12 cell lysates by a 5-HT$_{2C}$ receptor antibody, and the existence of PTEN in the complex was confirmed via Western blotting using a PTEN antibody (Ji et al. 2006). In parallel, 5-HT$_{2C}$ receptor was immunoprecipitated from the PC12 cell lysates by the PTEN antibody (Ji et al. 2006). These results demonstrate a physical interaction between PTEN and 5-HT$_{2C}$ receptors.

Then, we tried to determine the exact domain of 5-HT$_{2C}$ receptor responsible for its interaction with PTEN employing a pull-down assay, which utilizes a purified and tagged or labeled bait protein to generate a specific affinity support able to bind and purify a prey protein from a lysate sample, or other protein-containing mixtures. In an attempt to determine the site of interaction, a fusion protein incorporating the carboxyl-terminal tail of the 5-HT2cR was synthesized. The belief that this particular section was responsible for the interaction between PTEN and 5-HT$_{2C}$ receptor was due to the knowledge that the C-terminal of GPCRs commonly contains protein residues that couple with other proteins (Lee et al. 2002, 2003). Unfortunately, our original pull-down assay excluded a direct binding of PTEN with the C-terminal domain of 5-HT$_{2C}$ receptors. We then focused on the third intracellular loop of the 5-HT$_{2C}$ receptor, which is prominently longer than the C-terminal region. A pull-down assay utilizing two fusion proteins (one containing the full sequence of the C-terminal, the other comprising the entire sequence of the third intracellular loop) showed that the third intracellular loop, but not the C-terminal or GST alone, precipitated PTEN from PC12 lysates (Ji et al. 2006), thereby suggesting the critical involvement of the third intracellular loop of 5-HT$_{2C}$ receptor in the physical interaction between 5-HT$_{2C}$ receptor and PTEN.

While a motif of several amino acids is usually required for protein–protein interaction, the third intracellular loop of 5-HT$_{2C}$ receptor consists of 87 amino acids. We therefore tried to identify the exact motif within the third intracellular loop of 5-HT$_{2C}$ receptor that is responsible for the physical interaction between 5-HT2cR and PTEN. The third intracellular loop was split into five segments: 3L1F (Leu237-Gly252), 3L2F (His253-Asn267), 3L3F (Cys268-Asn282), 3L4F (Pro283-Arg297), and 3L5F (Pro298-Lys313). Each segment was integrated into a fusion protein and a pull-down assay using PC12 cell cultures demonstrated that 3L4F, but not 3L1F, 3L2F, 3L3F, 3L5F, or GST alone, precipitated PTEN (Ji et al. 2006). The above experiments provide the first evidence that 5-HT$_{2C}$ receptor physically interact with PTEN through the third intracellular loop of 5-HT$_{2C}$ receptor, but these experiments did not answer the question whether 5-HT$_{2C}$ receptor interact with PTEN directly or indirectly though additional protein(s). To answer this question, we conducted an in vitro pull-down assay in which only purified PTEN and testing fusion proteins were added, so that if PTEN is precipitated by one or more fusion proteins originating from 5-HT$_{2C}$ receptor, it is strongly suggestive that PTEN directly interact with 5-HT$_{2C}$ receptor. We showed that PTEN was selectively precipitated by the fusion proteins containing the third loop and

3L4F, suggesting the necessity of the 3L4F motif in mediating the direct interaction between PTEN and 5-HT$_{2C}$ receptor (Ji et al. 2006). All the data collected so far just indicate the importance of 3L4F in mediating 5-HT$_{2C}$ receptor coupling PTEN, but it is still unknown whether 3L4F is the only motif in 5-HT$_{2C}$ receptor that makes major contributions to protein–protein interaction between PTEN and 5-HT$_{2C}$ receptor. This is because we have not examined whether other domains of 5-HT$_{2C}$ receptor, including the first or second intracellular loop, transmembrane domains, extracellular loops, or extracellular N-terminal, may also contain motif(s) responsible for 5-HT$_{2C}$ receptor coupling with PTEN. Our further experiment demonstrated that addition of synthesized 3L4F to PC12 cell culture lysate before the conduction of Co-IP completely blocked Co-IP between 5-HT$_{2C}$ receptor and PTEN (Ji et al. 2006). Not only do these data provide convincing evidence indicating the importance of 3L4F in the interaction of 5-HT$_{2C}$R with PTEN, but they also suggest that 3L4F can be used to competitively disrupt the interaction between PTEN and 5-HT2cR.

In summary, we have established using PC12 cells that 5-HT$_{2C}$ receptor interacts directly with PTEN through the 3L4F motif of its third intracellular loop. These results led us to further explore whether 5-HT2cR also forms heterodimer with PTEN through 3L4F in the rat VTA, a brain region critically involved in drug reward. We found that similar to PC12 cell culture, addition of synthesized 3L4F to VTA lysate before the conduction of Co-IP completely blocked Co-IP between 5-HT$_{2C}$ receptor and PTEN (Ji et al. 2006), suggesting that 5-HT$_{2C}$ receptor:PTEN heterodimers exist in the rat VTA and could be disrupted by the 3L4F peptide.

16.3   PTEN Modulates 5-HT$_{2C}$ Receptor Through Its Phosphatase Activity

While PTEN is able to dephosphorylate proteins through its protein phosphastase activity (Leslie and Downes 2004), phosphorylation can occur to the intracellular loops of 5-HT$_{2C}$ receptor as described above. Thus, given that PTEN forms direct protein–protein interaction with 5-HT$_{2C}$ receptor shown above, it is plausible to hypothesize that PTEN may modulate the phosphorylation of 5-HT$_{2C}$ receptor through its protein phosphatase activity. To critically test this hypothesis, we prepared two types of stable PC12 cell lines. The first type of cell line demonstrated stable downregulation of PTEN after transfection of the cell line with vectors overexpressing PTEN-targeted small interfering RNA (siRNA), which results in prominently knocking down the expression of PTEN. The other cell line was transfected with wild-type PTEN constructs, which are able to overexpress PTEN. The 5-HT$_{2C}$ receptor agonist RO6001675 induced phosphorylation in PTEN stable knockdown cells but not in wild-type PTEN-overexpressing cells

(Ji et al. 2006), suggesting that PTEN is able to dephosphorylate 5-HT$_{2C}$ receptor to counteract RO600175-induced phosphorylation.

16.3.1   PTEN Dephosphorylates 5-HT$_{2C}$ Receptor via Its Protein Phosphatase but Not Its Lipid Phosphatase Activity

The above results did not show, however, the detailed mechanism underlying PTEN dephosphorylation of 5-HT$_{2C}$ receptor. In an attempt to discover whether PTEN utilizes its protein phosphatase or lipid phosphatase to regulate 5-HT$_{2C}$ receptor phosphorylation, two additional types of stable PC12 cell lines were prepared. G129R, a PTEN mutant eliminating both protein and lipid phosphatase functions (Furnari et al. 1997), was overexpressed in one type of cell line. G129E, another PTEN mutant without the lipid phosphatase activity only (Myers et al. 1998), was overexpressed in the other cell line. RO600175 was able to phosphorylate 5-HT$_{2C}$ receptor in the G129E cells but not in G129R cells (Ji et al. 2006), indicating that PTEN dephosphorylates 5-HT2cR via its protein phosphatase but not its lipid phosphatase activity.

Next, we examined whether disruption of protein–protein coupling between 5-HT2cR and PTEN could block the ability of PTEN to dephosphorylate phosphorylated 5-HT2cR. In order to do so, 3L4F peptide was rendered cell permeable by fusing it to the cell-membrane transduction domain of an human immunodeficiency virus (HIV) Tat protein (YGRKKRRQRRR) to obtain the Tat-3L4F peptide. Bath application of the Tat-3L4F into wild-type PTEN-overexpressing cells before pretreatment with Ro600175 abolished the capacity of PTEN to dephosphorylate Ro600175-induced phosphorylation of 5-HT2cR (Ji et al. 2006), suggesting that the Tat-3L4F peptide is effective in blocking PTEN from dephosphorylating 5-HT2cR. Considering that a previous study (Backstrom et al. 2000) showed that the site of phosphorylation of 5-HT2cR was on its C-terminal, how PTEN dephosphorylates 5-HT2cR by binding to its third intracellular loop is not clear. One possibility is that 5-HT$_{2C}$ receptor may interact with PTEN through additional contacting sites on its C-terminal (Ji et al. 2006).

16.4   Disruption of PTEN Coupling to 5-HT$_{2C}$ Receptor Reduces VTA Dopamine Activity

While phosphorylation of 5-HT$_{2C}$ receptor blocks its desensitization, enhances its resensitization, and is necessary for efficient signaling (Backstrom et al. 2000; Muller and Carey 2006), direct protein–protein coupling between PTEN and 5-HT$_{2C}$ receptor provides PTEN with the capacity to dephosphorylate 5-HT$_{2C}$ receptor. PTEN may therefore be able to regulate the function of 5-HT$_{2C}$ receptor, which could be blocked by Tat-3L4F-mediated disruption of PTEN coupling with 5-HT$_{2C}$ receptor. Previous research has demonstrated that the activity or firing rate of VTA dopamine neurons innervating the NAc was increased by blockade of the 5-HT$_{2C}$ receptor and decreased by activation of the 5-HT$_{2C}$ receptor (Di Matteo et al. 2002; Higgins and Fletcher 2003). To determine whether PTEN can regulate the firing rate of VTA dopamine neurons through its interaction with 5-HT$_{2C}$ receptor, we conducted in vivo recording of the activity of VTA dopamine neurons directly innervating the NAc. We first confirmed that selective 5-HT$_{2C}$ receptor antagonist and agonist decreased and increased the firing rate of VTA dopamine neurons, respectively (Ji et al. 2006), thereby supporting the previous findings that VTA dopamine neurons are tonically inhibited by 5-HT$_{2C}$ receptor. Since drug reward is associated with increased firing of VTA dopamine neurons resulting in elevated levels of dopamine in the NAc, controlling the firing rate through the action of the 5-HT$_{2C}$ receptor holds great therapeutic potential. Because PTEN interacts with 5-HT$_{2C}$ receptor by dephosphorylating the phosphorylation induced by the 5-HT$_{2C}$ receptor agonist R0600175, disrupting this coupling may inhibit the firing rate of the VTA dopamine neurons by maintaining the activation of 5-HT$_{2C}$ receptor. In agreement with this hypothesis, both a systemic and an intra-VTA injection of Tat-4L4F interfering peptide significantly diminished the firing rate of VTA dopamine neurons, which was effectively reversed by the 5-HT$_{2C}$ antagonist SB 242084 (Ji et al. 2006). These results suggest that disruption of the interaction between PTEN and 5-HT$_{2C}$ receptor using Tat-3L4F can inhibit the VTA dopamine neurons and reduce dopamine release in the NAc.

16.5   Tat-3L4F Can Diminish the Rewarding Effects of Abused Drugs

16.5.1   Electrophysiological Evidence

The next endeavor was to examine if Tat-3L4F interfering peptide could suppress the rewarding effects of abused drugs that lead to addiction. Marijuana is one of many abused drugs that induce gratifying effects by enhancing dopamine levels in the NAc after stimulating VTA dopamine neurons (Zangen et al. 2006). With the in vivo recording of the activity of VTA dopamine neurons again, we observed that when Δ9-tetrahydrocannabinol (THC), the major psychoactive ingredient of marijuana, was systemically injected alone, dopamine neuronal activity significantly increased as expected (Ji et al. 2006). However, when THC and Tat-3L4F were systemically administered together, the firing rate of VTA dopamine neurons was significantly suppressed, resembling injection by the 5-HT$_{2C}$ agonist RO600175, and reversed by the 5-HT$_{2C}$ antagonist SB 242084 (Ji et al. 2006). These results imply that Tat-3L4f interfering peptide may be able to block the rewarding effects of marijuana.

16.5.2 Behavioral Evidence

16.5.2.1 Conditioned Place Preference Paradigm

In order to test further the above argument, we employed conditioned place preference (CPP) paradigm, which consists of preconditioning, conditioning, and postconditioning phases. In the preconditioning phase, rats were allowed free access to two distinct compartments for 15 min, and the time spent in each compartment was recorded. During the conditioning phase, the same rats were placed in the conditioned compartment for 30 min after receiving an abused drug such as THC on days 1, 3, 5, and 7 and on alternate days (i.e., days 2, 4, 6, and 8) were placed in the opposite compartment for 30 m after receiving vehicle (Braida et al. 2004). In the postconditioning phase, rats were allowed free access to the two compartments for 15 min, and the motivational responses were evaluated by comparing the time spent in the conditioned compartment before and after conditioning on day 9, during which neither drug nor vehicle was injected. The hypothesis was that Tat-3L4F would reduce the CPP generated by THC. Rats showed significant preference for the drug-paired compartment, which is in agreement with previous study. However, rats that were injected with Tat-3L4F prior to receiving THC in the conditioning phase did not display a preference for either compartment (Ji et al. 2006), suggesting the possibility that Tat-3L4F interfering peptide effectively suppresses the rewarding effects produced by THC. Most drugs of abuse may be likely to produce rewarding effects by activating VTA dopamine neurons (Koob and Moal 1997; Nestler 2004), and thus, the capability of Tat-3L4F to suppress the firing rate of VTA dopamine neurons shown above indicates the potential of Tat-3L4F in suppressing the rewarding effects produced by most drugs of abuse. In support of this idea, we further found that systemic injection of Tat-3L4F prior to daily nicotine injection in the conditioning phase effective suppressed the CPP produced by nicotine (Ji et al. 2006).

The current understanding of drug addiction is a tight association of compulsive drug use with molecular and cellular mechanisms underlying long-term associative learning and memory (Hyman et al. 2006; Kauer 2004; Robbins and Everitt 1999, 2002). The CPP test is highly reliant on classical conditioning, which rests on learning and memory capabilities. Accordingly, the suppression of THC- and nicotine-induced CPP by Tat-3L4F interfering peptide may be produced by the suppression of drug reward, the learning and memory associated with drug reward, or both.

16.5.2.2 Standard Morris Water Maze

To determine whether Tat-3L4F interfering peptide is able to suppress the capacity of rats to learn and memorize, we recently examined the effects of this interfering peptide on the requisition and retrieval of spatial memory. A standard Morris water maze was conducted as previously described (Ferrarl et al. 1999; Hannesson and Corcoran 2000; Remondes and Schuman 2004). Briefly, the water maze was a circular plastic pool (200 cm diameter) filled with $22 \pm 1°C$ water made opaque so that rats could locate only a submerged platform for escape with the help of distinctive distal visual cues surrounding the pool. The escape latency and distance, i.e., the time and distance each rat required to locate the hidden platform, were used as an index of hippocampal-dependent spatial memory. Two experiments were conducted. Experiment 1 was designed to examine the effects of Tat-3L4F on spatial learning and memory with injection of the potent cannabinoid HU210 as positive control (Ferrarl et al. 1999; Hill et al. 2004). Three groups of rats received systemic injections of vehicle, Tat-3L4F, and HU210 once daily for 4 days and were trained on four consecutive daily sessions. Experiment 2 was designed to examine the effects of Tat-3L4F on the retrieval of spatial memory alone. Rats received four daily sessions of training with four trials in each session, and 1 day after the last training trial, two groups of rats were injected with vehicle and Tat-3L4F, respectively, followed 1 h later by a probe trial. Then, the same rats were given the same dose of vehicle or Tat-3L4F injection once every other day, and 28 days after the last training trial, rats were tested for long-term memory with one probe trial.

Rats receiving vehicle injection showed significant spatial learning during the four trials per session (Li et al. 2008), indicating that our training protocol is sufficient to produce spatial learning. This idea is further supported by our findings that daily treatment with HU210 at the dosage of 100 µg/kg prominently suppressed spatial learning (Li et al. 2008), because similar results have recently been reported by two groups (Ferrarl et al. 1999; Hill et al. 2004). Having once acquired the spatial memory, vehicle-treated animals were able to retrieve the spatial memory 1 and 28 days after training (Li et al. 2008), suggesting that our training protocol is sufficient to produce both short- and long-term memory. Vehicle injection before the retrieval of spatial memory did not significantly affect both short- and long-term memory, further suggesting that our Morris water maze paradigm may allow us to study the retrieval of spatial memory.

Animals that received four daily injections of Tat-3L4F showed gradual spatial learning and both short- and long-term memory that were indistinguishable from vehicle-treated controls (Li et al. 2008), suggesting that Tat-3L4F treatment does not significantly affect rats' acquisition of spatial memory. After requisition of spatial memory, during which no treatment was given, Tat-3L4F treatment before probe trial did not produce significant inhibitory effects on both short- and long-term memory (Li et al. 2008), thus supporting the view that Tat-3L4F treatment does not significantly affect rats in retrieval of spatial memory. Because we also gave four daily injections of Tat-3L4F at the dosage of 1 µmol/ kg to suppress the conditioned place preference induced by THC and nicotine (Ji et al. 2006), these results together suggest that Tat-3L4F is able to suppress the rewarding effects of drugs of abuse without significant effects on learning and memory mechanism associated with conditioned place preference.

16.5.3   Tat-3L4F Does Not Produce Significant Behavioral Side Effects

Previous studies have utilized nonselective treatments focused on enhancing 5-HT neurotransmission, such as L-tryptophan (Fletcher et al. 2004; McGregor et al. 1993) and fluoxetine (Carroll et al. 1990; Fletcher et al. 2004; Richardson and Roberts 1991), which were shown to reduce self-administration of cocaine and amphetamine. It was suggested that this occurred due to stimulation of 5-HT$_{2C}$ receptor in the VTA (Fletcher et al. 2004). However, nonselective treatments typically have more side effects. While the selective 5-HT$_{2C}$ agonist RO600175 inhibits the VTA dopamine pathway, RO600175 has also been shown to cause significant side effects in rats such as hypophagia, hypolocomotion, penile erection, motor functional suppression, and anxiety (Alves et al. 2004; Grottick et al. 2000; Higgins and Fletcher 2003; Wood 2003).

Although PTEN inhibitors may reduce the dephosphorylation effects of PTEN on 5-HT$_{2C}$ receptor to suppress the rewarding effects of abused drugs, the role of PTEN as a tumor suppressor may make the use of PTEN inhibitors disadvantageous (Ji et al. 2006). Tat-3L4F interfering peptide is able to suppress the dephosphrylation effects of 5-HT$_{2C}$ receptor while, theoretically, it exerts no significant effects on PTEN itself. However, both Tat-3L4F and RO600175 are able to suppress the firing rate of VTA dopamine neurons and CPP induced by THC, and Tat-3L4F may also share the "side effects" of RO600175. To test this hypothesis, we conducted a battery of behavioral tests showing that RO600175, but not Tat-3L4F, caused anxiety, penile erection, hypophagia, hypolocomotion, and motor functional suppression (Ji et al. 2006).

A possible explanation for the occurrence and absence of those side effects produced by RO600175 and Tat-3L4F, respectively, is that PTEN may regulate an intracellular signaling pathway that is different from that regulated by RO600175. Thus, despite the fact that RO600175 has the strong ability to suppress the rewarding effects of THC, Tat-3L4F may be a safer alternative.

References

Alex K, Pehek E (2007) Pharmacologic mechanisms of serotonergic regulation of dopamine neurotransmission. Pharmacol Ther 113:296–320.

Alves SH, Pinheiro G, Motta V, et al (2004) Anxiogenic effects in the rat elevated plus-maze of 5-HT(2C) agonists into ventral but not dorsal hippocampus. Behav Pharmacol 15:37–43.

Backstrom JR, Price RD, Reasoner DT, et al (2000) Deletion of the serotonin 5-HT2c receptor PDZ recognition motif prevents receptor phosphorylation and delays resensitization of receptor responses. J Biol Chem 275:23620–23626.

Braida D, Iosue S, Pegorini S, et al (2004) Delta9-tetrahydrocannabinol-induced conditioned place preference and intracerebroventricular self-administration in rats. Eur J Pharmacol 506:63–69.

Bubar MJ, Cunningham KA (2006) Serotonin 5-HT2A and 5-HT2C receptors as potential targets for modulation of psychostimulant use and dependence. Curr Top Med Chem 6:1971–1985.

Carroll ME, Lac ST, Asencio M, et al (1990) Fluoxetine reduces intravenous cocaine self-administration in rats. Pharmacol Biochem Behav, 35:237–244.

Di Matteo V, Di Giovanni G, Di Mascio M, et al (2000) Biochemical and electrophysiological evidence that Ro 60-0175 inhibits mesolimbic dopaminergic function through serotonin2c receptors. Brain Res 865:85–90.

Di Matteo V, Cacchio M, Di Giulio C, et al (2002) Role of serotonin(2C) receptors in the control of brain dopaminergic function. Pharmacol Biochem Behav 71:727–734.

Everitt BJ, Robbins TW (2005) Neural systems of reinforcement for drug addiction: from actions to habits to compulsion. Nat Neurosci 8:1481–1489.

Ferrarl FA, Ottani R, Vivoli, D, et al (1999) Learning impairment produced in rats by the cannabinoid agonist HU 210 in a water-maze task, Pharmacol Biochem Beh 64:555–561.

Fletcher PJ, Chintoh AR, Sinyard J, et al (2004) Injection of the 5-HT2c receptor agonist Ro60-0175 into the ventral tegmental area reduces cocaine-induced locomotor activity and cocaine self-administration. Neuropsychopharmacology 29:308–318.

Flomen R, Knight J, Sham P, et al (2004) Evidence that RNA editing modulates splice site selection in the 5-HT2C receptor gene. Nucleic Acids Res 32:2113–2122.

Furnari FB, Lin H, Huang HS, et al (1997) Growth suppression of glioma cells by PTEN requires a functional phosphatase catalytic domain. Proc Natl Acad Sci USA 94:12479–12484.

Grottick AJ, Fletcher PJ, Higgins GA (2000) Studies to investigate the role of 5-HT(2C) receptors on cocaine- and food-maintained behavior. J Pharmacol Exp Ther 295:1183–1191.

Grottick AJ, Corrigall WA, Higgins GA (2001) Activation of 5-HT2c receptors reduces the locomotor and rewarding effects of nicotine. Psychopharmacology 157:292–298.

Halliday G, Tork I (1989) Serotonin-like immunoreactive cells and fibres in the rat ventromedial mesencephalic tegmentum. Brain Res Bull 22:725–735.

Hannesson DK, Corcoran ME (2000) The effects of kindling on mnemonic function. Neurosci Biobehav Rev 24:725–751.

Higgins GA, Fletcher PJ (2003) Serotonin and drug reward: focus on 5-HT2C receptors. Eur J Pharmacol 480:151–162.

Hill MN, Froc DJ, Fox CJ, et al (2004) Prolonged cannabinoid treatment results in spatial working memory deficits and impaired long-term otentiation in the CA1 region of the hippocampus in vivo. Eur J Neurosci 20:859–863.

Hyman SE, Malenka RC, Nestler EJ (2006) Neural mechanisms of addiction: the role of reward-related learning and memory. Ann Rev Neurosci 29:565–598.

Ji SP, Zhang Y, Van Cleemput J, et al (2006) Disruption of PTEN coupling with 5-HT2C receptors suppresses behavioral responses induced by drugs of abuse. Nat Med 12:324–329.

Kalivas PW, Volkow ND (2005) The neural basis of addiction: a pathology of motivation and choice. Am J Psychiatry 162:1403–1413.

Kauer JA (2004) Learning mechanisms in addiction: synaptic plasticity in the ventral tegmental area as a result of exposure to drugs of abuse. Annu Rev Physiol 66:447–475.

Koob GF, Moal ML (1997) Drug abuse: hedonic homeostatic dysregulation. Science 278:52–58.

Lachyankar MB, Sultana N, Schonhoff CM, et al (2000) A role for nuclear PTEN in neuronal differentiation. J Neurosci 20:1404–1413.

Laviolette SR, van der Kooy D (2004) The neurobiology of nicotine addiction: bridging the gap from molecule to behavior. Nat Rev Neurosci 5:55–65.

Lee FS, Rajagopal R, Chao MV (2002) Distinctive features of Trk neurotrophin receptor transactivation by G protein-coupled receptors. Cytokine Growth Factor Rev 13:11–17.

Lee SP, O'Dowd BF, George SR (2003) Homo- and hetero-oligomerization of G protein-coupled receptors. Life Sci 74:173–180.

Leslie NR, Downes CP (2004) PTEN function: how normal cells control it and tumour cells lose it. Biochem J 382:1–11.

Li X, Zhang Y, Zhang X (2008) Tat-3L4F does not significantly affect spatial learning and memory. Behav Brain Res 193:170–173.

McGregor A, Lacosta S, Roberts DC (1993) L-tryptophan decreases the breaking point under a progressive ratio schedule of intravenous cocaine reinforcement in the rat. Pharmacol Biochem Behav 44:651–655.

Muller CP, Carey RJ (2006) Intracellular 5-HT2c receptor dephosphorylation: a new target for treating drug addiction. Trends Pharmacol Sci 27:455–458.

Musatov S, Roberts J, Brooks AI, et al (2004) Inhibition of neuronal phenotype by PTEN in PC12 cells. Proc Natl Acad Sci USA 101:3627–3631.

Myers MP, Pass I, Batty IH, et al (1998) The lipid phosphatase activity of PTEN is critical for its tumor supressor function. Proc Natl Acad Sci USA 95:13513–13518.

Nestler EJ (2004) Historical review: molecular and cellular mechanisms of opiate and cocaine addiction. Trends Pharmacol Sci 25:210–218.

Ning K, Pei L, Liao M, et al (2004) Dual neuroprotective signaling mediated by downregulating two distinct phosphatase activities of PTEN. J Neurosci 24:4052–4060.

Remondes M, Schuman EM. (2004) Role for a cortical input to hippocampal area CA1 in the consolidation of a long-term memory. Nature 431:699–703.

Richardson NR, Roberts DC (1991) Fluoxetine pretreatment reduces breaking points on a progressive ratio schedule reinforced by intravenous cocaine self-administration in the rat. Life Sci 49:833–840.

Robbins TW, Everitt BJ (1999) Drug addiction: bad habits add up. Nature 398:567–570.

Robbins TW, Everitt BJ (2002) Limbic–striatal memory systems and drug addiction. Neurobiol Learn Memory 78:625–636.

Simpson L, Parsons R (2001) PTEN: Life as a tumor suppressor. Exp Cell Res 264:29–41.

Waite K, Eng C (2002) Protean PTEN: form and function. Am J Hum Genet 70:829–844.

Wood MD (2003) Therapeutic potential of 5-HT2C receptor antagonists in the treatment of anxiety disorders. Curr Drug Targets CNS Neurol Disord 2:383–387.

Yamaguchi T, Sheen W, Morales M (2007) Glutamatergic neurons are present in the rat ventral tegmental area. Eur J Neurosci 25:106–118.

Zangen A, Solinas M, Ikemoto S, et al (2006) Two brain sites for cannabinoid reward. J Neurosci 26:4901–4907.

Chapter 17
The Role of Serotonin in Eating Behavior: Focus on 5-HT$_{2C}$ Receptors

Jason C.G. Halford

17.1   Introduction

An understanding of the role of serotonin in eating behavior is inextricably linked to the history of and continued search for appetite suppressing drugs. Appetite suppressants are characterized as drugs that alter energy balance to produce an energy deficit and thus weight loss by changing eating behavior to reduce caloric intake. Nonserotonergic drugs, particularly amphetamines in its various forms, have been used to control hunger and cause weight loss from the 1930s despite their psychological effects, effects on blood pressure, and abuse potential. These were the first true "appetite suppressants." Other monoaminergic drugs with lower abuse potential such as phentermine, diethylpropion, and phenylpropanolamine were also used for weight control, but cardiovascular stimulation, insomnia, anxiety, and irritability remained an issue with many of these drugs. However, from the late 1960s and the 1970s the beneficial effects of the amphetamine-related compound fenfluramine on body weight were noted.

17.1.1   Early Research into the Role of Endogenous 5-HT in Appetite Regulation

Fenfluramine appeared to be equally as effective as amphetamine and lacked its side-effect profile and abuse potential. It was also a serotonin (5-hydroxytryptamine or 5-HT)-acting drug. It was the combination of this drugs specific pharmacological and behavioral properties that triggered much of the early research into the role of endogenous 5-HT in appetite regulation. In 1997, all forms of fenfluramine

J.C.G. Halford (✉)
Kissileff Laboratory for the Study of Human Ingestive Behaviour, School of Psychology, University of Liverpool, Liverpool, UK, L69 7ZA
e-mail: j.c.g.halford@liverpool.ac.uk

G. Di Giovanni et al. (eds.), *5-HT$_{2C}$ Receptors in the Pathophysiology of CNS Disease*,   339
The Receptors 22, DOI 10.1007/978-1-60761-941-3_17,
© Springer Science+Business Media, LLC 2011

were withdrawn from the global market due to serious side-effect issues of valvular heart disease and pulmonary hypertension (Abeniam et al. 1996) an effect associated with peripheral 5-HT$_{2B}$ receptors. However, this withdrawal coincided with the introduction of sibutramine, a noradrenergic and serotonergic reuptake inhibitor, for the treatment of obesity.

The monoamine neurotransmitter serotonin has been intimately linked with appetite expression of over 40 years. Moreover, serotonin-acting drugs constitute the first class of appetite suppressing drugs to clearly modulate human appetite expression without contaminant behavioral effects or abuse potential. For no other peripheral or central target does the wealth of data on the effects of drugs on human appetite exist (Halford et al. 2007). These extensive behavioral data demonstrate that the endogenous hypothalamic serotonin plays an important part in within-meal satiation and postmeal satiety processes. Selectively targeting 5-HT receptors particularly involved in appetite control provides still provides an opportunity for the development of safe, tolerable, and more effective antiobesity agents (Halford et al. 2007).

17.1.2   Appetite Regulation: Episodic and Tonic Signals

To understand the role of 5-HT in appetite control, it is essential to understand the nature of appetite regulation and the episodic and tonic signals critical to appetite expression. Episodic signals are a crucial in the meal-by-meal regulation of energy intake and are critical to both the appetite fluctuations and patterns of eating behavior we undertake throughout the day (Halford and Blundell 2000). They are generated by the anticipation and ingestion of food and the digestion, absorption, and initial metabolism of nutrients. In contrasts, tonic inhibitory signals are generated by the storage and general metabolism of energy. Whilst episodic and tonic factors are composed of distinct processes, they both act to inhibit food intake via common hypothalamic circuitry (Halford and Blundell 2000). Serotonin plays an important part in the episodic regulation of appetite linking peripheral generated signals to hypothalamic circuitry essential to long-term energy balance.

17.1.3   Linking Between 5-HT and Control of Both Food Intake and of Feeding Behavior

5-HT was linked to the control of both food intake and feeding behavior for the first time 40 years ago (Blundell 1977). Blundell (1977) proposed that the 5-HT system had not only an inhibitory role in feeding but also was a key satiety factor. Specifically, he noted that the changes in caloric intake produced by the 5-HT interventions did not disrupt feeding behavior but altered in it in manner consistent with the behavioral effects of ingestion. At the time of this theoretical notion, little was know about the pharmacology of the 5-HT system. However, over the next 20 years significant advances in the identification and classification of numerous novel 5-HT receptors occurred (Hoyer and Martin 1997). The main focus for appetite research became the 5-HT$_1$ and 5-HT$_2$ receptor families, and eventually, postsynaptic 5-HT$_{1B}$ and the 5-HT$_{2C}$ receptors were established to be those criticial to satiety (see later and Chap. 4) (Blundell and Halford 1998). 5-HT neurons involved in appetite control project up towards the hypothalamic region, an area rich in orexigenic and anorexigenic systems critical in energy regulation. These systems include the hypothalamic anorexigenic neuropeptide melanocortin (MC) system.

The involvement of MC system in the regulation of food intake and body weight is well established, and recently it has become evident that the MC system may modulate the effects of 5-HT drugs on rodent feeding behavior (Heisler et al. 2002, 2003). Heisler et al. (2006) demonstrated the role of downstream melanocortin 4 receptors (MC4R) in mediating 5-HT-induced hypophagia. Activation of both 5-HT$_{2C}$ and 5-HT$_{1B}$ receptors produces hypophagia by promoting the release of the endogenous agonist and inhibiting the release of the endogenous antagonist of the MC4R receptor. This effect appears dependent on three mechanisms. First, 5-HT inhibits orexigenic agouti-related peptide (AgRP) neurons in the arcuate nucleus via 5-HT$_{1B}$ receptor activation. This inhibits the releases of the MC4R antagonist AgRP. Second, activation of axonal 5HT$_{1B}$ receptors on AgRP neuronal projections also decreases their inhibitory effect on adjacent anorexigneic pro-opiomelanocortin (POMC) neurons. This promotes the release of the MC4R agonist alpha-melanocyte stimulating hormone (α-MSH). Finally, as previous studies have shown, 5-HT also activates these anorexignic POMC neurons via activation of 5-HT$_{2C}$ receptors again promoting the release α-MSH. These data appear to confirm that the hypothalamus, specifically the MC system, is a key site of integration between the episodic signals via 5-HT and tonic signals via leptin (Halford and Blundell 2000). A recent study has uncovered a novel central 5-HT$_{2C}$ receptor pathway regulating physiologic fasted and fed motor activities, which may represent an integrated mechanism linking feeding behavior and gastrointestinal (GI) motor activities through the gut–brain axis (Fujitsuka et al. 2009). Indeed, 5-HT$_{2C}$ receptor seems to have also a major role through changes of endogenous ghrelin that are closely associated with feeding behavior. Therefore, a functional divergence of central 5-HT$_{2C}$ receptor pathway regulating physiologic GI motor activities that is different from that controlling food intake is likely to exist (Fig. 17.1).

17.2   5-HT Receptors: Feeding Behavior and Body Weight in Rodents

The role of 5-HT in the control of appetite on the structure of rodent feeding behavior were first examined in a series of studies using the 5-HT releasing and reuptake-inhibiting drug fenfluramine (Blundell 1977). Fenfluramine produced reductions in caloric intake that corresponded to changes in feeding behavior normally brought about by ingestion, rather than hyperactivity, sedation, or malaise. Measurements

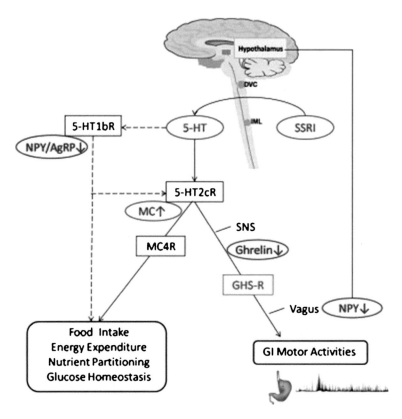

Fig. 17.1 Functional divergence in 5-HT$_{2C}$ receptor pathway. Serotonin (5-HT) inhibits feeding-stimulatory NPY/AgRP neurons through 5-HT$_{1B}$ receptor, while activating feeding-inhibitory *POMC*/CART neurons through 5-HT$_{2C}$ receptors. The primary effect by 5-HT receptor on energy metabolism requires downstream activation of MC4R. A novel 5-HT$_{2C}$ receptor pathway that regulates physiologic gastrointestinal (*GI*) motility, requiring sympathetic activity and leading to inhibition of ghrelin but not MC4R. 5-HT indirect agonist-selective serotonin reuptake inhibitors activate both 5-HT$_{2C}$ receptor pathways, but not the 5-HT$_{1B}$ pathway, for the anorexia and GI motor effects produced. *AgRP* agouti-related peptide, *CART* cocaine- and amphetamine-regulated transcript, *DVC* dorsal vagal complex, *IML* intermediolateral nucleus of the spinal cord, *NPY* neuropeptide Y, *POMC* pro-opiomelanocortin, *SNS* sympathetic nervous system (Reproduced from Fujitsuka et al. 2009. With permission)

or meal number, meal duration, eating rate, and activity demonstrated that fenfluramine (and then latter the more selective isomer d-fenfluramine) reduced food intake without disrupting feeding behavior (Blundell 1977). Fenfluramine and d-fenfluramine were also found to adjust feeding behavior in a manner consistent with the operation of satiety (Blundell and Latham 1978, 1980; Blundell and McArthur 1981; Halford and Blundell 1993).

With the availability of selective serotonin reuptake inhibitors (SSRIs) it became clear that drugs that inhibited the reuptake of 5-HT such as the fluoxetine and sertraline and as well as the serotoninegeric and noradrenergic reuptake inhibitors

(SNRI) sibutramine, produced similar changes in rodent feeding behavior. All of these drugs enhanced the behavioral satiety sequence (BSS) in rodents (a biobehavioral assay of drug action) indicating drug-induced hypophagic was associated with the natural development of satiety (Halford and Blundell 1996a, b; Halford et al. 1998; Clifton et al. 1989; Simansky and Viadya 1990; McGuirk et al. 1992). However, the 5-HT receptors underlying these satiety indicating behavioral effects remained to be elucidated.

5-HT receptor subtype agonists, antagonists, and knockout rodent strains have over time allowed researchers to identify the 5-HT receptors critical for satiety. To summarize these data, selective agonists of 5-HT$_{1B}$ and the 5-HT$_{2C}$ receptors have be shown to produce changes in feeding behavior consistent with the operation of satiety, notably the 5-HT$_{1B}$ receptor agonist CP-94 253, the preferential 5-HT$_{1B/2C}$ receptor agonists meta-chlorophenylpiperazine (mCPP) and trifluromethylphenylpiperazine (TFMPP), and the selective 5-HT$_{2C}$ receptor agonists (Halford and Blundell 1996a, b; Halford et al. 1998; Simansky and Viadya 1990; Lee and Simansky 1997; Lee et al. 2002). Importantly, drugs that have also been shown to agonize other 5-HT receptor subtypes, for example, DOI (5-HT$_{2A/2C}$) or RU-24 969 (5-HT$_{1A/1B}$), have been shown to disrupt the BSS by inducing hyperactivity, a critical side effect (Halford et al. 1998; Kennett and Curzon 1988a; Kitchener and Dourish 1994; Hewitt et al. 2002).

The importance of 5-HT$_{1B}$ and 5-HT$_{2C}$ in the control of rodent feeding behavior has also examined through the pharmacological challenge of 5-HT releasing and reuptake inhibiting drugs using selective receptor subtype antagonists. For d-fenfluramine in particular, strong evidence exists that drug-induced hypophagia is mediated by 5-HT$_{1B}$ and/or 5-HT$_{2C}$ (Clifton 1994; Neill and Cooper 1989; Samanin et al. 1989; Neill et al. 1990; Simansky and Nicklous 2002; Vickers et al. 2001). Similarly, whilst the effects of reuptake inhibitors are notorious difficult to block with antagonists (Wong et al. 1988; Grignaschi and Samanin 1992; Lee and Clifton 1992; Lightowler et al. 1996), the effects of fluoxetine are blocked by 5-HT$_1$ and 5-HT$_2$ antagonism, which also reverses the effect of fluoxetine on the BBS confirming the behavioral as well as the pharmacological specificity of the SSRIs' effect (Halford and Blundell 1996b). Antagonist studies have also demonstrated the importance of and 5-HT$_{2C}$ receptor subtypes in the hypophagic effects of the anti-obesity drug sibutramine (Halford et al. 2007; Jackson et al. 1997). In conclusion, antagonist studies demonstrate that the effects of 5-HT promoting drugs on both intake and behavior are mediated by 5-HT$_1$ and 5-HT$_2$ receptors, the effects of d-fenfluramine specifically by 5-HT$_{1B}$ and 5-HT$_{2C}$ receptor subtypes.

Finally, the key role of 5-HT$_{1B}$ and the 5-HT$_{2C}$ receptors in the hypophagic effects of endogenous 5-HT has also been demonstrated using knockout mice models (see Chap. 4). Mice lacking functional 5-HT$_{2C}$ receptors display marked hyperphagia leading to the development of obesity (Tecott et al. 1995; Nonogaki et al. 2003). These mice were also partly resistant to d-fenfluramine-induced hypophagia (Vickers et al. 1999), confirming the role of this receptor subtype in fenfluramine-induced hypophagia. Additionally, 5-HT$_{1B}$ receptor knockout mice are significantly heavier than wild types (Bouwknecht et al. 2001), an effect associated with significantly greater food consumption.

17.3 5-HT Drugs, Food Intake, and Feeding Behavior in Humans

As stated previously the effects of 5-HT drugs on human appetite expression are also well documented. Rogers and Blundell (Rogers and Blundell 1979) demonstrated that a single dose of fenfluramine (60 mg), when given to lean healthy males, could reduce food intake of a lunchtime meal by 789 kJ (26%), an effect accompanied by significant decreases in eating rate and desire to eat. These changes in eating behavior contrasted with the effects of amphetamine, which increased eating rate instead, despite reducing overall intake. It was later shown that both d-fenfluramine and fluoxetine produced similar changes to human appetite expression as fenfluramine (Goodall and Silverstone 1988; McGuirk and Silverstone 1990). In addition acute doses of sibutramine, a 5-HT and noradrenaline reuptake inhibitor, have been shown to reduce food intake and appetite in lean male participants (Hansen et al. 1998; Chapelot et al. 2000). Fenfluramine, d-fenfluramine, fluoxetine, and sibutramine also bring about similar changes in appetite in the obese (Blundell and Hill 1990; Drent et al. 1995; Pijl et al. 1991; Lawton et al. 1995; Ward et al. 1999; Rolls et al. 1998; Halford et al. 2010). With regard to sibutramine, it should also be noted that Barkeling et al. (2003) have demonstrated a link between sibutramine-induced hypophagia and subsequent sibutramine-induced weight loss in humans.

With regard to the direct agonism of 5-HT receptors, mCPP (a 5-HT $_{1B/2C}$ receptor preferential agonist) has been shown to reliably reduce food intake in humans. The effect of an acute dose of mCPP (0.4 mg/kg) was initially investigated in a double blind, placebo controlled, crossover design, study conducted with lean healthy female volunteers (Walsh et al. 1994). Each participant was dosed orally with either mCPP or a placebo 150 m prior to the presentation of a buffet lunch. At this ad libitum meal, food intake in those receiving mCPP treatments was reduced by 30% (approximately 1,000 kJ). This study also showed that the effect of mCPP on food intake was significantly associated with premeal hunger ratings being reduced. The study was replicated in a larger group of both male and female lean and healthy volunteers (Cowen et al. 1995). The drug was again effective at reducing food intake in women (28% reduction, 1,205 kJ), but also in men (20% reduction, 1,219 kJ). The drug produced significantly reduced hunger ratings prior to the meal in both men and women (150 min post dosing). This effect occurred marginally after peak plasma levels of mCPP (120 min post dosing), and just prior to the lunch.

The effects of mCPP on appetite and body weight have also been examined in the obese (Sargent et al. 1997). In this placebo controlled, double-blinded crossover trial; participants were treated for 14 days with mCPP (20 mg twice daily for women, 25 mg twice daily for men). Significantly more weight (0.8 kg) was lost from baseline with mCPP compared with placebo (0.04 kg). Participants were invited to the laboratory on the penultimate day of dosing and blood samples were taken to assess mCPP level and prolactin response. During this procedure, hunger-rating scales were also completed by the participants. Analysis of the scales subsequently showed that drug treatment produced a significant decrease in hunger ratings. It is important to note that mCPP can also produce transient but significant increases in the self-reported subjective ratings of light-headedness, anxiety, and nausea (Walsh et al. 1994; Cowen et al. 1995). It has also been observed that in response to acute doses of mCPP, transient increases in blood pressure and heart rate can occur (Ghaziuddin et al. 2003).

17.4   Food Intake and Body Weight: Experimental Findings in Rodents

The effect of chronic alterations in caloric intake produced by 5-HT drugs must result in changes in body weight and composition. Thus, a general increase in 5-HT function or selective activation of 5-HT receptors involved in satiety must produce losses in body mass or at least prevent the body weight gain associated with hyperphagia produced by obesity inducing diets. In rodent models of obesity it appears that direct agonism of 5-HT$_{2C}$ receptors does indeed affect weight gain. Vickers et al. (Vickers et al. 2000, 2003), for example, have shown the inhibitory effects of the preferential 5-HT$_{2C}$ receptor agonist mCPP on rodent body weight gain. As mCPP also agonizes several other 5-HT receptors (e.g., 5-HT$_{2A}$ and 5-HT$_{2B}$ receptors), therefore it is theoretically possible that the drug-induced attenuation of body weight gain could be due to activation of any of these receptors. However, using selective antagonists it has been demonstrated that mCPP-induced hypophagia results specifically from activation of the 5-HT$_{2C}$ receptor (Kennett and Curzon 1988b, 1991; Kennett et al. 1997).

The effects of several more selective 5-HT$_{2C}$ receptor agonists on rodent body weight have recently been studied. Vickers et al. (Vickers et al. 2000) infused Ro 60-0175 (26 mg/kg/day) into the animals via implanted mini-pumps for 14 days, which produced a significant reduction in body weight gain. The animals treated with Ro 60-0175 weighed 10% less than controls at the end of the study. Similarly, YM348, another potent and highly selective 5-HT$_{2C}$ receptor agonist, also produced an attenuation of body weight gain over a 2-week treatment period (at doses of 3 and 20 mg/kg/day) (Hayashi et al. 2004). Those animals treated with the higher dose of YM348 weighed 21.5% less than controls at the end of the study. Multiple doses of the novel selective 5-HT$_{2C}$ receptor agonist lorcaserin–APD356 (4.5, 9, 18, and 36 mg/kg) have recently been shown to inhibit the development of dietary-induced obesity (Bjenning et al. 2004). Lorcaserin–APD356 significantly reduced body weight gain in both male and female rats. The reduction in body weight gain was associated with robust hypophagia in the early phase of dosing.

17.5   5-HT Drugs and Weight Loss: Clinical Data

Few drugs, 5-HT or otherwise, are robustly tested in clinical trails for weight control. D-fenfluramine (Redux) was voluntarily withdrawn in 1997 due to primary pulmonary hypertension, a move that dealt a blow to the development of serotoninergic antiobesity compounds (Abeniam et al. 1996). Subsequently, the noradrenergic and 5-HT reuptake inhibitor sibutramine (Reductil, Meridia) was approved for the treatment of obesity. This has been in the market nearly 10 years. Since these, no new 5-HT antiobesity drugs have been licensed (Halford 2006a, b). From the metaanalysis of Haddock et al. (Haddock et al. 2002), fenfluramine produces average placebo subtracted of 2.41 kg and D-fenfluramine one of 3.82 kg. With regard to selective reuptake inhibitors, the noradrenergic and serotoninergic reuptake inhibitor sibutramine has been shown to produce placebo subtracted weight loss of 4.45 or 4.3 kg, in two metaanalyses of clinical data (Arterburn et al. 2004; Padwal et al. 2003).

Considerable effort has been made to develop a new generation of side-effect-free 5-HT drugs (Vickers and Dourish 2004). Some of the compounds have passed into phase 1 and (in the case of BVT-933 from Biovitrum and GlaxoSmithKline) phase 2 trials, it is regrettable that their effects of human appetite, food intake, and body weight remain largely unknown. One issue may have been drug affinity to 5-HT receptors other than 5-HT$_{2C}$, causing unwanted side effects during the clinical trial studies. However, lorcaserin–APD356 (Arena Pharmaceuticals) is currently undergoing clinical trials (Smith et al. 2005a). The structure of lorcaserin is undisclosed, but it is likely to have come from a series of novel 3-benzazepine derivatives (Smith et al. 2005a). The other 5-HT$_{2C}$ agonist under development is ATHX-105 from Athersys. Although approval for clinical testing was given in 2006, and phase 1 testing started a year later and finished January 2008, little is known about the effects of this drug on food intake or appetite expression. However, a 12-week double-blind, placebo controlled weight loss trial in the obese was scheduled to start in September 2008, potentially completing in April 2009.

Lorcaserin (ADP356) has successfully completed as series of clinical trials. In a phase 1b safety dose-escalation study no effect on heart values or pulmonary artery pressure was observed (Smith et al. 2005b). In phase 2a trials, a 15-mg dose produced a statistically significant mean weight loss of 1.3 kg (compared with 0.4 kg in the placebo group) over a 28-day treatment period. Phase 2b trials show that treatment with lorcaserin was associated with a highly significant average weight loss of 1.8, 2.6, and 3.6 kg at daily doses of 10, 15, and 20 mg, respectively, over the 12-week treatment period. In comparison, those in the placebo group lost just 0.3 kg in that time.

In 2006 lorcaserin entered the first of three phase 3 clinical trials [behavioral modification and lorcaserin for overweight and obesity management (BLOOM), behavioral modification and lorcaserin second study for obesity management (BLOSSOM) and behavioral modification and lorcaserin for overweight and obesity management in diabetes mellitus (BLOOM-DM)]. However, it has been commented that the effects of lorcaserin on weight loss in the preceding 12-week phase 2b trial appeared to be no greater than its "antecedent" d-fenfluramine and possible less than is currently achieved with sibutramine (Heal et al. 2008). Despite commonality of design, it is difficult to compare phase 2 trials directly, particular as unusually no dietary advice was given to participants during lorcaserin trails. Whether Heal et al. (Heal et al. 2008) are proved correct in their assertion will be shortly be demonstrated with the imminent completion of these lorcaserin phase 3 trials.

17.6   Conclusions

Drugs which either directly or indirectly stimulate hypothalamic 5-HT$_{2C}$ receptors in rodents produce both changes in the structure of feeding behavior and reductions in food intake that are consistent with the satiety process. These drugs cause an enhancement of the postmeal satiety potency of fixed caloric loads and reduce premeal appetite and food intake at ad libitum meals in both lean and obese humans. 5-HT drugs also produce sustained reductions in body weight gain in rodents, effects that mirror the weight loss–inducing effects of fenfluramine, d-fenfluramine, and sibutramine in humans. Antiobesity treatments have and will continue to target the endogenous 5-HT satiety system. A new generation of selective 5-HT$_{2C}$ agonists have been developed and some have passed into clinical testing. The selectivity of these compounds should ensure that they avoid the negative side effects associated with their predecessors. Currently, the effects on appetite expression of 5-HT drugs currently under development remain at least unpublished if not completely unknown. However, results are imminent. To conclude, it is over 40 years since Blundell linked the effects of 5-HT drugs on rodent feeding behavior with satiety. Despite the intervening years, and the identification of many other systems critical to energy regulation, endogenous 5-HT remains central to our understanding of human appetite expression.

References

Abeniam L, Moride Y, Brenot F, et al (1996) Appetite suppressant drugs and the risk of primary pulmonary hypertension. N Engl J Med 335:609–616.

Arterburn DE, Crane PK, Veenstra DL (2004) The efficacy and safety of sibutramine for weight loss: a systematic review. Arch Intern Med 164:994–1003.

Barkeling B, Elfhag K, Rooth P, et al (2003) Short-term effects of sibutramine (Reductil™) on appetite and eating behavior and the long-term therapeutic outcome. Int J Obes 27:693–700.

Bjenning C, Williams J, Whelan K, et al (2004) Chronic oral administration of APD356 significantly reduces body weight and fat mass in obesity-prone (DIO) male and female rats. Int J Obes 28 (1 suppl):214s

Blundell JE (1977) Is there a role for serotonin (5-hydroxytryptamine) in feeding? Int J Obes 1:15–42.

Blundell JE, Halford JCG (1998) Serotonin and appetite regulation: implications for the treatment of Obesity. CNS Drugs 9:473–495.

Blundell JE, Hill AJ (1990) Sensitivity of the appetite control system in obese subjects to nutritional and serotoninergic challenges. Int J Obes 14:219–233.

Blundell JE, Latham CJ (1978) Pharmacological manipulation of feeding behavior: possible influences of serotonin and dopamine on food intake. In: Garattini S, Samanin R, editors. Central mechanisms of anorectic drugs. New York: Raven, 78:83–109.

Blundell JE, Latham CJ (1980). Characteristic adjustments to the structure of feeding behavior following pharmacological treatments: effects of amphetamine and fenfluramine and the antagonism by pimozide and metergoline. Pharmacol Biochem Behav 12:717–722.

Blundell JE, McArthur RA (1981) Behavioural flux and feeding: continuous monitoring of food intake and food selection, and the video-recording of appetitive and satiety sequences for the analysis of drug action. In: Samanin R, Garattini S, editors. Anorectic agents: Mechanisms of action and tolerance. New York: Raven, 78:19–43.

Bouwknecht JA, van der Guten J, Hijsenm TH, et al (2001) Male and female 5-HT$_{1B}$ receptor knockout mice have higher body weights than wild types. Physiol Behav 74:507–516.

Chapelot D, Mamonier C, Thomas F, et al (2000) Modalities of the food intake-reducing effect of sibutramine in humans. Physiol Behav 68:299–308.

Clifton PG (1994) The neuropharmacology of meal patterning. In: Cooper SJ, editor. Ethology and Psychopharmacology, vol 94. Chichester: Wiley, pp. 313–328.

Clifton PG, Barnfield AMC, Philcox L (1989) A behavioural profile of fluoxetine induced anorexia. Psychopharmacology 97:89–95.

Cowen PJ, Sargent PA, Williams C, et al (1995) Hypophagic, endocrine and subjective responses to m-chlorophenylpiperazine in healthy men and women. Hum Psychopharmacol 10:385–391.

Drent ML, Zelissen PMJ, Kopperchaar HPF, et al (1995) The effect of dexfenfluramine on eating habits in a Dutch ambulatory android overweight population with an over-consumption of snacks. Int J Obes 19:299–304.

Fujitsuka N, Asakawa A, Hayashi M, et al (2009) Selective serotonin reuptake inhibitors modify physiological gastrointestinal motor activities via 5-HT$_{2C}$ receptor and acyl ghrelin. Biol Psychiatry 65:748–759.

Ghaziuddin N, Welch K, Greden J (2003) Central serotonergic effects of m-chlorophenylpiperazine (mCPP) among normal control adolescents. Neuropsychopharmacology 28:133–139.

Goodall E, Silverstone T (1988) Differential effect of d-fenfluramine and metergoline on food intake in human subjects. Appetite 11:215–288.

Grignaschi G, Samanin R (1992) Role of serotonin and catecholamines in brain in feeding suppressant effects of fluoxetine. Neuropharmacology 31:445–449.

Haddock CK, Poston WSC, Dill PL, et al (2002) Pharmacotherapy for obesity: a quantitative analysis of four decades of published randomized clinical trials. Int J Obes 26:262–273.

Halford JCG (2006) Pharmacotherapy for obesity. Appetite 45:6–10.

Halford JCG (2006) Obesity drugs in clinical development. Curr Opin Investig Drugs 7:312–318.

Halford JCG, Blundell JE (1993) 5-Hydroxytryptaminergic drugs compared on the behavioural sequence associated with satiety. Br J Pharmacol 100:95P.

Halford JCG, Blundell JE (1996) The 5-HT$_{1B}$ receptor agonist CP-94,253 reduces food intake and preserves the behavioural satiety sequence. Physiol Behav 60:933–939.

Halford JCG, Blundell JE (1996) Metergoline antagonizes fluoxetine induced suppression of food intake but not changes in the behavioural satiety sequence. Pharmacol Biochem Behav 54:745–751.

Halford JCG, Blundell JE (2000) Separate systems for serotonin and leptin in appetite control. Ann Med 32:222–232.

Halford JCG, Wanninayake SCD, Blundell JE (1998) Behavioural satiety sequence (BSS) for the diagnosis of drug action on food intake. Pharmacol Biochem Behav 61:159–168.

Halford JCG, Harrold JA, Boyland EJ, et al (2007) Serotonergic drugs: effects on appetite expression and use for the treatment of obesity. Drugs 67:27–55.

Halford JCG, Boyland EJ, Cooper SJ, et al (2010) The effects of sibutramine on the microstructure of eating behavior and energy expenditure in obese women. J Psychopharmacol 24:99–109.

Hansen DL, Toubro S, Stock MJ, et al (1998) Thermogenic effects of sibutramine in humans. Am J Clin Nutr 1998; 68: 1180–1186

Hayashi A, Sonoda R, Kimura Y, et al (2004) Antiobesity effect of YM348, a novel 5-HT$_{2C}$ receptor agonist, in Zucker rats. Brain Res 1011:221–227.

Heal DJ, Smith SL, Fisas A, et al (2008) Selective 5-HT6 receptor ligands: progress in the development of a novel pharmacological approach to the development of obesity and related metabolic disorders. Pharmacol Ther 117:207–234.

Heisler LK, Cowley MA, Tecott LH, et al (2002) Activation of central melanocortin pathways by fenfluramine. Science 297:609–611.

Heisler LK, Cowley MA, Kishi T, et al (2003) Central serotonin and melanocortin pathways regulating energy homeostasis. Ann N Y Acad Sci 994:169–174.

Heisler LK, Jobst EE, Sutton GM, et al (2006) Serotonin reciprocally regulates melanocortin neurons to modulate food intake. Neuron 51:239–249.

Hewitt KN, Lee MD, Dourish CT, et al (2002) Serotonin 2C receptor agonists and the behavioural satiety sequence in mice. Pharmacol Biochem Behav 71:691–700.

Hoyer D, Martin G (1997) 5-HT receptor classification and nomenclature: towards a harmonization with the human genome. Neuropharmacology 36:419–428.

Jackson, HC, Bearham MC, Hutchins, LJ, et al (1997) Investigation of the mechanisms underlying the hypophagic effects of the 5-HT and noradrenaline reuptake inhibitor, sibutramine, in the rat. Br J Pharmacol 121:1613–1618.

Kennett GA, Curzon G (1988) Evidence that the hypophagia induced by mCPP and TFMPP requires 5-HT$_{1C}$ and 5-HT$_{1B}$ receptors; hypophagia induced by RU-24969 only requires 5-HT$_{1B}$ receptors. Psychopharmacology 96:93–100.

Kennett GA, Curzon G (1988) Evidence that mCPP may have behavioural effects mediated by central 5-HT1C receptors. Br J Pharmacol 94:137–147.

Kennett GA, Curzon G (1991) Potencies of antagonists indicate that 5-HT$_{1C}$ receptors mediate 1-3(chlorophenyl)piperazine-induced hypophagia. Br J Pharmacol 10:2016–2020.

Kennett GA, Wood MD, Bright F, et al (1997) SB 242084, a selective and brain potent 5-HT$_{2C}$ receptor. Neuropharmacology 36:609–620.

Kitchener SJ, Dourish CT (1994) An examination of the behavioural specificity of hypophagia induced by 5-HT1B, 5-HT1C and 5-HT2 receptor agonists using the post-prandial sequence in rats. Psychopharmacology 113:368–377.

Lawton CL, Wales JK, Hill AJ, et al (1995) Serotoninergic manipulation, meal-induced satiety and eating patterns. Obes Res 3:345–356.

Lee MD, Clifton PG (1992) Partial reversal of fluoxetine anorexia by the 5-HT antagonist metergoline. Psychopharmacology 107:359–364.

Lee MD, Simansky KJ (1997) CP-94,253: a selective serotonin$_{1B}$ (5-HT$_{1B}$) agonist that promotes satiety. Psychopharmacology 131:264–270.

Lee MD, Kennett GA, Dourish CT, et al (2002) 5-HT$_{1B}$ receptors modulate components of satiety in the rat: behavioural and pharmacological analyses of the selective serotonin$_{1B}$ agonist CP-94,253. Psychopharmacology 164:49–60.

Lightowler S, Wood M, Brown T, et al (1996) An investigation of the mechanism responsible for fluoxetine-induced hypophagia in rats. Eur J Pharmacol 296:137–143.

McGuirk J, Silverstone T (1990) The effect of 5-HT re-uptake inhibitor fluoxetine on food intake and body weight in healthy male subjects. Int J Obes 14:361–372.

McGuirk J, Muscat R, Willner P (1992) Effects of the 5-HT uptake inhibitors femoxetine and parpexetine, and the 5-HT1A agonist cltoprazine, on the behavioural satiety sequence. Pharmacol Biochem Behav 41:801–805.

Neill JC, Cooper SJ (1989) Evidence that d-fenfluramine anorexia is mediated by 5-HT$_1$ receptors. Psychopharmacology 97:213–218.

Neill JC, Bendotti C, Samanin R (1990) Studies on the role of 5-HT receptors in satiation and the effect of d-fenfluramine in the runway test. Eur J Pharmacol 190:105–112.

Nonogaki K, Abdullah L, Goulding EH, et al (2003) Hyperactivity and Reduced Energy Cost of Physical Activity in Serotonin 5-HT$_{2C}$ Receptor Mutant Mice. Diabetes 52:315–320.

Padwal R, Li SK, Lau DCW (2003) Long-term pharmacotherapy for overweight and obesity: a systematic review and meta-analysis of randomized controlled trials. Int J Obes 27:1437–1446.

Pijl H, Koppeschaar HPF, Willekens FLA, et al (1991) Effect of serotonin re-uptake inhibition by fluoxetine on body weight and spontaneous food choice in obesity. Int J Obes 1991; 15:237–242

Rogers PJ, Blundell JE (1979) Effect of anorexic drugs on food intake and the micro-structure of eating in human subjects. Psychopharmacology 66:159–165.

Rolls BJ, Shide DJ, Thorward ML, et al (1998) Sibutramine reduces food intake in non-dieting women with obesity. Obed Res 6:1–11.

Samanin R, Mennini T, Bendotti C, et al (1989) Evidence that central 5-HT$_{2C}$ receptors do not play an important role in anorectic activity of d-fenfluramine in the rat. Neuropharmacology 28:465–469.

Sargent PA, Sharpley AL, Williams C, et al (1997) 5-HT$_{2C}$ receptor activation decreases appetite and body weight in obese subjects. Psychopharmacology 133:309–312.

Simansky JJ, Nicklous DM (2002) Parabrachial infusion of D-fenfluramine reduces food intake: blockade by the 5-HT1B antagonist SB-216641. Pharmacol Biochem Behav 71:681–690.

Simansky KJ, Viadya AH (1990) Behavioural mechanisms for the anorectic actions of the serotonin (5-HT) uptake inhibitor sertraline in rats: comparison with directly acting agonists. Brain Res Bull 25:953–960.

Smith S, Anderson J, Frank A, et al (2005) The effects of APD356, a selective 5-HT$_{2C}$ agonist, on weight loss in a 4 week study in healthy obese patients. Obes Res 13(suppl):101-OR

Smith BM, Smith JM, Tsai JH, et al (2005) Discovery and SAR of new benzazapines and potent and selective 5-HT$_{2C}$ receptor agonist for the treatment of obesity. Bioorg Med Chem Lett 12:1467–1470.

Tecott LH, Sun LM, Akanna SF, et al (1995) Eating disorder and epilepsy in mice lacking 5-HT$_{2C}$ serotonin receptors. Nature 374:542–546.

Vickers SP, Dourish CT (2004) Serotonin receptor ligands and the treatment of obesity. Curr Opin Invest Drug 5:377–388.

Vickers SP, Clifton PG, Dourish CT, et aln (1999) Reduced satiating effect of d-fenfluramine in serotonin 5-HT$_{2C}$ receptor mutant mice. Psychopharmacology 143:309–314

Vickers SP, Benwell KR, Porter RH, et al (2000) Comparative effects of continuous infusion of mCPP, Ro 60-0175 and d-fenfluramine on food intake, water intake, body weight and locomotor activity in rats. Br J Pharmacol 130:1305–1314.

Vickers SP, Dourish CT, Kennett GA (2001) Evidence that hypophagia induced by d-fenfluramine and d-norfenfluramine in the rat is mediated by 5-HT$_{2C}$ receptors. Neuropharmacology 41:200–209.

Vickers SP, Easton N, Webster LJ, et al (2003) Oral administration of the 5-HT$_{2C}$ receptor agonist, mCPP, reduces body weight gain in rats over 28 days as a result of maintained hypophagia. Psychopharmacology 167:274–280.

Walsh AE, Smith KA, Oldman AD (1994) M-Chlorophenylpiperazine decreases food intake in a test meal. Psychopharmacology 116:120–122.

Ward AS, Comer SD, Haney M, et al (1999) Fluoxetine-maintained obese humans: effect on food intake and body weight. Physiol Behav 66:815–821.

Wong DT, Reid LR, Threlkeld PG (1988) Suppression of food intake in rats by fluoxetine: comparison of anantiomers and effects of serotonin antagonists. Pharmacol Biochem Behav 31:475–479.

Chapter 18
Physiological and Pathophysiological Aspects of 5-HT$_{2c}$ Receptors in Basal Ganglia

**Philippe De Deurwaerdère, Laurence Mignon,
and Marie-Françoise Chesselet**

18.1 Introduction

The basal ganglia comprise a group of subcortical regions involved in motor and cognitive behavior. The effects of 5-HT in the basal ganglia are mediated by a variety of 5-HT receptors (Di Matteo et al. 2008). Among them, the 5-HT$_{2C}$ receptors are likely to have an important role in view of their widespread distribution and the specific neuronal types they control. This chapter summarizes the different facets of 5-HT$_{2C}$-mediated transmission in the basal ganglia. Particular attention will be given to its functional implications in normal and pathological states associated with basal ganglia dysfunction such as Parkinson's disease and antipsychotic-induced tardive dyskinesia.

18.1.1 Basal Ganglia Contain 5-HT Fibers and 5-HT$_{2C}$ Receptors

18.1.1.1 Functional Organizations of the Basal Ganglia

The basal ganglia are subcortical structures involved in the control of motor and cognitive functions, including sensorimotor integration, procedural memory, formation of habits, and behavioral automatisms (Graybiel 2005; Mink 1996). Based on their functional connections, modern definitions of the basal ganglia include the dorsal striatum (caudate nucleus and putamen), internal and external segments of the globus pallidus (GPi–entopeduncular nucleus, EPN, in rodents and GPe,

P. De Deurwaerdère (✉)
Unité Mixte de Recherche-Centre National de la Recherche Scientifique
(UMR-CNRS) 5227, Université Victor Segalen Bordeaux 2, Bordeaux, France
e-mail: deurwaer@u-bordeaux2.fr

G. Di Giovanni et al. (eds.), *5-HT$_{2C}$ Receptors in the Pathophysiology of CNS Disease*,   351
The Receptors 22, DOI 10.1007/978-1-60761-941-3_18,
© Springer Science + Business Media, LLC 2011

respectively), substantia nigra pars compacta and reticulata (SNc and SNr), and subthalamic nucleus (STN) (Alexander and Crutcher 1990) (Fig. 18.1). Limbic structures such as the nucleus accumbens (sometimes referred to as the *ventral striatum*) and ventral pallidum are closely associated with the basal ganglia and are involved in neuronal circuits parallel to those linking regions of the basal ganglia. The basal ganglia receive inputs from most cortical areas and convey this information back to premotor and motor cortical areas by way of the thalamus. A main organizational principle of the basal ganglia network is that cortical inputs reach the basal ganglia directly by monosynaptic projections to the striatum and the STN. In turn, STN and striatal neurons send glutamatergic excitatory or γ-aminobutyric acid (GABA)-ergic inhibitory fibers, respectively, to GPe, GPi, and SNr neurons. GPi and SNr neurons are GABA-ergic neurons that send inhibitory projections to the thalamus (Fig. 18.1) (Obeso et al. 2000). A second organizational principle emerges from the topographical organization of the cortical inputs, which determine functional territories in the basal ganglia. These territories represent sensorimotor, associative, and limbic regions, in which information is conveyed in parallel and in a partially segregated manner across the various anatomical nuclei that form the basal ganglia (Mink 1996; Kolomiets et al. 2003; Parent and Hazrati 1995; Sgambato et al. 1997).

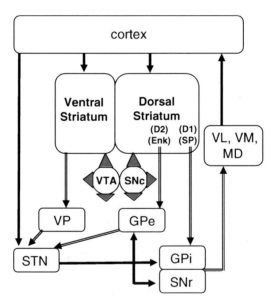

Fig. 18.1 Anatomofunctional organization of the basal ganglia. GPe and GPi external and internal globus pallidus, VP ventral pallidum, SNr and SNc substantia nigra pars reticulata and pars compacta, STN subthalamic nucleus, MD, VL and VM mediodorsal, ventrolateral and ventromedial thalamus, VTA ventral tegmental area, Enk enkephalin, SP substance P, D1 and D2 dopaminergic receptors. Filled arrows correspond to excitatory neurons; the other arrows correspond to inhibitory neurons. Dopaminergic neurons innervate most regions of the basal ganglia, as symbolized here by *arrows* (Adapted from Alexander and Crutcher 1990)

The activity of the basal ganglia is critically dependent on the functional integrity of dopaminergic (DA) neurons. The cell bodies of these DA neurons are located in the SNc and the ventral tegmental area (VTA), from where they innervate all basal ganglia regions. By far the highest density of DA fibers, receptors, and neurotransmitters, however, is found in the striatum and the nucleus accumbens. Abnormal DA transmission in the basal ganglia is involved in devastating diseases such as Parkinson's disease, iatrogenic dyskinesia induced by antiparkinsonian and antipsychotic medication, and drug addiction. Alteration in DA neuronal activity is associated with profound, molecular, cellular, and behavioral modifications within the basal ganglia and in connected regions (Albin et al. 1989; DeLong 1990; Gerfen et al. 1990; Chesselet and Delfs 1996). 5-HT neurons are particularly sensitive to changes of DA transmission as shown by the sprouting of 5-HT neurons observed after destruction of DA neurons in rodents and nonhuman primates (Boulet et al. 2008; Zhou et al. 1991).

18.1.2 5-HT Innervation of the Basal Ganglia

The 5-HT innervation of the basal ganglia comes from midbrain raphe nuclei (Azmitia and Segal 1978). The most rostral raphe nuclei, i.e., the dorsal raphe nucleus (DRN) and the caudal linear nucleus predominantly innervate the basal ganglia. Fibers ascend to innervate all areas of the basal ganglia, including the striatum, GP, SN, and STN. The 5-HT ascending pathway gives collaterals to each region as it travels to the frontal cortex (Van der Kooy and Hattori 1980). The medial raphe nuclei (MRN) also send fibers to the basal ganglia, in particular to the SN and the nucleus accumbens. However, despite the existence of some 5-HT cell bodies in the MRN, the projection of MRN to the basal ganglia does not seem to be serotonergic. High densities of 5-HT terminals, studied by anterograde labeling, immunohistochemistry (Steinbush 1984), postmortem tissue content, or 5-HT transporter binding, have been found in the SN, the GPi, the STN, the GPe, the nucleus accumbens, and the striatum. However, the 5-HT axon terminals are not homogeneously distributed in the basal ganglia. Gradients of innervations can be observed even in a single region. Notably, the ventral striatum presents higher densities of 5-HT fibers compared with the medial and lateral striatum (Soghomonian et al. 1989). Overall, the 5-HT system is present in all regions of the basal ganglia with a higher density found in associative and limbic territories compared with sensorimotor areas. Ultrastructurally, the morphology of nerve endings differs among regions. In the striatum, varicosities predominate and make en passant synapses, whereas classical nerve endings are less often observed, particularly in the lateral striatum (Soghomonian et al. 1989). Conversely, a higher proportion of true nerve endings have been described in the SNr and SNc, and these make synapses on dentrites and cell bodies of various neuronal types (Corvaja et al. 1993; Moukhles et al. 1997).

18.1.3   Distribution of 5-HT$_{2C}$ Receptors in the Basal Ganglia

18.1.3.1   General Overview

The 5-HT$_{2C}$ receptor (termed *5-HT$_{1C}$ receptors* before 1994; see Hoyer et al. 1994) belongs to the seven transmembrane G-protein-coupled receptor family. It is located exclusively within the central nervous system (CNS) with no messenger ribonucleic acid (mRNA) detected in heart, lungs, intestine, or kidney (Julius et al. 1988). Numerous studies have demonstrated a widespread and similar distribution of the 5-HT$_{2C}$ receptor subtype in rat, monkey, and human brains. The highest concentration is observed in choroid plexus and moderate levels have been reported in the different regions of the basal ganglia. Like the 5-HT axon terminals, the 5-HT$_{2C}$ receptors display a heterogeneous distribution in the basal ganglia (Mengod et al. 1990; Pompeiano et al. 1994). First of all, the mRNA has been observed in all regions except the GPe in rats, monkeys, and humans (Eberle-Wang et al. 1997; López-Giménez et al. 2001; Pasqualetti et al. 1999). In addition, rostral levels of the SNr are devoid of 5-HT$_{2C}$ receptor mRNA whereas the highest expression in this region has been reported in central and caudal parts, preferentially medially (Eberle-Wang et al. 1997). The highest concentration of mRNA has been found in the STN, SNc, SNr, VTA, and EPN with virtually all neurons in the latter region being labeled (Mengod et al. 1990; Eberle-Wang et al. 1997). In the striatum, 5-HT$_{2C}$ receptor mRNA shows a topographical gradient with higher levels medially and anteroventrally (Pompeiano et al. 1994; Eberle-Wang et al. 1997; Ward and Dorsa 1996). High levels are also seen within the nucleus accumbens (Wright et al. 1995).

Autoradiographic studies have reported moderate levels of 5-HT$_{2C}$ receptors in the basal ganglia. It should be noted, however, that the radioligands used in early studies, including [^{3}H]lysergic acid diethylamide (LSD), [^{3}H]5HT, [^{125}I]LSD, or [^{125}I]SCH 23982 are not selective for 5-HT$_2$ receptors (Hoyer and Karpf 1988; Meibach et al. 1980; Young and Kuhar 1980). The radioligands used nowadays are [^{3}H]mesulergine, and to a lesser extent, [^{3}H]1-(2,5-dimethoxy-4-iodophenyl)-2-aminopropane (DOI) that is also commonly used to label 5-HT$_{2A}$ receptors. Mesulergine is an ergot derivative with a high and preferential affinity for 5-HT$_{2C}$ receptors but also binds to 5-HT$_{2A}$, 5-HT$_{2B}$, 5-HT$_6$, and 5-HT$_7$ and D2 receptors. Furthermore, the cold ligands used to displace unwanted sites bound by [^{3}H]mesulergine including 5-HT itself, mianserine (which displays equal affinity for 5-HT$_{2A}$ and 5-HT$_{2C}$ receptors), or spiperone (which displaces mesulergine from D$_2$ and 5-HT$_{2A}$ receptors) are not selective. New ligands that are selective for 5-HT$_{2C}$ exist, but their use has been sporadic (Fox and Brotchie 2000a). Despite these limitations, ligand binding studies of 5-HT$_{2C}$ receptors within the basal ganglia show a high correlation with in situ hybridization for 5-HT$_{2C}$ mRNA in the SNr and GPi (high binding) and in the SNc, putamen, and caudate (low binding) (Mengod et al. 1990; López-Giménez et al. 2001; Wright et al. 1995; Hoyer et al. 1986; Pazos et al. 1987; Hoffman and Mezey 1989; Molineaux et al. 1989). Exceptions to a good match with mRNA distribution include the STN and the GPe. For example, Mengod et al. (1990)

have reported relatively low binding sites in the STN compared with the high levels of mRNA (Mengod et al. 1990). Furthermore, binding studies in rat and humans indicate equal levels of 5-HT$_{2C}$ receptors binding in the GPe and GPi, whereas GPe neurons do not express 5-HT$_{2C}$ mRNA (Eberle-Wang et al. 1997; Fox and Brotchie 2000a; Pazos et al. 1987; Pazos and Palacios 1985). This distribution of 5-HT$_{2C}$ receptors has been confirmed using immunohistochemistry (Clemett et al. 2000). These data suggest that 5-HT$_{2C}$ receptors in the basal ganglia are mostly somato-dentritic except in the GPe, where they are likely located presynaptically, on axon terminals of afferent neurons (Fig. 18.2).

In addition to the limited selectivity of the ligands, an important limitation of the radioligand binding studies to date is their inability to distinguish among various forms of the receptor. Indeed, the 5-HT$_{2C}$ receptor mRNA undergoes an editing process by RNA adenosine deaminases at various sites of the region encoding for the second intracellular loop of the receptor, which is critical for G-protein coupling (Burns et al.

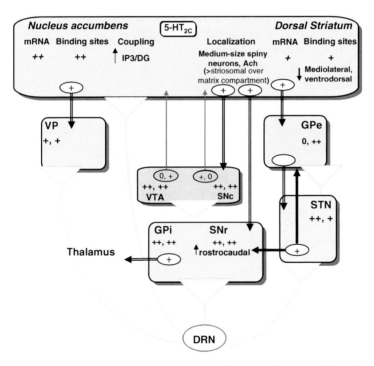

Fig. 18.2 Distribution of 5-HT$_{2C}$ receptors mRNA and binding sites in the basal ganglia. In each region, the density of mRNA is presented first by ++, +, or 0, relative to their level of expression, while the relative density of binding sites is presented second. In some occasions, neuronal populations have been labeled to illustrate their ability to express the receptor. 5-HT neurons from the dorsal raphe nucleus (DRN) have been added. GPe and GPi external and internal globus pallidus, VP ventral pallidum, SNr and SNc substantia nigra pars reticulata and pars compacta, STN subthalamic nucleus, VTA ventral tegmental area. Filled *arrows* correspond to excitatory neurons; the other *arrows* correspond to inhibitory neurons. Dopaminergic neurons diffusely innervate all basal ganglia regions

1997; Niswender et al. 1999; Berg et al. 2001). This process results in multiple 5-HT$_{2C}$ transcripts and isoforms that differ in their pharmacological properties. Mesulergine is not a pure neutral antagonist and displays inverse agonist properties. This implies that its binding is also dependent on the isoform considered. Therefore, other radioligands are needed to confirm the pattern of expression described earlier.

18.1.3.2   Cellular Types Expressing the 5-HT$_{2C}$ Receptors

Studies using in situ hybridization, chemical lesions of neuronal pathways, reverse transcription coupled to polymerase chain reaction (RT-PCR) and immuno-histochemistry have helped identify some of the cell types that express 5-HT$_{2C}$ receptors. In the striatum, most neurons expressing the receptor mRNA are medium-sized spiny neurons, the efferent GABA-ergic neurons innervating the GPe ("indirect" pathway), or the SNr/GPi ("direct" pathway). 5-HT$_{2C}$ receptor mRNA is similarly expressed in projection neurons forming both pathways because it colocalizes equally with enkephalin and substance P/dynorphin, respectively (Ward and Dorsa 1996). Recent RT-PCR studies have shown that the 5-HT$_{2C}$ mRNA is present in cholinergic interneurons. Interestingly, 5-HT$_{2C}$ mRNA showed a patchy distribution (Ward and Dorsa 1996) that could correspond to the patchy distribution of the 5-HT$_{2C}$ receptor binding of [^{3}H]DOI reported by Waeber and Palacios (Waeber and Palacios 1994). This finding suggests a preferential localiza-tion in the striosomal versus the extrastriosomal matrix compartment. The strio-somes, or "patches," and the extrastriosomal "matrix" represent an anatomofunctional organization of the striatum, with the striosomes being low in acetylcholinesterase, rich in mu opioid receptors, and containing the cell bodies of the GABA-ergic neurons projecting to the SNc (Gerfen 1984, 1985; Graybiel 1991).

In the STN, the cells expressing 5-HT$_{2C}$ receptors are presumably the gluta-matergic neurons innervating the GPe, the GPi, and the SN. The terminals carrying the presynaptic 5-HT$_{2C}$ receptors in the GPe have not been identified; they could include afferents from striatal and/or STN neurons. The cells expressing 5-HT$_{2C}$ receptors in the GPi and SNr likely include the GABA-ergic efferent neurons that innervate the thalamus and superior colliculus. Importantly, two populations of GABA-ergic neurons can be distinguished in the SNr; some neurons express a high level of 5-HT$_{2C}$ mRNA, whereas others seem devoid of the mRNA (Eberle-Wang et al. 1997). It is interesting to note that some authors consider that the SNr is com-posed of a mixed population of neurons equivalent to GPe and GPi neurons; the heterogenous distribution of the 5-HT$_{2C}$ receptor mRNA may in fact identify these two populations since GPe neurons do not express the mRNA, whereas GPi neu-rons do. Furthermore, a rostrocaudal gradient of expression of the mRNA has been detected in this region (Eberle-Wang et al. 1997).

The cellular location of 5-HT$_{2C}$ receptors in the ventral mesencephalon (SNc/VTA) may be more complex. In situ hybridization data suggest that, in the SNc, the receptor mRNA is selectively expressed by GABA-ergic neurons that are inter-mingled with DA neurons in this region. Indeed, 5-HT$_{2C}$ mRNA is colocalized with the mRNA for glutamic acid decarboxylase (GAD), the enzyme responsible for GABA synthesis, but not with tyrosine hydroxylase, the rate-limiting enzyme of DA biosynthesis in this region (Eberle-Wang et al. 1997). However, this apparently differential distribution may reflect large differences in levels of expression of the mRNA rather than a selective expression excluding the DA neurons. Indeed, more recent techniques of laser capture microdissection allow for the determination of mRNA expression in selected cell populations with quantitative polymerase chain reaction (Q-PCR), a technique that is much more sensitive than in situ hybridization histochemistry. With this approach, the 5HT$_{2C}$ mRNA can be detected in tyrosine hydroxylase positive neurons of the SNc (Mortazavi and Chesselet, unpublished observations). These data are compatible with results from a recent study reporting immunostaining with an antibody raised against 5-HT$_{2C}$ receptor in both GABA-ergic and DA neurons of the VTA (Bubar and Cunningham 2007). Even though such a colocalization appears to exist in the SN based on the PCR data, it may be limited because 5-HT$_{2C}$ binding sites tend to increase rather than decrease in the SNc after loss of nigrostriatal DA neurons both in rats and in humans (Fig. 18.2) (Fox and Brotchie 2000a; Radja et al. 1993).

18.1.3.3 Conclusions

5-HT$_{2C}$ receptors are heterogeneously expressed in the basal ganglia. A comparison between mRNA and binding sites suggests that this receptor is mostly a somatodendritic receptor, except in the GPe, where it may be located on axons. Numerous cell types express the receptor including GABA-ergic, glutamatergic and cholinergic neurons. The DA neuron may express very low levels of 5-HT$_{2C}$ receptors in the SNc but greater levels in the VTA. In general, the density of 5-HT$_{2C}$ receptors follows the density of the 5-HT innervation, the ventromedial parts of the basal ganglia being enriched in both. This pattern of expression of 5-HT$_{2C}$ receptors suggests that its functional influences may be stronger on associative and limbic circuits than on the sensorimotor pathways.

18.2 5-HT$_{2C}$ Receptor-Dependent Control of Neuronal Activity in the Basal Ganglia

18.2.1 Preamble: 5-HT$_{2C}$ Receptor and Second Messenger Coupling

In most cases, stimulation of the 5-HT$_{2C}$ receptor excites neurons. Indeed, stimulation of the 5-HT$_{2C}$ receptor enhances phosphoinositol turnover via Gq-protein coupling and activation of phospholipase C (Conn et al. 1986; Berg et al. 1998). This effect leads to a marked increase in intracellular Ca^{2+} (Julius et al. 1988).

In addition, the 5-HT$_{2C}$ receptor has been functionally linked to guanylyl cyclase in the porcine choroid plexus (Hartig et al. 1990) and its stimulation increases cyclic guanosine monophosphate (GMP) formation (Kaufman et al. 1995). The receptor may also couple to phosplipase A2 or a Gi protein when expressed in heterologous cell lines (Berg et al. 2001). More recently, a calmodulin–beta-arrestin pathway, independent from G proteins, has been reported (Berg et al. 2008; Labasque et al. 2008). Interestingly, in most cases, the 5-HT$_{2C}$ receptor appears constitutively active in heterologous systems. For example, 5-HT$_{2C}$ receptor activates phospho-inositol hydrolysis in the absence of agonists when expressed in NIH-3T3 fibro-blasts (Westphal et al. 1995) and could contribute to the basal activity of cells in the choroid plexus (Barker et al. 1994) and basal ganglia (De Deurwaerdère et al. 2004). In line with these data, several studies have shown that 5-HT$_{2C}$ receptors display a high propensity to couple to Gq protein. This property could account for the functional difference in the response of 5-HT$_{2A}$ and 5-HT$_{2C}$ receptors to 5-HT. Indeed, the ability of 5-HT$_{2C}$ receptors to stimulate phosphoinositol turnover is ten-fold greater than that of 5-HT$_{2A}$ receptors (Leonhardt et al. 1992). This difference has been observed in native rat striatal tissue, despite the four times lower expression of 5-HT$_{2C}$ receptors compared with 5-HT$_{2A}$ receptors in this region (Wolf and Schutz 1997). Yet, the higher coupling of 5-HT$_{2C}$ receptors to intracellular pathways is criti-cally dependent on the isoforms that result from the editing process. Indeed, editing dramatically lowers the coupling efficiency compared with the nonedited form of the receptor (Berg et al. 2001). This molecular mechanism is probably an important component of 5-HT$_{2C}$ receptor-dependent transmission in the brain, including in the basal ganglia. As discussed earlier, studies are still necessary to determine the distribution of the different isoforms of the receptor in these regions.

18.2.2   5-HT$_{2C}$ Receptor Influence in Basal Ganglia Regions

In agreement with the distribution of 5-HT$_{2C}$ receptors in the basal ganglia, a grow-ing number of electrophysiological and biochemical studies indicate that 5-HT$_{2C}$ receptors may affect neuronal function in most regions of the basal ganglia. Some of these results have been summarized in Fig. 18.3.

18.2.2.1   5-HT$_{2C}$ Receptor Modulation of Neuronal Activity in the Striatum

Effect of 5-HT$_{2C}$ Receptors in the Whole Striatum

Because many neuronal cell types express 5-HT$_{2C}$ receptors, it is likely that they will affect the activity of a variety of neuronal populations and the release of several neurotransmitters in this brain region. Early studies of the influence of 5-HT neurons from the DRN on striatal cell activity have reported mostly inhibitory effects that were blocked by the nonselective 5-HT$_{1/2}$ antagonist methysergide (Olpe and Koella

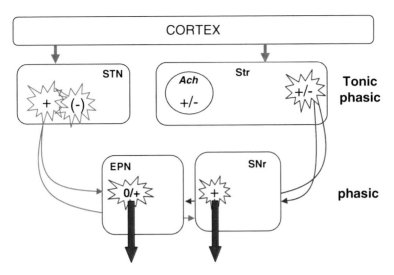

Fig. 18.3 Drawing summarizes the effects of 5-HT$_{2C}$ receptors in the basal ganglia. The in vitro data indicate that 5-HT$_{2C}$ receptors stimulate firing rate of STN and SNr neurons, while the data in the striatum are not very clear. On the other hand, the in vivo data indicate the existence of phasic and tonic controls in the striatum and putatively in the STN, whereas only phasic controls have been observed in the SNr of naive rats. The control exerted by 5-HT$_{2C}$ receptors in EPN, despite the presence of the receptor, would not be prominent in naive rats. EPN entopeduncular nucleus, SNr substantia nigra pars reticulate, STN subthalamic nucleus, Str striatum, Ach cholinergic interneuron; +,- or 0, relative to the influence of 5-HT$_{2C}$ agonists or antagonists on neuronal activity

1977). Mixed inhibitory/excitatory responses have also been reported (Olpe and Koella 1977; Davies and Tongroach 1978). This may explain why the stimulation of DRN does not induce any overall change in 2-deoxyglucose in the striatum (Cudennec et al. 1988), a finding at odds with electrophysiological and biochemical data indicating the presence of 5HT terminals and receptors in this region. Evidence that the release of endogenous 5-HT stimulates 5-HT$_{2C}$ receptors to affect neuronal activity in the striatum and/or nucleus accumbens is still lacking.

Preferential agonists such as meta-chlorophenylpiperazine (mCPP) or non selective agonists such as 1-(3-trifluoromethylphenyl) piperazine (TFMPP), RU29469 or DOI have been used to examine the influence of 5-HT receptors on indirect indexes of neuronal activity, in particular the expression of the protooncogene c-Fos by immunohistochemistry and the blood oxygen level dependent (BOLD) signal by functional magnetic resonance imaging in anesthetized rodents. All these compounds have been shown to affect c-Fos expression in the striatum (De Deurwaerdère and Chesselet 2000a; Leslie et al. 1993; Moorman and Leslie 1996; Lucas et al. 1997; Rouillard et al. 1996; Stark et al. 2006; Tremblay et al. 1998; Wirtshafter and Cook 1998). However, some studies did not find any effect of mCPP or DOI (Singewald et al. 2003; Tilakaratne and Friedman 1996). The pattern of Fos expression differs after administration of these various drugs. The mixed 5-HT$_{2A/2B/2C}$ agonist DOI enhances Fos in the dorsomedial striatum mainly, whereas RU-29469 and TFMPP induce a homogeneous increase in Fos expression in the anterocentral striatum and a patchy distribution in centrocaudal striatum (Lucas et al. 1997; Wirtshafter and Cook 1998). The effects of RU-29469 and TFMPP are likely related to their action on 5-HT_{1B} rather than 5HT_{2C} receptors because they are absent in 5-HT_{1B} knockout mice and blocked by preferential 5-HT_{1B} antagonists (Lucas et al. 1997). The effect of DOI has been related in part to both 5-HT_{2C} and 5-HT_{2A} receptors because it is abolished by the nonselective 5-HT_{2} antagonist ritanserin and partly reduced by the $5\text{-HT}_{2A}/D2$ antagonist spiperone (Leslie et al. 1993).

Meta-chlorophenylpiperazine (mCPP) induces a preferential increase in Fos expression in the medial striatum at a moderate dose and a substantial increase in both medial and lateral striata? at higher dose (De Deurwaerdère and Chesselet 2000a; Stark et al. 2006). According to some unpublished data, the effect elicited by 5 mg/kg mCPP is not related to 5-HT_{2C} receptors whereas part of the effect elicited by 1 mg/kg mCPP involves 5-HT_{2C} receptors, except in the dorsolateral striatum (Cook and Wirtshafter 1995; De Deurwaerdère and Chesselet 2000b). The BOLD signal elicited by 3 mg/kg mCPP is also higher in the medial striatum compared with lateral sectors (Stark et al. 2006, 2008; Hackler et al. 2007). At variance with Fos studies, the effect of mCPP on BOLD signal is abolished by the selective 5-HT_{2C} antagonist SB 242084 in the dorsolateral striatum and partly reduced in the medial striatum (Stark et al. 2008).

Single-unit recordings coupled to microiontophoresis have detected an inhibitory effect of 5-HT, DOI, and mCPP in vivo on the depolarization of striatal neurons produced by quisqualate in the head of the caudate nucleus (El Mansari et al. 1994; El Mansari and Blier 1997). Less quisqualate was required to activate neurons in the caudate nucleus of 5-HT_{2C} receptor mutant mice than in wild-type mice, suggesting that 5-HT_{2C} receptors play a tonic inhibitory role in membrane excitability (Rueter et al. 2000). However, the phasic inhibitory response induced by DOI and mCPP was not affected by a 5-HT_{2A} antagonist or by clozapine, an atypical antipsychotic that blocks not only DA receptors but also several 5-HT receptors including the 2A and 2C subtypes (Meltzer et al. 2003). It has been proposed that DOI and mCPP could bind an atypical 5-HT_{2} receptor in the head of the caudate nucleus (El Mansari and Blier 1997). In other regions of the striatum, while iontophoretic application of DOI still had an inhibitory effect on excitatory responses in some striatal neurons, 5-HT had an excitatory effect (Wilms et al. 2001). It is difficult to conclude from these data whether 5-HT_{2C} receptors mediate an inhibitory control on striatal medium-sized spiny neurons. One source of confusion could be the well-known heterogeneity of striatal efferent neurons, with neurons of the direct and indirect pathways presenting many molecular differences. Further studies in recently available mice expressing different fluorescent proteins in classes of striatal neurons should help clarify this issue (Lobo et al. 2006).

To summarize these in vivo data, it is interesting to note that the metabolic and genomic effects elicited by the preferential 5-HT_{2C} agonist mCPP follow in part the mediolateral gradient of 5-HT_{2C} receptors in the striatum. However, the lack of selectivity of these compounds makes it difficult to accurately determine the role of 5-HT_{2C} receptors. Indeed, other receptors, in particular 5HT_{1B}, may have been stimulated in these studies. These receptors may even be primarily responsible for the effects on Fos expression. This could explain why 5-HT$_2$ receptors appear not to be involved in the increase in Fos expression induced by drugs that enhance 5-HT release such as d-fenfluramine (Rouillard et al. 1996; Gardier et al. 2000; Javed et al. 1998).

Effect of 5-HT$_{2C}$ Receptors on Identified Neurons of the Striatum

As previously mentioned, the cellular types involved in increased Fos expression mediated by 5-HT$_{2C}$ receptors have not been characterized. To the best of our knowledge, no data are available on the effects of mCPP on striatal expression of the mRNAs encoding enkephalin or substance P in the striatum. In an unpublished study, we found that a subchronic treatment with mCPP decreased levels of dynorphin mRNA, which is expressed by neurons of the striatal patch compartment, without affecting enkephalin mRNA levels (Mignon et al. 2002). In contrast, DOI enhances both enkephalin and substance P mRNAs presumably via 5-HT$_{2A}$ receptors (Walker et al. 1991; Basura and Walker 2001). Thus, stimulation of 5-HT$_{2C}$ receptors may directly affect medium-sized spiny neurons but experimental support for this effect remains incomplete.

Various populations of interneurons have been described in the striatum, and these cells represent about 5–10% of striatal neurons (Kawaguchi et al. 1995). Cholinergic interneurons, characterized by their large cell body and their aspiny dendrites, have been the most extensively studied. An effect of 5-HT on striatal cholinergic cells has been reported (Euvrard et al. 1977; Gillet et al. 1985; Bianchi et al. 1989), and studies have shown either inhibitory or biphasic (inhibitory and excitatory) effects. The role of 5-HT$_{2C}$ receptors is not clear because the pharmacological studies relied upon nonselective compounds. More recently, 5-HT has been reported to strongly increase spontaneous firing rates of large aspiny neurons (presumably cholinergic interneurons) in vitro. Specifically, 5-HT, acting via 5-HT$_2$ receptors, reduced two pharmacologically and kinetically distinct after hyperpolarizations (AHPs) that play an important role in limiting the excitability of cholinergic interneurons (Blomeley and Bracci 2005). The selective 5-HT$_{2C}$ antagonist RS 102221 reduced the 5-HT-induced depolarization of cholinergic interneurons (Bonsi et al. 2007) suggesting that 5-HT$_{2C}$ receptors are responsible for this depolarization. The reduction of AHP by 5-HT$_{2C}$ receptor stimulation is also compatible with findings reported by North et al. (North and Uchimura 1989) in the nucleus accumbens and in other brain regions including the STN. The control exerted by cholinergic neurons on neighboring medium-sized spiny neurons may be inhibitory via M2 and M3 receptors located on presynaptic glutamatergic inputs (Pakhotin and Bracci 2007).

In conclusion, no clear picture of the influence of 5-HT$_{2C}$ receptors in the striatum emerges from the available studies. This likely reflects the existence of various mechanisms and numerous distinct cellular populations controlled by 5-HT$_{2C}$ receptors. Overall, stimulation of these receptors may indirectly inhibit efferent GABA-ergic neurons through an activation of cholinergic cells.

18.2.2.2   5-HT$_{2C}$ Receptor Modulation of SNr and EPN Neuronal Activity

Endogenous 5-HT has been shown to affect the electrophysiological activity of SNr neurons (Dray 1981). The responses are both excitatory and inhibitory. In mesencephalic brain slices containing the SNr, 5-HT also exerts mixed responses (Rick et al. 1995). 5-HT directly stimulates these cells, and these excitatory effects are mimicked by the nonselective 5-HT$_2$ agonist α-methyl-5-HT. In addition, the effect of 5-HT is suppressed by ketanserin, ritanserin, and several nonselective 5-HT$_{2C}$ antagonists (Rick et al. 1995; Gongora-Alfaro et al. 1997; Stanford and Lacey 1996). On the other hand, agonist and/or antagonist directed toward 5-HT$_{1B}$, 5-HT$_{2A}$, and 5-HT$_4$ receptors had no effect (Rick et al. 1995). In contrast, the inhibitory responses evoked by 5-HT in mesencephalic slices are not affected by nonselective 5-HT$_2$ antagonists and could be mediated by 5-HT$_{1B}$ receptors (Gongora-Alfaro et al. 1997; Stanford and Lacey 1996).

A role for 5-HT$_{2C}$ receptors in excitatory responses on SNr neurons has been further demonstrated in vivo by using the agonists mCPP and Ro-60-0175 as well as selective 5-HT$_{2C}$ antagonists. Thus, mCPP enhances the firing rate of some SNr neurons. The effect occurs in a subpopulation of neurons that are not sensitive to a noxious stimulus, presumably the efferent GABA-ergic neurons but not the GABA-ergic interneurons (Di Giovanni et al. 2001). Similarly, Ro-60-0175 excites some SNr neurons, and the excitatory effects of both agonists are suppressed by the systemic administration of the selective 5-HT$_{2C}$ antagonists SB 242084 and/or SB 243213 (Di Giovanni et al. 2001; Invernizzi et al. 2007). The effects can also be observed using iontophoretic application of 5-HT drugs, confirming the above-mentioned in vitro data that 5-HT$_{2C}$ receptors are involved in the stimulation of some SNr neurons. It is interesting to recall that in situ hybridization studies clearly demonstrated that only a subpopulation of SNr neurons expressed detectable levels of 5HT$_{2C}$ mRNA, thus supporting the possibility of a heterogeneous response to 5-HT$_{2C}$ agonists in this region (Eberle-Wang et al. 1997).

The excitatory effects of 5-HT agonists on the firing rate of SNr neurons are paralleled by an increase in nigral GABA, but not glutamate, release detected by in vivo intracerebral microdialysis (Invernizzi et al. 2007). Indeed, the effect elicited by Ro-60-0175 is prevented by the systemic administration of SB 243213. Interestingly, the intranigral administration of the 5-HT$_{2C}$ antagonist only partially reduced the effect of Ro-60-0175, thus suggesting that the enhancement of GABA release by the 5-HT$_{2C}$ agonist involves both intra- and extranigral mechanisms.

It is noteworthy that the response of EPN cells to 5-HT$_{2C}$ agonists differ slightly compared with SNr neurons. First, iontophoretic application of DOI in the EPN does not affect EPN neuron activity in vivo (Kita et al. 2007). Conversely, a moderate dose of mCPP (1 mg/kg) slightly enhances Fos expression in the EPN but not in the central or caudal SNr (De Deurwaerdère and Chesselet 2000a). The regulatory influence of 5-HT$_{2C}$ receptors on the output pathways does not seem to be equivalent in the SNr and the EPN, which may parallel the differential distribution of 5-HT$_{2C}$ mRNA in these regions (Eberle-Wang et al. 1997).

18.2.2.3   5-HT$_{2C}$ Receptor Modulation of GPe Neuronal Activity

To the best of out knowledge, no data support an effect of 5-HT$_{2C}$ receptors on neuronal activity in the GPe. In slices of GPe, 5-HT produces a depolarization leading to an increase in firing rate (Chen et al. 2008). This excitation occurs via 5-HT$_4$ and 5-HT$_7$ receptors but not via 5-HT$_{2C}$ or 5-HT$_3$ receptors (Kita et al. 2007; Chen et al. 2008; Bengtson et al. 2004; Hashimoto and Kita 2008). The lack of effect of 5-HT$_{2C}$ agents in vitro is in line with the lack of Fos induction seen with mCPP in the GPe (De Deurwaerdère and Chesselet 2000a) and the absence of expression of 5HT$_{2C}$ receptor mRNA in this region (Eberle-Wang et al. 1997).

18.2.2.4   5-HT$_{2C}$ Receptor Modulation of STN Neuronal Activity

Studies employing extracellular single-unit recordings or whole-cell patch clamp in rat brain slices have provided evidence that 5-HT induces both excitatory and inhibitory influences on STN neuron. As noted above, 5-HT$_{2C}$ receptors are expressed by some STN neurons and they likely mediate part of the excitatory action of 5-HT (Flores et al. 1995; Shen et al. 2007; Stanford et al. 1995; Xiang et al. 2005). Indeed, the effect of 5-HT is mimicked by nonselective 5-HT$_2$ agonists and is reduced by relatively selective antagonists that share the ability to block 5-HT$_{2C}$ receptors (Xiang et al. 2005). The direct stimulation of these neurons is characterized by an increase in STN neuron firing rate resulting from a reduction of potassium conductance, including AHP currents (Flores et al. 1995; Xiang et al. 2005). In some cases, burst firing can be facilitated by 5-HT$_{2C}$ receptor-dependent mechanism (Shen et al. 2007). This suggests that 5-HT$_{2C}$ receptors could contribute to the abnormal pattern of discharge of STN neurons reported in animal models of Parkinson's disease (Shen et al. 2007). Excitatory responses in the STN are mediated by both 5-HT$_{2C}$ and 5-HT$_4$ receptors, whereas inhibitory responses are not sensitive to 5-HT$_{2C}$ antagonists (Flores et al. 1995; Stanford et al. 1995). 5-HT may also indirectly affect neuronal activity in the STN by reducing the amplitude of glutamatergic excitatory postsynaptic currents or GABA-ergic inhibitory postsynaptic currents. However, these indirect responses are related to 5-HT$_{1B}$ and not 5-HT$_{2C}$ receptors (Shen and Johnson 2008).

Extracellular recordings in vivo have provided evidence that both the systemic and intra-STN administration of mCPP enhances the firing rate of STN neurons, an effect abolished by the selective 5-HT$_{2C}$ antagonist SB 242084 (Zhang et al. 2009). This finding is consistent with our earlier study that mCPP enhances Fos expression in the STN (De Deurwaerdère and Chesselet 2000a). In conclusion, a growing amount of concordant data indicates that stimulation of STN 5-HT$_{2C}$ receptors stimulates STN neuronal activity.

18.3   5-HT$_{2C}$ Receptors and Basal Ganglia Function

How does the stimulation of 5-HT$_{2C}$ receptors control basal ganglia function? As noted earlier, the basal ganglia consist of highly interconnected regions, each characterized by a specific organization and neuronal circuits. The influence of 5-HT$_{2C}$ receptors on this network will depend on the electrophysiological and metabolic features of each group of neurons. For example, it is likely that 5-HT$_{2C}$ receptors will differentially affect silent striatal neurons compared to tonically active SNr neurons. The challenge is to understand how the role of 5-HT$_{2C}$ receptors in one region will eventually impact the whole network and to determine whether common rules govern the effect of their stimulation.

A first evidence that the control exerted by 5-HT$_{2C}$ receptors in the basal ganglia depends on network information comes from the control they exert on the activity of mesencephalic DA neuron. 5-HT$_{2C}$ agonists and antagonists inhibit and enhance, respectively, the activity of DA neurons and the release of DA from striatal nerve terminals in rats (Di Giovanni et al., Chap. 11). It is noteworthy that the effects of 5-HT$_{2C}$ compounds vary with the state of the animal (anesthesia, awake) and are greater in the VTA than in the SNc (Navailles and De Deurwaerdère, Chap. 10). Few studies have been devoted to the brain regions involved in this regulation. Intrastriatal or intraccumbal administration of 5-HT$_{2C}$ antagonists may enhance DA release (Alex et al. 2005; Alex and Pehek 2007; Navailles et al. 2006a). Conversely, administration of these antagonists into the VTA does not affect DA release in the nucleus accumbens (Navailles et al. 2006a). Strikingly, neither intra-VTA nor intraaccumbal administrations of Ro-60-0175 inhibit DA release in the nucleus accumbens, but these regions may become involved when the agonist is administered systemically (Navailles et al. 2006a). Altogether, these results suggest that the tonic inhibitory control exerted by 5-HT$_{2C}$ receptors involves receptors located in terminal areas of DA neurons. This is consistent with the data mentioned earlier showing that 5-HT$_{2C}$ receptors are involved in the tonic inhibition of neuronal excitability in the striatum (Rueter et al. 2000). Thus, this phasic inhibitory control likely involves 5-HT$_{2C}$ receptors located in areas containing cell bodies and terminals of DA neurons but depends on information provided by other brain regions.

The second evidence for a role of neuronal circuits in the effects resulting from 5-HT$_{2C}$ receptor stimulation comes from Fos studies that revealed a greater effect of systemic injections of mCPP in the striatum and the STN than in the EPN or SNr. The mixed 5-HT$_{2C/2B}$ antagonist SDZ SER-082 and the selective 5-HT$_{2C}$ antagonist SB 243213 increased Fos expression only in the STN and the striatum (De Deurwaerdère et al. 2010). Thus, 5-HT$_{2C}$ receptors appear to exert a tonic (perhaps constitutive) influence in the striatum and the STN cells. In both regions, the

5-HT$_{2C}$ receptors could, in addition, mediate phasic responses. It is interesting that the higher responsiveness of basal ganglia structures to 5-HT$_{2C}$ receptor modulation occurs in the two regions receiving direct inputs from the cerebral cortex (De Deurwaerdère and Chesselet 2000a, b; De Deurwaerdère et al. 2010).

A third evidence comes from behavioral experiments reporting that alterations in 5-HT$_{2C}$ transmission in the STN or the striatum lead to rotational behavior in rats. Thus, unilateral injection of 5-HT, quipazine (a nonselective 5-HT$_{2C}$ agonist), or MK 212 (a preferential 5-HT$_{2C}$ agonist) into the STN induced a contralateral and dose-dependent turning behavior (Belforte and Pazo 2004). The contraversive turning behavior induced by 5-HT, blocked by the nonselective 5-HT$_2$ antagonist mianserine, is attributed to a decreased excitatory input from the STN to the SNr, which in turn enhances the activity of the ipsilateral motor thalamus. Indeed, kainic acid lesion of the SNr suppressed the contralateral rotations elicited by the stimulation of STN 5-HT$_{2B/2C}$ receptors (Belforte and Pazo 2004). Other data, however, indicate that blockade of subthalamic 5-HT$_{2C}$ receptors suppresses the stereotypic behavior induced by apomorphine administration (Barwick et al. 2000), which would support the opposite idea that 5-HT$_{2C}$ receptors activate subthalamonigral activity. In the striatum, local injections of 5-HT provoked contraversive turning, while the nonselective 5-HT$_{1/2}$ antagonist methysergide induced ipsiversive circling (James and Starr 1980). Although these data favor the existence of both tonic and phasic controls exerted by the 5-HT system on striatal cells, the interpretation of these data is limited by the lack of specificity of the pharmacological tools employed in these studies.

Altogether, the in vivo data support the idea that 5-HT$_{2C}$ transmission is heterogeneously organized in the basal ganglia. The main areas under the control of 5-HT$_{2C}$ receptors are the striatum and the STN, the two entries of cortical inputs to the basal ganglia. In contrast, the output regions of basal ganglia are less sensitive to 5-HT$_{2C}$ modulation in naive rats (Di Matteo et al. 2008). In addition, both tonic and phasic controls coexist in the STN and the striatum.

18.4   5-HT$_{2C}$ Receptors and the Motor Control of Orofacial Responses

The available, nonselective, 5-HT$_{2C}$ agonists elicit various alterations of motor responses including grooming, penile erection, hypolocomotor activity, forms of stereotypies, and purposeless oral movements. This latter response is interesting because it is observed at low doses of agonists and, at variance with some of the other behaviors evoked above, is only dependent on 5-HT$_{2C}$ receptors. Stewart and colleagues (1989) observed an increase in purposeless oral movements elicited by the nonselective 5-HT agonists mCPP and TFMPP (Stewart et al. 1989). The oral bouts consisted of vacuous chewing, jaw tremor, and tongue darting occurring without any physical object. The intensity of oral bouts induced by the 5-HT agonist was dose dependent and decreased when very high doses of mCPP were used

(Stewart et al. 1989; Gong and Kostrzewa 1992; Gong et al. 1992). The oral bouts induced by mCPP correspond to 5-HT$_{2C}$ receptor-dependent mechanisms because a variety of drugs able to block 5-HT$_{2C}$ receptors including mianserin, mesulergine, SDZ SER-082, or SB 206553 suppresses the oral responses to mCPP (Stewart et al. 1989; Gong and Kostrzewa 1992; Gong et al. 1992; Eberle-Wang et al. 1996; Wolf et al. 2005). Conversely, 5-HT$_{1B}$, 5-HT$_{2A}$, or 5-HT$_3$ antagonists do not affect mCPP-induced abnormal oral movements, while 5-HT$_{1A}$, 5-HT$_{1B}$, 5-HT$_3$ agonists do not elicit oral dyskinesia (Stewart et al. 1989; Gong et al. 1992; Kostrzewa et al. 1993). Although these latter data do not exclude a role of these receptors in the control of orofacial activity, these results emphasize a strong link between oral motor control and 5-HT$_{2C}$ receptors.

Intracerebral microinjections of drugs provided evidence that the abnormal oral response to mCPP is due to effects of the drug on receptors located within the basal ganglia, specifically in the STN and the striatum. Indeed, either unilateral or bilateral administration of low doses mCPP into the STN elicits oral bouts, and the effect is blocked by mesulergine (De Deurwaerdère and Chesselet 2000a; Eberle-Wang et al. 1996). Furthermore, the oral bouts elicited by the systemic administration of mCPP are abolished by the intra-STN administration of mesulergine (Eberle-Wang et al. 1996). Oral movements observed after systemic injection of mCPP are enhanced by a 5,7-dihydroxytryptamine lesion of 5-HT neurons, which indicates that these receptors are normally stimulated by endogenous 5HT (Mehta et al. 2001).

While the above studies favor the idea of an almost exclusive influence of STN 5-HT$_{2C}$ receptor in mCPP's effects, Plech et al. (1995) have also reported that intrastriatal administration of mCPP induced purposeless oral movements that are abolished by the intrastriatal administration of mianserin (Plech et al. 1995). Interestingly, despite the role of EPN in mediating abnormal movements (Adachi et al. 2002), we have found that the stimulation of EPN 5-HT$_{2C}$ receptors via the local administration of mCPP does not produce oral bouts. On the other hand, high doses of mCPP or TFMPP directly administered into the SNr have been shown to elicit abnormal oral movements (Liminga et al. 1993). For these authors, the fact that the nonselective 5-HT$_2$ agonist DOB did not induce vacuous chewing suggests a role for 5-HT$_{1B}$ receptors in these effects. Altogether these data suggest that striatal and STN 5-HT$_{2C}$ receptors are involved in the abnormal oral responses induced by 5-HT$_{2C}$ agonists in naive rats.

Nonselective 5-HT$_{2C}$ receptor antagonists also reduce oral bouts elicited by DA and muscarinic agonists (Gong et al. 1992; Carlson et al. 2003). Rosengarten and colleagues (1999) have shown that the effect of DA agonists and mCPP on vacuous chewing movements are additive. Nonetheless, mianserin is able to slightly reduce DA agonists-induced abnormal orofacial movements (Gong et al. 1992). In another study, mianserin blocked tacrine-induced vacuous chewing when injected into the dorsolateral part of the SNr (Carlson et al. 2003), a zone involved in facial motor control (Deniau et al. 1996). This study suggests that modification of basal ganglia activity may trigger phasic/tonic responses involving 5-HT$_{2C}$ receptors in the output regions.

18.5   Increased 5-HT$_{2C}$ Transmission in Animal Models of Diseases Related to Chronic DA Neuron Impairment

The link between 5-HT$_{2C}$ receptors and the control of orofacial movement is of particular importance in the context of the chronic use of antipsychotics and the impairment of DA function in Parkinson's disease (Kostrzewa et al. 2007). This link is further established in humans by data showing that the propensity of developing dyskinesia after neuroleptic use is correlated to polymorphisms of the 5-HT$_{2C}$ receptor (Gunes et al. 2007, 2008). In the last part of this chapter, we will review the consequences of chronic alterations in DA transmission on the role of 5-HT$_{2C}$ receptors in the basal ganglia.

18.5.1   Neuroleptic-Induced Dyskinesia and Increase of 5-HT$_{2C}$ Receptor Function

Classical antipsychotics like haloperidol cause motor side effects including acute parkinsonims and dystonia (often referred to as *extrapyramidal side effects*, or EPS) and late-emerging tardive dyskinesias (Janno et al. 2004; Tarsy and Baldessarini 1984; Tarsy et al. 2002). Despite the advent of newer antipsychotic drugs with reduced liability to induce EPS, these side effects remain a clinical concern, especially in vulnerable patient populations such as the elderly. Numerous reports demonstrate that a variety of haloperidol treatment paradigms induce vacuous chewing movements in rats (Rupniak et al. 1985; Waddington 1990). The emergence of vacuous chewing movements (VCMs) within days to weeks of initiating haloperidol treatment has been called *early onset* VCMs and appears to be associated with mechanisms related to EPS (Egan et al. 1996; Steinpreis et al. 1993). Late-emerging VCMs that are associated with weeks to months of treatment can persist for months after the drug is withdrawn and have been considered to model tardive dyskinesias (Waddington 1990). The precise mechanisms underlying VCMs after long-term treatment with antipsychotics is likely related to a primary mechanism of action of all antipsychotic drugs at the D2 receptor (Wadenberg et al. 2001).

Chronic treatment with haloperidol increases oral responses to mCPP (Wolf et al. 2005; Ikram et al. 2007). Furthermore, the dyskinesia measured after weeks of treatment with haloperidol are reduced by the nonselective 5-HT$_2$ antagonists ritanserin, seganserin, or ketanserin (Naidu and Kulkarni 2001). In another study, oral dyskinesia induced by haloperidol were reduced by concomitant daily administration of ritanserin, and the dyskinesias persisting after haloperidol withdrawal were also reduced by ritanserin administration (Marchese et al. 2004). The finding that the 5-HT$_{1A}$ agonists 8-OH-DPAT and buspirone also lower oral dyskinesia induced by long-term haloperidol suggests that 5-HT tone is altered by the neuroleptic (Haleem et al. 2007a, b). Indeed, these authors found that classical responses dependent on somatodendritic 5-HT$_{1A}$ receptors such as the control of 5-HT metabolism or locomotor responses to 8-OH-DPAT were enhanced by the neuroleptic (Samad et al. 2007). Using chronic administration of buspirone concomitantly to haloperidol, they reported a progressive suppression of oral dyskinesia induced by the neuroleptic. However, Wolf et al. (2005) have reported a slight increase in mCPP-induced inositol phosphate accumulation in striatal tissue of rats chronically treated by haloperidol. Moreover, Ikram and colleagues (2007) have shown a higher increase in 5-HIAA tissue? content induced by mCPP in the dorsolateral striatum of chronically haloperidol-treated rats (Ikram et al. 2007). It is obvious that haloperidol affects central 5-HT transmission, perhaps by modifying both the responses of 5-HT neurons, leading to alteration in 5-HT release, and the responses of 5-HT$_{2C}$ receptors, as suggested by modification of coupling efficiency of the receptor with intracellular second messenger pathways.

To the best of our knowledge, no data are available regarding the responses to 5-HT in the other basal ganglia regions in animals chronically treated with haloperidol. These studies would be important to provide a comprehensive explanation of the greater responsiveness to 5-HT$_{2C}$ agonists in these conditions. For instance, the blockade of 5-HT$_{2A/2C}$ receptors is thought to underline the inhibitory influence of clozapine and risperidone, two antipsychotic drugs that minimally induce oral dyskinesia in rats and tardive dyskinesia in humans, on the discharge of SNr neurons (Bruggeman et al. 2000). Indeed, these authors have reported that concurrent 5-HT$_{2A/2C}$ and moderate DA D2 receptor antagonism reproduce the in vivo effects of these atypical antipsychotics on the firing rate of SNr neurons. Therefore, it is possible that clozapine and risperidone, by blocking the excitatory response exerted by 5-HT$_{2C}$ receptors on SNr neurons activity, favor inhibitory effects. These data support the idea that alteration in 5-HT$_{2C}$ receptor transmission may occur in brain areas other than the striatum. Studies in animal models of Parkinson's disease are in fact supporting this hypothesis.

18.5.2 Chronic impairment of DA Neurons Induced by the Lesion of DA Neurons

The progressive destruction of the nigrostriatal DA neurons is the hallmark of Parkinson's disease. The loss of DA induces profound changes in the functional anatomy of the basal ganglia leading to the symptoms of parkinsonism, including bradykinesia, rigidity, and tremor at rest. As evoked earlier, the destruction of DA neurons in animal models of Parkinson's disease, the 1-methyl-4-phenyl-1,2,3,6-tetrahydropyridine treated monkeys or the hemiparkinsonian rats generated by a unilateral injection of 6-hydroxydopamine (6-OHDA), leads to a sprouting of 5-HT fibers in the striatum and the SN, with a corresponding increase in 5-HT tissue content and release in the striatum (Boulet et al. 2008; Zhou et al. 1991). In rodents, the increase has been more frequently observed in rats with a bilateral lesion as neonates compared with rats with a unilateral lesion as adults (Kostrzewa 1995; Mrini et al. 1995). Such modifications have not been observed in humans where some but not all data would rather support a damage of 5-HT fibers in the brain of parkinsonian patients (Kish et al. 2008; Scholtissen et al. 2006). This may represent a fundamental difference between toxin-induced models in which nigrostriatal neurons are selectively destroyed versus Parkinson's disease, in which multiple neuronal populations are affected, including the raphe nuclei (Braak et al. 2003). Of course, the interpretation in humans is complicated by the presence of treatments, and one cannot exclude that the 5-HT damages could have been higher if no sprouting had occurred. Interestingly, the levels of 5-HT$_{2C}$ receptor mRNA do not follow the increase in 5-HT innervation (Gong et al. 1994). Numan et al. (1995) have shown that 6-OHDA lesion of DA neurons did not affect 5-HT$_{2C}$ mRNA in the striatum, at variance with the 5-HT$_{2A}$ mRNA (Numan et al. 1995). In humans, 5-HT$_{2C}$ receptor binding is not affected in the brain of parkinsonian patients (Waeber and Palacios 1989), although an increased sensitivity to mesulergine binding has been reported specifically in the SNr (Fox and Brotchie 2000a).

Oral responses in the context of the rat model of Parkinson's disease have been described in detail by Salamone et al. (1998). Here we will report the results of experimental studies irrespective of their possible link to the iatrogenic effects of DA agents. In rats lesioned as adults, the purposeless oral response to peripheral mCPP is dramatically enhanced (De Deurwaerdère and Chesselet 2000a). The selective destruction of DA neurons in neonates also increases the sensitivity of adult rats to oral dyskinesia when challenged with DA drugs (Gong et al. 1992). This exaggerated response to DA agonists can be reduced by the nonselective 5-HT$_{2C}$ antagonist mianserin as well as by a lesion of 5-HT neurons with 5,7-dihyroxytryptamine. Moreover, DA neuronal loss also increases oral dyskinesia induced by mCPP (Gong and Kostrzewa 1992; Gong et al. 1992). These authors have concluded that the hypersensitivity of the oral responses to DA agonists in DA lesioned rats as neonates involves 5-HT neurons and 5-HT$_{2C}$ receptors.

The mechanisms whereby oral responses to 5-HT$_{2C}$ agonism are dramatically enhanced are not fully understood and could be related to multiple changes in 5-HT$_{2C}$-mediated neurotransmission in restricted areas of the basal ganglia in DA lesioned rats. By examining the expression of the protooncogene c-Fos, we have found that the lesion of nigrostriatal DA neurons does not modify the increase in Fos expression induced by peripheral administration of mCPP in the STN (Fig. 18.4). Three electrophysiological or behavioral data have confirmed the lack of changes of STN 5-HT$_{2C}$ receptors after a DA lesion. First, we have reported in the same study that the intra-STN administration of low doses mCPP similarly stimulates oral bouts in sham or 6-OHDA-lesioned rats (De Deurwaerdère and Chesselet 2000a). Second, the ability of mCPP to stimulate the firing rate of STN is equal in naive or lesioned rats (Zhang et al. 2009). Finally, the contraversive turning behavior induced by the intra-STN administration of 5-HT, attributed in part to 5-HT$_{2C}$ receptors, is not affected by the lesion of DA neurons (Belforte and Pazo 2004). Thus, 5-HT$_{2C}$ receptors of the STN are an important locus to generate oral dyskinesia elicited by mCPP, but they are not directly responsible for the higher oral response observed in 6-OHDA-lesioned rats.

The striatum might be one important locus. Indeed, the Fos response to mCPP is decreased in the medial, but not the lateral, striatum of 6-OHDA rats (Fig. 18.4). These data are difficult to interpret due to the multiple mechanisms possibly involved in the striatal effects elicited by mCPP (Stark et al. 2008). More directly, Plech et al. (1995) have reported that oral dyskinesia induced by the intrastriatal administration of mCPP are increased in 6-OHDA-lesioned rats (Plech et al. 1995). This increase could not be related to an increase in 5-HT$_{2C}$ mRNA (Gong et al. 1994). Together with the striatum, the most spectacular changes of 5-HT$_{2C}$ transmission in 6-OHDA-lesioned rats occur in output structures of the basal ganglia. Indeed, we found that the lesion of DA neurons enhances the ability of peripheral mCPP to enhance Fos expression in the EPN on the lesioned side only (Fig. 18.4). Interestingly, we have found that the administration of mCPP in the EPN of the lesioned side elicits purposeless oral movements (Mignon et al. 2002). An alteration of 5-HT$_{2C}$ receptor transmission in basal ganglia output is supported by data showing that intranigral infusion of the antagonist SB 206553 elicits contraversive turning behavior per se when administered on the lesioned side in 6-OHDA-lesioned rats (Fox et al. 1998). This locus would be responsible in part for the ability of 5-HT$_{2C}$ antagonists normethylclozapine, SB 200646, or SB 206553 to increase the contralateral rotations elicited by the D2 agonist quinpirole or the D1 agonist SKF 82958 in 6-OHDA-lesioned rats (Fox et al. 1998; Fox and Brotchie 1996, 1996b).

18.6   Conclusion and Clinical Perspectives

18.6.1   Role and Function of 5-HT$_{2C}$ Receptors in the Basal Ganglia

From a physiological point of view, 5-HT$_{2C}$ receptors may exert a general inhibitory control on membrane excitability, and this hypothesis has been extended to the control of functional network in the brain (Tecott et al. 1995; Trent and Tepper 1991). This idea corresponds to the observations in the basal ganglia, where stimulation of 5-HT$_{2C}$ receptors can limit membrane excitability in some neuronal populations (Rueter et al. 2000; Trent and Tepper 1991) and reduce behavioral output (Grottick et al. 2000; Rocha et al. 2002). This control over the excitability of cells in the basal ganglia may indirectly affect the impact of other 5-HT effects as already suggested by (Dalton et al. 2004). It is interesting to note that the 5-HT$_{2C}$ receptor exerts tonic, phasic, and constitutive controls over basal ganglia cells (De Deurwaerdère et al. 2004). These effects are not evenly distributed in the basal ganglia network. Indeed, tonic/constitutive controls mainly impact cellular activity in STN and striatum, whereas phasic controls affect all structures with the exception of the GPe. This peculiar organization is likely to contribute to the control of excitability of the whole network and to affect the response of the basal ganglia to incoming cortical information.

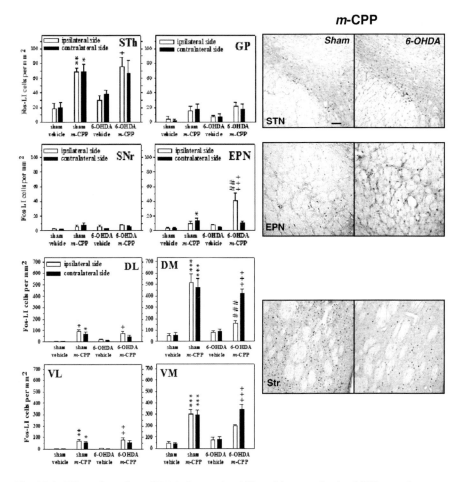

Fig. 18.4 Effect of a unilateral DA lesion on the ability of the nonselective 5-HT$_{2C}$ agonist meta-chlorophenylpiperazine (mCPP) to stimulate Fos expression in the basal ganglia. The data represent the number of Fos-immunoreactive cells/mm^2 in the subthalamic nucleus (STh), the globus pallidus (GP), the substantia nigra pars reticulate (SNr), the entopeduncular nucleus (EPN) and the four striatal quadrants (DL dorsolateral, DM dorsomedial, VL ventrolateral, VM ventromedial). Note the heterogeneity of the responses in the basal ganglia and the modifications of the responses by the lesion in the EPN and the striatum. (*or +, $p < 0.05$; **or ++, $p < 0.01$; *** or +++, $p < 0.001$ with respect to the same side in sham-lesioned rats receiving vehicle; #, $p < 0.05$ and ##, $p < 0.01$ with respect to the other side (Tuckey test).The photomicrographs on the left illustrate the effect of mCPP on Fos expression in the subthalamic nucleus (STN), the EPN, the medial striatum in sham or 6-hydroxydopamine (6-OHDA)-lesioned rats (Reproduced From De Deurwaerdère and Chesselet 2000a. With the authorization of Society for Neuroscience)

The link between molecular changes in 5-HT$_{2C}$ receptor-mediated signaling pathways and abnormal motor responses in the damaged basal ganglia remains unclear. At the cellular level, most of the little data available have focused on the Gq pathway, the main intracellular pathway coupled to 5-HT$_{2C}$ receptors in the brain. However, 5-HT$_{2C}$ receptors may also couple to other second intracellular messenger pathways that could play a role in pathological states. Additionally, where the cellular alterations in 5-HT$_{2C}$-mediated transmission occur is not clear. Several studies have focused on the striatum, but most other regions of the basal ganglia may be potential targets. In line with this thought, it is intriguing that aberrant motor responses to 5-HT$_{2C}$ receptor stimulation are related to changes of activity in associative/limbic regions of the basal ganglia (Beyeler et al. 2010). This observation is compatible with the preferential distribution of 5-HT fibers and 5-HT$_{2C}$ receptors in these regions of the basal ganglia (Beyeler et al. 2010). Consequently, it will be important to examine all parts of the basal ganglia to understand the neurobiological basis of the exaggerated oral responses to 5-HT$_{2C}$ receptor stimulation after chronic DA neuron impairment. To some extent, the situation is similar regarding DA transmission and its involvement in orofacial dyskinesia where numerous brain areas and circuits can be involved in this aberrant motor output (Parry et al. 1994; Prinssen et al. 1994; Uchida et al. 2005).

18.6.2 Clinical Perspectives with 5-HT$_{2C}$ Agents

The role of 5-HT$_{2C}$ receptors in the basal ganglia deserves further study in view of their therapeutic potential in neuropsychiatric disorders. Indeed, antagonists have been proposed as potential treatments for various pathologies including schizophrenia and Parkinson's disease.

Several atypical antipsychotics including clozapine, olanzapine, or risperone are associated with minimal occurrence of extrapyramidal symptoms and tardive dyskinesia (Meltzer et al. 2003; Meltzer and Huang 2008; Meltzer and Nash 1991). Their ability to block 5-HT$_{2C}$ receptors and/or to behave as antagonists/inverse agonists at 5-HT$_{2C}$ receptors might contribute to their clinical superiority compared with typical neuroleptics (Deniau et al. 1996; Canton et al. 1990; Di Matteo et al. 2002; Herrick-Davis et al. 2000; Kuoppamaki et al. 1993; Navailles et al. 2006b). The few clinical trials using nonselective 5-HT$_{2C}$ antagonists such as ritanserin in combination with typical antipsychotics have given positive results (Meltzer et al. 2003; Bersani et al. 1986). The use of 5-HT$_{2C}$ antagonists is therefore considered in the treatment of psychosis and schizophrenia to limit adverse motor effects including tardive dyskinesia (Gunes et al. 2008).

5-HT$_{2C}$ antagonists have been proposed in the treatment of Parkinson's disease. Previous clinical data have reported that the nonselective 5-HT$_{2C}$ antagonists mianserin or mirtazapine as well as treatments thought to reduce 5-HT tone may reduce dyskinesia in Parkinson's disease (Bonifati et al. 1994; Fox et al. 2008; Ikeguchi and Kuroda 1995; Nicholson and Brochtie 2002; Pact and Giduz 1999). The mechanisms underlying this efficacy are not yet understood. First, these compounds could ameliorate the efficacy of DA agonists, leading to the use of lower doses and consequently decreasing their side effects (Fox et al. 1998, 2008; Fox and Brotchie 2000b).

Indeed, antagonists would counteract the increase in nigrothalamic neurons induced by the combined stimulatory action exerted by 5-HT$_{2C}$ receptors upon subthalamonigral and nigrothalamic neurons. According to functional models of the basal ganglia, the reduction of nigrothalamic neuronal activity by 5-HT$_{2C}$ antagonists would release thalamocortical neurons from the tonic inhibition exerted by nigrothalamic neurons. Second, 5-HT$_{2C}$ antagonists could limit the occurrence of L-DOPA- or DA agonists-induced dyskinesia by limiting the hypersensitive responses to 5-HT$_{2C}$ receptor stimulation. Thus, 5-HT$_{2C}$ antagonists would lower an excessive stimulation of 5-HT$_{2C}$ receptors in the STN, the striatum, and/or the EPN. Moreover, dyskinesias are not specifically related to alterations within the sensorimotor circuit, and studies have reported that cognitive and limbic circuits may also participate (Guigoni et al. 2005; Beyeler et al. 2010). 5-HT$_{2C}$ receptor stimulation is known to enhance anxiety-like behaviors in rodents and humans (Curzon and Kennett 1990; Wood et al. 2001). The anxiolytic property of 5-HT$_{2C}$ antagonists could reduce the component of L-DOPA-induced dyskinesias putatively related to anxiety (Kuan et al. 2008).

Unfortunately, the use of 5-HT$_{2C}$ antagonists in the clinic may be limited by their facilitating effects on the reward circuit and feeding behavior, which may lead to addictive behavior and obesity (Tecott et al. 1995; Grottick et al. 2000; Fletcher et al. 2008; Marquis et al. 2007). Furthermore, the disinhibitory action on membrane excitability of some 5-HT$_{2C}$ antagonists could lead to seizure (see Fox et al. 2008), although some antagonists do not modify the threshold for seizure in rodents (Wood et al. 2001). On the other hand, the use of 5-HT$_{2C}$ agonists has been proposed in various pathologies including psychosis, depression, obesity, or addiction to drugs of abuse (Grottick et al. 2000; Fletcher et al. 2008; Marquis et al. 2007). The use of these agonists is in fact limited by their adverse oromotor effects, as already stressed (Simansky et al. 2004). In conclusion, 5-HT$_{2C}$ agonists and antagonists could be promising tools in various neuropsychiatric disorders but their use appears to be limited by potential side effects (Di Giovanni et al. 2006). Newer drugs and a better understanding of 5-HT$_{2C}$ receptors in the basal ganglia may help resolve these issues.

18.7 Conclusions

5-HT$_{2C}$ receptors influence the basal ganglia in a manner that is related to the resting and activated states of this neuronal network. Their action in the control of motor behavior can not be restricted to one area of the basal ganglia and their functional impact in each region is different. In particular, 5-HT$_{2C}$ receptors influence DA neuron activity and contribute to 5-HT transmission in the basal ganglia in both physiological and pathological conditions. A more in-depth knowledge of their role in these brain areas will permit to better understand iatrogenic motor and nonmotor side effects induced by current DA treatments of schizophrenia and Parkinson's disease.

References

Adachi K, Hasegawa M, Fujita S, et al (2002) Dopaminergic and cholinergic stimulation of the ventrolateral striatum elicit rat jaw movements that are funnelled via distinct efferents. Eur J Pharmacol 442:81–92.

Albin RL, Young AB, Penney JB (1989) The functional anatomy of basal ganglia disorders. Trends Neurosci 12:366–375.

Alex KD, Pehek EA (2007) Pharmacologic mechanisms of serotonergic regulation of dopamine neurotransmission. Pharmacol Ther 113:296–320.

Alex KD, Yavanian GJ, McFarlane HG, et al (2005) Modulation of dopamine release by striatal 5-HT$_{2C}$ receptors. Synapse 55:242–251.

Alexander GE, Crutcher MD (1990) Functional architecture of basal ganglia circuits: neural substrates of parallel processing. Trends Neurosci 13:266–271.

Azmitia EC, Segal L (1978) An autoradiographic analysis of the differential ascending projections of the midbrain dorsal and median raphe nuclei in the rat. J Comp Neurol 179:641–668.

Barker EL, Westphal RS, Schmidt D, et al (1994) Constitutively active 5-hydroxytryptamine2C receptors reveal novel inverse agonist activity of receptor ligands. J Biol Chem 269: 11687–11690.

Barwick VS, Jones DH, Richter JT, et al (2000) Subthalamic nucleus microinjections of 5-HT$_2$ receptor antagonists suppress stereotypy in rats. Neuroreport 11:267–270.

Basura GJ, Walker PD (2001) Serotonin 2A receptor regulation of striatal neuropeptide gene expression is selective for tachykinin, but not enkephalin neurons following dopamine depletion. Brain Res Mol Brain Res 92:66–77.

Belforte JE, Pazo JH (2004) Turning behavior induced by stimulation of the 5-HT receptors in the subthalamic nucleus. Eur J Neurosci 19:346–355.

Bengtson CP, Lee DJ, Osborne PB (2004) Opposing electrophysiological actions of 5-HT on noncholinergic and cholinergic neurons in the rat ventral pallidum in vitro. J Neurophysiol 92:433–443.

Berg KA, Maayani S, Goldfarb J, et al (1998) Effector pathway-dependent relative efficacy at serotonin type 2A and 2C receptors: evidence for agonist-directed trafficking of receptor stimulus. Mol Pharmacol 541:94–104.

Berg KA, Cropper JD, Niswender CM, et al (2001) RNA-editing of the 5-HT$_{2C}$ receptor alters agonist-receptor-effector coupling specificity. Br J Pharmacol 134:386–392.

Berg KA, Dunlop J, Sanchez T, et al (2008) A conservative, single-amino acid substitution in the second cytoplasmic domain of the human Serotonin2C receptor alters both ligand-dependent and -independent receptor signaling. J Pharmacol Exp Ther 324:1084–1092.

Bersani G, Grispini A, Marini S, et al (1986) Neuroleptic-induced extrapyramidal side effects: clinical perspective with Ritanserin (R 55667), a new selective 5-HT$_2$ receptor blocking agent. Curr Ther Res 40:492–499.

Beyeler A, Kadiri N, Navailles S, et al (2010) Stimulation of serotonin2c receptors elicits abnormal oral movements by acting on pathways other than the sensorimotor one in the rat basal ganglia. Neuroscience 169:158–170.

Bianchi C, Siniscalchi A, Beani L (1989) Effect of 5-hydroxytryptamine on [3H]-acetylcholine release from guinea-pig striatal slices. Br J Pharmacol 97:213–221.

Blomeley C, Bracci E (2005) Excitatory effects of serotonin on rat striatal cholinergic interneurones. J Physiol (Lond) 569:715–721.

Bonifati V, Fabrizio E, Cipriani R, et al (1994) Buspirone in levodopa-induced dyskinesias. Clin Neuropharmacol 17:73–82.

Bonsi P, Cuomo D, Ding J, et al (2007) Endogenous serotonin excites striatal cholinergic interneurons via the activation of 5-HT$_{2C}$, 5-HT$_6$, and 5-HT$_7$ serotonin receptors: implications for extrapyramidal side effects of serotonin reuptake inhibitors. Neuropsychopharmacology 32:1840–1854.

Boulet S, Mounayar S, Poupard A, et al (2008) Behavioral recovery in MPTP-treated monkeys: neurochemical mechanisms studied by intrastriatal microdialysis. J Neurosci 28:9575–9584.

Braak H, Del Tredici K, Rüb U, et al (2003) Staging of brain pathology related to sporadic Parkinson's disease. Neurobiol Aging 24:197–211.

Bruggeman R, Heeringa M, Westerink BH, et al (2000) Combined 5-HT$_2$/D2 receptor blockade inhibits the firing rate of SNR neurons in the rat brain. Prog Neuropsychopharmacol Biol Psychiatry 24:579–593.

Bubar MJ, Cunningham KA (2007) Distribution of serotonin 5-HT$_{2C}$ receptors in the ventral tegmental area. Neuroscience 146:286–297.

Burns CM, Chu H, Reuter SM, et al (1997) Regulation of serotonin-2C receptor G-protein coupling by RNA editing. Nature 387:303–308.

Canton H, Verriele L, Colpaert FC (1990) Binding of typical and atypical antipsychotics to 5-HT$_{1C}$ and 5-HT$_2$ sites: clozapine potently interacts with 5-HT1C sites. Eur J Pharmacol 191:93–96.

Carlson BB, Wisniecki A, Salamone JD (2003) Local injections of the 5-hydroxytryptamine antagonist mianserin into substantia nigra pars reticulata block tremulous jaw movements in rats: studies with a putative model of Parkinsonian tremor. Psychopharmacology (Berl) 165:229–237.

Chen L, Yung KK, Chan YS, et al (2008) 5-HT excites globus pallidus neurons by multiple receptor mechanisms. Neuroscience 151:439–451.

Chesselet MF, Delfs J (1996) Basal ganglia and movement disorders: an update. Trends Neurosci 19:417–422.

Clemett DA, Punhani T, Duxon MS, et al (2000) Immunohistochemical localisation of the 5-HT$_{2C}$ receptor protein in the rat CNS. Neuropharmacology 39:123–132.

Conn PJ, Sanders-Bush E, Hoffman BJ, et al (1986) A unique serotonin receptor in choroid plexus is linked to phosphatidylinositol turnover. Proc Natl Acad Sci USA 83:4086–4088.

Cook DF, Wirtshafter D (1995) Serotonin agonist-induced c-fos expression in the rat striatum. Soc Neurosci 21, Abstr 1424.

Corvaja N, Doucet G, Bolam JP (1993) Ultrastructure and synaptic targets of the raphe-nigral projection in the rat. Neuroscience 55:417–427.

Cudennec A, Duverger D, Serrano A, et al (1988) Influence of ascending serotonergic pathways on glucose use in the conscious rat brain. II. Effects of electrical stimulation of the rostral raphe nuclei. Brain Res 444:227–246.

Curzon G, Kennett GA (1990) m-CPP: a tool for studying behavioral responses associated with 5-HT1c receptors. Trends Pharmacol Sci 11:181–182.

Dalton GL, Lee MD, Kennett GA, et al (2004) mCPP-induced hyperactivity in 5-HT2C receptor mutant mice is mediated by activation of multiple 5-HT receptor subtypes. Neuropharmacology 46:663–671.

Davies J, Tongroach P (1978) Neuropharmacological studies on the nigro-striatal and raphe-striatal system in the rat. Eur J Pharmacol 51:91–100.

De Deurwaerdère P, Chesselet M-F (2000) Nigrostriatal lesions alter orofacial dyskinesia and c-fos expression induced by the serotonin agonist 1-(m-cholophenyl)piperazine in adult rats. J Neurosci 20:5170–5178.

De Deurwaerdère P, Chesselet MF (2000) Nigrostriatal lesions alter Serotonin 2C receptor stimulation-induced oro-facial dyskinesia and Fos expression in the basal ganglia. In: Serotonin club: from molecules to clinic: November 1–3, 2000, New Orleans, LA, Abstr 66.

De Deurwaerdère P, Navailles S, Berg KA, et al (2004) Constitutive activity of the serotonin2C receptor inhibits in vivo dopamine release in the rat striatum and nucleus accumbens. J Neurosci 24:3235–3241.

De Deurwaerdère P, Le Moine C and Chesselet M-F (2010) Selective blockade of Serotonin2C receptor enhances Fos expression specifically in the striatum and the subthalamic nucleus within the basal ganglia. Neurosci Lett 469:251–255.

DeLong MR (1990) Primate models of movement disorders of basal ganglia origin. Trends Neurosci 13:281–285.

Deniau JM, Menetrey A, Charpier S (1996) The lamellar organization of the rat substantia nigra pars reticulata: segregated patterns of striatal afferents and relationship to the topography of corticostriatal projections. Neuroscience 73:761–781.

Di Giovanni G, Di Matteo V, La Grutta V, et al (2001) m-Chlorophenylpiperazine excites non-dopaminergic neurons in the rat substantia nigra and ventral tegmental area by activating serotonin-2C receptors. Neuroscience 103:111–116.

Di Giovanni G, Di Matteo V, Pierucci M, et al (2006) Serotonin involvement in the basal ganglia pathophysiology: could the 5-HT$_{2C}$ receptor be a new target for therapeutic strategies? Curr Med Chem 13:3069–3081.

Di Matteo V, Cacchio M, Di Giulio C, et al (2002) Biochemical evidence that the atypical antipsychotic drugs clozapine and risperidone block 5-HT$_{2C}$ receptors in vivo. Pharmacol Biochem Behav 71:607–613.

Di Matteo V, Pierucci M, Esposito E, et al (2008) Serotonin modulation of the basal ganglia circuitry: therapeutic implication for Parkinson's disease and other motor disorders. Prog Brain Res 172:423–463.

Dray A (1981) Serotonin in the basal ganglia : functions and interactions with other neuronal pathways. J Physiol 77:393–403.

Eberle-Wang K, Lucki I, Chesselet MF (1996) A role for the subthalamic nucleus in 5-HT$_{2C}$-induced oral dyskinesia. Neuroscience 72:117–128.

Eberle-Wang K, Mikeladze Z, Uryu K, et al (1997) Pattern of expression of the Serotonin$_{2C}$ receptor messenger RNA in the basal ganglia of adults rats. J Comp Neurol 384:233–247.

Egan MF, Hurd Y, Ferguson J, et al (1996) Pharmacological and neurochemical differences between acute and tardive vacuous chewing movements induced by haloperidol. Psychopharmacology (Berl) 127:337–345.

El Mansari M, Blier P (1997) In vivo electrophysiological characterization of 5-HT receptors in the guinea pig head of caudate nucleus and orbitofrontal cortex. Neuropharmacology 36:577–588.

El Mansari M, Radja F, Ferron A, et al (1994) Hypersensitivity to serotonin and its agonists in serotonin-hyperinnervated neostriatum after neonatal dopamine denervation. Eur J Pharmacol 261:171–178.

Euvrard C, Javoy F, Herbet A, et al (1977) Effect of quipazine, a serotonin-like drug, on striatal cholinergic interneurones. Eur J Pharmacol 41:281–289.

Fletcher PJ, Rizos Z, Sinyard J, et al (2008) The 5-HT$_{2C}$ receptor agonist Ro60-0175 reduces cocaine self-administration and reinstatement induced by the stressor yohimbine, and contextual cues. Neuropsychopharmacology 33:1402–1412.

Flores G, Rosales MG, Hernandez S, et al (1995) 5-Hydroxytryptamine increases spontaneous activity of subthalamic neurons in the rat. Neurosci Lett 192:17–20.

Fox S, Brotchie J (1996) Normethylclozapine potentiates the action of quinpirole in the 6-hydroxydopamine lesioned rat. Eur J Pharmacol 301:27–30.

Fox SH, Brotchie JM (2000a) 5-HT$_{2C}$ receptor binding is increased in the substantia nigra pars reticulata in Parkinson's disease. Mov Disord 15:1064–1069.

Fox SH, Brotchie JM (2000b) 5-HT$_{2C}$ receptor antagonists enhance the behavioral response to dopamine D1 receptor agonists in the 6-hydroxydopamine-lesioned rat. Eur J Pharmacol 398:59–64.

Fox SH, Moser B, Brotchie JM (1998) Behavioral effects of 5-HT$_{2C}$ receptor antagonism in the substantia nigra zona reticulata of the 6-hydroxydopamine-lesioned rat model of Parkinson's disease. Exp Neurol 151:35–49.

Fox SH, Chuang R, Brotchie JM (2008) Parkinson's disease – opportunities for novel therapeutics to reduce the problems of levodopa therapy. Prog Brain Res 172:479–494.

Gardier AM, Moratalla R, Cuéllar B, et al (2000) Interaction between the serotoninergic and dopaminergic systems in d-fenfluramine-induced activation of c-fos and jun B genes in rat striatal neurons. J Neurochem 74:1363–1373.

Gerfen CR (1984) The neostriatal mosaic: compartmentalization of corticostriatal input and striatonigral output systems. Nature 311:461–464.

Gerfen CR (1985) The neostriatal mosaic. I. Compartmental organization of projections from the striatum to the substantia nigra in the rat. J Comp Neurol 236:454–476.

Gerfen CR, Engber TM, Mahan LC, et al (1990) D1 and D2 dopamine receptor-regulated gene expression of striatonigral and striatopallidal neurons. Science 250:1429–1432.

Gillet G, Ammor S, Fillion G (1985) Serotonin inhibits acetylcholine release from rat striatum slices: evidence for a presynaptic receptor-mediated effect. J Neurochem 45:1687–1691.

Gong L, Kostrzewa RM (1992) Supersensitized oral responses to a serotonin agonist in neonatal 6-OHDA-treated rats. Pharmacol Biochem Behav 41:621–623.

Gong L, Kostrzewa RM, Fuller RW, et al (1992) Supersensitization of the oral response to SKF 38393 in neonatal 6-OHDA-lesioned rats is mediated through a serotonin system. J Pharmacol Exp Ther 261:1000–1007.

Gong L, Kostrzewa RM, Li C (1994) Neonatal 6-hydroxydopamine and adult SKF 38393 treatments alter dopamine D1 receptor mRNA levels: absence of other neurochemical associations with the enhanced behavioral responses of lesioned rats. J Neurochem 63:1282–1290.

Gongora-Alfaro JL, Hernandez-Lopez S, Flores-Hernandez J, et al (1997). Firing frequency modulation of substantia nigra reticulata neurons by 5-hydroxytryptamine. Neurosci Res 29:225–231.

Graybiel AM (1991) Basal ganglia–input, neural activity, and relation to the cortex. Curr Opin Neurobiol 1:644–651.

Graybiel AM (2005) The basal ganglia: learning new tricks and loving it. Curr Opin Neurobiol 15:638–644.

Grottick AJ, Fletcher PJ, Higgins GA (2000) Studies to investigate the role of 5-HT$_{2C}$ receptors on cocaine- and food-maintained behavior. J Pharmacol Exp Ther 295:1183–1191.

Guigoni C, Li Q, Aubert I, et al (2005) Involvement of sensorimotor, limbic, and associative basal ganglia domains in L-3,4-dihydroxyphenylalanine-induced dyskinesia. J Neurosci 25:2102–2107.

Gunes A, Scordo M, Jaanson P, et al (2007) Serotonin and dopamine receptor gene polymorphisms and the risk of extrapyramidal side effects in perphenazine-treated schizophrenic patients. Psychopharmacology (Berl) 190:479–484.

Gunes A, Dahl M-L, Spina E, et al (2008) Further evidence for the association between 5-HT2C receptor gene polymorphisms and extrapyramidal side effects in male schizophrenic patients. Eur J Clin Pharmacol 64:477–482.

Hackler EA, Turner GH, Gresch PJ, et al (2007) 5-Hydroxytryptamine2C receptor contribution to m-chlorophenylpiperazine and N-methyl-beta-carboline-3-carboxamide-induced anxiety-like behavior and limbic brain activation. J Pharm Exp Ther 320:1023–1029.

Haleem DJ, Samad N, Haleem MA (2007). Reversal of haloperidol-induced extrapyramidal symptoms by buspirone: a time-related study. Behav Pharmacol 18:147–153.

Haleem DJ, Samad N, Haleem MA (2007) Reversal of haloperidol-induced tardive vacuous chewing movements and supersensitive somatodendritic serotonergic response by buspirone in rats. Pharmacol Biochem Behav 87:115–121.

Hartig PR, Hoffman BJ, Kaufman MJ, et al (1990) The 5-HT1C receptor. Ann N Y Acad Sci 600:149–66.

Hashimoto K, Kita H (2008) Serotonin activates presynaptic and postsynaptic receptors in rat globus pallidus. J Neurophysiol 99:1723–1732.

Herrick-Davis K, Grinde E, Teitler M (2000) Inverse agonist activity of atypical antipsychotic drugs at human 5-hydroxytryptamine2C receptors. J Pharmacol Exp Ther 295:226–232.

Hoffman BJ, Mezey E (1989) Distribution of serotonin 5-HT1C receptor mRNA in adult rat brain. FEBS Lett 247:453–462.

Hoyer D, Karpf A (1988) [^{125}I]SCH 23982, a "selective" D-1 receptor antagonist, labels with high affinity 5-HT$_{1C}$ sites in pig choroid plexus. Eur J Pharmacol 150:181–184.

Hoyer D, Pazos A, Probst A, et al (1986) Serotonin receptors in the human brain. II. Characterization and autoradiographic localization of 5-HT1C and 5-HT2 recognition sites. Brain Res 376:97–107.

Hoyer D, Clarke DE, Fozard JR, et al (1994). VII. International union of pharmacology classification of receptors for 5-hydroxytryptamine (serotonin). Pharmacol Rev 46:157–203.

Ikeguchi K, Kuroda A (1995) Mianserin treatment of patients with psychosis induced by antiparkinsonian drugs. Eur Arch Psychiatry Clin Neurosci 244:320–324.

Ikram H, Samad N, Haleem DJ (2007) Neurochemical and behavioral effects of m-CPP in a rat model of tardive dyskinesia. Pak J Pharm Sci 20:188–195.

Invernizzi RW, Pierucci M, Calcagno E, et al (2007) Selective activation of 5-HT$_{2C}$ receptors stimulates GABA-ergic function in the rat substantia nigra pars reticulata: a combined in vivo electrophysiological and neurochemical study. Neuroscience 144:1523–1535.

James TA, Starr MS (1980) Rotational behavior elicited by 5-HT in the rat: evidence for an inhibitory role of 5-HT in the substantia nigra and corpus striatum. J Pharmacol 32:196–200.

Janno S, Holi M, Tuisku K, et al (2004) Prevalence of neuroleptic-induced movement disorders in chronic schizophrenia inpatients. Am J Psychiatry 161:160–163.

Javed A, Van de Kar LD, Gray TS (1998) The 5-HT$_{1A}$ and 5-HT$_{2A/2C}$ receptor antagonists WAY 100635 and ritanserin do not attenuate D-fenfluramine-induced fos expression in the brain. Brain Res 791:67–74.

Julius D, MacDermott AB, Jessel TM, et al (1988) Functional expression of the 5-HT1c receptor in neuronal and nonneuronal cells. Cold Spring Harb Symp Quant Biol 53 Pt 1:385–393.

Kaufman MJ, Hartig PR, Hoffman BJ (1995) Serotonin 5-HT$_{2C}$ receptor stimulates cyclic GMP formation in choroid plexus. J Neurochem 64:199–205.

Kawaguchi Y, Wilson CJ, Auggod SJ, et al (1995) Striatal interneurones: chemical, physiological and morphological characterization. Trends Neurosci 12:527–535.

Kish SJ, Tong J, Hornykiewicz O, et al (2008) Preferential loss of serotonin markers in caudate versus putamen in Parkinson's disease. Brain 131:120–131.

Kita H, Chiken S, Tachibana Y, et al (2007) Serotonin modulates pallidal neuronal activity in the awake monkey. J Neurosci 27:75–83.

Kolomiets BP, Deniau JM, Glowinski J, et al (2003) Basal ganglia and processing of cortical information: functional interactions between trans-striatal and trans-subthalamic circuits in the substantia nigra pars reticulata. Neuroscience 117:931–938.

Kostrzewa RM (1995) Dopamine receptor supersensitivity. Neurosci Biobehav Rev 1:1–17.

Kostrzewa RM, Gong L, Brus R (1993) Serotonin (5-HT) systems mediate dopamine (DA) receptor supersensitivity. Acta Neurobiol Exp (Wars) 53:31–41.

Kostrzewa RM, Huang NY, Kostrzewa JP, et al (2007) Modeling tardive dyskinesia: predictive 5-HT$_{2C}$ receptor antagonist treatment. Neurotox Res 11:41–50.

Kuan WL, Zhao JW, Barker RA (2008) The role of anxiety in the development of levodopa-induced dyskinesias in an animal model of Parkinson's disease, and the effect of chronic treatment with the selective serotonin reuptake inhibitor citalopram. Psychopharmacology (Berl) 197:279–293.

Kuoppamaki M, Syvalahti E, Hietala J (1993) Clozapine and N-desmethylclozapine are potent 5-HT$_{1C}$ receptor antagonists. Eur J Pharmacol 245:179–182.

Labasque M, Reiter E, Becamel C, et al (2008) Physical interaction of calmodulin with the 5-hydroxytryptamine2C receptor C-terminus is essential for G protein-independent, arrestin-dependent receptor signaling. Mol Biol Cell 19:4640–4650.

Leonhardt S, Gorospe E, Hoffman BJ, et al (1992) Molecular pharmacological differences in the interaction of serotonin with 5-hydroxytryptamine1C and 5-hydroxytryptamine2 receptors. Mol Pharmacol 42:328–335.

Leslie RA, Moorman JM, Coulson A, et al (1993) Serotonin2/1C receptor activation causes a localized expression of the immediate-early gene c-fos in rat brain: evidence for involvement of dorsal raphe nucleus projection fibres. Neuroscience 53:457–463.

Liminga U, Johnson AE, Andrén PE, et al (1993) Modulation of oral movements by intranigral 5-hydroxytryptamine receptor agonists in the rat. Pharmacol Biochem Behav 46:427–433.

Lobo MK, Karsten SL, Gray M, et al (2006) FACS-array profiling of striatal projection neuron subtypes in juvenile and adult mouse brains. Nat Neurosci 9:443–452.

López-Giménez JF, Mengod G, Palacios JM, et al (2001) Regional distribution and cellular localization of 5-HT2C receptor mRNA in monkey brain: comparison with [^{3}H]mesulergine binding sites and choline acetyltransferase mRNA. Synapse 42:12–26.

Lucas JJ, Segu L, Hen R (1997) 5-Hydroxytryptamine1B receptors modulate the effect of cocaine on c-fos expression: converging evidence using 5-hydroxytryptamine1B knockout mice and the 5-hydroxytryptamine1B/1D antagonist GR127935. Mol Pharmacol 51:755–763.

Marchese G, Bartholini F, Ruiu S, et al (2004) Ritanserin counteracts both rat vacuous chewing movements and nigro-striatal tyrosine hydroxylase-immunostaining alterations induced by haloperidol. Eur J Pharmacol 483:65–69.

Marquis KL, Sabb AL, Logue SF, et al (2007) WAY 163909 [(7bR,10aR)-1,2,3,4,8,9,10, 10a-octahydro-7bH-cyclopenta-[b][1,4]diazepino[6,7,1hi]indole]: a novel 5-hydroxytryptamine 2C receptor-selective agonist with preclinical antipsychotic-like activity. J Pharmacol Exp Ther 320:486–496.

Mehta A, Eberle-Wang K, Chesselet MF (2001) Increased m-CPP-induced oral dyskinesia after lesion of serotonergic neurons. Pharmacol Biochem Behav 68:347–353.

Meibach RC, Maayani S, Green JP (1980) Characterization and radioautography of [³H]LSD binding by rat brain slices in vitro: the effect of 5-hydroxytryptamine. Eur J Pharmacol 67:371–382.

Meltzer HY, Huang M (2008) In vivo actions of atypical antipsychotic drug on serotonergic and dopaminergic systems. Prog Brain Res 172:177–197.

Meltzer HY, Nash JF (1991) Effects of antipsychotic drugs on serotonin receptors. Pharmacol Rev 43:587–604.

Meltzer HY, Li Z, Kaneda Y, et al (2003) Serotonin receptors: their key role in drugs to treat schizophrenia. Prog Neuropsychopharmacol Biol Psychiatry 27:1159–1172.

Mengod G, Nguyen H, Le H, et al (1990) The distribution and cellular localisation of the serotonin 1C receptor mRNA in the rodent brain examined by in situ hybridization hystochemistry. Comparison with receptor binding distribution. Neuroscience 35:577–591.

Mignon L, De Deurwaerdère P, Chesselet MF (2002) Behavioral and molecular effects of the 5-HT$_{2C}$ agonist mCPP in an animal model of Parkinson's Disease. Portland: Gordon conference, June 2002.

Mink JW (1996) The basal ganglia: focused selection and inhibition of competing motor programs. Prog Neurobiol 50:381–425.

Molineaux SM, Jessell TM, Axel R, et al (1989) 5-HT1c receptor is a prominent serotonin receptor subtype in the central nervous system. Proc Natl Acad Sci USA 86:6793–6797.

Moorman JM, Leslie RA (1996) P-chloroamphetamine induces c-fos in rat brain: a study of serotonin2A/2C receptor function. Neuroscience 72:129–139.

Moukhles H, Bosler O, Bolam JP, et al (1997) Quantitative and morphometric data indicate precise cellular interactions between serotonin terminals and postsynaptic targets in rat substantia nigra. Neuroscience 76:1159–1171.

Mrini A, Soucy J-P, Lafaille F, et al (1995) Quantification of the serotonin hyperinnervation in adult rat neostriatum after neonatal 6-hydroxydopamine lesion of nigral dopamine neurons. Brain Res 669:303–308.

Naidu PS, Kulkarni SK (2001) Effect of 5-HT$_{1A}$ and 5-HT$_{2A/2C}$ receptor modulation on neuroleptic-induced vacuous chewing movements. Eur J Pharmacol 428:81–86.

Navailles S, Moison D, Ryczko D, et al (2006) Region-dependent regulation of mesoaccumbens dopamine neurons in vivo by the constitutive activity of central serotonin$_{2C}$ receptors. J Neurochem 99:1311–1319.

Navailles S, De Deurwaerdère P, Spampinato U (2006) Clozapine and haloperidol differentially alter the constitutive activity of central serotonin2C receptors in vivo. Biol Psychiatry 59:568–575.

Nicholson SL, Brochtie JM (2002) 5-Hydroxytryptamine and Parkinson's disease – opportunities for novel therapeutics to reduce the problems of levodopa therapy. Eur J Neurol 3:1–6.

Niswender CM, Copeland SC, Herrick-Davis K, et al (1999) RNA editing of the human serotonin 5-hydroxytryptamine 2C receptor silences constitutive activity. J Biol Chem 274:9472–9478.

North A, Uchimura N (1989) 5-Hydroxytryptamine acts at 5-HT$_2$ receptors to decrease potassium conductance in rat nucleus accumbens neurons. J Physiol 417:1–12.

Numan S, Lundgren KH, Wright DE, et al (1995) Increased expression of 5HT$_2$ receptor mRNA in rat striatum following 6-OHDA lesions of the adult nigrostriatal pathway. Brain Res Mol Brain Res 29:391–396.

Obeso JA, Rodríguez-Oroz MC, Rodríguez M, et al (2000) Pathophysiology of the basal ganglia in Parkinson's disease. Trends Neurosci 23(10 suppl):S8–S19.

Olpe HR, Koella WP (1977) The response of striatal cells upon stimulation of the dorsal and median raphe nuclei. Brain Res 122:357–360.

Pact V, Giduz T (1999) Mirtazapine treats resting tremor, essential tremor, and levodopa-induced dyskinesias. Neurology 53:1154.

Pakhotin P, Bracci E (2007) Cholinergic interneurons control the excitatory input to the striatum. J Neurosci 27:391–400.

Parent A, Hazrati LN (1995) Functional anatomy of the basal ganglia. I. The cortico-basal ganglia-thalamo-cortical loop. Brain Res Brain Res Rev 20:91–127.

Parry TJ, Eberle-Wang K, Lucki I, et al (1994) Dopaminergic stimulation of subthalamic nucleus elicits oral dyskinesia in rats. Exp Neurol 128:181–190.

Pasqualetti M, Ori M, Castagna M, et al (1999) Distribution and cellular localization of the serotonin type 2C receptor messenger RNA in human brain. Neuroscience 92:601–611.

Pazos A, Palacios JM (1985) Quantitative autoradiographic mapping of serotonin receptors in the rat brain. I. Serotonin-1 receptors. Brain Res. 3462:205–230.

Pazos A, Probst A, Palacios JM (1987) Serotonin receptors in the human brain. III. Autoradiographic mapping of serotonin-1 receptors. Neuroscience 21:97–122.

Plech A, Brus R, Kalbfleisch JH, et al (1995) Enhanced oral activity responses to intrastriatal SKF 38393 and m-CCP are attenuated by intrastriatal mianserin in neonatal 6-OHDA-lesionned rats. Psychopharmacology (Berl) 119:466–473.

Pompeiano M, Palacios JM, Mengod G (1994) Distribution of the serotonin 5-HT$_2$ receptor family mRNAs: comparison between 5-HT$_{2A}$ and 5-HT$_{2B}$ receptors. Mol Brain Res 23:163–178.

Prinssen EP, Balestra W, Bemelmans FF, et al (1994) Evidence for a role of the shell of the nucleus accumbens in oral behavior of freely moving rats. J Neurosci 14:1555–1562.

Radja F, Descarries L, Dewar KM, et al (1993). Serotonin 5-HT$_1$ and 5-HT$_2$ receptors in adult rat brain after neonatal destruction of nigrostriatal dopamine neurons: a quantitative autoradiographic study. Brain Res 606:273–285.

Rick CE, Stanford IM, Lacey MG (1995) Exitation of rat substantia nigra pars reticulata neurons by 5-Hydroxytryptamine in vitro: evidence for a direct action mediated by 5-hydroxytryptamine2C receptors. Neuroscience 69:903–913.

Rocha BA, Goulding EH, O'Dell LE, et al (2002) Enhanced locomotor, reinforcing, and neurochemical effects of cocaine in serotonin 5-hydroxytryptamine 2C receptor mutant mice. J Neurosci 22:10039–10045.

Rosengarten H, Schweitzer JW, Friedhoff AJ (1999) The effect of novel antipsychotics in rat oral dyskinesia. Prog Neuropsychopharmacol Biol Psychiatry 23:1389–1404.

Rouillard C, Bovetto S, Gervais J, et al (1996) Fenfluramine-induced activation of the immediate-early gene c-fos in the striatum: possible interaction between serotonin and dopamine. Brain Res Mol Brain Res 37:105–115.

Rueter LE, Tecott LH, Blier P (2000) In vivo electrophysiological examination of 5-HT$_2$ responses in 5-HT$_{2C}$ receptor mutant mice. Naunyn Schmiedebergs Arch Pharmacol 361:484–491.

Rupniak NM, Jenner P, Marsden CD (1985) Pharmacological characterisation of spontaneous or drug-associated purposeless chewing movements in rats. Psychopharmacology (Berl) 85:71–79.

Salamone JD, Mayorga AJ, Trevitt JT, et al (1998) Tremulous jaw movements in rats: a model of parkinsonian tremor. Prog Neurobiol 56:591–611.

Samad N, Khan A, Perveen T, et al (2007) Increase in the effectiveness of somatodendritic 5-HT$_{1A}$ receptors in a rat model of tardive dyskinesia. Acta Neurobiol Exp 67:389–397.

Scholtissen B, Verhey FR, Steinbusch HW, et al (2006) Serotonergic mechanisms in Parkinson's disease: opposing results from preclinical and clinical data. J Neural Transm 113:59–73.

Sgambato V, Abo V, Rogard M, et al (1997) Effect of electrical stimulation of the cerebral cortex on the expression of the Fos protein in the basal ganglia. Neuroscience 81:93–112.

Shen KZ, Johnson SW (2008) 5-HT inhibits synaptic transmission in rat subthalamic nucleus neurons in vitro. Neuroscience 151:1029–1033.

Shen KZ, Kozell LB, Johnson SW (2007) Multiple conductances are modulated by 5-HT receptor subtypes in rat subthalamic nucleus neurons. Neuroscience 148:996–1003.

Simansky KJ, Dave KD, Inemer BR, et al (2004) A 5-HT$_{2C}$ agonist elicits hyperactivity and oral dyskinesia with hypophagia in rabbits. Physiol Behav 82:97–107.

Singewald N, Salchner P, Sharp T (2003) Induction of c-Fos expression in specific areas of the fear circuitry in rat forebrain by anxiogenic drugs. Biol Psychiatry 53:275–283.

Soghomonian JJ, Descarries L, Watkins K (1989) Serotonin innervation in adult rat neostriatum. II. Ultrastructural features: a radioautographic and immunocytochemical study. Brain Res 481:67–86.

Stanford IM, Lacey MG (1996) Differential actions of serotonin, mediated by 5-HT$_{1B}$ and 5-HT$_{2C}$ receptors, on GABA-mediated synaptic input to rat substantia nigra pars reticulata neurons in vitro. J Neurosci 16:7566–7573.

Stanford IM, Kantaria MA, Chahal HS, et al (1995) 5-Hydroxytryptamine induced excitation and inhibition in the subthalamic nucleus: action at 5-HT$_{2C}$, 5-HT$_4$ and 5-HT$_{1A}$ receptors. Neuropharmacology 49:1228–1234.

Stark JA, Davies KE, Williams SR, et al (2006) Functional magnetic resonance imaging and c-Fos mapping in rats following an anorectic dose of m-chlorophenylpiperazine. Neuroimage 31:1228–1237.

Stark JA, McKie S, Davies KE, et al (2008) 5-HT$_{2C}$ antagonism blocks blood oxygen level-dependent pharmacological-challenge magnetic resonance imaging signal in rat brain areas related to feeding. Eur J Neurosci 27:457–465.

Steinbush HW (1984) Serotonin-immunoreactive neurons and their projections in the CNS. In: Björklund A, Hökfelt T, Kuhar MJ, eds. Handbook of Chemical Neuroanatomy. Vol. 3, Part II. Amsterdam: Elsevier, pp. 68–125.

Steinpreis RE, Baskin P, Salamone JD (1993) Vacuous jaw movements induced by sub-chronic administration of haloperidol: interactions with scopolamine. Psychopharmacology (Berl) 111:99–105.

Stewart BR, Jenner P, Marsden CD (1989) Induction of purposeless chewing behavior in rats by 5-HT agonists drugs. Eur J Pharmacol 162:101–107.

Tarsy D, Baldessarini RJ (1984) Tardive dyskinesia. Annu Rev Med 35:605–623.

Tarsy D, Baldessarini RJ, Tarazi FI (2002) Effects of newer antipsychotics on extrapyramidal function. CNS Drugs 16:23–45.

Tecott LH, Sun LM, Akana SF, et al (1995) Eating disorder and epilepsy in mice lacking 5-HT2c serotonin receptors. Nature 374:542–546.

Tilakaratne N, Friedman E (1996) Genomic responses to 5-HT$_{1A}$ or 5-HT$_{2A/2C}$ receptor activation is differentially regulated in four regions of rat brain. Eur J Pharmacol 307:211–217.

Tremblay PO, Gervais J, Rouillard C (1998) Modification of haloperidol-induced pattern of c-fos expression by serotonin agonists. Eur J Neurosci 10:3546–3555.

Trent F, Tepper JM (1991) Dorsal raphé stimulation modifies striatal-evoked antidromic invasion of nigral dopaminergic neurons in vivo. Exp Brain Res 84:620–630.

Uchida T, Adachi K, Fujita S, et al (2005) Role of GABA(A) receptors in the retrorubral field and ventral pallidum in rat jaw movements elicited by dopaminergic stimulation of the nucleus accumbens shell. Eur J Pharmacol 510:39–47.

Van der Kooy D, Hattori T (1980) Bilaterally situated dorsal raphe cell bodies have only unilateral forebrain projections in rat. Brain Res 192:550–554.

Waddington JL (1990) Spontaneous orofacial movements induced in rodents by very long-term neuroleptic drug administration: phenomenology, pathophysiology and putative relationship to tardive dyskinesia. Psychopharmacology (Berl) 101:431–447.

Wadenberg ML, Soliman A, VanderSpek SC, et al (2001) Dopamine D2 receptor occupancy is a common mechanism underlying animal models of antipsychotics and their clinical effects. Neuropsychopharmacology 25:633–641.

Waeber C, Palacios JM (1989) Serotonin-1 receptor binding sites in the human basal ganglia are decreased in Huntington's chorea but not in Parkinson's disease: a quantitative in vitro auoradiography study. Neuroscience 32:337–347.

Waeber C, Palacios JM (1994) Binding sites for 5-hydroxytryptamine-2 receptor agonists are predominantly located in striosomes in the human basal ganglia. Mol Br Res 24:199–209.

Walker PD, Riley LA, Hart RP, et al (1991) Serotonin regulation of neostriatal tachykinins following neonatal 6-hydroxydopamine lesions. Brain Res 557:31–36.

Ward RP, Dorsa DM (1996) Colocalization of serotonin receptor subtypes 5-HT_{2A}, 5-HT_{2C} and 5-HT_6 with neuropeptides in rat striatum. J Comp Neurol 370:405–414.

Westphal RS, Backstrom JR, Sanders-Bush E (1995) Increased basal phosphorylation of the constitutively active serotonin 2C receptor accompanies agonist-mediated desensitization. Mol Pharmacol 48:200–205.

Wilms K, Vierig G, Davidowa H (2001) Interactive effects of cholecystokinin-8S and various serotonin receptor agonists on the firing activity of neostriatal neuronesin rats. Neuropeptides 35:257–270.

Wirtshafter D, Cook DF (1998) Serotonin-1B agonists induce compartmentally organized striatal Fos expression in rats. Neuroreport 9:1217–1221.

Wolf WA, Schutz LJ (1997) The serotonin 5-HT_{2C} receptor is a prominent serotonin receptor in basal ganglia:evidence from functional studies on serotonin-mediated phosphoinositide hydrolysis. J Neurochem 69:1449–1458.

Wolf WA, Bieganski GJ, Guillen V, et al (2005) Enhanced 5-HT_{2C} receptor signalling is associated with haloperidol-induced "early onset" vacuous chewing in rats: implications for antipsychotic drug therapy. Psychopharmacology (Berl) 182:84–94.

Wood MD, Reavill C, Trail B, et al (2001) SB-243213; a selective 5-HT_{2C} receptor inverse agonist with improved anxiolytic profile: lack of tolerance and withdrawal anxiety. Neuropharmacology 41:186–199.

Wright DE, Seroogy KB, Lundgren KH, et al (1995) Comparative localization of serotonin1A, 1C, and 2 receptor subtype mRNAs in rat brain. J Comp Neurol 351:357–373.

Xiang Z, Wang L, Kitai ST (2005) Modulation of spontaneous firing in rat subthalamic neurons by 5-HT receptor subtypes. J Neurophysiol 93:1145–1157.

Young WS, Kuhar MJ (1980) Serotonin receptor localization in rat brain by light microscopic autoradiography. Eur J Pharmacol 62:237–239.

Zhang QJ, Liu X, Liu J, et al (2009) Subthalamic neurons show increased firing to 5-HT_{2C} receptor activation in 6-hydroxydopamine-lesioned rats. Brain Res 1256:180–189.

Zhou FC, Bledsoe S, Murphy J (1991) Serotonergic sprouting is induced by dopamine-lesion in substantia nigra of adult rat brain. Brain Res 556:108–116.

Chapter 19
Modeling Tardive Dyskinesia: Predictive 5-HT$_{2C}$ Receptor Antagonist Treatment

Richard M. Kostrzewa

19.1 Tardive Dyskinesia: An Overview

19.1.1 Definition

From the Website of the National Institute of Neurological Disorders and Stroke (NINDS) (http://www.ninds.nih.gov/disorders/tardive/tardive.htm) tardive dyskinesia (TD) is classified as a neurological syndrome produced by neuroleptics, typically "conventional" antipsychotics. Although different muscle groups can be involved (e.g., fingers, arms, legs, trunk), the classic presentation is repetitive, involuntary, purposeless (i.e., vacuous) chewing movements with or without tongue protrusion and lip smacking (Casey 1987; Jeste and Caligiuri 1993).

19.1.2 Incidence, Persistence, Treatment of TD

The overall incidence of TD is ~5% per treatment year (Kane et al. 1982). Thereby, in a population of 100 patients treated with conventional antipsychotics for 3-years, ~15 patients would present with TD. When symptoms of TD initially appear, they can sometimes be suppressed by abruptly elevating the dose of the antipsychotic. However, symptoms of TD can reappear. Typically, TD is long-lived, often irreversible with or without continued treatment with an antipcychotic drug. Unfortunately, there is no recognized treatment for TD.

R.M. Kostrzewa (✉)
Department of Pharmacology, Quillen College of Medicine, East Tennessee State University, Johnson City, TN 37614-1708, USA
e-mail: kostrzew@etsu.edu

G. Di Giovanni et al. (eds.), *5-HT$_{2C}$ Receptors in the Pathophysiology of CNS Disease*, The Receptors 22, DOI 10.1007/978-1-60761-941-3_19,

19.2 Dopamine Systems and TD

Conventional antipsychotics are dopamine (DA) D_2 receptor antagonists. All dopaminergic tracts in brain are affected, but it appears that the nigrostriatal tract is most intimately associated with TD, since this tract exerts fine motor control. At an effective dose of antipsychotic, ~70–80% of the D_2 receptors are blocked. The acute effect of D_2 receptor block does not produce TD, but prolonged D_2 block can result in TD – hence the term *tardive dyskinesia* (i.e., delayed induction of a dyskinesia).

19.2.1 Acutely Administered DA Agents and Associated Oral Activity

The acute treatment of rats with either a DA D_1 agonist or DA D_2 antagonist results in the production of vacuous (i.e., purposeless) chewing movements (VCMs) (Rosengarten et al. 1983a, b, 1986; Rupniak et al. 1983; Arnt et al. 1987; Koshikawa et al. 1987; Molloy and Waddington 1988; Levin et al. 1989; Murray and Waddington 1989). Accordingly, the promotion of VCMs was thought to be related to an imbalance in D_1/D_2 receptor stimulation.

19.2.2 Initial Animal Modeling of TD

Although the association of DA receptor activation and VCM induction is a rational one, there is the realization that TD is not an acute phenomenon; TD is a movement disorder that develops gradually and technically only after long-term treatment with an antipsychotic. To this end, Waddington et al. (Waddington et al. 1983; Waddington 1990) first showed that long-term treatment with a neuroleptic resulted in spontaneous orofacial dyskinesias in rats. Later it was shown that the form and periodicity of these oral dyskinesias were similar to those of humans with tardive dyskinesia (Ellison and See 1989). However, when neuroleptic treatment was discontinued, the orofacial dyskinesias disappeared over an approximate 1-month period – unlike the persistence of TD following discontinuation of neuroleptics in humans (Waddington et al. 1983; Gunne et al. 1982; Mithani et al. 1987; Gunne and Haggstron 1983). Therefore, this rodent model of TD lacked some replicability with the human condition.

19.2.3 Presumed DA Receptor Supersensitivity in TD

From the 1970s DA receptor supersensitivity (DARSS) had been invoked as the underlying neural basis for development of TD (Klawans 1973; Chiu et al. 1981).

Because acute or long-term D$_2$ antagonist treatment produced orofacial dyskinesias, and because D$_1$ agonists acutely had that effect in rats, it was reasonable to conclude that VCMs and TD might be the product of endogenous DA acting at supersensitized DA D$_1$ receptors (Waddington 1990; Kostrzewa 1995). However, as we later showed, a DA D$_1$ receptor antagonist failed to attenuate spontaneous orofacial dyskinesias in rodents being treated for 1 year with a D$_2$ receptor antagonist (Kostrzewa et al. 2007).

19.2.4   Incidence of D$_2$ DARSS

Prolonged block of DA D$_2$ receptors over a period of weeks, as per antipsychotic treatment, is associated with an increase in D$_2$ receptor number – so that there is a "supersensitized" response to a D$_2$ agonist after the block is discontinued. D$_2$ DARSS, however, is also observed after repeated treatments with a direct or indirect agonist. For example, repeated treatments with amphetamine result in an exaggerated behavioral effect even if the interval between doses is great enough to account for full metabolism of the previous dose(s). This effect is most notable when treatments are initiated during postnatal ontogenetic development. For example, a single daily low dose of the D$_2$ agonist quinpirole results in exaggerated behavioral responses to quinpirole into adulthood. Rats display enhanced quinpirole-induced yawning, vertical jumping, locomotor activity, changes in reactivity to pain, and other effects (Kostrzewa et al. 1990, 1991, 1993a, 2004; Kostrzewa and Brus 1991; Brown et al. 2002). Moreover, in rats "primed" with daily quinpirole treatments, an acute dose of amphetamine produced a fivefold increase in DA exocytosis, as indicated by the rise in DA levels in the in vivo microdialysate levels of such rats (Nowak et al. 2001, 2007).

19.2.5   Evolution from DARSS Hypothesis to Multineuronal Associations with TD

Eventually the DARSS hypothesis was recognized as being too simplistic, and a multineuronal association was invoked to explain the underlying basis of TD (Casey 1987; Jeste and Caligiuri 1993; Waddington 1990; Gunne and Haggstron 1983; Kostrzewa 1995; Gong et al. 1992; Knable et al. 1994; Egan et al. 1995). Cholinergic (Rupniak et al. 1983, 1985; Salamone et al. 1990), γ-aminobutyric acid (GABA)-ergic (Mithani et al. 1987; Tamminga et al. 1979; Lloyd et al. 1985; Gunne et al. 1988), and serotoninergic systems (Gong et al. 1992; Gong and Kostrzewa 1992) appeared to also be involved in the production of orofacial dyskinesia. Nevertheless, DA systems were still recognized as crucial in the development of TD. Most intriguing was a finding by Gunne et al that neuroleptic-evoked VCMs in rats were enhanced after partial damage to the frontal cortex (Gunne et al. 1982).

19.2.6 DA D₁ Receptors and VCMs

In the 1980s Breese and colleagues demonstrated that a perinatal 6-hydroxydopamine (6-OHDA) lesion produced not only the expected DA denervation of neostriatum and nucleus accumbens but also life-long DA D_1 receptor supersensitization of rats. Moreover, this effect was unaccompanied by an increase in DA D_1 receptor number in the forebrain regions (Breese et al. 1985, 1987; Criswell et al. 1989). To better explore the association of DA D_1 receptors with perioral movements, a series of studies was undertaken on rats so lesioned perinatally with 6-OHDA. These rats, in adulthood, were found to be 100–1,000 times more sensitive to D_1 agonist-induced VCMs. Whereas 1- to 10-mg/kg dose of the D_1 agonist SKF 38393 [(±)1-phenyl-2,3,4,5-tetrahydro-1H-3-benzazepine-7,8-diol] was needed to produce VCMs in intact rats (Rosengarten et al. 1983a), a 0.01- to 0.1-mg/kg dose was sufficient in producing VCMs in the 6-OHDA-lesioned rats (Kostrzewa and Gong 1991; Gong et al. 1994). The enhancement of D_1 agonist-induced VCMs was evident as long as the 6-OHDA lesioning occurred within the first week after birth (Kostrzewa et al. 1993b) and as long as the 6-OHDA dose was sufficient to produce 98.5% DA depletion of neostriatum (Gong et al. 1993). D_1 receptor sensitization was unaccompanied by a change in the B_{max} (i.e., number) and K_d (i.e., affinity) of D_1 receptors, and there was no increase in the expression of high-affinity D_1 receptors in neostriatum of those rats with enhancement of VCMs (Gong et al. 1994). Clearly, however, D_1 receptor supersensitivity was manifest in 6-OHDA-lesioned rats in relation to VCM induction (Kostrzewa 1995; Kostrzewa et al. 1998, 1999, 2003, 2008a).

19.3 Serotonin (5-HT) Systems and TD

19.3.1 5-HT Receptors and VCMs

When the $5\text{-}HT_2$ agonist meta-chlorophenylpiperazine (mCPP) was administered to adult rats that had been lesioned neonatally with 6-OHDA there was enhanced induction of VCMs, such that a 1- to 3-mg/kg dose produced three to four times the number of VCMs observed in intact control rats (Gong et al. 1992). This effect was not replicated by the $5\text{-}HT_{1A}$ agonist 8-OH-DPAT [(±) 8-hydroxydipropylaminotetralin] or by the $5\text{-}HT_{1B}$ agonist CGS-12066B (7-trifluoromethyl-4(4-methyl-1-piperazinyl)-pyrrolo[1,2-alquinoxaline]. Moreover, the mCPP effect was not attenuated by the $5\text{-}HT_{1A/1B}$ antagonist pindolol, by the prodiminate $5\text{-}HT_{2A}$ antagonist ketanserin, or by the $5\text{-}HT_3$ antagonist MDL 72222 (3-tropanyl-3,5-dichlorobenzoate). However, the $5\text{-}HT_2$ antagonist mianserin did attenuate the mCPP effect (Gong et al. 1992). These findings demonstrated that $5\text{-}HT_2$ receptors also could play a role in the induction of VCMs in rats in which DA innervations of striatum and nucleus was compromised.

19.3.2 5-HT System Influence on D$_1$ Agonist Induction of VCMs

The above series of studies demonstrated that both DA systems and 5-HT systems have a prominent influence on the induction of perioral movements, and that both systems exert enhanced effects when DA denervation of neostriatum is produced perinatally. Many years earlier, Zigmond's group demonstrated that perinatal 6-OHDA lesioning of the nigrostriatal DA tract was accompanied by 5-HT fiber hyperinnervation of neostriatum (Berger et al. 1985; Snyder et al. 1986). Therefore, it was natural to explore the influence on one system on the other in relation to VCM induction.

Initially we found that the 5-HT$_2$ antagonist mianserin attenuated D$_1$ agonist-induced VCMs in 6-OHDA-lesioned rats (Gong et al. 1992; Plech et al. 1995). In contrast, the D$_1$ antagonist SCH 23390 [R-(+)-7-chloro-8-hydroxy-3-methyl-1-phenyl-2,3,4,5-tetrahydro-1H-3-benzazepine] did not attenuate mCPP induction of VCMs (Gong et al. 1992). This finding indicated that enhanced D$_1$ agonist induction of VCMs was mediated through a 5-HT system, not vice versa (Kostrzewa et al. 1992). Moreover, when 5-HT innervation in brain was largely destroyed in perinatal 6-OHDA-lesioned rats by perinatal 5,7-dihydroxytryptamine (5,7-DHT) treatment, there was failure of development of the enhanced D$_1$ agonist induction of VCMs in rats in adulthood (Brus et al. 1994). Similarly, in neonatally 6-OHDA-lesioned rats that already displayed enhanced D$_1$ agonist induction of VCMs in adulthood, a subsequent 5-HT lesion with 5,7-DHT eliminated the enhanced D$_1$ agonist effect (unpublished).

The series of findings implicate the 5-HT system as a regulator of the sensitivity status of DA D$_1$ receptors at the receptor level (5-HT$_2$) and as presumably neural mediators of D$_1$ modulation of oral activity or oral dyskinesia.

19.4 A New Rodent Model of TD

19.4.1 Development of the Model

To replicate the essential feature of TD, an animal model must display regularity of spontaneous orofacial dyskinesias, i.e., perioral movements occurring in the absence of any treatment. Occasional sporadic VCMs in rodents do not fulfill this criterion. Therefore, the behavioral outcome (VCMs) produced by long-term administration of haloperidol was the first true animal model of TD (Waddington et al. 1983; Waddington 1990). Long-term haloperidol administration did ultimately increase the number of VCMs in rats, but only after ~4 months; thus, the characteristic tardive feature was fulfilled. Additionally, the high level of perioral movements persisted for as long as haloperidol continued to be administered. However, as already noted, VCMs disappeared within 4–6 weeks after haloperidol treatment was discontinued.

Taking into account the finding of Gunne et al. (Gunne et al. 1982) regarding enhancement of neuroleptic-evoked perioral dyskinesias after brain injury, as well as the generally accepted notion of DA D_1 receptor supersensitivity in TD and in animal models of TD, we undertook the combined approach of first DA-denervating neostriatum in neonates with 6-OHDA, then starting the daily haloperidol treatment regimen when rats attained adulthood (Huang and Kostrzewa 1994). Advantages of this animal model were as follows.

An increase in spontaneous oral dyskinesia occurred after 3 months of haloperidol administration (in drinking water) in 6-OHDA-lesioned rats versus 4 months in intact controls that received haloperidol.

Numbers of vacuous chewing movements were approximately fivefold greater in haloperidol-treated 6-OHDA-lesioned rats than in haloperidol-treated intact controls, beginning from its initiation and through the remainder of the 1-year haloperidol treatment period.

After discontinuing haloperidol as a 1-year treatment, spontaneous oral dyskinesias persisted at the same high level in 6-OHDA-lesioned rats as during the haloperidol treatment phase – the first time that this effect was obtained in chronic haloperidol-treated rats.

In intact and in 6-OHDA-lesioned rats terminated 1 week after discontinuing haloperidol, the B_{max} for raclopride binding in neostriatum was elevated, while the K_d was unaltered. This animal model thus replicated the effects of haloperidol in initiating TD. Moreover, the persistence of a high level of VCMs following withdrawal of haloperidol as a treatment in 6-OHDA-lesioned rats, more closely replicated the effects of human TD. Also, the biochemical parameters relating to haloperidol-induced proliferation of D_2 receptor number in 6-OHDA-lesioned rats similarly mirrored findings of human TD (Huang et al. 1997).

One other finding revealed another aspect of TD. In 6-OHDA-lesioned rats terminated 8 months after discontinuing haloperidol, both the B_{max} and K_d for raclopride binding to neostriatal homogenates were at the same level as in intact control rats. The finding that VCMs persist at a high level in 6-OHDA-lesioned rats following withdrawal from long-term haloperidol treatment, indicates that changes in D_2 receptor number appear to be irrelevant in terms of sustenance of the high level of oral dyskinesia in the rats (Huang et al. 1997).

19.4.2   DA RSS, 5-HT RSS, and VCMs in the Haloperidol-Withdrawal Phase

6-OHDA-lesioned rats were tested during the haloperidol-withdrawal phase at intervals, with a series of agonists or antagonists for multiple neurotransmitter systems. The persistent high level of VCMs was not attenuated by a DA D_1 antagonist, DA D_2 antagonist, alpha-adrenoceptor antagonist, beta-adrenoceptor antagonist, NMDA antagonist, opioid mu antagonist, muscarinic antagonist, $GABA_A$ antagonist, adenosine A_{2A} antagonist, or $H_1/5$-HT antagonist. Similarly, a 5-HT_{1A} agonist and 5-HT$_{2A}$ antagonist failed to alter VCM number. However, on testing and retesting, several different 5-HT$_2$ antagonists attenuated VCM number. Because at least two of the 5-HT$_2$ antagonists (i.e., mesulergine, mianserin) have preferential 5-HT$_{2C}$ antagonist activity, we believe the effectiveness of the 5-HT$_2$ antagonists relates to their ability to block 5-HT$_{2C}$ receptors (Kostrzewa et al. 2007).

19.5   5-HT Systems in Other DA-Associated Disorders

In rats lesioned as neonates with 6-OHDA and lesioned again in adulthood with the serotoninergic toxin 5,7-DHT, there was an inordinate increase in hyperlocomotor activity. This combined treatment in rats was proposed to model attention deficit hyperactivity disorder (Kostrzewa et al. 1994, 2008b). Moreover, the hyperlocomotor activity was abated by 5-HT$_2$ receptor antagonists (Nowak et al. 2007; Brus et al. 2004). 5-HT systems are also proposed as an alternate phenotypic system beneficially preserving function in Parkinson disease (Kostrzewa and Moratalla, submitted)).

19.6   Summary on 5-HT$_{2C}$ Receptors and TD

The rodent model, in which there was perinatal destruction on the dopaminergic nigrostriatal tract along with administration of a conventional antipsychotic drug (i.e., haloperidol) for 1 year, appears to fulfill the most significant criteria for an animal model of TD – namely, gradual development of perioral dyskinesia, persistence of the dyskinesia after discontinuing the antipsychotic treatment, reproducible changes in D$_2$ receptor number in brain, and failure of D$_1$ and D$_2$ receptor antagonists to attenuate the dyskinesia. However, it is particularly relevant from a therapeutic perspective, that 5-HT$_2$ antagonists, and more specifically presumed 5-HT$_{2C}$ receptor antagonists, attenuate the oral dyskinesia in the model (particularly after antipsychotic treatment withdrawal). The rodent model is viable, and the prospect of 5-HT$_{2C}$ antagonists as treatements for TD is encouraging.

References

Arnt J, Hyttel J, Perregard J (1987) Dopamine D-1 receptor agonists combined with the selective D-2 agonist, quinpirole, facilitate the expression of oral stereotyped behavior in rats. Wur J Pharmacol 133:137–145.

Berger TW, Kaul S, Stricker EM, et al (1985) Hyperinnervation of the striatum by dorsal raphe afferents after dopamine-depleting brain lesions in neonatal rats. Brain Res 366:354–358.

Breese GR, Baumeister AA, Napier TC, et al (1985) Evidence that D$_1$ dopamine receptors contribute to the supersensitive behavioral responses induced by L-dihydroxyphenylalanine in rats treated neonatally with 6-hydroxydopamine. J Pharmacol Exp Ther 235:287–295.

Breese GR, Duncan GE, Napier TC, et al (1987) 6-Hydroxydopamine treatments enhance behavioral responses to intracerebral microinjection of D_1 and D_2-dopamine agonists into the nucleus accumbens and striatum without changing dopamine antagonist binding. J Pharmacol Exp Ther 240:167–176.

Brown RW, Gass JT, Kostrzewa RM (2002) Ontogenetic quinpirole treatments produce spatial memory deficits and enhance skilled reaching in adult rats. Pharmacol Biochem Behav 72:591–600.

Brus R, Kostrzewa RM, Perry KW, et al (1994) Supersensitization of the oral response to SKF 38393 in neonatal 6-hydroxydopamine-lesioned rats is eliminated by neonatal 5,7-dihydroxytryptamine treatment. J Pharmacol Exp Ther 268:231–237.

Brus R, Nowak P, Szkilnik R, et al (2004) Serotoninergics attenuate hyperlocomotor activity in rats. Potential new therapeutic strategy for hyperactivity. Neurotox Res 6, 317–325.

Casey DE (1987) Tardive dyskinesia. In: Meltzer HY, ed. Psychopharmacology. The Third Generation of Progress. New York: Raven, pp 1411–1419.

Chiu S, Paulose CS, Mishra RK (1981) Neuroleptic drug-induced dopamine receptor supersensitivity: antagonism by L-prolyl-L-loeucyl-glycinamide. Science 214:1261–1262.

Criswell HE, Mueller RA, Breese GR (1989) Priming of D_1-dopamine receptor responses: long-lasting behavioral supersensitivity to a D_1-dopamine agonist following repeated administration to neonatal 6-OHDA-lesioned rats. J Neurosci 9:125–133.

Egan MF, Hyde TM, Lleinman JE, et al (1995) Neuroleptic-induced vacuous chewing movements in rodents: incidence and effects of long-term increases in haloperidol dose. Psychopharmacology 117:74–81.

Ellison GD, See RE (1989) Rats administered chronic neuroleptics develop oral movements which are similar in form to those in humans with tardive dyskinesia. Psychopharmacology (Berl) 98:564–566.

Gong L, Kostrzewa RM (1992) Supersensitized oral response to a serotonin agonist in neonatal 6-OHDA treated rats. Pharmacol Biochem Behav 41:621–623.

Gong L, Kostrzewa RM, Fuller RW, et al (1992) Supersensitization of the oral response to SKF 38393 in neonatal 6-OHDA-lesioned rats is mediated through a serotonin system. J Pharmacol Exp Ther 261:1000–1007.

Gong L, Kostrzewa RM, Perry KW, et al (1993) Dose-related effects of a neonatal 6-OHDA lesion on SKF 38393- and m-chlorophenylpiperazine-induced oral activity responses of rats. Dev Brain Res 76:233–238.

Gong L, Kostrzewa RM, Li C (1994) Neonatal 6-OHDA and adult SKF 38393 treatments alter dopamine D_1 receptor mRNA levels: absence of other neurochemical associations with the enhanced behavioral responses of lesioned rats. J Neurochem 63:1282–1290.

Gunne LM, Haggstron, JE (1983) Reduction in nigral glutamic acid decarbosylase in rats with neuroleptic-induced oral dyskinesia. Psychopharmacology (Berl) 81:191–194.

Gunne LM, Growdon J, Glaeser B (1982) Oral dyskinesia in rats following brain lesions and neuroleptic drug administration. Psychopharmacology (Berl) 81:134–139.

Gunne LM, Bachus SE, Gale K (1988) Oral movements induced by interference with nigral GABA neurotransmission: relationship to tardive dyskinesias. Exp Neurol 100:459–469.

Huang N-Y, Kostrzewa RM (1994) Persistent oral dyskinesias in haloperidol-withdrawn neonatal 6-OHDA-lesioned rats. Eur J Pharmacol 271:433–437.

Huang N-Y, Kostrzewa RM, Li C, et al (1997) Persistent spontaneous oral dyskinesias in haloperidol-withdrawn rats neonatally lesioned with 6-hydroxydopamine: absence of an association with the B_{max} for [^{3}H]raclopride binding to neostriatal homogenates. J Pharmacol Exp Ther 280:268–276.

Jeste DV, Caligiuri MP (1993) Tardive dyskinesia. Schizophr Bull 19:303–315.

Kane JM, Woerner M, Weinhold P, et al (1982) A prospective study of tardive dyskinesia development: preliminary results. J Clin Psychopharmacol 2:345–349.

Klawans HL (1973) The pharmacology of tardive dyskinesia. Am J Psychiatry 130:82–86.

Knable MB, Hyde TM, Egan MF, et al (1994) Quantitative autoradiography of striatal dopamine D1, D2 and re-uptake sites in rats with vacuous chewing movements. Brain Res 646:217–222.

Koshikawa N, Aoki S, Tomiyama M, et al (1987) Sulpiride injection into the dorsal striatum increases methamphetamine-induced gnawing in rats. Eur J Pharmacol 133:119–125.

Kostrzewa RM (1995) Dopamine receptor supersensitivity. Neurosci Biobehav Rev 19:1–17.

Kostrzewa RM, Moratalla R (submitted) Serotonin and Parkinson disease.

Kostrzewa RM, Brus R (1991) Ontogenic homologous supersensitization of quinpirole-induced yawning in rats. Pharmacol Biochem Behav 39:517–519.

Kostrzewa RM, Gong L (1991) Supersensitized D$_1$ receptors mediate enhanced oal activity after neonatal 6-OHDA. Pharmacol Biochem Behav 39:677–682.

Kostrzewa RM, Hamdi A, Kostrzewa FP (1990) Production of prolonged supersensitization of dopamine D$_2$ receptors. Eur J Pharmacol 183:1411–1412.

Kostrzewa RM, Brus R, Kalbfleisch J (1991) Ontogenetic homologous sensitization to the antinociceptive action of quinpirole in rats. Eur J Pharmacol 209:157–161.

Kostrzewa RM, Gong L, Brus R (1992) Serotonin (5-HT) systems mediate dopamine (DA) receptor supersensitivity. Acta Neurobiol Exp 53:31–41.

Kostrzewa RM, Brus R, Rykaczewska M, et al (1993b) Low dose quinpirole ontogenically sensitizes to quinpirole-induced yawning in rats. Pharmacol Biochem Behav 44:487–489.

Kostrzewa RM, Brus R, Perry KW, et al (1993a) Age-dependence of a 6-hydroxydopamine lesion on SKF 38393- and m-chlorophenylpiperazine-induced oral activity responses of rats. Dev Brain Res 76:87–93.

Kostrzewa RM, Brus R, Kalbflesich JH, et al (1994) Proposed animal model of attention deficit hyperactivity disorder. Brain Res Bull 34:161–167.

Kostrzewa RM, Reader TA, Descarries L (1998) Serotonin neural adaptations to ontogenetic loss of dopamine neurons in rat brain. J Neurochem 70:889–898.

Kostrzewa RM, Brus R, Perry KW (1999) Interactive modulation by dopamine and serotonin neurons of receptor sensitivity of the alternate neurochemical system. Pol J Pharmacol 5:39–47.

Kostrzewa RM, Kostrzewa JP, Brus R (2003) Dopamine receptor supersensitivity: an outcome and index of neurotoxicity. Neurotox Res 5:111–118.

Kostrzewa RM, Kostrzewa JP, Nowak P, et al (2004) Dopamine D$_2$ agonist priming in intact and dopamine-lesioned rats. Neurotox Res 6:457–462.

Kostrzewa RM, Huang N-Y, Kostrzewa JP, et al (2007) Modeling tardive dyskinesia: predictive 5-HT$_{2C}$ receptor antagonist treatment. Neurotox Res 11:41–50.

Kostrzewa RM, Kostrzewa JP, Brown R, et al (2008a) Dopamine receptor supersensitivity: development, mechanisms, presentation, and clinical applicability. Neurotox Res 14:121–128.

Kostrzewa RM, Kostrzewa JP, Kostrzewa RA, et al (2008) Pharmacological models of ADHD. J Neural Transm 115, 287–298.

Levin ED, See RE, South D (1989) Effects of dopamine D$_1$ and D$_2$ receptor antagonists on oral activity in rats. Pharmacol Biochem Behav 34:43–48.

Lloyd KG, Willigens MT, Goldstein M (1985) Induction and reverse of dopamine dyskinesia in rat, cat, and monkey. Psychopharmacology (Berl) 85(suppl 2):200–210.

Mithani S, Atmadja S, Baimbridge KG, et al (1987) Neuroleptic-induced oral dyskinesias: effects of progabide and lack of correlation with regional changes in glutamic acid decarboxylase and choline acetyltransferase activities. Psychopharmacology (Berl) 93:94–100.

Molloy AG, Waddington JL (1988) Behavioral responses to the selective D$_1$-dopamine receptor agonist R-SK&F 38393 and the selective D$_2$-agonist RU 24213 in young compared with aged rat. Br J Pharmacol 95:335–342.

Murray AM, Waddington JL (1989) The induction of grooming and vacuous chewing by a series of selective D-1 dopamine receptor agonists: two directions of D-1:D-2 interaction. Eur J Pharmacol 160:377–387.

Nowak P, Brus R, Kostrzewa RM (2001) Amphetamine-induced enhancement of neostriatal in vivo microdialysate dopamine content in rats, quinpirole-primed as neonates. Pol J Pharmacol 53:319–329.

Nowak P, Bortel A, Dąbrowska J, et al (2007) Amphetamine and mCPP effects on dopamine and serotonin striatal in vivo microdialysates in an animal model of hyperactivity. Neurotox Res 11(2):131–144.

Plech A, Brus R, Kalbfleisch JH, et al (1995) Enhanced oral activity responses to intrastriatal SKF 38393 and m-CPP are attenuated by intrastriatal mianserin in neonatal 6-OHDA-lesioned rats. Psychopharmacology 119:466–473.

Rosengarten H, Schweitzer JW, Friedhoff AJ (1983) Induction of oral dyskinesias in naïve rats by D_1 stimulation. Life Sci 33:2479–2482.

Rosengarten H, Schweitzer JW, Egawa J, et al (1983) Diminished D_2 dopamine receptor function and the emergence of repetitive jaw movements. Adv Exp Med Biol 235:159–169.

Rosengarten H, Schweitzer JW, Egawa J, et al (1986) Diminished D_2 dopamine receptor function and the emergence of repetitive jaw movements. Adv Exp Med Biol 235:159–167.

Rupniak NMJ, Jenner P, Marsden CD (1983) Cholinergic manipulation of perioral behavior-induced by chronic neuroleptic administration to rats. Psychopharmacology (Berl) 79:226–230.

Rupniak NMJ, Jenner P, Marsden CD (1985) Pharmacological characterization of spontaneous or drug-associated purposeless chewing movements in rats. Psychopharmacology (Berl) 85:71–79.

Salamone JD, Johnson CJ, McCullough LD, et al (1990) Lateral striatal cholinergic mechanisms involved in oral motor activities in the rat. Psychopharmacology (Berl) 102:529–534.

Snyder AM, Zigmond MJ, Lund RD (1986) Sprouting of serotonergic afferents into striatum after dopamine depleting lesions in infant rats: a retrograde transport and immunocytochemical study. J Comp Neurol 245:274–281.

Tamminga CA, Crayton W, Chase TN (1979) Improvement in tardive dyskinesia after muscimol therapy. Arch Gen Psychiatry 36:595–598.

Waddington JL (1990) Spontaneous orofacial movements induced in rodents by very long-term neuroleptic drug administration: phenomenology, pathophysiology and putative relationship to tardive dyskinesia. Psychopharmacology (Berl) 101:431–447.

Waddington JL, Cross AJ, Gamble SJ, et al (1983) Spontaneous orofacial dyskinesia and dopaminergic function in rats after 6 months of neuroleptic treatment. Science 220:530–532.

Chapter 20
The Role of 5-HT$_{2A/2C}$ Receptors in Sleep and Waking

Jaime M. Monti and Héctor Jantos

20.1 Introduction

The presence of serotonin (5-HT) in the brain was reported more than 50 years ago (Amin et al. 1954). Soon after, it was realized that 5-HT participates in the control of functions as diverse as sleep and waking (W), circadian rhythms, mood, motor activity, analgesia, thermoregulation, food intake, and sexual behavior (Jacobs and Azmitia 1992; Wang and Nakai 1994; Jacobs and Fornal 1999; Monti and Monti 2000a; Portas et al. 2000). Moreover, several psychotropic drugs used for the treatment of anxiety disorders, major depression, and schizophrenia act, at least partly, through serotonergic mechanisms.

Neurotransmitters control of sleep and W is an extremely complex process and 5-HT represents only a facet of monoamines, acetylcholine (ACh), peptides and aminoacids control of the behavioral state. Serotonin shares with other neurotransmitters the ability to promote W and to suppress rapid-eye-movement (REM) sleep. The proposal that 5-HT is involved in arousal mechanisms comes mainly from electrophysiological and pharmacological studies. In this respect, it has been shown that dorsal raphe nucleus (DRN) 5-HT neuron firing varies across the sleep–wake cycle, being highest during the waking state, slowing during slow wave sleep (SWS) and stopping during REM sleep (Trulson and Jacobs 1979). Moreover, systemic injection of 5-HT receptor agonists increases W and reduces SWS and REM sleep (Monti et al. 2008a). Why activation of 5-HT receptors raises W is a matter of debate. There are at least two views: It could be a direct effect; on the other hand, it could be an indirect action related to the activation of other neurotransmitter systems. Presently available evidence tends to indicate that 5-HT receptor agonist-induced suppression of REM sleep depends on either the direct inhibition of neurons responsible for the induction and maintenance of the behavioral

J.M. Monti (✉)
Department of Pharmacology and Therapeutics, School of Medicine Clinics Hospital,
Montevideo 11600, Uruguay
e-mail: jmonti@mednet.org.uy

G. Di Giovanni et al. (eds.), *5-HT$_{2C}$ Receptors in the Pathophysiology of CNS Disease*,
The Receptors 22, DOI 10.1007/978-1-60761-941-3_20,
© Springer Science+Business Media, LLC 2011

state or the activation of γ-aminobutyric acid (GABA)-ergic cells located within or around the neuroanatomical structures responsible for REM sleep occurrence (Monti et al. 2008a).

More recently, a number of 5-HT receptors have been characterized at central sites that add to the complexity of the mechanisms involved in the functional activity of this indoleamine. They include seven different classes of receptors designated 5-HT_{1-7}. Except for the 5-HT_3 receptor that is a ligand-gated ion channel receptor, all other receptors are structurally related to the superfamily of G-protein-coupled receptors (Hannon and Hoyer 2008). The 5-HT_2 subfamily of 5-HT receptors is composed of three subtypes, the 5-HT_{2A}, 5-HT_{2B}, and 5-HT_{2C} receptors. All three receptors are G-protein coupled to the activation of the phospholipase C (PLC), functionally linked to phosphatidylinositol hydrolysis, and subsequent mobilization of intracellular Ca^{2+} and influx of extracellular Ca^{2+} with a resulting depolarization of the host cell (Leysen 2004).

Receptors corresponding to the 5-HT_2 subfamily are located within postsynaptic structures, mostly on proximal and distal dendritic shafts. Autoradiographic and immunohistochemical studies have demonstrated that each receptor subtype displays a particular distribution in the brain (Tables 20.1 and 20.2). 5-HT_{2A} and 5-HT_{2C} receptors are distributed predominantly in the cerebral cortex, olfactory system, septal nuclei, hippocampal formation, amygdala, basal ganglia, thalamus, hypothalamus, brainstem, cerebellum, and spinal cord of rat brain (Cornea-Hébert et al. 1999; Clemett et al. 2000). In rat cortex, the highest binding density has been localized to a continuous band that comprises layer IV and extends into layer III (Pazos et al. 1985). In man, the highest binding of 5-HT_2 receptors, predominantly 5-HT_{2A} receptors, has been found in the frontal and temporal cortices followed by the parietal cortex and motor regions, whereas intermediate and low levels have been described in the basal ganglia and the thalamus, respectively (Nichols and Nichols 2008). Of note, $5\text{-HT}_{2A/2C}$ receptor binding sites and $5\text{-HT}_{2A/2C}$ mRNA expression have been found in most neuroanatomical structures involved in the regulation of the behavioral state, including the DRN, median raphe nucleus (MRN), locus coeruleus (LC), tuberomammillary nucleus (TMN), lateral and dorsal hypothalamus, laterodorsal and pedunculopontine tegmental nuclei (LDP/PPT), medial pontine reticular formation (mPRF), preoptic area, anterior and lateral hypothalamic areas, thalamus (nonspecific nuclei), and basal forebrain (nucleus of the diagonal band of Broca, bed nucleus of the stria terminalis) (Figs. 20.1 and 20.2).

With respect to the 5-HT_{2B} receptor, only weak, limited expression is observed in the brain of laboratory animals and man (Duxon et al. 1997; Kitka and Bagdy 2008).

20.1.1   Role of 5-HT_{2A} and 5-HT_{2C} Receptors in the Regulation of the Behavioral State

Several approaches have been followed to characterize the role of the 5-HT_{2A} and 5-HT_{2C} receptors in the regulation of sleep and W. They include

Table 20.1 Distribution of the 5-HT$_{2A}$ receptor in rat brain (From Mengod et al. 1990; Abramowski et al. 1995; Duxon et al. 1997; Cornea-Hébert et al. 1999; Clemett et al. 2000)

Telencephalon	1. Cerebral cortex	→ Frontal
		→ Parietal
		→ Occipital
		→ Piriform
	2. Olfactory system	→Olfactory bulb
	3. Limbic system	→ Medial and lateral septal nuclei
		→ Hippocampal formation
		→ Amygdala
	4. Basal forebrain	→ Nucleus of the diagonal band of Broca
		→ Ventral pallidum
	Basal ganglia	→ Nucleus accumbens
		→ Caudate-putamen
		→ Globus pallidus
		→ Subthalamic nucleus
5-HT$_{2A}$ receptor		
Diencephalon	1. Thalamus	→ Medial, lateral, habenular, reticular, intralaminar, reuniens, anterior
	2. Hypothalamus	→ Medial and lateral preoptic area
		→ Anterior and lateral hypothalamic areas
		→ Ventromedial nucleus
		→ Mammillary nucleus
Mesencephalon		→ Central gray
		→ Substantia nigra – pars compacta
		→ Ventral tegmental area
Rhombencephalon		→ Laterodorsal and pedunculopontine tegmental nuclei
		→ Dorsal raphe nucleus and median raphe nuclei
		→ Locus coeruleus
		→ Medial pontine reticular formation
		→ Gigantocellular reticular nucleus

1. Characterization of the changes occurring in serotonin 5-HT$_{2A}$ and 5-HT$_{2C}$ receptor deficient mice
2. Determination of the changes that follow the administration of selective and nonselective 5-HT$_{2A}$ and 5-HT$_{2C}$ receptor agonists and antagonists to laboratory animals and man

20.1.1.1   Sleep Patterns in Mutant Mice That Do Not Express 5-HT$_{2A}$ or 5-HT$_{2C}$ Receptors

Popa et al. (2005) investigated the role of the 5-HT$_{2A}$ receptor on non-rapid-eye-movement (NREM) sleep and W regulation using wild-type and knockout mice that do not express the 5-HT$_{2A}$ receptor. 5-HT$_{2A}$ receptor knockout mice showed a significant increase of W and a reduction of NREM sleep. Values of REM sleep

Table 20.2 Distribution of the 5-HT$_{2C}$ receptor in the rat brain (From Mengod et al. 1990; Abramowski et al. 1995; Duxon et al. 1997; Cornea-Hébert et al. 1999; Clemett et al. 2000)

Telencephalon	1. Cerebral cortex	→ Frontal
		→ Parietal
		→ Occipital
		→ Piriform
		→ Cingulate
	2. Olfactory system	→ Olfactory bulb
	3. Limbic system	→ Lateral septal nucleus
		→ Hippocampal formation
		→ Amygdala
	4. Basal forebrain	→ Nucleus of the diagonal band of Broca
		→ Ventral pallidum
		→ Bed nucleus of the stria terminalis
	5. Basal ganglia	→ Nucleus accumbens
		→ Caudate-putamen
		→ Globus pallidus
		→ Subthalamic nucleus
5-HT$_{2C}$ receptor		
Diencephalon	1. Thalamus	→ Medial, lateral, reuniens, habenular, lateral, and geniculate nuclei
	2. Hypothalamus	→ Medial preoptic area
		→ Ventromedial nucleus
		→ Dorsomedial nucleus
		→ Mammillary nucleus
Mesencephalon		→ Central gray
		→ Substantia nigra pars compacta
Rhombencephalon		→ Laterodorsal and pedunculopontine tegmental nuclei
		→ Dorsal raphe nucleus and median raphe nucleus

were not significantly different from those found in the wild-type mice. Systemic administration of the selective 5-HT$_{2A}$ receptor antagonist MDL 100907 {R-(+)-α-(2,3-dimethoxyphenyl)-1-[2-(4-fluorophenylethyl)]-4-piperidine-methanol} significantly augmented NREM sleep and decreased W and REM sleep in the wild-type mice only.

Sleep has been characterized also in 5-HT$_{2C}$ receptor knockout mice (Frank et al. 2002). Mice lacking the 5-HT$_{2C}$ receptor had significantly less NREM sleep and greater amounts of W compared with the wild-type animals. In contrast, REM sleep values remained unchanged in the mutant mice. Since 5-HT$_{2A}$ and 5-HT$_{2C}$ receptors are positively coupled to PLC via $G_{Q/11}$ proteins and mobilize intracellular Ca^{2+}, a reduction of W and an increase of NREM sleep should have been expected in the mutants. In other words, opposite effects of gene deletion versus acute pharmacological activation of the same protein with selective 5-HT$_{2A}$ and 5-HT$_{2C}$ receptor agonists (to be dealt with in the next section) should have been expected on W and NREM sleep. Adrian (2008) has proposed that the discrepancy might depend, at

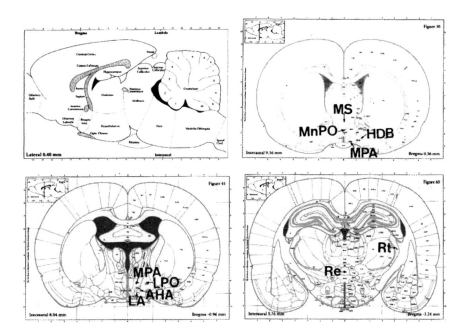

Fig. 20.1 5-HT$_{2A/2C}$ receptor binding sites in neuroanatomical structures involved in the regulation of sleep and waking: basal forebrain, thalamus, preoptic area, and hypothalamus. *AHA* anterior hypothalamic area, *HDB* nucleus of the horizontal limb of the diagonal band, *LA* lateroanterior hypothalamic nucleus, *LPO* lateral preoptic area, *MnPO* median preoptic nucleus, *MPA* median preoptic area, *MS* medial septal nucleus, *Re* reuniens thalamic nucleus, *Rt* reticular thalamic nucleus (Modified from Paxinos and Watson 2005)

least in part, on compensatory mechanisms in constitutive mutants. On the other hand, Frank et al. (2002) have posed that the greater amounts of W in the 5-HT$_{2C}$ receptor knockout mice could be related to the increase of catecholaminergic neurotransmission involving mainly the noradrenergic and dopaminergic systems.

It has been established that acetylcholine (ACh), 5-HT, noradrenaline (NA), dopamine (DA), histamine (HA), orexin (OX), and glutamate function to promote W (Pace-Schott and Hobson 2002; Jones 2003). The results obtained from studies aimed at determining the role of the 5-HT$_{2A}$ and 5-HT$_{2C}$ receptors in the modulation of ACh release are inconsistent. Thus, 5-HT and the 5-HT$_{2A/2C}$ receptor agonist 2,5-dimethoxy-4-bromo-amphetamine (DOB) inhibited ACh release of superfused rat neocortical slices, and this effect was attenuated by ketanserin, a predominantly 5-HT$_{2A}$ receptor antagonist (Muramatsu et al. 1990). On the other hand, systemic administration of the 5-HT$_{2A/2C}$ agonist 1-(2,5-dimethoxy-4-iodophenyl)-2-aminopropane (DOI) increased ACh release in the prefrontal cortex and the hippocampus of the rat, and this effect was prevented by LY-53857 {4-isopropyl-7-methyl-9-(2-hydroxy-1-methyl-propoxycarbonyl)-4,6A,7,8,9,10,10A-octahydro-indolo[4,3-FG]quinolone maleate}, a 5-HT$_{2A/2C}$ receptor antagonist (Nair and Gudelsky 2004). Fink and Göthert (2007) have

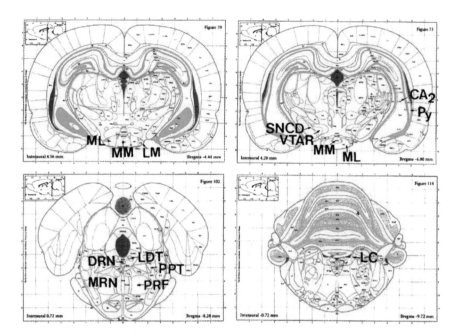

Fig. 20.2 5-HT$_{2A/2C}$ receptor binding sites in neuroanatomical structures involved in the regulation of sleep and waking: hippocampus, preoptic area, hypothalamus, and pons. *CA2* CA2 field of the hippocampus, *DRN* dorsal raphe nucleus, *LC* locus coeruleus, *LDT* laterodorsal tegmental nucleus, *LM* lateral mammillary nucleus, *ML* medial mammillary nucleus, lateral part, *MM* medial mammillary nucleus, medial part, *MRN* median raphe nucleus, *PPT* pedunculopontine tegmental nucleus, *PRF* pontine reticular formation, *Py* pyramidal cells, *SNCD* substantia nigra pars compacta, *VTAR* ventral tegmental area (Modified from Paxinos and Watson 2005)

proposed that the 5-HT$_2$ receptors mediating ACh release in the prefrontal cortex of the rat are located on cholinergic axon terminals.

As regards the involvement of 5-HT$_{2A}$ and 5-HT$_{2C}$ receptors in the release of DA, it has been found that systemically injected or locally applied MDL 100907 increases DA release in the medial prefrontal cortex (mPFC) of the rat (Schmidt and Fadayel 1995). On the other hand, administration of the 5-HT$_{2C}$ receptor agonist RO 60-0175 [(S)-2-(chloro-5-fluoro-indol-1-yl)-1-methylethylamine] by the intraperitoneal (IP) route decreases DA release in the nucleus accumbens of the rat, and this effect is prevented by the selective 5-HT$_{2C}$ antagonist SB 242084 [1-(1-methylindol-5-yl)-3-(3-pyridyl) urea] (Di Matteo et al. 2000). Fink and Göthert (2007) have suggested that the 5-HT$_{2A}$ and 5-HT$_{2C}$ receptors responsible for the inhibition of DA release are expressed by inhibitory GABA-ergic interneurons and that activation of these receptors indirectly inhibits the neurotransmitter release. With respect to the participation of 5-HT$_{2A}$ and 5-HT$_{2C}$ receptors in the release of NE, it has been reported that systemic administration of DOB and DOI inhibits the release of the catecholamine in the rat hippocampus and that this effect is prevented by ketanserin (Done and Sharp 1994). The inhibitory effect of the 5-HT$_2$ receptors on NE release has been ascribed also to the activation of GABA-ergic interneurons. The evidence in favor of the involvement of 5-HT$_{2A}$ and 5-HT$_{2C}$ receptors in the release of HA is scanty. It has been reported that 5-HT excites histaminergic tubero-mammillary neurons by activation of 5-HT$_{2C}$ receptors and Na$^+$/Ca^{2+} exchange (Eriksson et al. 2001). However, no attempts have been made to quantify HA levels at postsynaptic sites in animals treated with 5-HT$_{2A}$ and 5-HT$_{2C}$ ligands. Muraki et al. (2004) examined the effect of 5-HT on orexin neurons using hypothalamic slices from orexin-enhanced green fluorescent protein transgenic mice in which the protein was expressed exclusively in orexin neurons. 5-HT hyperpolarized orexin neurons in a concentration-dependent manner. In this study the authors proposed that this inhibitory 5-HT input to the orexin cells was predominantly dependent on the activation of the 5-HT$_{1A}$ receptor. However, no attempts have been made to elucidate the role of 5-HT$_2$ receptors in the functional activity of orexin neurons. It has been observed that most glutamatergic cells in layers II to IV of human and monkey prefrontal cortex express 5-HT$_{2A}$ receptor messenger ribonucleic acid (mRNA) (de Almeida and Mengod 2007). In addition, it has been established that amino-hydroxy-methyl-isoxazole propionic acid (AMPA) and kainate receptors mediate the 5-HT-induced excitatory postsynaptic potentials recorded from layer V pyramidal cells in the rat mPFC and that this effect is suppressed by a selective AMPA receptor antagonist (Zhang and Marek 2008). However, further studies are needed to determine whether glutamate participates in the 5-HT$_{2A}$ receptor-induced increase of W and decrease of NREM sleep. Thus, the limited available evidence tends to suggest that the increase of W and reduction of NREM sleep in 5-HT$_{2A}$ and 5-HT$_{2C}$ receptor knockout mice is related, at least in part, to the increased release of NA and DA. With respect to the mechanisms involved, it can be speculated that the reduction of GABA release at critical sites in the central nervous system (CNS) of 5-HT$_{2A}$ and 5-HT$_{2C}$ receptor knockout mice would be indirectly responsible for the increased availability of NA and DA. However, further studies are needed to resolve this issue.

20.2   Sleep Patterns in Laboratory Animals and Man Administered 5HT$_2$ Receptor Ligands

20.2.1   Systemic Administration of Selective and Nonselective 5-HT$_2$ Receptor Ligands to Laboratory Animals

Systemic administration of the 5-HT$_{2A/2C}$ receptor agonists DOI or 1-(2,5-dime-thoxy-4-methylphenyl)-2-aminopropane (DOM) has been shown to reduce SWS and REM sleep and to augment W in the rat (Dugovic and Wauquier 1987; Dugovic et al. 1989; Monti et al. 1990). In addition, systemic or intrathalamic injection of DOI decreased the neocortical high-voltage spindle activity that occurs during relaxed W in the rat (Jäkälä et al. 1995). The intraperitoneal (IP) and oral (p.o.)

Table 20.3 The effect of selective and nonselective 5-HT$_2$ receptor agonists administration on sleep and waking in laboratory animals

Compound	Route of administration	W	SWS	REMS	Reference
DOI (5-HT$_{2A/2C}$ agonist)	i.p	↑	↓	↓	(Monti et al. 1990)
DOM (HT$_{2A/2C}$ agonist)	i.p	↑	↓	↓	(Dugovic and Wauquier 1987; Dugovic et al. 1989)
RO 60-0175 (5-HT$_{2C}$ agonist)	i.p. and p.o	↑	n.s	↓	(Martin et al. 1998)
RO 60-0332 (5-HT$_{2C}$ agonist)	i.p. and p.o	↑	n.s.	↓	(Martin et al. 1998)
DOI (5-HT$_{2A/2C}$ agonist)	Microinjection into the DRN	n.s.	n.s	↓	(Monti and Jantos 2006b)
DOI (5-HT$_{2A/2C}$ agonist)	Microinjection into the LDT	n.s.	n.s.	↓	(Amici et al. 2004)

DRN dorsal raphe nucleus, *LDT* laterodorsal tegmental nucleus, *W* waking, *SWS* slow wave sleep, *REMS* REM sleep, *n.s.* nonsignificant, ↑ increased, ↓ decreased

{[S]-2-(chloro-5-fluoro-indol-1-yl)-1-methyletylamine} administration of the selective 5-HT$_{2C}$ receptor agonists RO 60-0175/ORG 35030 or RO-60-0332/ORG 35035 {[S]-2-[4,4,7-trimethyl-1,4-dihydro-indeno(1,2-b)pyrrol-1-yl]-1- methylethylamine} induced also an increase of W and a reduction of REM sleep in the rat (Martin et al. 1998) (Table 20.3). Injection of the 5-HT$_{2A/2C}$ receptor antagonists ritanserin, ketanserin, ICI 170,809 [2(2-dimethylamino-2-methylpropylthio)-3-phenylquinoline], or sertindole at the beginning of the light period induced a significant increase of SWS and a reduction of REM sleep in the rat. The 5-HT$_{2A/2C}$ antagonist ICI 169,369 [2-(2-dimethylaminoethylthio)-3-phenylquinoline] suppressed REM sleep; however, SWS remained unchanged. Waking was also diminished in most of these studies (Dugovic et al. 1989; Monti et al. 1990; Tortella et al. 1989; Silhol et al. 1991; Coenen et al. 1995; Kirov and Moyanova 1998a, b) (Table 20.4).

More recently, the action of subtype-selective 5-HT$_2$ receptor antagonists on sleep variables was assessed in rats and mice. Subcutaneous administration of the 5-HT$_{2A}$ antagonist EMD 281014 {7-[4-(2-(4-fluorophenyl)ethyl)-piperazine-1-carbonyl]-1H-indole-3-carbonitrile} significantly reduced REM sleep (Monti and Jantos 2006a). On the other hand, administration of the 5-HT$_{2A}$ receptor antagonist MDL 100907 by IP route during the light period augmented NREM sleep and reduced W and REM sleep in adult male mice (Popa et al. 2005). Moreover, oral administration of the 5-HT$_{2C}$ antagonist SB 243213 {5-methyl-1-[[-2-[(-2-methyl-3-pyridyl)oxy]-5-pyridyl] carbamoyl]-6-trifluoromethylindoline} significantly increased SWS and reduced REM sleep during the light period in the rat (Smith et al. 2002). However, REM sleep suppression was the only noticeable effect when the compound was given by the subcutaneous (SC) route (Monti and Jantos 2006a) (Table 20.4). Pretreatment with ritanserin prevented the enhancement of W and the deficit of SWS induced by DOI and DOM, but not the REM sleep suppression

Table 20.4 The effect of selective and nonselective 5-HT$_{2A}$ and 5-HT$_{2C}$ receptor antagonists on sleep and waking in laboratory animals

Compound	Route of administration	W	SWS	REMS	Reference
Ritanserin (5-HT$_{2A/2C}$ antagonist)	i.p.	n.s.	↑	↓	(Dugovic and Wauquier 1987)
Ritanserin (5-HT$_{2A/2C}$ antagonist)	i.p.	↓	↑	↓	(Dugovic et al. 1989)
Ritanserin (5-HT$_{2A/2C}$ antagonist)	i.p.	n.s.	↑	↓	(Monti et al. 1990)
Ritanserin (5-HT$_{2A/2C}$ antagonist)	i.p.	↓	↑	↓	(Kirov and Moyanova 1998a)
Ketanserin (5-HT$_{2A/2C}$ antagonist)	i.p.	↓	↑	↓	(Kirov and Moyanova 1998b)
ICI 169,369 (5-HT$_{2A/2C}$ antagonist)	p.o.	↑	n.s.	↓	(Tortella et al. 1989)
ICI 170,809 (5-HT$_{2A/2C}$ antagonist)	p.o.	↑	(delayed increase of SWS)	↓	(Tortella et al. 1989)
Sertindole (5-HT$_{2A/2C}$ antagonist)	i.p.	n.s.	↑	↓	(Coenen et al. 1995)
MDL 100907 (5-HT$_{2A}$ antagonist)	i.p.	↓	↑	↓	(Popa et al. 2005)
EMD 281014 (5-HT$_{2A}$ antagonist)	s.c.	n.s.	n.s.	↓	(Monti and Jantos 2006a)
SB-243213 (5-HT$_{2C}$ antagonist)	s.c.	n.s.	n.s.	↓	(Monti and Jantos 2006a)
SB-243213 (5-HT$_{2C}$ antagonist)	p.o.	n.s.	↑	↓	(Smith et al. 2002)

W waking, *SWS* slow wave sleep, *REMS* REM sleep, *n.s.* nonsignificant, ↑ increase, ↓ decrease

(Dugovic et al. 1989; Monti et al. 1990). Ritanserin also antagonized the DOI-induced decrease of neocortical high-voltage spindle activity (Jäkälä et al. 1995). In order to gain further insight into the roles of 5-HT$_{2A}$ and 5-HT$_{2C}$ receptors in the DOI-induced disruption of the sleep-wake cycle, animals were pretreated with either EMD 281014 or SB 243213, which selectively block the 5-HT$_{2A}$ or the 5-HT$_{2C}$ receptor, respectively. EMD 281014 prevented the increase of W and the reduction of SWS induced by DOI. However, REM sleep remained suppressed. In contrast SB 243213 failed to reverse the DOI-induced disruption of sleep and W (Monti and Jantos 2006a).

Thus on the basis of these results it appears that 5-HT$_{2A}$ mechanisms predominate following the systemic administration of DOI. However, the role of the 5-HT$_{2C}$ receptor cannot be excluded, and further studies using additional 5-HT$_{2C}$ antagonists are warranted. It should be stressed that the failure of EMD 281014 to prevent the suppression of REM sleep tends to indicate that the effect of DOI is not restricted to the 5-HT system.

In conclusion, systemic administration of selective 5-HT$_{2C}$ or nonselective 5-HT$_{2A/2C}$ receptor agonists to laboratory animals induced a consistent increase of W and a reduction of SWS and/or REM sleep. In contrast, systemic injection of selective and nonselective 5-HT$_{2A}$ and 5-HT$_{2C}$ receptor antagonists produced in almost all instances an increase of SWS and a suppression of REM sleep. Waking was found also to be reduced in a number of studies. Differences in species (mouse, rat), route (p.o., SC, IP), time of drug administration during the light phase, drug concentration, and the use of different approaches to analyze data could tentatively explain the discrepancies observed among studies. Is serotonin directly responsible for the increase of W induced by the selective 5-HT$_{2C}$ and nonselective 5-HT$_{2A/2C}$ agonists? It has been shown that systemic administration of DOI inhibits the firing of serotonergic neurons in the DRN and noradrenergic neurons in the LC. In addition, 5-HT and NA levels are diminished at postsynaptic sites. The reduction of the firing rate of serotonergic and noradrenergic cells is reverted by MDL 100907 and ketanserin, respectively (Garratt et al. 1991; Chiang and Aston-Jones 1993). As mentioned earlier, the IP injection of DOI significantly increases the extracellular concentration of ACh in the medial prefrontal cortex and the hippocampus of the rat, and the effect is blocked by the 5-HT$_{2A/2C}$ receptor antagonist LY53-857 (Nair and Gudelsky 2004). Thus, indirect evidence tends to indicate that ACh is involved in the increase of W after systemic DOI. However, to elucidate the mechanisms underlying the 5HT$_{2A/2C}$ receptor agonist-induced increase of W, further investigations are needed.

20.2.2   Local Administration of Selective and Nonselective 5-HT$_2$ Receptor Ligands to Laboratory Animals

Intraraphe or microiontophoretic administration of DOI inhibits the firing of serotonergic neurons in the DRN and reduces the extracellular concentration of 5-HT (Garratt et al. 1991). Notwithstanding the above, microinjection of DOI into the DRN results in the suppression of REM sleep (Table 20.3). Pretreatment with EMD 281014 or SB 243213 prevents the DOI-induced suppression of REM sleep, which indicates that it is mediated by the 5-HT$_{2A}$ and 5-HT$_{2C}$ receptors located in the DRN (Monti and Jantos 2006b). From a functional point of view, two types of GABA-ergic cells are found in the DRN: GABA-ergic interneurons that project to 5-HT cells and contribute to the decrease of their activity during SWS and long-projection GABA-ergic interneurons that are responsible for REM sleep suppression (Ford et al. 1995). Activation of long-projection GABA-ergic interneurons by DOI would inhibit the activity of cholinergic cells in the LDT/PPT nuclei and decrease REM sleep.

Amici et al. (2004) locally microinjected DOI, or the 5-HT$_2$ receptor antagonist ketanserin, into the LDT of rats. DOI significantly decreased the number of REM sleep episodes, whereas ketanserin induced the opposite effect (Table 20.3). The finding by Fay and Kubin (2000) that 5-HT$_{2A/2C}$ receptors are located not on cholinergic cells but on GABA-ergic interneurons intermingled with mesopontine cholinergic cells, tends to explain the inhibitory effect of DOI on cholinergic LDT neurons and the reduction of REM sleep episodes (Monti et al. 2008a).

20.2.3 Administration of Nonselective 5-HT$_2$ Receptor Ligands to Man

Idzikowski et al. (1986) characterized for the first time the effect of acute administration of ritanserin 10 mg on nocturnal sleep in healthy subjects. Administration of the 5-HT$_{2A/2C}$ receptor antagonist in the morning (8:00 am) significantly increased SWS (stages 3 and 4) and reduced stage 2 sleep. REM sleep was also suppressed when the compound was given in the evening (10:30 pm). Idzikowski et al. (1987) investigated also the effect of repeated morning administration of ritanserin 10 mg on sleep of healthy volunteers. After 2 weeks ritanserin treatment SWS remained increased. The increase of SWS was coupled with a reduction of stage 2 sleep. All other whole-night measures remained unchanged.

During the comparison of 1, 3, 10, and 30 mg ritanserin, a clear dose–response relationship was observed for the 5-HT$_2$ antagonist with greater doses inducing increased duration of SWS (Idzikowski et al. 1991). Ritanserin 10 mg discontinuation following daily administration for 8 weeks was not associated with the occurrence of withdrawal symptoms (Kamali et al. 1992). Sharpley et al. (1994) compared the effects of ritanserin 5 mg and ketanserin 20 mg and 40 mg on the sleep of healthy volunteers. Ritanserin and ketanserin 40 mg significantly increased SWS and reduced stage 2 sleep. REM sleep was also suppressed following ritanserin administration. The effects of the 5-HT$_{2A/2C}$ receptor antagonists seganserin, ICI 169,369, and SR 46349B {4-[(3Z)3-(2-dimethylaminoethyl)oxyimino-3 (2-flurophenyl) propen-1-yl] phenol hemifumarate} were studied also on the sleep EEG of subjects with normal sleep. Seganserin induced an increase of SWS and an enhacement of the power density in the delta and theta frequencies during NREM sleep. In addition, intermittent W showed a reduction after drug administration (Dijk et al. 1989). Compound ICI 169,369 induced also an increase of SWS (Sharpley et al. 1990) whereas SR 46349B increased SWS and reduced stage 2 sleep. Analysis of the EEG power spectra showed an increase of power within 0.75–4.5 Hz (delta activity) and a decrease of power within 12.25–15 Hz (spindle frequency activity) after SR 46349B administration (Landolt et al. 1999). Ritanserin has been administered also to poor sleepers, patients with chronic insomnia, and psychiatric patients with a generalized anxiety disorder (GAD) or a mood disorder. Ritanserin 5 mg taken by poor sleepers for 20 days caused a large and significant increase of SWS during the early and the late drug period compared with baseline placebo nights. Concomitantly with the increase of SWS there was a reduction of stage 2 and of the frequencies of awakenings. REM sleep was not affected (Adam and Oswald 1989).

The administration of ritanserin 10 mg during morning time for 5 days to a group of patients with chronic insomnia increased the duration of SWS without modifying stage 2 or REM sleep. The increase of SWS was related to greater amounts of stage 4 (Ruiz-Primo et al. 1989). The effect of ritanserin was characterized also in abstinent alcoholic patients with comorbid insomnia. The 5-HT$_2$ antagonist was given at a daily dose of 10 mg for 28 days. Ritanserin reduced wake time after sleep onset. The increase of total sleep time (TST) was associated with significantly greater amounts of NREM sleep. Slow wave sleep and REM sleep were not significantly modified (Monti et al. 1993).

da Roza Davis et al. (1992) determined the acute effects of ritanserin 5 mg on sleep variables in patients with GAD and matched healthy controls. Ritanserin produced a significant increase of SWS together with a reduction of stage 1 and wake time after sleep onset. Unexpectedly, the 5-HT$_2$ receptor antagonist increased REM sleep. Polysomnographic recordings of dysthymia patients (DSM-III) who received ritanserin 10 mg for 4 weeks showed a significant increase of SWS. No other variables were modified by the drug (Paiva et al. 1988). Moreover, acute administration of ritanserin 5 mg in patients with a diagnosis of major depression induced a significant increase of SWS without changing stage 2 or REM sleep duration. The increase of SWS was related to greater amounts of stage 3 (Staner et al. 1992).

In conclusion, the 5-HT$_{2A/2C}$ antagonists ritanserin, ketanserin, seganserin, ICI 169,369, and SR 46349B consistently increased SWS in subjects with normal sleep. In addition, ritanserin was shown to augment SWS in poor sleepers, chronic insomniacs, and patients with GAD or a mood disorder.

20.3 Effects of Typical and Atypical Antipsychotics on Sleep in Schizophrenia Patients

Typical and atypical antipsychotic drugs bind to a wide variety of CNS receptors including the 5-HT$_{2A}$ and 5-HT$_{2C}$ receptors. Thus, it seems pertinent to analyze the effects of antipsychotic drugs on sleep variables and the mechanisms involved in their actions on sleep and W. It should be noted that Monti and Monti (2004) and Winokur and Kamath (2008) have provided detailed reviews of work carried out in this area. Our analysis will be limited to the effects on sleep variables of the typical antipsychotics haloperidol, thiothixene, and flupentixol and the atypical antipsychotics olanzapine, risperidone, clozapine, quetiapine, and ziprasidone. Various populations were included in the studies to be described: healthy volunteers, outpatients, and hospitalized patients.

Typical antipsychotic drugs including haloperidol, thiothixene, and flupentixol have been shown to reduce stage 2 sleep latency and to increase total sleep time and sleep efficiency. Stage 4 sleep and SWS remained unchanged, whereas REM latency was significantly increased (Taylor et al. 1991; Nofzinger et al. 1993; Wetter et al. 1996; Maixner et al. 1998). Olanzapine given to schizophrenia patients or healthy subjects reduced stage 2 sleep latency and wake time after sleep onset, whereas total sleep time and sleep efficiency were enhanced. Concerning sleep architecture, stage 1 sleep was decreased, whereas stage 2 and SWS were augmented. On the other hand, olanzapine tended to disrupt REM sleep as judged by the reduction of REM sleep duration and the increase of REM latency (Salin-Pascual et al. 1999; Sharpley et al. 2000; Muller et al. 2004).

The limited information on risperidone tends to indicate that the compound improves sleep maintenance and increases SWS in schizophrenia patients (Dursun et al. 1999; Haffmans et al. 2001; Yamashita et al. 2002). In addition, in a study that involved healthy control subjects, risperidone produced a reduction of REM sleep time compared with placebo (Sharpley et al. 2003). The administration of clozapine to schizophrenia patients increased total sleep time, sleep efficiency, and stage 2 sleep. Stage 4 sleep and SWS tended to be reduced, whereas REM sleep was not significantly affected (Wetter et al. 1996; Hinze-Selch et al. 1997; Lee et al. 2001). In the study by Touyz et al. (1978) clozapine was given at a relatively low dose to normal young adults; the effects of the compound were limited to a decrease of stage 1 sleep and REM sleep duration. Cohrs et al. (2004) and Cohrs et al. (2005) examined the effects of quetiapine and ziprasidone on sleep in healthy subjects. The antipsychotic drugs significantly reduced sleep onset latency, wake time after sleep onset, and REM sleep and increased total sleep time, sleep efficiency, stage 2 sleep, and REM latency. To date no studies have been carried on the effect of quetiapine and ziprasidone on sleep of schizophrenia patients.

Typical and atypical antipsychotic drugs produce their effects by blocking, among others, dopamine, serotonin, α-adrenergic, histamine, and acetylcholine (muscarinic) receptors. Irrespective of their chemical structure, antipsychotics show low (olanzapine, quetiapine), intermediate (risperidone, clozapine, ziprasidone) to high (haloperidol, flupentixol, thiothixene) affinity for the D$_2$ receptor (Table 20.5). In contrast to the classical antipsychotics, the newer antipsychotics show moderate (quetiapine) to high (clozapine, olanzapine, risperidone, ziprasidone) affinity for the serotonin 5-HT$_{2A}$ receptor and to a lesser extent for the 5-HT$_{2C}$ receptor. The atypical antipsychotics bind with moderate (quetiapine, ziprasidone) and high affinity (olanzapine, risperidone, clozapine) to the α$_1$ adrenoceptor, whereas olanzapine and clozapine display high affinity for both the histamine H$_1$ and the acetylcholine (muscarinic) receptor (Table 20.5). The blockade of D$_2$ receptors could be partly responsible for the improvement of sleep in schizophrenia patients. On the other hand, the increase of SWS and the suppression of REM sleep induced by some of the atypical antipsychotics could be related to the blockade of

Table 20.5 Receptor binding profile of typical and atypical antipsychotics (From Horácek 2000; Reynolds 2004)

	D$_2$	5-HT$_{2A}$	5-HT$_{2C}$	α$_1$	Ach (M)	H$_1$
Haloperidol	+++	+	–	+	+	+
Flupentixol	+++	+	?	+	–	–
Olanzapine	+	+++	++	++	+++	+++
Risperidone	++	+++	+	+++	–	–
Clozapine	++	+	++	+++	+++	+++
Quetiapine	+	+	?	++	–	+
Ziprasidone	++	+++	?	++	–	+

D$_2$ dopamine receptor, *5-HT$_{2A}$* 5-HT$_{2C}$, serotonin receptors, *α$_1$* adrenergic receptor, *ACh (M)* cholinergic (muscarinic) receptors, *H$_1$* histamine receptor. Affinity: (–) absent; (+) low; (++) intermediate; (+++) high; (?) unknown

5-HT$_{2A}$ receptors. The blockade of α_1-adrenergic, histamine H$_1$, and muscarinic acetylcholine receptors could participate also in the amelioration of sleep provoked by the antipsychotic drugs. This is based on the premise that they produce somnolence, an increased likelihood of falling asleep and reduced concentration (Monti 1987; Heller-Brown and Taylor 1996; Monti and Monti 2000b).

20.4 Conclusions

Receptors corresponding to the 5-HT$_2$ subfamily are located within postsynaptic structures. They are coupled to the activation of the phospholipase C with a resulting depolarization of the host cell. 5-HT$_{2A/2C}$ receptor binding sites and 5-HT$_{2A/2C}$ mRNA expression have been found in most neuroanatomical structures involved in the regulation of the behavioral state. They include:

1. Serotonin (dorsal raphe nucleus, median raphe nucleus)-, noradrenaline (locus coeruleus)-, histamine (tuberomammillary nucleus)-, dopamine (ventral tegmental area, substantia nigra pars compacta)-, orexin (lateral and dorsal hypothalamus)-, acetylcholine (midbrain tegmentum, basal forebrain)-, and glutamate (medial pontine reticular formation)-containing neurons involved in the regulation of W.
2. γ-Aminobutyric acid/galanin (preoptic area, anterior and lateral hypothalamic areas)-containing cells that constitute the sleep-inducing system.
3. Cholinergic neurons (laterodorsal and pedunculopontine tegmental nuclei) that act to promote REM sleep.

The role of 5-HT$_{2A}$ and 5-HT$_{2C}$ receptors in the regulation of the behavioral state has been characterized: (1) in mutant mice that do not express 5-HT$_{2A}$ or 5-HT$_{2C}$ receptors and (2) following the administration of selective and nonselective 5-HT$_{2A}$ and 5-HT$_{2C}$ receptor agonists and antagonists to laboratory animals and man. 5-HT$_{2A}$ receptor knockout mice show a significant increase of W and a reduction of NREM sleep. Values corresponding to REM sleep remain unaltered. Similar changes have been observed in 5-HT$_{2C}$ receptor knockout mice. The limited available evidence tends to indicate that the increase of W and reduction of NREM sleep in the mutant mice is related, at least in part, to the increased release of NA and DA at central sites.

It is worth mentioning that sleep has been studied also in 5-HT$_{1A}$ and 5-HT$_{1B}$ receptor knockout mice (Adrian 2008). Mutant mice that do not express 5-HT$_{1A}$ or 5-HT$_{1B}$ receptor exhibit greater amounts of REM sleep than their wild-type counterparts. On the other hand, W and SWS remain unchanged. It has been shown that microinjection of 5-HT$_{1A}$ receptor agonists into areas in which the cholinergic REM sleep induction neurons are located results in the suppression of REM sleep(Monti and Jantos 2004, 2008). Thus, the increase of REM sleep in the 5-HT$_{1A}$ receptor knockout mice could be related to the absence of a postsynaptic 5-HT$_{1A}$ receptor inhibitory effect on REM-on neurons of the LDT/PPT (Table 20.6). 5-HT$_{1B}$ receptor activation facilitates the occurrence of W and negatively influences REM sleep (Monti and

Table 20.6 Influence of 5-HT$_{1A}$, 5-HT$_{1B}$, 5-HT$_{2A}$, 5-HT$_{2C}$, 5-HT$_3$, and 5-HT$_7$ receptor on sleep and waking in laboratory animals

Experimental procedure	W	SWS	REMS	Reference
5-HT$_{1A}$ receptor knockout mice (5-HT$_{1A}$ agonists: 8-OH-DPAT, flesinoxan)	n.s.	n.s.	+	(Boutrel et al. 2002)
Somatodendritic: microinjection into the DRN	n.s.	n.s.	+	(Monti et al. 2002)
Postsynaptic: systemic injection	+	–	–	(Monti and Jantos 2004)
5-HT$_{1B}$ receptor knockout mice (5-HT$_{1B}$ agonists: CGS 12066B, CP-94253)	n.s.	n.s.	+	(Boutrel et al. 1999)
Microinjection into the DRN	n.s.	n.s.	–	(Monti et al. 2010)
Systemic injection	+	–	–	(Monti et al. 1995)
5-HT$_{2A}$ receptor knockout mice	+	–	n.s.	(Popa et al. 2005)
5-HT$_{2C}$ receptor knockout mice (5-HT$_{2A/2C}$ agonist: DOI)	+	–	n.s.	(Frank et al. 2002)
Microinjection into the DRN	n.s.	n.s.	–	(Monti and Jantos 2006b)
Systemic injection	+	–	–	(Monti and Jantos 2006a)
(5-HT$_3$ receptor agonist: m-chlorophenylbiguanide)				
Microinjection into the DRN	n.s.	n.s.	–	(Monti and Jantos 2008)
i.c.v. injection	+	–	–	(Ponzoni et al. 1993)
5-HT$_7$ receptor knockout mice (5-HT$_7$ receptor agonist: LP-44)	n.s.	n.s.	–	(Hedlund et al. 2005)
Microinjection into the DRN	n.s.	n.s.	–	(Monti et al. 2008b)

DRN dorsal raphe nucleus, *W* waking, *SWS* slow wave sleep, *REMS* REM sleep, *n.s.* nonsignificant, + increased, – decreased

Jantos 2008). Accordingly, the increase of REM sleep in mice that do not express 5-HT$_{1B}$ receptor could be related also to the absence of an inhibitory effect on cholinergic REM-on cells. Systemic administration of selective 5-HT$_{2C}$ agonists and nonselective 5-HT$_{2A/2C}$ agonists induces an increase of W and a reduction of SWS and REM sleep in laboratory animals. On the other hand, injection of selective 5-HT$_{2A}$ and 5-HT$_{2C}$ antagonists or nonselective 5-HT$_{2A/2C}$ antagonists increases SWS and reduces REM sleep. Waking has been shown to be decreased in most studies.

Studies aimed at determining the effect of 5-HT$_2$ agonists on neurotransmitter release have shown that they increase ACh release and reduce NA and DA availability at central sites. Notwithstanding this, further studies are needed to determine whether other neurotransmitter systems are involved in the 5-HT$_{2A}$ and 5-HT$_{2C}$ receptor-induced increase of W and reduction of sleep. Interestingly, administration of nonselective 5-HT$_{2A/2C}$ antagonists increases SWS in subjects with normal sleep, poor sleepers, patients with chronic insomnia, and psychiatric patients with GAD or a mood disorder. Presently, benzodiazepine and nonbenzodiazepine derivatives (zolpidem, zopiclone, eszopiclone) are the predominant hypnotic drugs. All these compounds improve the induction and maintenance of sleep in patients with primary and comorbid insomnia. However, benzodiazepine agonists suppress SWS sleep and REM sleep. On the other hand, SWS and REM sleep tend to remain unchanged during the administration of nonbenzodiazepine hypnotics. The finding that ritanserin and ketanserin promote SWS in man has led to the development of new 5-HT$_2$ receptor antagonists targeted towards improving sleep architecture in insomniac patients. In other words, their administration together with benzodiazepine or nonbenzodiazepine hypnotic drugs would be expected to improve sleep and to increase SWS values (Wafford and Ebert 2008).

References

Abramowski D, Rigo M, Duc D, et al (1995) Localization of the 5-hydroxytryptamine$_{2C}$ receptor protein in human and rat brain using specific antisera. Neuropharmacology 34:1635–1645.

Adam K, Oswald I (1989) Effects of repeated ritanserin on middle-aged poor sleepers. Psychopharmacology 99:219–221.

Adrian J (2008) Sleep and waking in mutant mice that do not express various proteins involved in serotonergic neurotransmission such as the serotonergic transporter, monoamine oxidase A, and 5-HT$_{1A}$, 5-HT$_{1B}$, 5-HT$_{2A}$, 5-HT$_{2C}$ and 5-HT$_7$ receptors. In: Monti JM, Pandi-Perumal SR, Jacobs BL, et al, eds. Serotonin and Sleep: Molecular, Functional and Clinical Aspects. Basel: Birkhäuser, pp. 457–475

Amici R, Sanford LD, Kearney K, et al (2004) A serotonergic (5-HT$_2$) receptor mechanism in the laterodorsal tegmental nucleus participates in regulating the pattern of rapid-eye-movement sleep occurrence in the rat. Brain Res 996:9–18.

Amin AH, Crawford TBB, Gaddum JH (1954) The distribution of substance P and 5-hydroxytryptamine in the central nervous system of the dog. J Physiol 125:596–618.

Boutrel B, Franc B, Hen R, et al (1999) Key role of 5-HT$_{1B}$ receptors in the regulation of paradoxical sleep as evidenced in 5-HT$_{1B}$ knock-out mice. J Neurosci 19:3204–3212.

Boutrel B, Monaca C, Hen R, et al (2002) Involvement of 5-HT$_{1A}$ receptors in homeostasis and stress-induced adaptive regulations of paradoxical sleep: studies in 5-HT$_{1A}$ knock-out mice. J Neurosci 22:4686–4692.

Chiang C, Aston-Jones G (1993) A 5-hydroxytryptamine-2 agonist augments gamma-aminobutyric acid and excitatory amino acid inputs to noradrenergic locus coeruleus neurons. Neuroscience 54:409–420.

Clemett DA, Punhani T, Duxon MS, et al (2000) Immunohistochemical localisation of the 5-HT$_{2C}$ receptor protein in the rat CNS. Neuropharmacology 39:123–132.

Coenen AML, Ates N, Skarsfeldt T, et al (1995) Effects of sertindole on sleep-wake states, electroencephalogram, behavioral patterns and epileptic activity in rats. Pharmacol Biochem Behav 51:353–357.

Cohrs S, Rodenbeck A, Guan Z, et al (2004) Sleep-promoting properties of quetiapine in healthy subjects. Psychopharmacology 174:421–429.

Cohrs S, Meier A, Neumann A-C, et al (2005) Improved sleep continuity and increased slow wave sleep and REM latency during ziprasidone treatment: a randomized controlled crossover trial of 12 healthy male subjects. J Clin Psychiatry 66:989–996.

Cornea-Hébert V, Riad M, Wu C, et al (1999) Cellular and subcellular distribution of the serotonin 5-HT$_{2A}$ receptor in the central nervous system of adult rat. J Comp Neurol 409:187–209.

da Roza Davis JM, Sharpley AL, Cowen PJ (1992) Slow wave sleep and 5-HT$_2$ receptor sensitivity in generalised anxiety disorder. Psychopharmacology 108:387–389.

de Almeida J, Mengod G (2007) Quantitative analysis of glutamatergic and GABAergic neurons expressing 5-HT(2A) receptors in human and monkey prefrontal cortex. J Neurochem 103:475–486.

Di Matteo V, Di Giovanni G, Di Mascio M, et al (2000) Biochemical and electrophysiological evidence that RO 60-0175 inhibits mesolimbic dopaminergic function through serotonin$_{2C}$ receptors. Brain Res 865:85–90.

Dijk DJ, Beersma DGM, Daan S, et al (1989) Effects of seganserin, a 5-HT$_2$ antagonist, and temazepam on human sleep stages and EEG power spectra. Eur J Pharmacol 171:207–218.

Done CJ, Sharp T (1994) Biochemical evidence for the regulation of central noradrenergic activity by 5-HT$_{1A}$ and 5-HT$_2$ receptors: microdialysis studies in the awake and anaesthetized rat. Neuropharmacology 33:411–421.

Dugovic C, Wauquier A (1987) 5-HT$_2$ receptors could be primarily involved in the regulation of slow wave sleep in the rat. Eur J Pharmacol 137:145–146.

Dugovic C, Wauquier A, Leysen JE, et al (1989) Functional role of 5-HT$_2$ receptors in the regulation of sleep and wakefulness in the rat. Psychopharmacology 97:436–442.

Dursun SM, Patel JKM, Burke JG, et al (1999) Effects of typical antipsychotic drugs and risperidone on the quality of sleep in patients with schizophrenia: a pilot study. J Psychiatry Neurosci 24:333–337.

Duxon MS, Flanigan TP, Reavley AC, et al (1997) Evidence for expression of the 5-hydroxytryptamine-2B receptor protein in the rat central nervous system. Neuroscience 76:323–329.

Eriksson KS, Stevens DR, Haas HL (2001) Serotonin excites tuberomammillary neurons by activation of Na$^{(+)}$/Ca$^{(2+)}$-exchange. Neuropharmacology 40:345–351.

Fay R, Kubin L (2000) Pontomedullary distribution of 5-HT$_{2A}$ receptor-like protein in the rat. J Comp Neurol 418:323–345.

Fink KB, Göthert M (2007) 5-HT receptor regulation of neurotransmitter release. Pharmacol Rev 59:360–417.

Ford B, Holmes CJ, Mainville L, et al (1995) GABAergic neurons in the rat pontomesencephalic tegmentum: codistribution with cholinergic and other tegmental neurons projecting to the posterior lateral hypothalamus. J Comp Neurol 363:177–196.

Frank MG, Stryker MP, Tecott LH (2002) Sleep and sleep homeostasis in mice lacking the 5-HT$_{2C}$ receptor. Neuropsychopharmacology 27:869–873.

Garratt JC, Kidd EJ, Wright IK, et al (1991) Inhibition of 5-hydroxytryptamine neuronal activity by the 5-HT agonist DOI. Eur J Pharmacol 199:349–355.

Haffmans PMJ, Oolders JM, Hoencamp E, et al (2001) The effect of risperidone versus haloperidol on sleep patterns of schizophrenic patients: results of a double-blind, randomised pilot trial. Eur Neuropsychopharm 11 (suppl 3):S260.

Hannon J, Hoyer D (2008) Molecular biology of 5-HT receptors. In: Monti JM, Pandi-Perumal SR, Jacobs BL, et al., eds. Serotonin and Sleep: Molecular, Functional and Clinical Aspects. Basel: Birkhäuser, pp 155–182.

Hedlund PB, Huitron-Resendiz S, Henriksen SJ, et al (2005) 5-HT$_7$ receptor inhibition and inactivation induced antidepressantlike behavior and sleep pattern. Biol Psychiatry 58:831–837.

Heller-Brown J, Taylor P (1996) Muscarinic receptor agonists and antagonists. In: Hardman JG, Limbird LE, eds. The Pharmacological Basis of Therapeutics. New York: McGraw-Hill, pp. 141–160.

Hinze-Selch D, Mullington J, Orth A, et al (1997) Effects of clozapine on sleep: a longitudinal study. Biol Psychiatry 42:260–266.

Horácek J (2000) Novel antipsychotics and extrapyramidal side effects. Theory and reality. Pharmacopsychiatry 33 (suppl):34–42.

Idzikowski C, Mills FJ, Glennard R (1986) 5-Hydroxytryptamine-2 antagonist increases human slow wave sleep. Brain Res 378:164–168.

Idzikowski C, Cowen PJ, Nutt D, et al (1987) The effects of chronic ritanserin treatment on sleep and the neuroendocrine response to L-tryptophan. Psychopharmacology 93:416–420.

Idzikowski C, Mills FJ, James RJ (1991) A dose-response study examining the effects of ritanserin on human slow wave sleep. Br J Clin Pharm 31:193–196.

Jacobs BL, Azmitia EC (1992) Structure and function of the brain serotonin system. Physiol Rev 72:165–229.

Jacobs BL, Fornal CA (1999) Activity of serotonergic neurons in behaving animals. Neuropsychopharmacology 21 (suppl 2):9S–15S.

Jäkälä P, Sirvio J, Koivisto E, et al (1995) Modulation of rat neocortical high-voltage spindle activity by 5-HT$_1$/5-HT$_2$ receptor subtype specific drugs. Eur J Pharmacol 82:39–55.

Jones FJ (2003) Arousal systems. Frontiers Biosci 8:438–451.

Kamali F, Stansfield SC, Ashton CH, et al (1992) Absence of withdrawal effects of ritanserin following chronic dosing in healthy volunteers. Psychopharmacology 108:213–217.

Kirov R, Moyanova S (1998) Age-dependent effect of ketanserin on the sleep-waking phases in rats. Int J Neurosci 93:257–264.

Kirov R, Moyanova S (1998) Age-related effect of ritanserin on the sleep-waking phases in rats. Int J Neurosci 93:265–278.

Kitka T, Bagdy G (2008) Effect of 5-HT$_{2A/2B/2C}$ receptor agonists and antagonists on sleep and waking in laboratory animals and humans. In: Monti JM, Pandi-Perumal SR, Jacobs BL, et al, eds. Serotonin and Sleep: Molecular, Functional and Clinical Aspects. Basel: Birkhäuser, pp. 387–414

Landolt HP, Viola M, Burgess HJ, et al (1999) Serotonin-2 receptors and human sleep: Effect of a selective antagonist on EEG power spectra. Neuropsychopharmacology 21:455–466.

Lee JH, Woo JI, Meltzer HY (2001) Effects of clozapine on sleep measures and sleep-associated changes in growth hormone and cortisol in patients with schizophrenia. Psychiatry Res 103:157–166.

Leysen JE (2004) 5-HT$_2$ receptors. Curr Drug Targets CNS Neurol Disord 3:11–26.

Maixner S, Tandon R, Eiser A, et al (1998) Effects of antipsychotic treatment on polysomnographic measures in schizophrenia: a replication and extension. Am Psychiatry 155: 1600–1602.

Martin JR, Bos M, Jenck F, et al (1998) 5-HT$_{2C}$ receptor agonists: Pharmacological characteristics and therapeutic potential. J Pharmacol Exp Ther 286:913–924.

Mengod G, Nguyen H, Le H, et al (1990) The distribution and cellular localization of the serotonin 1C receptor in the rodent brain examined by in situ hybridization histochemistry. Comparison with receptor binding distribution. Neuroscience 35:577–591.

Monti JM (1987) Disturbances of sleep and wakefulness associated with the use of antihypertensive agents. Life Sci 41: 1979–1988.

Monti JM, Jantos H (2004) Effects of the 5-HT$_{1A}$ receptor ligands flesinoxan and WAY 100635 given systemically or microinjected into the laterodorsal tegmental nucleus on REM sleep in the rat. Behav Brain Res 151:159–166.

Monti JM, Jantos H (2006a) Effects of the serotonin 5-HT$_{2A/2C}$ receptor agonist DOI and of the selective 5-HT$_{2A}$ or 5-HT$_{2C}$ receptor antagonists EMD 281014 and SB-243213, respectively, on sleep and waking in the rat. Eur J Pharmacol 553:163–170.

Monti JM, Jantos H (2006b) Effects of activation and blockade of 5-HT$_{2A/2C}$ receptors in the dorsal raphe nucleus on sleep and waking in the rat. Prog Neuropsychopharmacol Biol Psychiatry 30:1189–1195.

Monti JM, Jantos H (2008) Mechanisms involved in the inhibition of REM sleep by serotonin. In: Monti JM, Pandi-Perumal SR, Jacobs BL, et al, eds. Serotonin and Sleep: Molecular, Functional and Clinical Aspects. Basel: Birkhäuser, pp. 371–385

Monti JM, Monti D (2000a) Role of dorsal raphe nucleus serotonin 5-HT$_{1A}$ receptor in the regulation of REM sleep. Life Sci 66:1999–2012.

Monti JM, Monti D (2000b) Histamine H$_1$ receptor antagonists in the treatment of insomnia. Is there a rational basis for use? CNS Drugs 13:87–96.

Monti JM, Monti D (2004) Sleep in schizophrenia patients and the effects of antipsychotic drugs. Sleep Med Rev 8:133–148.

Monti JM, Orellana C, Boussard M, et al (1990) 5-HT receptor agonists 1-(2,5-dimethoxy-4-iodophenyl)-2-aminopropane (DOI) and 8-OH-DPAT increase wakefulness in the rat. Biogen Amines 7:145–151.

Monti JM, Alterwain P, Estévez F, et al (1993) The effects of ritanserin on mood and sleep in abstinent alcoholic patients. Sleep 16:647–654.

Monti JM, Monti D, Jantos H, et al (1995) Effects of selective activation of the 5-HT$_{1B}$ receptor with CP-94,253 on sleep and wakefulness in the rat. Neuropharmacology 34:1647–1651.

Monti JM, Jantos H, Monti D (2002) Increased REM sleep after intra-dorsal raphe nucleus injection of flesinoxan or 8-OHDPAT: prevention with WAY 100635. Eur Neuropsychopharmacol 12:47–55.

Monti JM, Jantos H, Monti D (2008a) Serotonin and sleep-wake regulation. In: Monti JM, Pandi-Perumal SR, Sinton CM Eds., Neurochemistry of Sleep and Wakefulness. Cambridge: Cambridge University Press, pp. 244–279.

Monti JM, Leopoldo M, Jantos H (2008b) The serotonin 5-HT$_7$ receptor agonist LP-44 microinjected into the dorsal raphe nucleus suppresses REM sleep in the rat. Behav Brain Res 191:184–189.

Monti JM, Jantos H, Lagos P (2010) Activation of serotonin 5-HT$_{1B}$ receptor in the dorsal raphe nucleus affects REM sleep in the rat. Behav Brain Res 206:8–16.

Muller MJ, Rossbach W, Mann K, et al (2004) Subchronic effects of olanzapine on sleep EEG in schizophrenic patients with predominantly negative symptoms. Pharmacopsychiatry 37:157–162.

Muraki Y, Yamanaka A, Tsujino N, et al (2004) Serotonergic regulation of the orexin/hypocretin neurons through the 5-HT$_{1A}$ receptor. J Neurosci 24:7159–7166.

Muramatsu M, Chaki S, Usuki-Ito C, et al (1990) Attenuation of serotonin-induced suppression of [^{3}H]acetylcholine release from rat cerebral cortex by minaprine: possible involvement of the serotonin-2 receptor and K$^+$ channel. Neurochem Int 16:301–307.

Nair SG, Gudelsky GA (2004) Activation of 5-HT$_2$ receptors enhances the release of acetylcholine in the prefrontal cortex and hippocampus of the rat. Synapse 53:202–207.

Nichols DE, Nichols CD (2008) Serotonin receptors. Chem Rev 108:1614–1641.

Nofzinger EA, van Kammen DP, Gilbertson MW, et al (1993) Electroencephalographic sleep in clinically stable schizophrenic patients: two-weeks versus six-weeks neuroleptic free. Biol Psychiatry 33:829–835.

Pace-Schott EF, Hobson J (2002) Basic mechanisms of sleep: New evidence on the neuroanatomy and neuromodulation of the NREM-REM cycle. In: Charney D, Nemeroff C, eds. Neuropsychopharmacology – The Fifth Generation of Progress. Philadelphia: Lippincott Williams & Wilkins, pp. 1859–1877.

Paiva T, Arriaga F, Wauquier A, et al (1988) Effects of ritanserin on sleep disturbances in dysthymic patients. Psychopharmacology 96:395–399.

Paxinos G, Watson C (2005) The Rat in Stereotaxic Coordinates – The New Coronal Set, 5th edn. Sydney: Academic.

Pazos A, Cortés R, Palacios JM (1985) Quantitative autoradiographic mapping of serotonin receptors in the rat brain. II. Serotonin-2 receptors. Brain Res 346:231–249.

Ponzoni A, Monti JM, Jantos H (1993) The effects of selective activation of the 5-HT$_3$ receptor with m-chlorophenylbiguanide on sleep and wakefulness in the rat. Eur J Pharmacol 249:259–264.

Popa D, Lena C, Fabre V, et al (2005) Contribution of 5-HT$_2$ receptor subtypes to sleep-wakefulness and respiratory control, and functional adaptations in knock-out mice lacking 5-HT$_{2A}$ receptors. J Neurosci 25:11231–11238.

Portas CM, Bjorvatn B, Ursin R (2000) Serotonin and the sleep/wake cycle: special emphasis on microdialysis studies. Prog Neurobiol 60:13–35.

Reynolds GF (2004) Receptor mechanisms in the treatment of schizophrenia. J Psychopharmacol 18:340–345.

Ruiz-Primo E, Haro R, Valencia M (1989) Polysomnographic effects of ritanserin in insomniacs: a crossed double-blind controlled study. Sleep Res 18:72.

Salin-Pascual RJ, Herrera-Estrella M, Galicia-Polo L, et al (1999) Olanzapine acute administration in schizophrenic patients increases delta sleep and sleep efficiency. Biol Psychiatry 46:141–143.

Schmidt CJ, Fadayel GM (1995) The selective 5-HT$_{2A}$ receptor antagonist, MDL 100,907, increases dopamine efflux in the prefrontal cortex of the rat. Eur J Pharmacol 273:273–279.

Sharpley AL, Solomon RA, Fernando AI, et al (1990) Dose-related effects of selective 5-HT$_2$ receptor antagonists on slow wave sleep in humans. Psychopharmacology 101:568–569.

Sharpley AL, Elliott JM, Attenburrow MJ, et al (1994) Slow wave sleep in humans: role of 5-HT$_{2A}$ and 5-HT$_{2C}$ receptors. Neuropharmacology 33:467–471.

Sharpley AL, Vassallo CM, Cowen PJ (2000) Olanzapine increases slow-wave sleep: evidence for blockade of central 5-HT$_{2C}$ receptors in vivo. Biol Psychiatry 47:468–470.

Sharpley AL, Bhagwagar Z, Hafizi S, et al (2003) Risperidone augmentation decreases rapid eye movement sleep and decreases wake in treatment-resistant depressed patients. J Clin Psychiatry 64:192–196.

Silhol S, Glin L, Gottesmann C (1991) Study of the 5-HT$_2$ antagonist ritanserin on sleep-waking cycle in the rat. Physiol Behav 41:241–243.

Smith MI, Piper DC, Duxon MS, et al (2002) Effect of SB-243213, a selective 5-HT$_{2C}$ receptor antagonist, on the rat sleep profile: a comparison to paroxetine. Pharmacol Biochem Behav 71:599–605.

Staner L, Kempenaers C, Simonnet MP, et al (1992) 5-HT2 receptor antagonism and slow-wave sleep in major depression. Acta Psychiatr Scand 86:133–137.

Taylor SF, Tandon R, Shipley JE, et al (1991) Effect of neuroleptic treatment on polysomnographic measures in schizophrenia. Biol Psychiatry 30:904–912.

Tortella FC, Echevarría E, Pastel RH, et al (1989) Suppressant effects of selective 5-HT$_2$ antagonists on rapid eye movement sleep in rats. Brain Res 485:294–300.

Touyz SW, Saayman GS, Zabow T (1978) A psychophysiological investigation of the long-term effects of clozapine upon sleep patterns of normal young adults. Psychopharmacology 56:69–73.

Trulson ME, Jacobs BL (1979) Raphe unit activity in freely moving cats: correlation with level of behavioral arousal. Brain Res 163:135–50.

Wafford KA, Ebert B (2008) Emerging anti-insomnia drugs: tackling sleeplessness and the quality of wake time. Nature Rev Drug Discov. doi:10.1038/nrd2464.

Wang QP, Nakai Y (1994) The dorsal raphe: an important nucleus in pain modulation. Brain Res Bull 34:575–585.

Wetter TC, Lauer CJ, Gillich G, et al (1996) The electroencephalographic sleep pattern in schizophrenic patients treated with clozapine or classical antipsychotic drugs. J Psychiat Res 30:411–419.

Winokur A, Kamath J (2008) The effect of typical and atypical antipsychotic drugs on sleep of schizophrenic patients. In: Monti JM, Pandi-Perumal SR, Jacobs BL, et al, eds. Serotonin and Sleep: Molecular, Functional and Clinics Studies. Basel: Birkhäuser, pp. 587–610

Yamashita H, Morinobu S, Yamakawi S, et al (2002) Effect of risperidone on sleep in schizophrenia: a comparison with haloperidol. Psychiatry Res 109:137–142.

Zhang C, Marek GJ (2008) AMPA receptor involvement in 5-hydroxytryptamine2A receptor-mediated prefrontal cortical excitatory synaptic currents and DOI-induced head shakes. Prog Neuropsychopharmacol Biol Psychiatry 32:62–71.

Chapter 21
Role of Alternative Splicing of the 5-HT$_{2C}$ Receptor in the Prader–Willi Syndrome

Shivendra Kishore and Stefan Stamm

21.1 Editing

The 5-HT$_{2C}$ receptor pre-messenger ribonucleic acid (pre-mRNA) undergoes both editing and alternative pre-mRNA splicing. Both events occur in the cell nucleous and are summarized below. RNA editing is the chemical modification of RNA bases. Common editing events include 2′-O-methylation, the addition of a methyl group on the ribose; conversion of cytidine to uridine; and conversion from adenine to inosine. The modification of the bases is catalyzed by deaminating enzymes that hydrolyze specific amino groups of the bases. The cytidine-to-uridine editing is catalyzed by cytidine deaminase, and the adenine-to-inosine editing is catalyzed by adenosine deaminase acting on RNAs (ADARs) (reviewed in Jepson and Reenan 2008 and Mehler and Mattick 2007). Editing of RNA has multiple effects on the resulting RNAs. It can lead to alteration of coding capacity, altered microRNA (miRNA) or small inhibitory RNA (siRNA) target populations, heterochromatin formation, nuclear sequestration, cytoplasmic sequestration, inhibition of miRNA and siRNA processing, and altered alternative splicing patterns (Nishikura 2006). The 5-HT$_{2C}$ receptor pre-mRNA undergoes adenine-to-inosine editing on at least five editing sites. The combination of these sites could generate 32 isoforms theoretically, but not all the predicted mRNA forms have been identified (reviewed in Werry et al. 2008). As the translational machinery interprets an inosine as a guanosine the adenine-to-guanine editing changes the protein sequence that is encoded by the edited pre-mRNA. Due to the degeneracy of the genetic code, some of the edited mRNAs encode the same protein, which reduced the number to proteins generated by editing to 24.

S. Stamm (✉)
Department of Molecular and Cellular Biochemistry, University of Kentucky, College of Medicine, B283 Biomedical Biological Sciences Research Building, 741 South Limestone, Lexington, KY 40536-0509, USA
e-mail: stefan@stamms-lab.net

G. Di Giovanni et al. (eds.), *5-HT$_{2C}$ Receptors in the Pathophysiology of CNS Disease*, 413
The Receptors 22, DOI 10.1007/978-1-60761-941-3_21,
© Springer Science+Business Media, LLC 2011

21.2   Pre-mRNA Splicing

21.2.1   General Mechanisms

In addition to editing, the 5-HT$_{2C}$ receptor pre-mRNA undergoes alternative splicing. Alternative splicing affects an estimated 95% of human intron containing genes and is one of the most important mechanisms to increase the use of information encoded in eukaryotic genomes (Pan et al. 2008; Wang et al. 2008). The mechanism of splicing catalysis has been studied in considerable detail (Jurica and Moore 2003; Wahl et al. 2009). Critical for the catalysis are transient interactions between the pre-mRNA and five small nuclear ribonucleoprotein (snRNPs) (reviewed in Biamonti and Caceres 2009; Stark and Luhrmann 2006). The interaction is based on imperfect base complementarity between the snRNPs and the pre-mRNA (Sharp 1994).

In contrast to the constitutive splicing mechanism, it is not fully understood how splice sites, especially the alternative ones, are selected. Currently, it is not possible to accurately identify alternative spliced exons from genomic DNA sequences. The problem is that splice sites exhibit a large degree of sequence variations and only the four GU-AG nucleotides flanking the intron are conserved (Stamm et al. 2006; Thanaraj and Clark 2001). The splicing machinery needs therefore additional signals that define an exon (Robberson et al. 1990). This signal is provided by transient complexes of splicing regulatory proteins and pre-mRNA. Once these complexes have formed, they interact with components of the core spliceosome, which allows the correct identification of splice sites.

Serine-arginine-rich proteins (SR-proteins) and heterogenous ribonuclearproteins (hnRNPs) are the major classes of proteins identified in complexes forming on pre-mRNA. These proteins bind to short degenerate sequences on the pre-mRNA. The degeneracy of the sequences allows the coding requirements of the pre-mRNA to be independent from splicing requirements. Depending on which proteins they bind, sequence elements on the pre-mRNA can act as either enhancers or silencers, which either promote or antagonize exon usage. Splicing regulatory proteins generally possess RNA-binding and protein-interaction domains that allow weak, transient binding between protein and pre-mRNA as well as among proteins that assemble on the pre-mRNA. The combination of these multiple weak interactions ultimately leads to the accurate recognition of exons by the spliceosome (Smith and Valcarcel 2000; Maniatis and Tasic 2002; Hertel 2008).

21.2.2   Control of Alternative Splicing

The formation of the exon-recognition complexes is subject to numerous controls. The first level of control is the variation of regulatory factor concentrations that often differ between cell types (Hanamura et al. 1998). In addition to this regulation, the activity of splicing regulatory proteins is regulated by post-translational modifications, especially reversible phosphorylation. Phosphorylation influences the binding affinity between splicing regulatory proteins and can therefore control the formation of exon-recognition complexes. It is not fully understood what controls the phosphorylation of splicing factors, but numerous studies indicate that well-established signaling routes, such as mitogen activated protein (MAP)-kinase, Ca^{2+}-dependent kinase, and cyclic adenosine monophosphate (cAMP)-dependent pathways are involved. This model explains why numerous cellular stimuli, such as receptor activation or membrane depolarization can influence splice site selection. This paradigm also implies that tissue- and cell-specific differences in alternative splicing could be due to different signals that cells receive (reviewed in Stamm 2008; Shin and Manley 2004; Stamm 2002).

21.3   Gene Structure and Processing of the 5-HT$_{2C}$ Pre-mRNA

21.3.1   Gene Structure

The 5-HT$_{2C}$ receptor pre-mRNA is composed of at least seven exons (Fig. 21.1a). The entire human 5-HT$_{2C}$R gene from exon I to exon VI spans around 326 kilobase (kb) of DNA. Current databases annotate six exons and five introns. Recently, a human 5-HT$_{2C}$ receptor mRNA was identified that contains a novel 91 nucleotide long alternatively spliced exon in the 5′ untranslated region (UTR) between exon II and III (accession numbers M81778, DR003480). In addition, in mouse, a not previously annotated exon is located between Exon II and III of the mouse 5-HT$_{2C}$R gene (mRNA accession number BC098327). This exon is unrelated to the human one. All introns of the serotonin receptor are larger than the mammalian average of 3365 nt (Thanaraj and Stamm 2003). The largest one is intron IV that spans about 117 kb. The 5′ UTR of the gene comprises exons I and II and a part of exon III. The coding region of the 5-HT$_{2C}$R cDNA spans from part of exon III to exon VI (Fig. 21.1b).

In contrast to many G-protein-coupled receptors that do not contain introns in their coding regions, the coding sequence of the human 5-HT$_{2C}$R gene is interrupted by three introns. The long 3′ UTR of the receptor is generated by exon VI. Key features of the gene structure, such as the positions of the intron–exon junctions as well as the promoter regions have been conserved between rodents and humans. This suggests that similar *cis*- and *trans*-acting elements regulate gene expression in both species. In its 5′ UTR, the 5-HT$_{2C}$ receptor gene hosts at least one snoRNA (HBI-36) between exons II and III and two putative miRNAs hsa-mir-1264 and hsa-mir-1298. However, the function of these RNAs remains to be determined. Theoretically, the three alternative exons can be combined to generate nine mRNA isoforms. Combining these nine mRNA isoforms with 32 variants generated by pre-mRNA editing predicts that the 5-HT$_{2C}$R gene can generate 288 mRNA isoforms. It is not clear whether all of these isoforms are actually generated. Detailed reverse transcription-polymerase chain reaction (RT-PCR) studies produced evidence for fragments of most of the isoforms

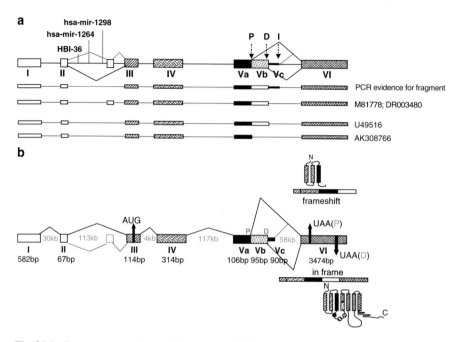

Fig. 21.1 Gene structure of serotonin receptors. (**a**) Overview of pre-mRNAs generated from the serotonin receptor. Exons are indicated by boxes or a thick line and roman numerals. Introns are indicated as lines. Exons that contribute to the open reading frame are shaded. Splicing patterns are indicated by lines. Annotated and predicted RNA isoforms are indicated underneath the gene structure. Exon V undergoes alternative splicing. The splice sites I to III are indicated. The proximal, distal, and intronic splice sites I (P, D, and I) are indicated. (**b**) Protein isoforms and splicing events. The structure of the human 5-HT$_{2C}$R receptor is indicated as in panel (**a**), and the intron and exon lengths are indicted. P and D indicate the location of proximal and distal splice sites, respectively. An arrow pointing towards AUG indicates the start codon of the longest open reading frame. UAA(P) and UAA(D) are the stop codons resulting from usage of the proximal and distal splice site. Exons are indicated as boxes with roman numerals. The mRNA isoforms and the resulting proteins are schematically shown. The shading reflects contribution of the different exons to the protein composition. Exon Vb encodes the second intracellular loop, and due to RNA editing, three amino acids, which are indicated as dots, are variable

(Burns et al. 1997; Niswender et al. 1999; Wang et al. 2000; Fitzgerald et al. 1999; Hackler et al. 2006). As the expression of these isoforms changes due to environmental stress and is altered in disease processes (Englander et al. 2005), their regulation has been studied in detail.

21.3.2 *Alternative Pre-mRNA Processing of Exon Vb*

Exon V of the 5-HT$_{2C}$ pre-mRNA has three alternative 5′ splice sites, proximal P, distal D, and intronic I. Different authors use different nomenclatures for these sites, for example, splice site I, II, and III. Their location is shown in Fig. 21.1a. The usage of these sites define exon Vb and Vc (Fig. 21.1a). There is RT-PCR evidence for the usage of the intronic site I (also named *donor site III*), but a full-length mRNA containing exon Vc has not been described. However, the RT-PCR data suggest that RNAs with this exon exist (Flomen et al. 2004). Alternative usage of exon Vb is documented in mRNA databases. Exon Vb is of special importance for the regulation of the 5-HT$_{2C}$ pre-mRNA, since it is located in the coding region of the protein and is targeted by both RNA editing and alternative splicing. Exon Vb encodes the part of the protein that composes its second intracellular loop. This loop couples to the G protein and is therefore essential for signaling. The exon is 95 nucleotides long and, thus, can not accommodate an integer number of the three-nucleotide long codons. Therefore, skipping of this exon causes a frameshift and leads to the generation of a truncated receptor mRNA. It is not clear whether this mRNA is translated into a nonfunctional receptor that lacks the G-protein-coupling ability or undergoes nonsense-mediated RNA decay. Nonsense mediated decay is a posttranscriptional surveillance mechanism that can degrade mRNA with premature stop codons (Neu-Yilik and Kulozik 2008). Since there is no published evidence for the expression of short 5-HT$_{2C}$ mRNA forms, it is likely that skipping of exon Vb leads to the degradation of the resulting mRNA. Exon Vb is localized in a predicted extended secondary structure that harbors editing sites described above (Fig. 21.2). In addition to the five edited adenine residues in exon Vb, a sixth site in exon Vc has been described (Flomen et al. 2004). Editing of these nucleotides changes the encoded protein in the second intracellular loop that is involved in receptor signaling.

In addition to influencing the encoded protein, editing influences the splicing of exon Vb when tested in cell culture based assays (Flomen et al. 2004; Kishore and Stamm 2006a). The distal splice site (TAGgtaaat) deviates on two positions from the consensus, "optimal" 5′ splice site (AAGgtaagt). The splice site is not used when analyzed in reporter gene assays. In these types of assays, a fragment of the gene is transfected into cells and the splicing pattern is analyzed by subsequent RT-PCR (reviewed in Tang et al. 2005; Stoss et al. 1999). These analyses showed that when the splice site is mutated into the consensus sequence, the exon is included into the mRNA (Kishore and Stamm 2006a). However, even after this splice site is mutated into a perfect mammalian consensus, exon Vb is still predominantly skipped. This suggested the existence of a splicing silencer element in the exon. Such a silencing element was bioinformatically predicted in the exon (Kishore and Stamm 2006a). This splicing silencing element partially overlaps with the adenine editing sites. Their conversion from adenine to inosine in the editing process weakens the splicing silencer, and as a result, exon Vb is now included (Flomen et al. 2004; Kishore and Stamm 2006a). These experiments were performed in transfected cell lines that have a different set of splicing regulatory proteins than differentiated neurons that express the 5-HT$_{2C}$ receptor under physiological conditions. When the effect of editing of exon Vb was studied in knock-in mouse models no effect on

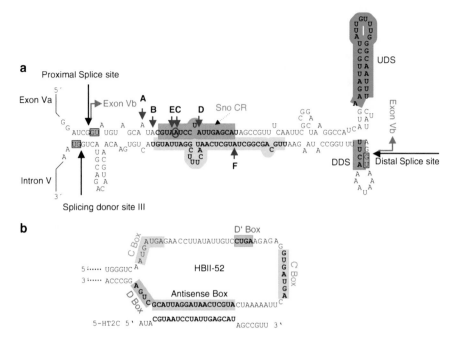

Fig. 21.2 Predicted RNA structure of exon Vb and complementarity to the snoRNA HBII-52. (**a**) RNA structure: The sequence of the serotonin receptor 5-HT$_{2C}$ exon V is indicated. *Arrows* point to the five nucleotides that are edited from A to I (A–E editing sites). Structural elements are indicated by shading: UDS upstream distal splice site, DDS downstream distal splice site, *sno-CR* snoRNA complementarity region. The base at editing site C fulfills the requirements to be 2'-O-methylated (circle) by the snoRNA. Exon Vb is indicated by *arrows*; the GU nucleotides of the proximal, distal, and donor site III are boxed. (**b**) Complementarity between exon Vb and the snoRNA HBII-52. The snoRNA is shown in 3'–5' orientation to illustrate the base pairing with the sno-CR of the serotonin receptor. The structural elements of the snoRNA, the C, C' box, D, D' box, and antisense box are indicated

exon inclusion was found. In these mice, the wild-type exon Vb was substituted with an exon that had adenine-to-guanine mutations at the five editing sites (Kawahara et al. 2008). This suggests the presence of an activity in neurons that promotes exon Vb inclusion and can overwrite the influence of a splicing silencer.

Exon Vb harbors an 18 nt sequence that exhibits full complementarity to the antisense box of a small nucleolar RNA (snoRNA), HBII-52. The official name of this snoRNA is SNORD115, but the historical name HBII-52 for "second human brain library, clone number 52," is widely used in the literature. The snoRNA HBII-52 is expressed only in neurons (Cavaille et al. 2000), and its coexpression with 5 HT$_{2C}$ splicing reporter constructs promotes exon Vb inclusion in cell culture based assays (Kishore and Stamm 2006a). This finding suggested that the snoRNA HBII-52 is involved in splice site selection of the 5-HT$_{2C}$ receptor pre-mRNA.

21.4   Small Nucleolar Rnas

21.4.1   Traditional View of Snornas

Small nucleolar RNAs are small, noncoding RNAs. Based on their sequence, they can be subdivided into C/D and H/ACA snoRNAs. C/D box snoRNAs have C and D boxes as characteristic sequence elements that help form the snoRNA particle, or snoRNP. Small nucleolar RNAs reside in introns from which they are released during pre-mRNA processing of the hosting gene. During the splicing reaction, the intron is released as a lariat structure that contains a 2′ to 5′ phosphodiester bond at the adenosine branch point. The lariat is opened by a debranching enzyme, and the intron is typically rapidly removed by nuclease action. If the intron contains snoRNAs, proteins associate with the snoRNA sequences and prevent their further degradation. As a result, the snoRNA that "resides" in an intron is released as a snoRNP.

A major function attributed to C/D box snoRNAs is their guiding of 2′-O-methylation in ribosomal, transfer, and snRNAs. The guiding activity of the snoRNAs is achieved by the formation of a specific RNA:RNA duplex between the snoRNA and its target. Most snoRNAs contain two regions to interact with other RNAs, the antisense boxes. Each antisense box exhibits sequence complementarity to its target and forms a short, transient double strand with it. On the target RNA, the nucleotide base pairing with the snoRNA nucleotide positioned five nucleotides downstream of the snoRNA D box is methylated on the 2′-O-hydroxyl group (reviewed in Matera et al. 2007). Several snoRNAs show complementarity towards pre-rRNA, but the rRNA is not 2′-O-methylated at the predicted positions (Steitz and Tycowski 1995). Recently, numerous C/D box snoRNAs were discovered that show no sequence complementarity to other RNAs, suggesting that C/D box snoR-NAs might have function other than 2′-O-methylation (Filipowicz and Pogacic 2002). Furthermore, bioinformatics analysis of high-throughput sequencing data provided evidence for shorter forms of snoRNAs, suggesting that snoRNAs could be precursors for miRNA-like nuclear RNAs (Taft et al. 2009; Scott et al. 2009).

21.4.2   Small Nucleolar Rnas Missing in the Prader–Willi Syndrome

HBII-52 is a neuron-specific C/D box snoRNA (Cavaille et al. 2000). The snoRNA resides in the SNURF-SNRPN locus (Fig. 21.3a). Loss of expression from this locus is the most likely cause for Prader–Willi syndrome (PWS) (Butler et al. 2006).

In contrast to most other C/D box snoRNAs, HBII-52 contains only a single antisense box. This antisense box exhibits sequence complementarity to exon Vb of the 5-HT$_{2C}$ receptor and is phylogenetically highly conserved (Nahkuri et al. 2008). In most species that express HBII-52, clusters of this snoRNA are expressed.

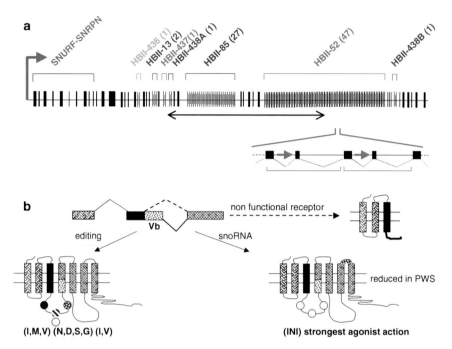

Fig. 21.3 5-HT$_{2C}$ receptor and HBII-52 snoRNA. (**a**) Small nucleolar RNAs are generated from the SNURF-SNRPN locus. The *SNURF-SNRPN* gene, located in the Prader–Willi critical region is schematically shown. Exons are shown as boxes, introns as horizontal lines. The bracket labeled SNURF-SNRPN indicates the protein-coding part of the pre-mRNA. Small nucleolar RNAs are located between noncoding exons and are shown as shorter and lighter vertical lines. Their names are indicated on top of the gene structure. A magnification of two snoRNAs (arrows) from the HBII-52 cluster is shown as an enlargement. A thick line with double *arrows* shows the microdeletion that causes PWS in one patient (Sahoo et al. 2008). (**b**) Model for HBII-52 action on the serotonin receptor pre-mRNA. Exon Vb is alternatively spliced. Skipping of the exon leads to a premature stop codon and is predicted to generate a nonfunctional receptor. However, it is not clear whether this nonfunctional protein is formed, since the mRNA is predicted to undergo nonsense mediated decay. Editing of the pre-mRNA promotes exon Vb inclusion but changes the amino acid composition in the second intracellular loop that couples to the G protein. The three positions that are changed by editing are indicated with circles, and the possible amino acids are shown below. The snoRNA promotes inclusion of the exon without editing and leads to a receptor that has the amino acids INI (isoleucine, asparagine, isoleucine) at the positions. This receptor has the strongest agonist response, and its expression is reduced in people with Prader–Willi syndrome

For example, humans posses 47 HBII-52 copies flanked by noncoding exons. The exon–snoRNA–exon structure is arranged in tandem (Fig. 21.3a). Each of the copies exhibits sequence complementarity to exonVb of the Serotonin receptor $_{2C}$ mRNA. The corresponding mouse MBII-52 snoRNAs are expressed throughout the mouse brain. They are most abundant in hippocampus, but absent in choroid plexus and some thalamic nuclei (Rogelj et al. 2003). The expression of MBII-52 is upregulated during early memory consolidation in the hippocampus (Rogelj et al. 2003). This indicates that snoRNAs could convey a "memory signal."

ExonVb of the serotonin receptor is expressed throughout the brain, but it is mostly absent in the choriod plexus. In contrast, the snoRNA MBII-52 is expressed throughout the brain, but is absent in the choroid plexus, indicating a correlation between HBII-52 expression and exon Vb usage. We therefore analyzed the influence of HBII-52 on 5-HT$_{2C}$R pre-mRNA processing. In these experiments, the snoRNA and a 5-HT$_{2C}$ receptor splicing reporter were transiently expressed in cell lines. This allowed to analyze the influence of HBII-52 on exon Vb inclusion of reporter. We found that HBII-52 promotes usage of exon Vb. In addition, a transient binding of the snoRNA to the 5 HT$_{2C}$ receptor RNA could be detected by UV cross-link followed by RT-PCR (Kishore and Stamm 2006a). These experiments suggested that the snoRNA HBII-52 can promote inclusion of exon Vb and could be a factor that "overwrites" the splicing silencing element in this exon.

21.4.3   Mechanism of Snorna Acting on the Serotonin Receptor

The mechanism used by snoRNPs to change splice site selection is not obvious, as snoRNPs are mainly located in the nucleolus and splicing takes place in the nucleoplasma. However, snoRNAs are generated in the nucleoplasma during the splicing reaction and intron release. They can therefore contact pre-mRNA during their generation. In addition, snoRNAs share some proteins with the splicing machinery. For example, the 15.5 K protein that was originally identified as part of a the C/D box snoRNP complex where it binds to a conserved kink turn binds also to a similar structure in the U4 snRNA where it interacts with the splicing factor hPrp31 (Liu et al. 2007). This raises the possibility that the 15.5 K protein bound to snoRNPs interferes with the U4/U6 rearrangement during the splicing reaction by interacting with hPrp31.

Insight into the mechanism came from experiments that analyzed the RNAs from a single HBII-52 expression unit by RNase protection analysis, which directly quantifies the expressed RNAs. The data indicated that the HBII-52 expression unit generates several RNAs. Mutation studies showed that these shorter RNAs are only made when their precursor snoRNA contains intact C and D boxes. This indicates that they are most likely generated by further processing of the snoRNA and were therefore termed *psnoRNAs* (for processed snoRNAs) (Kishore et al. 2010). The main product of the 48 HBII-52 expressing units that are missing in PWS is therefore not a C/D box snoRNA but a psnoRNA that lacks several nucleotides at the ends. This shorter version lacks the stem of the snoRNA that is crucial for the assembly of a functional snoRNPs but still contains the antisense box needed for targeting to pre-mRNA. In addition to this form, three other shorter RNAs (60 to 37 nt) could be detected. The psnoRNAs were present in the nucleoplasma, where they could interact with pre-mRNA. The analysis of the protein composition showed that the RNAs associate with hnRNPs commonly implicated in splice site regulation but not with the known structural C/D box snoRNA proteins or the 2'-*O*-methylase (Kishore et al. 2010). This strongly suggest that HBII-52 has a role different from C/D box snoRNAs that function in 2'-*O*-methylation of RNA.

As the major HBII-52 psnoRNA form still contains the antisense box that targets the serotonin receptor exon Vb sequence, it is likely that this RNA form brings processing factors to this exon, similar to a bifunctional oligonucleotide. Studies of miRNAs, Dscam selector RNA or U1 snRNAs showed that RNA:RNA interactions can tolerate multiple mismatches towards their targets. This indicates that HBII-52 could also regulate other splicing events.

A bioinformatic analysis predicted about 220 alternative exons that have evolutionary conserved sites that exhibit limited complementarity to the antisense box. Five of these exons were regulated by HBII-52 expression (Kishore et al. 2010). In each of the identified exons there were three mismatches between the 18 nt antisense element and the target RNA, which is reminiscent of U1, where the majority of 5' splice sites has a mismatch in two of the nine possible bases.

Together, these data indicate that HBII-52 derived RNAs promote exonVb inclusion either by recruiting other pre-mRNA processing factors to 5-HT$_{2C}$ pre-mRNA or by forming novel small RNAs that interfere with the splicing process.

21.5   Prader–Willi Syndrome

21.5.1   *Organization of Snornas in the Prader–Willi Critical Region*

The HBII-52 snoRNA resides in the SNURF-SNRPN locus (for small RNP in neurons [SmN] upstream reading frame). The SNURF-SNRPN locus whose loss of expression causes PWS spans more than 460 kb and contains at least 148 exons (Runte et al. 2001) (Fig. 21.3a). The locus is maternally imprinted, meaning that only the allele from the father is expressed. The loss of expression from this paternal allele, most frequently through genomic deletions causes PWS. The SNURF-SNRPN locus has a complex architecture. Ten exons in the 5' part of the gene are transcribed into a bicistronic mRNA that encodes the SNURF (SmN upstream reading frame) and the SmN (small RNP in Neurons) protein. Adjacent to the SNURF-SNRPN gene is a bipartite imprinting center (IC) that silences most maternal genes of the PWS critical region. The large 3' UTR region of the SNURF-SNRPN gene harbors clusters of the C/D box snoRNAs HBII-85 and HBII-52 that are present in at least 24 and 47 copies, respectively. In addition, the region harbors single copies of other C/D box snoRNAs: HBII-13, HBII-436, HBII-437 HBII-438A, and HBII-438B. The snoRNAs are flanked by noncoding exons and show a large degree of conservation between mammalian species. Their flanking, noncoding exons are only poorly conserved, suggesting that the snoRNAs are important, not the flanking exons (Fig. 21.3a). The snoRNAs in this locus show a tissue-specific expression. Expression of HBII-52 could be detected only in brain, whereas other snoRNAs are expressed also in nonbrain tissues (reviewed in Kishore and Stamm 2006b). A link between PWS and snoRNAs was supported by the recent finding that a microdeletion containing only snoRNAs, including all the HBII-85 and most of the HBII-52 cluster leads to the PWS phenotype (Sahoo et al. 2008) (Fig. 21.3a). The accumulated data strongly suggest that the loss of snoRNA expression from the SNURF-SNRNP region plays a decisive role in PWS.

21.5.2 Features of Prader–Willi Syndrome

Prader–Willi syndrome is a congenital disease with an incidence of about 1 in 8,000 to 20,000 live births. Prader–Willi syndrome is the most common genetic cause of marked obesity in humans. The excess weight causes type II diabetes as a major complication. This makes PWS the most frequent genetic cause for type II diabetes (Butler et al. 2006). Early PWS is characterized by a failure to thrive, feeding difficulties and hypogonadism. Later, the patients are characterized by short stature and develop mild to moderate mental retardation, behavioral problems and hyperphagia that leads to severe obesity. Children with PWS show low levels of growth hormone, Insulin-like growth factor 1 (IGF-I), and insulin as well as elevated levels of ghrelin (Eiholzer et al. 1998a, b; Cummings et al. 2002) and often exhibit central adrenal insufficiency (de Lind van Wijngaarden et al. 2008). Subsequently, growth hormone substitution was approved for treatment of children with PWS (Carrel et al. 2006). The growth hormone substitution represents to date the only pharmaceutical therapy and is successful in weight management.

21.5.3 Serotonin Receptor in Patients with Prader–Willi Syndrome

HBII-52 snoRNA expression has not been detected in people with Prader–Willi syndrome (Kishore and Stamm 2006a; Cavaille et al. 2000). This raises the question whether these patients also exhibit an imbalance in the serotonin mRNA isoforms. Therefore, brain samples from patients with Prader–Willi syndrome were analyzed by RT-PCR. By using primers that specifically recognize the five editing sites, the 5-HT$_{2C}$ mRNA isoforms could be compared between Prader–Willi patients and age-matched controls. The experiments indicated a significantly reduced editing in three of the four tested sites when the same brain regions are compared. This suggests a reduced expression of the nonedited pre-mRNA in Prader–Willi patients. However, due to the intrinsic problems with human tissues, the protein composition could not be analyzed. These data support a model where HBII-52 promotes exon inclusion of the nonedited exon Vb (Fig. 21.3b).

Exon Vb encodes the second intracellular loop of the receptor that couples to G proteins. Editing changes the amino acids in this loop and alters the receptor properties. The nonedited version of the receptor shows the highest efficacy towards serotonin. Changing the amino acids through editing generates multiple receptor isoforms with 10- to 100-fold lower efficacy (Wang et al. 2000). These conclusions have been derived from studies performed cell culture. The studies have been recently recapitulated in a knock-in mouse model (Kawahara et al. 2008). Mouse lines were engineered where the wild-type 5-HT_{2C} receptor allele was exchanged with an allele generating only the fully edited receptor version. In the mutant allele all five adenosine residues were replaced by guanine residues, which are similar to the inosine residues generated by editing. The mice harboring the fully edited VGV allele of serotonin receptor 2C showed growth retardation, an increased energy expenditure, and a constitutively activated sympathetic nervous system, as well as hyperphagia (Kawahara et al. 2008). These mice did not express any nonedited INI allele. These findings were confirmed by a second, similar mouse model (Morabito et al. 2010). Together, these data indicate the importance of the physiological balance of 5-HT_{2C} splice variants. Some aspects of the phenotype, such as hyperphagia and growth retardation, correlate with the PWS phenotype (Morabito et al. 2010).

21.5.4   Is There a Link Between HBII-52 Expression and Hunger Control?

A molecular link between a defect in the 5-HT_{2C} production and PWS is an attractive hypothesis, as the 5-HT_{2C} receptor plays a crucial role in hunger control and satiety, which is the major problem in PWS. Since HBII-52 promotes the generation of the most active receptor, it acts like a "genetic agonist" of the serotonin receptor. The administration of selective $5\text{-HT}_{2C}R$ agonists, such as d-fenfluramine has a strong appetite-suppressing effect (Vickers et al. 2001). Underlining the importance of the 5-HT_{2C} receptor for hunger control, the mouse knockout of $5\text{-HT}_{2C}R$ is hyperphagic and develops obesity. Expression of the $5\text{-HT}_{2C}R$ in the arcuate nucleus, a major hunger control center reverses the hyperphagic phenotype (Xu et al. 2008). As mentioned above, when a mutant of the receptor that represents the fully edited $5\text{-HT}_{2C}R$ is expressed in knockout mice, the resulting mice remain hyperphagic (Kawahara et al. 2008; Morabito et al. 2010). Collectively, the data strongly support a model where the loss of HBII-52 causes a loss of the mRNA isoform that encodes the most active form of the receptor, which is necessary for proper hunger control.

21.6   Conclusion

Together, the data indicate that the snoRNA HBII-52 contributes to alternative splicing regulation of the 5-HT_{2C} receptor. The exact molecular mechanism of the regulation is currently unveiled. The HBII-52 snoRNA is processed into smaller snoRNA fragments (Kishore et al. 2010) that most likely directly influence the 5-HT$_{2C}$ receptor pre-mRNA. Finally, it is likely HBII-52 controls other pre-mRNAs, and it is therefore expected that the 5-HT$_{2C}$ receptor mRNA is not the only deregulated RNA in PWS. However, since mouse models of the 5-HT$_{2C}$ receptor mutants recapitulate some aspects of PWS, we expect that the dysregulation of the receptor plays a decisive role in this disease.

Acknowledgments   The laboratory is supported by the Deutsche Forschungsgemeinschaft (DFG), the European Union (EURASNET), and the NIH.

References

Biamonti G, Caceres JF (2009) Cellular stress and RNA splicing. Trends Biochem Sci 34:146–153.

Burns CM, Chu H, Rueter SM, et al (1997) Regulation of serotonin-2C receptor G-protein coupling by RNA editing. Nature 387:303–308.

Butler MG, Hanchett JM, Thompson TE (2006) Clinical Findings and Natural History of Prader-Willi Syndrome. In: Butler MG, Lee PDK, Whitman BY, eds. Management of Prader–Willi Syndrome. New York: Springer, pp. 3–48

Carrel AL, Lee PDK, Mogul HR (2006) Growth Hormone and Prader-Willi Syndrome. In: Butler MG, Lee PDK, Whitman BY, eds. Management of Prader-Willi Syndrome. New York: Springer, pp. 201–241

Cavaille J, Buiting K, Kiefmann M, et al (2000) Identification of brain-specific and imprinted small nucleolar RNA genes exhibiting an unusual genomic organization. Proc Natl Acad Sci USA 97:14311–14316.

Cummings DE, Clement K, Purnell JQ, et al (2002) Elevated plasma ghrelin levels in Prader–Willi syndrome. Nat Med 8:643–644.

de Lind van Wijngaarden RF, Otten BJ, Festen DA, et al (2008) High prevalence of central adrenal insufficiency in patients with Prader–Willi syndrome. J Clin Endocrinol Metab 93:1649–1654.

Eiholzer U, Stutz K, Weinmann C, et al (1998) Low insulin, IGF-I and IGFBP-3 levels in children with Prader–Labhart–Willi syndrome. Eur J Pediatr 157:890–893.

Eiholzer U, Gisin R, Weinmann C, et al (1998) Treatment with human growth hormone in patients with Prader-Labhart–Willi syndrome reduces body fat and increases muscle mass and physical performance. Eur J Pediatr 157:368–377.

Englander MT, Dulawa SC, Bhansali P, et al (2005) How stress and fluoxetine modulate serotonin 2C receptor pre-mRNA editing. J Neurosci 25:648–651.

Filipowicz W, Pogacic V (2002) Biogenesis of small nucleolar ribonucleoproteins. Curr Opin Cell Biol 14:319–327.

Fitzgerald LW, Iyer G, Conklin DS, et al (1999) Messenger RNA editing of the human serotonin 5-HT2C receptor. Neuropsychopharmacology 21:82 S–90 S.

Flomen R, Knight J, Sham P, et al (2004) Evidence that RNA editing modulates splice site selection in the 5-HT2C receptor gene. Nucleic Acids Res 32:2113–2122.

Hackler EA, Airey DC, Shannon CC, et al (2006) 5-HT(2C) receptor RNA editing in the amygdala of C57BL/6 J, DBA/2 J, and BALB/cJ mice. Neurosci Res 55:96–104.

Hanamura A, Caceres JF, Mayeda A, et al (1998) Regulated tissue-specific expression of antagonistic pre-mRNA splicing factors. RNA 4:430–444.

Hertel KJ (2008) Combinatorial control of exon recognition. J Biol Chem 283:1211–1215.

Jepson JE, Reenan RA (2008) RNA editing in regulating gene expression in the brain. Biochim Biophys Acta 1779:459–470.

Jurica MS, Moore MJ (2003) Pre-mRNA splicing: awash in a sea of proteins. Mol Cell 12:5–14.

Kawahara Y, Grimberg A, Teegarden S, et al (2008) Dysregulated editing of serotonin 2C receptor mRNAs results in energy dissipation and loss of fat mass. J Neurosci 28:12834–12844.

Kishore S, Stamm S (2006) The snoRNA HBII-52 regulates alternative splicing of the serotonin receptor 2C. Science 311:230–232.

Kishore S, Stamm S (2006) Regulation of alternative splicing by snoRNAs. Cold Spring Harb Symp Quant Biol LXXI:329–334

Kishore S, Khanna A, Zhang Z, et al (2010) The snoRNA MBII-52 (SNORD 115) is processed into smaller RNAs and regulates alternative splicing. Hum Mol Genet 19:1153–1164.

Liu S, Li P, Dybkov O, et al (2007) Binding of the human Prp31 Nop domain to a composite RNA-protein platform in U4 snRNP. Science 316:115–120.

Maniatis T, Tasic B (2002) Alternative pre-mRNA splicing and proteome expansion in metazoans. Nature 418:236–243.

Matera AG, Terns RM, Terns MP (2007) Non-coding RNAs: lessons from the small nuclear and small nucleolar RNAs. Nat Rev Mol Cell Biol 8:209–220.

Mehler MF, Mattick JS (2007) Noncoding RNAs and RNA editing in brain development, functional diversification, and neurological disease. Physiol Rev 87:799–823.

Morabito MV, Abbas AI, Hood JL, et al (2010) Mice with altered serotonin 2C receptor RNA editing display characteristics of Prader-Willi Syndrome. Neurobiol Dis 39:169–180.

Nahkuri S, Taft RJ, Korbie DJ, et al (2008) Molecular evolution of the HBII-52 snoRNA cluster. J Mol Biol 381:810–815.

Neu-Yilik G, Kulozik AE (2008) NMD: multitasking between mRNA surveillance and modulation of gene expression. Adv Genet 62:185–243.

Nishikura K (2006) Editor meets silencer: crosstalk between RNA editing and RNA interference. Nat Rev Mol Cell Biol 7:919–931.

Niswender CM, Copeland SC, Herrick-Davis K, et al (1999) RNA editing of the human serotonin 5-hydroxytryptamine 2C receptor silences constitutive activity. J Biol Chem 274:9472–9478.

Pan Q, Shai O, Lee LJ, et al (2008) Deep surveying of alternative splicing complexity in the human transcriptome by high-throughput sequencing. Nat Genet 40:1413–1415.

Robberson BL, Cote GJ, Berget SM (1990) Exon definition may facilitate splice site selection in RNAs with multiple exons. Mol Cell Biol 10:84–94.

Rogelj B, Hartmann CE, Yeo CH, et al (2003) Contextual fear conditioning regulates the expression of brain-specific small nucleolar RNAs in hippocampus. Eur J Neurosci 18:3089–3096.

Runte M, Huttenhofer A, Gross S, et al (2001) The IC-SNURF-SNRPN transcript serves as a host for multiple small nucleolar RNA species and as an antisense RNA for UBE3A. Hum Mol Genet 10:2687–2700.

Sahoo T, del Gaudio D, German JR, et al (2008) Prader–Willi phenotype caused by paternal deficiency for the HBII-85 C/D box small nucleolar RNA cluster. Nat Genet 40:719–721.

Scott MS, Avolio F, Ono M, et al (2009) Human miRNA precursors with box H/ACA snoRNA features. PLoS Comput Biol 5:e1000507.

Sharp PA (1994) Split genes and RNA splicing. Cell 77:805–815.

Shin C, Manley JL (2004) Cell signalling and the control of pre-mRNA splicing. Nat Rev Mol Cell Biol 5:727–738.

Smith CW, Valcarcel J (2000) Alternative pre-mRNA splicing: the logic of combinatorial control. Trends Biochem Sci 25:381–388.

Stamm S (2002) Signals and their transduction pathways regulating alternative splicing: a new dimension of the human genome. Hum Mol Genet 11:2409–2416.

Stamm S (2008) Regulation of alternative splicing by reversible phosphorylation. J Biol Chem 283:1223–1227.

Stamm S, Riethoven JJ, Le Texier V, et al (2006) ASD: a bioinformatics resource on alternative splicing. Nucleic Acids Res 34:D46–D55.

Stark H, Luhrmann R (2006) Cryo-electron microscopy of spliceosomal components. Annu Rev Biophys Biomol Struct 35:435–457.

Steitz JA, Tycowski KT (1995) Small RNA chaperones for ribosome biogenesis. Science 270:1626–1627.

Stoss O, Stoilov P, Hartmann AM, et al (1999) The in vivo minigene approach to analyze tissue-specific splicing. Brain Res Prot 4:383–394.

Taft RJ, Glazov EA, Lassmann T, et al (2009) Small RNAs derived from snoRNAs. Rna 15:1233–1240.

Tang Y, Novoyatleva T, Benderska N, et al (2005) Analysis of Alternative Splicing In Vivo using Minigenes. In: Westhof E, Bindereif A, Schön A, Hartmann E, eds. Handbook of RNA Biochemistry. Weinheim: Wiley-VCH, pp. 755–782.

Thanaraj TA, Clark F (2001) Human GC-AG alternative intron isoforms with weak donor sites show enhanced consensus at acceptor exon positions. Nucleic Acids Res 29:2581–2593.

Thanaraj TA, Stamm S (2003) Prediction and statistical analysis of alternatively spliced exons. Prog Mol Sub Biol 31:1–31.

Vickers SP, Dourish CT, Kennett GA (2001) Evidence that hypophagia induced by d-fenfluramine and d-norfenfluramine in the rat is mediated by 5-HT2C receptors. Neuropharmacology 41:200–209.

Wahl MC, Will CL, Luhrmann R (2009) The spliceosome: design principles of a dynamic RNP machine. Cell 136:701–718.

Wang Q, O'Brien PJ., Chen C-X, et al (2000) Altered G protein-coupling functions of RNA editing isoform and splicing variant serotonin 2C receptors. J Neurochem 74:1290–1300.

Wang ET, Sandberg R, Luo S, et al (2008) Alternative isoform regulation in human tissue transcriptomes. Nature 456:470–476.

Werry TD, Loiacono R, Sexton PM, et al (2008) RNA editing of the serotonin 5HT2C receptor and its effects on cell signalling, pharmacology and brain function. Pharmacol Ther 119:7–23.

Xu Y, Jones JE, Kohno D, et al (2008) 5-HT2CRs expressed by pro-opiomelanocortin neurons regulate energy homeostasis. Neuron 60:582–589.

Chapter 22
The Role of 5-HT$_{2C}$ Receptor in Epilepsy

Rita Jakus and Gyorgy Bagdy

22.1 Types and Mechanisms of Epilepsy

Various types of seizure can be recognized on the basis of the nature and distribution of the abnormal discharge. Epileptic seizures are generally classified into focal (partial) and generalized, though there is some overlap and many varieties of each. Partial seizures are those in which the discharges begin locally and often remain localized. The clinical manifestation of focal seizures varies depending on the origin of epileptic discharges (so-called epileptic focus); the attack may involve motor, sensory, autonomic and psychic symptoms. They can be divided into two categories, simple (if consciousness is not lost) or complex (if consciousness is lost, due to the involvement of the reticular formation). The commonest form of focal epilepsies (50–60%) originates from the temporal lobe (TLE), on the basis of mesial temporal (hippocampal) sclerosis, associated with previous complex febrile seizures, or developmental abnormalities and dysembryoplastic neuroepithelioma. Generalized seizures involve the whole brain, including the reticular system, thus producing abnormal electrical activity throughout both hemispheres. Immediate loss of consciousness is characteristic of generalized seizures. Two important categories are tonic–clonic (grand mal) seizures and absence epilepsy. According to the casual aetiology, epilepsy is also classified into idiopathic and symptomatic (Browne and Holmes 2003; Engel 2004; Holmes 2004).

In general, excitation will naturally tend to spread throughout a network of interconnected neurons but is normally prevented from doing so by inhibitory mechanisms. Thus, epileptogenesis can arise if excitatory transmission is facilitated or inhibitory transmission is reduced. In certain respects epileptogenesis resembles long-term potentiation, and similar types of use-dependent synaptic plasticity may be involved (Kullmann et al. 2000). The highly interconnected networks of the mammalian forebrain can generate a wide variety of synchronized activities, including those underlying epileptic seizures, which often appear

---

G. Bagdy (✉)

Department of Pharmacology and Pharmacotherapy, Semmelweis University, Budapest, Hungary

G. Di Giovanni et al. (eds.), *5-HT$_{2C}$ Receptors in the Pathophysiology of CNS Disease*,   429
The Receptors 22, DOI 10.1007/978-1-60761-941-3_22,
© Springer Science+Business Media, LLC 2011

as a transformation of otherwise normal brain rhythms. The simplest form of epileptiform activity in these structures is the interictal spike, a synchronized burst of action potentials generated by recurrent excitation, followed by a period of hyperpolarization, in a localized pool of pyramidal neurons. Seizures can also be generated in response to a loss of balance between excitatory and inhibitory influences and can take the form of either tonic depolarizations or repetitive, rhythmic burst discharges, either as clonic or spike wave activity, again mediated by both intrinsic membrane properties and synaptic interactions.

22.2 Animal Models for Epilepsies

22.2.1 Focal Epilepsy Models

An important early work to elucidate the neuronal disturbances underlying ictal discharge, and interictal electroencephalogram (EEG) spikes, used the acute cat neocortical penicillin focus (Matsumoto and Ajmone-Marsan 1964). In vivo extracellular recording within the experimental epileptic focus revealed normally firing neurons, except for abnormal burst discharges during the EEG spike, and cessation of firing during the aftercoming slow wave. Intracellular recordings demonstrated that the bursting was caused by membrane depolarization of unusually high amplitude and prolonged duration, which was called a *paroxysmal depolarization shift* (Matsumoto and Ajmone-Marsan 1964). This was followed by a prolonged high amplitude after hyperpolarization during which normal action potentials were inhibited. In most cases, the paroxysmal depolarization shift appeared to reflect an abnormal Ca^{2+} current of dentrites and soma, associated with continuous Na^+ action potentials at the axon hillock for as long as the depolarization persisted. A high percentage of neurons within the experimental epileptic focus participated synchronously in these transient events, producing the negative EEG deflection characteristic of the interictal spike. It could be concluded that ictal onset in the penicillin focus model appears to be due to dysinhibition.

Subsequent in vivo studies on patients with mesial temporal lobe epilepsy, using depth electrodes, including microelectrodes, revealed that only 5% of the recorded neurons in humans demonstrate this behavior (Babb et al. 1973), compared with over 90% in the experimental penicillin focus (Matsumoto and Ajmone-Marsan 1964). Consequently, synchrony in bursting neurons is difficult to demonstrate in patients. Furthermore, the EEG pattern of ictal onset in the human epileptogenic hippocampus typically does not consist of recruiting rhythm but rather pronounced repetitive high amplitude spike wave discharges (Velasco et al. 2000), sometimes resembling the EEG pattern of absence seizures, where the prominent slow wave represents enhanced inhibition (Giaretta et al. 1987).

The kindling model may approximate the human condition more closely than directly evoked seizure models. Since its discovery by Goddard et al. (1969), the kindling phenomenon has been used as a chronic animal model of TLE. In limbic kindling, low-intensity electrical stimulation of certain regions of the limbic system, such as the amygdala, with implanted electrodes normally produces no seizure response. If a brief period of stimulation is repeated daily for several days, the response gradually increases until very low levels of stimulation will evoke a full seizure and eventually seizures begin to occur spontaneously (Goddard et al. 1969). Once produced, the kindled state persists indefinitely. Kindling is still widely accepted as a functional model in which the altered neuronal response develops in the absence of gross morphological damage, such as that seen in many other epilepsy models. High doses of neurotoxins such as kainate or pilocarpine are administered systematically to produce status epilepticus (continual recurrent seizures). In that case, it is not the status epilepticus that is of interest, but the delayed appearance of spontaneous seizures. Thus, this model has been named "post-status epilepticus models of TLE" (Morimoto et al. 2004). Unilateral lesions more similar to unilateral human mesial temporal lobe epilepsy can be produced with intrahippocampal injections of kainic acid (Bragin et al. 1999). After transient intense stimulation, spontaneous seizures begin to occur 2–4 weeks later, and again continue indefinitely. The hippocampal lesions in all of these models consist of the same cell loss, axon sprouting, synaptic reorganization, and gliosis seen in human hippocampal sclerosis with mesial temporal lobe epilepsy, but the maximal cell loss in patients is in the CA1 region and in the CA3 region in rats. Dube et al. (2006) directly address the causal relationship of long febrile seizures and development of TLE. Focal neocortical epilepsy has been modeled using topical application of toxic metals such as cobalt, aluminium, and iron (Ward 1972). These lesions produce focal seizures for prolonged periods of time and may mimic human focal epilepsies due to scars and hemosiderin deposits caused by trauma, stroke, and vascular malformations.

Kharatishvili and coworkers studied the electrophysiological, behavioral, and structural features of posttraumatic epilepsy induced by severe, nonpenetrating lateral fluid-percussion brain injury in rats (D'Ambrosio et al. 2005; Kharatishvili et al. 2006).

There is also evidence that neurotrophins, particularly brain-derived neurotrophic factor (BDNF), may play a role in epileptogenesis. Brain-derived neurotrophic factor, which acts on a membrane receptor tyrosine kinase, enhances membrane excitability and also stimulates synapse formation. Production and release of BDNF is increased in the kindling models, and there is also evidence for its involvement in human epilepsy. Models of focal epilepsies are summarized in Table 22.1.

22.2.2 Generalized Epilepsy and Seizure Models

There are many different animal models of generalized epilepsy (see Table 22.2). Electrical stimulation (maximal electroshock [MES]), chemoconvulsants (kainic acid, pilocarpine, pentylentatrazol, bicuculline, picrotoxin, flurothyl), and genetic models have been used to generate generalized seizures.

Table 22.1 Focal epilepsy models (Modified from Engel 2004; Bagdy et al. 2007; Löscher 2002)

Electrical stimulation – acute seizures and chronic (kindling) models

Topical convulsants that block inhibition (penicillin, bicuculline, picrotoxin, pentylentetrazol, strychnine) – acute or chronic seizures

Topical convulsants that enhance excitation (carbachol, kainic acid) – acute seizures and chronic (kindling) models

Freeze lesion or partially isolated cortical slab (with intact vascularization) – chronic seizures

Metals (Al_2O_3, cobalt) – chronic seizures

Kindling (electrical or chemical) – chronic model

Experimental febrile seizures – (acute and) chronic model

Posttraumatic epilepsy (PTE) induced by lateral fluid percussion brain injury – chronic model

Hippocampal sclerosis (kainic acid, pilocarpine, poststatus epilepticus models of temporal lobe epilepsy (TLE) – chronic models

Focal dysplasia (neonatal freeze, prenatal radiation, methylazoxymethanol) – neonatal, prenatal treatment models

Table 22.2 Generalized epilepsy models (Modified from Holmes 2004; Bagdy et al. 2007)

Genetics

Genetically epilepsy prone rats (GEPRs)

Mongolian gerbil

DBA/2 J mouse

Chromosome 4 congenic mice

Photosensitive baboon

$5\text{-}HT_{2C}$ receptor knock out mice

$5\text{-}HT_{1A}$ receptor knock out mice

Generalized tonic-clonic seizure

Maximal electroshock (MES)

Chemoconvulsant

Glutamate agonists

• Kainic acid

• Pilocarpine

GABA antagonists

• Pentylentetrazol

• Bicuculline

• Picrotoxin

Other

Absence

Genetic absence rats from Strasburg (GAERS)

Wistar Albino Glaxo/Rijswijk (WAG/Rij)

Low-dose penthylenetetrazol

Cholesterol biosynthesis inhibitor (AY-994) – atypical

Mice

• Tottering

• Stargazer

• Lethargic

• Slow-wave epilepsy mice

• Mocha mouse

• Ducky mouse

Maximal electroshock and chemoconvulsants are useful in generating acute seizures but are not adequate models for studying epilepsy. While spontaneous recurrent seizures can occur following status epilepticus induced by chemoconvulsants, the seizures usually are partial with secondary generalization.

The basic underlying mechanism in absence seizure, characterized by the generation of intermittent synchronized bursting of neurons separated by periods of normal function, arises from thalamus–cortex interaction. During spike and wave discharges, a large number of neurons oscillate between short periods of excitation, corresponding to the spike, and longer periods of inhibition, corresponding to the slow wave component of the spike and wave complex (Gloor 1978). Both in vivo and in vitro studies have demonstrated the neuronal circuit that generates the oscillatory thalamocortical burst firing observed during absence seizures (Snead 1995). Within the thalamus, sleep spindles are generated as a recurrent interaction between thalamocortical and thalamic reticular cells (Steriade et al. 1993). It has been suggested, based on the resemblance in the EEG and the similar circadian pattern, that spike-wave discharges (SWD) are modified sleep spindles (Steriade et al. 1993; McCormick 2002). Spike-wave discharges never develop in genetic rat absence models with lesions in their thalamic reticular nucleus, which is considered the primary pacemaker of spindle rhythm. In idiopathic generalized epilepsy, spindles transform to SWD pattern; in other words, SWD represent the epileptic variant of the complex thalamocortical system function, which is the substrate of non-REM sleep EEG phenomena (Halasz et al. 2002). The circuit comprises only three neuronal populations: cortical pyramidal neurons, thalamocortical relay neurons, and neurons of nucleus reticularis thalami (NRT). The principal synaptic connections of the thalamocortical circuit include glutamatergic fibers between neocortical pyramidal cells and the NRT, γ-aminobutyric acid (GABA)-ergic fibers from NRT neurons that activate GABA$_A$ and GABA$_B$ receptors on thalamic relay neurons, and recurrent collateral GABA-ergic fibers from NRT neurons that activate GABA$_A$ receptors on adjacent NRT neurons. Thalamic relay neurons and NRT neurons possess low-threshold, transient Ca^{2+} channels (T-type Ca2+ channels) that allow them to exhibit a burst firing mode, followed by an inactive mode. Mild depolarization of these neurons is sufficient to activate these T-type Ca2+ channels and to allow the influx of extracellular Ca^{2+}. Further depolarization produced by Ca^{2+} inflow often exceeds the threshold for firing a burst of action potentials. After T-type Ca2+ channels are activated, they become inactivated quite quickly; hence the name transient. T-type Ca2+ channels require a long, intense hyperpolarization to remove their activation (deinactivation). The required hyperpolarization can be provided by GABA$_B$ receptors that are present on thalamic relay neurons. The interplay between GABA$_B$-mediated inhibition and the low threshold T-type calcium channel therefore plays a critical role in generating the oscillating hyperpolarization/depolarization activity seen in the thalamus. In animal absence models, GABA$_B$ receptor agonists produce an increase in seizure frequency (by facilitating deinactivation of T-type Ca2+ channels), whereas GABA$_B$ receptor antagonists reduce seizure frequency.

As noted earlier, collateral GABA-ergic fibers from the NRT neurons activate GABA$_A$ receptors on adjacent NRT neurons. Activating GABA$_A$ receptors in NRT, therefore, results in reduction of GABA-ergic output to the thalamic relay neurons. α-Amino-3-hydroxy-5-methyl-4-isoxazole-propionate (AMPA) receptors appear to mediate fast transmission in the thalamus. According to data derived from the computational models that produce spike-wave oscillations, both AMPA and $GABA_B$ receptors interact during the generation and propagation of the oscillations (Destexhe et al. 1996). Thalamocortical relay cells can elicit AMPA receptor-mediated excitatory postsynaptic potentials in NRT, while the latter neurons elicit $GABA_A$ and $GABA_B$-related inhibitory postsynaptic potentials, generating continuous oscillations. Kaminski et al. (2001) found that $GABA_B$ and AMPA receptors-mediated neurotransmission regulates the occurrence of SWD in an additive manner in WAG/Rij rats.

22.3 Serotonin Concentration and Release in Epilepsy

The functions of 5-HT on the central nervous system (CNS) are numerous and appear to involve control of appetite, sleep, memory and learning, temperature regulation, mood, behavior, cardiovascular function, muscle contraction, endocrine regulation, maturation of neuronal and glial cells and synaptic connections, and, as discussed here, epilepsy (Barnes and Sharp 1999; Hoyer et al. 1994; Azmitia and Whitaker-Azmitia 1999). The idea that there may be a link between serotonin and the inhibition of epilepsy was suggested as early as 1957 by Bonnycastle et al. (1957). In their study they demonstrated that a series of anticonvulsants, including phenytoin, elevated brain serotonin levels.

Serotonergic neurotransmission modulates a wide variety of experimentally induced seizures and is involved in the enhanced seizure susceptibility observed in rodents genetically prone to epilepsy (Kilian and Frey 1973; Buterbaugh 1978; Przegalinski 1985; Hiramatsu et al. 1987; Dailey et al. 1992; Gerber et al. 1998; Filakovszky et al. 1999). Generally, agents that elevate extracellular 5-HT levels, such as 5-hydroxytryptophan and 5-HT reuptake blockers, inhibit both focal (limbic) and generalized seizures (Löscher 1984; Prendiville and Gale 1993; Yan et al. 1994). Conversely, depletion of brain 5-HT lowers the threshold to audiogenically, chemically, and electrically evoked convulsion (Browning et al. 1978; Statnick et al. 1996).

Seizure models in which 5-HT appears to play a prominent role include audiogenic seizures in genetically epilepsy prone rats (GEPR3 and 9) (Daily et al. 1989; Jobe et al. 1973), audiogenic seizures in DBA/2 J mice (Brennan et al. 1997; Tecott et al. 1995), limbic and thalamic seizures in cats (Wada et al. 1993), sensory-induced seizures in El mice (Hiramatsu et al. 1987), seizures induced by focal injection of bicuculline into area tempestas of deep prepiriform cortex of rats (Prendiville and Gale 1993), maximal electroshock seizures in mice and rats (Buterbaugh 1978; Browning et al. 1978), and in WAG/Rij rats, an accepted genetic model of human absence epilepsy (Gerber et al. 1998; Filakovszky et al. 1999; Löscher and Schmidt 1988; Coenen et al. 1992; Van Luijtelaar et al. 2002; Jakus et al. 2003; Graf et al. 2004), handling-induced convulsion in chromosome 4 congenic mice (Reilly et al. 2008). The rat model of myoclonic epilepsy is associated with a profound loss of serotonin throughout the brain (except in the striatum) (Welsh et al. 2002). In a recent study, the effect of 5-HT in an atypical absence model of AY-9944-treated rats was studied (Bercovici et al. 2006). The increased levels of 5-hydroxyindoleactic acid and 5-HT, as well as the altered rates of serotonin turnover, suggest that serotonergic transmission may be perturbed in the AY-9944-treated rats. Mazarati et al. (2006) reported that serotonin depletion by parachloroamphetamine increased the severity of limbic status epilepticus (summarized in Table 22.3).

22.4 5-HT$_{2C}$ Receptors and Epilepsy

22.4.1 Localization and Functions

Several neurons involved in excitatory and inhibitory neurotransmission, like GABA-ergic or glutamaterg neuronal cells, express 5-HT$_{2C}$ receptors. Thus, 5-HT$_{2C}$ receptors regulate the activity of neuronal networks involved either in epileptogenesis or in propagation (Table 22.4). The highest concentration of 5-HT$_{2C}$ receptors is in the choroid plexus, but 5-HT$_{2C}$ binding sites are widely distributed and present in areas of the cortex (olfactory nucleus, pyriform, cingulated, and retrospenial), limbic system (nucleus accumbens, hippocampus, amygdala,), and the basal ganglia (caudate nucleus, substantia nigra). They are present in the pyriform cortex and substantia nigra, demonstrated by 5-HT$_{2C}$ receptor-mediated electrophysiological response in these regions (Sheldon and Aghajanian 1991). Activation of the 5-HT$_{2C}$ receptor increases phospholipase PLC activity in the choroid plexus of various species, but these receptors function also via phospholipase A$_2$ (PLA) and phospholipase D, even more, agonist-independent association with β-arrestins was reported for nonedited and partially edited 5-HT$_{2C}$ receptors (Millan et al. 2008) (Table 22.4).

22.4.2 Pharmacological Studies

There is evidence for the 5-HT$_{2C}$ receptor-mediated excitation of neurons in several brain regions (Sheldon and Aghajanian 1991). Extracellular and intracellular recordings revealed that 5-HT-induced inhibition of burst firing in NRT is mediated through 5-HT$_2$ receptors with the possible involvement of 5-HT$_{2C}$ (formerly 5-HT$_{1C}$) receptors (Pape and McCormick 1989; McCormick and Wang 1991; McCormick 1992) as measured by neocortical high-voltage spindle activity (Jäkälä et al. 1995). In a recent study it was demonstrated that activation of 5-HT$_{2C}$ receptors by receptor

Table 22.3 Effects of activation of 5-HT$_{2C}$ receptors on epilepsy in various animal models

Agonist	Effect	Model	Reference
mCPP (5-HT$_{2C,2B}$)	↑ Seizure threshold	**Mouse electroshock test, pentylenetetrazole test in rat and mouse**	Upton et al. (1998)
	↓ SWD	**WAG/Rij**	Jakus et al. (2003)
	↑ Survival rate and duration	**Pentylenetetrazole-induced seizures in mice**	Kecskemeti et al. (2005)
	↑ Seizure latency	**Pentylenetetrazole-induced seizures in mice**	Kecskemeti et al. (2005)
DOI (5-HT$_{2A,B,C}$)	↑ Seizure development	*AM kindling*	Wada et al. (1996)
TFMPP (5-HT$_{1A,B,C,}$ 5-HT$_{2C}$)	↓ Threshold	**MES in mice**	Hoyer and Middlemiss (1989)
TFMPP (5-HT$_{1A,B,C,}$ 5-HT$_{2C}$)	↓ Motor seizure	*Pilocarpine model of TLE*	Hernandez et al. (2002)

Models are distinguished by typeface: *focal* (italic) and **generalized** (boldface).

agonists or by increase in endogenous 5-HT by administration of selective serotonin reuptake inhibitors (SSRIs), inhibits SWD, although this latter inhibitory effect is not significant at basal 5-HT levels. In contrast, activation of postsynaptic 5-HT$_{1A}$ receptors by receptor agonists or increase in extracellular 5-HT concentration promotes SWD, especially when inhibitory 5-HT$_{2C}$ receptors are inactivated by subtype-selective receptor antagonists (Jakus et al. 2003).

Watanabe et al. (2000) assessed the effects of the acute administration of various 5-HT receptor antagonists on hippocampal partial seizures generated by low-frequency electrical stimulation in rats. The results in this model suggest that stimulation of 5-HT$_{1A}$, 5-HT$_{2A,C}$, and 5-HT$_3$ receptors does not alter the seizure threshold or severity and that the blockade of 5-HT uptake produced by low dose of fluoxetine appears to increase seizure threshold and decrease seizure severity. Upton et al. (1998) investigated the role of 5-HT$_{2C}$ and 5-HT$_{2B}$ receptors in the generation of pentylenetetrazol and electroshock-evoked seizures in rodents. The results indicate that the observed anticonvulsant effects of 1-(*m*-chlorophenyl)-piperazine (mCPP), a 5-HT$_{2C/2B}$ receptor agonist, are likely to be mediated by activation of 5-HT$_{2C}$ receptors. The effect of the subtype-selective 5-HT$_{2C}$ receptor antagonist SB 242084 on mCPP-induced inhibition of pentylenetetrazol-induced seizure provides further support for this conclusion (see Table 22.5) (Bagdy et al. 2007).

However, blockade of these receptors in mice (or rats) by SB 206553 or SB 242084 did not result in reduced seizure threshold characteristic of mutant mice deficient in 5-HT$_{2C}$ receptors, suggesting that in normal adult animals this receptor subtype may usually be subjected to only a low level of 5-HT tone. These findings are consistent with experiments demonstrating that the blockade of 5-HT$_{2C}$ receptors with the subtype-selective antagonist SB 242084 was not

Table 22.4 Distribution, cellular, electrophysiological, and neurochemical responses associated with the activation of 5-HT$_{2C}$ receptor-related epilepsy (Modified from Barnes and Sharp 1999; Bagdy et al. 2007)

Distribution/brain areas	Level	Response	Mechanism	References
Chorioid plexus, cortex (olfactory nucleus, pyriform, cingulate and retrosplenial), limbic system (nucleus accumbens, hippocampus, amygdala), basal ganglia (globus pallidus, substantia nigra) (Sheldon and Aghajanian 1991; Palacios et al. 1991; Radja et al. 1991; Rick et al. 1995)	Cellular/subcellular	PLC/Ca^{2+}/PKC (+), PLA$_2$/AA (+), PLD (+), NOS (+), gK$^+$ (-), gCl$^-$ (+), pERK (+), pAkt (+), gNa$^+$ (-)	Postsynaptic	Millan et al. (2008), Sanders-Bush et al. (1988), Julius et al. (1988)
	Electro physiological	Neuronal depolarization	Postsynaptic	Sheldon and Aghajanian (1991), Rick et al. (1995), Aghajanian (1995)
	Neuro chemical	GABA release (+)	Postsynaptic	Aghajanian (1995),
		Glutamate release (+)	Postsynaptic	Bobker (1994), Done
		NA/dopamine release (-)	Postsynaptic	and Sharp (1994)

associated with significant changes in the epileptiform activity, but pretreatment with this compound reversed the effects of mCPP on SWD in WAG/Rij rats (Jakus et al. 2003). Furthermore, SSRI antidepressants like fluoxetine and citalopram increase SWD when 5-HT_{2C} receptors are blocked by subtype-selective SB 242084. The inability of 5-HT_{2C} receptor antagonists to reduce seizure threshold in adult rodents is in contrast to the observed characteristics of mutant mice lacking the 5-HT_{2C} receptor (Tecott et al. 1995). Although other epilepsy models in addition to WAG/Rij rats have to be tested, these results together (Jakus et al. 2003; Upton et al. 1998) suggest that pharmacological blockade of the receptor and knockout of the receptor gene may result in somewhat different effects. This might be explained by developmental or neuroadaptive changes in the brain. In contrast, activation of 5-HT_{2C} receptors potentiates cocaine-induced seizures (O'Dell et al. 2000).

Other experiments using pharmacological probes have also implicated 5-HT_2 receptors in the development of amygdala-kindled limbic seizures (Wada et al. 1997) and the expression of electrically induced generalized seizures (Przegalinski et al. 1994) in rodents. Morita et al. (2005) examined whether the inhibition of serotonin transporter (SERT) contributes to cocaine- and other local anesthetics-induced convulsions, together with subtypes of 5-HT receptor subtypes are involved in the convulsion. For this porpuse cocaine, meprylcaine and lidocaine, all of which have different effects on SERT, were used as convulsants and the effects of SSRIs, specific agonists and antagonists of 5-HT receptor subtypes were evaluated in mice. Incidence of cocaine- and meprylcaine-induced convulsions was significantly reduced by $5\text{-HT}_{2A,2B,2C}$ antagonist, LY-53857, and 5-HT_{2C} antagonist, RS 102221. The threshold of cocaine and meprylcaine was significantly increased by both antagonists. 5-HT_{2A} antagonists (MDL 11939 and ketanserin) and 5-HT_{2B} antagonist (SB 204741), except at high doses, had little effect on cocaine- and meprylcaine-induced convulsions. None of these antagonists altered the parameters of lidocaine-induced convulsions. They also found, that pretreatment with fluoxetine, but not citalopram,

Table 22.5 Effect of the subtype-selective 5-HT_{2C} receptor antagonist SB 242084 (5 mg/kg) alone and on mCPP (1 mg/kg)-induced inhibition of pentylenetetrazol-induced seizure

Pretreatment	Treatment	Duration of survival[a] (s)	
		Mean	SEM
Vehicle	Vehicle	1,073	229
Vehicle	m-CPP	2,597[b]	263
SB 242084	Vehicle	569	85
SB 242084	m-CPP	1,080	278

[a]Duration of survival was measured from the administration of pentylenetetrazol (100 mg/kg, sc) until the death of the mice (male CFLP mice, 23 ± 1.5 g) (Bagdy et al. 2007)
[b]Significant compared with vehicle + vehicle, $p < 0.05$

increased the plasma concentration of lidocaine. Their results suggest that increase of serotonergic neuronal activity through 5-HT$_{2C}$ receptor stimulation was responsible for increased activity of local anesthetics-induced convulsions and support the involvement of this mechanism in cocaine-meprylcaine – but not in lidocaine-induced convulsions – through their direct inhibitory action on central SERT. The possible use of ligands of the 5-HT$_{2C}$ receptor is in the focus of interest and many molecular tools are available to further prove the consequences of 5-HT$_{2C}$ activation in epilepsy (see Table 22.6). In this regard, Isaac (2005) gave a summary of 5-HT$_{2C}$ ligands as possible tools and candidates for preclinical and clinical evaluation in epilepsies.

22.4.3   Genetically Modified Animals

Several knockout mouse models suggest a relationship between 5-HT and epilepsy. The audiogenic seizures syndrome is the first known defect caused by genetic manipulation of a 5-HT receptor subtype; this provides a robust model for examination of serotonergic mechanism in epilepsy. Mutant mice lacking the 5-HT$_{2C}$ receptor subtype are extremely susceptible to audiogenic seizures and are prone to spontaneous death from seizures, suggesting that serotonergic neurotransmission, mediated by 5-HT$_{2C}$ receptors, suppresses neuronal network hyperexcitability and seizure activity (Brennan et al. 1997; Tecott et al. 1995; Applagate and Tecott 1998). Fehr et al. (2002) developed a congenic strain that possesses a segment of chromosome 4 from the C57BL/6 J(B6) donor strain superimposed on a genetic background estimated to be >99% from the DBA/2 J (D2) mouse strain. The introduced segment spans the Mpdz gene. Mpdz encodes the multi-PDZ domain protein (MPDZ). Multi-PDZ domain protein's interaction with 5-HT$_{2C}$ receptors has been confirmed in vivo and is mediated via a PDZ domain in MPDZ (Becamel et al. 2001; Parker et al. 2003; 2002). This congenic strain exhibits significantly less severe alcohol withdrawal hyperexcitabilty than background strain mice. Reilly et al. (2008) compared the chromosome 4 congenic and background strains for their susceptibilities to CNS hyperexcitability in response to the selective 5-HT$_{2C}$ receptor antagonist SB 242084. They found that chromosome 4 congenic mice show significantly fewer severe convulsions in response to SB 242084 than background strain mice show. This is consistent with the hypothesis that allelic variation in Mpdz or a linked gene affects 5-HT$_{2C}$ receptor function and thereby influences convulsion severity (Tables 22.6 and 22.7).

In conclusion, genetic models, pharmacological studies, and all other data provide massive evidence for the role of 5-HT$_{2C}$ receptors in epileptogenesis and/or propagation. Actions of currently available drugs with significant affinity to this receptor or potential new molecules under development may provide effective tools in the treatment of a wide variety of different epilepsies.

Table 22.6 Effects of inhibition of 5-HT$_{2C}$ receptors on epilepsy in various animal models (Modified from Bagdy et al. 2007)

Antagonist	Effect	Model (generalized)	References
Mesulergine *(DA)/* (5-HT $_{2A,C}$)	↓ Effect of TFMPP	MES in mice	Hoyer and Middlemiss (1989), Van Wijngaarden et al. (1990)
Ritanserin	↓ Effect of TFMPP	MES in mice	Hoyer and Middlemiss (1989), Van Wijngaarden et al. (1990)
Ketanserin (5-HT$_{2A,B,C}$)	↓ Effect of TFMPP	MES in mice	Hoyer and Middlemiss (1989), Van Wijngaarden et al. (1990)
SB 206553 (5-HT$_{2B,C}$)	↓ Effect of mCPP	Mouse electroshock test, penthylenetetrasol test in rat and mouse	Upton et al. (1998)
SB 242084 (5-HT$_{2C}$)	↓ Effect of mCPP on SWD	WAG/Rij rat	Jakus et al. (2003)
SB 242084 (5-HT$_{2C}$)	↓ Effect of mCPP	Pentylenetetrazole-induced seizures in mice	Bagdy et al. (2007)
SB 242084 (5-HT$_{2C}$)	↓ Severity of convulsion	Chromosome 4 congenic mice	Reilly et al. (2008)

MES maximal electroshock, *WAG/Rij* Wistar albino Glaxo rats, bred in Rijswijk, *SWD* spike-wave discharges, *AM* amygdale, *TLE* temporal lobe epilepsy, *DOI* 1-(2,5-dimethoxy-4-iodophenyl)-2-aminopropane, *mCPP* *meta*-chlorophenylpiperazine HCl, *SB* 242084, 6-chloro-5-methyl-1-(2-[2-methylpyrid-3-yloxy]pyrid-5-yl)carbamoyl]indoline dihydrochloride, *TFMPP* 3-trifluoromethylphenylpiperazine monohydrochloride, *SB206553* 5-methyl-1-(3-pyridylcarbamoyl)-1,2,3,5-tetrahydropyrrolo [2,3-*f*]indole

Table 22.7 Effects of the Lack of 5-HT$_{2C}$ receptors on epilepsy-related parameters in mice

Effect	Model	References
↑ Audiogenic seizure	5-HT$_{2C}$ receptor knockout mice	Tecott et al. (1995)
↑ Spontaneous death from seizure	5-HT$_{2C}$ receptor knockout mice	Brennan et al. (1997), Applagate and Tecott (1998)

References

Aghajanian GK (1995) Electrophysiology of serotonin receptor subtypes, signal transduction pathways. In: Bloom FR, Kupfer DJ, eds. Psychopharmacology: The Fourth Generation of Progress. New York: Raven, pp. 1451–1459

Applagate CD, Tecott LH (1998) Global increases in seizure susceptibility in mice lacking 5-HT2C receptors: a behavioral analysis. Exp Neurol 154(2):522–530

Azmitia EC, Whitaker-Azmitia, EC (1999) Development and adult plasticity of serotoninergic neurons and their target. In: Baumgarten HG, Göther M, eds. Serotonergic Neurons and 5-HT Receptors in the CNS. Springer: New York, pp. 1–39

Babb TL, Carr E, Crandall PH (1973) Analysis of extracellular firing patterns of deep temporal lobe structures in man. Electroencephalogr Clin Neurophysiol 34(3):247–257.

Bagdy G, Kecskemeti V, Riba P, Jakus R (2007) Serotonin and epilepsy. J Neurochem 100:857–873.

Barnes NM, Sharp T (1999) A review of central 5-HT receptors and their function. Neuropharmacology 38:1083–1152.

Becamel C, Figge A, Poliak S, et al (2001) Interaction of serotonin 5-hydroxytryptamine type 2C receptors with PDZ10 of the multi-PDZ domain protein MUPP1. J Biol Chem 276:12974–12982

Bercovici E, Cortez MA, Wang X, Snead OC 3rd (2006) Serotonin depletion attenuates AY-9944-mediated atypical absence seizures. Epilepsia 47(2):240–246.

Bobker DH (1994) A slow excitatory postsynaptic potential mediated by 5-HT$_2$ receptors in nucleus prepositus hypoglossi. J Neurosci 14:2428–2434.

Bonnycastle DD, Giarman NJ, Paasonen MK (1957) Anticonvulsant compounds and 5-hydroxytryptamine in rat brain. Br J Pharmacol 12(2):228–231.

Bragin A, Engel Jr J, Wilson CL, Vizentin E, Mathern GW (1999) Electrophysiologic analysis of a chronic seizure model after unilateral hippocampal KA injection. Epilepsia 40(9):1210–1221.

Brennan TJ, Seeley WW, Kilgard M, Schreiner CE, Tecott LH (1997) Sound-induced seizures in serotonin 5-HT2C receptor mutant mice. Nat Genet 16:387–390.

Browne TR, Holmes GL (2003) Handbook of Epilepsy, 3rd ed. Philadelphia, PA: Lippinkott Williams &Wilkins.

Browning RA, Hoffman WE, Simonton RL (1978) Changes in seizure susceptibility after intracerebral treatment with 5,7-dihydroxy-tryptamine: role of serotonergic neurones. Ann NY Acad Sci 305:437–456.

Buterbaugh CG (1978) Effects of drugs modifying central serotonergic function on the response of extensor and non extensor rats to maximal electroshock. Life Sci 23:2393–2904.

Coenen AM, Drinkenburg WH, Inoue M, van Luijtelaar EL (1992) Genetic models of absence epilepsy, with emphasis on the WAG/Rij strain of rats. Epilepsy Res 12(2):75–86.

D'Ambrosio R, Fender JS, Fairbanks JP, et al (2005) Progression from frontal-parietal to mesial-temporal epilepsy after fluid percussion injury in the rat. Brain 128:174–188.

Dailey JW, Yan QS, Mishra PK, Burger RL, Jobe PC (1992) Effects of fluoxetine on convulsions and brain serotonin as detected by microdialysis in genetically epilepsy-prone rats. J Pharmacol 260:533–540.

Daily JW, Reigel CE, Mishura PK, Jobe PC (1989) Neurobiology seizure predisposition in the genetically epilepsy prone rat (Review). Epilepsy Res 3(1), 3–17.

Destexhe A, Bal T, McCormick DA, Sejnowski TJ (1996) Ionic mechanisms underlying synchronized oscillations and propagating waves in a model of ferret thalamic slices. J Neurophysiol 76(3):2049–2070.

Done CJ, Sharp T (1994) Biochemical evidence for regulation of central noradrenergic activity by 5-HT$_{1A}$ and 5-HT$_2$ receptors: microdialysis studies in the awake and anaesthetized rat. Neuropharmacology 33:411–421.

Dube C, Richichi C, Bender RA, Chung G, Litt B, Baram TZ (2006) Temporal lobe epilepsy after experimental prolonged febrile seizures: prospective analysis. Brain 129:911–922.

Engel J Jr (2004) Models of focal epilepsy. In: Hallett M, Phillips LH, Schomer DL II, Massey JM, eds. Advances in Clinical Neurophysiology. Clin Neurophysiol 57:392–399.

Fehr C, Shirly RL, Belknap JK, Crabbe JC, Buck KJ (2002) Congenic mapping of alcohol and pentobarbital withdrawal liability loci to a <1 centimorgan interval of murine chromosome 4: identification of Mpdz as a candidate gene. J Neurosci 22:3730–3738

Filakovszky J, Gerber K, Bagdy G (1999) A serotonin-1A receptor agonist and N-methyl-D-aspartate receptor antagonist oppose each others effects in a genetic rat epilepsy model. Neurosci Lett 261:89–92.

Gerber K, Filakovszky J, Halasz P, Bagdy G (1998) The 5-HT1A agonist 8-OH-DPAT increases the number of spike-wave discharges in a genetic rat model of absence epilepsy. Brain Res 807:243–245.

Giaretta D, Avoli M, Gloor P (1987) Intracellular recordings in pericruciate neurons during spike and wave discharges of feline generalized penicillin epilepsy. Brain Res 405(1):68–79.

Gloor P (1978) Generalized epilepsy with bilateral synchronous spike and wave discharge. New findings concerning its physiological mechanisms. Electroencephalogr Clin Neurophysiol Suppl 34:245–249.

Goddard GV, McIntyre DC, Leech CK (1969) A permanent change in brain function resulting from daily electrical stimulation. Exp Neurol 25(3):295–330.

Graf M, Jakus R, Kantor S, Levay G, Bagdy G (2004) Selective 5-HT$_{1A}$ and 5-HT$_7$ antagonists decrease epileptic activity in the WAG/Rij rat model of absence epilepsy. Neurosci Lett 359:45–48.

Halasz P, Terzano MG, Parrino L (2002) Spike-wave discharge and the microstructure of sleep-wake continuum in idiopathic generalised epilepsy, Neurophysiol Clin 32: 38–53.

Hernandez EJ, Williams PA, Dudek PA (2002) Effects of fluoxetine and TFMPP on spontaneous seizures in rats with pilocarpine-induced epilepsy. Epilepsia 43(11):1337–1345.

Hiramatsu MK, Kawanaga K, Kabuto H, Mori A (1987) Reduced uptake and release of 5-hydroxytryptamine and taurine in the cerebral cortex of epileptic El mice. Epilepsy Res 1:40–44.

Holmes GL (2004) Models for generalized seizures. In: Hallett M, Phillips LH, Schomer DL II, Massey JM, eds. Advances in Clinical Neurophysiology (Clinical Neurophysiology Supplement, vol 57). The Netherlands: Elsevier BV, pp. 415–438.

Hoyer D, Middlemiss DN (1989) Species differences in the pharmacology of terminal 5-HT autoreceptors in mammalian brain (Review). Trends Pharmacol Sci 10(4):130–132.

Hoyer D, Clarke DE, Fozrd JR, Harting PR, Mylecharane EJ, Saxena PR, Humpherey PPA (1994) VII. International union of pharmacology classification of receptors for 5-hydroxytryptamine (serotonin). Pharmacol Rev 46:157–203.

Isaac M (2005) Serotonergic 5-HT$_{2C}$ receptors as a potential therapeutic target for the design antiepileptic drugs. Curr Top Med Chem 5(1):59–64

Jäkälä P, Sirviö J, Koivisto E, Björklund M, Kaukua J, Riekkinen PJ (1995) Modulation of rat neocortical high-voltage spindle activity by 5-HT1/5-HT2 receptor subtype specific drugs. Eur J Pharmacol 282:39–55.

Jakus R, Graf M, Juhasz G, Gerber K, Levay G, Halasz P, Bagdy G (2003) 5-HT$_{2C}$ receptors inhibit, 5-HT$_{1A}$ receptors activate the generation of spike-wave discharges in a genetic rat model of absence epilepsy. Exp Neurol 182, 964–972.

Jobe PC, Picchioni AL, Chin L (1973) Role of brain 5-hydroxytryptamine in audiogenic seizure in the rat. Life Sci 13(1):1–1.

Julius D, MacDermott AB, Axel R, et al (1988) Molecular characterisation of a functional cDNA encoding the serotonin$_{1C}$ receptor. Science 244:558–564

Kamiński RM, Van Rijn CM, Turski WA, Czuczwar SJ, Van Luijtelaar G (2001) AMPA and GABA(B) receptor antagonists and their interaction in rats with a genetic form of absence epilepsy. Eur J Pharmacol 430(2–3):251–259.

Kecskemeti V, Rusznák Z, Riba P, et al (2005) Norfluoxetine and fluoxetine have similar anticonvulsant and Ca^{2+} channel blocking potencies. Brain Res Bull 67:112–132.

Kharatishvili I, Nissinen JP, McIntosh TK, Pitkanen A (2006) A model of posttraumatic epilepsy induced by lateral fluid-percussion brain injury in rats. Neuroscience 140:685–697.

Kilian M, Frey HH (1973) Central monoamines and convulsive thresholds in mice and rats. Neuropharmacology 12:681–692.

Kullmann DM, Asztely F, Walker MC (2000) The role of mammalian ionotropic receptors in synaptic plasticity: LTP, LTD and epilepsy (Review). Cell Mol Life Sci 57(11):1551–1561.

Löscher W (1984) Genetic animal models of epilepsy as a unique resource for the evaluation of anticonvulsant drugs (Review). Methods Find Exp Clin Pharmacol 6(9):531–547

Löscher W (2002) Animal models of epilepsy for the development of antiepileptogenic and disease modifying drugs. A comparison of the pharmacology of kindling and post-status epilepticus models of temporal lobe epilepsy. Epilepsy Res 50:105–123.

Löscher W, Schmidt D (1988) Which animal models should be used in the search for new antiepileptic drugs? A proposal based on experimental and clinical considerations. Epilepsy Res 2:145–181.

Matsumoto H, Ajmone-Marsan C (1964) Cellular mechanisms in experimental seizures. Science 144:193–194.

Mazarati AM, Baldwin RA, Shinmei S, Sankar R (2006) In vivo interaction between serotonin and galanin receptors types 1 and 2 in the dorsal raphe: implication for limbic seizures. J Neurochem 95:1495–1503.

McCormick DA (1992) Neurotransmitter actions in the thalamus and cerebral cortex. J Clin Neurophysiol 9:212–223

McCormick DA (2002) Cortical and subcortical generators of normal and abnormal thythmicity. Int Rev Neurobiol 49:99–114.

McCormick DA, Wang Z (1991) Serotonin and noradrenaline excite GABAergic neurones of guinea-pig and cat nucleus reticularis thalami. J Physiol 442:235–255.

Millan MJ, Marin P, Bockaert J, la Cour CM (2008) Signaling at G-protein-coupled serotonin receptors: recent advances and future research directions. Trends Pharmacol Sci 29(9):454–64.

Morimoto K, Fahnestock M, Racine RJ (2004) Kindling and status epilepticus models of epilepsy: rewiring the brain. Prog Neurobiol 73:1–60.

Morita K, Hamamoto M, Arai S, et al (2005) Inhibition of serotonin transporters by cocaine and meprylcaine through 5-HT$_{2C}$ receptor stimulation facilitates their seizure activities. Brain Res 1057(1–2):153–160

O'Dell LE, Li R, George FR, Ritz MC (2000) Molecular serotonergic mechanisms appear to mediate genetic sensitivity to cocaine-induced convulsions. Brain Res 863(1–2):213–224.

Palacios JM, Waeber C, Mengod G, et al (1991) Autoradiography of 5-HT receptors: a critical appraisal. Neurochem Int 18:17–25

Pape HC, McCormick DA (1989) Noradrenaline and serotonin selectively modulate thalamic burst firing by enhancing a hyperpolarisation-activated cation current. Nature 340:715–718.

Parker LL, Backstrom JR, Sanders-Bush E, Shiieh BH (2003) Agonist-induced phosphorylation of the serotonin 5-HT2C receptor regulates its interaction with multiple PDZ protein 1. J Biol Chem 278:21576–21583

Prendiville S, Gale K (1993) Anticonvulsant effect of systemic fluoxetine on focally-evoked limbic motor seizures in rats. Epilepsia 34:381–384.

Przegalinski E (1985) Monoamines and the pathophysiology of seizure disorders. In: Frey HH, Janz D, eds. Handbook of Experimental Pharmacology. Berlin: Springer-Verlag, pp. 101–137.

Przegalinski E, Baran J, Siwanowicz J (1994) Role of 5-hydroxytryptamine receptor subtypes in the 1-[3-(trifluoromethyl)phenyl] piperazine-induced increase in threshold for maximal electroconvulsions in mice. Epilepsia 35:889–894

Radja F, Laporte AM, Daval G, et al (1991) Autoradiography of serotoin receptor subtypes in the central nervous system. Neurochem Int 18:1–15

Reilly MT, Milner LC, Shirley RL, Crabbe JC, Buck KJ (2008) 5-HT$_{2C}$ and GABA$_B$ receptors influence handling-induced convulsion severity in chromosome 4 congenic and BBA/2 J background strain mice. Brain Res 1198:124–131.

Rick CE, Stanford IM, Lacey MG (1995) Excitation of rat substantia nigra pars reticulate neurons by 5-hydroxytryptamine in vitro: evidence for a direct action mediated by 5-hydroxytryptamine$_{2C}$ receptors. Neuroscience 69:903–913

Sanders-Bush E, Burris KD, Knoth K (1988) Lysergic acid diethylamide and 2,5-dimethoxy-4-methylamphetamine are partial agonists at serotonin receptors linked to phosphoinositide hydrolysis. J Pharmacol Exp Ther 246:924–928

Sheldon PW, Aghajanian GK (1991) Excitatory responses to serotonin (5-HT) in neurons of the rat piriform cortex: evidence for mediation by 5-HT1C receptors in pyramidal cells and 5-HT2 receptors in interneurons. Synapse 9(3):208–218.

Snead OC 3rd (1995) Basic mechanisms of generalized absence seizures. Ann Neurol 37:146–157.

Statnick MA, Maring-Smith ML, Clough RW, et al (1996) Effects of 5,7-dihydroxy-tryptamine on audiogenic seizures in genetically epilepsy-prone rats. Life Sci 59:1763–1771.

Steriade M, McCormick DA, Sejnowski J (1993) Thalamocortical oscillacions in sleeping and arousal brain. Science 262:679–685.

Tecott LH, Sun LM, Akana SF, Strack AM, Lowenstein DH, Dallman MF, Julius D (1995) Eating disorder and epilepsy in mice lacking 5-HT$_{2C}$ serotonin receptors. Nature 374:542–546.

Upton N, Stean T, Middlemiss D, Blackburn T, Kennett G (1998) Studies on the role of 5-HT2C and 5-HT2B receptors in regulating generalised seizure threshold in rodents. Eur J Pharmacol 359:33–40.

Van Luijtelaar EL, Drinkenburg WH, van Rijn CM, Coenen AM (2002) Rat models of genetic absence epilepsy: what do EEG spike-wave discharges tell us about drug effects? Methods Find Exp Clin Pharmacol 24(suppl D):65–70.

Van Wijngaarden I, Tulp MT, Soudijn W (1990) The concept of selectivity in 5-HT receptor research (Review). Eur J Pharmacol 188(6):301–312.

Velasco M, Velasco F, Velasco AL, Jimenez F, Brito F, Marquez I (2000) Acute and chronic electrical stimulation of the centromedian thalamic nucleus: modulation of reticulo-cortical systems and predictor factors for generalized seizure control (Review). Arch Med Res 31(3): 304–315.

Wada Y, Nakamura M, Hasegawa H, Yamaguchi N (1993) Intra-hippocampal injection of 8-hydroxy-2-(di-n-propylamino)tetralin (8-OH-DPAT) inhibits partial and generalized seizures induced by kindling stimulation in cats. Neurosci Lett 159(1–2):179–182.

Wada Y, Shiraishi J, Nakamura M, Koshino Y (1996) Biphasic action of the histamine precursor L-histidine in the rat kindling model of epilepsy. Neurosci Lett 204(3):205–208.

Wada Y, Shiraishi J, Nakamura M, Koshino Y (1997) Role of serotonin receptor subtypes in the development of amygdaloid kindling in rats. Brain Res 747:338–342

Ward AA Jr (1972) Topical convulsant metals. In: Purpura DP, Penry JK, Tower DB, Woodbury DM, Walter RD, eds. Experimental Models of Epilepsy: A Manual for the Laboratory Worker., New York: Raven, pp. 13–35

Watanabe K, Ashby Jr CR, Katsumori H, Minabe Y (2000) The effect of the acute administration of various selective 5-HT receptor antagonists on focal hippocampal seizures in freely-moving rats. Eur J Pharmacol 398:239–246.

Welsh JP, Placantonakis DG, Warsetsky SI, Marquez RG, Bernstein L, Aicher SA (2002) The serotonin hypothesis of myoclonus from the perspective of neuronal rhythmicity. Adv Neurol 89:307–329.

Yan QS, Jobe PC, Cheong JH, Ko KH, Dailey JW (1994) Role of serotonin in the anticonvulsant effect of fluoxetine in genetically epilepsy prone rats. Naunyn-Schmiedebergs. Arch Pharmacol 350:149–152.

Chapter 23
The Role of Serotonin on Attentional Processes and Executive Functioning: Focus on 5-HT$_{2C}$ Receptors

Eleftheria Tsaltas and Vasileios Boulougouris

23.1 Introduction

Attention refers to the processes determining an organism's receptivity to external or internal excitation and hence the probability that it will engage in the processing of that excitation (Parasuraman 1998). Although it is often related as a cognitive function, it is distinct in encompassing a multitude of manifestations that underlie and sustain the activity of the other cognitive and behavioral performance in several ways: through the selection and integration of sensory inputs, which is essential for efficient learning and remembering as well as for the organization of appropriate responses. Impaired attentional processing may therefore become manifested as inattention, distractibility, memory impairment, confusion, perseveration, or disinhibition. Recognition of the diversity of attention has led to the identification of three distinct fundamental qualities: selection, enabling the allocation of priority to certain informational elements to the exclusion of others; vigilance, referring to the capacity for attentional persistence over time; and control, which optimizes performance, for example, by inhibition of concurrent activities (Parasuraman 1998; Robbins 2002, 2005).

Attempts to uncover neural mechanisms through which brain serotonin systems influence attentional processes as well as other executive functions are complicated by the heterogeneity of the receptors through which serotonin acts. At least 14 distinct subtypes of serotonin (5-hydroxytryptamine, or 5-HT) receptors are expressed within the central nervous system (Barnes and Sharp 1999). They are highly diverse in respect to their structures, gene regulation, primary effect or mechanisms, regional and subcellular expression patterns and physiological actions. However, the multiplicity of 5-HT receptors provides an opportunity for a fine functional dissection of brain serotonin systems, one receptor at a time.

V. Boulougouris (✉)
Experimental Psychology Laboratory, Department of Psychiatry, Athens University Medical School, Eginition Hospital, 74, Vas. Sofias Ave., 11528 Athens, Greece
e-mail: vboulougouris@googlemail.com

G.Di Giovanni et al. (eds.), *5-HT$_{2C}$ Receptors in the Pathophysiology of CNS Disease*, 445
The Receptors 22, DOI 10.1007/978-1-60761-941-3_23,
© Springer Science + Business Media, LLC 2011

Progress in this area has been facilitated by the development of relatively selective pharmacological tools and by molecular genetic techniques enabling the generation of animals with planned 5-HT receptor gene mutations. In order to ascertain neuroanatomical and neurochemical specificity of experimental interventions, it is necessary to resort to the use of experimental animal models. This endeavor has been facilitated by the current availability of comparable cross-species tests of cognitive function. These enable the identification of common neural substrates that subserve similar functions across species, increasing the likelihood that the same cognitive functions are being studied in each species.

In this chapter, the contribution of the serotonergic system to basic operations such as vigilance, shifting, and executive control are surveyed with emphasis to a prominent central serotonin receptor subtype – the 5-HT_{2C} receptor. Following a brief description of the anatomy of 5-HT_{2A} and 5-HT_{2C} receptor, the survey is focused on evidence from experimental animals. It encompasses data generated by four different experimental conditions, each of which centers on a specific aspect of the attentional and executive function.

The first paradigm is the five-choice serial reaction time task, which provides a direct measure of sustained attention and bears good analogy to the human continuous performance test (CPT), a traditional index of human vigilance. The second paradigm is attentional set shifting and reversal learning, which has been used to decompose the types of processes engaged by tests of attentional flexibility such as the Wisconsin Card Sort Test (WCST). The third paradigm is the reinforced spatial alternation measuring memory, cognitive flexibility, as well as persistent behavior. Finally, the signal attenuation paradigm models certain components of executive control, including attention and inhibition. In each case, the types of operation that are measured by the given paradigm will be defined. Then, the role of the serotonergic systems in the neurochemical modulation of its behavioral output will be examined, focusing on the contribution of the 5-HT_{2C} receptor subtype. In conclusion, reference to clinical implications for neurological and neuropsychiatric disorders will be made.

23.2   5-HT Receptor Subtypes

The true complexity of the serotonergic system is revealed when it is considered that over 14 different types of 5-HT receptor, assigned to one of seven families (5-HT_{1-7}), have currently been identified and that the number is set to rise (Barnes and Sharp 1999; Hoyer et al. 2002). Investigation of the possibility that these different receptors mediate different functions within the 5-HT system has only begun more recently with the advent of selective pharmaceutical compounds that can distinguish between the different receptor subtypes, and this avenue of research is constantly growing with the continuous development of more selective agents. Given the increasing interest in the serotonergic system in relation to psychiatric disorders and the escalating number of drug targets available, a comprehensive review of this work would be lengthy undertaking. Discussion that is more detailed will therefore be limited to 5-HT_{2A} and 5-HT_{2C} receptor subtypes, with the caveat that this is not a comprehensive delineation of the 5-HT receptors that may be implicated in the regulation of attentional processes and executive functions.

23.2.1 5-HT_{2A} Receptor Subtypes

There are currently three members of the 5-HT_2 receptor family (5-HT_{2A}, 5-HT_{2B}, 5-HT_{2C}) all of which are coupled positively to phospholipase C and mobilize intracellular calcium. Investigation of the function of individual members of the 5-HT_2 family has only proved possible quite recently with the development of selective 5-HT2A and 5-HT_{2C} receptor antagonists (5-HT_{2A} receptor antagonist: M100907, formerly MDL 100907; 5-HT_{2C} receptor antagonists: SB 242084 and RS 102221). However, there is still a need for more selective agonists, particularly at the 5-HT_{2A} receptor. Receptor autoradiography studies using tritiated ligands demonstrate high levels of 5-HT_{2A} receptors in many forebrain regions, and particularly in cortical areas, including frontal cortex, the nucleus accumbens, caudate nucleus, and HPC (Lopez-Gimenez et al. 1997; Pazos et al. 1985, 1987).

5-HT$_{2A}$ receptors are located postsynaptically to serotonergic neurons and have been found on both γ-aminobutyric acid (GABA)-ergic interneurons as well as glutamatergic cortical pyramidal cells (Burnet et al. 1995; Francis et al. 1992; Morilak et al. 1993, 1994; Wright et al. 1995). Activation of the 5-HT_{2A} receptors depolarizes the cell membrane, potentially through a decrease in K^+ currents (Marek and Aghajanian 1995; Aghajanian and Marek 1997; Araneda and Andrade 1991). Although it has generally been reported that none of the more selective 5-HT_2 receptor agonists or antagonists alter levels of 5-HT (e.g., Gobert and Millan 1999; Gobert et al. 2000), it has been suggested that 5-HT_{2A} receptors are involved in a glutamatergic feedback loop originating from the prefrontal cortex (PFC) and terminating in the dorsal raphé nucleus (DRN). Therefore stimulation of 5-HT_{2A} receptors in this region may have the capacity to alter 5-HT release throughout the forebrain (Martin-Ruiz et al. 2001).

23.2.2 5-HT_{2C} Receptor Subtypes

The 5-HT_{2C} receptor was originally classified as a member of the 5-HT1 family (5-HT_{1C}) (Pazos et al. 1987) but has been reclassified following more extensive investigation of its structure and function (Humphrey et al. 1993). Unlike the 5-HT_{2A} receptors, 5-HT_{2C} receptors are only found within the central nervous system (Palacios et al. 1990). However, 5-HT_{2C} receptors are widely distributed within the brain, and high levels are found in the cortex, limbic system, and basal ganglia. The majority of studies point to a predominantly postsynaptic location for 5-HT_{2C}

receptors, but 5-HT$_{2C}$ receptor mRNA has been localized within the DRN, indicating that this receptor could be found at presynaptic sites as well.

In common with the 5-HT$_{2A}$ receptor, activation of the 5-HT$_{2C}$ receptor also appears to depolarize the cell membrane (Rick et al. 1995; Sheldon and Aghajanian 1991), although as with the majority of data regarding the 5-HT$_{2A}$ receptor, much has been inferred through observations that 5-HT-induced increases in activity are not blocked by a range of other 5-HT-receptor specific antagonists. Hopefully, the fact that there is now a commercially available 5-HT$_{2C}$ receptor agonist, WAY 161503, will enable clarification of both the physiological and behavioral effects of 5-HT$_{2C}$ receptor stimulation.

23.3   5-HT and the Five-Choice Serial Reaction Time Task (5CSRTT)

The 5CSRTT is an animal test widely used with rodents providing substantial validity as a direct measure of different components of attention (for details see Boulougouris and Tsaltas 2008; Carli et al. 1983). In brief, animals are trained to detect the location of a brief visual stimulus presented pseudorandomly in one of the five apertures over a large number of trials. The performance measures include choice accuracy, omissions, premature responses (responses made before the target stimulus), perseverative responses (additional nose pokes made postpresentation of the stimulus in any nose-poke aperture), perseverative panel pushes (additional responses made at the food magazine before or after food retrieval), correct response latency, and food collection latency.

Optimal performance on this apparently simple task requires the integration of several cognitive processes. Sustained attention to the goal area for the duration of the intertrial interval (ITI) is necessary in order not to miss the target, while divided attention across all five exposed holes is essential in order to scan the entire visual array. Other processes measured by this task include sensor, motor or motivational processes, decision making, and inhibitory control (for details see Boulougouris and Tsaltas 2008). Apart from aspects of attention and impulse control, the task is also capable of dissociating performance elements which usually covary, although they probably rely on processes that are under control of different neural mechanisms.

Neurochemically speaking, apart from the involvement of the dopaminergic system in the modulation of the 5CSRTT (discussed in Boulougouris and Tsaltas 2008), the serotonergic system is also heavily implicated. The 5CSRTT is demonstrably sensitive to serotonergic manipulations: Global, 5,7-dihydroxytryptamine (5,7-DHT) lesion-induced 5-HT depletion consistently appears to spare response accuracy, while it increases impulsivity as reflected by increased premature responding and decreased omissions as well as correct response latency (Harrison et al. 1997; Winstanley et al. 2003a, 2004; Koskinen et al. 2000). However, systemic administration of the 5-HT$_{1A}$ receptor agonist 8-hydroxy-2-(di-n-propylamino)tetraline

(8-OH-DPAT), which also decreases 5-HT release (Bonvento et al. 1992; Hajos et al. 1999; Celada et al. 2001), does not affect impulsive responding and improves attentional performance (Winstanley et al. 2003a). At higher doses, the selective 5-HT1A receptor agonist 8-OH-DPAT reportedly increased impulsivity, possibly by activating presynaptic 5-HT_{1A} receptors (Carli and Samanin 2000). There is an incongruence, then, between the effects of chronic lesion-induced global 5-HT decreases and the effects of acute global decreases such as those affected by systemic administration of a 5-HT_{1A} receptor agonist.

The apparent inconsistency is compounded by the observation that systemic and intra-cerebral administration of the 5-HT_{2A} receptor antagonist M100907 in the prefrontal cortex (PFC) decrease impulsive responding (Winstanley et al. 2003a). Moreover, infusions of M100907 in the medial prefrontal cortex (mPFC) counteracted the loss of executive control [impulsivity induced by the competitive N-methyl-D-aspartate (NMDA) receptor antagonist 3-(R)-2-carboxypiperazin-4-propyl-1-phosphonic acid (CPP)], while 8-OH-DPAT decreased compulsive perseveration (Carli et al. 2006). Thus, an antagonist of the 5-HT system effectively produces effects opposite of those of global decrease in 5-HT transmission. This paradox, along with the observation that 2,5-dimethoxy-4-iodoamphetamine (DOI), a $5\text{-HT}_{2A/2C}$ agonist does increase premature responding, probably through activation of the 5-HT_{2A} receptor (Koskinen et al. 2000), suggests dissociable behavioral contribution of 5-HT receptor subtypes in the 5CSRTT.

Indeed, evidence suggests that the 5-HT_{2A} and 5-HT_{2C} receptors have opposing neurochemical effects. 5-HT_{2C} receptor activation inhibits dopamine release, whereas 5-HT_{2A} activation enhances dopamine release (Di Matteo et al. 2000, 2001; Millan et al. 1998). Antagonism of 5-HT_{2C} and 5-HT_{2A} receptors has opposite effects on some behavioral effects of cocaine (Fletcher et al. 2002). Furthermore, it has been demonstrated that 5-HT_{2C} and 5-HT_{2A} receptors also have contrasting and dissociable behavioral contribution on impulsivity in the 5CSRTT. The selective 5-HT_{2C} antagonist SB 242084 increases premature responding and decreases correct response latency (Winstanley et al. 2004; Higgins et al. 2003) (Fig. 23.1a). This premature responding increase has recently been shown to be mediated by the nucleus accumbens (NAc) (Robinson et al. 2008) (Fig. 23.1c). When the antagonist was administered systemically to 5,7-DHT-lesioned animals, the increase in premature responding emerged over and above the similar effects of the 5,7-DHT lesion (Winstanley et al. 2004). In contrast, the selective 5-HT_{2A} antagonist M100907 had no effect on response latency and actually reduced premature responding, an effect mediated by the NAc (Robinson et al. 2008). The effect of M100907 (administered systemically) was abolished by 5,7-DHT lesions (Winstanley et al. 2004). This dissociation challenges the hypothesis that general decreases in 5-HT neurotransmission increase impulsivity. Furthermore, the fact that antagonism of the 5-HT_{2C} receptor produces a behavioral profile closer to 5,7-DHT lesions than any other receptor so far tested including the 5-HT_{2A} receptor suggests that the 5-HT_{2C} receptor is central in the serotonergic regulation of behavioral inhibition.

Compulsivity, another form of inhibition deficit, is also accessible by the 5CSRTT via the measure of repeated responding at the holes (perseverative responding), offering a putative index of compulsivity. Winstanley et al. (2004) demonstrated that 5,7-DHT lesions increased perseverative as well as impulsive responding, a finding consistent with increased perseverative errors during reversal in the marmoset after localized 5-HT depletion within the PFC (Clarke et al. 2004) and after orbitofrontal cortex (OFC) damage (Jones and Mishkin 1972; Rogers et al. 1999; Schoenbaum et al. 2002; Chudasama and Robbins 2003; Chudasama et al. 2003). Neither 5-HT$_{2A}$ antagonism (M100907) nor 5-HT$_{2C}$ antagonism (SB 242084) appears to affect perseverative responses (Winstanley et al. 2003a, 2004; Higgins et al. 2003; Chudasama and Robbins 2003; Chudasama et al. 2003). These data suggest that different kinds of motor disinhibition differ in their neurobiological bases, as impulsivity and compulsivity appear to be differentially regulated by the 5-HT system.

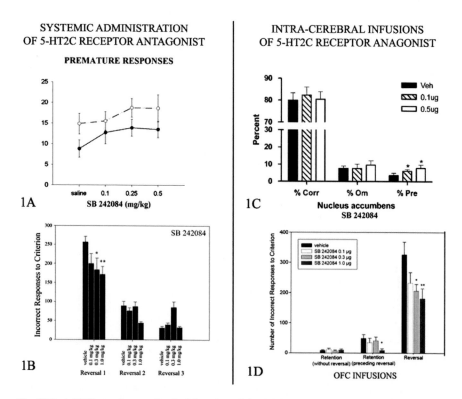

Fig. 23.1 (**a**) Effects of systemic administration of the 5-HT$_{2C}$ receptor antagonist SB 242084 on the percentage of premature responses performed during the 5CSRTT in ICV 5,7-DHT-lesioned animals and sham-operated controls (Reproduced from Winstanley et al. 2004. With permission). (**b**) Effects of systemic administration of SB 242084 on incorrect responses during spatial reversal learning (Adapted from Boulougouris et al. 2008. With permission). (**c**) Effects of intra-NAc infusions of SB 242084 on the percent correct, omissions, and premature responses during the 5CSRTT (Reproduced from Robinson et al. 2008. With permission). (**d**) Effects of intra-OFC infusions of SB 242084 on incorrect responses during spatial reversal learning (Adapted from Boulougouris and Robbins 2010. With permission)

23.4   5-HT and Reversal Learning

Tests such as the WCST, which index cognitive flexibility, in fact address several similar yet distinct forms of attentional shifts. For example, if we consider discrimination learning based on compound stimuli involving two perceptual dimensions (e.g., shapes and lines), where exemplars of these dimensions occur in combination with one another on successive trials, one exemplar of one particular dimension being correct (e.g., vertical but not skewed line correct), then (1) when the relevant stimulus dimension (i.e., lines) stays constant but novel stimuli are used (e.g., straight but not curly line correct), this is an intradimensional (ID) shift; (2) when an exemplar from the previously irrelevant dimension (shapes) becomes correct (square but not triangle), then an extradimensional (ED) shift is demanded; finally (3) when the stimuli remain the same, but the previously correct exemplar is now incorrect (triangle but not square), then we refer to *reversal learning*, a shift which can occur either at the compound discrimination learning stage or after the ID or ED shift.

Different tests of attentional flexibility involving ID or ED shifts and reversal are used translationally. Such procedures by necessity engage other processes besides switching attention (e.g., ability to utilize feedback denoting that a shift is necessary, ability to overcome "learned irrelevance" of a previously nonoperative perceptual dimension). However, the precise nature of any failure to make a required shift can be further analyzed (see, e.g., Owen et al. 1993).

Accumulating evidence implicated the serotonergic system in reversal learning but not in attentional shifting. Selective 5-HT depletion in the marmoset had no effect on ED or serial ID shifting, but it produced a large deficit in reversal learning due to perseverative responding to the previously rewarded object (Clarke et al. 2004, 2008, 2005, 2007).

In human volunteers, transient depletion of central 5-HT by the tryptophan depletion technique produced effects on discrimination learning that were especially evident in reversal learning (Park et al. 1994). Another study (Rogers et al. 1999) also reported that tryptophan depletion led to relatively selective effects on human reversal learning (but see also reference Talbot et al. 2006) with no effect on ED shifting. Evers et al. (2005) showed that behavioral reversal was accompanied by significant signal change in the right ventrolateral and dorsomedial PFC of healthy volunteers performing a probabilistic reversal task. Tryptophan depletion enhanced reversal-related signal change in the dorsomedial PFC only, affecting the blood oxygen level-dependent (BOLD) signal specifically associated with negative feedback. These data indicate that the 5-HT system has a modulatory role in reversal learning specifically.

On the receptor level, recent evidence suggests that different 5-HT receptor subtypes have distinct roles in the modulation of reversal learning. Boulougouris et al. (2008) established a double dissociation in the role of 5-HT_{2C} and 5-HT_{2A} receptor subtypes in serial spatial reversal learning. Specifically, systemic administration of the 5-HT_{2C} receptor antagonist SB 242084 facilitated spatial reversal learning in a dose-dependent manner (Fig. 23.1b). Selective infusions into the orbitofrontal cortex (OFC) of SB 242084 also promoted reversal learning, whereas infusions in the mPFC or nucleus accumbens did not. The facilitation of reversal learning therefore appears to be mediated by 5-HT$_{2C}$ receptors within the OFC (Boulougouris and Robbins 2010) (Fig. 23.1d). In contrast, systemic treatment with the 5-HT$_{2A}$ receptor antagonist M100907 dose-dependently impaired reversal learning, on the first reversal of the series in particular. This deficit emerged as increased perseveration of the previously correct response, reproducing the effects observed after selective orbitofrontal 5,7-DHT lesions (Clarke et al. 2004, 2005, 2007) as well as orbitofrontal cortical lesions in rats and nonhuman primates (Chudasama and Robbins 2003; Dias et al. 1996; Boulougouris et al. 2007).

The finding that the enhancement of spatial reversal learning via 5-HT$_{2C}$ receptor blockade is actually mediated by the OFC is apparently at odds with the above-mentioned lesion studies. This is not the only instance where contrasting effects between 5-HT depletion and 5-HT receptor antagonism have been reported. For example, recent studies showed no effect of 5-HT depletion on the delayed discounting task (Winstanley et al. 2003b), while the 5-HT$_{1A}$ receptor agonist 8-OH-DPAT (shown to turn off 5-HT release at autoreceptors) produces impulsive choice (Winstanley et al. 2005). Therefore, although the discrepancy could be attributed to task differences between lesion and antagonist studies (e.g., differences in the modalities of the reversal learning task used here and by Roberts and colleagues: object versus spatial response reversal), such explanations would appear rather superficial. A more interesting hypothesis is that the discrepancy between the lesion and antagonist studies may reflect incomplete 5-HT depletion from OFC resulting in 5-HT$_{2C}$ receptor supersensitivity (as may occur in OCD) (Graf et al. 2003; Yamauchi et al. 2004). This possibility could perhaps be investigated through infusions of 5-HT$_{2C}$ and 5-HT$_{2A}$ receptor antagonists on 5-HT depleted animals.

These findings are of considerable theoretical and clinical importance. At a theoretical level, the opposing effects of 5-HT$_{2A}$ and 5-HT$_{2C}$ antagonism on perseverative responding in spatial reversal learning task (increase and decrease, respectively) contrast with the also reverse effects of these agents on impulsive responding in the 5CSRTT (see Sect. 23.3 on 5CSRTT). Specifically, intra-PFC 5-HT$_{2A}$ antagonism decreases impulsive responding (Winstanley et al. 2003a; Higgins et al. 2003), whereas 5-HT$_{2C}$ antagonism increases it (Winstanley et al. 2004). These observations are relevant to the concept of an impulsivity–compulsivity spectrum in obsessive–compulsive spectrum disorders (Hollander and Rosen 2000). At a clinical level, these data also bear on the issue of whether 5-HT$_{2C}$ receptor antagonists might be expected to be useful in the treatment of human obsessive–compulsive disorder (OCD).

23.5   5-HT and the Reinforced Spatial Alternation Task

The reinforced spatial alternation task is a behavioral procedure used in animals to measure memory, executive functions, as well as persistent behavior (Tsaltas et al. 2005; Rawlins and Olton 1982; Rawlins and Tsaltas 1983; Givens and Olton 1995).

Each alternation trial includes two runs through the T-maze, with both food cups baited. The animal is placed on the start point with its back toward the closed guillotine door. In the first run (forced, information run), one arm of the maze is blocked. As soon as the animal reaches the goal and eats the reinforcer, it is moved on the start point, the obstacle is removed, and the second run (free direction choice run) begins. The choice run is completed when all paws of the animal are in the lateral arm. Thereafter, change in choice is prevented. Choice of the arm opposite to the preceding forced arm is rewarded and of the same resulted in nonreward.

A model of compulsive behavior based on a spontaneous behavioral persistence tendency in the framework of spatial reward alternation in the T-maze was recently developed by Tsaltas et al. (2005). This model's focal behavioral criterion is directional persistence. It has been established that this model responds isomorphically with clinical compulsive behavior to a number of serotonergic manipulations, while it is not influenced by nonserotoninergic antidepressants or benzodiazepines. Specifically, it has been demonstrated that meta-chlorophenylpiperazine (mCPP), a nonspecific 5-HT_{2C} agonist, -induced directional persistence is blocked by chronic treatment with the selective serotonin reuptake inhibitor (SSRI) fluoxetine but not with diazepam (benzodiazepine, anxiolytic) or desipramine (tricyclic antidepressant); these data support the reliability and predictive validity of the model's target behavior. Moreover, the focal obsessive behavior of the model has been demonstrated to (1) be controlled by 5-HT_{2C} and not by 5-HT_{1D} receptors, since the specific 5-HT_{1D} agonist naratriptan had no effect on mCPP-induced persistence (Tsaltas et al. 2005), (2) exhibit cross-tolerance between SSRI fluoxetine and the nonspecific serotonin agonist mCPP, suggesting a possible common path of action of the two substances, and (3) be sensitive to the administration of the agonist of the dopaminergic D_2/D_3 receptors of quinpirole, supporting the hypothesis of the serotonin–dopamine interaction and contributing to its construct validity (Kontis et al. 2008).

The initial finding implicating the 5-HT_{2C} receptor in the mediation of persistent behavior in this model was further investigated with the use of specific 5-HT_{2A} and 5-HT_{2C} receptor antagonists, M100907 and SB 242084, respectively. Systemic blockade of the 5-HT_{2C}, but not the 5-HT_{2A}, receptor offered protection against the mCPP-induced directional persistence, thus strengthening 5-HT_{2C} receptor involvement in compulsive behavior (Papakosta submitted) (Fig. 23.2). It should be noted that the above findings constitute new evidence in the understanding of OCD etiopathogenesis, as well as other psychiatric afflictions where inflexible behavior is a feature.

23.6 5-HT and the Signal Attenuation Task

Another sophisticated task used to measure both attention and inhibitory control is that of "signal attenuation." The signal attenuation model, developed by Joel et al. (Joel and Avisar 2001; Joel et al. 2001, 2005a, b), is based on the hypothesis that compulsive behavior results from deficient feedback associated with the completion

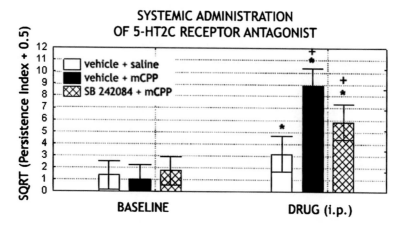

Fig. 23.2 Effects of systemic administration of the 5-HT$_{2C}$ receptor antagonist SB 242084. on mCPP-induced directional persistence in the reinforced spatial alternation task in the T-maze (Adapted from Papakosta et al. unpublished observations)

of goal-directed responses: Normal functioning of such feedback prevents pointless repetitions of responses once their goal has been attained. The goal-directed behavior of this model is instrumental lever pressing for food. The feedback for a successful response is a compound stimulus of light and tone. The "feedback deficit" assumed to underlie compulsive behavior is induced in the model by means of attenuation of the "signaling property" of this compound stimulus (repeated presentation without food and without lever-press opportunity). The behavioral control condition for this attenuation process is called *regular extinction,* and it is an identical training and testing sequence, apart from the omission of the stimulus devaluation stage. The effects of signal attenuation on lever-press responding are assessed under extinction conditions through comparisons to the effects of regular extinction. Regular extinction and, to a lesser extent, extinction after signal attenuation both produce excessive lever presses (ELP), followed by magazine entry (ELP completed, or ELP-C) extinction after signal attenuation additionally produces ELP not followed by magazine entry (ELP uncompleted, or ELP-U). According to the authors, ELP-C reflects rats' responses to nonreward, while ELP-U reflects response to the encounter of an attenuated signal and constitutes the model's focal behavior (compulsive lever pressing).

It has been demonstrated that the model is sensitive to serotonergic manipulations since administration of SSRIs (paroxetine and fluvoxamine) had an "anticompulsive" effect on "compulsive" lever pressing (Joel et al. 2004; Joel and Doljansky 2003). Although there are no studies investigating the contribution of distinct 5-HT receptor subtypes, it has been recently reported that the 5-HT$_{2C}$, but not the 5-HT$_{2A}$, receptor subtype is implicated in inhibitory control. Specifically, 5-HT$_{2C}$ receptor blockade following administration of the selective 5-HT$_{2C}$ receptor antagonist RS

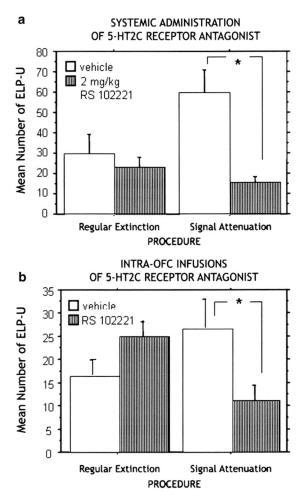

Fig. 23.3 (**a**) Effects of systemic administration of the 5-HT$_{2C}$ receptor antagonist RS 102221 on excessive uncompleted lever presses (ELP-U) in the posttraining signal attenuation and regular extinction procedures. (**b**) Effects of intra-OFC infusions of RS 102221 on the same behavioral measure in both procedures (Adapted from Flaisher-Grinberg et al. 2008. With permission)

102221 reduced compulsivelever pressing, an effect mediated within the orbitofrontal cortex (Flaisher-Grinberg et al. 2008) (Fig. 23.3).

23.7   5-HT$_2$ Receptors and Clinical Implications

Studies on the involvement of 5-HT$_2$ receptors in attentional processes and executive control may be relevant to various neuropsychiatric disorders. The evidence emerging from studies in the 5CSRTT suggests that serotonergic modulation in the mPFC and the NAc can increase attentional selectivity and decrease impulsivity via 5-HT$_{1A}$ and 5-HT$_{2A}$ receptors. These findings bear clinical relevance, given that some atypical antipsychotics have 5-HT$_{2A}$ receptor antagonist actions that may potentially contribute to a procognitive effect in schizophrenia (Meltzer et al. 2003). The opposing effects of 5-HT$_{2A}$ and 5-HT$_{2C}$ receptor antagonism on premature responding in the 5CSRTT (Winstanley et al. 2004) indicate that selective 5-HT$_{2A}$ receptor antagonists and/or 5-HT$_{2C}$ receptor agonists may have beneficial effects in psychiatric disorders where coexisting impulsivity is often present, including attention deficit hyperactivity disorder, schizophrenia, and substance abuse.

Finally, the Boulougouris et al. (2008) data may be useful in relieving reversal deficits such as those noted in Huntington disease. In fact, they may deserve consideration as a means of controlling compulsivity in the context of OCD. The latter is strengthened by the anticompulsive effect of 5-HT$_{2C}$ receptor antagonism on the reinforced spatial alternation and signal attenuation tasks.

23.8 Conclusions

This survey provides an integrative account of the contribution of serotonin, with emphasis on the 5-HT$_{2C}$ receptor subtype, to specific aspects of attentional processes and executive functioning as they emerge from experimental animal work. Four tasks allowing translational study have been used to that purpose:

1. The 5CSRTT, an analogue of the human CPT, is designed to measure several attentional operations with an emphasis on sustained attention or vigilance.
2. Attentional set shifting including reversal, intra- and intradimensional shifts, as the human WCST, tap attentional flexibility, that is the ability of humans and animals to develop and maintain higher-order rules, and shift attention according to changing reward contingencies.
3. The reinforced spatial alternation assesses working memory and other executive functions as well as persistent behavior.
4. Finally, the signal attenuation task addresses the issue of behavioral control by means of inhibition of activities that no longer serve environmental demands.

Taken together, the findings detailed above highlight the specificity of influences that the serotonin system has on overall prefrontal executive control, acting to promote distinct components of prefrontal processing in a context-dependent manner. Future directions must focus toward the definition of the specific aspects of attentional functions in which the serotonergic system is acting to influence prefrontal processing.

Acknowledgements VB would like to thank the Bodossaki Foundation for funding.

References

Aghajanian GK, Marek GJ (1997) Serotonin induces excitatory postsynaptic potentials in apical dendrites of neocortical pyramidal cells. Neuropharmacology 36:589–599.

Araneda R, Andrade R (1991) 5-Hydroxytryptamine2 and 5-hydroxytryptamine 1A receptors mediate opposing responses on membrane excitability in rat association cortex. Neuroscience 40:399–412.

Barnes NM, Sharp T (1999) A review of central 5-HT receptors and their function. Neuropharmacology 38:1083–1152.

Bonvento G, Scatton B, Claustre Y, et al (1992) Effect of local injection of 8-OH-DPAT into the dorsal or median raphe nuclei on extracellular levels of serotonin in serotonergic projection areas in the rat brain. Neurosci Lett 137:101–104.

Boulougouris V, Robbins TW (2010) Enhancement of spatial reversal learning by 5-HT2C receptor antagonism is neuroanatomically specific. J Neurosci 30:930–938.

Boulougouris V, Tsaltas E (2008) Serotonergic and dopaminergic modulation of attentional processes. Prog Brain Res 172:517–42.

Boulougouris V, Dalley JW, Robbins TW (2007) Effects of orbitofrontal, infralimbic and prelimbic cortical lesions on serial spatial reversal learning in the rat. Behav Brain Res 179:219–228.

Boulougouris V, Glennos JC, Robbins TW (2008) Dissociable effects of selective 5-HT2A and 5-HT2C receptor antagonists on serial spatial reversal learning in rats. Neuropsychopharmacology 33:2007–2019.

Burnet PW, Eastwood SL, Lacey K, et al (1995) The distribution of 5-HT1A and 5-HT2A receptor mRNA in human brain. Brain Res 676:157–168.

Carli M, Samanin R (2000) The 5-HT(1A) receptor agonist 8-OH-DPAT reduces rats' accuracy of attentional performance and enhances impulsive responding in a five-choice serial reaction time task: role of presynaptic 5-HT(1A) receptors. Psychopharmacology (Berl) 149: 259–268.

Carli M, Robbins TW, Evenden JL, et al (1983) Effects of lesions to ascending noradrenergic neurones on performance of a 5-choice serial reaction task in rats; implications for theories of dorsal noradrenergic bundle function based on selective attention and arousal. Behav Brain Res 9:361–380.

Carli M, Baviera M, Invernizzi RW, et al (2006) Dissociable contribution of 5-HT1A and 5-HT2A receptors in the medial prefrontal cortex to different aspects of executive control such as impulsivity and compulsive perseveration in rats. Neuropsychopharmacology 31:757–767.

Celada P, Puig MV, Casanovas JM, et al (2001) Control of dorsal raphe serotonergic neurons by the medial prefrontal cortex: involvement of serotonin-1A, GABA(A), and glutamate receptors. J Neurosci 21:9917–9929.

Chudasama Y, Robbins TW (2003) Dissociable contributions of the orbitofrontal and infralimbic cortex to pavlovian autoshaping and discrimination reversal learning: further evidence for the functional heterogeneity of the rodent frontal cortex. J Neurosci 23:8771–8780.

Chudasama Y, Passetti F, Rhodes SE, et al (2003) Dissociable aspects of performance on the 5-choice serial reaction time task following lesions of the dorsal anterior cingulate, infralimbic and orbitofrontal cortex in the rat: differential effects on selectivity, impulsivity and compulsivity. Behav Brain Res 146: 105–119.

Clarke HF, Dalley JW, Crofts HS, et al (2004) Cognitive inflexibility after prefrontal serotonin depletion. Science 304:878–880.

Clarke HF, Walker SC, Crofts HS, et al (2005) Prefrontal serotonin depletion affects reversal learning but not attentional set shifting. J Neurosci 25:532–538.

Clarke HF, Walker SC, Dalley JW, et al (2007) Cognitive inflexibility after prefrontal serotonin depletion is behaviorally and neurochemically specific. Cereb Cortex 17: 18–27.

Clarke HF, Robbins TW, Roberts AC (2008) Lesions of the medial striatum in monkeys produce perseverative impairments during reversal learning similar to those produced by lesions of the orbitofrontal cortex. J Neurosci 28:10972–10982.

Di Matteo V, Di Giovanni G, Esposito E (2000) SB 242084: a selective 5-HT(2C) receptor antagonist. CNS Drug Rev 6:195–205.

Di Matteo V, De Blasi A, Di Giulio C, et al (2001) Role of 5-HT(2C) receptors in the control of central dopamine function. Trends Pharmacol Sci 22:229–232.

Dias R, Robbins TW, Roberts AC (1996) Dissociation in prefrontal cortex of affective and attentional shifts. Nature 380, 69–72.

Evers EA, Cools R, Clark L, et al (2005). Serotonergic modulation of prefrontal cortex during negative feedback in probabilistic reversal learning. Neuropsychopharmacology 30: 1138–1147.

Flaisher-Grinberg S., Klavir O, Joel D (2008) The role of 5-HT2A and 5-HT2C receptors in the signal attenuation rat model of obsessive-compulsive disorder. Int J Neuropsychopharmacol 11:811–825.

Fletcher PJ, Grottick AJ, Higgins GA (2002) Differential effects of the 5-HT(2A) receptor antagonist M100907 and the 5-HT(2C) receptor antagonist SB242084 on cocaine-induced locomotor activity, cocaine self-administration and cocaine-induced reinstatement of responding. Neuropsychopharmacology 27:576–586.

Francis PT, Pangalos MN, Pearson RC, et al (1992) 5-Hydroxytryptamine1A but not 5-hydroxytryptamine2 receptors are enriched on neocortical pyramidal neurones destroyed by intrastriatal volkensin. J Pharmacol Exp Ther 261:1273–1281.

Givens B, Olton DS (1995) Bidirectional modulation of scopolamine-induced working memory impairments by muscarinic activation of the medial septal area. Neurobiol Learn Mem 63:269–276.

Gobert A, Millan MJ (1999) Serotonin (5-HT)2A receptor activation enhances dialysate levels of dopamine and noradrenaline, but not 5-HT, in the frontal cortex of freely-moving rats. Neuropharmacology 38:315–317.

Gobert A, Rivet JM, Lejeune F, et al (2000) Serotonin(2C) receptors tonically suppress the activity of mesocortical dopaminergic and adrenergic, but not serotonergic, pathways: a combined dialysis and electrophysiological analysis in the rat. Synapse 36: 205–221.

Graf M, Kantor S, Anheuer ZE, et al (2003) m-CPP-induced self-grooming is mediated by 5-HT2C receptors. Behav Brain Res 142:175–179.

Hajos M, Hajos-Korcsok E, Sharp T (1999) Role of the medial prefrontal cortex in 5-HT1A receptor-induced inhibition of 5-HT neuronal activity in the rat. Br J Pharmacol 126:1741–1750.

Harrison AA, Everitt BJ, Robbins TW (1997) Central 5-HT depletion enhances impulsive responding without affecting the accuracy of attentional performance: interactions with dopaminergic mechanisms. Psychopharmacology (Berl) 133:329–342.

Higgins GA, Enderlin M, Haman M, et al (2003) The 5-HT2A receptor antagonist M100,907 attenuates motor and "impulsive-type" behaviors produced by NMDA receptor antagonism. Psychopharmacology (Berl) 170:309–319.

Hollander E, Rosen J (2000) Impulsivity. J Psychopharmacol, 14: S39–S44.

Hoyer D, Hannon JP, Martin GR (2002) Molecular, pharmacological and functional diversity of 5-HT receptors. Pharmacol Biochem Behav 71:533–554.

Humphrey PP, Hartig P, Hoyer D (1993) A proposed new nomenclature for 5-HT receptors. Trends Pharmacol Sci 14:233–236.

Joel D, Avisar A (2001) Excessive lever pressing following post-training signal attenuation in rats: a possible animal model of obsessive compulsive disorder? Behav Brain Res 123:77–87.

Joel D, Doljansky J (2003) Selective alleviation of compulsive lever-pressing in rats by D1, but not D2, blockade: possible implications for the involvement of D1 receptors in obsessive-compulsive disorder. Neuropsychopharmacology 28:77–85.

Joel D, Avisar A, Doljansky J (2001) Enhancement of excessive lever-pressing after post-training signal attenuation in rats by repeated administration of the D1 antagonist SCH 23390 or the D2 agonist quinpirole, but not the D1 agonist SKF 38393 or the D2 antagonist haloperidol. Behav Neurosci 115:1291–1300.

Joel D, Ben-Amir E, Doljansky J, et al (2004) "Compulsive" lever-pressing in rats is attenuated by the serotonin re-uptake inhibitors paroxetine and fluvoxamine but not by the tricyclic antidepressant desipramine or the anxiolytic diazepam. Behav Pharmacol 15:241–252.

Joel D, Doljansky J, Roz N, et al (2005a) Role of the orbital cortex and of the serotonergic system in a rat model of obsessive compulsive disorder. Neuroscience 130:25–36.

Joel D, Doljansky J, Schiller D (2005b) Compulsive' lever pressing in rats is enhanced following lesions to the orbital cortex, but not to the basolateral nucleus of the amygdala or to the dorsal medial prefrontal cortex. Eur J Neurosci 21:2252–2262.

Jones B, Mishkin M (1972) Limbic lesions and the problem of stimulus–reinforcement associations. Exp Neurol 36:362–377.

Kontis D, Boulougouris V, Papakosta VM, et al (2008) Dopaminergic and serotonergic modulation of persistent behavior in the reinforced spatial alternation model of obsessive-compulsive disorder. Psychopharmacology (Berl) 200:597–610.

Koskinen T, Ruotsalainen S, Puumala T, et al (2000) Activation of 5-HT2A receptors impairs response control of rats in a five-choice serial reaction time task. Neuropharmacology 39:471–481.

Lopez-Gimenez JF, Mengod G, Palacios JM, et al (1997) Selective visualization of rat brain 5-HT2A receptors by autoradiography with [3H]MDL 100,907. Naunyn Schmiedebergs Arch Pharmacol 356:446–454.

Marek GJ, Aghajanian GK (1995) Protein kinase C inhibitors enhance the 5-HT2A receptor-mediated excitatory effects of serotonin on interneurons in rat piriform cortex. Synapse 21:123–130.

Martin-Ruiz R, Puig MV, Celada P, et al (2001) Control of serotonergic function in medial prefrontal cortex by serotonin-2A receptors through a glutamate-dependent mechanism. J Neurosci 21:9856–9866.

Meltzer HY, Li Z, Kaneda Y, Ichikawa J (2003) Serotonin receptors: their key role in drugs to treat schizophrenia. Prog Neuropsychopharmacol Biol Psychiatry 27:1159–1172.

Millan MJ, Dekeyne A, Gobert A (1998) Serotonin (5-HT)2C receptors tonically inhibit dopamine (DA) and noradrenaline (NA), but not 5-HT, release in the frontal cortex in vivo. Neuropharmacology 37:953–955.

Morilak DA, Garlow SJ, Ciaranello RD (1993) Immunocytochemical localization and description of neurons expressing serotonin2 receptors in the rat brain. Neuroscience 54:701–717.

Morilak DA, Somogyi P, Lujan-Miras R, et al (1994) Neurons expressing 5-HT2 receptors in the rat brain: neurochemical identification of cell types by immunocytochemistry. Neuropsychopharmacology 11:157–166.

Owen AM, Roberts AC, Hodges JR, et al (1993) Contrasting mechanisms of impaired attentional set-shifting in patients with frontal lobe damage or Parkinson's disease. Brain 116: 1159–1175.

Palacios JM, Waeber C, Hoyer D, et al (1990) Distribution of serotonin receptors. Ann N Y Acad Sci 600:36–52.

Papakosta VM, Boulougouris V, Kalogerakou S, et al. 5-HT2C but not 5-HT2A receptor blockade modulates pharmacologically induced persistence in the spatial alternation model of obsessive-compulsive disorder (submitted manuscript).

Parasuraman R (1998) The attentive brain: issues and concepts. In: Parasuraman R, ed. The Attentive Brain. Cambridge, MA: MIT Press, pp. 3–15.

Park SB, Coull JT, McShane RH, et al (1994). Tryptophan depletion in normal volunteers produces selective impairments in learning and memory. Neuropharmacology 33:575–588.

Pazos A, Cortes R, Palacios JM (1985) Quantitative autoradiographic mapping of serotonin receptors in the rat brain. II. Serotonin-2 receptors. Brain Res 346:231–249.

Pazos A, Probst A, Palacios JM (1987) Serotonin receptors in the human brain – IV. Autoradiographic mapping of serotonin-2 receptors. Neuroscience 21:123–139.

Rawlins JN, Olton DS (1982) The septo-hippocampal system and cognitive mapping. Behav Brain Res 5:331–358.

Rawlins JN, Tsaltas E (1983) The hippocampus, time and working memory. Behav Brain Res 10:233–262.

Rick CE, Stanford IM, Lacey MG (1995) Excitation of rat substantia nigra pars reticulata neurons by 5-hydroxytryptamine in vitro: evidence for a direct action mediated by 5-hydroxytryptamine2C receptors. Neuroscience 69:903–913.

Robbins TW (2002) The 5-choice serial reaction time task: behavioral pharmacology and functional neurochemistry. Psychopharmacology (Berl) 163:362–380.

Robbins TW (2005) Chemistry of the mind: neurochemical modulation of prefrontal cortical function. J Comp Neurol 493:140–146.

Robinson ES, Dalley JW, Theobald DE, et al (2008) Opposing roles for 5-HT2A and 5-HT2C receptors in the nucleus accumbens on inhibitory response control in the 5-choice serial reaction time task. Neuropsychopharmacology 33:2398–2406.

Rogers RD, Blackshaw AJ, Middleton HC, et al (1999) Tryptophan depletion impairs stimulus-reward learning while methylphenidate disrupts attentional control in healthy young adults: implications for the monoaminergic basis of impulsive behavior. Psychopharmacology (Berl), 146:482–491.

Schoenbaum G, Nugent SL, Saddoris MP, et al (2002) Orbitofrontal lesions in rats impair reversal but not acquisition of go, no-go odor discriminations. Neuroreport 13:885–890.

Sheldon PW, Aghajanian GK (1991) Excitatory responses to serotonin (5-HT) in neurons of the rat piriform cortex: evidence for mediation by 5-HT1C receptors in pyramidal cells and 5-HT2 receptors in interneurons. Synapse 9:208–218.

Talbot PS, Watson DR, Barrett SL, et al (2006) Rapid tryptophan depletion improves decision-making cognition in healthy humans without affecting reversal learning or set shifting. Neuropsychopharmacology 31:1519–1525.

Tsaltas E, Kontis D, Chrysikakou S, et al (2005) Reinforced spatial alternation as an animal model of obsessive-compulsive disorder (OCD): investigation of 5-HT2C and 5-HT1D receptor involvement in OCD pathophysiology. Biol Psychiatry 57:1176–1185.

Winstanley CA, Chudasama Y, Dalley JW, et al (2003a) Intra-prefrontal 8-OH-DPAT and M100907 improve visuospatial attention and decrease impulsivity on the five-choice serial reaction time task in rats. Psychopharmacology (Berl) 167:304–314.

Winstanley CA, Dalley JW, Theobald DE, et al (2003b) Global 5-HT depletion attenuates the ability of amphetamine to decrease impulsive choice on a delay-discounting task in rats. Psychopharmacology (Berl) 170: 320–331.

Winstanley CA, Theobald DE, Dalley JW, et al (2004) 5-HT2A and 5-HT2C receptor antagonists have opposing effects on a measure of impulsivity: interactions with global 5-HT depletion. Psychopharmacology (Berl) 176:376–385.

Winstanley CA, Theobald DE, Dalley JW, et al (2005) Interactions between serotonin and dopamine in the control of impulsive choice in rats: therapeutic implications for impulse control disorders. Neuropsychopharmacology 30:669–682.

Wright DE, Seroogy KB, Lundgren KH, et al (1995) Comparative localization of serotonin1A, 1C, and 2 receptor subtype mRNAs in rat brain. J Comp Neurol 351:357–373.

Yamauchi M, Tatebayashi T, Nagase K, et al (2004) Chronic treatment with fluvoxamine desensitizes 5-HT2C receptor-mediated hypolocomotion in rats. Pharmacol Biochem Behav 78:683–689.

Chapter 24
5-HT$_{2C}$ Receptors in Learning

López-Vázquez Miguel Ángel, Gutiérrez-Guzmán Blanca Érika,
Cervantes Miguel, and Olvera-Cortés María Esther

24.1 Serotonin and Cognition: General Insights

Serotonin (5-hydroxytryptamine, or 5-HT) is involved in multiple cerebral functions
including learning, memory, mood, anxiety, and impulsiveness (Aouizerate et al.
2005; Coccaro 1989; Cools et al. 2008). Serotonin acts through multiple receptors
that are classified as 5-HT$_1$ to 5-HT$_7$, each one showing subclasses, based on their
affinity to agonists and based on their cellular signaling pathways (Roth 1994). 5-HT$_2$
receptors are all coupled to Gq protein and are members of the rhodopsin family of
G-protein-coupled receptors (Aghajanian and Marek 1997). Their activation results
in stimulation of phospholipase C and the production of inositol phosphate (Conn and
Sanders-Bush 1986). 5-HT$_2$ receptors are located in 5-HT terminal regions and are
particularly dense in deeper cortical layers, including the prefrontal and anterior cingulated cortices, and in the hippocampus and the thalamus, all of which are brain
regions known to be involved in learning (Fischette et al. 1987; Pazos et al. 1987).

Subclasses of 5-HT$_2$ receptors include 5-HT$_{2A}$, 5-HT$_{2B}$, and 5-HT$_{2C}$; the latter
exhibits the unique mechanism of generating multiple functional receptor variants
through messenger ribonucleic acid (mRNA) editing (Burns et al. 1997).

5-HT$_{2C}$ mRNA is restricted almost exclusively to the central nervous system
(Julius et al. 1988), where it is distributed in areas including the cerebral cortex,
limbic system, and basal ganglia (Rick et al. 1995; Sheldon and Aghajanian 1991).
5-HT$_{2C}$ receptors have a widespread distribution in humans, monkeys, and rats and
exist both pre- and postsynaptically (Lopez-Gimenez et al. 2001). In the monkey

M.E. Olvera-Cortés (✉)
Centro de Investigación Biomédica de Michoacán, Instituto Mexicano del Seguro Social
and
División de Estudios de Posgrado, Facultad de Ciencias Médicas y Biológicas "Dr. Ignacio
Chávez", Universidad Michoacana de San Nicolás de Hidalgo
and
Instituto de Física y Matemáticas de física y matemáticas, Universidad Michoacana de San
Nicolás de Hidalgo
e-mail: maesolco@yahoo.com; maria.olverac@imss.gob.mx

G.Di Giovanni et al. (eds.), *5-HT$_{2C}$ Receptors in the Pathophysiology of CNS Disease*,
The Receptors 22, DOI 10.1007/978-1-60761-941-3_24,
© Springer Science+Business Media, LLC 2011

brain, 5-HT$_{2C}$ receptors show an extensive and heterogeneous distribution, with the strongest concentration in the choroid plexus (Lopez-Gimenez et al. 2001). In the neocortex, 5-HT$_{2C}$ receptor mRNA has been detected in layer V of all cortical regions except in the calcarine sulcus. The striatum and basal forebrain show strong staining for the receptor mRNA (autoradiographic study) in the nucleus accumbens, ventral aspects of caudate–putamen (striatum), septal nuclei, Broca diagonal band, ventral striatum, and amygdale, as well as in several thalamic, midbrain, and brainstem nuclei. The location of receptors is primarily somatodendritic, possibly, on axon terminals in several regions including septal nuclei and the horizontal limb of the diagonal band and the interpeduncular nucleus (presence of mRNA without binding sites). Especially abundant 5-HT$_{2C}$ mRNA has been found in some regions with a high population of cholinergic cells (Lopez-Gimenez et al. 2001). In the rat, high levels of 5-HT$_{2C}$ receptors have been observed in the prefrontal cortex (Pompeiano et al. 1994), mainly located on the apical dendrites of pyramidal neurons and less on parvalbumin-labeled γ-aminobutyric acid (GABA) interneurons (Miner et al. 2003). A widespread distribution of 5-HT$_{2C}$ mRNA has also been found in the rat. Clemett (Clemett et al. 2000) observed the most abundant mRNA content in the neuronal bodies in the anterior olfactory nucleus, medial and intercalated amygdaloid nuclei, hippocampus subfields CA1 to CA3, laterodorsal and lateral geniculated thalamic nuclei, caudate–putamen, and several cortical areas, including piriform and frontal. The 5-HT$_{2C}$ receptors are present in the CA1 subfield of the rat hippocampus at all developmental stages (Garcia-Alcocer et al. 2006). Immunopositive neurons are also located in the dorsal raphe, where a role as autoreceptors has been proposed (Clemett et al. 2000).

The presence of 5-HT$_{2C}$ receptors in the cell bodies or terminal regions of the three dopaminergic systems (Alex and Pehek 2007) suggests that this receptor is a modulator of dopaminergic system functioning and of the learning processes in which dopamine neurotransmission is involved. An indirect inhibition of tonic dopamine release by 5-HT$_{2C}$ receptors through the activation of GABA-ergic cells and the consequent increase in GABA release occurs in either the substantia nigra pars reticulata or the striatum (nigrostriatal pathway). The 5-HT$_{2C}$ receptors appear to tonically inhibit the release of dopamine in the cortical targets of the mesocortical pathway, but this effect does not appear to be exerted at the cortical level but at the ventral tegmental area (Alex and Pehek 2007; Di Matteo et al. 2001).

In the ventral tegmental area, 5-HT$_{2C}$ receptors are located on subpopulations of both GABA-ergic and dopaminergic neurons, its density (of 5-HT$_{2C}$ receptors) showing higher levels in the middle region and a lower expression in more caudal regions (Bubar and Cunningham 2007; Ji et al. 2006), whereas in the striatum and nucleus accumbens, 5-HT$_{2C}$ receptors are only located on GABA-ergic neurons (Eberle-Wang et al. 1997).

Thus, in view of the presence of 5-HT$_{2C}$ receptors on dopaminergic neurons, it is possible that these receptors exert a modulatory effect on learning involving the cerebral areas that are targets of the dopaminergic system, in which the dopamine exerts a principal role in organizing learning processes, as is the case in the prefrontal cortex and striatum.

24.2 Striatum-Related Learning

The basal ganglia are a group of subcortical nuclei controlling voluntary movement, which in vertebrates include the striatum, globus pallidus (globus pallidus external segment), entopeduncular nucleus (globus pallidus internal segment), subthalamic nucleus, substantia nigra, and the pedunculopontine nucleus (Wichmann and DeLong 1998).

The striatum is the major input structure of the basal ganglia and comprises two functionally similar nuclei: the caudate and the putamen. Inputs arrive to the striatum from many cortical areas (Kemp and Powell 1970), at different subregions of both nuclei, and as a result, five parallel information processing circuits have been identified: a motor circuit, an oculomotor circuit, a dorsolateral prefrontal circuit, a lateral orbitofrontal circuit, and an anterior cingulated circuit (Alexander et al. 1986). The output structures in the basal ganglia circuit are the globus pallidus internal segment and the substantia nigra pars reticulata (DeLong 1990).

The basal ganglia are affected in diseases such as Parkinson and Huntington, in which a loss of motor coordination is the main clinical symptom. However, both diseases are associated with dementia because the cognitive functions sustained by these cerebral nuclei are altered by the degeneration of their neurons (Packard and Knowlton 2002). Experimental evidence supporting the role of basal ganglia in learning is abundant. The procedural learning, sequential motor learning, conditioning, and egocentric learning have all been extensively shown to require the participation of striatal activity, particularly because they involve the acquisition of stimulus–response associations (McDonald and White 1993, 1994). These cognitive processes imply a basal ganglia interaction with the cerebral cortex through afferent information (Packard and Knowlton 2002; Wilson 1998).

Lesion and inactivation of the striatum have been shown to impede the acquisition of instrumental tasks, in which a discrete stimulus predicts the availability of reinforcement (McDonald and White 1993), as well as stimulus–response learning and temporal expectation in rats (McDonald and White 1993; Florio et al. 1999; Hudzik et al. 2000). More specifically, the dorsolateral striatum has been associated with performance in procedural tasks and habit learning (Featherstone and McDonald 2004a, b), whereas the dorsomedial striatum supports the flexible shifting of response patterns (Ragozzino 2003; Ragozzino et al. 2002a, b).

On the other hand, the role of serotonin in striatal-dependent learning has not been well defined until now. The main origin of the serotonin arriving to basal ganglia is the dorsal raphe nucleus, which sends projections to the striatum, globus pallidus, subthalamic nucleus, substantia nigra, and pedunculopontine nucleus, as well as to the prefrontal cortex (Dahlstroem et al. 1964; Herve et al. 1987; Van Bockstaele et al. 1994). The selective destruction of serotoninergic neurons through injection of 5,7-dihydroxytryptamine (5,7-DHT) into the raphe increases responses for a conditioned reward, and it has been proposed that the removal of inhibitory serotoninergic influence on the mesolimbic dopamine system might underlie this effect (Fletcher et al. 1999).

As previously stated, 5-HT$_{2C}$ mRNA is differentially distributed among basal ganglia subregions (Hoffman and Mezey 1989; Mengod et al. 1990). Moreover, a higher density of 5-HT$_{2C}$ receptor mRNA is observed on medium-sized neurons, probably projection neurons, preferentially localized in ventrolateral subareas of the caudate–putamen. However, there are no differences between substance P-containing neurons and enkephalin-containing neurons (Ward and Dorsa 1996). Conversely, the preferential presence of 5-HT$_{2C}$ receptors almost exclusively in the patch compartment (Ward and Dorsa 1996) implies differential modulation of the projection onto the substantia nigra pars compacta and thus the modulation of dopaminergic input to the striatum (Gerfen 1984).

In spite of the abundant presence of 5-HT$_{2C}$ receptors, few pharmacological studies through locally administering compounds with 5-HT$_{2C}$ affinity have been done to assess the 5-HT$_{2C}$ receptor participation in striatal-dependent learning. Thus, little is known about the participation of this receptor in the modulation of striatal-dependent cognition processes, although systemic application of these drugs is the general rule. 5-HT$_{2C}$ receptor involvement in learning tasks requiring participation of the striatum is summarized below.

24.2.1 Passive Avoidance

There is evidence that the dorsal striatum is involved in aversive conditioning. Lesions of this cerebral structure produce deficits in active avoidance (Winocur 1974) and passive avoidance learning (Winocur 1974; Prado-Alcala et al. 1975). Passive avoidance learning is based on the innate preference of rodents for darkness. An inescapable shock is given to the animals when they enter the dark compartment of an apparatus containing a dark and an illuminated compartment. The suppression of the innate preference as consequence of the shock serves as a measure of learning (retention of this memory is tested 24 h later). The serotoninergic modulation of this passive avoidance has been observed in studies in which modulation of 24-h post-training retention occurs after application of serotonin agents. Specifically, the post-training application of the 5-HT$_{1A}$ receptor agonist 8-hydroxy-2-(di-n-propylamino) tetralin (8-OH-DPAT) (0.1–0.3 mg/kg subcutaneously [SC]) produced a dose-dependent deficit in the 24-h retention test (Misane and Ogren 2000).

Misane and Ogren (2000) evaluated the involvement of multiple serotonin receptors in passive avoidance using different compounds including the serotonin-releasing compound p-chloroamphetamine (PCA) (0.3, 1.0, and 3.0 mg/kg intraperitoneal IP]) in rats, showing PCA impairment in this task, which has been related to the PCA-induced release of serotonin, when evaluated 24 h after training. This impairment was antagonized by the selective 5-HT-reuptake inhibitor zimelidine, whereas no effect was observed after desipramine administration (inhibitor of the noradrenaline reuptake), showing the involvement of serotonin on the detrimental effect of PCA (Ogren 1985, 1986). Similarly, Santucci et al. (1996) showed that increased serotoninergic activity after p-chlorophenylalanine (PCPA) administration in rats induces deficits in passive avoidance, related to the acute increase of serotonin but not to the subsequent serotonin depletion.

In accordance with this finding, the intrastriatal infusion of serotonin produces retention deficits in inhibitory avoidance tests (Prado-Alcala et al. 2003a).

Concerning the specific role of 5-HT$_{2C}$ receptors, Martin et al. (1998) administered the 5-HT$_{2C}$ agonist (S)-2-(4,4,7-trimethyl-1,4,-dihydro-indeno[1,2-b]pyrrol-1-yl)-1-methylethylamine (Ro 60-0332) (5 mg/kg SC) and observed an amelioration of the passive avoidance deficits caused by bulbectomy in the rat. Bulbectomy is used as a model for depression in the rat because it produces disturbed emotive behavior and impaired passive avoidance acquisition that is ameliorated by antidepressant treatment (Broekkamp et al. 1980). Misane and Ogren (2000) also administered m-chlorophenylpiperazine (mCPP) (3.0 and 5.0 mg/kg IP), a 5-HT$_{2A/2C/1B}$ receptor agonist, before the passive avoidance test and observed that mCPP produced a dose-related impairment of memory in the 24-h retention test. Apparently, the effect was mediated through 5-HT$_{1A}$ receptor activation because the impairment was reversed by the application of 5-HT$_{1A}$ receptor antagonists, whereas the authors did not observe any effect mediated by 5-HT$_{2C}$ receptors. However, the simultaneous application of the 5-HT$_{2C}$ receptor antagonist N-(2-naphthyl)-N'-(3-pyridyl)-urea (Ro 60-0491) (3.0 mg/kg IP) with mCPP, was able to block the inhibitory effect of mCPP, suggesting an indirect modulatory role for 5-HT$_{2C}$ receptors. Recently, Marquis (Marquis et al. 2007) tested the 5-HT$_{2C}$ receptor agonist (7bR,10aR)-1,2,3,4,8,9,10,10a-octahydro-7bH-cyclopenta-[b](Aouizerate et al. 2005; Roth 1994)diaxepino[6,7,1hi]indole (WAY 163909) and observed a dose-dependent reduction in the avoidance response (0.3–3 mg/kg IP) at doses that had little or no effect on the number of failures to escape. The effect was blocked by the application of the 5-HT$_{2B/2C}$ receptor antagonist 5-methyl-1-(3-pyridylcarbamoyl)-1,2,3,5-tetrahydropyrrolo[2,3-f]indole (SB 206553) (10.0 mg/kg p.o.).

From previous findings, we can summarize that agonists of 5-HT$_{2C}$ (mCPP and WAY 163909) administered systemically produced impairment of passive avoidance tasks that were reversed by the administration of specific 5-HT$_{2C}$ receptor antagonists (Ro 60-0491 and SB 206553), supporting an inhibitory role for 5-HT$_{2C}$ receptor activation on memory in retention tests. Di Giovanni et al. (2000) have reported that mCPP IP administration decreased in a dose-dependent manner the firing rate of ventral tegmental area dopaminergic neurons and the basal firing rate of dopaminergic neurons from the substantia nigra pars compacta. These changes were associated with a decrease in dopamine release in the nucleus accumbens. All effects were blocked by the application of the 5-HT$_{2C}$ receptor antagonist 6-chloro-5-methyl-N-[6-[(2-methylpyridin-3-yl)oxy]pyridin-3-yl]indoline-1-carboxamide (SB 242084). Thus, the inhibitory effect of 5-HT$_{2C}$ receptors on mesolimbic and nigrostriatal dopaminergic function could underlie the effects of 5-HT$_{2C}$ activation or inactivation in passive avoidance tests. The 5-HT$_{2C}$ receptor antagonist SB 206553, dose-dependently increases dopamine levels in the prefrontal cortex, accumbens, and striatum (Gobert et al. 2000) similarly to SB 242084, whereas the 5-HT$_{2C}$ receptor agonist (S)-2-(chloro-5-fluoro-indol-1-yl)-1-methylethylamine (Ro 60-0175) reduced dopamine levels in the same areas. Thus, the increase and decrease of dopamine function could be related to the deficit of retention produced by 5-HT$_{2C}$ agonists in passive avoidance tests.

Blocking of 5-HT$_2$ receptors, through the bilateral intrastriatal infusion of ketanserin (5-HT$_{2A/2C}$ receptor antagonist; 0.5, 1.0, 2.0, and 4.0 ng) immediately after inhibitory avoidance training, induced a dose-dependent retention deficit when tested 24 h later (Prado-Alcala et al. 2003b). Lucas and Spampinato (2000) reported that the intrastriatal infusion of SB 206553 reduced the dopamine efflux in freely moving rats, whereas the 5-HT$_{2A}$ receptor antagonist 1(Z)-[2-(dimethylamino) ethoxyimino]-1(2-fluorophenyl)-3-(4-hydroxyphenyl)-2(E)-propene (SR 46349B) had no effect on dopamine efflux. Thus, the impairment in passive avoidance observed by these authors could be related to a decrease in dopamine efflux in the striatum caused by ketanserin acting through 5-HT$_{2C}$ antagonism. However, this interpretation could be an over simplification of striatal neurotransmitter interactions. For example, the role of striatal acetylcholine on consolidation processes in passive avoidance has been extensively demonstrated. The posttraining intrastriatal infusion of atropine (acetylcholine antagonist) and scopolamine (cholinergic muscarinic antagonist) impairs passive avoidance in a dose-dependent manner (Prado-Alcala et al. 1984, 1985; Quirarte et al. 1993). Moreover, experiments evaluating passive avoidance under dopaminergic manipulation showed a detrimental effect mediated through the activation of D2 receptors (Ichihara et al. 1992). So, it is clear that studies that specifically evaluate 5-HT$_{2C}$ receptor activity in specific cerebral areas are required to gain a more complete understanding of their role in learning processes, including the 5-HT$_{2C}$ modulation (and modulation of all other receptors) of dopamine and acetylcholine, acting either locally in the striatum or on dopaminergic systems to influence the organization of passive avoidance learning.

24.2.2 Differential Reinforcement of Low Rate

As was previously mentioned, the establishment of stimulus–response associations requires the striatal processing of stimulus–response contingencies, as demonstrated by McDonald and White, among others, in their studies of the dissociation of three memory systems (for a complete review see McDonald and White 1993; White and McDonald 2002). In the differential reinforcement of a low rate 72-s operant task (DRL-72), the rat is trained in an operant chamber and must learn to wait a minimum of 72 s between lever presses in order to obtain food reinforcement. Any early response resets the clock and is not reinforced (Andrews et al. 1994). Thus, this task implies stimulus–response association as well as temporal discrimination. Early work demonstrated that the neurotoxic lesions produced by either kainic acid or 6-hydroxydopamine (6-OHDA) injection into the ventrolateral striatum impair DRL-20 performance (Dunnett and Iversen 1982). Likewise, the dopaminergic modulation of differential reinforcement of low rate (DRL) schedules through action on the ventral striatum was evaluated by Neill and Herndon (Neill and Herndon 1978). They observed that the direct application of dopamine, D-amphetamine, and scopolamine into the ventral anterior region of the neostriatum of the rat decreased the efficiency of the animals in a DRL-10 schedule of reinforcement.

Serotoninergic modulation of conditioned reward responses has been addressed leading to the paradigm that an increase in serotoninergic activity attenuates conditioned reward responses, whereas selective reduction of cerebral serotoninergic activity enhances conditioned responses. Fletcher et al. (1993) reported that intraraphe injection 8-OHDPAT (where it acts on autoreceptors and inhibits serotonin release) is sufficient to induce conditioned place preference, serving as an unconditioned stimulus. In a later work, Fletcher et al. (1999) observed that the destruction of serotoninergic raphe bodies with 5,7-DHT increases response in conditioned rewarded tasks. However, drugs that inhibit serotonin uptake (such as fluoxetine), improve the performance of rats in a DRL-72 schedule, presumably through indirect stimulation of 5-HT1 receptors (Marek et al. 1989a), whereas the nonselective serotonin antagonist methysergide, blocks the increase in reinforcement seen under DRL schedules (Marek and Seiden 1988). Accordingly, McGuire and Seiden (1980) demonstrated that tricyclic antidepressant drugs may reduce the response rate and increase the reinforcement rate of rats under a DRL-18 operant task, depending on the doses. In this operant schedule responses are reinforced only when they occurred with a delay of at least 18 s, after a previous response, resulting in interresponse intervals (IRTs) longer than 18 s. Interresponse-interval frequency histogram typically showed an inverted U-shaped distribution that changes by shifting to the right as a consequence of lengthening of IRT induced by imipramine or desipramine. However, the effect of increased serotonin could be mediated through diverse 5-HT receptors.

Marek et al. (1989b) did not observe a consistent effect of mCPP, or other 5-HT$_2$ agonists, on DRL-72. However, when the more specific 5-HT$_{2C}$ agonists Ro 60-0175 and Ro 60-0332 where tested, the rate of lever pressing (10 mg/kg IP) significantly decreased (without observable sedation), although only Ro 60-0175 significantly increased the number of reinforcements obtained, whereas a nonsignificant increase (1.7-fold vehicle baseline) in this parameter was induced by Ro 60-0332 (Martin et al. 1998). Thus, 5-HT$_{2C}$ activation apparently modulates the accuracy, evaluated by the increase in the number of reinforcements obtained, in this task.

Striatal 5-HT$_{2C}$ receptor activation or inactivation under DRL schedules has not been evaluated; hence, the assumption that the effect observed in these previous experiments occurs through the action of agonists on striatal 5-HT$_{2C}$ receptors could be speculative.

It can be stated that the effect of an increase of serotonin on the reinforcement rate may be exerted through 5-HT$_{2C}$ receptors. However, specific experimental designs aimed to elucidate a principal role of 5-HT$_{2C}$ receptors must be done since several results show an inhibitory influence of these receptors on reward or reinforcement responses (Di Giovanni et al. 2000; Di Matteo et al. 2000a). The relevant role of dopamine in reward and reinforcement regulation (Bardo 1998; Wise 2002) through projections from the ventral tegmental area to the nucleus accumbens (Wise 2002) is widely accepted. Moreover, 5-HT$_{2C}$ receptors are located in the mesocorticolimbic system affecting the activity of dopamine-dependent neurons, through an inhibitory effect on dopamine release (Di Giovanni et al. 2000; Di Matteo et al. 2000a; De Deurwaerdere et al. 2004), whereas the systemic application of 5-HT$_{2C}$ agonists has been reported to decrease the dopamine efflux into the striatum

(Gobert et al. 2000; De Deurwaerdere et al. 2004; Alex et al. 2005). The effect of the reduction of dopamine after 5-HT$_{2C}$ receptor stimulation and the increase in rewarded responses in the DRL task seems difficult to interpret in view of evidence indicating a relevant participation of dopamine in sustaining the task. It has been reported that the application of the D1 antagonist R(+)-7-chloro-8-hydroxy-3-methyl-1-phenyl-2,3,4,5,-tetrahydro-1H-3-benzazepine (SCH-23390) or the D2 receptor antagonist raclopride caused a severe deficit on DRL-10 schedule behavior in rats, evident in a flattening of the IRT curve after the administration of these antagonists (Cheng and Liao 2007). In agreement with these above data, dopamine depletion in the prefrontal cortex by local injection of 6-OHDP, produced an impairment in DRL-30 schedules evidenced by an increase in the number of responses at shorter IRTs and a fewer number of the obtained reinforcers; in absence of an increase in spontaneous locomotor activity, suggesting a deficit in inhibitory control of these responses (Sokolowski and Salamone 1994). The possibility that the observed effect after the application of a 5-HT$_{2C}$ receptor agonist was due to influences on temporal discrimination seems improbable, since the effects 5-HT$_2$ receptor stimulation on this ability are mediated through action on 5-HT$_{2A}$ receptors, whose activation causes impairment in temporal discrimination in rats (Asgari et al. 2006; Body 2003).

Thus, an indirect involvement of the 5-HT$_{2C}$ receptors on DRL tasks may be inferred from these results.

24.2.3 Auto-shaping Learning Test

Auto-shaping learning tests consist of a Pavlovian instrumental test, which requires striatal (stimulus–response habit formation) participation (Meneses 2002a), and prefrontal interaction, particularly of the orbitofrontal cortex (Chudasama and Robbins 2003). The auto-shaping learning test is conducted in an operant chamber in which ten trials are given in a first test and 20 trials are given in a second test to food-deprived animals. A trial consists of the presentation of an illuminated retractable lever for 8 s (conditioned stimulus) followed by delivery of a food pellet (unconditioned stimulus) with a 60-s intertrial interval. When the animals press the lever the trial is shortened, the lever retracted, the light turned off, and the food pellet is immediately delivered. The response during the lever presentation is regarded as a conditioned response. The increase or decrease of conditioned responses is considered as enhancement or impairment of learning, respectively. The first session (training) occurs 24 h before the retention test and the treatments are administered immediately after the first session of training (Meneses 2002b). In rats, 5-HT depletions produce an increase in the speed and number of responses in this task (Winstanley et al. 2004a), whereas the administration of serotonin uptake inhibitors such as fluoxetine (10 mg/kg IP) increases the proportion of conditioned responses, when tested shortly (1.5 h) after drug administration or longer (24 h) (Meneses 2007a).

Meneses and Hong (1997) tested the effects of several 5HT receptor antagonists as well as an uptake inhibitor and 5-HT depleters on the auto-shaping task. The results showed that both postsynaptic stimulation of 5-HT$_{2A/2C/1B}$ (through mCPP administration, 5 and 10 mg/kg IP) and postsynaptic blockade of 5-HT$_{2B/2C}$ receptors through [1-(naphthyl)piperazine, 1-NP; 0.5–1.0 mg/kg IP] impaired the memory. Moreover, an improvement of this cognitive function occurred after presynaptic activation of 5-HT$_{2C}$ receptors by administration of (±)-2,5-dimethoxy-4-iodoamphetamine (DOI), a 5-HT$_{2A/2C}$ receptor agonist (0.01 and 0.1 mg/kg IP), or presynaptic blockade of 5-HT$_{2C}$ receptors through 6-(2-[4-[bis(4-fluorophenyl)methylene]-1-piperidinyl]-7-methyl-5H-thiazolo[3,2-a]pyrimidin-5-one (ketanserin), a 5-HT$_{2A/2C}$ receptor antagonist (0.001, 0.01, and 0.1 mg/kg IP) and ritanserin (0.5 mg/kg), result in improvement in the retention test 24 h later. In addition, the injection of 5 and 10 mg/kg of 1-(3-trifluoromethyl)phenyl)piperazine (TFMPP), a 5-HT$_{2A/2B/2C}$ agonist, or mesulergine (0.4 mg/kg) decreased the rate of conditioned responses, indicating impairment of memory (Meneses and Hong 1997). In a later work, Meneses (Meneses 2002b) analyzed the participation of 5-HT$_2$ receptors in the auto-shaping learning task and observed that the posttraining administration of N-(1-methyl-indolyl)-N'-(3-pyridyl)urea (SB 200646; 2–20 mg/kg IP), a selective 5-HT$_{2B/2C}$ receptor antagonist, had no effect on consolidation; however, it was able to antagonize the memory impairment induced by mCPP, 1-NP, and mesulergine and to attenuate the impairment produced by TFMPP administration even though SB 200646 was unable to block the memory enhancement caused by DOI and ketanserin. Recently, Meneses evaluated the effect of different agonists and antagonists of 5-HT receptors in short-term memory (STM), tested 1.5 h after the training session, and long-term memory (LTM), testing 24 h after the training session in the auto-shaping test (Meneses 2007a). Both STM and LTM were impaired by mCPP and mesulergine, a 5-HT$_{2C}$ antagonist (although it displays affinity for 5-HT$_{2C/2A/7}$ receptors), at doses of 1.0, 5.0, and 10.0 mg/kg IP, while DOI also impaired STM (0.1, 0.5, and 1.0 mg/kg) and LTM (1.0 mg/kg). In other work, it was found that the administration of SB 200646 (2 mg/kg IP) significantly increased the number of conditioned responses in rats in the LTM test but not in the STM test (Meneses 2007b).

Studies aimed to evaluate auto-shaping learning after the administration of highly specific 5-HT$_{2C}$ agonists or antagonist compounds have not been carried out. However, the results summarized here serve to highlight the fact that the nonselective agonists mCPP and TFMPP impair animal performance, whereas the agonist DOI leads to either impairment or improvement, depending on the dose. The detrimental effect of mCPP and TFMPP could be mediated through 5-HT$_B$ receptors because, as Meneses (Meneses 2007a) showed, 7-trifluoromethyl-4(4-met-hyl-1-piperazinyl)-pyrrolo[1,2-a]quinoxaline maleate (CGS-12066A), a 5-HT$_{1B}$ agonist, causes impairment in this task when administered at doses of 10 mg/kg. Accordingly, Meneses (2003) reported that deficits in learning have been observed after the presynaptic overstimulation of 5-HT$_{1B}$ receptors by drugs displaying high affinity for 5-HT$_{1B}$ receptors, including mCPP and TFMPP. Moreover, antagonism of 5-HT$_{1B/1D}$ receptors through 2'-methyl-4'-(5-methyl[1,2,4]oxadiazol-3-yl)-biphenyl]-amide (GR 127935) increases the consolidation (Meneses et al. 1997).

In line with this interpretation, 1-NP has been reported to act as a 5-HT$_{1B}$ agonist (Hoyer 1988).

The 5-HT$_{2A/2B/2C}$ receptor agonist TFMPP causes impairment in consolidation that could be related to the agonism on 5-HT$_{2B}$ receptors. In fact, DOI, an agonist of 5-HT$_{2A/2C}$ receptors, improves the consolidation through 5-HT$_{2A}$ receptors, as can be inferred from the effects of the antagonist of 5-HT$_{2A}$ (+)-(2,3-dimethoxyphenyl)-1-[2-(4-fluorophenylethyl)]-4-piperidinemethanol (MDL-100907), which abolishes the improving effect induced by DOI, that is not affected by SB 200646 (Meneses et al. 1997). The 5-HT$_{2B/2C}$ receptor antagonist SB 200646 has no effect when administered alone (2 mg/kg IP) but is able to prevent the impairment in memory consolidation caused by mCPP and TFMPP. This may support an effect of SB 200646 through 5-HT$_{B}$ receptor blockade since, as mentioned previously, TFMPP also has affinity for 5-HT$_{1B}$ receptors. However, recently, Meneses (2007a, b) also reported that SB 200646 was able to increase the proportion of conditioned responses in LTM but not in STM. The possible influence of extra training provided by the STM session in which the antagonists had no effect and the proposed better efficacy of SB 200646 as a 5-HT$_{2B}$ antagonist (Hannon and Hoyer 2008) in these effects deserves further research.

In this sense, some apparently contradictory results show that DOI, acting as agonist, as well as ketanserin and ritanserin, acting as antagonists of 5-HT$_{2A/2C}$ receptors, improved the consolidation, while mesulergine caused impairment in memory consolidation by acting through 5-HT$_{2A/2C/7}$ receptor antagonism. Since it has been proposed that the DOI effect was mediated by 5-HT$_{2A}$ receptors and that 5-HT$_{2A}$ and 5-HT$_{7}$ antagonists were inactive in this task (Meneses 2007a; Meneses et al. 1997), it seems possible that both impairment and improvement would be observed after 5-HT$_{2C}$ activation. However, experimental data concerning the effects of highly specific 5-HT$_{2C}$ agonists and antagonists have still to be obtained in order to gain information to clarify these apparently conflicting findings.

24.2.4   Place Conditioning

Place conditioning is a paradigm that involves the preference of a place in which the animal had received any reinforcement (or rewarding stimulus). Using the 5-HT$_{2C}$ receptor agonist 8,9-dichloro-2,3,4,4a-tetrahydro-1H-pyrazino[1,2-a]quinoxalin-5(6H)-one (WAY 161503), Mosher et al. (2005) evaluated place conditioning and did not find any evidence of conditioning either for WAY 161503 (3.0 mg/kg SC) or TFMPP (3.0 mg/kg SC), but found a decrease in locomotor activity that was blocked by the 5-HT$_{2C}$ receptor antagonist SB 242084. Hayes et al. (2009) evaluated the role of 5-HT$_{2C}$ receptors in nicotine-induced place conditioning through the administration of the 5-HT$_{2C}$ receptor agonist WAY 161503 (1.0 and 3.0 mg/kg) and the 5-HT$_{2C}$ receptor antagonist SB 242084 (1 mg/kg). WAY 161530 had no effects on place conditioning, although it decreased basal and nicotine-induced locomotor activity and the effect was blocked by SB 242084.

From the preceding data, it can be observed that agonists having affinity for 5-HT$_{2C}$ receptors produce deficits in passive avoidance learning, whereas antagonists reverted the effects because the activation of 5-HT$_{2C}$ receptors apparently causes amnesic effects in the retention test (24 h after training). Thus, serotonin, through this receptor, may modulate the passive avoidance learning process. Until now, a detrimental effect by activation and improvement by inactivation of 5-HT$_{2C}$ receptors in striatal-dependent learning appeared to occur. It is known that mesocorticolimbic dopamine activity also plays a central role in mediating motivation- and reward-related behavior (Ikemoto and Wise 2004) and that dopamine is a principal neurotransmitter accounting for the organization of striatal-related learning (Badgaiyan et al. 2007); thus, the effects observed in those learning processes sustained by the striatum after serotonin 5-HT$_{2C}$ manipulations could be related to modulation of dopamine release or utilization in this cerebral region. Systemic application of 5-HT$_{2C}$ receptor agonists (such as Ro 60-0175) does not significantly decrease basal firing of substantia nigra pars compacta neurons (Di Giovanni et al. 2000), though it decreases the dopamine efflux in the striatum (Alex et al. 2005). By contrast, the administration of 5-HT$_{2C}$ receptor antagonists (SB 242084) or inverse agonists (SB 206553) result in increase of dopamine efflux in this region (De Deurwaerdere et al. 2004). This effect could be related to the impairment of striatal-dependent learning observed after administration of compounds acting as 5-HT$_{2C}$ receptor agonists because striatal dopamine plays a relevant role in all learning processes sustained by this memory system (White et al. 1993). However, more complex neurochemical mechanisms could underlie these cognitive functions in view of the results obtained in DRL-72 and in auto-shaping learning tests, where a specific 5-HT$_{2C}$ receptor agonist produces improvement of learning (see Table 24.1).

It seems that reaching of a better understanding of the complex interaction of 5-HT receptors in general and 5-HT$_{2C}$ receptors in particular in the regulation of this learning process requires further research through experimental designs including highly specific 5-HT$_{2C}$ receptor agonists and antagonists.

24.3   Prefrontal-Dependent Learning

The prefrontal cortex receives a dense serotonergic innervation from the raphe nuclei (Smiley and Goldman-Rakic 1996), which constitutes a serotonergic system. Prefrontal pyramidal neurons possess several serotonin receptor subtypes, with a particularly high density of 5-HT$_{1A}$ and 5-HT$_{2A}$ receptors (Feng et al. 2001; Jakab and Goldman-Rakic 1998). Although there is comparatively less evidence on the presence of 5-HT$_{2C}$ receptor mRNA in the prefrontal cortex, several studies have shown the presence of 5-HT$_{2C}$ receptor mRNA in the cingulate cortex or in the anterior cingulate cortex (Pompeiano et al. 1994; Hoffman and Mezey 1989). Additionally, a study in human brain showed a similar distribution of 5-HT$_{2C}$ receptor mRNA as has been found in rat brain, including its expression in the anterior cingulate cortex (Pasqualetti et al. 1999). 5-HT$_{2C}$ mRNA is expressed in the same

Table 24.1 Modulation of striatal dependent learning tasks by 5-HT$_{2C}$ Receptors

Compound	Dose, via	Task	Effect	Main result (References)
Ro 60-0332 5-HT$_{2C}$ agonist	5.0 mg/kg, s.c.	PA	Revert deficits	Ameliorate deficits caused by bulbectomy (Martin et al. 1998)
mCPP 5-HT$_{1B/2A/2C}$ agonist	3.0 and 5.0 mg/kg, i.p.	PA	Impairment	Probably acting through 5-HT$_{1A}$ receptors Blocked by Ro 60-0491 (Misane and Ogren 2000)
WAY-163,909 5-HT$_{2C}$ agonist	0.3–3.0 mg/kg, i.p.	PA	Impairment	Dose-dependent deficit (Marquis et al. 2007)
Ro 60-0491 5-HT$_{2C}$ antagonist	3.0 mg/kg i.p.	PA	Block deficit	Block impairment caused by mCPP (Misane and Ogren 2000)
SB-206,553 5-HT$_{2B/2C}$ antagonist	10.0 mg/kg, p.o.	PA	Block deficit	Block the effect of WAY-163909 (Marquis et al. 2007)
Ketanserine 5-HT$_{2A/2C}$ antagonist	0.5, 1.0, 2.0, 4.0 ng Intra-striatum	PA	Deficit	Dose-dependent deficit (Prado-Alcala et al. 2003b)
Ro 60-0175 5-HT$_{2C}$ agonist	10.0 mg/kg i.p.	DRL-72	Improvement	Increased number of reinforcements obtained (Martin et al. 1998)
WAY-161,503 5-HT$_{2C}$ agonist	3.0 mg/kg s.c.	PC	No effect	Hypolocomotion (Mosher et al. 2005)
TFMPP 5-HT$_{1B/2C}$ Agonist	3.0 mg/kg s.c.	PC	No effect	Hypolocomotion (Mosher et al. 2005)
mCPP 5-HT$_{1B/2A/2C}$ agonist	1–10 mg/kg i.p.	ALT	Impairment	Decrease the rate of conditioned responses (Meneses and Hong 1997)
Mesurgeline 5-HT$_{2C/2A7}$ antagonist	(a) 0.4 mg/kg i.p. (b) 1–10 mg/kg i.p.	ALT	a and b. Impairment	a and b. Decrease the rate of conditioned responses (Meneses and Hong 1997; Meneses 2007b)
DOI	(a) 0.01 and 0.1 mg/kg i.p.	ALT	(a) Improvement	(a) Increase of the rate of conditioned responses (Meneses and Hong 1997)
5-HT$_2$ agonist	(b) 0.1, 0.5 and 1.0 mg/kg i.p.		(b) Impairment	(b) Decrease of the rate of conditioned responses (Meneses 2007b)

Ketanserine 5-HT$_{2A/2C}$ antagonist	0.5 mg/kg i. p.	ALT	Improvement	Increase of the rate of conditioned responses (Meneses and Hong 1997)
Ritanserin 5-HT$_{2A/2C}$ antagonist	0.001–0.1 mg/kg i.p.	ALT	Improvement	Increase of the rate of conditioned responses (Meneses and Hong 1997)
SB-200646 5-HT$_{2C/2B}$ antagonist	2.0 mg/kg i.p.	ALT	(a) No effect	(a) Antagonize the impairment induced by mCPP, 1-NP (Meneses 2002a)
			(b) Improvement	(b) Increase of the rate of conditioned responses (Meneses 2007b)

ALT auto-shaping learning task, *PA* passive avoidance, *DRL-72* differential reinforcement of low rate 72 s operant task

brain regions where 5-HT$_{2C}$ receptors are located, and they could be predominately postsynaptic receptors rather than autoreceptors on presynaptic terminals (Clemett et al. 2000). However, the possibility that 5-HT$_{2C}$ receptors act as presynaptic heteroreceptors in some regions cannot be excluded (Alex and Pehek 2007).

5-HT$_{2C}$ receptors also appear to tonically inhibit dopamine release from the mesocortical pathway (Alex and Pehek 2007). Nevertheless, some studies suggest that 5-HT$_{2C}$ receptors localized in the prefrontal cortex do not modulate dopamine release in this region, either tonically (Alex et al. 2005; Pozzi et al. 2002) or phasically (Alex et al. 2005; Pozzi et al. 2002; Pehek et al. 2006). Thus, infusions of SB 206553 directly into the prefrontal cortex did not alter basal, K$^+$-stimulated, or stress-induced cortical dopamine release (Alex et al. 2005; Pehek et al. 2006). However, cortical 5-HT$_{2C}$ receptors modulate dopamine-mediated behaviors because intracortical infusions of a 5-HT$_{2C}$ antagonist potentiated the hyperlocomotion induced by a systemic injection of cocaine (Filip and Cunningham 2003), an effect that is not mediated by alterations in extracellular dopamine in the prefrontal cortex (Alex et al. 2005). Nevertheless, serotonin can influence the excitability of pyramidal neurons through several pathways.

24.3.1 Serotonergic Modulation of Cortical Neuronal Excitability

Serotonin, by activating different receptor subtypes, regulates membrane excitability in the central nervous system in a complex manner (Andrade 1998). The involvement of various voltage-gated ion channels in this action of serotonin receptors has been demonstrated (Carr et al. 2002; Colino and Halliwell 1987; Haj-Dahmane and Andrade 1996; Penington and Kelly 1990). However, application of a low concentration of a 5-HT$_{1A}$ or 5-HT$_{2A/2C}$ agonist alone fails to alter the action potential firing elicited by somatic current injection in prefrontal pyramidal neurons. These results had suggested that the low level of 5-HT$_{1A}$ or 5-HT$_{2A/C}$ activation is not sufficient to trigger the change in voltage-gated channels that affects neuronal excitability. In spite of this, 5HT$_{1A}$ receptor activation can induce a membrane hyperpolarization in cortical neurons (Araneda and Andrade 1991) and in hippocampal neurons (Andrade et al. 1986); this last probably by opening inwardly rectifying potassium channels (Colino and Halliwell 1987) and inhibiting voltage-dependent calcium channels (Penington and Kelly 1990). In contrast, 5-HT$_{2A/2C}$ receptor activation can induce a membrane depolarization in cortical neurons (Aghajanian and Marek 1997; Araneda and Andrade 1991), probably by inhibiting an inwardly rectifying potassium conductance (North and Uchimura 1989) or activating a nonselective cationic current (Haj-Dahmane and Andrade 1996).

Regarding 5-HT$_{2C}$ receptor participation on cellular excitability, it has been observed that iontophoretic application of 5-HT ligands suppresses spontaneous firing of prefrontal cortical neurons (Berg et al. 2008; Bergqvist et al. 1999), suggesting that the 5-HT$_{2C}$ receptor limits the excitability of cortical pyramidal neurons

(Carr et al. 2002). Moreover, it has been observed that the 5HT$_{2C}$ receptor agonist 6-chloro-2-(1-piperazinyl)-pyrazine (MK-212) completely prevents the expression of behavioral sensitization in animals pretreated with 3,4-methylenedioxymetham-phetamine (MDMA) (Ramos et al. 2005). Accordingly, previous studies by Pan and Wang (1991a, b) showed that the inhibition of pyramidal cells in the prefrontal cortex produced by MDMA is mediated mainly through the serotoninergic system. Moreover, it has been hypothesized that lesions of serotonergic terminals induced by repeated MDMA injections could increase the inhibitory effect of serotonin through a hypersensitization mechanism, similarly to the effect of 5,7-DHT lesions (Ashby et al. 1994; Conn et al. 1987; Conn and Sanders-Bush 1987). In this manner, these cells would show higher sensitivity to the inhibitory effects of 5-HT$_{2C}$ receptor stimulation by SCH-23390, which shows high affinity for 5-HT$_{2C}$ receptors (Aguirre et al. 1995). It is also noteworthy that a similar MDMA treatment produces a long-term increase of 5-HT$_{2C}$ receptor mRNA expression in the hippocampus (Yau et al. 1994).

Recently, it was shown that the 5-HT$_{2C}$ receptor is predominantly expressed in the deep layers (layers V/VI) in comparison with those in superficial layers (layers I/II/III) of the medial prefrontal cortex (Liu et al. 2007). The 5-HT$_{2C}$ receptor immunoreactivity was found to be intense in the periphery of cell bodies and processes proximal to cell bodies in primarily round- or fusiform-shaped cells in the rat medial prefrontal cortex. Moreover, 50%t of the 5-HT$_{2C}$ receptor immunoreactivity in this region is colocalized with glutamate decarboxylase (GAD) isoform 67 immunoreactivity, a marker of GABA interneurons (Liu et al. 2007). These findings are coincident with previous data demonstrating that 5-HT$_{2C}$ receptor mRNA is present in a subpopulation of GABA-ergic interneurons in the prefrontal cortex (Vysokanov et al. 1998). In addition, serotonergic axons in the monkey prefrontal cortex predominantly synapse on interneurons (Smiley and Goldman-Rakic 1996). Studies have also shown that the local administration of the 5-HT$_{2A/2C}$ receptor agonist DOI within the medial prefrontal cortex increased extracellular GABA levels in this area (Abi-Saab et al. 1999), an effect mediated by the 5-HT$_{2C}$ receptors and not through the 5-HT$_{2A}$ receptor (Liu et al. 2007).

The 5-HT$_{2C}$ receptor immunoreactivity in the prelimbic cortex was localized primarily to parvalbumin-positive interneurons, which include basket and chandelier neurons that innervate the cell body and the initial segment, respectively, of pyramidal neurons (Conde et al. 1994; Gabbott and Bacon 1997). In contrast, some studies have demonstrated that 25–50% of pyramidal neurons in the prefrontal cortex contain mRNA for 5-HT$_{2C}$ receptors (Vysokanov et al. 1998). Thus, the inhibitory synapses formed by parvalbumin-expressing basket and chandelier neurons are proximal to the initial axon segment of pyramidal cell, the sites at which action potentials are generated. As such, a primary function of parvalbumin-positive GABA neurons is to modulate the efferent signaling of pyramidal neurons (Lewis et al. 2005; Markram et al. 2004; Miles et al. 1996). The 5-HT$_{2C}$ receptor-mediated influence on GABA interneurons would be expected to primarily target the basket and chandelier cells that provide the strongest inhibitory effects on the output of cortical pyramidal cells, compared with interneurons that innervate distal dendrites of the pyramidal cells, which probably function to regulate incoming afferent signals to the pyramidal neurons (Miles et al. 1996). The localization of 5-HT$_{2C}$ receptors at the deep layers of the rat medial prefrontal cortex (layers V/VI) suggests that these receptors act to modulate the output of neurons in these layers. Thus, release of GABA via activation of the 5-HT$_{2C}$ receptors on parvalbumin-positive medial prefrontal cortex GABA interneurons would be expected to reduce excitatory glutamate output as well as subsequent dopamine neurotransmission within the mesoaccumbens pathway (Liu et al. 2007).

With respect to serotonin–glutamate interaction, previous studies have shown that the N-methyl-D-aspartate (NMDA) channel receptor is an important target of 5-HT$_2$ and 5-HT$_{1A}$ receptor modulation (Arvanov et al. 1999; Blank et al. 1996; Yuen et al. 2005). In the presence of NMDA, application of low concentrations of 5-HT$_{1A}$ or 5-HT$_{2A/2C}$ agonists exerts a reducing or enhancing effect, respectively, on the action potential firing, suggesting that NMDA receptor activation provides a "gate" to facilitate the opposing regulation of neuronal excitability by the low level of 5-HT$_{1A}$ or 5-HT$_{2A/2C}$ agonists (Ping et al. 2008). The reason why NMDA facilitates serotonergic regulation of neuronal excitability is likely to be the change of intracellular signaling molecule(s) downstream of Ca^{2+} flow through NMDA channel receptors (Zhong et al. 2008). Evidence shows that the opposing effects of 5-HT$_{1A}$ or 5-HT$_{2A/2C}$ on neuronal excitability are mediated by differential regulation of a converging target, extracellular signal-regulated kinase (ERK). Extracellular signal-regulated kinase can be regulated by the protein kinase A (PKA) or the protein kinase C (PKC) cascade (Roberson et al. 1999) downstream of 5-HT$_{1A}$ or 5-HT$_{2A/C}$ receptors since ERK activation can increase the amplitude of backpropagating action potentials by phosphorylating dendritic A-type K$^+$ channel Kv4.2 subunits (Yuan et al. 2000). In this way, one possible mechanism underlying the regulation of neuronal excitability by 5-HT–NMDA interactions is through the ERK modification of dendritic K$^+$ channels (Zhong et al. 2008).

Recently, the possible involvement of the regulation of neuronal excitability by 5-HT–NMDA interactions in cognitive and emotional processes has been addressed by examining animals exposed to acute stress, since many mental illnesses are exacerbated by stressful conditions (Arnsten and Goldman-Rakic 1998; Mazure et al. 1995). Accordingly, several lines of evidence have shown that stress interferes with serotonin neurotransmission by changing serotonin release or serotonin receptor functions (Adell et al. 1997; Lowry et al. 2000; Maswood et al. 1998; Tan et al. 2004). For example, in the forced swim test, a behavioral paradigm often used to evaluate antidepressant/anxiolytic efficacy, the effect of 5-HT$_{2A/2C}$, but not 5-HT$_{1A}$, on action potential firing is lost in stressed animals, which is associated with the selective loss of 5-HT$_{2A/2C}$-induced increase of ERK activity. This suggests that 5-HT$_{2A/2C}$ receptors are probably desensitized and inactivated by elevated levels of serotonin in response to stress stimulation (Zhong et al. 2008). The other substrate through which 5-HT$_{2C}$ receptors affect psychological processes is the mesoaccumbens pathway, central in psychological processes including motivation, reward, and mood (Nestler and Carlezon 2006; Pierce and Kumaresan 2006; Salamone 1996) and an important site for the actions of psychostimulants such as cocaine (Filip and Cunningham 2003; Pierce and Kumaresan 2006). Thus, alterations in accumbal dopamine or glutamate levels consequent to stimulation of 5-HT$_{2C}$ receptors in the medial prefrontal cortex may be one mechanism by which 5-HT$_{2C}$ receptors modulate the neurochemical and behavioral effects of psychostimulants (Alex and Pehek 2007).

Intramedial prefrontal infusion of 5-HT$_{2C}$ receptor agonists has been shown to block the hypermotive and discriminative stimulus effects of cocaine (Filip and Cunningham 2003) and block the expression of sensitization (the progressive enhancement of the hypermotive effects of a drug following repeated drug exposure) to MDMA (Ramos et al. 2005). On the other hand, intramedial prefrontal cortical infusion of 5-HT$_{2C}$ receptor antagonists enhanced cocaine-induced hyperactivity and increased recognition of the stimulus effects of cocaine (Filip and Cunningham 2003).

Conversely, a growing body of evidence suggests that 5-HT$_{2A}$ and 5-HT$_{2C}$ receptors have opposing functional roles. For example, 5-HT$_{2C}$ receptors appear to inhibit dopamine release, whereas activation of 5-HT$_{2A}$ receptors enhances it (Di Matteo et al. 2001, 2002; Millan et al. 1998). Moreover, antagonism of 5-HT$_{2C}$ receptors potentiates some of the behavioral effects of cocaine, whereas antagonism of 5-HT$_{2A}$ receptors attenuates both cocaine-induced hypermotility and reinstatement of cocaine-seeking behavior (Cunningham et al. 1992; Fletcher et al. 2002a).

In the next section, the 5-HT$_{2C}$ effects on sustained prefrontal learning tasks will be summarized.

24.3.2 Reversal Learning

The serotonergic system regulates the main functions of the prefrontal cortex including emotional control, cognitive behaviors, and working memory (Buhot 1997; Williams et al. 2002). As a result, abnormalities in the serotonergic system have been implicated in the pathogenesis of mental disorders associated with prefrontal cortex dysfunction, such as depression, anxiety, obsessive–compulsive disorder (OCD), and schizophrenia (Doris et al. 1999; Gross et al. 2002; Lemonde et al. 2003). A common characteristic associated with these disorders is cognitive inflexibility, that is, an inability to spontaneously withhold, modify, or sustain adaptive behavior in response to changing situational demands. Thus, in order to study the role of the prefrontal cortex in sustaining behavioral flexibility, reversal learning tasks have been designed for humans (Fellows and Farah 2003; Murphy 2002; Rogers et al. 2000; Rolls et al. 1994), nonhuman primates (Butter et al. 1969; Clarke et al. 2004, 2005, 2007; Dias and Segraves 1996; Lee et al. 2007), and rats (Birrell and Brown 2000; Boulougouris et al. 2007; Idris et al. 2005; McAlonan and Brown 2003; van der Meulen et al. 2007). In these tasks, efficient reversal learning calls upon specific operations such as (1) detection of the shift in contingency; (2) inhibition of a prepotent, learned response; (3) overcoming "learned irrelevance"; and (4) new associative learning (Boulougouris et al. 2008).

The monoamine neurotransmitter serotonin has been strongly implicated in behavioral flexibility (Boulougouris et al. 2008), possibly through its influence on the mentioned functions. In accordance with this hypothesis, selective serotonin depletions in the marmoset prefrontal cortex induced by the neurotoxin 5,7-DHT impaired performance on a serial visual discrimination reversal learning task, which was mainly involved with perseverative responses to a previously rewarded stimulus (Clarke et al. 2004). Subsequent work has established that this deficit was specific to reversal learning and not attentional set shifting (Clarke et al. 2005). More recently, it has been demonstrated that this deficit in reversal learning was specific to serotonin and not dopamine depletion in the orbitofrontal cortex (Clarke et al. 2007). Similarly, systemic administration of the $5\text{-}HT_{1A}$ receptor agonist 8-OH-DPAT impaired serial reversal learning by enhancing perseverative tendencies, an effect that was reversed by the selective $5\text{-}HT_{1A}$ receptor antagonist N-[2-[4-(2-methoxyphenyl)-1-piperazinyl]ethyl]-N-(2-pyridinyl)   cyclohexane carboxamide trihydrochloride (WAY 100635) (Clarke et al. 2003). Similar deficits were observed in rats after 5HT-reduction by administration of a tryptophan-deficient diet and in monkeys with administration of the $5\text{-}HT_3$ receptor antagonist ondansetron in high doses (Barnes et al. 1990; Domeney et al. 1991).

While the involvement of 5-HT systems in reversal learning is well established, the particular 5-HT receptor subtypes that underlie these effects are not well understood.

With respect to inhibitory response control, recent reports indicate that the $5\text{-}HT_{2C}$ receptor antagonist SB 242084 increases premature responding on the five-choice serial reaction time (5-CSRT) task, whereas the $5\text{-}HT_{2A}$ receptor antagonist MDL-100907 decreases it by the same measure (Higgins and Fletcher 2003; Winstanley et al. 2004b). In this sense, a dissociable behavioral effect of the selective $5\text{-}HT_{2A}$ antagonist MDL-100907 and the $5\text{-}HT_{2C}$ antagonist SB 242084 on serial spatial reversal learning has been observed (Boulougouris et al. 2008). MDL-100907 impaired initial reversal learning by increasing the number of trials (highest dose, 0.1 mg/kg IP) and incorrect responses to the criterion (two highest doses, 0.03 and 0.1 mg/kg IP). This impairment, perseverative in nature, occurred in the absence of significant effects on retention of previous stimulus–reward contingencies. In contrast, SB 242084 improved reversal learning by decreasing the same measures (the two highest doses, 0.3 and 1.0 mg/kg). Moreover, the analysis of type of errors revealed that the opposing effects of the $5\text{-}HT_{2A}$ and $5\text{-}HT_{2C}$ receptor antagonists were specific to early reversal I stages (when more reversal tests were made, no effects were induced by the drugs) affecting perseverative but not learning errors (Boulougouris et al. 2008).

However, the study of the effects of $5\text{-}HT_{2C}$ receptor agonists on compulsive behavior has led to contradictory results. It has been observed that $5\text{-}HT_{2C}$ receptor activation induced "compulsive" grooming (Graf 2006; Graf et al. 2003) and directional persistence in spatial alternation (Tsaltas et al. 2005); whereas, in other models such as marble burying and schedule-induced polydipsia, $5\text{-}HT_{2C}$ agonists attenuated compulsive behavior and blockade of $5\text{-}HT_{2C}$ receptors increased compulsive drinking (in the polydipsia model) (Martin et al. 1998, 2002), though the anticompulsive effects of 5HT$_{2C}$ agonists have been attributed to their sedative effects (Kennett et al. 2000). On the other hand, it has been demonstrated that 5-HT$_{2C}$ antagonism mimics some of the effects of psychostimulant drugs such as D-amphetamine, which increases dopamine release in the nucleus accumbens (Cole and Robbins 1987, 1989); D-amphetamine causes a similar pattern of behavioral effects on the 5-CSRT tasks as SB 242084, increasing the number of premature responses (Cole and Robbins 1987; Harrison et al. 1997). Thus, there could be an interaction with dopamine modulation by 5-HT$_{2C}$ receptors.

Effects of 5-HT$_{2C}$ receptor antagonism have also been reported to enhance the stimulant effects of several drugs of abuse such as phencyclidine and MDMA (Fletcher et al. 2002a; b, 2001, Hutson et al. 2000). A number of studies attribute the proaddictive effects of 5-HT$_{2C}$ antagonists to an increase of the dopaminergic activity in the ventral tegmental area and the consequent increased release of dopamine in the nucleus accumbens (Gobert et al. 2000; Di Matteo et al. 1999, 2000a, b; Millan et al. 1998; Higgins and Fletcher 2003; Di Giovanni et al. 2001). In contrast, MDL-100907 neither influences the spontaneous firing rate of dopaminergic neurons nor alters basal levels of dopamine or norepinephrine (noradrenaline) release (Kehne et al. 1996), but it does attenuate amphetamine-induced hyperactivity (Sorensen et al. 1993) and amphetamine, DOI, or MDMA-induced dopamine release (Gobert and Millan 1999; Porras et al. 2002; Schmidt et al. 1994).

The finding that SB 242084 reduced perseverative responding in spatial reversal learning, a task dependent on the orbitofrontal cortex (Boulougouris et al. 2008), suggests that this facilitatory effect of 5-HT$_{2C}$ antagonists is possibly mediated by the orbitofrontal cortex (Boulougouris et al. 2008). Nevertheless, Clarke et al. (2007) showed that selective 5-HT depletion of the marmoset orbitofrontal cortex impaired performance of a visual serial reversal learning task; this deficit was due to a failure to inhibit responding to the previously rewarded stimulus. On the other hand, mutant mice devoid of 5-HT$_{2C}$ receptors display enhanced exploration of a novel environment (Rocha et al. 2002), as well as perseverative behavior, a component of OCDs. Likewise, it has been reported that 5-HT$_{2C}$ receptor knockout mice may display compulsive-like behavior (Chou-Green et al. 2003).

It is known that mCPP causes exacerbation of OCD symptoms in patients (Zohar and Insel 1987) and it appears that selective serotonin reuptake inhibitors produce anti-OCD effects through 5-HT$_2$ receptors (Erzegovesi et al. 1992). Since several lines of evidence show any participation of 5-HT$_{2A}$ receptors in OCD, it has been suggested that the effects of mCPP on the persistence are mediated by mCPP acting via 5-HT$_{2C}$ receptors. Meta-chlorophenylpiperazine acts as a nonselective agonist for 5-HT$_{1A/1B/2C}$ receptors and antagonist of 5-HT$_3$ receptors. Since MK-212, which has a high affinity for 5-HT$_{1A}$ and 5-HT$_{2C}$ receptor subtypes, had no effect on OCD symptom intensity, this apparently excludes the participation of 5-HT$_{2C}$ in OCD behavioral expression (Gross et al. 1998). However, in monkeys, compulsive whole-body scratching is produced by 8-OH-DPAT, and the effect can be fully reversed by subsequent treatment with 5-HT$_{2C}$ receptor agonists more potently than with fluoxetine (Martin et al. 1998).

As observed from the previous data, a complex modulation of perseveration could be occurring through 5-HT$_2$ serotonin receptors, and the specific role of 5-HT$_{2C}$ is beginning to be elucidated.

24.4 Hippocampal-Dependent Learning

The hippocampus has been widely related to spatial processing, as well as to long-term memory establishment, among other cognitive processes. It has been proposed that the hippocampus is part of a memory system involved in the processing of relationships between configurations of stimuli such as those required in the integration of spatial maps (cognitive mapping) (Jarrard 1993; O'Keefe and Nadel 1978; Poucet 1993). Based on this assumption, the hippocampus has been considered crucial in those learning tasks that involve spatial information such as spatial reference memory and spatial working memory, acting in concert with the prefrontal cortex (Jones and Wilson 2005; Puryear et al. 2006; van Asselen et al. 2006; Wang and Cai 2006).

Studies addressing the serotonergic modulation of place learning have showed that cerebral serotonin depletion has no effect on the learning ability of rats evaluated using the Morris water maze (Murtha and Pappas 1994; Richter-Levin and Segal 1989), whereas other works have reported the enhancement of spatial discrimination as a consequence of cerebral serotonin depletion (Altman et al. 1990; Normile et al. 1990). However, when rats are submitted to double lesion including cholinergic and serotonergic afferents or neurotransmitter depletion, the deficiencies observed are more severe that those caused by a cholinergic lesion alone (Murtha and Pappas 1994; Richter-Levin and Segal 1989; Nilsson et al. 1988; Richter-Levin et al. 1993). Thus, a modulation of place navigation by serotonin has been proposed from the results of early studies, as indicated.

However, only a few works addressed the participation of 5-HT$_{2C}$ receptors in hippocampal-dependent learning, in spite of the diverse evidence supporting its participation in the regulation of hippocampal physiology. In this sense, it has been observed that both hippocampus and medial septum/Broca diagonal band (MS/BDB) complex express a variety of serotonin receptors, including the 5-HT$_{2C}$ receptor (Pompeiano et al. 1994; Clemett et al. 2000) located exclusively on GABA-ergic interneurons (Gulyas et al. 1999; Leranth and Vertes 1999). It has been observed that mCPP application (0.1 mg/kg IP) to anesthetized rats inhibits the firing of MS/BDB vertical limb complex cells and that this parallels the abolition of theta oscillations registered in dentate gyrus and CA1. The 5-HT$_{2C}$ receptor mediation of this effect was tested after the administration of Ro 60-0175 (1.0 mg/kg), which produced the same effect (Hajos et al. 2003). Conversely, the application of the selective antagonists SB 242084 (1.0 mg/kg IP) and SB 206553 (1.0 mg/kg IP) induced an increase of neuronal oscillations of the medial septum and hippocampus, as well as increases in hippocampal theta activity.

These findings support the participation of 5-HT$_{2C}$ in the tonic regulation of hippocampal electrical theta activity and the modulation of several physiological and pathological processes sustained by the hippocampus (Hajos et al. 2003). In another study, the administration of the selective 5-HT$_{2C}$ receptor antagonist SB 242084 (1.0 mg/kg IP) increased the theta power in the frequency range 5–8 Hz in awake rats (Kantor et al. 2005). Evidence that theta oscillations are related to memory formation has been obtained (Hasselmo et al. 2002; Seager et al. 2002; Winson 1978). Particularly in the Morris water maze, increases in CA1 theta activity at frequencies of 6.5–9.5 Hz are associated with learning (Olvera-Cortes et al. 2002, 2004). In addition, pharmacological manipulation has shown that drugs that reduce theta activity also block learning (Hasselmo et al. 2002; Givens 1995; Givens and Olton 1995) and that drugs that increase the expression of theta activity facilitate the induction of long-term potentiation (LTP) and learning (Staubli and Xu 1995; Wu et al. 2000).

Thus, the ability of 5-HT$_{2C}$ agonists and antagonists to modulate theta activity supports the possibility that these receptors are modulating the hippocampal-dependent learning process. However, as mentioned, few works have evaluated the effect of compounds with selective 5-HT$_{2C}$ receptor affinity in hippocampal-dependent learning.

24.4.1   Place Learning

Place learning in the Morris water maze is a test extensively used in the evaluation of hippocampal learning and memory (McNamara and Skelton 1993).

Tecott et al. (1998) did not observe differences during the training between mutant mice, lacking 5-HT$_{2C}$ receptors, and wild-type mice in the Morris water maze. However, in the probe trial effected the last day of training, mutant mice did not show any preference for the target quadrant where the escape platform had been located, whereas the wild-type mice preferred this quadrant. This lack of preference for the escape platform quadrant indicates that the mutant mice probably used a nonhippocampal strategy to solve the training tasks (Tecott et al. 1998). In fact, animals can solve these tasks by using cue learning or egocentric learning as efficiently as place learning (Olvera-Cortes et al. 2004). In the same study, Tecott et al. (1998) used cerebral slices from mutant mice to evaluate the induction of LTP in different hippocampal regions and observed that only the LTP in the perforant path–dentate gyrus synapse was attenuated. This finding indicates a perturbation in dentate gyrus functioning that could impede the animals to use a hippocampal-dependent strategy, since it has been reported that the selective lesion of granular dentate cells impairs early phases of acquisition in hippocampal-dependent learning (Schuster et al. 1997) and that LTP has been proposed as a mechanism underlying memory formation (Lynch 2004). On the other hand, 5-HT$_{2C}$ deficient mice are prone to spontaneous convulsions, and it has been suggested that these receptors mediate the tonic inhibition of neuronal network excitability (Heisler et al. 1998;

Tecott et al. 1995). However, in the previously described work, the 5-HT$_{2C}$ receptor mutation was generalized, and therefore functional alterations in other regions relevant to spatial learning such as the medial septum, which is the pacemaker of hippocampal electrical activity and the source of acetylcholine release in the hippocampus, cannot be discarded. Acetylcholine modulation of learning has been extensively established (Altman et al. 1990; McNamara and Skelton 1993; Albert 1996; Lamberty and Gower 1991).

With regard to the evaluation of the effect of compounds with affinity for 5-HT$_{2C}$ receptors, Kant et al. (1996) evaluated the effect of the 5-HT$_{2C}$ receptor agonist TFMPP (5.0 mg/kg IP) on rat learning by using a water maze test in which the rats are challenged to learn to swim through alleyways and doors to reach a platform, in order to evaluate working spatial memory. Both the time required to find the platform and the number of errors made were the variables considered. The different tests in rats receiving TFMPP showed an increase of the time required by the rats to solve the task, but the number of errors were not affected, though these effects seem to be better explained by alteration of motor functions instead of learning process, as was suggested by the same authors in other work in which DOI (0.1 and 0.25 mg/kg IP) had a slowing effect on movement and the latencies to find the platform increased) without an increase in the number of errors (Kant et al. 1998). Thus, hypolocomotion has been observed after acute administration of mCPP (2.5 mg/kg IP) in rats evaluated in a novel open field test (exploration-associated locomotion), where the animals displayed a low number of turns and rears (Fone et al. 1998). Similar observations were made after the administration of TFMPP and DOI (Lucki et al. 1989; Pranzatelli et al. 1992). Therefore, it is important to mention that 5-HT$_{2C}$ receptors seem not to be involved in the working spatial memory.

Recently, Khaliq et al. (2008) tested place learning of rats after the administration of mCPP and observed a negative correlation between memory function and 5-HT$_{2C}$ receptor stimulation. The mCPP (1.0, 3.0, and 5.0 mg/kg IP) impaired memory function in a dose-dependent manner, increasing the latency to find a sunken platform in the water maze in a 24-h posttraining memory test. The authors stated that this detrimental effect was not related to hypolocomotion because passive avoidance tests produced similarly impaired retention at all doses tested.

However, passive avoidance implicates an absence of response, whereas the swimming speed in any water maze could be influenced by hypolocomotion; hence, a measure of pathway lengths must be conducted to assert the influence of motor alterations, which may be relevant because hypolocomotion has been consistently observed when locomotor activity was evaluated. Further experimental work must be done to eliminate motor influences and to obtain conclusive results about 5-HT$_{2C}$ receptor regulation in place learning ability.

In a spatial working memory test with delay (delayed nonmatching to position task [DNMTP]), cerebral serotonin depletion induced by the administration of 5,7-DHT in rats had no effects. After the authors applied DOI (100 and 300 mg/kg IP), no effects on choice accuracy were observed (Ruotsalainen et al. 1998). Thus, evidence on 5-HT$_{2C}$ receptor directly affecting hippocampal-dependent test was not obtained. Although tests of working memory are prefrontally driven, the spatial component of this function implies hippocampal–prefrontal interactive relationships, and no effects after serotonin manipulation were observed.

24.4.2   Novel Object Recognition

Rodents naturally tend to approach and explore novel objects, which lack of significance for the animals and which have never been paired with a reinforcing stimulus; thus, they show innate preference for novel objects over familiar objects. The rodents approach novel objects and investigate them by touching and sniffing them (Aggleton et al. 1995). The novel object exploration can be evaluated experimentally by testing recognition memory. Briefly, animals are exposed to two identical objects placed in two opposite corners of an apparatus and exploration of the objects is allowed during a period of 2 min. Twenty-four hours later the animals are submitted to an exploration session with a known and a novel object, and the time employed in exploring each object is recorded (Ennaceur and Delacour 1988).

Recognition memory tests are dependent on normal hippocampal functioning when they are tested after longer delays (10 min, 1 h, and 24 h), as shown by the impairment of object recognition after lesions of the hippocampus (Clark et al. 2000). Likewise, inactivation of the dorsal hippocampus through locally infused lidocaine application impairs one-trial object recognition testing with 24 h delay in C57BL/6 mice (Hammond et al. 2004).

Acute tryptophan depletion is a pharmacological procedure to decrease cerebral serotonin levels (Lieben et al. 2004a). Early studies found that serotonin depletion caused by tryptophan depletion impairs object recognition (Lieben et al. 2004b; van Donkelaar et al. 2008). Moreover, dorsal raphe lesion with 5,7-DHT produced a pronounced impairment in object recognition (Lieben et al. 2006). However, chronic administration of fluoxetine also causes object recognition impairment in the rat (Valluzzi and Chan 2007).

Pitsikas and Sakellaridis (2005) evaluated the effect of the administration of 3.0 mg/kg Ro 60-0491, a 5-HT$_{2C}$ receptor antagonist, just after the training trial and evaluated the retention 24 h later, suggesting that the 5-HT$_{2C}$ receptor antagonist modulates the storage and retrieval of information because the drug-treated rats spent more time exploring the novel object than did the saline-treated animals. Moreover, effects on motor activity or exploration were not observed in the drug-treated animals. Siuciak et al. (2007) evaluated the effect of 2-(3-chlorobenzyloxy)-6-(piperazin-1-yl)pyrazine (CP-809101), a potent selective 5-HT$_{2C}$ receptor agonist, in a novel object recognition task and observed an enhancing effect of the drug compared with vehicle-treated mice, because mice treated with CP-809101 (1 mg/kg SC) spent significantly more time exploring novel objects. In this work, the authors applied the drug before the first exposure to the objects, in a manner such that the effects could influence the acquisition process, consolidation, or both.

At present, few works address the participation of 5-HT$_{2C}$ receptors in recognition memory and data from the previous studies presented here were obtained from experimental animals of different species and several schedules of drug administration; hence, well-supported proposals about the nature of 5-HT$_{2C}$ receptor participation in these cognitive phenomena could not easily be done.

In summary, experimental data from pharmacological studies concerning 5-HT$_{2C}$ receptor participation in the hippocampal-related learning process have not yielded consistent evidence. Few studies have suggested a role for 5-HT$_{2C}$ receptors in the modulation of place learning ability, and most of the obtained results indicate a hypolocomotion influence on escape latency, a parameter that has been used as the index of learning in the Morris water maze. It seems that path lengths must be analyzed in place learning evaluations under the actions of 5-HT$_{2C}$ agonist or antagonist compounds in order to discard the influence of drug-induced locomotor alterations. With regard to tests with spatial components such as DNMTP, compounds with affinity for 5-HT$_{2C}$ receptors have not elicited effects. Finally, inconsistent results have been obtained in object recognition memory tests in which antagonists or agonists improve recognition memory (see Table 24.2).

In view of the influence of 5-HT$_{2C}$ receptors on hippocampal physiology, we believe that exhaustive analysis at behavioral, biochemical, and electrophysiological levels addressing the role of 5-HT$_{2C}$ receptors must be conducted.

Table 24.2 Hippocampal dependent learning modulation by 5-HT$_{2C}$ receptors

Approach	Dose, Via	Task	Effect	Main result (Reference)
Mutant mice (no functional 5-HT$_{2C}$ receptors)		WM	Impairment	Used no hippocampal strategy (Tecott et al. 1998)
TFMPP 5-HT$_{2C}$ agonist	5.0 mg/kg, i.p.	WM: RM/ WM	No effect	Latencies increased by hypolocomotion (Kant et al. 1996)
DOI 5-HT$_{2A/2B/2C}$ agonist	0.1 and 0.25 mg/kg, i.p.	WM: RM/ WM	No effect	Latencies increased by hypolocomotion (Kant et al. 1998)
mCPP 5-HT$_{1B/2A/2C}$ agonist	1.0, 3.0, 5.0 mg/ kg, i.p.	WM	Impairment	Latencies increased. Possibly due to hypolocomotion (Khaliq et al. 2008)
DOI 5-HT$_{2A/2B/2C}$ agonist	100 and 300 µg/ kg, i.p.	DNMTP	No effect	Affect only reversal learning (Ruotsalainen et al. 1998)
Ro 60-0491 5-HT$_{2C}$ antagonist	3.0 mg/kg, i.p.	OR	Improvement	More time exploring novel object (Pitsikas and Sakellaridis 2005)
CP-809,101 5-HT$_{2C}$ agonist	1.0 mg/kg, s.c.	OR	Improvement	More time exploring the novel object (Siuciak et al. 2007)

DNMTP delayed not matching to place task, *WM* water maze, *RM* reference memory, *WM* working memory, *OR* object recognition task

24.5   5-HT$_{2C}$ Effects on Factors Influencing Learning Processes

24.5.1   Anxiety

The 5-HT$_2$ receptor family seems to be particularly involved in anxiety because several drugs effective for the treatment of anxiety disorders interact with this type of receptor (Mora et al. 1997; Peroutka 1995). Mora et al. (1997) tested mCPP, ritanserin and trans-4-[(3Z)3-(2-dimethylaminoethyl)oxyimino-3-(2-fluorophenyl) propen-1-yl]phenol hemifumarate (SR-463496A), a selective 5-HT$_{2A/2C}$ receptor antagonist (Rinaldi-Carmona et al. 1992) on two types of fear. The authors used a paradigm in which the same rat in one experimental session was exposed to two types of fear, in an elevated T-maze with the perpendicular arm closed and the other with arms open. By placing the animal in the closed arm, the training of inhibitory passive avoidance is evaluated by measurement of the time that the animal takes to leave the closed arm during three consecutive trials. This inhibitory avoidance task is assumed to represent conditioned fear. The same animal is placed in the open arm in a one-way escape task, which is assumed to represent unconditioned fear (Graeff et al. 1993). Using this task, Mora et al. (1997) observed a facilitatory effect of TFMPP on inhibitory avoidance, which was evident in an increase in the latency to leave the closed arm, whereas the latency to leave the open arm was increased in rats receiving 0.2 mg/kg IP Thus, the drug impaired the one-way escape task. Similarly, mCPP increased the avoidance latency in a dose-dependent manner, while it tended to impair one-way escape. SB 200646A, a 5-HT$_{2C}$ receptor antagonist, dose-dependently decreased the avoidance latency. However, the one-way escape task was not affected by the drug. Inhibitory avoidance was also impaired by the selective 5-HT$_{2C}$ antagonist (+)-*cis*-4,5,7a,8,9,10,11,11a-octahydro-7H-10-methylindolo(1,7-bC)(2,6) naphthyridine (SER-082; 0.1–1.0 mg/kg). Thus, the conditioned fear seems to be tonically facilitated through 5-HT$_{2C}$ receptor stimulation, whereas unconditioned fear might be phasically inhibited by activation of 5-HT$_{2C}$ receptors (Graeff et al. 1993).

Jones et al. (2002) evaluated the 5-HT$_{2C}$ receptor participation in unconditioned escape behavior using an unstable elevated exposed plus maze. Based on the evidence that mCPP induces panic in humans and increases the unconditioned escape behavior in a dose-dependent manner in the rat, the 5-HT$_{2C}$ antagonist SB 242084 was administered to rats previously treated with mCPP, and a dose-dependent inhibition of the increase in escape and hypolocomotor effects was observed. Moreover, Martin et al. (2002) tested the anxiolytic effects of SB 242084 administered in rats at doses that reverse the mCPP-induced hypolocomotor effect and observed that the 5-HT$_{2C}$ antagonist is able to cause anxiolytic effects as determined by the elevated plus maze task.

In other studies, local administration of TFMPP (0.75 and 1.5 µg) in the ventral hippocampus produced a reduction in the open arm exploration by rats exposed to the elevated plus maze, without effects on the number of entries in the closed arm, indicating a selective anxiogenic profile. A higher dose of TFMPP (3.0 µg) reduced both open and closed arm entries, suggesting hypolocomotion. Similar results were observed after application of MK-212, suggesting anxiogenic effects due to 5-HT$_{2C}$ agonism (Alves et al. 2004).

Moreover, in other studies in rats, it was observed that chronic unpredictable mild stress facilitates 5-HT$_{2C}$ receptor function, whereas antidepressant treatments reduced 5-HT$_{2C}$ receptor function (Jenck et al. 1993). The pharmacological characterization of the effects of the 5-HT$_{2C}$ agonist mCPP led to the hypothesis that activation of 5-HT$_{2C}$ receptors may mediate anxiety in humans and in rodents (Kantor et al. 2005). Earlier, it was suggested that the anxiety produced in rodents and possibly in humans after the administration of a selective serotonin reuptake inhibitor (SSRI), antidepressants, or mCPP could be mediated by activation of 5-HT$_{2C}$ receptors (Bagdy et al. 2001; Kennett et al. 1989). In addition, the 5-HT$_{2C}$ receptor antagonist SB 242084 has shown anxiolytic effects in several anxiety tests (Martin et al. 2002; Kennett et al. 1997).

The social environment can be, among others, an important source of ethological-based stress (File and Seth 2003; Koolhaas et al. 1997). Kantor et al. (2005) evaluated the effect of the administration of SB 242084 in a social interaction anxiety test using a highlighted unfamiliar arena in which a pair of rats was placed together in order to evaluate the social interaction of each rat with an unknown, similar sized (none differing more than 15 g in weight) test partner. Both the magnitude and the time of social interaction under the effects of SB 242084 were compared with those recorded after chlordiazepoxide (CDP), a widely used benzodiazepine, administration. SB 242084 increased the time of social interaction at doses of 0.3 and 1.0 mg/kg, while the magnitude of social interaction was similar to that observed under CDP effects.

As previously discussed, 5-HT$_{2C}$ receptors could influence learning processing through their effect on anxiety levels. Particularly, acute inescapable stress in rats dramatically affects synaptic plasticity in the hippocampus and inhibits LTP when stress occurred prior to the induction of this electrical phenomenon (Diamond et al. 2005; Shors et al. 1989). It is known that inescapable stress led to an increase of 5-HT output in several brain areas including the hippocampus (Vahabzadeh and Fillenz 1994), and in unstressed animals, drugs increasing serotonin activity such as fluoxetine (an SSRI) and fluvoxamine inhibit the induction of LTP (Kojima et al. 2003; Shakesby et al. 2002). In stressed animals, Ryan et al. (2008) evaluated the participation of 5-HT$_2$ receptors on the inhibition of CA1 high frequency stimulation-induced LTP, produced by unavoidable stress; they observed that fenfluramine (5 mg/kg IP) is able to reverse the stress-induced LTP inhibition, and the 5-HT$_2$ receptor antagonist cinanserin prevented this reversal, indicating a primarily 5-HT$_2$ effect of the fenfluramine. In addition, the preferential 5-HT$_{2C}$ receptor antagonist MK-212, enabled the induction of LTP in stressed rats (3 mg/kg IP). The same effect was observed after the application of α-methyl-5-(2-thienylmethoxy)-1H-indole-3-ethanamine hydrochloride (BW 723C86), a 5-HT$_{2B}$ receptor agonist, showing that the activation of both receptor subtypes is enough to revert the stress-induced inhibition of LTP recorded from the CA1 hippocampal subfield.

Thus, besides direct influences on learning processes, 5-HT$_{2C}$ receptors can be involved in the modulation of other processes, like anxiety, that may in turn influence learning ability.

24.5.2   Impulsiveness

Forebrain serotonin depletion induced by 5,7-DHT administration increases impulsive responses in different experimental paradigms including go/no-go and differential-reinforcement-of-low-rate schedule of reinforcement tasks (Fletcher 1995; Harrison et al. 1999).

Impulsiveness is defined as an action without adequate forethought; in addition, alterations in the impulse control or lack of behavioral inhibition are part of disorders such as OCD, attention deficit hyperactivity disorder (ADHD), schizophrenia, antisocial behavior, and addictive behavior (American Psychiatric Association 1994).

Five-choice serial reaction time is helpful in identifying animals with attentional deficits in conjunction with impulsiveness (Puumala et al. 1996). The task was adapted from Leonard reaction time test for humans (Wilkinson 1963) and requires a rat to detect and respond to brief flashes of light presented randomly in one of five spatial diverse locations in order to evaluate attention, impulsiveness, speed of response, and motivation (Carli et al. 1983). Puumala et al. reported that the measure of attention (percentage of correct responses) is inversely correlated with the probability of premature responses, a reflection of impulsiveness (Puumala et al. 1997). Further, Puumala and Sirvio (1998) demonstrated that poorly performing rats had a higher serotonin utilization ratio (5-HIAA/5-HT) in the frontal cortex than well-performing rats in the task. This suggests a serotonin influences against the accuracy of responses in the 5-CSRT task.

However, global serotonin depletion caused by intracerebroventricular administration of 5,7-DHT increases the number of premature responses in the 5-CSRT task in rats (Winstanley et al. 2004a). When the 5-HT$_{2A/2C}$ receptor agonist DOI was used, Koskinen et al. (2000a) observed that DOI decreased the number of completed trials (0.1 or 0.15 mg/kg IP), whereas only the administration of 1.0 mg/kg resulted in increase of the number of premature responses, without effect on choice accuracy.

Serotonin plays a role in OCD, a relatively common anxiety disorder characterized by recurrent intrusive thoughts and repetitive time-consuming behaviors (Antony et al. 1998). Thus, an effect on learning processes could arise in the ability to modulate perseveration in rats, that could be an opposing effect in learning tests in which the animals must learn to act depending on information provided and to inhibit perseverative responses.

Higgins and Fletcher (2003) reported that the 5-HT$_{2C}$ receptor antagonist SB 242084 increases premature responding on the 5-CSRT task, as well as decreases the latency for correct responses. Using this task, Winstanley and colleagues administered the 5-HT$_{2C}$ receptor antagonist SB 242084 to rats previously lesioned with 5,7-DHT and control rats. They reported an increase in premature responding both in control and in serotonin-depleted animals, antagonist treated. However, there was no effect on the accuracy of responses (Winstanley et al. 2004b). Talpos et al. (2006) compared the effects of ketanserin (5-HT$_{2A/C}$ receptor antagonist) and SER-082 (5-HT$_{2C/2B}$ receptor antagonist) by means of two tests of impulsiveness, the 5-SCRT task and the delayed reward task. The authors observed a decrease in premature (impulsive) responding after ketanserin administration in the 5-CSRT task, as was previously described (Koskinen et al. 2000b; Passetti et al. 2003), although no effect was caused by the treatment in the delayed reward task. SER-082 had no effect on the 5-CSRT task, whereas impulsive responding decreased in the delayed reward task. The authors explained the results of this work by considering the properties of SB 242084 as an antagonist, an inverse agonist, or as a weak agonist. However, similarly, the accuracy of responses as reported previously was not affected in any of the tests.

Fletcher et al. (2007) applied different compounds to rats and evaluated the animals using the 5-SCRT task. They observed that ketanserin (1.0 mg/kg IP) and the 5-HT$_{2A}$ receptor antagonist MDL100197 (0.01, 0.1, and 0.5 mg/kg IP) decreased the number of premature responses, using an intertrial interval (ITI) of 5 and 9 s, whereas the 5-HT$_{2B}$ receptor antagonist 6-chloro-5-methyl-1-(5-quinolylcarbam-oyl)indoline (SB 215505) had no effect, and the 5-HT$_{2C}$ receptor antagonist SB 242084 increased the number of premature responses (0.1 and 0.5 mg/kg IP) with an ITI of both 5 and 9 s. DOI (0.3 and 0.6 mg/kg) reduced the accuracy of responding, increased the number of omissions, and increased the latencies of responses and reinforcer collections, whereas it did not affect premature responding. The 5-HT$_{2C}$ receptor agonist Ro 60-0175 increased only the proportion of omitted trials, the latency to respond, and the latency to collect the reinforcer (0.6 mg/kg). The results replicate the extended result that 5-HT$_{2A}$ receptor blockade reduces and 5-HT$_{2C}$ blockade enhances premature responding (impulsiveness). In another study, Quarta et al. (2007) analyzed the modulation of 5-HT$_{2C}$ on behavioral nicotine effects and evaluated the influence of the 5-HT$_{2C}$ receptor agonist Ro 60-0175 on responses to nicotine in the 5-CSRT task. Nicotine positively affected the response indices as response latencies and omission errors and produced anticipatory responding. Ro 60-175 counteracted the effects of nicotine, increasing the response latencies and omission errors, but it had little effect on response accuracy.

It appears that 5-HT$_{2C}$ receptors actively participate in the regulation of impulsiveness, because the administration of antagonists increases the impulsiveness measured as an increase in premature responses. However, this increase in the impulsiveness does not act in detriment to the accuracy of responses and may be limited to a motor influence of 5-HT$_{2C}$. In accordance with this, the application of 5-HT$_{2C}$ agonists induces hypolocomotion as was mentioned previously.

Thus, both direct and indirect effects of 5-HT$_{2C}$ receptors activation or inactivation in learning process have been observed. It seems that several neurotransmitter systems (e.g.,dopaminergic, cholinergic, glutamatergic) and several cerebral systems (striatum, prefrontal cortex, and hippocampus) are involved as the neural substrate sustaining this modulation. Until now, the principal restraint in the experimental approach of 5-HT$_{2C}$ receptors is the unavailability of selective compounds that allow attribution of specific effects on learning processes to 5-HT$_{2C}$ receptors. Nevertheless, the results summarized in the review show that these receptors are important in learning modulation.

24.6   Clinical Implications

Data concerning 5HT-DA interaction in brain processes of learning and memory have been recently reviewed (Olvera-Cortes et al. 2008), showing the relevant regulatory role of serotonergic neurotransmission on dopaminergic-mediated phenomena in the central nervous system, including those accounting for cognition, under both physiological and pathological conditions.

Nowadays it is recognized that serotonin–dopamine interactions may have important clinical implications since they are involved in pathophysiological mechanisms of some neurological and psychiatric diseases (like schizophrenia, Parkinson and Huntinton diseases, ADHD, Alzheimer disease, anxiety, depression, and drug addiction) that may evolve with cognitive dysfunctions besides motor, mood, or behavioral disturbances and may able to be improved by drugs acting on dopamine or serotonin neurotransmission (Araki et al. 2006; Boulougouris and Tsaltas 2008; Cools 2006; Di Pietro and Seamans 2007; Gray and Roth 2007; Kostrzewa et al. 2005; Remington 2008; Scholes et al. 2007; Scholtissen et al. 2006). Further, even though derangement of different brain dopaminergic neurotransmission systems has been mainly implicated in these pathologies, they also affect other nondopaminergic neuronal structures, including neurotransmitter receptors underlying synaptic serotonergic activity.

Also, degeneration of serotonergic neurons in the medial raphe nucleus (Halliday et al. 1990; Paulus and Jellinger 1991), reduction of several key markers of serotonergic activity in the striatum (Chinaglia et al. 1993; Guttman et al. 2007; Kerenyi et al. 2003; Kim et al. 2003; Kish 2003; Kish et al. 2008), the cerebral cortex (Scatton et al. 1983), and the cerebrospinal fluid (Kuhn et al. 1996; Mayeux et al. 1984) as well as changes in the density and the activity of several types of serotonin receptors (Castro et al. 1998; Cheng et al. 1991; Fox and Brotchie 2000) have been demonstrated in Parkinson disease patients.

On the other hand, altered neurotransmission processes dealing with dopamine and serotonin uptake, synthesis, breakdown, and receptor activation have been described as related to symptoms shown by children and adults with ADHD, including cognitive compulsivity (Oades 2008).

Some pathophysiological phenomena in schizophrenia have also been related to dopamine–serotonin interactions. Thus, the better clinical profile of novel atypical antipsychotic drugs resulting in effective improvement of mood and behavioral alterations and less untoward effects in schizophrenic patients has been ascribed to their actions on both serotonin and dopamine systems (Alex and Pehek 2007; Meltzer and Huang 2008; Richtand et al. 2008; Stone and Pilowsky 2007; Werkman et al. 2006). It has also been shown that some cognitive expressions can be differentially affected in schizophrenic patients depending on the antipsychotic drug treatment they receive (Araki et al. 2006; Di Pietro and Seamans 2007; Gray and Roth 2007). Once experimental evidence showing the important regulatory role of serotonin on cerebral physiological and physiopathological dopamine-mediated processes had been obtained, efforts have been focused on finding out whether the different types of serotonin receptors are involved in motor, mood, behavioral, and cognitive functions mainly dependent on dopaminergic activity. Since alterations of these functions are implicated in some human neurological and psychiatric diseases, it could be assumed that the differential roles of the various serotonin receptor subtypes may result in the variants of clinical manifestations of these cerebral pathologies. In fact, activation of $5HT_{2A}$ or $5HT_{2C}$ receptors results in opposite neurochemical effects on dopamine release (Di Matteo et al. 2001), its inhibition being a $5HT_{2C}$ receptor-mediated phenomena.

However, extensive investigation aimed to obtain direct evidence on the role of $5HT_{2C}$ and other serotonin receptor subtypes in neural processes and behaviors that require a high cognitive demand, memory improvement, or recovery after an impaired cognitive performance, either under physiological or under physiopathological conditions, in human beings, has been handicapped until now by the lack of selective agonists or antagonists suitable for human use.

Nevertheless, it has been suggested that alterations of some $5HT_{2C}$-mediated neural phenomena could be a part of the pathophysiological mechanisms of cerebral diseases such as Parkinson disease, epilepsy, anxiety, depression, attention deficit, and hyperactivity disorder in human beings (Oades 2008; Di Giovanni et al. 2006; Isaac 2005; Scarpelli et al. 2008; Sodhi et al. 2001). A widespread distribution of the $5HT_{2C}$ receptors has been found in the brain (Pazos et al. 1987), and its functioning may depend on both its agonist-mediated activation and its constitutive active condition, meaning that it is activated even in the absence of an agonist (De Deurwaerdere et al. 2004). Thus, $5HT_{2C}$ receptors have been proposed as a feasible potential target for pharmacological interventions aimed to improve neurological, behavioral, emotional, or cognitive disorders in patients being affected by these diseases.

Accordingly, attention has been paid to the possible role of $5HT_{2C}$ receptors in the mechanisms of action of antipsychotic, mainly those atypical, drugs leading to effects on mood, behavioral, or cognitive alterations in schizophrenic patients (Meltzer and Huang 2008). It has been observed that the RNA editing of the $5HT_{2C}$ receptor in specific brain structures related to mood, behavior, and cognition is reduced in schizophrenia (Sodhi et al. 2001) and that ser23cis single nucleotide polymorphism of $5HT_{2C}$ receptor gene could be predictive of the clinical response to clozapine (Sodhi et al. 1995). It has also been suggested that the $5HT_{2C}$ receptor gene may be very important to epigenetic events that may influence the course of schizophrenia and response to treatment (Reynolds et al. 2005; Sodhi et al. 2005).

Differential binding affinity of typical and atypical antipsychotic drugs to cloned serotonin and dopamine receptor subtypes has been found to correlate with their clinical efficacy. Thus, an inverse correlation between clinically effective doses of some typical antipsychotics and 5HT$_{2C}$ receptor affinity has been shown, while effective doses of atypical antipsychotic drugs are strongly correlated to 5HT$_{2C}$/D$_2$ affinity ratio. However, correlations possibly accounting for antipsychotic effectiveness also have been shown between other serotonergic and dopamine receptor subtypes affinity ratios. These data have been interpreted as suggesting that interaction with constitutive 5HT$_{2C}$ receptor signaling facilitates the antipsychotic effects of typical antipsychotic drugs (Richtand et al. 2008).

An important finding regarding the 5HT$_{2C}$ receptor is the behavioral profile of the 5HT$_{2C}$ selective receptor agonist WAY 163909, similar to that of atypical antipsychotics, and its effectiveness to revert alterations in cognition induced by drugs in experimental animals (Dunlop et al. 2006). Data from experimental studies also suggest that the 5HT$_{2C}$ receptor antagonism appears to have useful effects on certain types of memory impairment. Thus, SB 200646 antagonizes memory impairment due to some 5HT$_{2A/2C}$ agonists and dizolcipine but not scopolamine (Meneses 2002b).

Experimental data dealing with anatomical and functional characteristics of 5HT$_{2C}$ and 5HT$_{2C}$ receptor-mediated serotonergic activity have shown that this receptor plays a pivotal role for basal ganglia physiology and pathophysiology. Thus, the involvement of 5HT$_{2C}$ receptors in the tonic and phasic regulation of mesencephalic, mesocorticolimbic, and nigrostriatal neuronal dopaminergic activity (Di Matteo et al. 2001; Fox and Brotchie 2000; Pierucci et al. 2004), as well as in the functional relationships between these brain regions underlying movement, mood, behavioral, and cognitive functions, supports the proposal of 5HT$_{2C}$ receptors as a target for novel therapeutic strategies in Parkinson disease (Di Giovanni et al. 2006; Di Matteo et al. 2008). Human postmortem tissue samples from Parkinson disease patients have revealed that dopamine depletion may result in adjustments of 5HT$_{2C}$ receptors that appear to be upregulated in the substantia nigra pars reticulata (Fox and Brotchie 2000) without change in its density in the striatum, while other adjustments as those in 5HT$_{2A}$ receptor density are seen in the striatum and substantia nigra.

An association between 5HT$_{2A/2C}$ receptors and increased hyperactivity and impulsivity components of the ADHD has been suggested (Oades 2007). Further, since 5HT$_{2C}$ receptors, among other serotonin receptor subtypes, have been identified on astrocytes (Hirst et al. 1998), their role in glial activities accounting for pathophysiological mechanisms and possible therapeutic proposals for the ADHD, remains to be clarified.

It has been suggested that 5HT$_{2C}$ receptors play an important role in the pathophysiology of anxiety and depression. Thus, beneficial effects can be expected from specific 5HT$_{2C}$ receptor antagonists on mood and cognitive functions that are reduced in depression, since these drugs increase the dopaminergic neurotransmission (Berg et al. 2008; Millan 2006). Density of 5HT$_{2C}$ receptors has been found to be increased in depressed patients, while enhancement of RNA editing leading to changes in the proportions of nonedited and partially edited receptor mRNA transcripts were found in depressed suicide patients (Gurevich et al. 2002; Iwamoto and Kato 2003). On the other hand, feelings of anxiety and panic can be induced by the $5HT_{2C}$ receptor agonist mCPP in humans (Charney et al. 1987; Gatch 2003; Klein et al. 1991) as well as an anxiogenic-like behavior in animals, which can be blocked by $5HT_{2C}$ receptor antagonists (Martin et al. 2002; Jones et al. 2002; Bagdy et al. 2001; Klein et al. 1991; Cornelio and Nunes-de-Souza 2007; Hackler et al. 2007).

Some experimental data point out the role of $5HT_{2C}$ among other 5HT receptors as key modulators of dopamine output in several brain structures, where dopamine is a primary mediator of the rewarding effects of psychostimulants (Bubar and Cunningham 2006). It is known that stimulation of $5HT_{2C}$ receptors attenuates the cocaine-induced release of dopamine from the rat's nucleus accumbens (Navailles et al. 2008). In addition, interference of agonist-induced $5HT_{2C}$ receptor phosphorylation in the ventral tegmental region results in blockade of the conditioned place preference induced by Δ9-tetrahydrocannabinol and nicotine, by suppressing the rewarding neural processes linked to learning and memory mechanisms associated with this conditioned response (Ji et al. 2006; Maillet et al. 2008).

Although the clinical significance of $5HT_{2C}$ receptors deserves further research, the above-mentioned data support a role of $5HT_{2C}$ receptors in cognitive functions in human beings, under both physiological and phatophysiological conditions. In addition, $5HT_{2C}$ receptors should be a target for drug development, as has been suggested. Accordingly, future clinical trials aimed to obtain direct evidence on the effectiveness of new drugs to improve cognitive alterations associated with some neurological and psychiatric diseases could be based on the use of selective $5HT_{2C}$ receptor ligands.

References

Abi-Saab WM, Bubser M, Roth RH, et al (1999) 5-HT2 receptor regulation of extracellular GABA levels in the prefrontal cortex. Neuropsychopharmacology 20:92–96.

Adell A, Casanovas JM, Artigas F (1997) Comparative study in the rat of the actions of different types of stress on the release of 5-HT in raphe nuclei and forebrain areas. Neuropharmacology 36:735–741.

Aggleton JP, Neave N, Nagle S, et al (1995) A comparison of the effects of anterior thalamic, mamillary body and fornix lesions on reinforced spatial alternation. Behav Brain Res 68:91–101.

Aghajanian GK, Marek GJ (1997) Serotonin induces excitatory postsynaptic potentials in apical dendrites of neocortical pyramidal cells. Neuropharmacology 36:589–599.

Aguirre N, Galbete JL, Lasheras B, et al (1995) Methylenedioxymethamphetamine induces opposite changes in central pre- and postsynaptic 5-HT1A receptors in rats. Eur J Pharmacol 281:101–105.

Albert MS (1996) Cognitive and neurobiologic markers of early Alzheimer disease. Proc Natl Acad Sci USA 93:13547–13551.

Alex KD, Pehek EA (2007) Pharmacologic mechanisms of serotonergic regulation of dopamine neurotransmission. Pharmacol Ther 113:296–320.

Alex KD, Yavanian GJ, McFarlane HG, et al (2005) Modulation of dopamine release by striatal 5-HT2C receptors. Synapse 55:242–251.

Alexander GE, DeLong MR, Strick PL (1986) Parallel organization of functionally segregated circuits linking basal ganglia and cortex. Annu Rev Neurosci 9:357–381.

Altman HJ, Normile HJ, Galloway MP, et al (1990) Enhanced spatial discrimination learning in rats following 5,7-DHT-induced serotonergic deafferentation of the hippocampus. Brain Res 518:61–66.

Alves SH, Pinheiro G, Motta V (2004) Anxiogenic effects in the rat elevated plus-maze of 5-HT(2C) agonists into ventral but not dorsal hippocampus. Behav Pharmacol 15:37–43.

American Psychiatric Association (1994): Diagnostic and Statistical Manual of Mental Disorders. 4th ed. Washington DC: American Psychiatric Press.

Andrade R (1998) Regulation of membrane excitability in the central nervous system by serotonin receptor subtypes. Ann N Y Acad Sci 861:190–203.

Andrade R, Malenka RC, Nicoll RA (1986) A G protein couples serotonin and GABAB receptors to the same channels in hippocampus. Science 234:1261–1265.

Andrews JS, Jansen JHM, Linders S, et al (1994) Effects of imipramine and mirtazepine on operant performance in the rat. Drug Dev Res 32:58–66.

Antony MM, Roth D, Swinson RP, et al (1998) Illness intrusiveness in individuals with panic disorder, obsessive-compulsive disorder, or social phobia. J Nerv Ment Dis 186:311–315.

Aouizerate B, Guehl D, Cuny E, et al (2005) Updated overview of the putative role of the serotoninergic system in obsessive-compulsive disorder. Neuropsychiatr Dis Treat 1:231–243.

Araki T, Yamasue H, Sumiyoshi T, et al (2006) Perospirone in the treatment of schizophrenia: effect on verbal memory organization. Prog Neuropsychopharmacol Biol Psychiatry 30:204–208.

Araneda R, Andrade R (1991) 5-Hydroxytryptamine2 and 5-hydroxytryptamine 1A receptors mediate opposing responses on membrane excitability in rat association cortex. Neuroscience 40:399–412.

Arnsten AF, Goldman-Rakic PS (1998) Noise stress impairs prefrontal cortical cognitive function in monkeys: evidence for a hyperdopaminergic mechanism. Arch Gen Psychiatry 55:362–368.

Arvanov VL, Liang X, Magro P, Roberts R, Wang RY, et al (1999) A pre- and postsynaptic modulatory action of 5-HT and the 5-HT2A, 2C receptor agonist DOB on NMDA-evoked responses in the rat medial prefrontal cortex. Eur J Neurosci 11:2917–2934.

Asgari K, Body S, Bak VK, et al (2006) Effects of 5-HT2A receptor stimulation on the discrimination of durations by rats. Behav Pharmacol 17:51–59.

Ashby CR, Jr., Zhang JY, Edwards E, Wang RY, et al (1994) The induction of serotonin3-like receptor supersensitivity and dopamine receptor subsensitivity in the rat medial prefrontal cortex after the intraventricular administration of the neurotoxin 5,7-dihydroxytryptamine: a microiontophoretic study. Neuroscience 60:453–462.

Badgaiyan RD, Fischman AJ, Alpert NM (2007) Striatal dopamine release in sequential learning. Neuroimage 38:549–556.

Bagdy G, Graf M, Anheuer ZE, et al (2001) Anxiety-like effects induced by acute fluoxetine, sertraline or m-CPP treatment are reversed by pretreatment with the 5-HT2C receptor antagonist SB-242084 but not the 5-HT1A receptor antagonist WAY 100635. Int J Neuropsychopharmacol 4:399–408.

Bardo MT (1998) Neuropharmacological mechanisms of drug reward: beyond dopamine in the nucleus accumbens. Crit Rev Neurobiol 12:37–67.

Barnes JM, Costall B, Coughlan J, et al (1990) The effects of ondansetron, a 5-HT3 receptor antagonist, on cognition in rodents and primates. Pharmacol Biochem Behav 35:955–962.

Berg KA, Harvey JA, Spampinato U, et al (2008) Physiological and therapeutic relevance of constitutive activity of 5-HT 2A and 5-HT 2C receptors for the treatment of depression. Prog Brain Res 172:287–305.

Bergqvist PB, Bouchard C, Blier P (1999) Effect of long-term administration of antidepressant treatments on serotonin release in brain regions involved in obsessive-compulsive disorder. Biol Psychiatry 45:164–174.

Birrell JM, Brown VJ (2000) Medial frontal cortex mediates perceptual attentional set shifting in the rat. J Neurosci 20:4320–4324.

Blank T, Zwart R, Nijholt I, et al (1996) Serotonin 5-HT2 receptor activation potentiates N-methyl-D-aspartate receptor-mediated ion currents by a protein kinase C-dependent mechanism. J Neurosci Res 45:153–160.

Body JJ (2003) Rationale for the use of bisphosphonates in osteoblastic and osteolytic bone lesions. Breast 2:S37–44.

Boulougouris V, Tsaltas E (2008) Serotonergic and dopaminergic modulation of attentional processes. Prog Brain Res 172:517–542.

Boulougouris V, Dalley JW, Robbins TW (2007) Effects of orbitofrontal, infralimbic and prelimbic cortical lesions on serial spatial reversal learning in the rat. Behav Brain Res 179:219–228.

Boulougouris V, Glennon JC, Robbins TW (2008) Dissociable effects of selective 5-HT2A and 5-HT2C receptor antagonists on serial spatial reversal learning in rats. Neuropsychopharmacology 33:2007–2019.

Broekkamp CL, Garrigou DL, KG L (1980) Serotonin-mimetic and antidepressant drugs on passive avoidance learning by olfactory bulbectomised rats. Pharmacol Biochem Behav 13:643–646.

Bubar MJ, Cunningham KA (2006) Serotonin 5-HT2A and 5-HT2C receptors as potential targets for modulation of psychostimulant use and dependence. Curr Top Med Chem 6:1971–1985.

Bubar MJ, Cunningham KA (2007) Distribution of serotonin 5-HT2C receptors in the ventral tegmental area. Neuroscience 146:286–297.

Buhot MC (1997) Serotonin receptors in cognitive behaviors. Curr Opin Neurobiol 7:243–254.

Burns CM, Chu H, Rueter SM, et al (1997) Regulation of serotonin-2C receptor G-protein coupling by RNA editing. Nature 387:303–308.

Butter CM, McDonald JA, Snyder DR (1969) Orality, preference behavior, and reinforcement value of nonfood object in monkeys with orbital frontal lesions. Science 164:1306–1307.

Carli M, Robbins TW, Evenden JL, et al (1983) Effects of lesions to ascending noradrenergic neurones on performance of a 5-choice serial reaction task in rats; implications for theories of dorsal noradrenergic bundle function based on selective attention and arousal. Behav Brain Res 9:361–380.

Carr DB, Cooper DC, Ulrich SL, et al (2002) Serotonin receptor activation inhibits sodium current and dendritic excitability in prefrontal cortex via a protein kinase C-dependent mechanism. J Neurosci 22:6846–6855.

Castro ME, Pascual J, Romon T, et al (1998) 5-HT1B receptor binding in degenerative movement disorders. Brain Res 790:323–328.

Charney DS, Woods SW, Goodman WK, et al (1987) Serotonin function in anxiety. II. Effects of the serotonin agonist MCPP in panic disorder patients and healthy subjects. Psychopharmacology (Berl) 92:14–24.

Cheng RK, Liao RM (2007) Dopamine receptor antagonists reverse amphetamine-induced behavioral alteration on a differential reinforcement for low-rate (DRL) operant task in the rat. Chin J Physiol 50:77–88.

Cheng AV, Ferrier IN, Morris CM, et al (1991) Cortical serotonin-S2 receptor binding in Lewy body dementia, Alzheimer's and Parkinson's diseases. J Neurol Sci 106:50–55.

Chinaglia G, Landwehrmeyer B, Probst A, et al (1993) Serotoninergic terminal transporters are differentially affected in Parkinson's disease and progressive supranuclear palsy: an autoradiographic study with [3H]citalopram. Neuroscience 54:691–699.

Chou-Green JM, Holscher TD, Dallman MF, et al (2003). Compulsive behavior in the 5-HT2C receptor knockout mouse. Physiol Behav 78:641–649.

Chudasama Y, Robbins TW (2003) Dissociable contributions of the orbitofrontal and infralimbic cortex to pavlovian autoshaping and discrimination reversal learning: further evidence for the functional heterogeneity of the rodent frontal cortex. J Neurosci 23:8771–8780.

Clark RE, Zola SM, Squire LR (2000) Impaired recognition memory in rats after damage to the hippocampus. J Neurosci 20:8853–8860.

Clarke HF, Dalley JW, Crofts HF Robbins TW, Roberts AC, Eds. (2003) Prefrontal serotonin and serial reversal learning: the effects of serotonin depletion and serotonin 1 A manipulation. Presentation at EBPS. Antwerp, Belgium.

Clarke HF, Dalley JW, Crofts HS, et al (2004) Cognitive inflexibility after prefrontal serotonin depletion. Science 304:878–880.

Clarke HF, Walker SC, Crofts HS, et al (2005) Prefrontal serotonin depletion affects reversal learning but not attentional set shifting. J Neurosci 25:532–538.

Clarke HF, Walker SC, Dalley JW, et al (2007) Cognitive inflexibility after prefrontal serotonin depletion is behaviorally and neurochemically specific. Cereb Cortex 17:18–27.

Clemett DA, Punhani T, Duxon MS, et al (2000) Immunohistochemical localisation of the 5-HT2C receptor protein in the rat CNS. Neuropharmacology 39:123–132.

Coccaro EF (1989) Central serotonin and impulsive aggression. Br J Psychiatry 115(suppl 8):52–62.

Cole BJ, Robbins TW (1987) Amphetamine impairs the discriminative performance of rats with dorsal noradrenergic bundle lesions on a 5-choice serial reaction time task: new evidence for central dopaminergic-noradrenergic interactions. Psychopharmacology (Berl) 91:458–466.

Cole BJ, Robbins TW (1989) Effects of 6-hydroxydopamine lesions of the nucleus accumbens septi on performance of a 5-choice serial reaction time task in rats: implications for theories of selective attention and arousal. Behav Brain Res 33:165–179.

Colino A, Halliwell JV (1987) Differential modulation of three separate K-conductances in hippocampal CA1 neurons by serotonin. Nature 328:73–77.

Conde F, Lund JS, Jacobowitz DM, et al (1994) Baimbridge KG, Lewis DA. Local circuit neurons immunoreactive for calretinin, calbindin D-28 k or parvalbumin in monkey prefrontal cortex: distribution and morphology. J Comp Neurol 341:95–116.

Conn PJ, Sanders-Bush E (1986) Biochemical characterization of serotonin stimulated phosphoinositide turnover. Life Sci 38:663–669.

Conn PJ, Sanders-Bush E (1987) Central serotonin receptors: effector systems, physiological roles and regulation. Psychopharmacology (Berl) 92:267–277.

Conn PJ, Janowsky A, Sanders-Bush E (1987) Denervation supersensitivity of 5-HT-1c receptors in rat choroid plexus. Brain Res 400:396–398.

Cools R (2006) Dopaminergic modulation of cognitive function-implications for L-DOPA treatment in Parkinson's disease. Neurosci Biobehav Rev 30:1–23.

Cools R, Roberts AC, Robbins TW (2008) Serotoninergic regulation of emotional and behavioral control processes. Trends Cogn Sci 12:31–40.

Cornelio AM, Nunes-de-Souza RL (2007) Anxiogenic-like effects of mCPP microinfusions into the amygdala (but not dorsal or ventral hippocampus) in mice exposed to elevated plus-maze. Behav Brain Res 178:82–89.

Cunningham KA, Paris JM, Goeders NE (1992) Serotonin neurotransmission in cocaine sensitization. Ann N Y Acad Sci 654:117–127.

Dahlstroem A, Fuxe K, Olson L, et al (1964) Ascending systems of catecholamine neurons from the lower brain stem. Acta Physiol Scand 62:485–486.

De Deurwaerdere P, Navailles S, Berg KA, et al (2004) Constitutive activity of the serotonin2C receptor inhibits in vivo dopamine release in the rat striatum and nucleus accumbens. J Neurosci 24:3235–3241.

DeLong MR (1990) Primate models of movement disorders of basal ganglia origin. Trends Neurosci 13:281–285.

Di Giovanni G, Di Matteo V, Di Mascio M, et al (2000) Preferential modulation of mesolimbic vs. nigrostriatal dopaminergic function by serotonin(2C/2B) receptor agonists: a combined in vivo electrophysiological and microdialysis study. Synapse 35:53–61.

Di Giovanni G, Di Matteo V, La Grutta V, et al (2001) m-Chlorophenylpiperazine excites non-dopaminergic neurons in the rat substantia nigra and ventral tegmental area by activating serotonin-2C receptors. Neuroscience 103:111–116.

Di Giovanni G, Di Matteo V, Pierucci M, et al (2006) Serotonin involvement in the basal ganglia pathophysiology: could the 5-HT2C receptor be a new target for therapeutic strategies? Curr Med Chem 13:3069–3081.

Di Matteo V, Di Giovanni G, Di Mascio M, et al (1999) SB 242084, a selective serotonin2C receptor antagonist, increases dopaminergic transmission in the mesolimbic system. Neuropharmacology 38:1195–1205.

Di Matteo V, Di Giovanni G, Di Mascio M, et al (2000) Biochemical and electrophysiological evidence that RO 60-0175 inhibits mesolimbic dopaminergic function through serotonin(2C) receptors. Brain Res 865:85–90.

Di Matteo V, Di Mascio M, Di Giovanni G, et al (2000) Acute administration of amitriptyline and mianserin increases dopamine release in the rat nucleus accumbens: possible involvement of serotonin2C receptors. Psychopharmacology (Berl) 150:45–51.

Di Matteo V, De Blasi A, Di Giulio C (2001) Role of 5-HT(2C) receptors in the control of central dopamine function. Trends Pharmacol Sci 22:229–232.

Di Matteo V, Cacchio M, Di Giulio C, et al (2002) Role of serotonin(2C) receptors in the control of brain dopaminergic function. Pharmacol Biochem Behav 71:727–734.

Di Matteo V, Pierucci M, Esposito E, et al (2008) Serotonin modulation of the basal ganglia circuitry: therapeutic implication for Parkinson's disease and other motor disorders. Prog Brain Res 172:423–463.

Di Pietro NC, Seamans JK (2007) Dopamine and serotonin interactions in the prefrontal cortex: insights on antipsychotic drugs and their mechanism of action. Pharmacopsychiatry 40 (suppl 1):S27–S33.

Diamond DM, Park CR, Campbell AM, et al (2005) Competitive interactions between endogenous LTD and LTP in the hippocampus underlie the storage of emotional memories and stress-induced amnesia. Hippocampus 15:1006–1025.

Dias EC, Segraves MA (1996) The primate frontal eye field and the generation of saccadic eye movements: comparison of lesion and acute inactivation/activation studies. Rev Bras Biol 2:239–255.

Domeney AM, Costall B, Gerrard PA, et al (1991) The effect of ondansetron on cognitive performance in the marmoset. Pharmacol Biochem Behav 38:169–175.

Doris A, Ebmeier K, Shajahan P (1999) Depressive illness. Lancet 354:1369–1375.

Dunlop J, Marquis KL, Lim HK, et al (2006) Pharmacological profile of the 5-HT(2C) receptor agonist WAY 163909; therapeutic potential in multiple indications. CNS Drug Rev 12:167–177.

Dunnett SB, Iversen SD (1982) Neurotoxic lesions of ventrolateral but not anteromedial neostriatum in rats impair differential reinforcement of low rates (DRL) performance. Behav Brain Res 6:213–226.

Eberle-Wang K, Mikeladze Z, Uryu K, et al (1997) Pattern of expression of the serotonin2C receptor messenger RNA in the basal ganglia of adult rats. J Comp Neurol 384:233–247.

Ennaceur A, Delacour J (1988) A new one-trial test for neurobiological studies of memory in rats. 1: behavioral data. Behav Brain Res 31:47–59.

Erzegovesi S, Ronchi P, Smeraldi E (1992) 5-HT2C receptor and fluvoxamine effect in obsessive-compulsive disorder. Human Psychopharmacology 7:287–289.

Featherstone RE, McDonald RJ (2004) Dorsal striatum and stimulus-response learning: lesions of the dorsolateral, but not dorsomedial, striatum impair acquisition of a simple discrimination task. Behav Brain Res 150:15–23.

Featherstone RE, McDonald RJ (2004) Dorsal striatum and stimulus-response learning: lesions of the dorsolateral, but not dorsomedial, striatum impair acquisition of a stimulus-response-based instrumental discrimination task, while sparing conditioned place preference learning. Neuroscience 124:23–31.

Fellows LK, Farah MJ (2003) Ventromedial frontal cortex mediates affective shifting in humans: evidence from a reversal learning paradigm. Brain 126(Pt 8):1830–1837.

Feng J, Cai X, Zhao J, et al (2001) Serotonin receptors modulate GABA(A) receptor channels through activation of anchored protein kinase C in prefrontal cortical neurons. J Neurosci 21:6502–6511.

File SE, Seth P (2003) A review of 25 years of the social interaction test. Eur J Pharmacol 463:35–53.

Filip M, Cunningham KA (2003) Hyperlocomotive and discriminative stimulus effects of cocaine are under the control of serotonin(2C) (5-HT(2C)) receptors in rat prefrontal cortex. J Pharmacol Exp Ther 306:734–743.

Fischette CT, Nock B, Renner K (1987) Effects of 5,7-dihydroxytryptamine on serotonin1 and serotonin2 receptors throughout the rat central nervous system using quantitative autoradiography. Brain Res 421:263–279.

Fletcher PJ (1995) Effects of combined or separate 5,7-dihydroxytryptamine lesions of the dorsal and median raphe nuclei on responding maintained by a DRL 20s schedule of food reinforcement. Brain Res 675:45–54.

Fletcher PJ, Ming ZH, Higgins GA (1993) Conditioned place preference induced by microinjection of 8-OH-DPAT into the dorsal or median raphe nucleus. Psychopharmacology (Berl) 113:31–36.

Fletcher PJ, Korth KM, Chambers JW (1999) Selective destruction of brain serotonin neurons by 5,7-dihydroxytryptamine increases responding for a conditioned reward. Psychopharmacology (Berl) 147:291–299.

Fletcher PJ, Robinson SR, Slippoy DL (2001) Pre-exposure to (+/-)3,4-methylenedioxy-methamphetamine (MDMA) facilitates acquisition of intravenous cocaine self-administration in rats. Neuropsychopharmacology 25:195–203.

Fletcher PJ, Grottick AJ, Higgins GA (2002) Differential effects of the 5-HT(2A) receptor antagonist M100907 and the 5-HT(2C) receptor antagonist SB242084 on cocaine-induced locomotor activity, cocaine self-administration and cocaine-induced reinstatement of responding. Neuropsychopharmacology 27:576–586.

Fletcher PJ, Korth KM, Robinson SR, et al (2002) Multiple 5-HT receptors are involved in the effects of acute MDMA treatment: studies on locomotor activity and responding for conditioned reinforcement. Psychopharmacology (Berl) 162:282–291.

Fletcher PJ, Tampakeras M, Sinyard J, et al (2007) Opposing effects of 5-HT(2A) and 5-HT(2C) receptor antagonists in the rat and mouse on premature responding in the five-choice serial reaction time test. Psychopharmacology (Berl) 195:223–234.

Florio T, Capozzo A, Nisini A, et al (1999) Dopamine denervation of specific striatal subregions differentially affects preparation and execution of a delayed response task in the rat. Behav Brain Res 104:51–62.

Fone KC, Austin RH, Topham IA, et al (1998) Effect of chronic m-CPP on locomotion, hypophagia, plasma corticosterone and 5-HT2C receptor levels in the rat. Br J Pharmacol 123:1707–1715.

Fox SH, Brotchie JM (2000) 5-HT2C receptor binding is increased in the substantia nigra pars reticulata in Parkinson's disease. Mov Disord 15:1064–1069.

Gabbott PL, Bacon SJ (1997) Calcineurin immunoreactivity in prelimbic cortex (area 32) of the rat. Brain Res 747:352–356.

Garcia-Alcocer G, Segura LC, Garcia Pena M, et al (2006) Ontogenetic distribution of 5-HT2C, 5-HT5A, and 5-HT7 receptors in the rat hippocampus. Gene Expr 13:53–57.

Gatch MB (2003) Discriminative stimulus effects of m-chlorophenylpiperazine as a model of the role of serotonin receptors in anxiety. Life Sci 73:1347–1367.

Gerfen CR (1984) The neostriatal mosaic: compartmentalization of corticostriatal input and striatonigral output systems. Nature 311:461–464.

Givens B (1995) Low doses of ethanol impair spatial working memory and reduce hippocampal theta activity. Alcohol Clin Exp Res 19:763–767.

Givens B, Olton DS (1995) Bidirectional modulation of scopolamine-induced working memory impairments by muscarinic activation of the medial septal area. Neurobiol Learn Mem 63:269–276.

Gobert A, Millan MJ (1999) Serotonin (5-HT)2A receptor activation enhances dialysate levels of dopamine and noradrenaline, but not 5-HT, in the frontal cortex of freely-moving rats. Neuropharmacology 38:315–317.

Gobert A, Rivet JM, Lejeune F, et al (2000) Serotonin(2C) receptors tonically suppress the activity of mesocortical dopaminergic and adrenergic, but not serotonergic, pathways: a combined dialysis and electrophysiological analysis in the rat. Synapse 36:205–221.

Graeff FG, Silveira MC, Nogueira RL (1993) Role of the amygdala and periaqueductal gray in anxiety and panic. Behav Brain Res 58:123–131.

Graf M (2006) 5-HT2c receptor activation induces grooming behavior in rats: possible correlations with obsessive-compulsive disorder. Neuropsychopharmacol Hung 8:23–28.

Graf M, Kantor S, Anheuer ZE, et al (2003) m-CPP-induced self-grooming is mediated by 5-HT2C receptors. Behav Brain Res 142:175–179.

Gray JA, Roth BL (2007) Molecular targets for treating cognitive dysfunction in schizophrenia. Schizophr Bull 33:1100–1119.

Gross R, Sasson, Y., Chopra, M, et al (1998) Biological models of obsessive-compulsive disorder. The serotonin Hypothesis. In: Swinson RP, Antony M.M., Rachman S., et al, Eds. Obsessive-compulsive disorder Theory, research and treatment. New York: Guilford. pp. 141–153

Gross C, Zhuang X, Stark K, et al (2002) Serotonin1A receptor acts during development to establish normal anxiety-like behavior in the adult. Nature 416:396–400.

Gulyas AI, Acsady L, Freund TF (1999) Structural basis of the cholinergic and serotonergic modulation of GABAergic neurons in the hippocampus. Neurochem Int 34:359–372.

Gurevich I, Tamir H, Arango V, et al (2002) Altered editing of serotonin 2C receptor pre-mRNA in the prefrontal cortex of depressed suicide victims. Neuron 3:349–356.

Guttman M, Boileau I, Warsh J, et al (2007) Brain serotonin transporter binding in non-depressed patients with Parkinson's disease. Eur J Neurol 14:523–528.

Hackler EA, Turner GH, Gresch PJ, et al (2007) 5-Hydroxytryptamine2C receptor contribution to m-chlorophenylpiperazine and N-methyl-beta-carboline-3-carboxamide-induced anxiety-like behavior and limbic brain activation. J Pharmacol Exp Ther 320:1023–1029.

Haj-Dahmane S, Andrade R (1996) Muscarinic activation of a voltage-dependent cation nonselective current in rat association cortex. J Neurosci 16:3848–3861.

Hajos M, Hoffmann WE, Weaver RJ (2003) Regulation of septo-hippocampal activity by 5-hydroxytryptamine(2C) receptors. J Pharmacol Exp Ther 306:605–615.

Halliday GM, Li YW, Blumbergs PC, et al (1990) Neuropathology of immunohistochemically identified brainstem neurons in Parkinson's disease. Ann Neurol 27:373–385.

Hammond RS, Tull LE, Stackman RW (2004) On the delay-dependent involvement of the hippocampus in object recognition memory. Neurobiol Learn Mem 82:26–34.

Hannon J, Hoyer D (2008) Molecular biology of 5-HT receptors. Behav Brain Res 195:198–213.

Harrison AA, Everitt BJ, Robbins TW (1997) Central 5-HT depletion enhances impulsive responding without affecting the accuracy of attentional performance: interactions with dopaminergic mechanisms. Psychopharmacology (Berl) 133:329–342.

Harrison AA, Everitt BJ, Robbins TW (1999) Central serotonin depletion impairs both the acquisition and performance of a symmetrically reinforced go/no-go conditional visual discrimination. Behav Brain Res 100:99–112.

Hasselmo ME, Hay J, Ilyn M, et al (2002) Neuromodulation, theta rhythm and rat spatial navigation. Neural Netw 15:689–707.

Hayes DJ, Mosher TM, Greenshaw AJ (2009) Differential effects of 5-HT(2C) receptor activation by WAY 161503 on nicotine-induced place conditioning and locomotor activity in rats. Behav Brain Res 197:323–330.

Heisler LK, Chu HM, Tecott LH (1998) Epilepsy and obesity in serotonin 5-HT2C receptor mutant mice. Ann N Y Acad Sci 861:74–78.

Herve D, Pickel VM, Joh TH, et al (1987) Serotonin axon terminals in the ventral tegmental area of the rat: fine structure and synaptic input to dopaminergic neurons. Brain Res 435:71–83.

Higgins GA, Fletcher PJ (2003) Serotonin and drug reward: focus on 5-HT2C receptors. Eur J Pharmacol 480:151–162.

Hirst WD, Cheung NY, Rattray M, et al (1998) Cultured astrocytes express messenger RNA for multiple serotonin receptor subtypes, without functional coupling of 5-HT1 receptor subtypes to adenylyl cyclase. Brain Res Mol Brain Res 61:90–99.

Hoffman BJ, Mezey E (1989) Distribution of serotonin 5-HT1C receptor mRNA in adult rat brain. FEBS Lett 247:453–462.

Hoyer D (1988) Functional correlates of serotonin 5-HT1 recognition sites. J Recept Res 8:59–81.

Hudzik TJ, Howell A, Georger M, et al (2000) Disruption of acquisition and performance of operant response-duration differentiation by unilateral nigrostriatal lesions. Behav Brain Res 114:65–77.

Hutson PH, Barton CL, Jay M, et al (2000) Activation of mesolimbic dopamine function by phencyclidine is enhanced by 5-HT(2C/2B) receptor antagonists: neurochemical and behavioral studies. Neuropharmacology 39:2318–2328.

Ichihara K, Nabeshima T, Kameyama T (1992) Effects of dopamine receptor agonists on passive avoidance learning in mice: interaction of dopamine D1 and D2 receptors. Eur J Pharmacol 213:243–249.

Idris NF, Repeto P, Neill JC, et al (2005) Investigation of the effects of lamotrigine and clozapine in improving reversal-learning impairments induced by acute phencyclidine and D-amphetamine in the rat. Psychopharmacology (Berl) 179:336–348.

Ikemoto S, Wise RA (2004) Mapping of chemical trigger zones for reward. Neuropharmacology 1:190–201.

Isaac M (2005) Serotonergic 5-HT2C receptors as a potential therapeutic target for the design antiepileptic drugs. Curr Top Med Chem 5:59–67.

Iwamoto K, Kato T (2003) RNA editing of serotonin 2C receptor in human postmortem brains of major mental disorders. Neurosci Lett 346:169–172.

Jakab RL, Goldman-Rakic PS (1998) 5-Hydroxytryptamine2 serotonin receptors in the primate cerebral cortex: possible site of action of hallucinogenic and antipsychotic drugs in pyramidal cell apical dendrites. Proc Natl Acad Sci USA 95:735–740.

Jarrard LE (1993) On the role of the hippocampus in learning and memory in the rat. Behav Neural Biol 60:9–26.

Jenck F, Moreau JL, Mutel V (1993) Evidence for a role of 5-HT1C receptors in the antiserotonergic properties of some antidepressant drugs. Eur J Pharmacol 231:223–229.

Ji SP, Zhang Y, Van Cleemput J, et al (2006) Disruption of PTEN coupling with 5-HT2C receptors suppresses behavioral responses induced by drugs of abuse. Nat Med 12:324–329.

Jones MW, Wilson MA (2005) Theta rhythms coordinate hippocampal-prefrontal interactions in a spatial memory task. PLoS Biol 3:e402.

Jones N, Duxon MS, King SM (2002) 5-HT2C receptor mediation of unconditioned escape behavior in the unstable elevated exposed plus maze. Psychopharmacology (Berl) 164:214–220.

Julius D, MacDermott AB, Jessel TM, et al(1988) Functional expression of the 5-HT1c receptor in neuronal and nonneuronal cells. Cold Spring Harb Symp Quant Biol 1:385–393.

Kant GJ, Meininger GR, Maughan KR, et al (1996) Effects of the serotonin receptor agonists 8-OH-DPAT and TFMPP on learning as assessed using a novel water maze. Pharmacol Biochem Behav 53:385–390.

Kant GJ, Wylie RM, Chu K, et al (1998) Effects of the serotonin agonists 8-OH-DPAT, buspirone, and DOI on water maze performance. Pharmacol Biochem Behav 59:729–735.

Kantor S, Jakus R, Molnar E, et al (2005) Despite similar anxiolytic potential, the 5-hydroxytryptamine 2C receptor antagonist SB-242084 [6-chloro-5-methyl-1-[2-(2-methylpyrid-3-yloxy)-pyrid-5-yl carbamoyl] indoline] and chlordiazepoxide produced differential effects on electroencephalogram power spectra. J Pharmacol Exp Ther 315:921–930.

Kehne JH, Ketteler HJ, McCloskey TC, et al (1996) Effects of the selective 5-HT2A receptor antagonist MDL 100,907 on MDMA-induced locomotor stimulation in rats. Neuropsychopharmacology 15:116–124.

Kemp JM, Powell TP (1970) The cortico-striate projection in the monkey. Brain 93:525–546.

Kennett GA, Whitton P, Shah K, et al (1989) Anxiogenic-like effects of mCPP and TFMPP in animal models are opposed by 5-HT1C receptor antagonists. Eur J Pharmacol 164:445–454.

Kennett GA, Wood MD, Bright F, et al (1997) SB 242084, a selective and brain penetrant 5-HT2C receptor antagonist. Neuropharmacology 36:609–620.

Kennett G, Lightowler S, Trail B, et al (2000) Effects of RO 60 0175, a 5-HT(2C) receptor agonist, in three animal models of anxiety. Eur J Pharmacol 387:197–204.

Kerenyi L, Ricaurte GA, Schretlen DJ, et al (2003) Positron emission tomography of striatal serotonin transporters in Parkinson disease. Arch Neurol 60:1223–1229.

Khaliq S, Irfan B, Haider S, et al (2008) m-CPP induced hypolocomotion does not interfere in the assessment of memory functions in rats. Pak J Pharm Sci 21:139–143.

Kim CH, Koo MS, Cheon KA, et al (2003) Dopamine transporter density of basal ganglia assessed with [123I]IPT SPET in obsessive-compulsive disorder. Eur J Nucl Med Mol Imaging 30:1637–1643.

Kish SJ (2003) Biochemistry of Parkinson's disease: is a brain serotonergic deficiency a characteristic of idiopathic Parkinson's disease? Adv Neurol 91:39–49.

Kish SJ, Tong J, Hornykiewicz O, et al (2008) Preferential loss of serotonin markers in caudate versus putamen in Parkinson's disease. Brain 131(Pt 1):120–131.

Klein E, Zohar J, Geraci MF, et al (1991) Anxiogenic effects of m-CPP in patients with panic disorder: comparison to caffeine's anxiogenic effects. Biol Psychiatry 30:973–984.

Kojima T, Matsumoto M, Togashi H, et al (2003) Fluvoxamine suppresses the long-term potentiation in the hippocampal CA1 field of anesthetized rats: an effect mediated via 5-HT1A receptors. Brain Res 959:165–168.

Koolhaas JM, De Boer SF, De Rutter AJ, et al (1997) Social stress in rats and mice. Acta Physiol Scand Suppl 640:69–72.

Koskinen T, Ruotsalainen S, Sirvio J (2000) The 5-HT(2) receptor activation enhances impulsive responding without increasing motor activity in rats. Pharmacol Biochem Behav 66:729–738.

Koskinen T, Ruotsalainen S, Puumala T, et al (2000). Activation of 5-HT2A receptors impairs response control of rats in a five-choice serial reaction time task. Neuropharmacology 39:471–481.

Kostrzewa RM, Nowak P, Kostrzewa JP, et al (2005) Peculiarities of L-DOPA treatment of Parkinson's disease. Amino Acids 28:157–164.

Kuhn W, Muller T, Gerlach M, et al (1996) Depression in Parkinson's disease: biogenic amines in CSF of "de novo" patients. J Neural Transm 103:1441–1445.

Lamberty Y, Gower AJ (1991) Cholinergic modulation of spatial learning in mice in a Morris-type water maze. Arch Int Pharmacodyn Ther 309:5–19.

Lee D, Rushworth MF, Walton ME, et al (2007) Functional specialization of the primate frontal cortex during decision making. J Neurosci 27:8170–8173.

Lemonde S, Turecki G, Bakish D, et al (2003) Impaired repression at a 5-hydroxytryptamine 1A receptor gene polymorphism associated with major depression and suicide. J Neurosci 23:8788–8799.

Leranth C, Vertes RP (1999) Median raphe serotonergic innervation of medial septum/diagonal band of broca (MSDB) parvalbumin-containing neurons: possible involvement of the MSDB in the desynchronization of the hippocampal EEG. J Comp Neurol 410:586–598.

Lewis DA, Hashimoto T, Volk DW (2005) Cortical inhibitory neurons and schizophrenia. Nat Rev Neurosci 6:312–324.

Lieben CK, Blokland A, Westerink B, et al (2004) Acute tryptophan and serotonin depletion using an optimized tryptophan-free protein-carbohydrate mixture in the adult rat. Neurochem Int 44:9–16.

Lieben CK, van Oorsouw K, Deutz NE (2004) Acute tryptophan depletion induced by a gelatin-based mixture impairs object memory but not affective behavior and spatial learning in the rat. Behav Brain Res 151:53–64.

Lieben CK, Steinbusch HW, Blokland A (2006) 5,7-DHT lesion of the dorsal raphe nuclei impairs object recognition but not affective behavior and corticosterone response to stressor in the rat. Behav Brain Res 168:197–207.

Liu S, Bubar MJ, Lanfranco MF, et al (2007) Serotonin2C receptor localization in GABA neurons of the rat medial prefrontal cortex: implications for understanding the neurobiology of addiction. Neuroscience 146:1677–1688.

Lopez-Gimenez JF, Mengod G, Palacios JM, et al (2001) Regional distribution and cellular localization of 5-HT2C receptor mRNA in monkey brain: comparison with [3H]mesulergine binding sites and choline acetyltransferase mRNA. Synapse 42:12–26.

Lowry CA, Rodda JE, Lightman SL, et al (2000) Corticotropin-releasing factor increases in vitro firing rates of serotonergic neurons in the rat dorsal raphe nucleus: evidence for activation of a topographically organized mesolimbocortical serotonergic system. J Neurosci 20:7728–7736.

Lucas G, Spampinato U (2000) Role of striatal serotonin2A and serotonin2C receptor subtypes in the control of in vivo dopamine outflow in the rat striatum. J Neurochem 74:693–701.

Lucki I, Ward HR, Frazer A (1989) Effect of 1-(m-chlorophenyl)piperazine and 1-(m-trifluoromethylphenyl)piperazine on locomotor activity. J Pharmacol Exp Ther 249:155–164.

Lynch MA (2004) Long-term potentiation and memory. Physiol Rev 84:87–136.

Maillet JC, Zhang Y, Li X, et al (2008) PTEN-5-HT2C coupling: a new target for treating drug addiction. Prog Brain Res 172:407–420.

Marek GJ, Seiden LS (1988) Effects of selective 5-hydroxytryptamine-2 and nonselective 5-hydroxytryptamine antagonists on the differential-reinforcement-of-low-rate 72-second schedule. J Pharmacol Exp Ther 244:650–658.

Marek GJ, Li AA, Seiden LS (1989) Evidence for involvement of 5-hydroxytryptamine1 receptors in antidepressant-like drug effects on differential-reinforcement-of-low-rate 72-second behavior. J Pharmacol Exp Ther 250:60–71.

Marek GJ, Li AA, Seiden LS (1989) Selective 5-hydroxytryptamine2 antagonists have antidepressant-like effects on differential-reinforcement-of-low-rate 72-second schedule. J Pharmacol Exp Ther 250:52–59.

Markram H, Toledo-Rodriguez M, Wang Y, et al (2004) Interneurons of the neocortical inhibitory system. Nat Rev Neurosci 5:793–807.

Marquis KL, Sabb AL, Logue SF, et al (2007) WAY 163909 [(7bR,10aR)-1,2,3,4,8,9,10,10a-octahydro-7bH-cyclopenta-[b][1,4]diazepino[6,7,1hi]indole]: A novel 5-hydroxytryptamine 2C receptor-selective agonist with preclinical antipsychotic-like activity. J Pharmacol Exp Ther 320:486–496.

Martin JR, Bos M, Jenck F, et al (1998) 5-HT2C receptor agonists: pharmacological characteristics and therapeutic potential. J Pharmacol Exp Ther 286:913–924.

Martin JR, Ballard TM, Higgins GA (2002) Influence of the 5-HT2C receptor antagonist, SB-242084, in tests of anxiety. Pharmacol Biochem Behav 71:615–625.

Maswood S, Barter JE, Watkins LR (1998) Exposure to inescapable but not escapable shock increases extracellular levels of 5-HT in the dorsal raphe nucleus of the rat. Brain Res 783:115–120.

Mayeux R, Stern Y, Cote L, et al (1984) Altered serotonin metabolism in depressed patients with parkinson's disease. Neurology 34:642–646.

Mazure CM, Kincare P, Schaffer CE (1995) DSM-III-R Axis IV: clinician reliability and comparability to patients' reports of stressor severity. Psychiatry 58:56–64.

McAlonan K, Brown VJ (2003) Orbital prefrontal cortex mediates reversal learning and not attentional set shifting in the rat. Behav Brain Res 146:97–103.

McDonald RJ, White NM (1993) A triple dissociation of memory systems: hippocampus, amygdala, and dorsal striatum. Behav Neurosci 107:3–22.

McDonald RJ, White NM (1994) Parallel information processing in the water maze: evidence for independent memory systems involving dorsal striatum and hippocampus. Behav Neural Biol 61:260–270.

McGuire PS, Seiden LS (1980) Differential effects of imipramine in rats as a function of DRL schedule value. Pharmacol Biochem Behav 13:691–694.

McNamara RK, Skelton RW (1993) The neuropharmacological and neurochemical basis of place learning in the Morris water maze. Brain Res Brain Res Rev 18:33–49.

Meltzer HY, Huang M (2008) In vivo actions of atypical antipsychotic drug on serotonergic and dopaminergic systems. Prog Brain Res 172:177–197.

Meneses A (2002) Tianeptine: 5-HT uptake sites and 5-HT(1-7) receptors modulate memory formation in an autoshaping Pavlovian/instrumental task. Neurosci Biobehav Rev 26:309–319.

Meneses A (2002) Involvement of 5-HT(2A/2B/2C) receptors on memory formation: simple agonism, antagonism, or inverse agonism? Cell Mol Neurobiol 22:675–688.

Meneses A (2003) A pharmacological analysis of an associative learning task: 5-HT(1) to 5-HT(7) receptor subtypes function on a pavlovian/instrumental autoshaped memory. Learn Mem 10:363–372.

Meneses A (2007) Stimulation of 5-HT1A, 5-HT1B, 5-HT2A/2C, 5-HT3 and 5-HT4 receptors or 5-HT uptake inhibition: short- and long-term memory. Behav Brain Res 184:81–90.

Meneses A (2007) Do serotonin(1-7) receptors modulate short and long-term memory? Neurobiol Learn Mem 87:561–572.

Meneses A, Hong E (1997) Role of 5-HT1B, 5-HT2A and 5-HT2C receptors in learning. Behav Brain Res 87:105–110.

Meneses A, Terron JA, Hong E (1997) Effects of the 5-HT receptor antagonists GR127935 (5-HT1B/1D) and MDL100907 (5-HT2A) in the consolidation of learning. Behav Brain Res 89:217–223.

Mengod G, Nguyen H, Le H, et al (1990) The distribution and cellular localization of the serotonin 1C receptor mRNA in the rodent brain examined by in situ hybridization histochemistry. Comparison with receptor binding distribution. Neuroscience 35:577–591.

Miles R, Toth K, Gulyas AI, et al (1996) Differences between somatic and dendritic inhibition in the hippocampus. Neuron 16:815–823.

Millan MJ (2006) Multi-target strategies for the improved treatment of depressive states: Conceptual foundations and neuronal substrates, drug discovery and therapeutic application. Pharmacol Ther 110:135–370.

Millan MJ, Dekeyne A, Gobert A (1998) Serotonin (5-HT)2C receptors tonically inhibit dopamine (DA) and noradrenaline (NA), but not 5-HT, release in the frontal cortex in vivo. Neuropharmacology 37:953–955.

Miner LA, Backstrom JR, Sanders-Bush E, et al (2003) Ultrastructural localization of serotonin2A receptors in the middle layers of the rat prelimbic prefrontal cortex. Neuroscience 116:107–117.

Misane I, Ogren SO (2000) Multiple 5-HT receptors in passive avoidance: comparative studies of p-chloroamphetamine and 8-OH-DPAT. Neuropsychopharmacology 22:168–190.

Mora PO, Netto CF, Graeff FG (1997) Role of 5-HT2A and 5-HT2C receptor subtypes in the two types of fear generated by the elevated T-maze. Pharmacol Biochem Behav 58:1051–1057.

Mosher T, Hayes D, Greenshaw A (2005) Differential effects of 5-HT2C receptor ligands on place conditioning and locomotor activity in rats. Eur J Pharmacol 515:107–116.

Murphy P (2002) Cognitive functioning in adults with attention-deficit/hyperactivity disorder. J Atten Disord 5:203–209.

Murtha SJ, Pappas BA (1994) Neurochemical, histopathological and mnemonic effects of combined lesions of the medial septal and serotonin afferents to the hippocampus. Brain Res 651:16–26.

Navailles S, Moison D, Cunningham KA, et al (2008) Differential regulation of the mesoaccumbens dopamine circuit by serotonin2C receptors in the ventral tegmental area and the nucleus accumbens: an in vivo microdialysis study with cocaine. Neuropsychopharmacology 33:237–246.

Neill DB, Herndon JG, Jr (1978) Anatomical specificity within rat striatum for the dopaminergic modulation of DRL responding and activity. Brain Res 153:529–538.

Nestler EJ, Carlezon WA, Jr. (2006) The mesolimbic dopamine reward circuit in depression. Biol Psychiatry 59:1151–1159.

Nilsson OG, Strecker RE, Daszuta A, et al (1988) Combined cholinergic and serotonergic denervation of the forebrain produces severe deficits in a spatial learning task in the rat. Brain Res 453:235–246.

Normile HJ, Jenden DJ, Kuhn DM, et al (1990) Effects of combined serotonin depletion and lesions of the nucleus basalis magnocellularis on acquisition of a complex spatial discrimination task in the rat. Brain Res 536:245–250.

North RA, Uchimura N (1989) 5-Hydroxytryptamine acts at 5-HT2 receptors to decrease potassium conductance in rat nucleus accumbens neurones. J Physiol 417:1–12.

O'Keefe J, Nadel L. The hippocampus as a cognitive map. Oxford: Clarendon; 1978.

Oades RD (2007) Role of the serotonin system in ADHD: treatment implications. Expert Rev Neurother 7:1357–1374.

Oades RD (2008) Dopamine-serotonin interactions in attention-deficit hyperactivity disorder (ADHD). Prog Brain Res 172:543–565.

Ogren SO (1985) Evidence for a role of brain serotonergic neurotransmission in avoidance learning. Acta Physiol Scand Suppl 544:1–71.

Ogren SO (1986) Analysis of the avoidance learning deficit induced by the serotonin releasing compound p-chloroamphetamine. Brain Res Bull 16:645–660.

Olvera-Cortes E, Cervantes M, Gonzalez-Burgos I (2002) Place-learning, but not cue-learning training, modifies the hippocampal theta rhythm in rats. Brain Res Bull 58:261–270.

Olvera-Cortes E, Guevara MA, Gonzalez-Burgos I (2004) Increase of the hippocampal theta activity in the Morris water maze reflects learning rather than motor activity. Brain Res Bull 62:379–384.

Olvera-Cortes ME, Anguiano-Rodriguez P, Lopez-Vazquez MA, et al (2008) Serotonin/dopamine interaction in learning. Prog Brain Res 172:567–602.

Packard MG, Knowlton BJ (2002) Learning and memory functions of the basal ganglia. Annu Rev Neurosci 25:563–593.

Pan HS, Wang RY (1991) The action of (+/-)-MDMA on medial prefrontal cortical neurons is mediated through the serotonergic system. Brain Res 543:56–60.

Pan HS, Wang RY (1991) MDMA: further evidence that its action in the medial prefrontal cortex is mediated by the serotonergic system. Brain Res 539:332–336.

Pasqualetti M, Ori M, Castagna M, et al (1999) Distribution and cellular localization of the serotonin type 2C receptor messenger RNA in human brain. Neuroscience 92:601–611.

Passetti F, Dalley JW, Robbins TW (2003) Double dissociation of serotonergic and dopaminergic mechanisms on attentional performance using a rodent five-choice reaction time task. Psychopharmacology (Berl) 165:136–145.

Paulus W, Jellinger K (1991) The neuropathologic basis of different clinical subgroups of Parkinson's disease. J Neuropathol Exp Neurol 50:743–755.

Pazos A, Probst A, Palacios JM (1987) Serotonin receptors in the human brain–IV. Autoradiographic mapping of serotonin-2 receptors. Neuroscience 21:123–139.

Pehek EA, Nocjar C, Roth BL, et al (2006) Evidence for the preferential involvement of 5-HT2A serotonin receptors in stress- and drug-induced dopamine release in the rat medial prefrontal cortex. Neuropsychopharmacology 31:265–277.

Penington NJ, Kelly JS (1990) Serotonin receptor activation reduces calcium current in an acutely dissociated adult central neuron. Neuron 4:751–758.

Peroutka SJ (1995) 5-HT receptors: past, present and future. Trends Neurosci 18:68–69.

Pierce RC, Kumaresan V (2006) The mesolimbic dopamine system: the final common pathway for the reinforcing effect of drugs of abuse? Neurosci Biobehav Rev 30:215–238.

Pierucci M, Di Matteo V, Esposito E (2004) Stimulation of serotonin2C receptors blocks the hyperactivation of midbrain dopamine neurons induced by nicotine administration. J Pharmacol Exp Ther 309:109–118.

Ping SE, Trieu J, Wlodek ME, et al (2008) Effects of estrogen on basal forebrain cholinergic neurons and spatial learning. J Neurosci Res 86:1588–1598.

Pitsikas N, Sakellaridis N (2005) The 5-HT2C receptor antagonist RO 60-0491 counteracts rats' retention deficits in a recognition memory task. Brain Res 1054:200–202.

Pompeiano M, Palacios JM, Mengod G (1994) Distribution of the serotonin 5-HT2 receptor family mRNAs: comparison between 5-HT2A and 5-HT2C receptors. Brain Res Mol Brain Res 23:163–178.

Porras G, Di Matteo V, Fracasso C, et al (2002) 5-HT2A and 5-HT2C/2B receptor subtypes modulate dopamine release induced in vivo by amphetamine and morphine in both the rat nucleus accumbens and striatum. Neuropsychopharmacology 26:311–324.

Poucet B (1993) Spatial cognitive maps in animals: new hypotheses on their structure and neural mechanisms. Psychol Rev 100:163–182.

Pozzi L, Acconcia S, Ceglia I, et al (2002) Stimulation of 5-hydroxytryptamine (5-HT(2C)) receptors in the ventrotegmental area inhibits stress-induced but not basal dopamine release in the rat prefrontal cortex. J Neurochem 82:93–100.

Prado-Alcala RA, Grinberg ZJ, Arditti ZL, et al (1975) Learning deficits produced by chronic and reversible lesions of the corpus striatum in rats. Physiol Behav 15:283–287.

Prado-Alcala RA, Signoret-Edward L, Figueroa M, et al (1984) Post-trial injection of atropine into the caudate nucleus interferes with long-term but not with short-term retention of passive avoidance. Behav Neural Biol 42:81–84.

Prado-Alcala RA, Fernandez-Samblancat M, Solodkin-Herrera M (1985) Injections of atropine into the caudate nucleus impair the acquisition and the maintenance of passive avoidance. Pharmacol Biochem Behav 22:243–247.

Prado-Alcala RA, Ruiloba MI, Rubio L, et al (2003) Regional infusions of serotonin into the striatum and memory consolidation. Synapse 47:169–175.

Prado-Alcala RA, Solana-Figueroa R, Galindo LE, et al (2003) Blockade of striatal 5-HT2 receptors produces retrograde amnesia in rats. Life Sci 74:481–488.

Pranzatelli MR, Murthy JN, Pluchino RS (1992) Identification of spinal 5-HT1C binding sites in the rat: characterization of [3H]mesulergine binding. J Pharmacol Exp Ther 261:161–165.

Puryear CB, King M, Mizumori SJ (2006) Specific changes in hippocampal spatial codes predict spatial working memory performance. Behav Brain Res 169:168–175.

Puumala T, Sirvio J (1998) Changes in activities of dopamine and serotonin systems in the frontal cortex underlie poor choice accuracy and impulsivity of rats in an attention task. Neuroscience 83:489–499.

Puumala T, Ruotsalainen S, Jakala P, et al (1996) Behavioral and pharmacological studies on the validation of a new animal model for attention deficit hyperactivity disorder. Neurobiol Learn Mem 66:198–211.

Puumala T, Bjorklund M, Ruotsalainen S, et al (1997) Lack of relationship between thalamic oscillations and attention in rats: differential modulation by an alpha-2 antagonist. Brain Res Bull 43:163–171.

Quarta D, Naylor CG, Stolerman IP (2007) The serotonin 2C receptor agonist Ro-60-0175 attenuates effects of nicotine in the five-choice serial reaction time task and in drug discrimination. Psychopharmacology (Berl) 193:391–402.

Quirarte GL, Cruz-Morales SE, Diaz del Guante MA, et al (1993) Protective effect of under-reinforcement of passive avoidance against scopolamine-induced amnesia. Brain Res Bull 32:521–524.

Ragozzino ME (2003) Acetylcholine actions in the dorsomedial striatum support the flexible shifting of response patterns. Neurobiol Learn Mem 80:257–267.

Ragozzino ME, Jih J, Tzavos A (2002) Involvement of the dorsomedial striatum in behavioral flexibility: role of muscarinic cholinergic receptors. Brain Res 953:205–214.

Ragozzino ME, Ragozzino KE, Mizumori SJ (2002) Kesner RP. Role of the dorsomedial striatum in behavioral flexibility for response and visual cue discrimination learning. Behav Neurosci 116:105–115.

Ramos M, Goni-Allo B, Aguirre N (2005) Administration of SCH 23390 into the medial prefrontal cortex blocks the expression of MDMA-induced behavioral sensitization in rats: an effect mediated by 5-HT2C receptor stimulation and not by D1 receptor blockade. Neuropsychopharmacology 30:2180–2191.

Remington G (2008) Alterations of dopamine and serotonin transmission in schizophrenia. Prog Brain Res 172:117–140.

Reynolds GP, Templeman LA, Zhang ZJ (2005) The role of 5-HT2C receptor polymorphisms in the pharmacogenetics of antipsychotic drug treatment. Prog Neuropsychopharmacol Biol Psychiatry 29:1021–1028.

Richtand NM, Welge JA, Logue AD, et al (2008) Role of serotonin and dopamine receptor binding in antipsychotic efficacy. Prog Brain Res 172:155–175.

Richter-Levin G, Segal M (1989) Spatial performance is severely impaired in rats with combined reduction of serotonergic and cholinergic transmission. Brain Res 477:404–407.

Richter-Levin G, Greenberger V, Segal M (1993) Regional specificity of raphe graft-induced recovery of behavioral functions impaired by combined serotonergic/cholinergic lesions. Exp Neurol 121:256–260.

Rick CE, Stanford IM, Lacey MG (1995) Excitation of rat substantia nigra pars reticulata neurons by 5-hydroxytryptamine in vitro: evidence for a direct action mediated by 5-hydroxytryptamine2C receptors. Neuroscience 69:903–913.

Rinaldi-Carmona M, Congy C, Santucci V, et al (1992) Biochemical and pharmacological properties of SR 46349B, a new potent and selective 5-hydroxytryptamine2 receptor antagonist. J Pharmacol Exp Ther 262:759–768.

Roberson ED, English JD, Adams JP, et al (1999) The mitogen-activated protein kinase cascade couples PKA and PKC to cAMP response element binding protein phosphorylation in area CA1 of hippocampus. J Neurosci 19:4337–4348.

Rocha BA, Goulding EH, O'Dell LE, et al (2002). Enhanced locomotor, reinforcing, and neurochemical effects of cocaine in serotonin 5-hydroxytryptamine 2C receptor mutant mice. J Neurosci 22:10039–10045.

Rogers RD, Andrews TC, Grasby PM (2000) Contrasting cortical and subcortical activations produced by attentional-set shifting and reversal learning in humans. J Cogn Neurosci 12:142–162.

Rolls ET, Hornak J, Wade D (1994) Emotion-related learning in patients with social and emotional changes associated with frontal lobe damage. J Neurol Neurosurg Psychiatry 57:1518–1524.

Roth BL (1994) Multiple serotonin receptors: clinical and experimental aspects. Ann Clin Psychiatry 6:67–78.

Ruotsalainen S, MacDonald E, Koivisto E, et al (1998) 5-HT1A receptor agonist (8-OH-DPAT) and 5-HT2 receptor agonist (DOI) disrupt the non-cognitive performance of rats in a working memory task. J Psychopharmacol 12:177–185.

Ryan BK, Anwyl R, Rowan MJ (2008) 5-HT2 receptor-mediated reversal of the inhibition of hippocampal long-term potentiation by acute inescapable stress. Neuropharmacology 55:175–182.

Salamone JD (1996) The behavioral neurochemistry of motivation: methodological and conceptual issues in studies of the dynamic activity of nucleus accumbens dopamine. J Neurosci Methods 64:137–149.

Santucci AC, Knott PJ, Haroutunian V (1996) Excessive serotonin release, not depletion, leads to memory impairments in rats. Eur J Pharmacol 295:7–17.

Scarpelli G, Alves SH, Landeira-Fernandez J, et al (2008) Effects of two selective 5-HT2C receptor-acting compounds into the ventral hippocampus of rats exposed to the elevated plus maze. Psychol Neurosci 1:87–96.

Scatton B, Javoy-Agid F, Rouquier L, et al (1983) Reduction of cortical dopamine, noradrenaline, serotonin and their metabolites in Parkinson's disease. Brain Res 275:321–328.

Schmidt CJ, Sullivan CK, Fadayel GM (1994) Blockade of striatal 5-hydroxytryptamine2 receptors reduces the increase in extracellular concentrations of dopamine produced by the amphetamine analogue 3,4-methylenedioxymethamphetamine. J Neurochem 62:1382–1389.

Scholes KE, Harrison BJ, O'Neill BV, et al (2007) Acute serotonin and dopamine depletion improves attentional control: findings from the stroop task. Neuropsychopharmacology 32:1600–1610.

Scholtissen B, Deumens R, Leentjens AF, et al (2006) Functional investigations into the role of dopamine and serotonin in partial bilateral striatal 6-hydroxydopamine lesioned rats. Pharmacol Biochem Behav 83:175–185.

Schuster G, Cassel JC, Will B (1997) Comparison of the behavioral and morphological effects of colchicine- or neutral fluid-induced destruction of granule cells in the dentate gyrus of the rat. Neurobiol Learn Mem 68:86–91.

Seager MA, Johnson LD, Chabot ES, et al (2002) Oscillatory brain states and learning: impact of hippocampal theta-contingent training. Proc Natl Acad Sci USA 99:1616–1620.

Shakesby AC, Anwyl R, Rowan MJ (2002) Overcoming the effects of stress on synaptic plasticity in the intact hippocampus: rapid actions of serotonergic and antidepressant agents. J Neurosci 22:3638–3644.

Sheldon PW, Aghajanian GK (1991) Excitatory responses to serotonin (5-HT) in neurons of the rat piriform cortex: evidence for mediation by 5-HT1C receptors in pyramidal cells and 5-HT2 receptors in interneurons. Synapse 9:208–218.

Shors TJ, Seib TB, Levine S, Thompson RF, et al (1989) Inescapable versus escapable shock modulates long-term potentiation in the rat hippocampus. Science 244:224–226.

Siuciak JA, Chapin DS, McCarthy SA, et al (2007). CP-809,101, a selective 5-HT2C agonist, shows activity in animal models of antipsychotic activity. Neuropharmacology 52:279–290.

Smiley JF, Goldman-Rakic PS (1996) Serotonergic axons in monkey prefrontal cerebral cortex synapse predominantly on interneurons as demonstrated by serial section electron microscopy. J Comp Neurol 367:431–443.

Sodhi MS, Arranz MJ, Curtis D, et al (1995) Association between clozapine response and allelic variation in the 5-HT2C receptor gene. Neuroreport 7:169–172.

Sodhi MS, Arranz MJ, Curtis D, et al (2001) RNA editing of the 5-HT2C receptor is reduced in schizophrenia. Mol Pharmacol 6:373–379.

Sodhi MS, Airey DC, Lambert W, et al (2005) A rapid new assay to detecting RNA editing revelas antipsychotic-induced changes in serotonin-2C transcripts. Mol Pharmacol 68:711–719.

Sokolowski JD, Salamone JD (1994) Effects of dopamine depletions in the medial prefrontal cortex on DRL performance and motor activity in the rat. Brain Res 642:20–28.

Sorensen SM, Kehne JH, Fadayel GM, et al (1993) Characterization of the 5-HT2 receptor antago-nist MDL 100907 as a putative atypical antipsychotic: behavioral, electrophysiological and neurochemical studies. J Pharmacol Exp Ther 266:684–691.

Staubli U, Xu FB (1995) Effects of 5-HT3 receptor antagonism on hippocampal theta rhythm, memory, and LTP induction in the freely moving rat. J Neurosci 15:2445–2452.

Stone JM, Pilowsky LS (2007) Novel targets for drugs in schizophrenia. CNS Neurol Disord Drug Targets 6:265–272.

Talpos JC, Wilkinson LS, Robbins TW (2006) A comparison of multiple 5-HT receptors in two tasks measuring impulsivity. J Psychopharmacol 20:47–58.

Tan H, Zhong P, Yan Z (2004) Corticotropin-releasing factor and acute stress prolongs serotoner-gic regulation of GABA transmission in prefrontal cortical pyramidal neurons. J Neurosci 24:5000–5008.

Tecott LH, Sun LM, Akana SF, et al (1995) Eating disorder and epilepsy in mice lacking 5-HT2c serotonin receptors. Nature 374:42–46.

Tecott LH, Logue SF, Wehner JM, et al (1998) Perturbed dentate gyrus function in serotonin 5-HT2C receptor mutant mice. Proc Natl Acad Sci USA 95:15026–15031.

Tsaltas E, Kontis D, Chrysikakou S, et al (2005) Reinforced spatial alternation as an animal model of obsessive-compulsive disorder (OCD): investigation of 5-HT2C and 5-HT1D receptor involvement in OCD pathophysiology. Biol Psychiatry 57:1176–1185.

Vahabzadeh A, Fillenz M (1994) Comparison of stress-induced changes in noradrenergic and sero-tonergic neurons in the rat hippocampus using microdialysis. Eur J Neurosci 6:1205–1212.

Valluzzi JA, Chan K (2007) Effects of fluoxetine on hippocampal-dependent and hippocampal-independent learning tasks. Behav Pharmacol 18:507–513.

van Asselen M, Kessels RP, Neggers SF, et al (2006) Brain areas involved in spatial working memory. Neuropsychologia 44:1185–1194.

Van Bockstaele EJ, Cestari DM, Pickel VM (1994) Synaptic structure and connectivity of sero-tonin terminals in the ventral tegmental area: potential sites for modulation of mesolimbic dopamine neurons. Brain Res 647:307–322.

van der Meulen JA, Joosten RN, de Bruin JP, et al (2007) Dopamine and noradrenaline efflux in the medial prefrontal cortex during serial reversals and extinction of instrumental goal-directed behavior. Cereb Cortex 17:1444–1453.

van Donkelaar EL, Rutten K, Blokland A (2008) Phosphodiesterase 2 and 5 inhibition attenuates the object memory deficit induced by acute tryptophan depletion. Eur J Pharmacol 600:98–104.

Vysokanov A, Flores-Hernandez J, Surmeier DJ (1998) mRNAs for clozapine-sensitive receptors co-localize in rat prefrontal cortex neurons. Neurosci Lett 258:179–182.

Wang GW, Cai JX (2006) Disconnection of the hippocampal-prefrontal cortical circuits impairs spatial working memory performance in rats. Behav Brain Res 175:329–336.

Ward RP, Dorsa DM (1996) Colocalization of serotonin receptor subtypes 5-HT2A, 5-HT2C, and 5-HT6 with neuropeptides in rat striatum. J Comp Neurol 370:405–414.

Werkman TR, Glennon JC, Wadman WJ, et al (2006) Dopamine receptor pharmacology: interactions with serotonin receptors and significance for the aetiology and treatment of schizophrenia. CNS Neurol Disord Drug Targets 1:3–23.

White NM, McDonald RJ (2002) Multiple parallel memory systems in the brain of the rat. Neurobiol Learn Mem 77:125–184.

White NM, Packard MG, Seamans J (1993) Memory enhancement by post-training peripheral administration of low doses of dopamine agonists: possible autoreceptor effect. Behav Neural Biol 59:230–241.

Wichmann T, DeLong MR (1998) Models of basal ganglia function and pathophysiology of movement disorders. Neurosurg Clin N Am 9:223–236.

Wilkinson RT (1963) Interaction of noise with knowledge of results and sleep deprivation. J Exp Psychol 66:332–337.

Williams GV, Rao SG, Goldman-Rakic PS (2002) The physiological role of 5-HT2A receptors in working memory. J Neurosci 22:2843–2854.

Wilson CJ (1998) Basal ganglia. In: Shepherd GM, ed. The synaptic organization of the brain, 4th ed. Oxford: Oxford University Press, pp. 329–375.

Winocur G (1974) Functional dissociation within the caudate nucleus of rats. J Comp Physiol Psychol 86:432–439.

Winson J (1978) Loss of hippocampal theta rhythm results in spatial memory deficit in the rat. Science 201:160–163.

Winstanley CA, Dalley JW, Theobald DE, et al (2004) Fractionating impulsivity: contrasting effects of central 5-HT depletion on different measures of impulsive behavior. Neuropsychopharmacology 29:1331–1343.

Winstanley CA, Theobald DE, Dalley JW, et al (2004) 5-HT2A and 5-HT2C receptor antagonists have opposing effects on a measure of impulsivity: interactions with global 5-HT depletion. Psychopharmacology (Berl) 176:376–385.

Wise RA (2002) Brain reward circuitry: insights from unsensed incentives. Neuron 36:229–240.

Wu M, Shanabrough M, Leranth C, et al (2000) Cholinergic excitation of septohippocampal GABA but not cholinergic neurons: implications for learning and memory. J Neurosci 20:3900–3908.

Yau JL, Morris RG, Seckl JR (1994) Hippocampal corticosteroid receptor mRNA expression and spatial learning in the aged Wistar rat. Brain Res 657:59–64.

Yuan A, Dourado M, Butler A, et al (2000) SLO-2, a K$^+$ channel with an unusual Cl$^-$ dependence. Nat Neurosci 3:771–779.

Yuen EY, Jiang Q, Chen P, et al (2005) Serotonin 5-HT1A receptors regulate NMDA receptor channels through a microtubule-dependent mechanism. J Neurosci 25:5488–5501.

Zhong P, Yuen EY, Yan Z (2008) Modulation of neuronal excitability by serotonin-NMDA interactions in prefrontal cortex. Mol Cell Neurosci 38:290–299.

Zohar J, Insel TR (1987) Drug treatment of obsessive-compulsive disorder. J Affect Disord 13:193–202.

Chapter 25
The Role of 5-HT$_{2C}$ Polymorphisms in Behavioral and Psychological Symptoms of Alzheimer Disease

Antonia Pritchard

25.1 Introduction to Alzheimer Disease

Alzheimer disease (AD) is a progressive neurodegenerative condition, which is clinically characterized by increasing memory loss, leading to dementia and eventually death. Alzheimer disease is the most common form of dementia, accounting for between 50% and 70% of all late onset dementias and affecting more than 25 million people worldwide. In the USA, the proportion of the population aged ≥65 years is projected to increase from 12.4% in 2000 to 19.6% in 2030 (US Census Burehttp: http://www.census.gov/population/www/projections/natdet-D1A.html). As the population lives longer, the prevalence of AD, which doubles every 5 years after age 65, is also expected to increase. The increasing healthcare demands associated with the rising aging population will pose a huge challenge for both personal and public resources.

Patients with AD usually first present with slowly progressive memory loss, and as the disease progresses, several areas of cognition are affected. These include a loss of ability to recognize objects, impaired visual–spatial skills, deterioration in the ability to perform daily tasks, difficulties with speech, and a decrease in the ability to reason the consequences of actions. Other neurological pathways can also become involved, including the extrapyramidal system, which can result in Parkinson-like movement difficulties. Patients commonly become less able to perform familiar daily tasks, such as cooking, washing, and dressing.

25.1.1 Pathological Features of Alzheimer Disease

There are several pathological features of AD, which can be used at autopsy for a definitive disease diagnosis (Fig. 25.1). The classic pathological hallmarks of AD

A. Pritchard (✉)
Molecular Psychiatry Group, G Floor, CBCRC Building, Queensland Institute of Medical Research, 300 Herston Road, Herston, Brisbane, Queensland, 4800, Australia
e-mail: a.pritchard2@uq.edu.au

G. Di Giovanni et al. (eds.), *5-HT$_{2C}$ Receptors in the Pathophysiology of CNS Disease*,
The Receptors 22, DOI 10.1007/978-1-60761-941-3_25,
© Springer Science+Business Media, LLC 2011

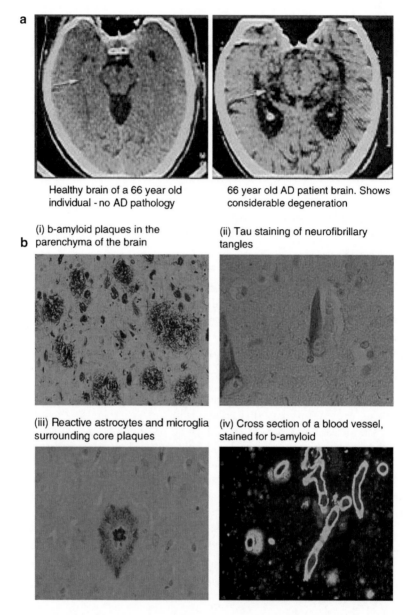

Healthy brain of a 66 year old individual - no AD pathology

66 year old AD patient brain. Shows considerable degeneration

(i) b-amyloid plaques in the parenchyma of the brain

(ii) Tau staining of neurofibrillary tangles

(iii) Reactive astrocytes and microglia surrounding core plaques

(iv) Cross section of a blood vessel, stained for b-amyloid

Fig. 25.1 The pathological changes in Alzheimer disease (*AD*) brains. (**a**) The overall changes to the structure of the brain in an AD patient compared with a healthy control. (**b**) Histopathological staining of β-amyloid and τ structures in AD brains (Pictures courtesy of Prof. D. Mann, University of Manchester, UK)

are abnormal clusters of protein in the brain. These accumulations occur in two forms: intracellularly within neurones, known as *neurofibrillary tangles* (NFT) (Fig. 25.1bii), and extracellularly between the cells of the brain, known as *β-amyloid plaques* (Aβ) (Fig. 25.1bi).

The NFT occur predominantly in the hippocampus and the temporal lobe cortex and are primarily composed of paired helical filaments (PHF), which consist of the hyperphosphorylated microtubule associated protein, tau.

The extracellular protein clusters are mainly composed of Aβ, a fragment of the amyloid precursor protein (APP). The microglia and astrocyte cells surrounding Aβ plaques are activated, which has been suggested to play a role in plaque development.

Other pathological changes include loss of synapses and synaptic markers and, on a more global level, neuronal loss and cerebral atrophy. The neurodegeneration in AD is restricted to the brain. The most vulnerable systems are those involved in the neural processing of memory, in particular the hippocampus, its connections to the cortex and temporal lobes, and the cortical cholinergic system.

25.2 The Behavioral and Psychological Symptoms of Alzheimer Disease

While cognitive impairments are often considered to be the defining feature of AD, many patients also display behavioral and psychiatric disturbances, known collectively as the *behavioral and psychological symptoms of dementia* (BPSD). These symptoms include personality change, depression, hallucinations, delusions, aggression, wandering, sexual disinhibition, and agitation. The BPSD are a major cause of distress to family members and caregivers and are a frequent contributor to patient institutionalization (Finkel 2003; Steele et al. 1990). It is clear that an individual patient does not experience every symptom. The BPSD do not progress regularly or exist continuously in a patient once they have occurred; for this reason, BPSD are described as an "episodic" phenomenon. However, the individual symptoms rarely appear in isolation during the course of the disease, and the presenting symptom(s) vary between patients.

25.2.1 What Causes the BPSD?

The individual behavioral or psychological symptoms do not occur in every patient; nor is it possible to predict which patients will suffer from a particular symptom. Speculation has therefore arisen as to the cause of BPSD: Why do some patients suffer from particular symptoms, but not others? Multiple factors are thought to contribute to the development of BPSD by individual patients. Studies have suggested that these include environmental, medical and brain neuropathological factors, the severity and progression of the neurodegenerative process of AD, and genetic susceptibility.

25.2.2 Genetic Influence for the Manifestation of BPSD

There are monogenetic familial forms of AD; however, these only account for approximately 5% of cases. The majority of AD cases are often called either *sporadic* or *complex inherited* as it is not inherited in families in a Mendelian fashion, i.e., not caused by a single gene mutation. This type of AD is indeed complex because no single gene or simple mode of inheritance accounts for its heritability or incidence and is thought to be caused by an intricate interaction between genetic and nongenetic factors. It is also clear that BPSD are not caused by single gene mutations inherited in a Mendelian fashion; however, a genetic influence to the manifestation of BPSD has been demonstrated.

25.2.2.1 Family Studies to Establish Inheritance of BPSD

In an ideal situation, a genetic component to BPSD would have been established by investigations using extended families or monozygotic twins and determining if there was an increase in the presence of certain symptoms in those affected by AD, greater than expected by chance. Such studies have been used to describe the heritability of disorders such as schizophrenia, bipolar disorder and unipolar depression, as well as AD itself (Porteous 2008). However, because AD is predominantly a late-onset disease, these possible routes of investigation are severely limited because there are few family members alive late in life. There have therefore been only a few studies of the heritability of behavioral or psychological symptoms in early onset familial or sporadic complex AD patients using these methods.

A more practical method, due to the late onset of the disorder, is the examination of whether there was an increase in family history in patients with certain symptoms compared with those who did not have that symptom. The basic hypothesis behind this theory is familial factors (most likely genetic) that predispose to a certain behavioral or psychiatric symptom in AD may be the same as those that predispose to that symptom in psychiatric illness earlier in life. Several lines of evidence using these methods support the involvement of genes in BPSD.

Pearlson et al. (1990), Strauss and Ogrocki (1996), and Lyketsos et al. (1996) showed a family history of depression was significantly associated with increased risk for major depression in Alzheimer patients. Sweet et al. (2002)found that the development of psychosis in AD is determined in part by genetic factors. Tunstall et al. (2000) found a significantly elevated pairwise concordance for agitation/aggression.

The conclusion reached from these investigations was a genetic component to the susceptibility to certain BPSD exists, and a number of researchers started the process of ascertaining of psychiatric information of patients in AD cohorts to enable the search for these genes.

25.2.3 Genetic Investigation of Susceptibility to BPSD

Many candidate genes have been proposed and investigated, the majority of which have been previously implicated in psychiatric disorders such as schizophrenia, bipolar disease, and major depression. Genes involved in neurotransmitter systems have been particularly implicated for two reasons (Rang et al. 2003; Barnes and Sharp 1999):

1. They are the pharmacological target of treatment for BPSD and other psychiatric disorders, such as schizophrenia, bipolar disorder, and major depression.
2. There is substantial evidence to suggest these systems play a role in many aspects of human and animal behavior and psychiatric symptoms, including aggression, hallucinations, delusions, depression, anxious behavior, and the regulation of appetite.

25.2.3.1 The Serotonergic System as a Candidate for BPSD Susceptibility

The serotonergic system includes the neurotransmitter 5-hydroxytryptamine (5-HT), enzymes that synthesise 5-HT, receptors of 5-HT, the serotonergic transporter (SERT), which clears 5-HT from the synaptic cleft, and the enzymes that break down 5-HT (Barnes and Sharp 1999). In the central nervous system (CNS), 5-HT is present in particularly high concentrations in localized regions of the midbrain and is known to be involved in mood, emotion, cognition, motor function, and circadian rhythms. To date, there have been seven families of 5-HT receptors identified, comprising 14 structurally and pharmacologically distinct subtypes (Barnes and Sharp 1999). These receptors control the different effects of 5-HT, in either an excitatory or an inhibitory manner. As a neurotransmitter, 5-HT is stored in and released from serotonergic neurones. The cell bodies of serotonergic neurones in the brain are found in nine clusters, most of which are located in the raphe nuclei of the midbrain, pons, and medulla. The receptors for 5-HT are also found in specific locations throughout the body. This chapter focuses on one of these receptors, namely, 5-HT$_{2C}$.

25.3 The 5-HT$_{2C}$ Serotonergic Receptor

The 5-HT$_{2C}$ receptor gene was originally cloned by Saltzman et al. in 1991. The receptor belongs to the G-protein-coupled receptor superfamily and exerts its cellular actions via inositol triphosphate (IP3) signaling to produce an excitatory effect (Saltzman et al. 1991). The CNS distribution of 5-HT$_{2C}$ has been extensively mapped by receptor autoradiography, and 5-HT$_{2C}$ has been found in high levels in the choroid plexus, areas of the cortex, the basal ganglia, and areas of the limbic system, including the hippocampus and the amygdala (Mengod et al. 1996; Lopez-Gimenez et al. 2001, 2002). It is thought that hypophagia and anxiety are behavioral responses associated with activation of the 5-HT$_{2C}$ receptor, especially given their locations within the brain. Certain hallucinogens have an agonistic effect on 5-HT$_{2C}$ to produce their effect, which additionally suggests a role for this receptor in psychosis (Barnes and Sharp 1999; Fiorella et al. 1995).

25.3.1 The 5-HT$_{2C}$ Protein

The 5-HT$_{2C}$ protein is a membrane spanning G-protein receptor (Barnes and Sharp 1999) (Fig. 25.2). A splice variant of the 5-HT$_{2C}$ protein has been detected in the brain tissue of humans, rats, and mice. This variant contains a 95-base-pair (bp) deletion in the region coding for the second intracellular loop and fourth transmembrane domain. This deletion at the RNA level leads to a frameshift and premature termination of the protein (Xie et al. 1996). However, the function of this variant is unclear, as it lacks a binding site for 5-HT (Barnes and Sharp 1999).

25.3.2 The Structure and Location of the 5-HT$_{2C}$ Gene

The gene for the human 5-HT$_{2C}$ gene was mapped to chromosome Xq24 and contains 6 exons, spanning more than 230 kilobase (kb) (Xie et al. 1996; Chen et al. 1992) (Fig. 25.3). No alternative transcripts at the genetic level have been described. The messenger ribonucleic acid (mRNA) product undergoes posttranscriptional editing, which yields multiple isoforms with a widespread distribution in the brain. This is potentially of great significance if the mRNA isoforms translate into proteins, because the predicted amino acid sequence suggests different regulatory and pharmacological properties (Burns et al. 1997).

25.3.3 5-HT$_{2C}$ as a Candidate Gene for Psychiatric Disorders

The locations and postulated function of the 5-HT$_{2C}$ receptor within the CNS have led geneticists to investigate variants within the gene for associations with psychiatric disorders. The fact that the dopaminergic system can be inhibited by serotonergic activity (Kapur and Remington 1996) and mediated by 5-HT$_{2C}$ in the

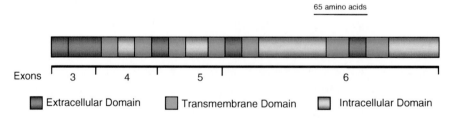

Fig. 25.2 Schematic diagram of the 5-HT$_{2C}$ protein showing the gene exons that correspond to the protein domains. The 5-HT$_{2C}$ protein domains and corresponding exons, as predicted by Swiss-prot (http://www.ebi.ac.uk/uniprot/) and Interpro (http://www.ebi.ac.uk/interpro/). The 5-HT$_{2C}$ protein is seven-transmembrane spanning, consists of 458 amino acids, and has a molecular weight of 51.8 kDa

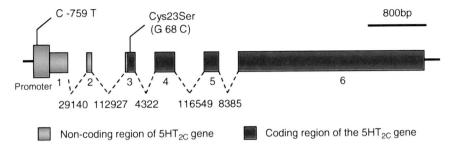

Fig. 25.3 Schematic diagram of the 5-HT$_{2C}$ gene showing the exons, size indication of the introns, and the location of the polymorphic variants discussed within this chapter. The 5-HT$_{2C}$ gene (NM_000868.1/gi:4504540) is located at Xq24, occupies approximately 326 kb of genomic DNA, and encodes for two different isoforms, with editing occurring at the RNA level. Other known variants can be found using the UCSC genome browser (http://genome.ucsc.edu/)

mesolimbic and mesostriatal pathways in rats (Di Matteo et al. 2001) further strengthened this gene as a good candidate.

A decrease in 5-HT$_{2C}$ gene expression has been reported in postmortem studies of schizophrenia (Castensson et al. 2005) and variants within 5-HT$_{2C}$, and the different patterns of RNA editing have been associated with schizophrenia; however the association studies have largely not revealed statistically significant associations (http://www.schizophreniaforum.org/res/sczgene/default.asp). Other psychiatric disorders investigated for association with different patterns of RNA editing or associations with polymorphic variants within the 5-HT$_{2C}$ gene include suicide, major depression (Gurevich et al. 2002), weight gain with antipsychotic medication (Ujike et al. 2008), and attention deficit hyperactivity disorder (Li et al. 2006); a comprehensive list of all these association studies can be found at the National Institutes of Health (NIH) genetic association database (http://geneticassociationdb.nih.gov/).

25.3.4   Polymorphisms Within the 5-HT$_{2C}$ Gene

One of the most extensively investigated variant within the 5-HT$_{2C}$ gene is a G-to-C polymorphism (rs6318) at nucleotide position 68 within exon 3 (the first coding exon) that results in an amino acid change from a cysteine to a serine, at amino acid 23, which falls in the first hydrophobic region of the protein (Lappalainen et al. 1995). There are also other variants within this gene that have been analyzed with psychiatric illnesses, particularly the promoter C-759T variant, with limited replication (http://geneticassociationdb.nih.gov/).

Analyses of 5-HT$_{2C}$ genotypes must be gender specific because the gene resides on the X chromosome; this means that males are usually hemizygous for the gene. In rare cases, there are males who appear as a heterozygote, implying that they possess more than one X chromosome. These unusual individuals are generally excluded from analyses.

25.3.5   Association of cys23ser with BPSD in Alzheimer Disease

The cys23ser (also known as G68C; rs6318) polymorphism has been examined in four studies for association with various BPSD, as summarized in Table 25.1 (Holmes et al. 1998, 2003; Assal et al. 2004; Pritchard et al. 2008). As can be seen from Table 25.1, there have been no consistent associations of the cys23ser variant with any of the individual BPSD examined.

Holmes et al. performed the first investigation of this polymorphism with BPSD in 1998, utilizing a cross-sectional study design. They found a nonsignificant association in both females ($\chi^2 = 5.2$, $p = 0.07$) and males ($\chi^2 = 3.4$, $p = 0.07$) for the presence of visual hallucinations in individuals with the C (ser23) allele. They also found statistically significant association in females for the presence of hyperphagia (abnormally increased appetite) with the C (ser23) allele ($\chi^2 = 7.6$, $p = 0.02$), which was not found in males (Holmes et al. 1998).

The second study of 5-HT$_{2C}$ in BPSD was performed by Holmes et al. (2003, 1998), who utilized a cross-sectional study design to investigate susceptibility to depression in AD. They did not report the analysis of separate genders in this paper, presumably due to the small number of patients with depression in the whole group ($n = 19$) compared with patients without depression ($n = 139$). However, they found an association between the CC (ser23/ser23) genotype with the presence of depression compared with patients with GG (cys23/cys23) genotype; OR = 12.4; 95% CI = 2.8–56.2; $p = 0.004$ (Holmes et al. 2003).

Assal et al. investigated whether the cys23ser polymorphic variant was associated with agitation/aggression, depression and anxiety in AD. In this cross-sectional study, they found no association with this variant and delusions, hallucinations, agitation/aggression, or anxiety (Assal et al. 2004).

Pritchard et al. (2008) sought to provide replication of the results from these three previous investigations. They utilized a large cohort, analysed longitudinally for BPSD using the NPI and investigated delusions, hallucinations, depression, anxiety, agitation/aggression, and appetite disturbances. An association was observed with the C allele and the presence of anxiety in females ($\chi^2 = 5.13$, $p = 0.02$) and a nonstatistically significant distortion in the C allele in the presence of appetite disturbance in females.

25.3.6   Discussion of Association Studies of 5-HT$_{2C}$ with BPSD

From these studies, there was a validation of the findings of no association between this variant in 5-HT$_{2C}$ and psychosis, agitation/aggression (Assal et al. 2004; Pritchard et al. 2008), or with delusions (Holmes et al. 1998; Assal et al. 2004; Pritchard et al. 2008). Despite the different functions of 5-HT$_{2C}$ that suggest it would be a good candidate gene for BPSD, the evidence currently suggests it is not implicated in the aetiology of these symptoms.

Table 25.1 Previous investigations of the cys23ser variant in 5-HT$_{2C}$ Polymorphisms for association with hallucinations, delusions, agitation/aggression, appetite disturbances, depression and anxiety, in Alzheimer disease patients

Authors, year	Ethnicity	AD Patients	Assessment	BPSD Symptoms Examined	Result
Holmes et al. (1998)	British	211	CAMDEX[a] and MOUSEPAD[b]	Visual hallucinations Auditory hallucinations Delusions Aggression Hyperphagia	C (ser) allele associated with visual hallucinations ($p = 0.07$) and hyperphagia in females ($p = 0.02$) but not males
Holmes et al. (2003)	British	158	CAMDEX and Cornell scale for depression in dementia	Depression	CC (ser/ser) genotype associated with depression ($p = 0.004$)
Assal et al. (2004))	American	96	NPI[c]	Delusions, hallucinations, agitation/aggression Anxiety	No association with delusions, hallucinations, or agitation/aggression
Pritchard et al. (2008)	British	394	NPI	Delusions Hallucinations Agitation/aggression Depression Anxiety Appetite disturbances	C (ser) allele associated with anxiety in females ($p = 0.02$)

[a]CAMDEX: The Cambridge Examination for Mental Disorders of Older People
[b]MOUSEPAD: Manchester and Oxford Universities Scale for the Psychopathological Assessment of Dementia
[c]NPI: Neuropsychiatric inventory

25.3.6.1   5-HT$_{2C}$ Associations with Visual Hallucinations in AD

Holmes et al. found an association with visual hallucinations; however, this was not replicated by either Assal et al. (2004) or Pritchard et al. (2008)). The investigation by Holmes et al. utilized an assessment tool that distinguishes the types of hallucinations observed, which the Neuropsychiatric Inventory, utilized by Assal et al. and Pritchard et al., does not (Holmes et al. 1998; Assal et al. 2004; Pritchard et al. 2008). This finding, therefore, clearly requires replication in a cohort with more detailed psychiatric assessment before a conclusion can be drawn over the role of the 5-HT$_{2C}$ cys23ser variant in the aetiology of visual hallucinations in AD patients.

25.3.6.2   5-HT$_{2C}$ Associations with Depression in AD

Holmes et al. found an association with depression, which was not replicated by either of the other two investigations examining depression (Holmes et al. 2003; Assal et al. 2004; Pritchard et al. 2008). Holmes et al. utilized an assessment tool that was developed specifically to examine depression in dementia patients, whereas Assal et al. and Pritchard et al. utilized the more general NPI. It is therefore possible that this specialized tool was more sensitive than the NPI. However, the low number of patients with depression and the lack of dichotomisation into males and females in the Holmes et al. cohort could also have lead to discrepancies between these investigations. Thus, despite there being a good case for a role of 5-HT$_{2C}$ in depression based on the actions of antidepressants on this receptor, its role in depressive symptomology of BPSD is still unclear.

25.3.6.3   5-HT$_{2C}$ Associations with Anxiety in AD

Assal et al. found no association with anxiety; however, Pritchard et al. found an association between the C allele and anxiety in females in their cohort. It was postulated by Pritchard et al. that the longitudinal nature of their investigation resulted in the identification of a large percentage of their cohort having anxiety, which could have an impact on the associations observed. As females and males are analyzed separately, the number of patients required to provide adequate power to detect true associations in a cohort with a high frequency of anxiety would be very large. 5-HT$_{2C}$ is known to be expressed in the amygdala, a region of the brain known to regulate anxiety (Barnes and Sharp 1999) and is therefore a good candidate gene for the regulation of anxiety. At this point, however, no conclusions can be drawn over the role of 5-HT$_{2C}$ in the aetiology of anxiety in AD without further replication investigations.

25.3.6.4   5-HT$_{2C}$ Associations with Appetite Disturbance in AD

While no association was reported by Assal et al. with appetite disturbance (Assal et al. 2004), Holmes et al. found an increase in the C allele with hyperphagia in females only (Holmes et al. 1998), and Pritchard et al. found a nonstatistically significant distortion in genotype and allele frequency in females (Pritchard et al. 2008), in the same direction as described by Holmes et al. It must be noted that the NPI does not distinguish the nature of appetite disturbances observed, so the results of Pritchard et al. are not directly comparable to those of Holmes et al. However, 5-HT$_{2C}$ has been implicated in the control of appetite, as demonstrated in 5-HT$_{2C}$ null mice, which exhibited an abnormal control of feeding behavior (Tecott et al. 1995) and by the use of 5-HT$_{2C}$ receptor agonists, which reduce feeding in rats (Heisler et al. 2002). The interesting findings of weight gain in schizophrenic patients on certain antipsychotics, obesity and weight gain with seasonal affective disorder, with the 759C/T promoter polymorphism (rs3813929) (De Luca et al. 2007; Ryu et al. 2007; Praschak-Rieder et al. 2005; Miller et al. 2005; Pooley et al. 2004; Zhang et al. 2002; Reynolds et al. 2002) suggest that this variant, as well as others within the gene, should be examined for changes in appetite in AD patients. These data suggest that there is evidence to support a role for genetic variation in the 5-HT$_{2C}$ receptor with appetite regulation in AD patients.

25.4 Conclusions

There have only been a few studies investigating a possible association between 5-HT$_{2C}$ and the BPSD occurring in AD. These investigations have focused on only one variant within the gene, and clearly more work needs to be performed before 5-HT$_{2C}$ can be excluded or verified as a susceptibility candidate. Discrepancies between the few previously reported studies could be due to the small sample size of some studies, especially given the need to dichotomize the sample for analysis into male and female. Other issues include differing selection criteria for AD patients, different sensitivities for detecting symptoms dependent on the neuropsychiatric test utilized, population or ethnic differences, which can alter the degrees of linkage disequilibrium between cohorts, and the discrepancy of symptoms detected between cross-sectional versus longitudinal study designs, and some findings will likely be spurious due to multiple testing and by chance.

The potentially interesting similarity in findings from investigations of appetite disturbances in AD patients (Holmes et al. 1998; Pritchard et al. 2008), combined with those previous findings in other disorders, utilizing a different variant in the 5-HT$_{2C}$ gene (De Luca et al. 2007; Ryu et al. 2007; Praschak-Rieder et al. 2005; Miller et al. 2005; Pooley et al. 2004; Zhang et al. 2002; Reynolds et al. 2002), suggests a potential role for the gene in appetite disturbances.

Genetic investigations of BPSD are an important area of research. If a risk factor is identified, the possibility of development of prophylactic treatment specific to the symptom in question becomes more realistic. Symptom specific treatment could make a large difference to patients and caregivers and relieve some of the burden on the health services, and therefore, continuation and expansion of these genetic investigations are vitally important.

Acknowledgements I thank the patients and carers who undertook the clinical trial that gave rise to the work of Pritchard et al. (2008) and the clinicians and nurses who were involved with the study. Funding was from the Birmingham and Solihull NHS, and work was carried out in Dr. Corinne Lendon's laboratory at the University of Birmingham, Birmingham, UK and at the Queensland Institute of Medical Research, Brisbane, Australia.

References

Assal F, Alarcon M, Soloman E, et al (2004) Association of the serotonin transporter and receptor gene polymorphisms in neuropsychiatric symptoms in Alzheimer's disease. Arch Neurol 61:1249–1253.

Barnes N, Sharp T (1999) A review of central 5HT receptors and their functions. Neuropharmacology 38:1083–1152.

Burns C, Chu H, Rueter S, et al (1997) Regulation of serotonin-2C receptor G-protein coupling by RNA editing. Nature 387:303–308.

Castensson A, Aberg K, McCarthy S, et al (2005) Serotonin receptor 2C (HTR2C) and schizophrenia: examination of possible medication and genetic influences on expression levels. Am J Med Genet B Neuropsychiatr Genet 134B:84–89.

Chen K, Yang W, Grimsby J, et al (1992) The human 5-HT2 receptor is encoded by a mulitple intron-exon gene. Mol Brain Res 14:20–26.

De Luca V, Mueller DJ, de Bartolomeis A, et al (2007) Association of the HTR2C gene and antipsychotic induced weight gain: a meta-analysis. Int J Neuropsychopharmacol 10:697–704.

Di Matteo V, De Blasi A, Di Giulio C, et al (2001) Role of 5-HT(2C) receptors in the control of central dopamine function. Trends Pharmacol Sci 22:229–232.

Finkel S (2003) Behavioral and psychological symptoms of dementia. Clin Geriatr Med 19:799–824.

Fiorella D, Rabin R, Winter J (1995) Role of the 5HT2A and 5HT2C receptors in the stimulus effects of hallucinogenic drugs II: Reassessment of LSD false positives. Psychopharmacology 121:357–363.

Gurevich I, Tamir H, Arango V, et al (2002). Altered editing of serotonin 2C receptor pre-mRNA in the prefrontal cortex of depressed suicide victims. Neuron 34:349–356

Heisler L, Cowley M, Tecott L, et al (2002) Activation of central melanocortin pathways by fenfluramine. Science 297:609–611.

Holmes C, Arranz M, Powell J, et al (1998) 5HT2A and 5HT2C receptor polymorphisms and psychopathology in late onset Alzheimer's disease. Hum Mol Genet 7:1507–1509.

Holmes C, Arranz M, Collier D, et al (2003) Depression in Alzheimer's disease: the effect of serotonin receptor gene variation. Am J Med Genet B Neuropsychiatr Genet 119B:40–43.

Kapur S, Remington G (1996) Serotonin-dopamine interaction and its relevance to schizophrenia. Am J Psychiatry 153:466–476.

Lappalainen J, Zhang L, Dean M, et al (1995) Identification, expression and pharmacology of a Cys23-Ser23 substitution in the human 5-HT2C receptor gene (HTR2C). Genomics 27:274–279

Li J, Wang Y, Zhou R, et al (2006) Association between polymorphisms in serotonin 2C receptor gene and attention-deficit/hyperactivity disorder in Han Chinese subjects. Neurosci Lett 407:107–111.

Lopez-Gimenez JF, Mengod G, Palacios JM, et al (2001) Regional distribution and cellular localization of 5-HT2C receptor mRNA in monkey brain: comparison with [3H]mesulergine binding sites and choline acetyltransferase mRNA. Synapse 42:12–26.

Lopez-Gimenez JF, Tecott LH, Palacios JM, et al (2002) Serotonin 5- HT (2C) receptor knockout mice: autoradiographic analysis of multiple serotonin receptors. J Neurosci Res 67:69–85.

Lyketsos CG, Tune L, Pearlson G, et al (1996) Major depression in Alzheimer's disease. An interaction between gender and family history. Psychosomatics 37:380–384.

Mengod G, Vilaro MT, Raurich A, et al (1996) 5-HT receptors in mammalian brain: receptor autoradiography and in situ hybridization studies of new ligands and newly identified receptors. Histochem J 28:747–758.

Miller DD, Ellingrod VL, Holman TL, et al (2005) Clozapine-induced weight gain associated with the 5HT2C receptor -759C/T polymorphism. Am J Med Genet B Neuropsychiatr Genet 133B:97–100.

Pearlson G, Ross C, Lohr W, et al (1990) Association between family history of affective disorder and the depressive syndrome of Alzheimer's disease. Am J Psychiatry 147:452–456.

Pooley EC, Fairburn CG, Cooper Z, et al (2004) A 5-HT2C receptor promoter polymorphism (HTR2C–759C/T) is associated with obesity in women, and with resistance to weight loss in heterozygotes. Am J Med Genet B Neuropsychiatr Genet 126B: 124–127.

Porteous D (2008) Genetic causality in schizophrenia and bipolar disorder: out with the old and in with the new. Curr Opin Genet Dev 18:229–234.

Praschak-Rieder N, Willeit M, Zill P, et al (2005) Cys 23-Ser 23 substitution in the 5-HT(2C) receptor gene influences body weight regulation in females with seasonal affective disorder: an Austrian-Canadian collaborative study. J Psychiatr Res 39:561–567.

Pritchard AL, Harris J, Pritchard CW, et al (2008) Role of 5HT 2A and 5HT 2C polymorphisms in behavioral and psychological symptoms of Alzheimer's disease. Neurobiol Aging 29:341–347.

Rang H, Dale M, Ritter J, et al (2003) Pharmacology. 5th Edition. London: Churchill Livingstone, pp. 244–253.

Reynolds GP, Zhang ZJ, Zhang XB (2002) Association of antipsychotic drug-induced weight gain with a 5-HT2C receptor gene polymorphism. Lancet 359:2086–2087.

Ryu S, Cho EY, Park T, et al (2007) -759 C/T polymorphism of 5-HT2C receptor gene and early phase weight gain associated with antipsychotic drug treatment. Prog Neuropsychopharmacol Biol Psychiatry 31:673–677.

Saltzman A, Morse B, Whitman M, et al (1991) Cloning of the human serotonin 5HT2 and 5HT1C receptor subtypes. Biochem Biophys Res Commun 18:1469–1478.

Steele C, Rovner B, Chase GA, et al (1990) Psychiatric symptoms and nursing home placement of patients with Alzheimer's disease. Am J Psychiatry 147:1049–1051.

Strauss ME, Ogrocki PK (1996) Confirmation of an association between family history of affective disorder and the depressive syndrome in Alzheimer's disease. Am J Psychiatry 153:1340–1342.

Sweet RA, Nimganokar V, Devlin B, et al (2002) Increased familial risk of the psychotic phenotype of Alzheimer's disease. Neurology 58:907–911.

Tecott L, Sun L, Akana S, et al (1995) Eating disorder and epilepsy in mice lacking 5HT2C serotonin receptors. Nature 374:542–546.

Tunstall N, Owen M, Williams J, et al (2000) Familial influence on variation in age of onset and behavioral phenotype in Alzheimer's disease. Br J Psychiatry 176:156–159.

Ujike H, Nomura A, Morita Y, et al (2008) Multiple genetic factors in olanzapine-induced weight gain in schizophrenia patients: a cohort study. J Clin Psychiatry 69(9):1416–1422.

Xie E, Zhu L, Zhao L, et al (1996) The human serotonin 5-HT2C receptor: Complete cDNA, genomic structure and alternatively spliced variant. Genomics 35:551–561.

Zhang Z, Zhang X, Yao Z, et al (2002) (Association of antipsychotic agent-induced weight gain with a polymorphism of the promotor region of the 5-HT2C receptor gene.) Zhonghua Yi Xue Za Zhi 82:1097–1101.

Chapter 26
Ocular Hypotension: Involvement of Serotonergic 5-HT$_2$ Receptors

Najam A. Sharif

26.1 Introduction

The eye is the window to the brain and is in fact connected to the latter via the optic nerve originating from the retina. The ease of accessibility to the eye has made it a model organ to study the effects of a variety of drugs in order to treat certain ocular diseases. The following simplified synopsis may be useful for readers who are unfamiliar with ocular anatomy and physiology (Fig. 26.1).

The eye is comprised of two chambers. The anterior chamber is bounded by the cornea on the outer surface and the lens on the inside and contains the aqueous humor. The posterior chamber is bounded by the lens and the retina choroid and contains the vitreous humor, a jelly-like substance that helps maintain the shape of the globe and cushions the retina. The aqueous humor, produced by the ciliary epithelium within the ciliary bodies located at the lateral edges of the lens, nourishes the cornea, lens, and trabecular meshwork (TM) from where the fluid exits the anterior chamber down the canal of Schlemm that in turn empties into the veinous circulation. The rates of production and egress of the aqueous humor are tightly controlled and are equivalent under normal physiological conditions.

However, if the aqueous humor cannot exit the eye fast enough and begins to accumulate, the pressure in the anterior chamber begins to rise, and if not reduced, eventually causes numerous pathological sequelae that lead to the death of retinal ganglion cells and thus loss of visual acuity leading eventually to blindness ("end-stage glaucoma"). Glaucoma is the second leading cause of blindness in the world (Quigley 1996; Congdon et al. 2004), and while it is composed of many separate subclasses of ocular/retinal diseases, it is now well accepted that ocular hypertension is a major risk factor associated with glaucoma (Quigley 1996; Congdon et al. 2004). Consequently, the reduction and control of intraocular pressure (IOP) has become the

N.A. Sharif (✉)
Pharmaceutical Research, Alcon Research, Ltd. (R6-19),
6201 South Freeway, Fort Worth, TX 76134, USA
e-mail: naj.sharif@alconlabs.com

G. Di Giovanni et al. (eds.), *5-HT$_{2C}$ Receptors in the Pathophysiology of CNS Disease*,   523
The Receptors 22, DOI 10.1007/978-1-60761-941-3_26,
© Springer Science+Business Media, LLC 2011

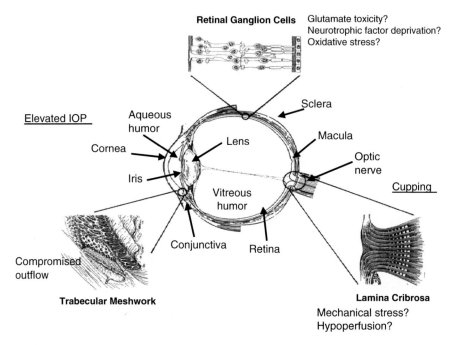

Retinal Ganglion Cells   Glutamate toxicity?
Neurotrophic factor deprivation?
Oxidative stress?

Sclera

Elevated IOP   Aqueous
humor

Cornea

Lens   Macula

Optic
nerve

Iris

Cupping

Vitreous
humor

Compromised
outflow

Conjunctiva   Retina

Trabecular Meshwork

Lamina Cribrosa
Mechanical stress?
Hypoperfusion?

Fig. 26.1 Anatomical and pathophysiological aspects of the eye

major treatment for glaucoma. While a number of pharmacotherapeutic agents have
been approved for treating the ocular hypertension, including β-blockers, carbonic
anhydrase inhibitors, $α_2$-adrenergic agonists, and prostaglandins that selectively activate the prostaglandin F (FP) receptor (Clark and Yorio 2003; Sharif and Klimko
2007), these agents have a variety of deficiencies that ultimately warrant the discovery and clinical utility of new drugs to achieve better compliance, greater potency and
efficacy, longer duration of action, and reduced off-target side effects or a combination of the latter attributes (Clark and Yorio 2003; Sharif and Klimko 2007).

Serotonergic innervation of the eye was demonstrated some years ago using
immunohistochemical/morphological methods (Ehinger and Floren 1980; Moro
et al. 1981; Tobin et al. 1988; Matsumoto et al. 1992; Pootanakit and Brunken 2002;
Pootanakit et al. 1999; Redburn and Churchill 1987) with subsequent detection of
free serotonin (5-HT) in both the anterior and posterior chambers and within certain
tissues of the eye (Cooper et al. 1984; Martin et al. 1988; Trope et al. 1987; Veglio
et al. 1998; Boerrigter et al. 1992). Receptor subtypes for 5-HT (including their messenger ribonucleic acid [mRNA]) have also been identified and pharmacologically
characterized in several ocular tissues and cells using assays and models involving
molecular biological (Turner et al. 2003; Chidlow et al. 1998, 2004; Sharif and
Senchyna 2006; Mallorga and Sugrue 1987), receptor binding (Barnes et al. 1993;
Chidlow et al. 1995; Neufeld et al. 1982), autoradiographical (Sharif et al. 2006),
second messenger (Blazynski et al. 1985; Akhtar 1987; Cutcliffe and Osborne 1987;

Barnett and Osborne 1993; Tobin and Osborne 1989; Osborne and Ghazi 1991; Crider et al. 2003; Inoue- Matsuhisa et al. 2003; Sharif et al. 2006a b, 2007, electrophysiological (Brunken and Jin 1993; Brunken et al. 1993) transmitter release (Harris et al. 2001, 2002), tissue contraction, (Lograno and Romano 2003) and other functional assays/techniques (Klyce et al. 1982; Brunken and Daw 1988).

Even though specific pathologies have not been ascribed to the various 5-HT receptors found in the eye, there has been some progress made with respect to their potential involvement in ocular hypertension and glaucoma. Early studies directed toward the effects of 5-HT uptake inhibitors on IOP suggested that 5-HT may regulate IOP, and indeed an elevation of IOP was observed following intravenous injection of fluoxetine (Costagliola et al. 2000, 2004; Krootila et al. 1987). However, topical ocular administration of 5-HT in rabbits produced either an elevation or decrease in IOP (Meyer-Bothling et al. 1993; Chu et al. 1999) leading to some confusion. Some investigators surmized that 5-HT$_{1A}$ receptor activation may be necessary to lower IOP by decreasing endogenous cAMP since 5-HT$_{1A}$ agonists, 8-hydroxy-DPAT (Chidlow et al. 1999, 2001; May et al. 2003a) and flesinoxan (May et al. 2003a), exhibited ocular hypotensive activity in normotensive rabbits. The presence of 5-HT$_{1A}$ receptor binding sites in rabbit iris-ciliary body (Chidlow et al. 1995) and functionally active 5-HT$_{1A}$ receptors in this tissue of rabbit and humans (Tobin and Osborne 1989; Osborne and Ghazi 1991) and cattle (Klyce et al. 1982) further corroborated the in vivo studies. Unfortunately, however, since 5-HT$_{1A}$ agonists failed to lower IOP of cynomolgus monkeys (both normotensive or ocular hypertensive made by laser ablation of the TM) (May et al. 2003a), it became clear that there are major species differences in the response to serotonergic compounds in terms of influencing ocular hypertension.

Another line of investigation centered around the observation that certain 5-HT$_2$ receptor antagonists, including ketanserin (Chang et al. 1985; Chiou and Li 1992; Costagliola et al. 1993; Mastropasqua et al. 1997; Takat et al. 2001; Takenaka et al. 1995), sarpogrelate (Takenake et al. 1995), and more recently with BVT-28949 and related compounds (Berthold et al. 2005; Cernerud et al. 2007), lowered IOP in normal rabbits and human subjects. Once again, since these results could not be reproduced in ocular hypertensive monkeys (May et al. 2003a; Sharif et al. 2007) it appears that species differences abound for the IOP reduction effects of this class of serotonergic agents. One confounding aspect, however, is the poly-pharmacology of 5-HT$_2$ antagonists (Hoyer et al. 1994, 2002; Oshika et al. 1991; Wang et al. 1997), thus making it difficult to conclude that the 5-HT$_2$ antagonism was responsible for initiating IOP lowering in rabbits and human subjects. Further detailed studies with a range of 5-HT$_2$ antagonists (ketanserin, M-100907, cinanserin, ritanserin) (Sharif et al. 2006a,b; Sharif et al. 2007; May et al. 2003a,b,c) and profiling of such compounds allowed us to conclude that the alpha-adrenergic antagonist activity (Wang et al. 1997; Yoshio et al. 2001; May et al. 2003a,b,c) of 5-HT$_2$ antagonists probably accounts for the ocular hypotensive actions of these and related compounds. With this background, we made a novel observation that 5-HT$_2$ agonists possessing activity at 5-HT$_{2A}$ (predominantly), 5-HT$_{2B}$, and 5-HT$_{2C}$ receptors consistently lower and control IOP in nonhuman primates with laser-induced ocular hypertension (May et al. 2003a,b,c; Sharif et al. 2006a,b, 2007 2009; May et al. 1992). A more detailed appraisal of the role of serotonergic agents in lowering and controlling IOP and the possible mechanisms of action of such compounds is presented below.

26.2 Ocular Distribution of 5-HT$_2$ Receptor mRNAs and Receptor Proteins

Since early studies had revealed a dense network of serotonergic nerves in and around the rabbit eye (Ehinger and Floren 1980; Moro et al. 1981; Tobin et al. 1988; Matsumoto et al. 1992), and free 5-HT had been found in the aqueous humor of rats (Boerrigter et al. 1992) rabbits (Martin et al. 1992) and humans (Martin et al. 1988; Trope et al. 1987; Veglio et al. 1998), it was of great interest to delineate the role of 5-HT and its multitude of receptors in ocular physiology and pharmacology. Messenger RNAs for many subtypes of 5-HT receptors have been detected in rabbit, porcine, and human ocular tissues by RT-PCR (Nash et al. 1993; Turner et al. 2003; Chidlow et al. 1998, 2004; Sharif and Senchyna 2006; Mallorga and Sugrue 1987) and in situ hybridization (Chidlow et al. 1988, 2004) techniques. The 5-HT$_2$ family of receptor mRNAs was found in tissues and cells involved in aqueous humor dynamics in the anterior segment of the eye, with specific enrichment in ciliary body (CB), ciliary epithelium (CE), and trabecular meshwork (TM) (Fig. 26.2, Table 26.1) (Chidlow et al. 2004; Sharif and Senchyna 2006; Mallorga and Sugrue 1987). Even though 5-HT$_{2A}$ receptor mRNAs predominated, 5-HT$_{2B}$ and 5-HT$_{2C}$ receptor mRNAs were also clearly visible in these tissues/cells (Table 26.1, Fig. 26.2). Using receptor binding techniques coupled with quantitative autoradiographic techniques, a high density of

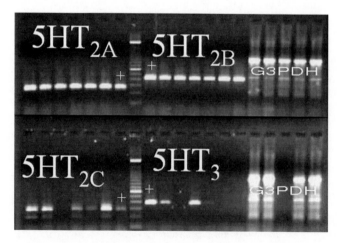

Fig. 26.2 5-HT$_2$ receptor subtype mRNAs in human ciliary body (Modified from Sharif et al. 2006b)

Table 26.1 5-HT receptor subtype mRNAs detected by RT-PCR in human ocular tissues and cells involved in aqueous humor dynamics and as targets for glaucomatous damage (Adapted from Chidlow et al. 1998, 2004; Sharif and Senchyna 2006)

Levels of 5-HT receptor subtype mRNAs in human ciliary body

Donor No.	5-HT$_{2A}$	5-HT$_{2B}$	5-HT$_{2C}$	5-HT$_3$	5-HT$_4$	5-HT$_5$	5-HT$_6$	5-HT$_7$
1	+++	+++	++	+++	+++	+++	−	+++
2	+++	+++	++	++	++	++	−	+++
3	+++	+++	−	−	++	++	−	+++
4	+++	+++	+	+++	++	++	−	+
5	+++	+++	+	+	++	−	−	+++
6	+++	+++	+++	−	+++	++	−	+++

Levels of 5-HT receptor subtype mRNAs in human ciliary epithelium

Donor No.	5-HT$_{2A}$	5-HT$_{2B}$	5-HT$_{2C}$	5-HT$_3$	5-HT$_4$	5-HT$_5$	5-HT$_6$	5-HT$_7$
1	+++	−	x	x	x	x	x	x
2	+++	+	+	x	x	x	x	−
3	+++	−	+	x	x	x	x	+
4	+++	+	+	x	x	x	x	+

Levels of 5-HT receptor subtype mRNAs in human TM cells

Donor	5-HT$_{2A}$	5-HT$_{2B}$	5-HT$_{2C}$	5-HT$_3$	5-HT$_4$	5-HT$_5$	5-HT$_6$	5-HT$_7$
Pooled TM Cells obtained from 8 donors	+++	+++	+	−	−	+	−	+

(continued)

Table 26.1 (continued)

Levels of 5-HT receptor subtype mRNAs in human retina

Donor No.	5-HT$_{2A}$	5-HT$_{2B}$	5-HT$_{2C}$	5-HT$_3$	5-HT$_4$	5-HT$_5$	5-HT$_6$	5-HT$_7$
1	++	++	+	–	+++	+++	–	–
2	+	++	–	–	+++	++	–	–
3	–	+	+	–	++	+	–	+
4	+++	++	++	–	+++	+++	–	+
5	+++	++	++	–	+++	+++	–	+
6	+++	++	++	–	+++	+++	–	+
7	+++	x	x	–	+++	+++	–	+
8	x	++	++	x	x	x	x	x

Qualitative grading of the relative optical density of the receptor mRNAs bands on gels used for RT-PCR of tissues or cells obtained from individual human donors' eyes: +++, very strong band; ++, easily visible band; +, extremely faint but present band; –, negative for presence of the band; x, incomplete data set due to lack of available tissue or total RNA

5-HT$_2$ receptors were localized in human ciliary processes, CB, and ciliary muscle (CM), in addition to other tissues (Sharif et al. 2006b) (Table 26.2). Likewise, 5-HT$_2$ agonists that were shown to stimulate functional responses via cloned human 5-HT$_2$ receptor subtypes and in a variety of relevant human ocular cells and lowered IOP (Mallorga and Sugrue 1987; Neufeld et al. 1982; Sharif et al. 2006a,b, 2007, 2009; May et al. 2003a,b,c, 2006; Gabelt et al. 2005) (see below) exhibited high nanomolar affinities for 5-HT$_{2A}$, 5-HT$_{2B}$, and 5-HT$_{2C}$ receptors (Table 26.3).

Table 26.2 Relative density of 5-HT receptor binding sites (and β-adrenoceptor types) in tissues involved in aqueous humor dynamics determined by quantitative autoradiographic techniques in human eye sections (Adapted from Sharif et al. 2006b with permission from Mary Ann Liebert, Inc. Publishers).

Ocular tissues	[³H]Ketaserin[a] binding (5-HT$_2$ receptors) Specific binding; DLU/mm² (% specific binding)	[³H]5-HT[b] binding (5-HT receptors) Specific binding; DLU/mm² (% specific binding)	[³H]Levobetaxolol binding (β-receptors) Specific binding; DLU/mm² (% specific binding)
Ciliary epithelium (process)	13,683 ± 5,870 (40%)	71,780 ± 2,725 (70%)	51,459 (76%)
Longitudinal ciliary muscle	14,459 ± 3,683 (47%)	14,232 ± 7,937 (48%	27,543 (83%)

[a][³H]Ketaserin (1 nmol/L) data are mean ± SEMs derived from >4 readings for each of three different human donor eyes
[b][³H]5-HT (4 nmol/L) data are mean ± SEM derived from >4 readings for each of two human donor eyes

Table 26.3 Human cloned 5-HT$_2$ receptor subtype binding affinities of selected serotonergic compounds with IOP-lowering activity in ocular hypertensive monkeys (Adapted from May et al. 2003a, 2006; Sharif et al. 2007, 2009 with permission from Mary Ann Liebert, Inc. Publishers).

Compound	Binding inhibition constant at 5-HT$_{2A}$ receptor (K$_i$; nmol/L)	Binding inhibition constant at 5-HT$_{2B}$ receptor (K$_i$; nmol/L)	Binding inhibition constant at 5-HT$_{2C}$ receptor (K$_i$; nmol/L)
(R)-DOI	1 ± 0.1	18 ± 3	4 ± 1
Cabergoline	5 ± 0.1	9 ± 0.3	6 ± 0.4
5-HT	8 ± 2	13 ± 5	8 ± 3
AL-34662 [(S)-isomer]	12 ± 3	8 ± 2	3 ± 0.4
α-methyl-5-HT	12 ± 2	13 ± 3	7 ± 1
5-methoxy-dimethyl tryptamine	15 ± 2	52 ± 2	42 ± 8
AL-34707 [(R)-isomer]	105 ± 38	35 ± 11	34 ± 5

Data are mean ± SEM from ≥3 experiments using displacement of [¹²⁵I]DOI binding to membranes of CHO cells expressing specific recombinant human 5-HT$_2$ receptor subtypes. K$_i$ values are inversely related to the relative affinity of the compound

26.3   Functional Responses to 5-HT$_2$ Agonists in Ocular Cells and Tissues

In order to functionally link the 5-HT$_2$ receptor binding sites characterized by autoradiography in human ocular tissues (Table 26.2), we next demonstrated that indeed numerous selective and nonselective 5-HT$_2$ receptor agonists (e.g., 5-HT, α-methyl 5-HT (R)-DOI, AL-34662, MK-212, mCPP, BW723C86, cabergoline) potently and efficaciously stimulated the production of inositol phosphates via phospholipase C-induced hydrolysis of phosphoinositides (PI) (Table 26.4) (Sharif et al. 2006a,b, 2007, 2009; May et al. 2003a,b,c) and mobilized intracellular Ca^{2+} ([Ca^{2+}]$_i$)(Table 26.5, Fig. 26.3) (Sharif et al. 2006a,b, 2007; May et al. 2003a,b,c) in human CM and TM cells with further confirmation in human recombinant 5-HT$_2$ receptor subtypes using GTP-γ-S^{35} binding (Table 26.5) and [Ca^{2+}]$_i$ mobilization assays (Table 26.5) (Sharif et al. 2006a,b, 2007; May et al. 2003a,b). In detailed studies using 5-HT$_2$ receptor-selective antagonists (Hoyer et al. 2002; Wainscott et al. 1996; Wood et al. 1997; Porter et al. 1999; Jerman et al. 2001; Kennett et al. 1996), we next demonstrated that the PI turnover and [Ca^{2+}]$_i$ mobilizing effects in h-TM and h-CM cells of various 5-HT$_2$ agonists (e.g., 5-HT, AL-34662) (Table 26.6, Fig. 26.4) were most potently blocked by the 5-HT$_{2A}$ antagonist (M-100970; IC$_{50}$s = 1–2 nmol/L), followed by 5-HT$_{2C}$ antagonists (RS-102221; SB 242084; IC$_{50}$s = 1–4 μmol/L) and then 5-HT$_{2B}$ antagonists (RS-127445; SDZSER082; IC$_{50}$s > 1–10 μmol/L) (Table 26.6) (Sharif et al. 2006a,b, 2007; May et al. 2003c). The consequence of 5-HT$_2$ receptor activation by cabergoline, for instance, in human CM cells appeared to be the synthesis and release of matrix metalloproteinases (sharif et al. 2009) that are known to be involved in remodeling of the extracellular space by digesting the extracellular matrix and other debris (Weinreb et al. 1997). Furthermore, such activities of cabergoline were perhaps responsible in whole or in part for the increased efflux of fluid from the anterior chamber of the isolated perfused pig eye (Fig. 26.5) (sharif et al. 2009) that in vivo would be reflected as an overall decrease in IOP (May et al. 2003c). The mechanical contraction/relaxation of CM and/or TM tissues by 5-HT$_2$ agonists to lower IOP is also possible (Wiederholt et al. 2000) with perhaps an additional involvement of matrix metallopeptidase (MMP) release during such tissue stretching (Keller et al. 2009; WuDunn 2009).

Taken together, these collective data indicated that functional 5-HT$_2$ receptors are present in human ocular cells that mediate many of the downstream biological effects of 5-HT$_2$ agonists known to have IOP-lowering activity in conscious ocular hypertensive cynomolgus monkeys (Sharif et al. 2006a,b, 2007, 2009; May et al. 2003a,b,c, 2006; Gabelt et al. 2005).

Table 26.4 Serotonergic compound-induced PI turnover in h-CM and h-TM cells and in cell expressing human cloned 5HT$_2$ receptor subtypes or in other cells/tissues (Adapted from Sharif et al. 2006a,b, 2007, 2009)

Serotonergic compound	Reported receptor selectivity	h-CM cells agonist potency (EC$_{50}$; nmol/L)	h-TM cells agonist potency (EC$_{50}$; nmol/L)	Human cloned 5-HT$_{2A}$ receptors agonist potency (EC$_{50}$; nmol/L)	Human cloned 5-HT$_{2B}$ receptors agonist potency (EC$_{50}$; nmol/L)	Human cloned 5-HT$_{2C}$ receptors agonist potency (EC$_{50}$; nmol/L)
α-Methyl-5-HT	5-HT$_{2A}$	63±17 (97%)	140±60 (128%)	146±29[a]	nd	nd
5-Methoxy tryptamine	Nonselective	69±18 (118%)	181±72 (95%)	nd	nd	nd
5-Methoxy-dimethyl tryptamine	Nonselective	80±10 (74%)	247±107 (62%)	462[b]	240±30[c]	78±6[d]
5-HT	Nonselective	85±16 (99%)	364±75 (107%)	105±15[a]	nd	nd
(R)-DOI	5-HT$_{2A}$	165±47 (117%)	13±5 (78%)	3±1[a]	nd	nd
5-Methoxy α-methyl tryptamine	Nonselective	1,200±270 (110%)	147±35 (110%)	48[b]	3±0.2[c]	3±1[d]
AL-34662 [(S) isomer]	Nonselective	±80 (83%)	254±50 (98%)	74±9#	2±1[c]	4±1[d]
AL-34707 [(R) isomer]	Nonselective	>18,000 (115%)	2,313±1,182 (55%)	236±34[b]	85±15[c]	87±11[d]
Cabergoline	Nonselective	76; 69	19±7	8	3	190

Data are mean ± SEM or from individual experiments

[a]From GTP-γ-S^{35} binding assays

[b]Rat 5-HT$_{2A}$

[c]Rat stomach fundus contraction

[d]Rat 5-HT$_{2C}$, nd, not determined

Table 26.5 Serotonergic compound-induced $[Ca^{2+}]_i$ mobilization induced in h-CM, h-TM, and human cloned 5-HT$_2$ receptor subtype-expressing cells (Adapted from Sharif et al. 2006a,b; Sharif et al. 2007, 2009)

Serotonergic agonist	Reported receptor selectivity	h-CM cells agonist potency (EC$_{50}$; nM)	h-TM cells agonist potency (EC$_{50}$; nM)	Human cloned 5-HT$_{2A}$ receptors agonist potency (EC$_{50}$; nM)	Human Cloned 5-HT$_{2B}$ receptors agonist potency (EC$_{50}$; nM)	Human cloned 5-HT$_{2C}$ receptors agonist potency (EC$_{50}$; nM)
α-Methyl-5-HT	5-HT$_{2A}$	36±11	22	3.9±0.7	3.8±0.8	2.3±1.0
5-Methoxy tryptamine	Nonselective	42±11	8	nd	nd	nd
(R)-DOI	5-HT$_{2A}$	120±68 (62±9%)	18	1.3±0.4	23.6±13.3	1.2±0.1
5-HT	Nonselective	130±36	40	4.1±0.6	4.1±0.7	0.24±0.04
5-Methoxy-n,n-dimethyl tryptamine	Nonselective	170±20 (57±7%)	64 (68%)	152±60[#]	nd	226±39[a]
BW-723C86	5-HT$_{2B}$	1,766±517 (51±9%)	1,213 (54%)	63±22	3±1	9±3
MK-212	5-HT$_{2C}$	470±350 (59±7%)	≥1,000 (25%)	529±198	230±77	8±3
mCPP	5-HT$_{2C}$	≥1,000 (46%)	≥1,000 (15%)	32±8	26±2	2±1
AL-34662 [(S) isomer]	Unknown	140±23 (87%)	38±8 (89%)	nd	nd	nd
AL-34707 [(R) isomer]	Unknown	1,300±240 (44%)	523±290 (62%)	nd	nd	nd
Cabergoline	Nonselective	900±320	570±83	63.4±10.3	871.2±85.8	>1,000

Data are mean±SEMs where shown. Intrinsic activity relative to 5-HT (100%) when >100% is shown. Majority of the compounds were full agonists at the human cloned 5-HT$_2$ receptor subtypes, except (R)-DOI, which approached 60%

nd not determined

[a]From GTP-γ-S^{35} binding to membranes of CHO cells expressing cloned human 5-HT$_2$ receptors

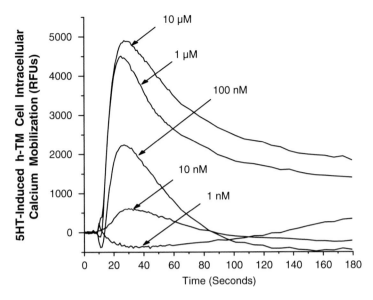

Fig. 26.3 [Ca²⁺]i mobilization by 5-HT in h-TM cells (Modified from Sharif et al. 2006a with permission from Assoc. Research in vision and opthalmology).

26.4 Ocular Hypotensive Activity of 5-HT₂ Agonists/Antagonists

Many contradictory publications have appeared over the years that described a variety of effects of different 5-HT agonists and antagonists tested in different animals and human subjects using various routes of administration (see Sect. 26.1). Consequently, no consensus was achieved as to the role of different 5-HT receptors in modulating aqueous humor dynamics and IOP, with further complications added by major species differences in the responses to various serotonergic compounds. Using a systematic approach, we conducted a detailed evaluation of numerous agonists and antagonists, some with well-defined pharmacological selectivities, for their ability to influence IOP in the laser-induced glaucomatous nonhuman primate eye in an effort to provide a framework for potential future testing of suitable serotonergic agents in humans to treat ocular hypertension and glaucoma (May et al. 2003a,b,c; Sharif et al. 2006a,b, 2007, 2009). The overall conclusion drawn from these extensive studies (Table 26.7) is that 5-HT₂ receptors are the primary mediators of IOP reduction in the cynomolgus monkey model of ocular hypertension and glaucoma and that 5-HT₂ agonists are potential drugs to lower and control IOP. Thus, use of selective 5-HT₂ receptor agonists and antagonists revealed that indeed activation of the 5-HT₂ₐ receptor, with some contribution(s) from 5-HT₂C and 5-HT₂B receptors, leads to IOP lowering in the monkey (Table 26.7) (Sharif et al. 2006a,b, 2007, 2009; May et al. 2003a,b,c, 2006) and rat (Sharif et al. 2009) but not in rabbits and cats (Sharif et al. 2009; May et al. 2003c).

Table 26.6 Inhibition of $[Ca^{2+}]_i$ mobilization induced by AL-34662 and 5-HT in h-CM and h-TM cells by 5-HT$_2$ receptor subtype-selective antagonists (Adapted from Sharif et al. 2007 with permission from Mary Ann Liebert Inc. Publishers)

5-HT$_2$ receptor antagonist	Receptor selectivity	h-CM cells IC$_{50}$ (nmol/L) vs AL-34662	h-CM cells IC$_{50}$ (nmol/L) vs 5-HT	h-TM cells IC$_{50}$ (nmol/L) vs AL-34662	h-TM cells IC$_{50}$ (nmol/L) vs 5-HT
M-100,970	5-HT$_{2A}$	1.8±0.7	0.9±0.1	1±0.2	1.1±0.2
RS-127445	5-HT$_{2B}$	>1,000	>1,000	>10,000	>1,000
RS-102221	5-HT$_{2C}$	>1,000	7,490±2,010	7,760±1,220	>1,000
SB-242084	5-HT$_{2C}$	nd	nd	1,060±282	1,760±776
SDZSER082	5-HT$_{2B/2C}$	2,080±719	2,306±815	nd	3,700±516

Data are mean±SEMs from 3 to 7 experiments

nd not determined

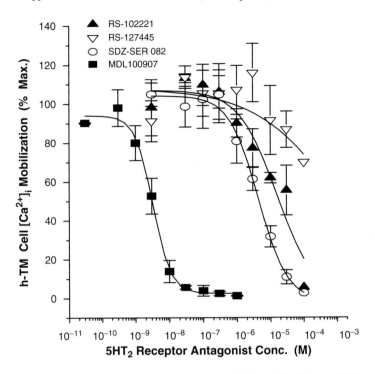

Fig. 26.4 Inhibition of [Ca^{2+}]i Mobilization induced by 5-HT in h-TM Cells by 5-HT$_2$ receptor antagonists (Modified from Sharif et al. 2006a) with permission from Assoc. Research in vision and opthalmology).

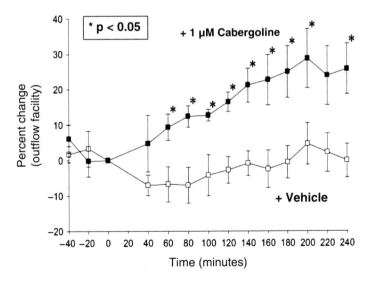

Fig. 26.5 Aqueous humor outflow induced by cabergoline in perfused porcine eye (Modified from Sharif et al. 2009 with permission from Mary Ann Liebert Inc. Publishers).

Table 26.7 IOP-lowering activities of selected compounds in conscious ocular hypertensive cynomolgus monkey eyes upon topical ocular administration (Adapted from Sharif et al. 2006a,b 2007, 2009; May et al. 2003c)

Compound	Reported receptor selectivity	Topical ocular dose (µg/eye)	Max. % IOP reduction
(R)-8-OH-DPAT	5-HT$_{1A}$ agonist	300	6.3±4.5
Flesinoxan	5-HT$_{1A}$ agonist; α$_1$ agonist	250	11.6±1.9
GR-43175C	5-HT$_{1D}$ agonist	250	2.8±2.0
		500	6.4±3.3
MDL-73005EF	5-HT$_{1A}$ antagonist	500	8.3±5.0
(R)-DOI	5-HT$_{2A}$ agonist	300	27.7±5.8*
		300	34.4±6.7*
α-Methyl-5-HT	5-HT$_{2A}$ agonist	150	30.9±3.8*
		250	30.8±7.7*
BW723C86	5-HT$_{2B}$ agonist	150	16.6±5.3
		300	18.0±5.0
mCPP	5-HT$_{2C}$ agonist	300	22.8±6.0*
		300	19.1±3.7
MK-212	5-HT$_{2C}$ agonist	150	13.0±3.0
		300	12.1±3.2
5-HT	Nonselective 5-HT agonist	250	18.0±5.1
5-Methoxy-*n,n*-dimethyl tryptamine	Nonselective 5-HT$_2$ agonist	300	30.2±4.4*
Clozapine	Mixed 5-HT agonist; muscarinic agonist	150	14.3±4.6
Ketanserin	5-HT$_2$ antagonist; α$_1$ antagonist	300	3.4±6.4
Cinanserin	5-HT$_2$ antagonist	100	9.2±2.3
		300	6.8±2.2
Ritanserin	5-HT$_2$ antagonist; α$_1$ antagonist	300	10.2±3.6
M-100907	5-HT$_{2A}$ antagonist	300	15.8±3.8
SB-206553	5-HT$_{2B/C}$ antagonist	150	15.1±4.5
		300	13.4±4.4
RS-102221	5-HT$_{2C}$ antagonist	300	7.9±2.6
SB-242084	5-HT$_{2C}$ antagonist	300	12.2±1.7*
Bufotenine	5-HT$_3$ agonist	100	8.9±5.0
		300	26.1±5.1*
Quipazine	5-HT$_3$ antagonist	300	12.9±5.1
R/S-Zacopride	5-HT$_3$ antagonist; 5-HT$_4$ agonist	500	8.6±1.8
SDZ-205557	5-HT$_4$ antagonist	300	14.7±3.4
SB-258510A	5-HT$_6$ antagonist	300	20.1±5.0*
(+)-SB-258719	5-HT$_7$ antagonist	300	12.8±3.3
Fluoxetine (caused intense discomfort)	5-HT uptake inhibitor	100	22.3±6.3*
Reference drugs			
Brimonidine	α$_2$ Agonist	150	31.2±6.3*
Levobetaxolol	β-Antagonist	150	25.9±3.2*
Pilocarpine	Muscarinic agonist	150	27.9±6.0*

(continued)

Table 26.7 (continued)

Compound	Reported receptor selectivity	Topical ocular dose (µg/eye)	Max. % IOP reduction
Travoprost	Prostaglandin FP agonist	0.1	22.7±5.8*
		0.3	25.7±5.1*
		1	28.7±3.5*
Latanoprost	Prostaglandin FP agonist	1	32.3±3.6*
		3	25.7±5.1*
		10	28.8±5.8*

Data are mean±SEMs from 8 to 9 animals per group per study. The baseline IOPs were ≥38.2 mmHg.
*$p < 0.05$ relative to baseline IOP for that dose using an unpaired Student's t test; ≥20% reduction is considered biologically significant

It was interesting to note that the pharmacological data obtained using human CM and TM cells correlated well with the ability of the 5-HT$_{2A}$ agonists to lower monkey IOP, indicating perhaps that the ocular hypertensive monkey model may be a useful surrogate for humans as far as the ocular effects of serotonergic agents are concerned. However, future studies with suitable serotonergic agents like AL-34662 and others (May et al. 2003b, 2006; Sharif et al. 2007, 2009) in ocular hypertensive human subjects will ultimately confirm or refute this apparent correlation. Unfortunately, due to the poor ocular bioavailability of selective 5-HT$_2$ receptor antagonists, it was not possible to conclusively delineate the relative proportion of contributory effects of the three 5-HT$_2$ receptor subtypes involved in mediating the IOP-lowering effects of serotonergic agonists possessing 5-HT$_{2A-C}$ receptor activities. Considering the relative lack of selectivity of the test agonist utilized in our studies across the human cloned 5-HT$_{2A}$, 5-HT$_{2B}$, and 5-HT$_{2C}$ receptor subtypes (Tables 26.3–26.5) and the unknown relative corneal permeability (and thus aqueous humor and anterior uveal tissue concentrations achieved of test agents), the overall conclusion is probably that all three 5-HT$_2$ receptors are involved in whole or in part in the IOP-lowering actions of compounds such as (R)-DOI, α-methyl-5-HT, cabergoline, and AL-34662 (Tables 26.7–26.9) (Sharif et al. 2006a,b, 2007, 2009; May et al. 2003a,b,c, 2006).

26.5 Mechanism of Action of 5-HT₂ Agonists to Lower IOP

Cabergoline is a polypharmacological agent (Sharif et al. 2009) that activated 5-HT$_2$ receptors in human CM, TM cells, and human cloned 5-HT$_2$ receptors to increase PI hydrolysis and mobilization of $[Ca^{2+}]_i$ (Sharif et al. 2009), activities that were most potently blocked by M-100907 (5-HT$_{2A}$ antagonist). Furthermore, cabergoline lowered IOP in conscious and ketamine-sedated ocular hypertensive cyno-

Table 26.8 Reduction of IOP by AL-34662 (S) and AL-34707 (R) in ocular hypertensive monkey eyes after topical ocular dosing (Adapted from Sharif et al. 2007 with permission from Mary Ann Liebert Inc. Publishers)

		Percent reduction in IOP[a] – hours post-dose		
	Topical ocular dose	1 h	3 h	6 h
AL-34662 (S)	300 µg	13.4±3.1[*]	31.4±3.2[**]	33.0±3.1[**]
AL-34707 (R)	300 µg	9.0±6.0	24.4±4.9[**]	16.6±6.8[***]

Data are mean±SEM of data from 8 to 9 monkeys per group. Baseline IOPs ranged between 35.8 and 41.8 mmHg in these studies

[*]$p<0.01$; [**]$p<0.001$; [***]$p<0.02$

[a]Statistically significant IOP reduction relative to the respective baseline

Table 26.9 Cabergoline-induced IOP lowering in conscious ocular hypertensive cynomolgus monkey eyes (Adapted from Sharif et al. 2009 with permission from Mary Ann Liebert Inc. Publishers)

		Percent reduction in IOP after topical ocular administration of cabergoline (relative to baseline)			
Treatment	Topical ocular dose	One hour postdose	Three hours postdose	Seven hours postdose	Twenty-four hours postdose
Vehicle		3.3±4.0	1.3±4.5	5.4±3.8	nd
Cabergoline	50 µg	16.0±2.4*	30.6±3.6*	28.8±5.5*	27.3±4.8*
Cabergoline	500 µg	18.9±4.5*	29.9±4.9*	30.4±4.5*	nd

Data are mean±SEM from nine monkeys per group. IOP reduction >25% is considered robust

nd not determined

*$p<0.05$ by Student's *t* test

molgus monkeys (Sharif et al. 2009). (R)-DOI is a potent and selective 5-HT$_{2A}$ agonist (Sharif et al. 2006a,b, 2007; May et al. 2003a; Hoyer et al. 2002) that also activated human CM and TM cells and that also lowered IOP in ocular hypertensive monkey eyes (Sharif et al. 2006a,b, 2007; May et al. 2003a,b,c; Gabelt et al. 2005). Both cabergoline (Table 26.10) (Sharif et al. 2009) and (R)-DOI (Gbdert et al. 2005) stimulated uveoscleral outflow of aqueous humor in the monkey eyes to lower IOP. While the cellular and molecular mechanism of action of these compounds has not been fully delineated, cabergoline increased ciliary muscle cell MMPs release (Sharif et al. 2009) that probably were responsible for remodeling the CM and the extracellular space to cause the IOP reduction as has been demonstrated for ocular hypotensive FP-class and DP-class prostaglandins Weinreb et al. 1997. It was interesting to note that cabergoline also stimulated aqueous humor outflow in perfused porcine eyes (Sharif et al. 2009). However, in view of the ability of 5-HT$_2$ agonists to contract bovine ciliary muscle in vitro (Lograno and Romano, 2003), a mechanical effect being responsible for IOP lowering via the 5-HT$_2$ receptors cannot be ruled out (Wiederholt et al. 2000; WuDun 2009). Yet, even the latter effect may ultimately involve MMPs since stretching of uveal tissues leads to release of MMPs (Keller et al. 2009; WuDunn 2009).

Table 26.10 Aqueous humor dynamics after topical ocular dosing with cabergoline (300 μg) in ketamine-sedated cynomolgus monkeys eyes (Adapted from Sharif et al. 2009) with permission from Mary Ann Liebert Inc. Publishers)

| | Normotensive eyes | | | | | Lasered (hypertensive) eyes | | | | |
	Baseline value	n	Treatment value	n	p*	Baseline value	n	Treatment value	n	p*
IOP^1(mmHg)	20.3±3.5	13	20.7±5.0	12	0.85	33.2±9.6	13	32.3±10.9	13	0.41
IOP^2 (mmHg)	19.3±5.5	12	20.1±4.3	12	0.80	25.7±11.2	12	21.7±9.1	13	0.02
IOP^3 (mmHg)	18.7±5.4	13	14.3±6.8	12	0.02	22.2±7.5	13	14.4±7.2	13	0.004
F$_a$ (:L/min)	1.65±0.45	13	1.69±0.68	12	0.79	1.49±0.44	13	1.39±0.57	12	0.66
C$_{fl}$ (:L/min/ mmHg)	0.23±0.27	12	0.15±0.20	12	0.46	0.09±0.06	13	0.27±0.42	11	0.15
F$_{us}$ (:L/min)	0.69±0.70	12	1.61±0.97	11	0.01	0.95±0.40	13	1.49±1.56	11	0.37

Data are mean±SD from 11 to 13 monkeys as shown above. C_{fl} fluorophotometric outflow facility, F_a aqueous flow, F_{us} uveoscleral outflow, IOP^1 intraocular pressure, taken at 5:00 PM just before the first drop of cabergoline, IOP^2 intraocular pressure taken at 9:00 AM just before the second drop of cabergoline, IOP^3 intraocular pressure taken at 11:30 AM, 2.5 h after the second drop of cabergoline. Times are ±0.5 h

Statistical analyses: *comparing baseline day with treatment day using Student's two-tailed, paired t test. Level of significance (p) values were as shown above

26.5.1 Future Directions

Even though it appears that the target serotonergic receptors involved in mediating IOP reduction in ocular hypertensive cynomolgus monkeys by increasing uveoscleral and conventional outflow are the 5-HT$_2$ receptors, the exact subtype(s) contributing to this effect remain unknown. Since 5-HT$_{2A}$ receptor mRNAs, proteins, and their signal transduction processes predominate in human CM and TM (cells and tissues) (Chidlow et al. 1987, 2004; Sharif and Senchyna, 2006; Sharif et al. 2006a,b), perhaps this subtype is the principle receptor subtype involved in reducing IOP. However, the participation of the 5-HT$_{2B}$ and 5-HT$_{2C}$ receptors cannot be discounted since the mRNAs for both these receptor subtypes are also present in the human ciliary body, ciliary epithelium, and TM, although at a lower level than that for 5-HT$_{2A}$ receptors (Table 26.1) (Sharif and Senchyna, (2006). Clearly, more potent, selective, and ocular bioavailable 5-HT$_2$ agonists (and antagonists) are needed to help tease out the relative contributions of the various 5-HT$_2$ receptor subtypes to the ocular hypotension observed. Perhaps 5-HT$_{2A-C}$ knockout mice and/or use of siRNAs and/or use of 5-HT$_2$ receptor antibodies will also prove useful in this endeavor in the near future. Since ultimately the retinal ganglion cells (RGCs) are compromized by the elevated IOP in glaucoma patients, and since RGCs (Table 26.1) (Sharif and Senchyna, 2006), retinal pigment epithelium cells (Osborne et al. 1993), and other cells of the retina (Mitchell and Redburn 1985) contain 5-HT$_2$ receptors, any potential or demonstrated neuroprotective activity of 5-HT$_2$ agonists may also prove very useful in addition to their ability to lower elevated IOP. The latter aspect warrants further research.

References

Akhtar RA (1987) Effects of norepinephrine and 5-hydroxytryptamine on phosphoinositide-PO$_4$ turnover in rabbit cornea. Exp Eye Res 44:849–862.

Barnes JM, Barnes NM, Brunken, WJ, et al (1993) Identification of 5-HT$_3$ receptor recognition sites in rabbit retina. 5-Hydroxytryptamine-CNS receptors and brain function. Proc Int Conf Serotonin, Vol. 1, Birmingham, UK, p. 53.

Barnett NL, Osborne NN (1993) The presence of serotonin (5-HT$_1$) receptor negatively coupled to adenylate cyclase in rabbit and human iris-ciliary process. Exp Eye Res 57:209–216.

Berthold M, Crossley R, Ward T (2005) Imidazol[1,5-a]pyridine or imadazo[1,5-a]piperidine derivatives and their uses for the preparation of medicaments against 5-HT$_{2A}$ receptor-related disorders. World Patent 2005/021545 A1.

Blazynski C, Ferrendelli JA, Cohen AI (1985) Indolamine-sensitive adenylate cyclase in rabbit retina:characterization and distribution. J Neurochem 45:440–447.

Boerrigter, RMM, Sietsema JV, Kema IP (1992) Serotonin (5-HT) and the rat's eye. Some pilot studies. Doc Ophthalmol 82:141–150.

Brunken WJ, Daw NW (1988) The effects of serotonin agonists and antagonists on the response properties of complex ganglion cells of the rabbit retina. Vis Neurosci 1:181–188.

Brunken WJ, Jin XT (1993) A role for 5-HT$_3$ receptors in visual processing in the mammalian retina. Vis Neurosci 10:511–522.

Brunken WJ, Jin XT, Pis-Lopez AM (1993) The properties of the serotonergic system in the retina. Prog Ret Res 16:75–99.

Cernerud M, Lundstrom H, Nilsson BM, et al (2007) Amino-substituted 1H-pyrazin-2-ones and 1H-quinoxalin-2-ones. US Patent 7244722.

Chang FW, Burke JA, Potter DE (1985) Mechanism of ocular hypotensive action of ketanserin. J Ocular Pharmacol 1:137–147.

Chidlow G, DeSantis LM, Sharif, NA, et al (1995) Characteristics of [^{3}H]-5-hydroxytryptamine binding to iris-ciliary body tissue of the rabbit. Invest Ophthalmol Vis Sci 36:2238–2245.

Chidlow G, Le Corre S, Osborne NN (1998) Localization of 5-hydroxytryptamine-1A and 5-hydroxytryptamine-7 receptors in rabbit ocular and brain tissues. Neurosci 87:675–689.

Chidlow G, Nash MS, DeSantis L, et al (1999) The 5-HT$_{1A}$ Receptor agonist 8-OH-DPAT lowers intraocular pressure in normotensive NZW rabbits. Exp Eye Res 69:587–593.

Chidlow G, Cupido A, Melena J, et al (2001) Flesinoxan, a 5-HT$_{1A}$ receptor agonist/α_1-adreno-ceptor antagonist, lowers intraocular pressure in NZW rabbits. Curr Eye Res 23:144–153.

Chidlow G, Hiscott PS, Osborne NN (2004) Expression of serotonin receptor mRNAs in human ciliary body:a polymerase chain reaction study. Graefe's Arch Clin Exp Ophthalmol 242:259–264.

Chiou GC, Li BH (1992) Ocular hypotensive actions of serotonin antagonist-ketanserin and ana-logs. J Ocular Pharmacol Ther 8:11–21.

Chu TC, Ogidigben MJ, Potter DE (1999) 8-OH-DPAT-Induced ocular hypotension:sites and mechanisms of action. Exp Eye Res. 69:227–238.

Clark AF, Yorio T (2003) Ophthalmic drug discovery. Nat Rev Drug Discov 2:448–459.

Congdon N, O'Colemain B, Klaver CC, et al (2004) Causes and prevalence of visual impairment among adults in the United States. Arch Ophthalmol 122:477–485.

Cooper RL, Constable IJ, Davidson L (1984) Catecholamines in aqueous humor of glaucoma patients. Aust J Ophthalmol 12:345–349.

Costagliola C, Iuliano G, Rinaldi M, et al (1993) Effect of topical ketanserin administration on intraocular pressure. Br J Pharmacol 77:344–348.

Costagliola C, Mastropasua L, Capone D, et al (2000) Effect of fluoxetine on intraocular pressure in the rabbit. Exp Eye Res 70:551–555.

Costagliola C, Parmeggiani F, Sebastiani A (2004) SSRIs and intraocular pressure modifications:evidence, therapeutic implications and possible mechanisms. CNS Drugs 18:475–484.

Crider JY, Williams GW, Drace CD, et al (2003) Pharmacological characterization of a serotonin receptor (5-HT$_7$) stimulating cAMP production in human corneal epithelial cells. Invest Ophthalmol Vis Sci 44:4837–4844.

Cutcliffe N, Osborne NN (1987) Serotonergic and cholinergic stimulation of inositol phosphate formation in the rabbit retina. Evidence for the presence of serotonin and muscarinic receptors. Brain Res 421:95–104.

Ehinger B, Floren I. (1980) Retinal indolamine accumulating neurons. Neurochem Int 1:209–229.

Gabelt BT, Okka M, Dean TR, et al (2005) Aqueous humor dynamics in monkeys after topical R-DOI. Invest Ophthalmol Vis Sci 46:4691–4696.

Harris LC, Awe SO, Opere CA, et al (2001) [^{3}H]Serotonin release from bovine iris-ciliary body:pharmacology of pre-junctional serotonin (5-HT$_7$) autoreceptors. Exp Eye Res 73:59–67.

Harris LC, Awe SO, Opere CA, LeDay AM, Ohia SE, Sharif NA. (2002) Pharmacology of sero-tonin receptors modulating electrically-induced [^{3}H]norepinephrine release from isolated mammalian iris-ciliary bodies. J Ocular Pharmacol Ther 18:339–348.

Hoyer D, Clarke DE, Fozard JR, et al (1994) VII. International union of pharmacology classifica-tion of receptors for 5-hydroxytryptamine (serotonin). Pharm Rev 46:157–203.

Hoyer D, Hannon JP, Martin GR (2002) Molecular, pharmacological and functional diversity of 5-HT receptors. Pharmacol Biochem Behav 71:533–554.

Jerman JC, Brough SJ, Gager T, et al (2001) Pharmacological characterization of human 5-HT$_2$ receptor subtypes. Eur J Pharmacol 414:23–30.

Kehne JH, Baron BM, Carr AA, et al (1996) Preclinical characterization of the potential of the putative atypical antipsychotic MDL 100,907 as a potent 5-HT$_{2A}$ antagonist with a favorable CNS safety profile. J Pharmacol Exp Ther 277:968–981.

Keller KE, Aga M, Bradley JM, et al (2009) Extracellular matrix turnover and outflow resistance. Exp Eye Res 88:676–682.

Kennett GA, Wood MD, Bright F, et al (1997) SB 242084, a selective and brain penetrant 5-HT$_{2C}$ receptor antagonist. Neuropharmacol 36:609–620.

Klyce SD, Palkama KA, Harkone M, et al (1982) Neural serotonin stimulates chloride transport in the rabbit corneal epithelium. Invest Ophthalmol Vis Sci 23:181–192.

Krootila K, Palkama A, Uusitalo H (1987) Effects of serotonin and its antagonist (ketanserin) on intraocular pressure in the rabbit. J Ocular Pharmacol 3:279–290.

Lograno, MS, Romano, MR (2003) Pharmacological characterization of the 5-HT$_{1A}$, 5-HT$_2$ and 5-HT$_3$ receptors in the bovine ciliary muscle. Eur J Pharmacol 464:69–74.

Mallorga P, Sugrue MF (1987) Characterization of serotonin receptors in the iris-ciliary body of the albino rabbit. Curr Eye Res 6:527–532.

Mangel SS, Brunken WJ (1992) The effects of serotonin drugs on horizontal and ganglion cells in the rabbit retina. Vis Neurosci 8:213–218.

Martin XD, Brennan MC, Lichter PR (1988) Serotonin in human aqueous humor. Ophthalmol 95:1221–1226.

Martin XD, Malina,H.Z, Brennan MC, et al (1992) The ciliary body – the third organ found to synthesize indoleamines in humans. Eur J Ophthalmol 2:67–72.

Mastropasqua L, Costagliola C, Ciancaglini M, et al (1997) Ocular hypotensive effect of ketanserin in patients with primary open angle glaucoma. Acta Ophthalmol Scand Suppl 224:24–25.

Matsumoto Y, Ueda S, Kawata M (1992) Morphological characterization and distribution of indolamine-accumulating cells in the rat retina. Acta Histochem Cytochem 25:45–51.

May JA. McLaughlin MA, Sharif NA, et al (2003a) Evaluation of the ocular hypotensive response of serotonin 5-HT$_{1A}$ and 5-HT$_2$ receptor ligands in conscious ocular hypertensive cynomolgus monkeys. J Pharmacol Exp Ther 306:301–309.

May JM, Chen H-H, Rusinko A, et al (2003b) A novel and selective 5-HT$_2$ receptor agonist with ocular hypotensive activity:(S)-(+)-1 – (2-aminopropyl)-8,9-dihydropyrano-[3,2-e]indole. J Med Chem 46:4188–4195.

May JA, Dean TR, Sharif NA, et al (2003c) Serotonergic 5-HT$_2$ agonists for treating glaucoma. US Patent 6664286 B1.

May JA, Dantanarayana AP, Zinke PW, et al (2006) 1-((S)-2-Aminopropyl)-1H-indazol-6-ol:A potent peripherally acting 5-HT$_2$ receptor agonist with ocular hypotensive activity. J Med Chem 49:318–328.

Meyer-Bothling U, Bron AJ, Osborne NN (1993) Topical application of serotonin or the 5-HT$_1$-agonist 5-CT on intraocular pressure in rabbits. Invest Ophthalmol Vis Sci 34:3035–3042.

Mitchell CK, Redburn DA (1985) Analysis of pre- and post-synaptic factors of the serotonin system in rabbit retina. J Cell Biol 100:64–73.

Moro F, Scapagnin U, Scaletta S, et al (1981) Serotonin nerve endings in the regulation of papillary diameter. Ann Ophthalmol 13:487–490.

Inoue-Matsuhisa E. Moroi SE, Takenaka H, et al (2003) 5-HT$_2$ receptor-mediated phosphoinositide hydrolysis in bovine ciliary epithelium. J Ocular Pharmacol Ther 19:55–62.

Nash M, Flanigan T, Leslie R, et al (1993) Serotonin-2A receptor mRNA expression in rat retinal pigment epithelial cells. Ophthalmic Res 31:1–4.

Neufeld AH, Ledgard SE, Jumblatt MM, et al (1982) Serotonin-stimulated cyclic AMP synthesis in the rabbit corneal epithelium. Invest Ophthalmol Vis Sci 23:193–198.

Osborne NN, Ghazi H (1991) 5-HT$_{1A}$ receptors positively coupled to cAMP formation in the rabbit retina. Neurochem Int 19:407–511.

Osborne NN, Fitzgibbon F, Nas M, et al (1993) Serotonergic, 5-HT$_2$, receptor-mediated phosphoinositide turnover and mobilization of calcium in cultured rat retinal pigment epithelium cells. Vis Res 33:2171–2179.

Oshika T, Araie Sugiyama T, Nakajima M, et al (1991) Effect of bunazosin hydrochloride on intraocular pressure and aqueous humor dynamics in normotensive human eyes. Arch Ophthalmol 109:1569–1574.

Pootanakit K, Brunken WJ (2002) 5-HT$_{1A}$ and 5-HT$_7$ receptor expression in the mammalian retina. Brain Res 875:152–156.

Pootanakit K, Prior KJ, Hunter DJ, et al (1999) 5-HT$_{2A}$ receptors in the rabbit retina: potential presynaptic modulators. Vis Neurosci 16:221–230.

Porter RHP, Benwell KR, Lamb H, et al (1999) Functional characterization of agonists at recombinant human 5-HT$_{2A}$, 5-HT$_{2B}$ and 5-HT$_{2C}$ receptors in CHO-K1 cells. Brit J Pharmacol 128:13–20.

Quigley HA (1996) Number of people with glaucoma worldwide. Br J Ophthalmol. 80:389–393.

Redburn DA, Churchill L (1987) An indoleamine system in photoreceptor cell terminals of the Long-Evans rat retina. J Neurosci 7:319–329.

Sharif NA, Klimko P (2007) CNS:ophthalmic agents, In: Taylor JB, Triggle DJ, eds. Comprehensive Medicinal Chemistry II, Vol. 6. Oxford: Elsevier, pp. 297–320.

Sharif NA, Senchyna M (2006) Serotonin receptor subtype mRNA expression in human ocular tissues determined by RT-PCR. Mol Vis 12:1040–1047.

Sharif NA, Kelly CR, Crider JY, et al (2006a) Serotonin-2 (5-HT$_2$) receptor-mediated signal transduction in human ciliary muscle cells:role in ocular hypotension. J Ocular Pharmacol Ther 22:389–401.

Sharif NA, Kelly C, McLaughlin MA (2006b) Human trabecular meshwork cells express functional serotonin-2A (5-HT$_{2A}$) receptors: role in IOP reduction. Invest Ophthalmol Vis Sci 47:4001–4010.

Sharif NA, McLaughlin MA, Kelly CR (2007) AL-34662:a potent, selective and efficacious ocular hypotensive serotonin-2 receptor agonist. J Ocular Pharmacol Ther 23:1–13.

Sharif NA, McLaughlin, MA, Kelly, et al (2009) Cabergoline:pharmacology, ocular hypotensive studies in multiple species, and aqueous humor dynamic modulation in cynomolgus monkey eyes. Exp Eye Res 88:386–397.

Takat D, Guler C, Arici M, et al (2001) Effect of ketanserin administration on intraocular pressure. Ophthalmologica 215:419–423.

Takenaka H, Mano T, Maeno T, et al (1995) The effect of anplag (sarpogrelate HCl), novel selective 5-HT$_2$ antagonist on intraocular pressure in glaucoma patients. Invest Ophthalmol Vis Sci 36:S734.

Tobin AB, Osborne NN (1989) Evidence for the presence of serotonin receptors negatively coupled to adenylate cyclase in the rabbit iris-ciliary body. J Neurochem 53:686–691.

Tobin AB, Unger W, Osborne NN (1988) Evidence for presence of serotonergic nerves and receptors in the iris-ciliary body complex of the rabbit. J Neurosci 8:3713–3721.

Trope GE, Sole M, Aedy L, et al (1987) Levels of norepinephrine, epinephrine, dopamine, serotonin and N-acetylserotonin in aqueous humor. Can J Ophthalmol 22:152–154.

Turner HC, Alvarez LJ, Candia OA, et al (2003) Characterization of serotonergic receptors in rabbit, porcine and human conjunctivae. Curr Eye Res 27:205–215.

Veglio F, De Sanctis U, Schiavone D, et al (1998) Evaluation of serotonin levels in human aqueous humor. Ophthalmologica 212:160–163.

Wainscott DB, Lucaites VL, Kursar JD, et al (1996) Pharmacologic characterization of the human 5-hydroxytryptamine$_{2B}$ receptor: evidence for species differences. J Pharmacol Exp Ther 276:720–726.

Wang RF, Lee PY, Mittag TW, et al (1997) Effect of 5-methyl-urapidil, an alpha-1a-adrenergic antagonist and 5-hydroxytrptamine-1A agonist, on aqueous humor dynamics in monkeys and rabbits. Curr Eye Res 16:769–775.

Weinreb RN, Kashiwagi K, Kashiwagi F, et al (1997) Prostaglandins increase matrix metalloproteinase release from human ciliary smooth muscle cells. Invest Ophthalmol Vis Sci 38:2772–2780.

Wiederholt M, Thieme H, Stumpff F (2000) The regulation of trabecular meshwork and ciliary muscle contractility. Prog Retinal Eye Res 19:271–295.

Wood MD, Thomas DR, Gager TL (1997) Pharmacological characterisation of the human 5-HT$_{2B}$ and 5-HT$_{2C}$ receptors in functional studies. Pharmacol Rev Commun 9:259–268.

WuDunn D (2009) Mechanobiology of trabecular meshwork cells. Exp Eye Res 88:718–723.

Yoshio R, Taniguchi T, Itoh H, et al (2001) Affinity of serotonin receptor antagonists and agonists to recombinant and native α1-adrenoceptor subtypes. Jpn J Pharmacol 86:189–195.

Index

Printed by Books on Demand, Germany